S0-DOR-608

# PRODUCT ENGINEERING DESIGN MANUAL

EDITED BY

**Douglas C. Greenwood**

EDITOR, *PRODUCT ENGINEERING DESIGN DIGEST*

McGRAW-HILL BOOK COMPANY, INC.

NEW YORK TORONTO LONDON

1959

 Library of Congress Catalog Card Number: 59-13681.

8 9 10 11 12 13 HL 9 8 7 6
24360

# PREFACE

Most advances in design engineering do not come about unaided. Even the most brilliant developments are based on the firm ground of existing, proven ideas. To this extent, therefore, improved mechanisms, better fastening and assembly methods, material savings, and design ingenuity in general are the result of an evolutionary process. One thing helps another and design takes a step forward.

But to trigger the inventive processes that yield the new idea, there must be a catalyst. Two or more ideas – even though quite different – can then combine to produce a completely new and better method, device, component, or complete product. The catalyst in all design-engineering problems is experience and recallable knowledge. That is one reason why engineering offices and individual engineers compile a design file. Another reason is simply to have quickly available various alternative ways of solving some specific design problems. Accumulated in these files is material of special, long-term reference value. Indeed it can almost be said that one mark of an experienced design engineer is the quality and size of his design file.

A treasure house of such material has appeared in the pages of *Product Engineering* magazine. One of the saddest reflections of an engineering editor is the realization that permanent-value material has been lost to many readers in the no-longer-available back issues of his magazine.

To my knowledge, only in the pages of *Product Engineering* can be found such specially edited design-file material. The pages that contain such material are never reprinted – except by special arrangement. They have consequently been much sought-after by engineers, many of whom have suggested their reproduction.

Now, at last, this stimulating material has been carefully sifted, condensed, and re-edited where necessary. The best and most practical *Product Engineering* design-file pages are presented in this manual. It contains material of the type for most of which you would search the usual engineering handbooks and textbooks in vain – material that is, nevertheless, the lifeblood of design engineering.

The illustrated mechanisms and other information have been supplemented, where necessary for design aid, with charts, formulas, and tables. To the younger engineer this represents the nearest thing to a ready-made experience that I can possibly imagine. For the experienced man it will be recognized immediately as an essential supplement to the fund of knowledge he already possesses.

For all levels of product-design engineers, it is intended that this will be a solid package of practical how-to's and the catalyst for further progress through new ideas.

I should like, at this point, to thank James B. Reswick, associate professor in charge of graphics and machine design at M.I.T., and James M. Aiken, assistant professor of the same department, whose suggestion that this book be compiled was supplemented with their own notebook of tear sheets as an example. Thanks are also due my associates on the staff of *Product Engineering* for data and helpful suggestions, and to former staff members who originally edited some of the earlier pages. Most of all, credit is due the engineers who prepared the material and submitted it to *Product Engineering* for publication.

DOUGLAS C. GREENWOOD
Editor
*Product Engineering Design Digest*

# CONTENTS


Preface iii

What Makes a Successful Design Engineer? ix

*Chapter 1. Accessories* 1

**Adhesive Applicators** for High-speed Machines 2
**Bead Chains** for Light Service 4
**Bellows:** Thirteen Ways to Use Metallic Bellows 6
**Detents:** Retaining and Locking Detents 8
**Frictional Supports** for Adjustable Parts 10
**Hoppers:** 7 Basic Hoppers for Parts 12
**Nibs and Catches** for Lids on Sheet-metal Containers 14
**Plugs,** Caps and Covers 16
**Ultrasonics:** Applications of Ultrasonic Transducers 20
**Vibration:** Selecting Vibration Isolators 22
Types of Mounts for Vibration Isolation 24

*Chapter 2. Assembly* 27

**Adhesives:** Properties and Uses of 43 Adhesives 28
Which Adhesives for What 32
How to Calculate Stresses in Adhesive Joints 34
**Aligning:** 16 Ways to Align Sheets and Plates with One Screw 38
**Circular Parts:** Joining Circular Parts without Fasteners 40
**Die Castings:** 23 Ways to Attach Small Die-cast Parts 42
**Fasteners:** Fundamentals of Selecting Locknuts 47
**Flexible Materials:** Fastening Methods for Flexible Sheet Materials 50
**Glass:** Methods of Attaching Glass to Metal Structures 52
**Inserts:** Applications of Helical Wire Inserts 54
**Locking:** Various Methods of Locking Threaded Members 56
**Quick-release** Fasteners 58
**Sheet Metal:** Design for Sheet-metal Boxes with Square and Round Corners 60
Top and Bottom Attachments for Sheet-metal Containers 62
Fastening Sheet-metal Parts by Tongues, Snaps and Clinching 64
Joining Sheet-metal Parts without Fasteners 66
Liquid-tight Riveted Joints 68
**Simplifying Assemblies** with Spring-steel Fasteners 70
**Tamper-proofing:** 20 Tamper-proof Fasteners 72
**Twisting:** How to Design Parts That Assemble and Lock by Twisting 74

*Chapter 3. Clutches* 77

**Basic Types** of Mechanical Clutches 78
**Over-riding:** Construction Details of Over-riding Clutches 80
10 Ways to Apply Overrunning Clutches 82
**Small** Mechanical Clutches for Precise Service 84
**Serrated** Clutches and Detents 86

*Chapter 4. Couplings* 89

**Flexible:** Typical Designs of Flexible Couplings – I 90
Typical Designs of Flexible Couplings – II 92
Typical Designs of Flexible Couplings – III 94
**Low-cost** Methods of Coupling Small Diameter Shafts 96
**Parallel Shafts:** Coupling of Parallel Shafts 98
**Splines:** Ten Different Types of Splined Connections 100
**Tubing:** How to Connect Tubing – Cross and Tee Joints 102
Different Mechanical Methods for Attaching Tubing 104

*Chapter 5. Bearings and Mounts* 107

**Lubrication:** Eleven Ways to Oil Lubricate Ball Bearings 108
Lubrication of Small Bearings 110
**Linear Motion:** Nine Types of Ball Slides for Linear Motion 112
**Miniature:** Unusual Applications of Miniature Bearings 114
**Mounting:** Rolling Contact Bearing Mounting Units 116

*Chapter 6. Control and Measurement Devices* 119

**Electric:** Methods of Electric Control 120
**Limit Switching** Mechanisms – Methods of Mechanical Actuation 122
**Liquid Level** Indicators and Controllers 124
Mechanisms of Indicators and Controllers 126
**Photoelectric** Controls 128
**Stop Mechanisms:** Automatic Stop Mechanisms Protect Machines and Work 130
Electric Automatic Stop Mechanisms 132
Mechanical Automatic Stop Mechanisms 134
**Speed:** How to Obtain Constant Speed Motion Below 10 RPM 136
12 Ways of Measuring Speed 140
**Temperature** Regulators 142
Thermocouple Details for Temperature Measurement 144
**Timers:** Types of Automatic Timers 146

*Chapter 7. Dimensions and Design* 149

**Automatic Assembly:** Check-chart for Single-part Complexity 150
**Continuous Beams:** Formulas and Curves Give Quick Answers for Partially Loaded Continuous Beams 152
Nomogram to Determine Parallel Axis Moment of Inertia 154
**Corrosion:** The 6 Types of Corrosion, a Preventive Guide 156
**Clad Steel:** Designing and Fabricating with Clad Steel Plate 158
**Circular Segments:** Universal Tables for Circular Segments and Sectors 160
**Knurls:** How to Specify Precision Knurls 164
**Piping Liquids:** How Much Horsepower to Pipe Liquids 166
**Plastics:** Design Fundamentals for Molded Plastic Parts 168
**Porcelain Enamel:** Recommended Practices for Designing Porcelain Enamel Products 172
**Press Fits:** How to Calculate Stresses in Press-fit Bushings 174
Increasing the Holding Power of Press Fits 176
**Ratchet Layout** Analyzed 178
**Rubber Parts:** Tips for Designing Rubber Parts and Assemblies 180
**Shrink-fit** Nomograph 182
**Stainless Steel:** The Right Finishes for Stainless Steel 184
**Stretch Forming:** How to Find Final Dimensions of Stretch-formed Parts 186
**Toothed Components:** Design of Toothed Mechanical Components 188
**Wire Stitching:** Design Data for Metal Wire Stitching 190

*Chapter 8. Drives* **193**

**Belt:** Ten Types of Belt Drives 194
Mechanisms for Adjusting Tension of Belt Drives 196
**Chain:** Methods of Reducing Pulsations in Chain Drives 198
**Horsepower Capacity:** Leather Belts – Hp Loss and Speeds 200
Power Capacity of Spur Gears 201
**Friction** Wheel Drives Designed for Maximum Torque 204
Accurate Solution for Disk-clutch Torque Capacity 207
Torque of Slip Couplings 208
**Shafts:** Torsional Strength of Shafts 210
**Variable Speed:** Basic Types of Variable Speed Friction Drives 212

*Chapter 9. Electrical, Electronic and Magnetic Components* **215**

**Electrical Symbols** and Standards 216
**Fundamental Electronic Circuits:** Power and Voltage Amplifiers 218
Tubes for Industrial Inspection and Control 220
**Heating Elements:** Designing Electric Heating Elements 222
Typical Industrial Uses of Electric Heating Elements 224
**Magnets:** Magnet Coil Design 226
Fundamental Types of Permanent Magnets 230
Applications for Permanent Magnets 232
Permanent Magnet Mechanisms, Their Design and Uses 234
**Motors:** Fractional Horsepower Motors 236
Torque Requirements of Various Motor Driven Loads 238
**Snap Switches:** Thirty-seven Ideas for the Application of
Precision Snap-acting Switches 241
**Terminals:** Electrical Terminal Connections 242
**Tubes:** 12 Ways to Retain Electron Tubes 244

*Chapter 10. Mechanical Movements and Linkages* **247**

**Air Actuated:** Mechanisms Actuated by Air or Hydraulic Cylinders 248
**Computing Mechanisms:** I & II 250
**Geneva Drives:** Modified Geneva Drives and Special Mechanisms 254
**Intermittent** Movements and Mechanisms 256
Mechanisms for Providing Intermittent Rotary Motion 258
Friction Devices for Intermittent Rotary Motion 260
**Kinematics of Intermittent Mechanisms:** I The External Geneva Wheel 262
II The Internal Geneva Wheel 265
III The Spherical Geneva Wheel 269
Kinematics of the Crank and Slot Drive 272
**Linear Error:** Linear to Angular Conversion of Gear-tooth Index Error 275
**Linear Motion:** Accelerated and Decelerated Linear Motion Elements 276
**Power Thrust** Linkages and Their Applications 278
**Short Motions:** Transmission Linkages for Multiplying Short Motions 280
**Special Purpose** Mechanisms 282
Mechanisms for Producing Specific Types of Motions 284
**Toggle Linkage** Applications in Different Mechanisms 286
**Traversing** Mechanisms Used on Winding Machines 288

*Chapter 11. Miscellaneous Design Aids* **291**

**Backlash:** How to Provide for Backlash in Threaded Parts 292
**Corrosion:** Recommended Design Details to Reduce Corrosion 294
**Electroplating:** For Better Electroplating, These Answers to Trouble Spots 296
For Better Electroplating – Better Mounting, Drainage, Radii 298

**Lubrication** of Roller Chains 300
**Variable Stress:** Design of Parts for Conditions of Variable Stress 302

*Chapter 12. Shaft Seals* 305

**O-rings:** Design Recommendations for O-ring Seals 306
**Non-rubbing** Seals for Oil Retention 308
**Rubbing** Seals for Oil Retention 310
**Sleeve Bearings:** Typical Methods of Sealing Rotating Shafts 312

*Chapter 13. Springs* 315

**Adjusting Methods:** Compression Spring Adjusting Methods – I & II 316
Adjustable Extension Springs 320
**Deflection:** How Much Force to Deflect a Spring Sideways? 322
**Overriding** Spring Mechanisms for Low-torque Drives 324
**Testing:** 17 Ways of Testing Springs 326
**Unusual Uses** for Helical Wire Springs 328

*Chapter 14. Welding and Brazing* 331

**Brazing:** Methods for Placing Brazing Materials, and Vent Locations 332
**Built-up Welded** Constructions 334
**Resistance Welding:** Preparation of Materials for Resistance Welding 338

# What Makes A Successful Design Engineer?

**GEORGE H. LOGAN**
Starr Engineering Company
North Hollywood, California

ENGINEERING DESIGN is creative work, the evolution by thought of things new and useful. No less than the artist or musician, the designer has to have the inner spark that urges him to produce entities out of void. His eye penetrates the opaque, his mind builds three-dimensional mechanisms of functional nature, he has an innate talent for *imagineering*. The spark need not be large; if it exists at all, it can be made to grow into a catalytic agent to the man's success in product engineering, to building the inner satisfaction linked with achievement.

For those who must invent and design mechanical devices, as other men to be content must paint or write music, how best is the spark nurtured? One highly important step in the development of creative capabilities is to think of the brain as a rubber band, compel the intellect to stretch in many directions when attacking a design problem. A good way to do this is to write down the hard-core parameters of the assignment and then think radially outward from the core, surrounding the hub with spokes of ideas for solution. Such cultivation of uninhibited thinking is a highly effective strategy in outmaneuvering design difficulties. The designer should urge his thoughts toward unorthodox approaches to the problem. Some of the attacks may well be impractical at the outset, but ideas beget ideas. Yet out of such thinking may come ingenious solutions of the kind that prompt people to say: "Why didn't I think of that myself!"

The sensation of dull labor is experienced at times when but one inadequate approach to a problem has been conceived, and it is a bitter struggle trying to make it alone provide a total answer. However, there is relief from such frustration once the "mind of rubber" concept is adopted and used. With it, the facility of design execution and the degree of inspired novelty are tremendously increased and the time for finishing the job reduced.

The conditioning of the mind to unfettered thinking, not bound by precedent or the obvious, cannot fail to develop singular versatility and sureness of performance. And once that control of the mind is completely held by a design engineer, he has taken a large step toward more recognition.

The belief in and practice of work organization should be closely associated with the acute-mind principle just described. To gain full advantage of a volatile imagination, the products of that imagination should be recorded in a manner easy to analyze. Explicitly, when the designer receives an assignment, he should start a notebook labeled "Original Data, Project ——————". The first sheet might carry the captions "Project Objective" and "Project Background." The definition under "Project Objective" can be based on discussions with the individual or individuals initiating the assignment. With this concise statement of what the project must accomplish, the engineer possesses a valuable springboard for his work. Lacking such definition, much time can be wasted through imperfect understanding of what the project is supposed to produce.

The data listed under the heading "Project Background" is, in general, anything that contributes to the expeditious handling of the problem. It can be a compilation of patents showing prior art in the field or a synopsis of sales department ideas on features that will give a product maximum marketability. This background material might also include any literature that describes competitive products. Such information is a valuable guide to setting design goals as they pertain to product cost, performance, and customer acceptance.

Once these basic facts are recorded, the engineer starts adding sheets describing his own work to the Data Book. These trace the progress of his thinking. The earliest of the pages are generally mere freehand sketches and notations on first ideas. But the very act of placing these visualizations on paper often suggests simplifications or combinations of preliminary ideas that later become the basis of the final product. Appropriate calculations are, of course, inserted as they occur.

Another step in cultivating imagineering is to make a habit of closely scanning both the editorial and advertising pages of industrial magazines and society journals. Each page has something to contribute to a designer's fund of knowledge. And, in many cases, design descriptions suggest the solution to an immediate problem.

Through the ages artists have borne the implication, whether justified or not, of being temperamental. The engineer cannot afford to be the target of such comment. He should discipline his mind, control his speech and actions so that he achieves full emotional balance. Regardless of how much brilliance he develops, how much information he has absorbed, unless he controls his reactions to disturbing stimuli, he builds a barrier to advancement. Emotional balance is an intangible will-o'-the-wisp, often more difficult to secure than technical skill itself; yet it is, at least, of equal importance.

Building a calm temperament is not always easy, but it can be done if *restraint* is practiced. Speech should be restrained when matters turn sharply controversial. At such times, it is wise to interject a cooling-off period before again trying to sell a sincere idea. Likewise, there is little value or reward for worrying about things over which one has absolutely no control. The emotionally balanced engineer never permits the factual approach to be warped by irritations, for it is the serene and objective mind that belongs to a man who can perform.

In consideration of an engineer's legitimate ambition to progress, the importance of broadening the knowledge of non-technical business topics should not be overlooked. It is largely true that this is an age of specialization, and that each individual must master his specialty. Therefore, it is the alert design engineer who channels into his available time the study of industrial subjects other than engineering. Consider men in management positions. Almost invariably they have stored information that equips them to make decisions on matters such as economics, accounting, law, or industrial relations. And the greater the responsibility, the broader the scope of knowledge needed to understand and evaluate the endless variety of subjects that thread industry. Therefore, it is also desirable to soak up information appearing in journals like the *Harvard Business Review*, which treats topics such as "Utilization of Older Manpower," "Basic Elements of a Free, Dynamic Society," and "An Economics for Administrators."

"Rubber band" thinking is required of an executive. His mind must blanket and swarm all over problems, exploring every pertinent facet. He must be able to make quick decisions that are correct most of the time. He is always ready to switch from objective thinking to abstract thinking, for it is by the latter thought process that sweeping, inspired plans are made. Thus, through rigorous discipline and actions ordered by a disciplined mind, the creative design engineer can make for himself an opportunity that leads to an outstanding position in his chosen field.

# 1
# ACCESSORIES


Adhesive Applicators for High Speed Machines 2
Bead Chains for Light Service 4
Thirteen Ways to Use Metallic Bellows 6
Retaining and Locking Detents 8
Frictional Supports for Adjustable Parts 10
7 Basic Hoppers for Parts 12
Nibs and Catches for Lids on Sheet-metal Containers 14
Plugs, Caps and Covers 16
Applications of Ultrasonic Transducers 20
Selecting Vibration Isolators 22
Types of Mounts for Vibration Isolation 24

# ADHESIVE APPLICATORS FOR

THE METHODS OF APPLYING LIQUID ADHESIVES that are illustrated here include rotary applicators on movable axes and otherwise movable between adhesive pick-up position and applying position, endless belt applicators, applicators in the form of moving daubers, plates, and the like, reciprocating dies exuding measured quantities of cement, and spray nozzles. All of these mechanisms are used or are applicable on production machines such as for making pasteboard boxes or cartons, pasting labels or envelopes, and making shoes or other products involving the use of liquid adhesives.

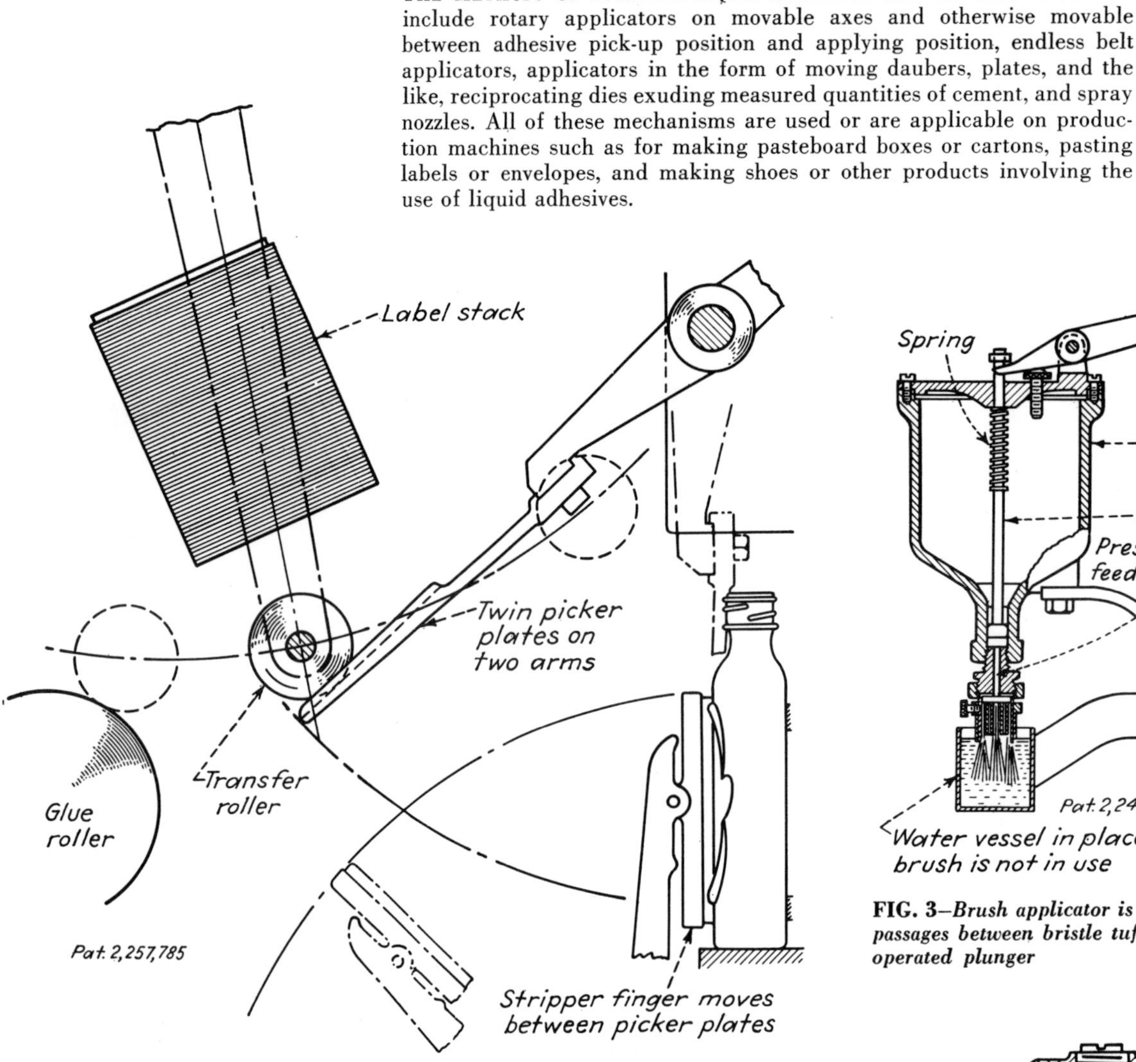


**FIG. 1** – *Bottom label is spread with glue by two abutting glue-coated picker plates, which separate during contact with label stack, then carry label to bottle*

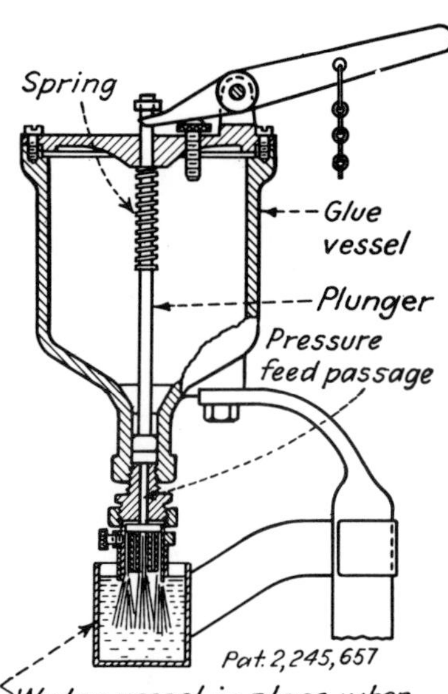


**FIG. 3**–*Brush applicator is fed through passages between bristle tufts by spring operated plunger*

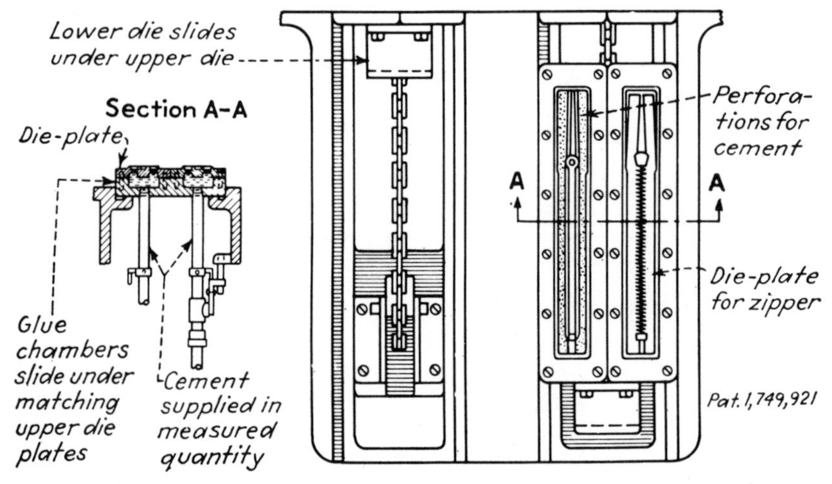


**FIG. 2** – *Measured quantities of cement are forced through perforations in specially designed upper and lower die plates, which are closed hydraulically over zippers. Lower die only is shown*

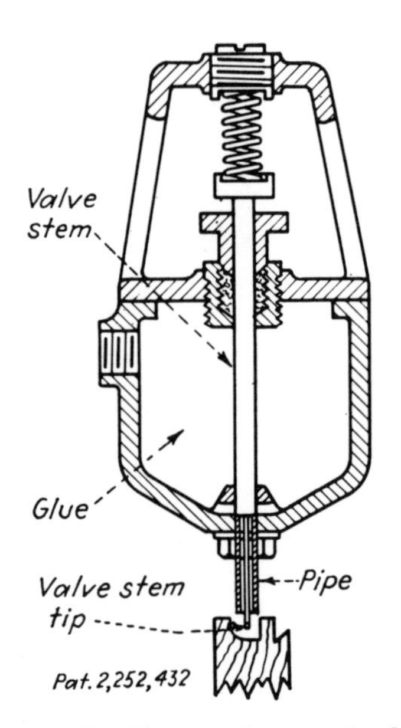


**FIG. 4**–*Shoulder on valve stem in glue chamber retains glue until pressure on tip opens bottom valve*

# HIGH-SPEED MACHINES

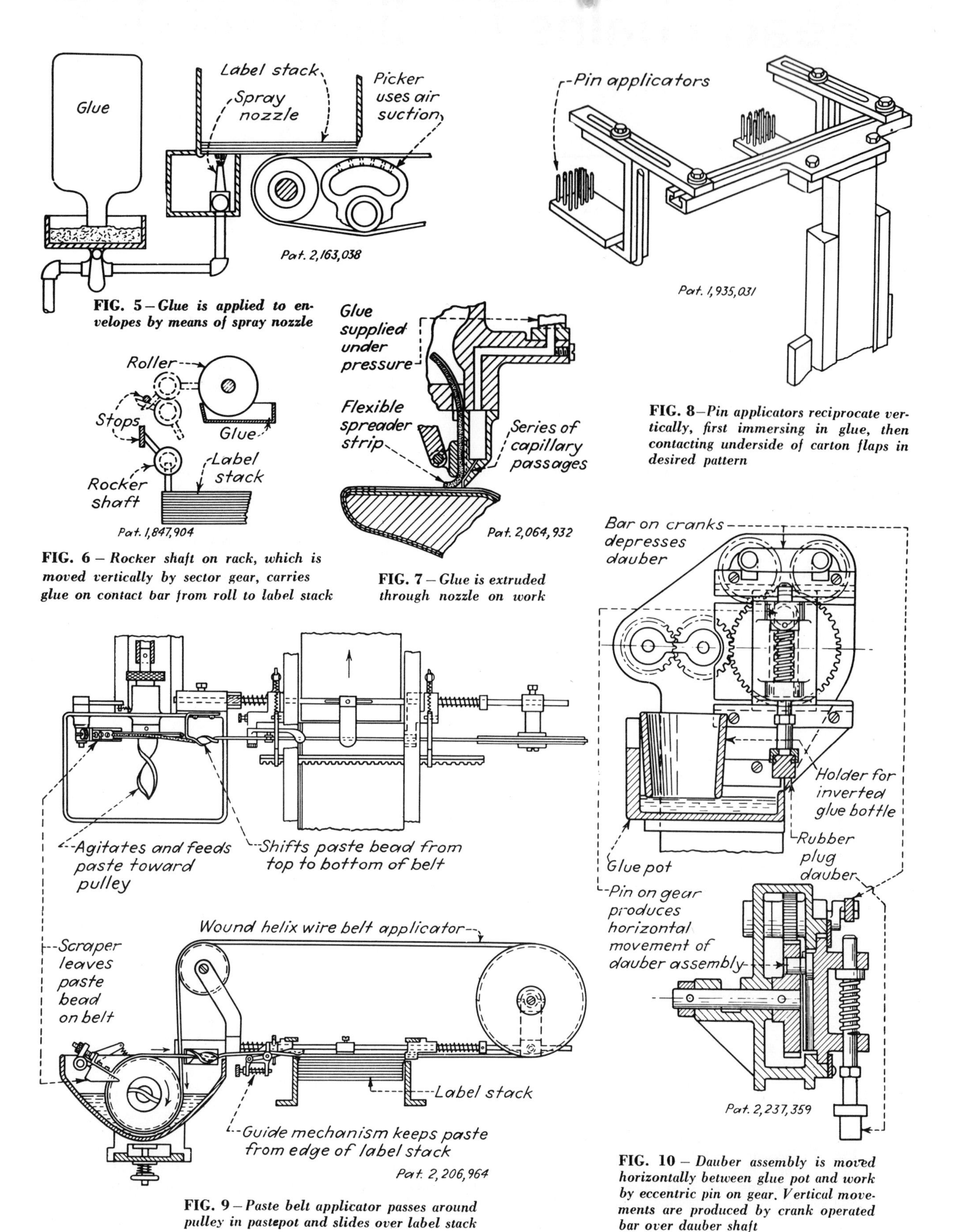


**FIG. 5** – *Glue is applied to envelopes by means of spray nozzle*

**FIG. 8** – *Pin applicators reciprocate vertically, first immersing in glue, then contacting underside of carton flaps in desired pattern*

**FIG. 6** – *Rocker shaft on rack, which is moved vertically by sector gear, carries glue on contact bar from roll to label stack*

**FIG. 7** – *Glue is extruded through nozzle on work*

**FIG. 9** – *Paste belt applicator passes around pulley in pastepot and slides over label stack*

**FIG. 10** – *Dauber assembly is moved horizontally between glue pot and work by eccentric pin on gear. Vertical movements are produced by crank operated bar over dauber shaft*

# Bead chains for light service

BERNARD WASKO, chief engineer, Voland and Sons, Inc.

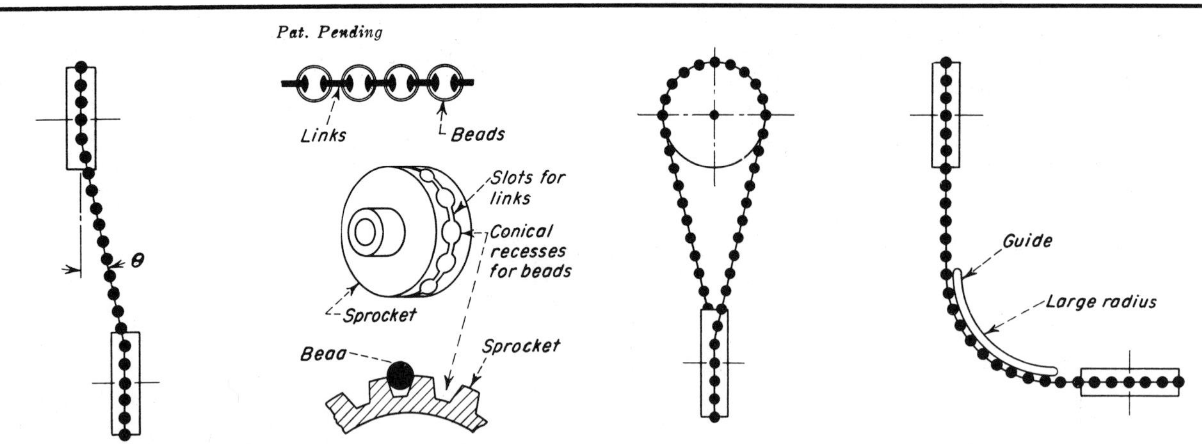


**Fig. 1—Misaligned sprockets. Nonparallel planes usually occur when alignment is too expensive to maintain. Bead chain can operate at angles up to $\theta = 20$ degrees.**

**Fig. 2—Details of bead chain and sprocket. Beads of chain seat themselves firmly in conical recesses in the face of sprocket. Links ride freely in slots between recesses in sprocket.**

**Fig. 3—Skewed shafts normally acquire two sets of spiral gears to bridge space between shafts. Angle misalignment does not interfere with qualified bead chain operation on sprockets.**

**Fig. 4—Right angle drive does not require idler sprockets to go around corner. Suitable only for very low torque application because of friction drag of bead chain against guide.**

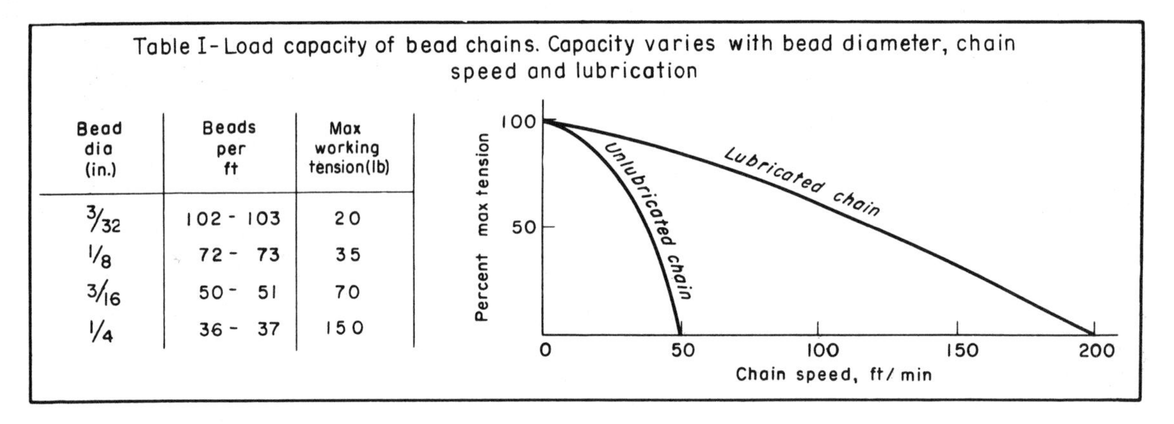

Table I - Load capacity of bead chains. Capacity varies with bead diameter, chain speed and lubrication

| Bead dia (in.) | Beads per ft | Max working tension (lb) |
|---|---|---|
| 3/32 | 102 - 103 | 20 |
| 1/8 | 72 - 73 | 35 |
| 3/16 | 50 - 51 | 70 |
| 1/4 | 36 - 37 | 150 |



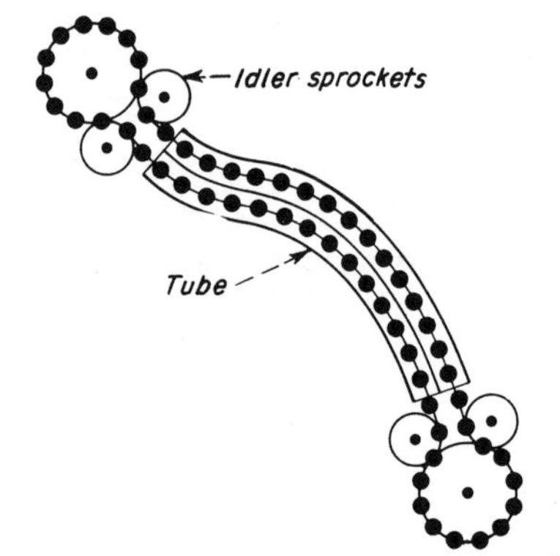


**Fig. 5—Remote control through rigid or flexible tube has almost no backlash and can keep input and output shafts synchronized.**

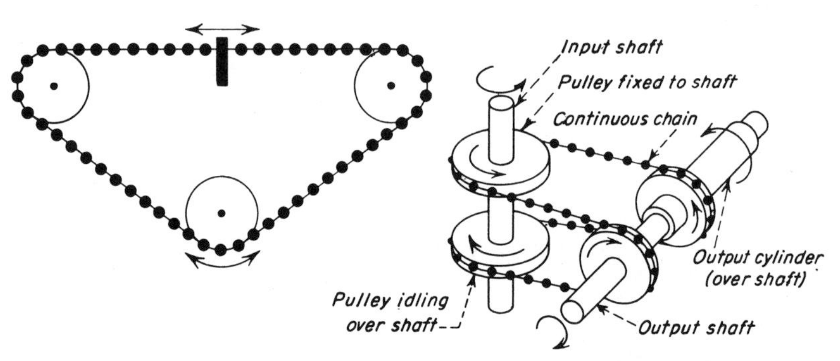


**Fig. 6—Linear output from rotary input. Beads prevent slippage and maintain accurate ratio between the input and output displacemonts.**

**Fig. 7—Counter-rotating shafts. Input shaft drives two counter-rotating outputs (shaft and cylinder) through a continuous chain.**

Where torque requirements and operating speeds are low, qualified bead chains offer a quick and economical way to: Couple misaligned shafts; convert from one type of motion to another; counter-rotate shafts; obtain high ratio drives and overload protection; control switches and serve as mechanical counters.

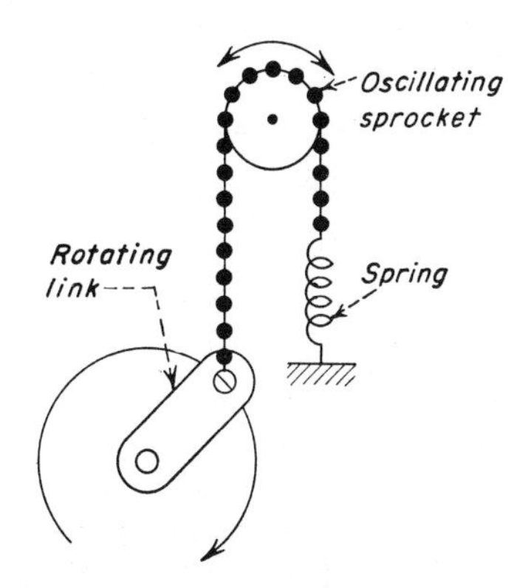


Fig. 8—Angular oscillations from rotary input. Link makes complete revolutions causing sprocket to oscillate. Spring maintains chain tension.

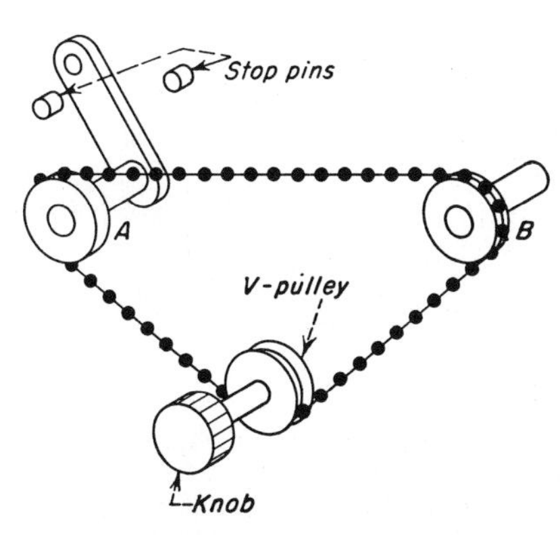


Fig. 9—Restricted angular motion. Pulley, rotated by knob, slips when limit stop is reached; shafts A and B remain stationary and synchronous.

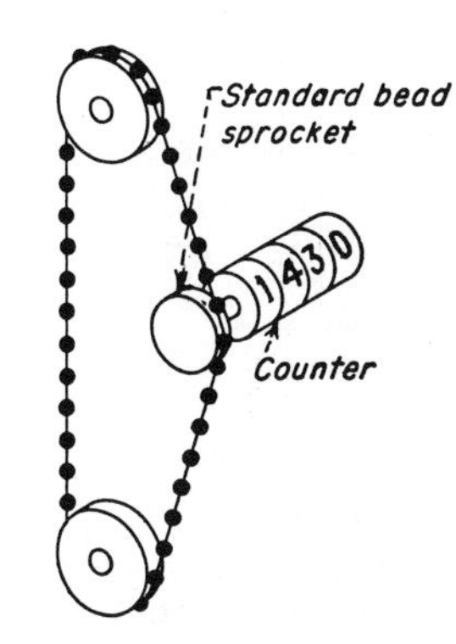


Fig. 10—Remote control of counter. For applications where counter cannot be coupled directly to shaft, bead chain and sprockets can be used.

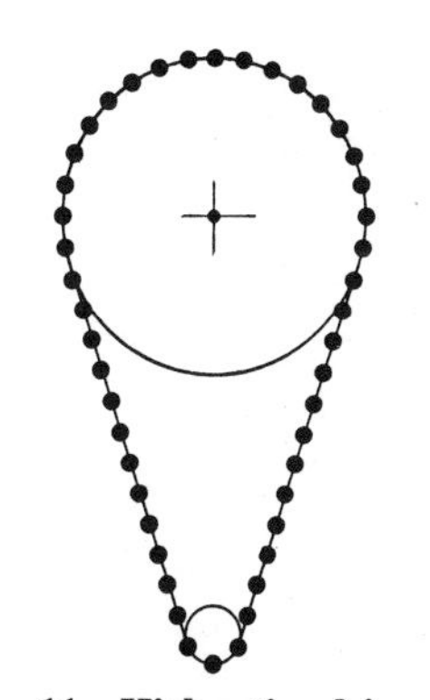

Fig. 11—High-ratio drive less expensive than gear trains. Qualified bead chains and sprockets will transmit power without slippage.

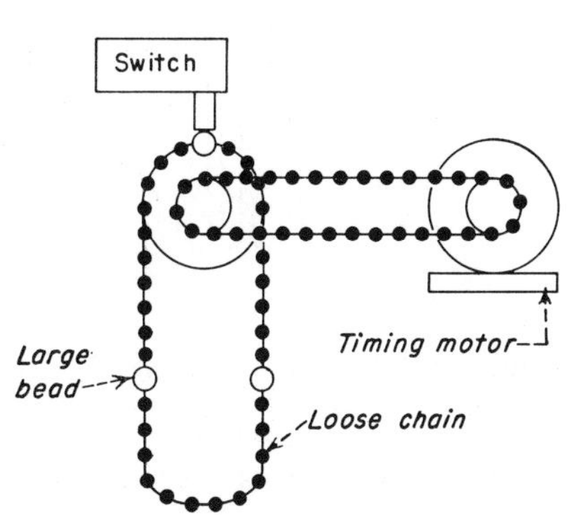


Fig. 12—Timing chain containing large beads at desired intervals operates micro-switch. Chain can be lengthened to contain thousands of intervals for complex timing.

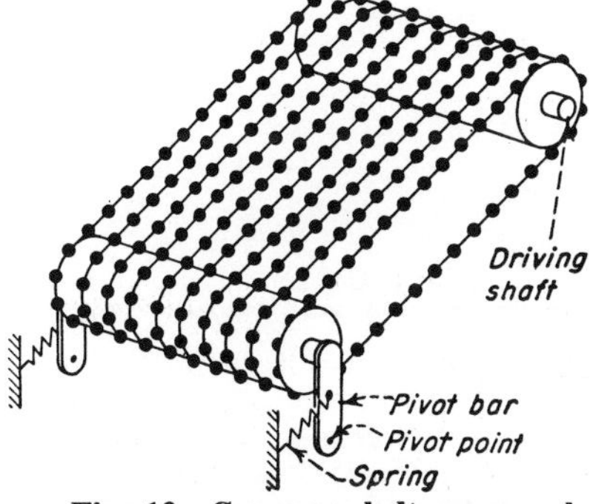


Fig. 13—Conveyor belt composed of multiple chains and sprockets. Tension maintained by pivot bar and spring. Width of belt easily changed.

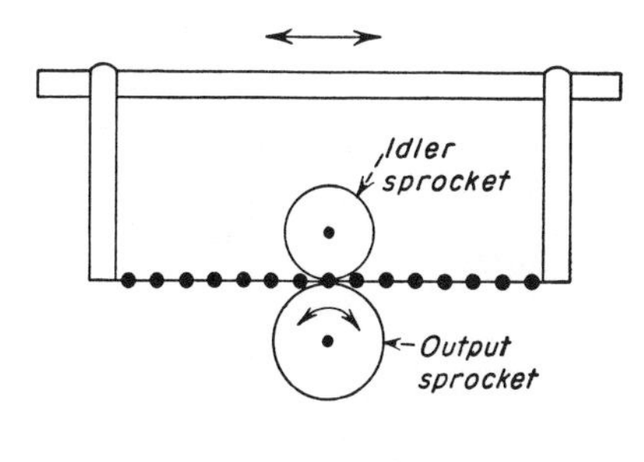


Fig. 14—Gear and rack duplicated by chain and two sprockets. Converts linear motion into rotary motion.

Standard sprocket

Sprocket with shallow recesses

Fig. 15 — Overload protection. Shallow sprocket gives positive drive for low loads; slips one bead at a time when overloaded.

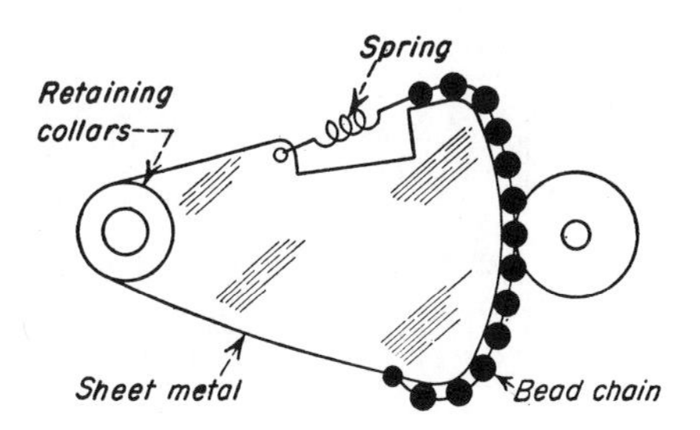


Fig. 16—Gear segment inexpensively made with bead chain and spring wrapped around edge of sheet metal. Retaining collars keep sheet metal sector from twisting on the shaft.

# Thirteen Ways to Use

Sketches serve two purposes: (1) Illustrate unique as well as typical applications;

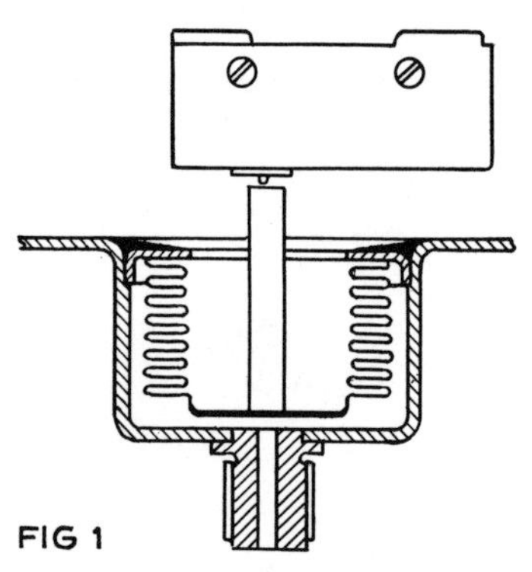
FIG 1

**ACTUATE GAGES** and switches. Pressures can be as high as 2,000 psi. Maximum value should exist when the bellows is near free-length.

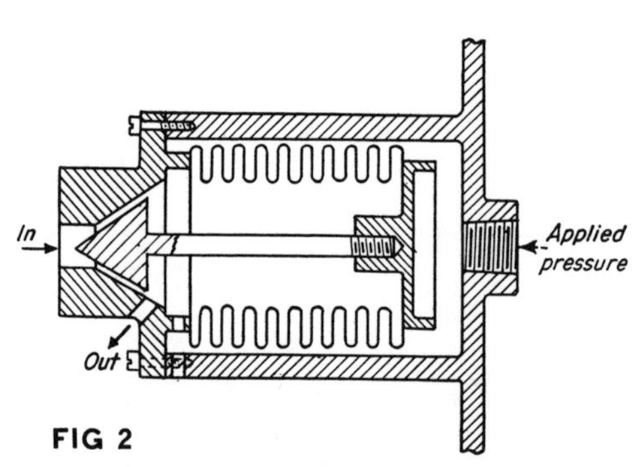

FIG 2

**FLOW CONTROL.** Variations in pressure adjust needle in flow valve. Also shows how bellows can be packless seals for valve stems and shafts.

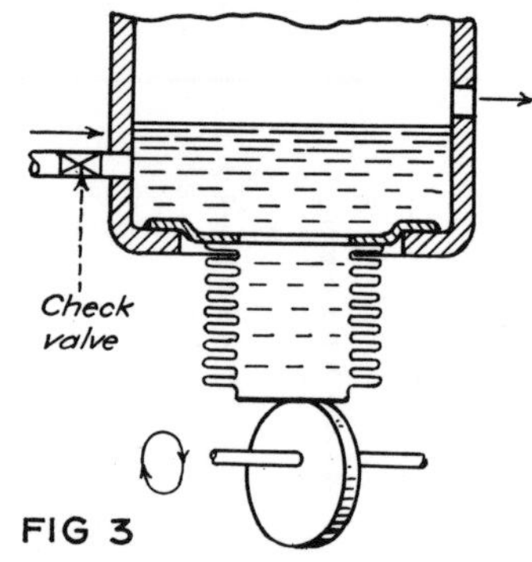

FIG 3

**METERING DEVICE.** Dispensing machines can use bellows as constant or variable displacement pump to measure and deliver predetermined amounts of liquids.

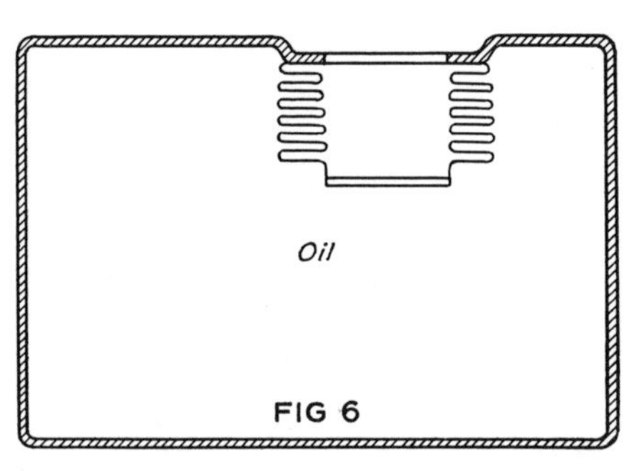

FIG 6

**ABSORB EXPANSION OF FLUIDS OR GASES.** Transformer (above) uses bellows to absorb increases in volume of oil caused by thermal expansion. Single controls of this type can operate from —70 to 250 F or from 0 to 650 F.

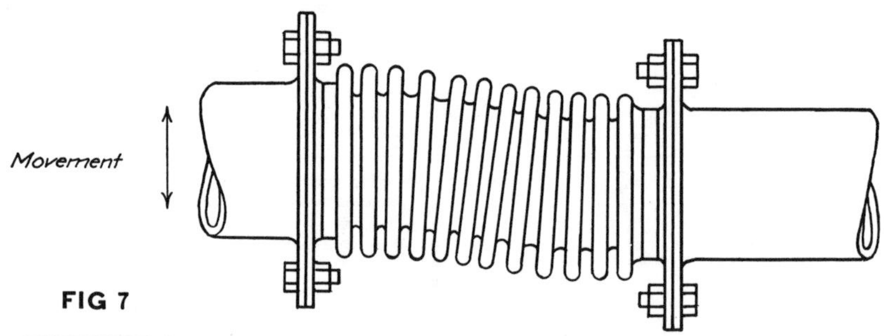

FIG 7

**FLEXIBLE CONNECTOR.** Suitable for wide range of applications from instruments to jet engines and large piping. Bellows absorb movement caused by thermal expansion, isolate vibration and noise as well as permit misalignment of mating elements. Wide variety of sizes and materials are now possible. Units are now in use from ¼ to 72 inches in diameter, made from such materials as brass, phosphor bronze, beryllium copper and stainless steel.

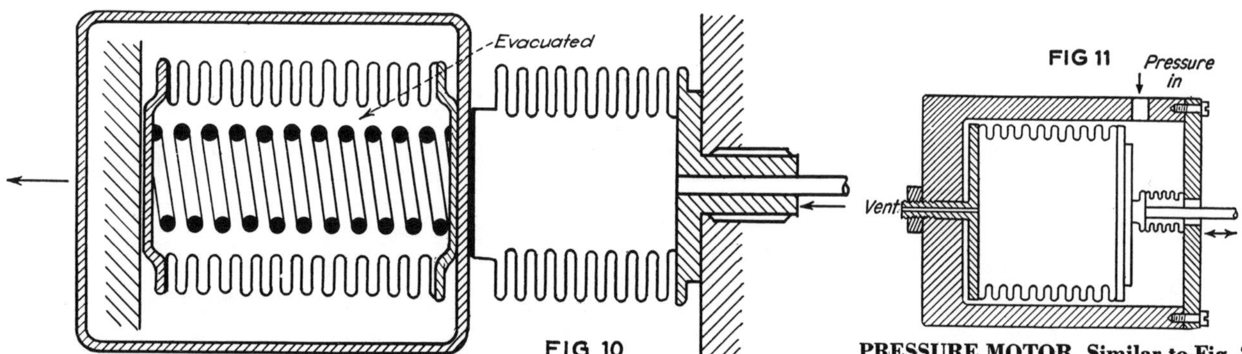

FIG 10

FIG 11

**PRESSURE COMPENSATOR.** Effect of ambient pressure can be eliminated in a pressure measuring system by matching the area of a pressure bellows with that of an aneroid and combining the two into a single assembly. Errors caused by ambient pressure can be held to a max of one percent. Present materials permit aneroid operation from —70 to 450 F.

**PRESSURE MOTOR.** Similar to Fig. 2. Bellows used instead of piston and cylinder arrangement. Eliminates effects of leakage and friction. Long stroke can be provided with sensitive response.

# Metallic Bellows

E. PERRY CUMMING
Bridgeport Thermostat Division
Robertshaw-Fulton Controls Company

(2) Show how the movement of bellows can be transferred to other elements.

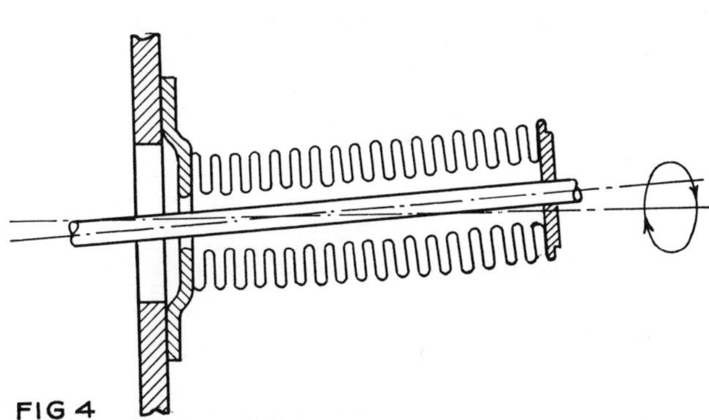

FIG 4
**FLEXIBLE COUPLING.** Bellows can transmit torque through oblique shafts with negligible amount of backlash or can be used to transmit circular motion through the wall of a sealed container as shown above.

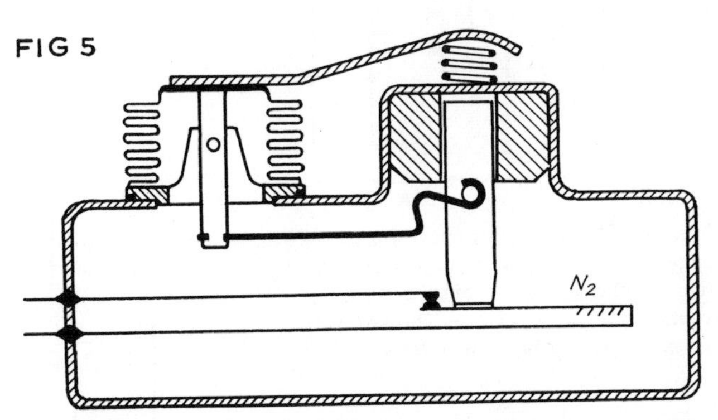


FIG 5
**HERMETIC SWITCH.** Bellows provide a gas tight flexible member through which motion can be transmitted into a sealed assembly. Flexibility and long life are important characteristics of these elements.

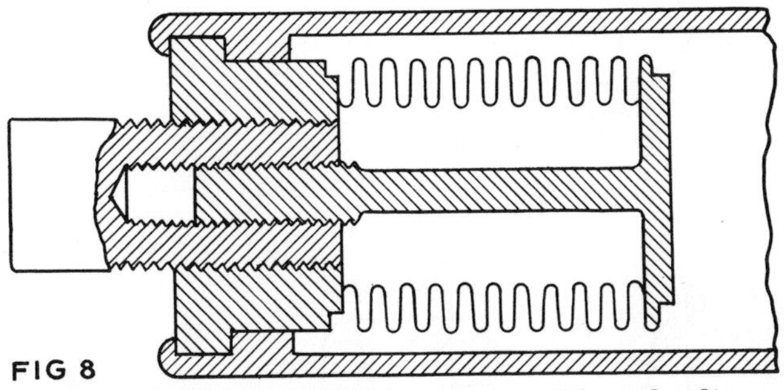

FIG 8
**SEALED ADJUSTMENT.** Accurately calibrated adjustments inside sealed instruments are possible by means of single or compound threads. To meet varied installation requirements, bellows are available with ends prepared for ring or disk end plates of standard or special design. These plates are fastened by brazing or welding techniques.

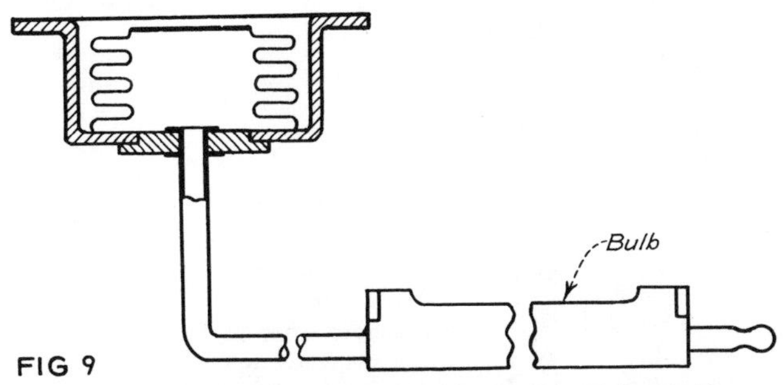


FIG 9
**VAPOR PRESSURE THERMOSTAT.** Small dia bellows offer large movement over a relatively small, adjustable temperature range. Can be filled so as to be unaffected by over-runs in temperature. Compensation for changes in ambient temperature is unnecessary whether this temperature is above or below the value selected for control purposes.

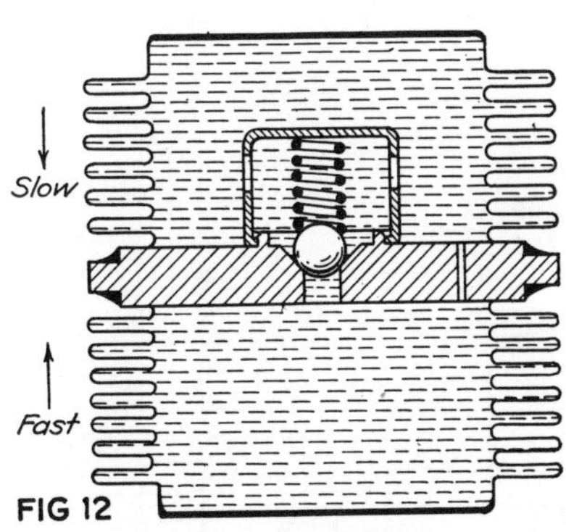


FIG 12
**TIME DELAY MECHANISM.** A check valve and proper size bleed hole between two liquid filled bellows allows fast motion in one direction and slow motion in the other direction.

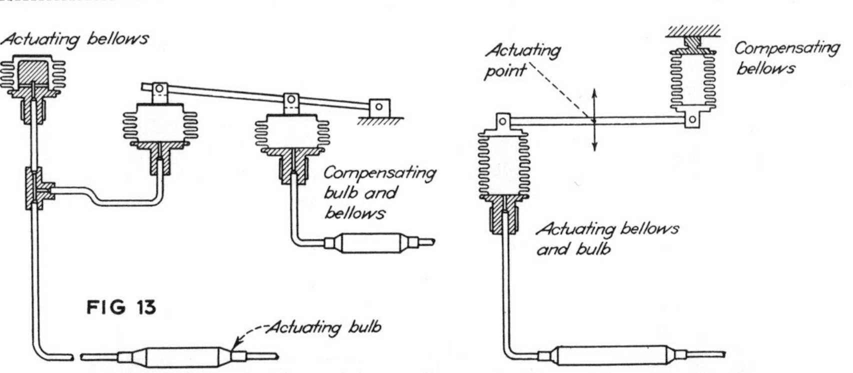


FIG 13
**AMBIENT TEMPERATURE COMPENSATION.** Two methods are shown. The left one uses two bellows in the actuating system. One is driven by the compensating assembly and correctly positions the actuating bellows as ambient temperature changes. The other method uses a floating lever whose mid-point is positioned by both the actuating and compensating bellows.

# Retaining and

Many forms of detents are used for positioning gears, levers, belts, covers and similar parts. Most of these embody some form of spring in varying degrees of tension, the working end of the detent being hardened to prevent wear

Fig. 1—Driving plunger, shown in engagement at *A* is pulled out, and given a 90-deg. turn, pin *X* slipping into the shallow groove as shown at *B*, thus disengaging both members.

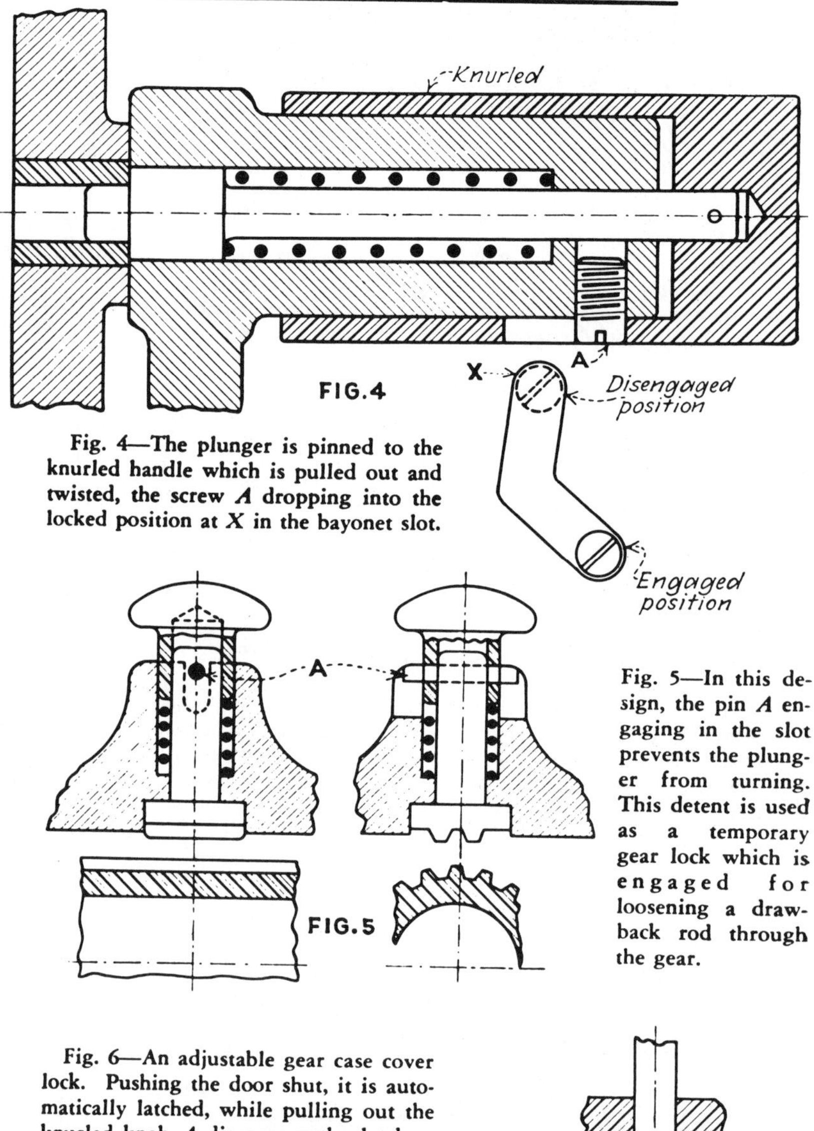


Fig. 4—The plunger is pinned to the knurled handle which is pulled out and twisted, the screw *A* dropping into the locked position at *X* in the bayonet slot.

Fig. 5—In this design, the pin *A* engaging in the slot prevents the plunger from turning. This detent is used as a temporary gear lock which is engaged for loosening a draw-back rod through the gear.

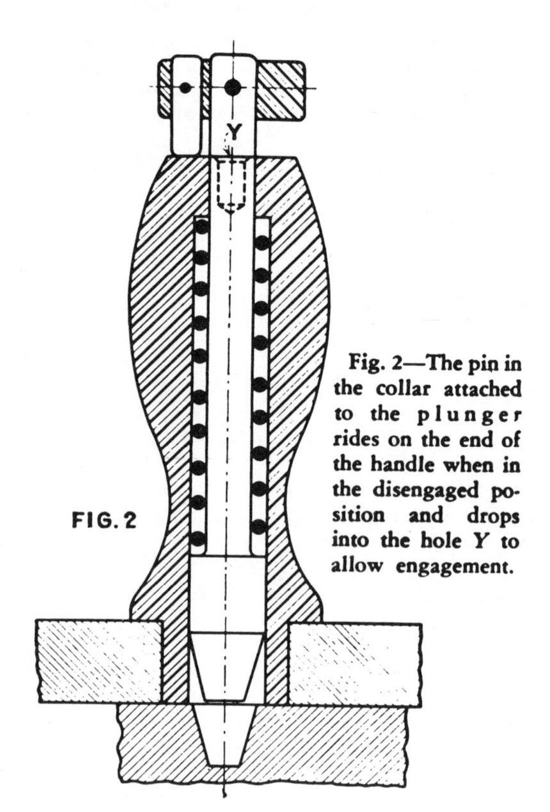


Fig. 2—The pin in the collar attached to the plunger rides on the end of the handle when in the disengaged position and drops into the hole *Y* to allow engagement.

Fig. 3—A long and a short slotted pin driven into the casting gives two plunger positions.

Fig. 6—An adjustable gear case cover lock. Pushing the door shut, it is automatically latched, while pulling out the knurled knob *A* disengages the latch.

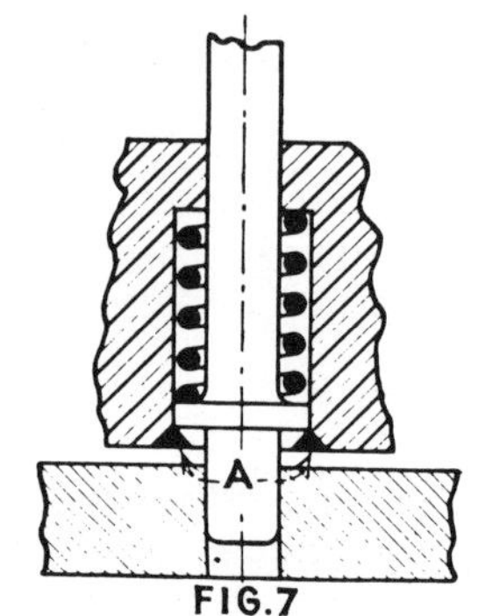


Fig. 7—In this design the plunger is retained by staking or spinning over the hole at *A*.

# Locking Detents

ADAM FREDERICKS

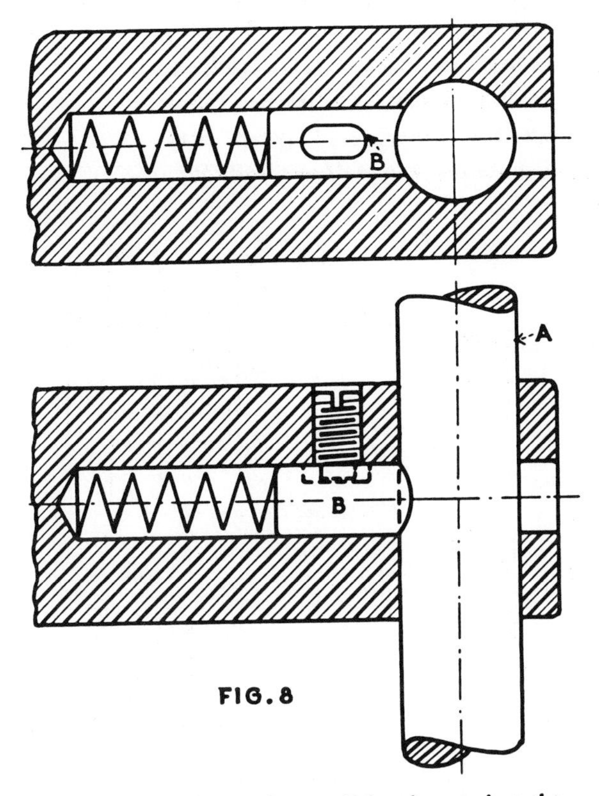


Fig. 8—End of the plunger *B* bearing against the hand lever *A* is concaved and prevented from turning by the dog point setscrew engaging the splined slot. Friction is the only thing that holds the adjustable hand lever *A* in position.

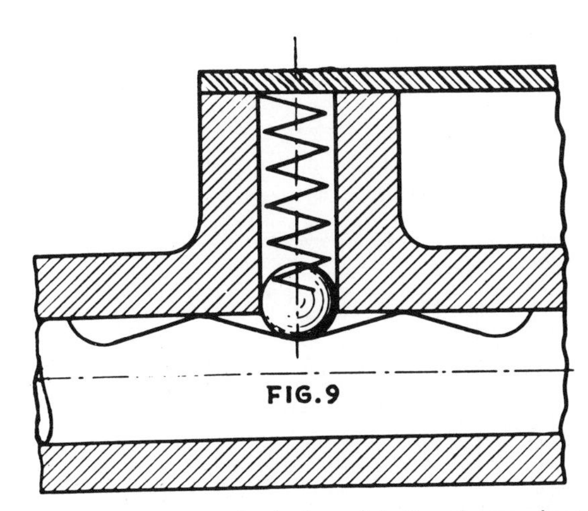


Fig. 9—A spring-backed steel ball makes a cheap but efficient detent, the grooves in the rod having a long, easy riding angle. For economy, rejected or undersized balls can be purchased from manufacturers.

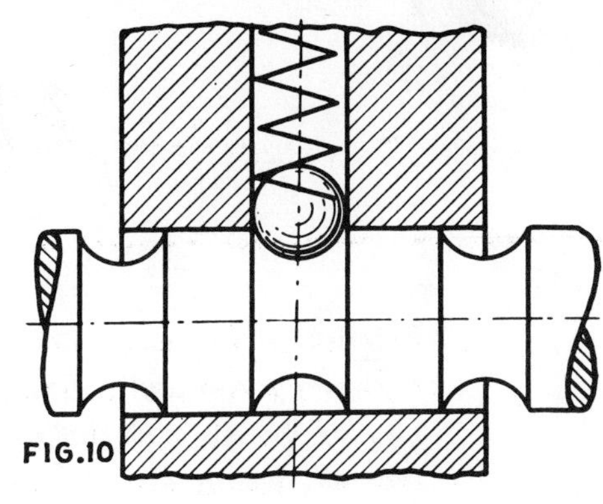


Fig. 10—Another form in which the grooves are cut all around the rod, which is then free to turn to any position.

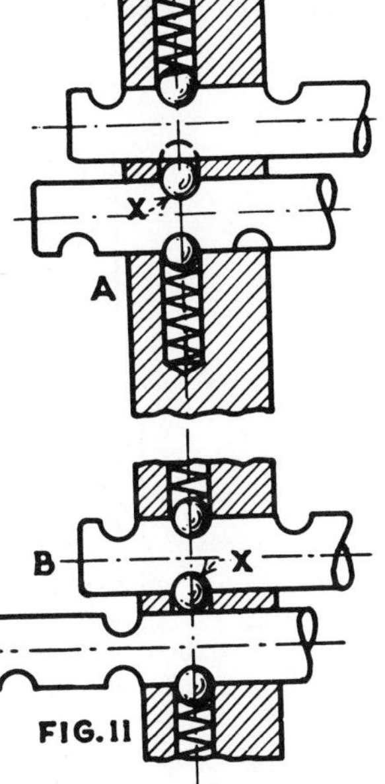


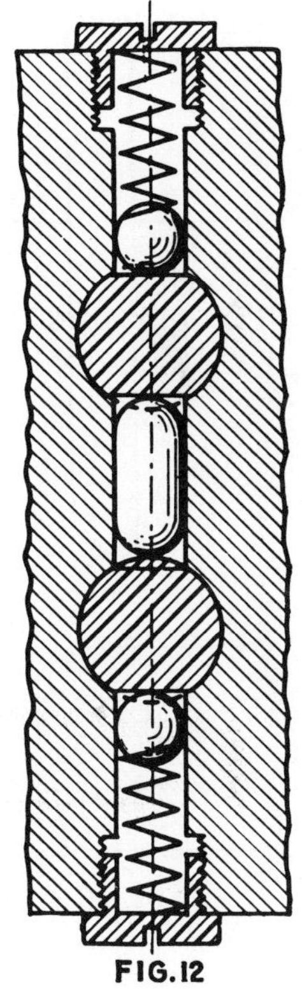


Figs. 11 and 12—Above is shown a double-locking device for gear shift yoke rods. At *A* the neutral position is shown with ball *X* free in the hole. At *B* the lower rod is shifted, forcing ball *X* upwards, retaining the upper rod in a neutral position. The lower rod must also be in neutral position before the upper rod can be moved. To the right is shown a similar design wherein a rod with hemispherical ends is used in place of ball *X*.

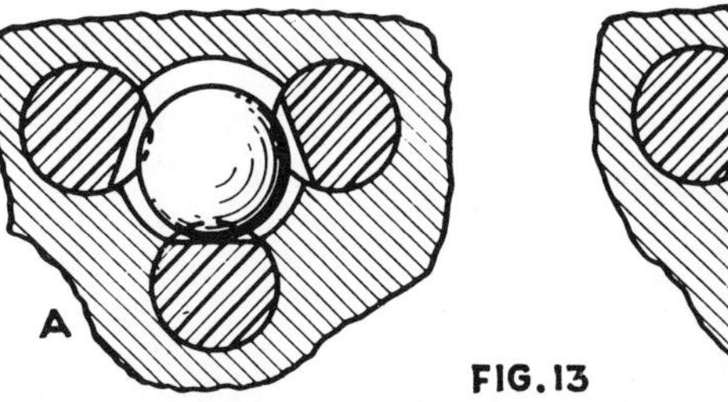


Fig. 13—Without using a spring of any kind, three gear-shifting rods are locked by a large steel ball. At *A*, the neutral position is shown. At *B*, the lower rod has been shifted, forcing the ball upwards, thereby locking the other two rods. The dashed circle shows the position of the ball when the right-hand rod has been shifted.

# FRICTIONAL SUPPORTS

L. KASPER

**Frictional supports permitting relative longitudinal and rotational adjustment between a rod and a clamping member have wide application because of their simplicity of design and the ease and rapidity with which they can be adjusted. Possibilities of design are endless, as indicated by the accompanying group of designs. Illustrations include types having slight and strong resistance to friction, types in which the frictional resistance can be varied to suit conditions, types that have greater resistance in one direction than the other, and types that have positive detents for certain positions. The sketches are self-explanatory.**

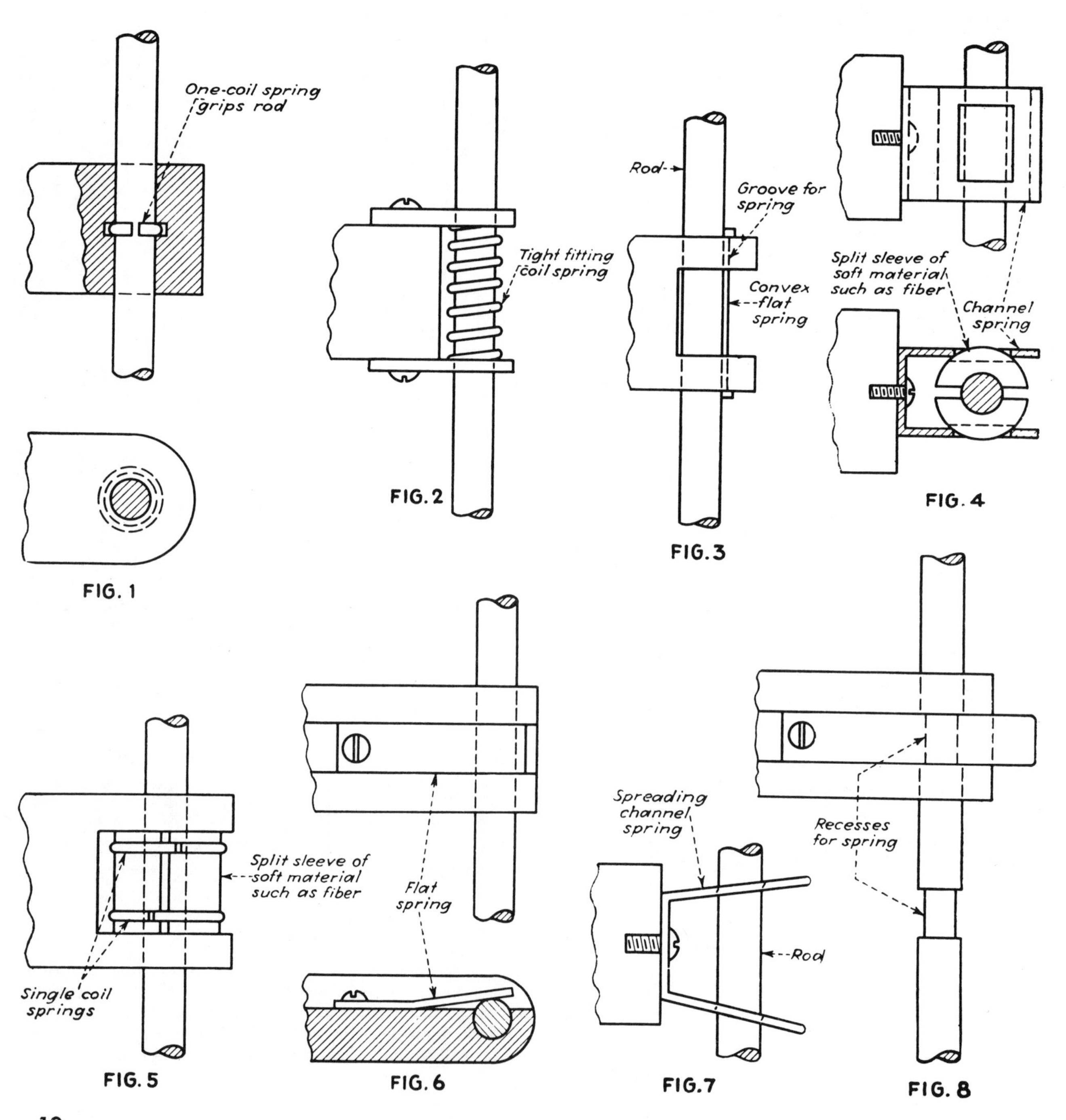

# FOR ADJUSTABLE PARTS

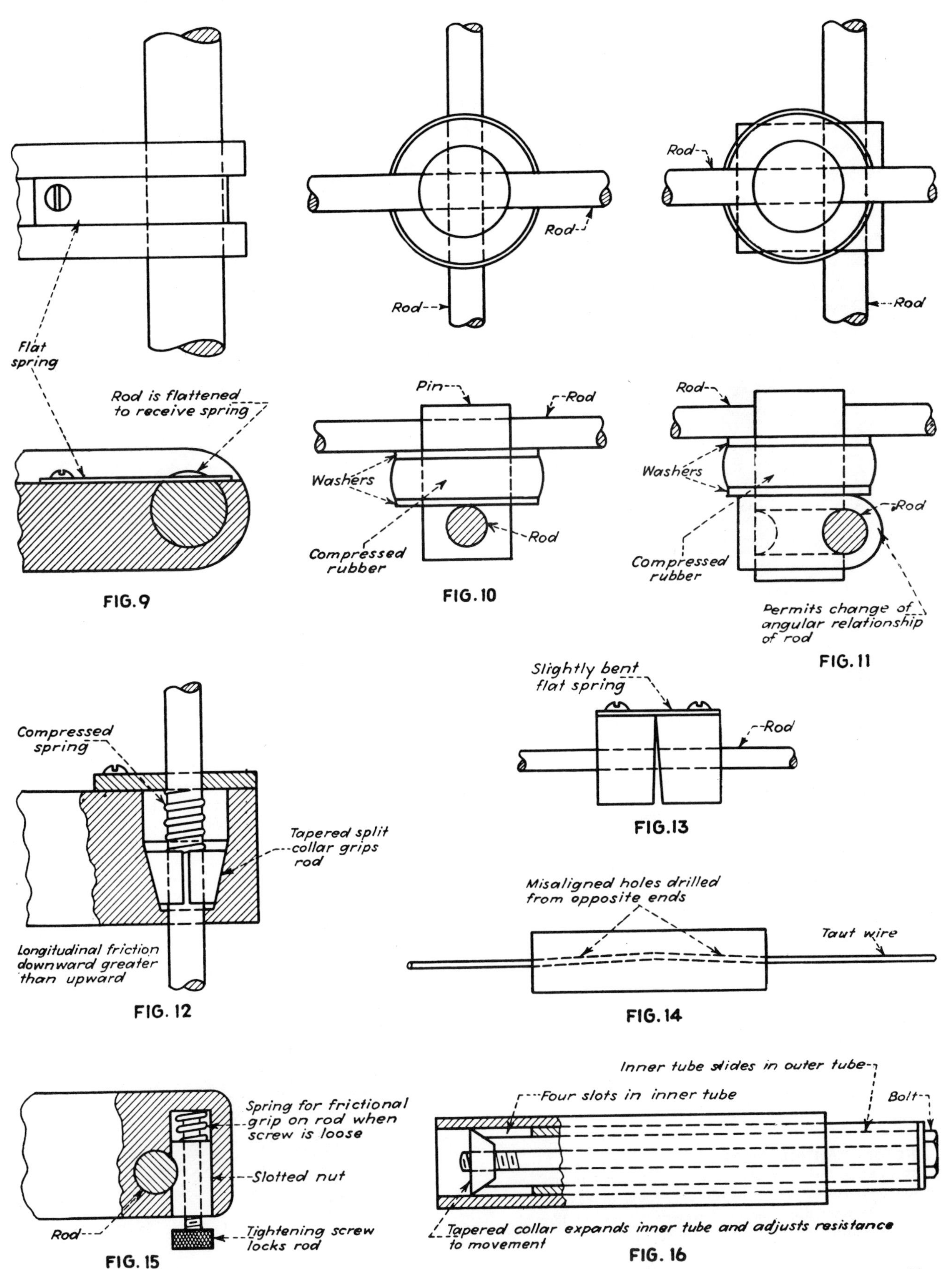

# 7 BASIC HOPPERS

**Hoppers, feeding single parts to an assembly station, speed up many operations. Reviewed here are some devices that may not be familiar to all design engineers.**

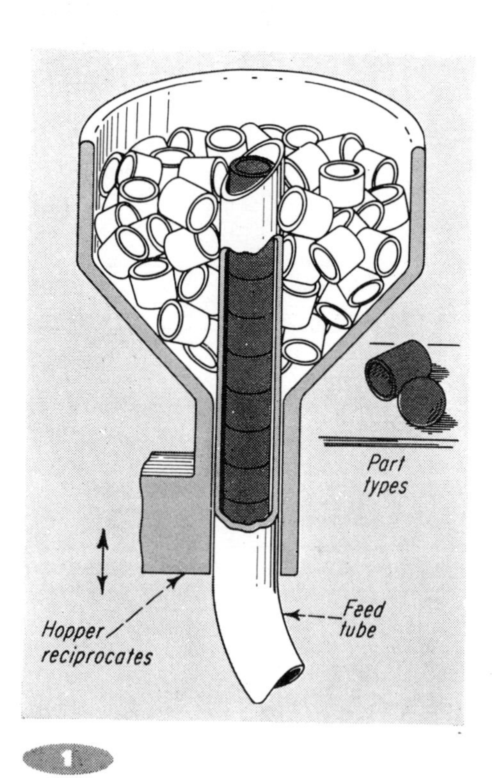


### Reciprocating feed . . .

for spheres or short cylinders is perhaps the simplest feed mechanism. Either the hopper or the tube reciprocates. The hopper must be be kept topped-up with parts unless the tube can be adjusted to the parts level.

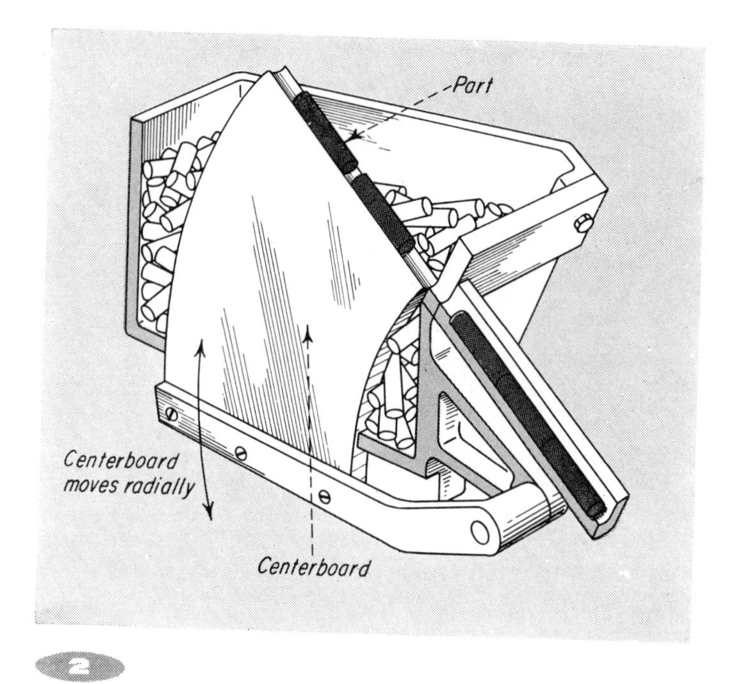


### Centerboard selector . . .

is similar to reciprocating feed. The centerboard top can be milled to various section shapes to pick up moderately complex parts. It works best, however, with cylinders too long to be fed with the reciprocating hopper. Feed can be continuous or as required.

### Rotary centerblades . . .

catch small U-shaped parts effectively if their legs are not too long. Parts must also be resilient enough to resist permanent set from displacement forces as blades cut through pile of parts. Feed is usually. continuous.

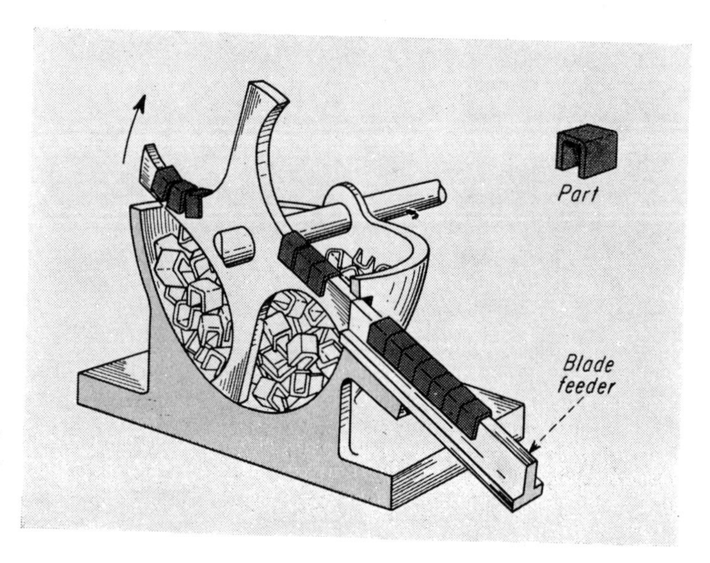

# for PARTS

**PETER C. NOY** *Manufacturing Engineer*
Canadian General Electric Co., Ltd., Barrie, Ont.

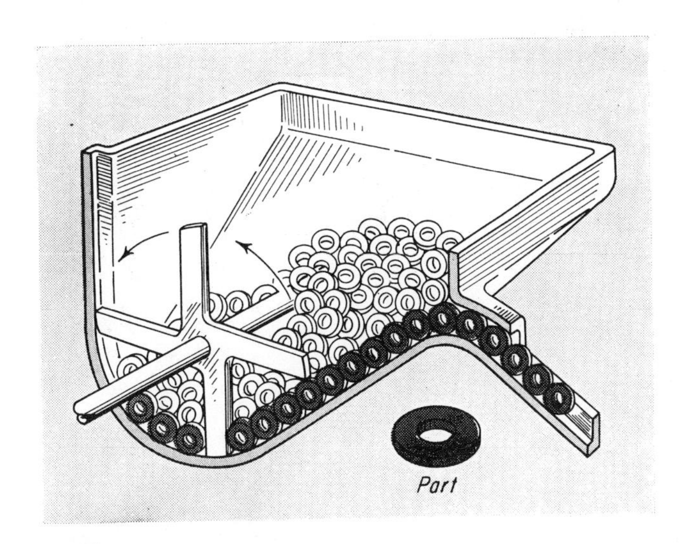


4

### Paddle wheel . . .

is effective for disk-shaped parts if they are stable enough. Thin, weak parts would bend and jam. Such designs must be avoided if possible—especially if automatic assembly methods will be employed. *(See pp. 150, 151.)*

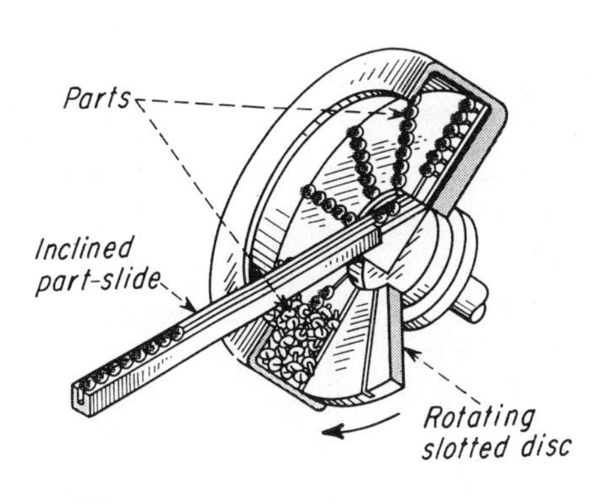


5

### Rotary screw-feed . . .

handles screws, headed pins, shouldered shafts and similar parts. In most hopper feeds, random selection of chance-orientated parts necessitates further machinery if parts must be fed in only one specific position. Here, however, all screws are fed in the same orientation (except for slot position) without separate machinery.

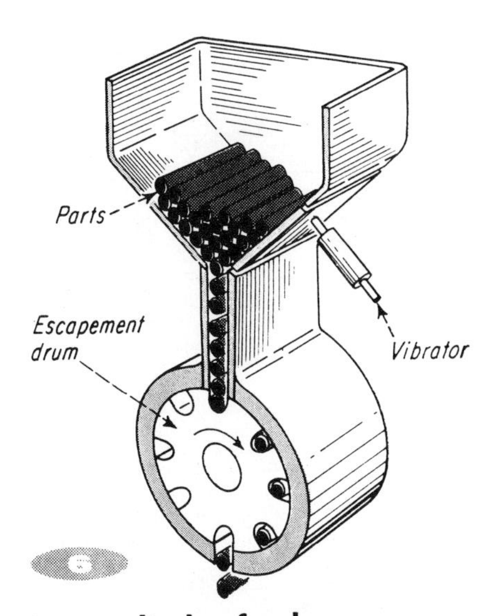


6

### Long-cylinder feeder . . .

is a variation of the first two hoppers. If the cylinders have similar ends, the part can be fed without pre-positioning, thus assisting automatic assembly. *(See pp. 150, 151.)* A cylinder with differently shaped ends requires extra machinery to orientate the part before it can be assembled.

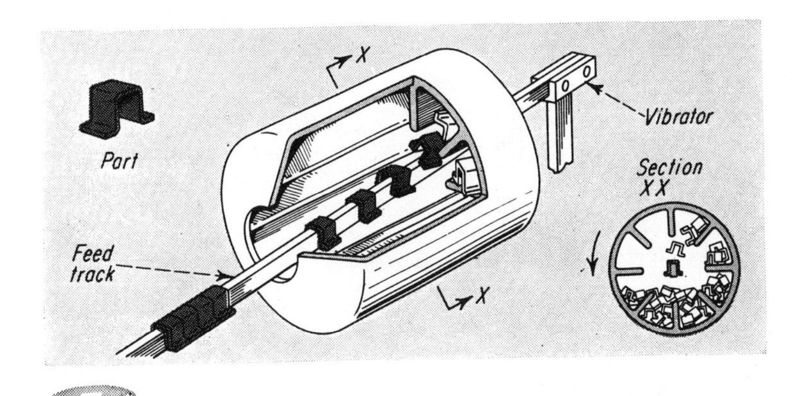


7

### Barrel hopper . . .

is most useful if parts tend to tangle. The parts drop free of the rotating-barrel sides. By chance selection some of them fall onto the vibrating rack and are fed out of the barrel. Parts should be stiff enough to resist excessive bending because the tumbling action can subject parts to relatively severe loads. The tumbling sometimes helps to remove sharp burrs.

# NIBS AND CATCHES FOR LIDS

**WALLACE C. MILLS**

Variety in the design of detents for removable or hinged lids on sheet metal containers such as tool boxes, cabinets, round cans, and boxes of all sizes and shapes has resulted from different service requirements and necessity for low cost production. A common and simple way of making catches is to form small projections or nibs in the metal of box and lid. The illustrations on these two pages show ways of using such projections as well as other ways of making catches.

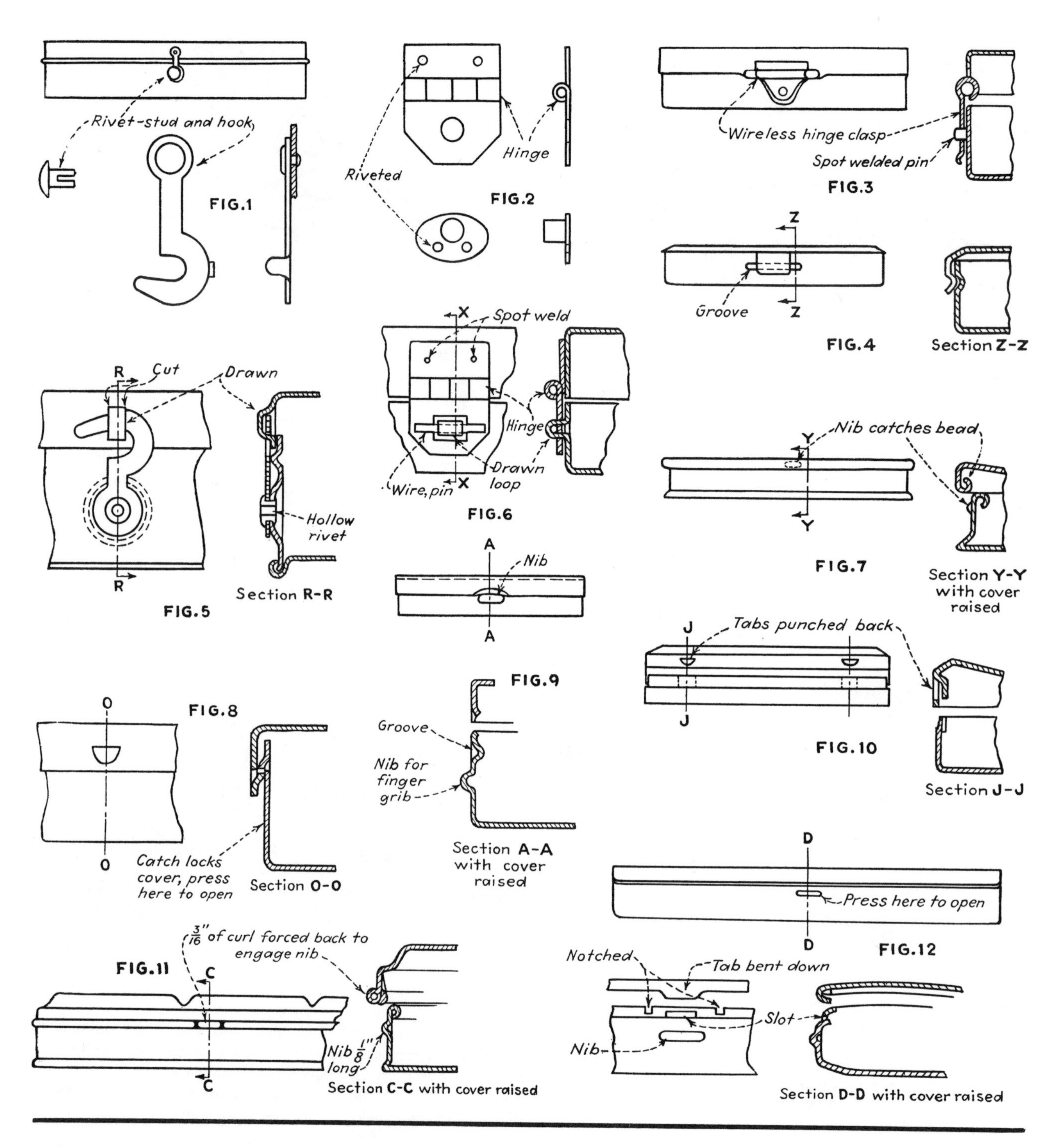

# ON SHEET METAL CONTAINERS

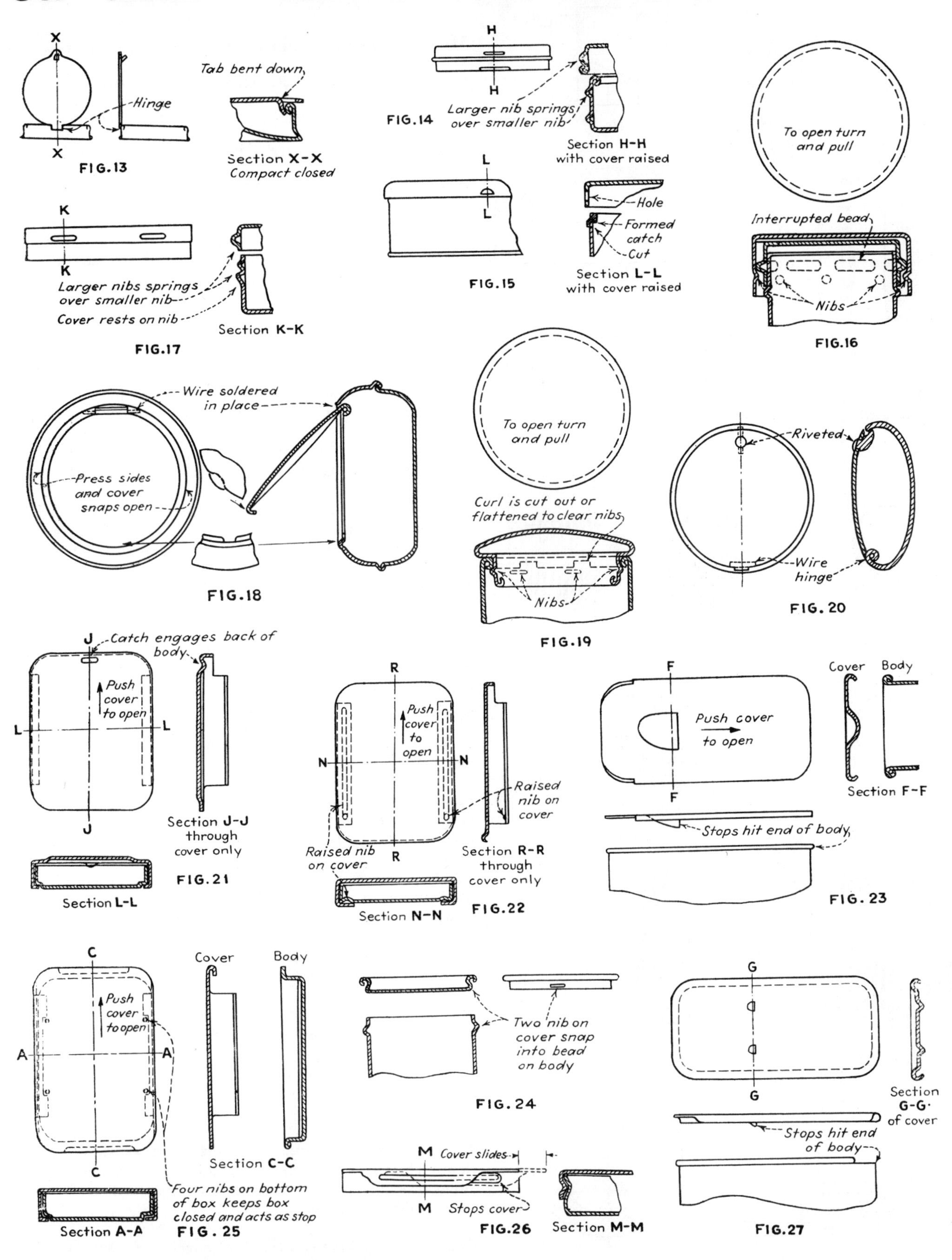

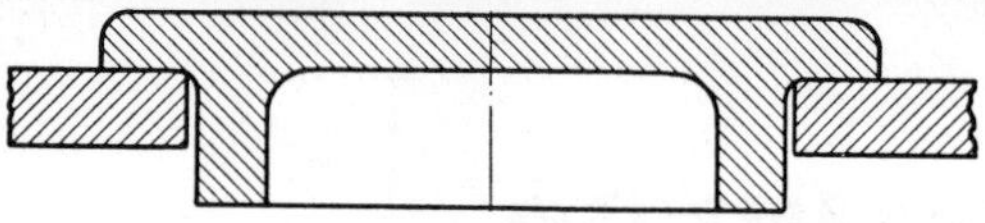

# PLUGS, CAPS

**Many instances occur in the design of machine tools, household appliances, automobiles and other mechanical contrivances, where holes must be covered or filled to keep out foreign materials, retain grease or oil, protect working parts against injury, or to improve appearance. Various methods of capping and plugging such holes are detailed below**

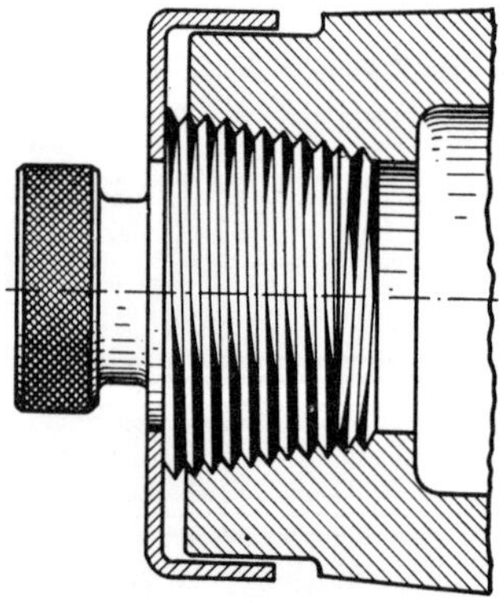

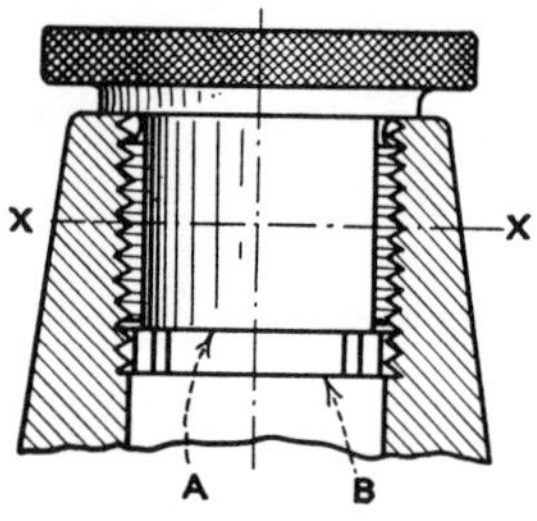

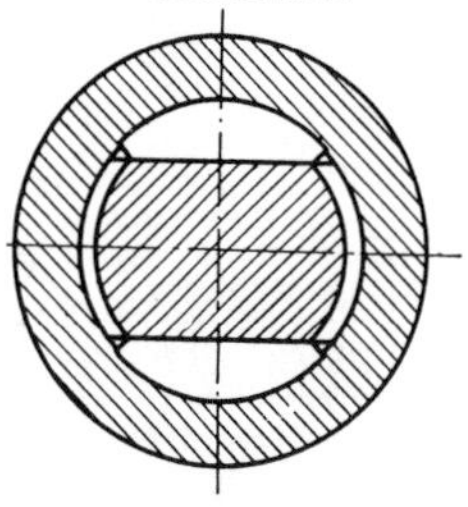

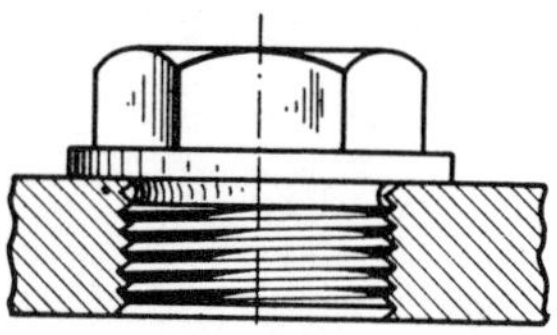

HENRY J. MARTIN

FIG. 1—Cover of sufficient weight to stay in place by gravity. If a flat finished surface is desirable, the cover may be recessed.

FIG. 2—To avoid chips falling through hold-down bolt holes in machine tools, taper-headed plugs can be inserted.

FIG. 3—A plain plug with end chamfered or rounded will enter a hole satisfactorily, but in this instance the plug end is as sheared, and the extra chamfering operation is unnecessary.

FIG. 4—When the hole is large, a plug of cast iron is inserted and held in place by a tangential headless screw.

FIG. 5—The "Welch" expansion plug is used universally. At *A* the original shape is shown. Flattening the center portion, as at *B*, expands it.

FIG. 6—A cupped plug made of sheet metal. Slight pressure at the center will expand plug sufficiently to hold. Similar type of hole cover has a seating flange.

FIG. 7—When oilhole cover must be frequently removed, a brass or steel plug with the end slit as shown will hold firmly when inserted into the hole.

FIG. 8—Common pipe plug of malleable iron or brass is used not only for oil drain holes, but in larger sizes can be used for inspection holes. Socket heads are available, as shown.

FIG. 9—For extra precautions against dust, the plug is pipe-threaded and shouldered for a sheet metal dust cap.

FIG. 10—Hole is finished with straight fine-pitch thread. The plug, having fairly loose fit, is held in place by cotter pin.

FIG. 11—A straight fine-threaded plug with hex. head has integral collar. Boss of casting is finished for proper contact with flange.

FIG. 12—Threaded portion is slabbed off as shown in end view. Threads in hole are also relieved so that 90 deg. turn disengages threads. At *A* is end of plug and at *B* end of relieved threads in hole.

FIG. 13—Plug taking end thrust of member *A*. Plug is slotted

FIG. 1

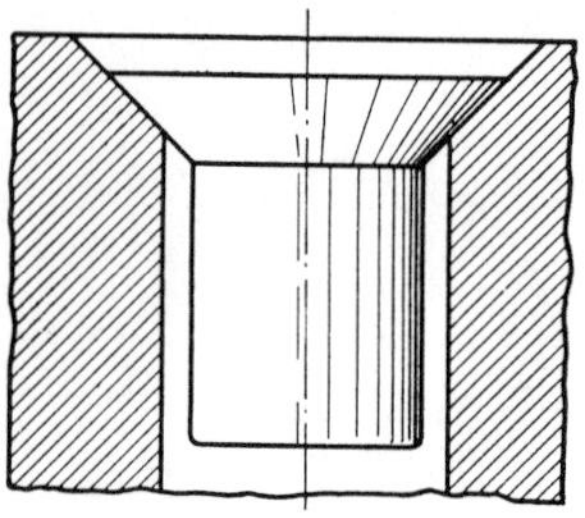

FIG. 2

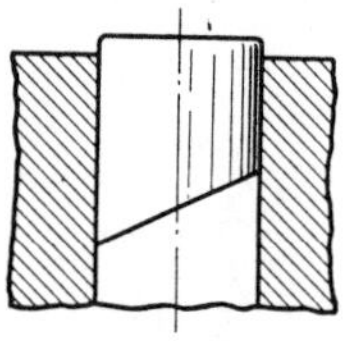

FIG. 3

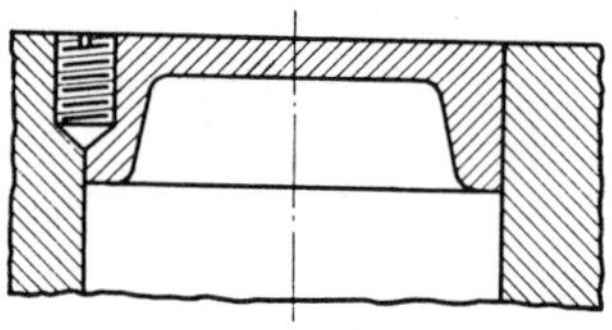

FIG. 4

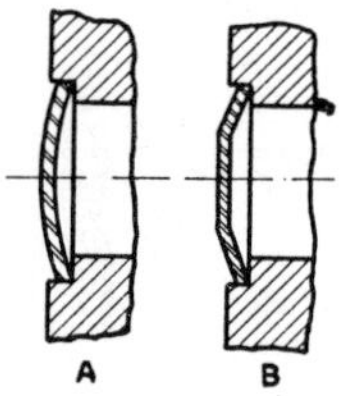

A B

FIG. 5

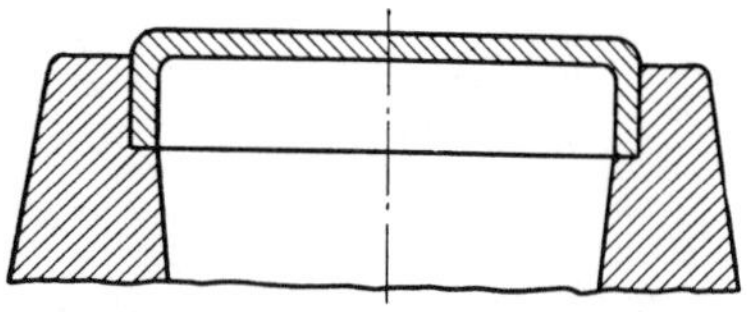

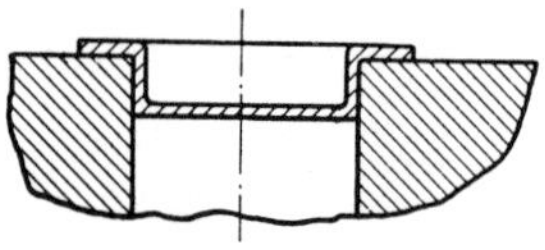

FIG. 6

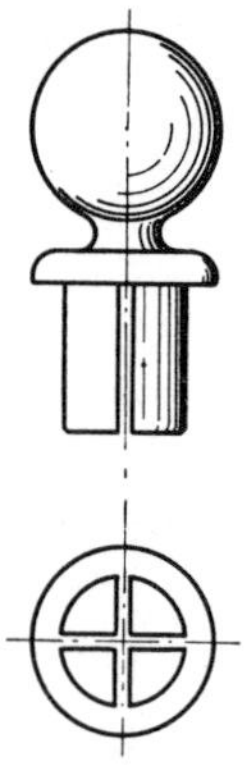

FIG. 7

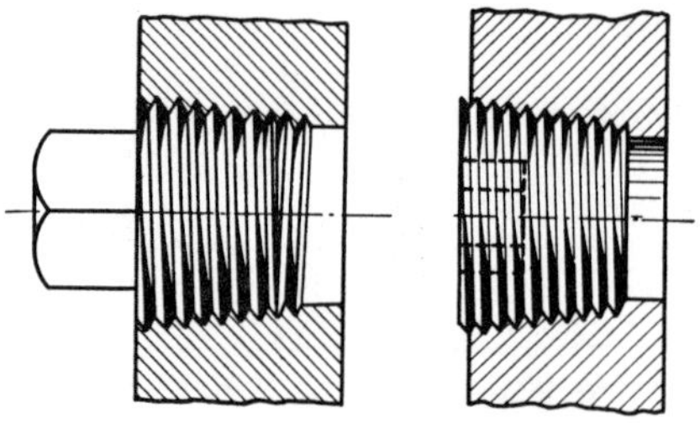

FIG. 8

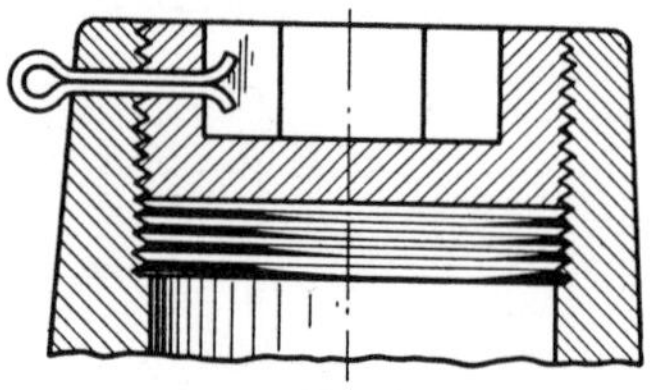

FIG. 9

FIG. 10

X X

A B

Section X-X

FIG. 11

FIG. 12

# AND COVERS

and fitted with tap bolt to expand plug threads in hole when proper adjustment is attained.

Fig. 14—Provisions for an oil retainer are made in the use of an asbestos-copper sealing ring or spark plug gasket. The hex. head of the plug is turned down for a washer effect to contact the gasket properly. These gaskets are procurable in numerous commercial sizes.

Fig. 15—Where an acorn-type of nut is used, sealing is effectively accomplished by use of a soft lead gasket. The casting, if fairly smooth, need not be machined as the lead will flow sufficiently for a proper joint. The integral washer portion of the nut can be larger than the hex. nut as shown or can be made as in Fig. 14, using a smaller lead gasket.

Fig. 16—Large cast iron or sheet metal covers with cork or composition gaskets for sealing. They can be fastened with a centrally-placed screw as illustrated or with a series of smaller screws near the periphery of the cover. In the method shown, the hand screw has a retaining ring so the screw cannot be separated from the cover.

Fig. 17—Top of sheet metal oil filling tube has a flange which is slotted diametrically opposite, as shown on the left of the center line. Into these slots two projections *A* of cap enter. Cap is then given a turn of approximately 135 deg. to lock. Two opposite projections *B* on cap form finger grips.

Fig. 18—Similar arrangement to that of Fig. 17, except that stiff flat spring *A*, 3/8 in. wide is riveted to cap. The spring ends enter the two slots as shown to right of centerline and cover is turned to locking position. The periphery of cap has straight knurl to facilitate turning.

Fig. 19—Another method of capping oil filling tube. Fiber washer is centered by round projection on formed locking piece *A*, both pieces being riveted to auxiliary cover *B*. Cap *C*, which has an enameled outer surface, is spun over the auxiliary cap.

Fig. 20—Oil level gage with hole-sealing washer is made of half-round wire bent in form of handle at upper end. End *A* is sheared off at an angle and is sprung outwards. Spring pressure against side of hole holds it in place.

Fig. 21—Similar method except that round wire, smaller than oil hole, is flattened and portion *A* in center of flat is extruded as shown to bear against side of hole. Four bearing surfaces, 90 deg. apart, keep gage from rattling. Steel-backed felt washer protects hole entrance.

Fig. 22—Die-cast or sheet steel oil hole cover in vertical or angular position is held closed by gravity. Top end is riveted to casting. May be hung at angle of 30 deg. to horizontal if necessary.

Fig. 23—Similar to Fig. 22 except that coiled spring under pivot screw keeps cover in any position desired. Handle *A* makes shifting possible. This type of oil hole cover has an additional advantage over the method illustrated in Fig. 22 as it can be used in horizontal positions.

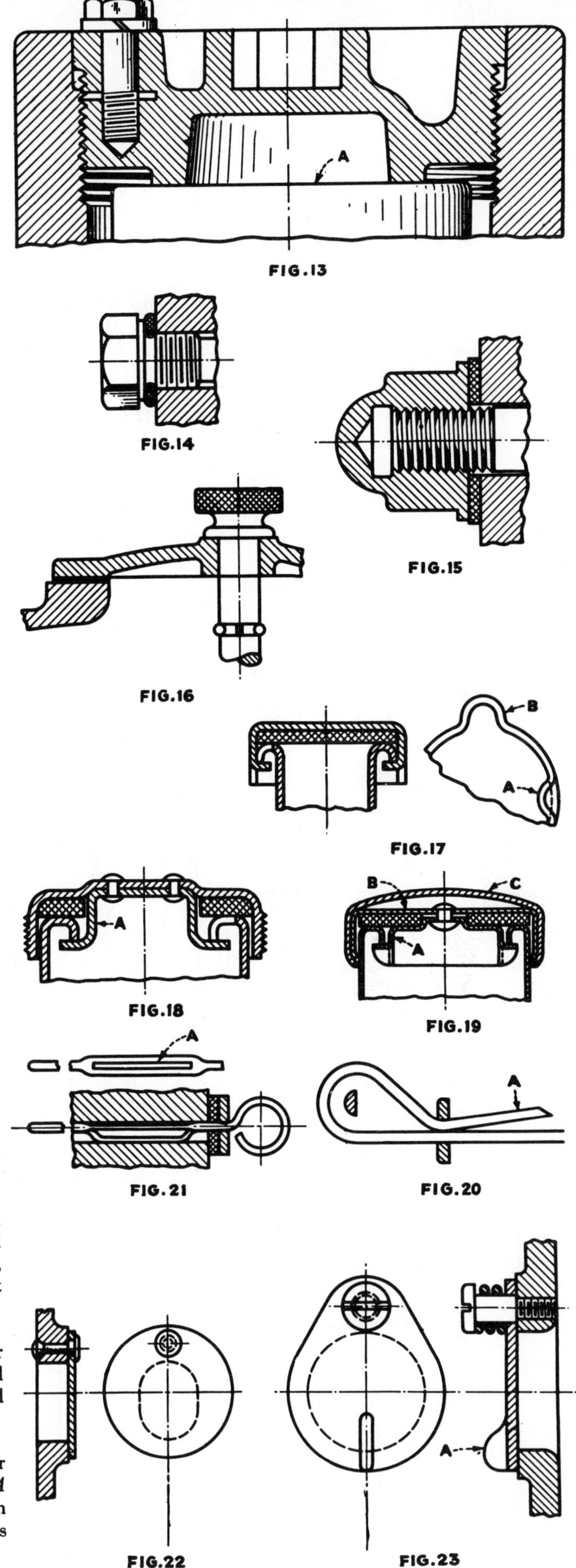

# PLUGS, CAPS AND

Fig. 24

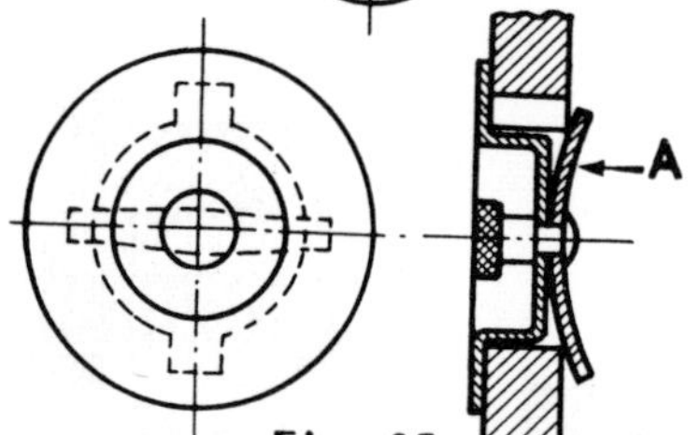

Fig. 25

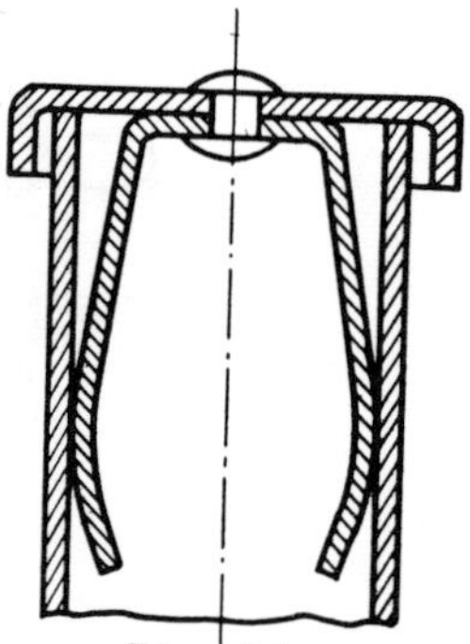

A

Fig. 26

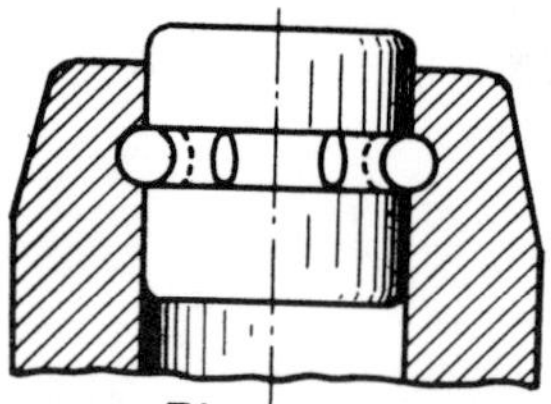

Fig. 27

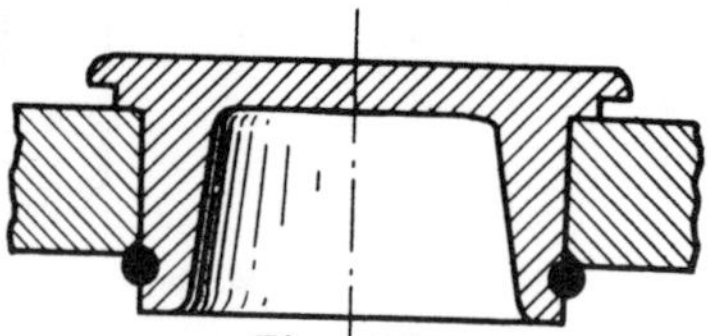

Fig. 28

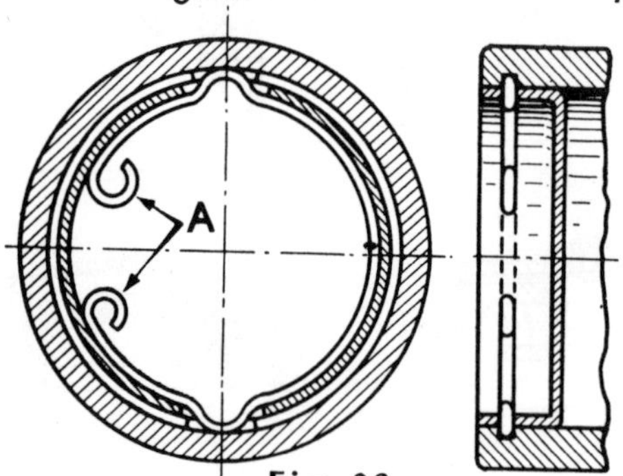

Fig. 29

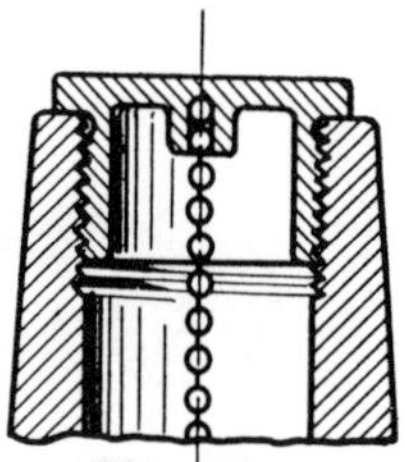

A

Fig. 30

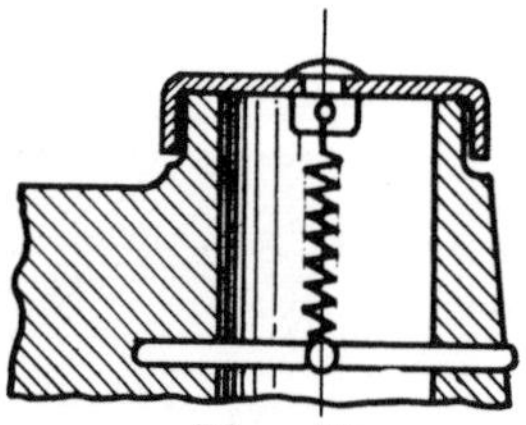

Fig. 31

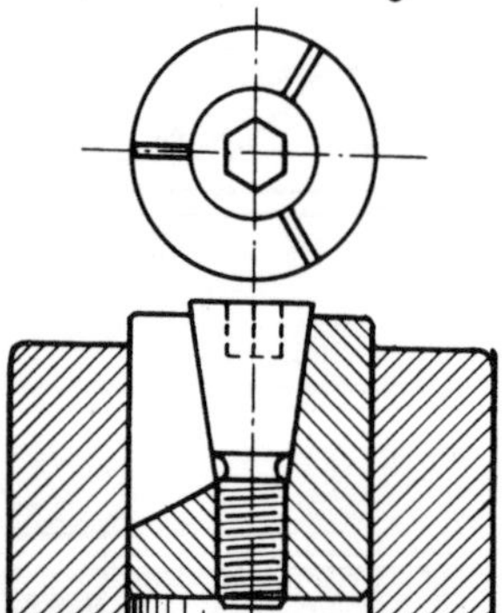

Fig. 32

Fig. 33

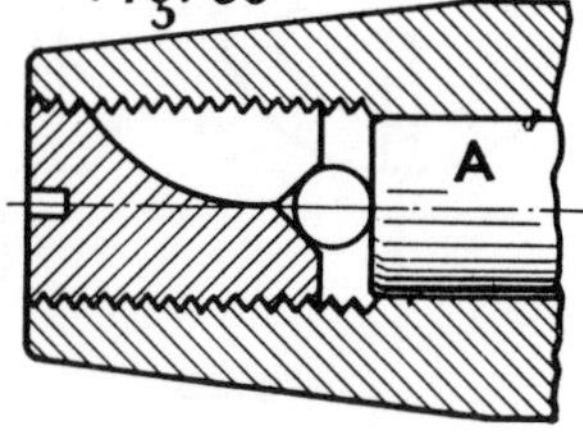

A

Fig. 34

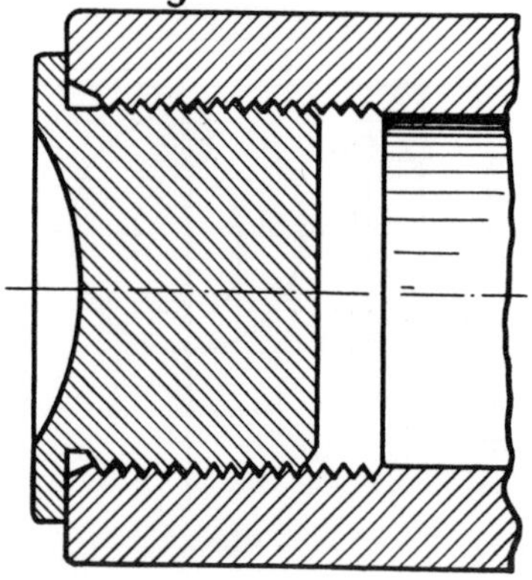

Fig. 24—Heavy cover for observation opening or cleaning hole. End opposite pivot screw is hook-shaped and engages thumb screw for locking.

Fig. 25—Another type of cover for large holes. Flat spring *A* of sufficient stiffness is keyed and riveted to knurled finger knob. Casting is slotted for entrance as shown in end view. Dished cover can be made to fit any shape of hole.

Fig. 26—Sheet steel cover for closing tube or pipe. A long piece of narrow flat stock, not necessarily spring steel, is shaped as shown and riveted to cover. Spring width is formed greater than inside diameter of tube and cams to retaining position when pushed into hole.

Fig. 27—Snap-spring lock for any size of hole plug. Depth of groove in casting is turned approximately 1/3 diameter of wire for disassembly. Groove in plug is sufficiently deep for contraction of snap-ring during assembly or removal of plug. Spring length is approximately 80 per cent on one complete coil.

Fig. 28—Method of retaining cast iron plug in hole. Inside of hole is chamfered into which snap ring partially fits. If ring is strong enough, it tends to creep up the chamfer incline, thereby assuring proper contact between surface and plug.

Fig. 29—Large dished cover held in place by wire snap-ring. Cover is slotted as shown in end view to accommodate the two ears formed into the snap-ring. These projections engage the continuous groove in the casting. To facilitate removal of retaining spring, the ends are looped at *A* for engagement of pliers.

Fig. 30—Screw plug has ball chain attached to center inside boss so plug cannot be misplaced. Cover can be turned without kinking ball chain, the latter being long enough to allow removal of plug. Other end of chain is fastened to inside casting.

Fig. 31—Pin with hole is riveted to sheet metal cover. Loop of helical spring engages hole in spring. Lower spring loop is slid over retaining pin in casting. Upon removal of cap, lower spring loop slides along pin towards wall of hole.

Fig. 32—The smooth plug with either three or four slots has the outer end taper-reamed for a special heat-treated socket head screw. Short plugs that cannot be slotted to sufficient depth should have a good fit in hole before expansion takes place.

Fig. 33—Straight fine-threaded plug is slotted as in Fig. 35. The inner end is taper spotted and acts as ball seat. The steel ball bears against member *A*, expanding plug in hole threads.

Fig. 34—Threaded plug with low flanged head for plugging end of spindle to prevent coolant drip. Flange is circular slotted so that head may be kept as short as possible.

Fig. 35—Unthreaded plug to take place of that shown in Fig. 37. Saves tapping. Plug is cross-drilled for steel pin, the end of which is beveled to engage cone-pointed headless set screw.

# COVERS (cont.)

Fig. 36—Another plug for same installation as in Fig. 38. Cross hole is not quite drilled through plug, and forms seat for ball. Pin-retained spring pushes ball into turned groove in spindle.

Fig. 37—Plunger pin is used instead of ball. Plug is drilled and reamed but not completely through. Plunger end is turned down and retained by inserted washer. Spindle has drill spot for engagement of plunger end. Hole in spindle is plugged at *A*.

Fig. 38—The flanged plug is tapped off center for a cone pointed headless set screw which pushes the locking ball into the groove in the casting. When removing plug, screw is backed out partially, the ball falling into the drilled pocket. Should be assembled with ball uppermost to prevent loss of ball when plug is removed.

Fig. 39—Two hemispherically headed drill rod pins with opposite ends beveled for screw engagement. Pins fit in V-groove in casting. When screw is backed off, pins will cam in towards center for removing plug from hole.

Fig. 40—Oil level sight gage in which steel plate *A* has two drilled holes for oil and air circulation. Two cork gaskets, one each side of glass disk, are retained by flanged cover.

Fig. 41—Similar arrangement as Fig. 43, except that circulation holes are drilled directly in casting wall, thereby eliminating one piece. Retaining collar is set into counterbored hole for better appearance.

Fig. 42—Oil drip gage inner member has oil-collecting channel *A* which catches oil running down inside wall of housing. Oil runs through inclined tube past glass window. Cork gaskets, glass and channel member are retained by collar fastened to outside of housing.

Fig. 43—Commercial type of oil sight gage is integral and purchased ready to assemble. The brass housing with hex.-shaped periphery to facilitate assembly is counterbored for two cork gaskets and glass disk. A metal washer *A* rests against outer cork gasket. The brass housing is then spun over onto this metal washer, making the assembly oil-tight.

Fig. 44—When a mechanism projects past the outer wall of a machine part, a tubular piece with closed end can be used. At *A*, the housing is expanded as shown in the end view, entering cross slots in hole and then turned 90 deg. to assembled position. Collar *B* is completely circular and acts as a locating flange.

Fig. 45—A similar type of tubular housing except that both flanges *A* and *B* are completely circular. The open end is slotted, and made of springy material, will contract sufficiently when entering the hole, and expand to a locking position when flange *A* reaches inner wall of housing.

Fig. 46—Sometimes a spring must extend through the wall of a casting. To house the spring, a commercial flanged tube with closed end is used. Assembly with flange against inner wall of housing is kept in place by resulting spring pressure.

Fig. 47—A copper bellows can be used when there is a variation of pressure from within. The threaded portion is spun over into the groove between the first two convolutions. When assembled, the threaded tube is screwed tightly against the first convolution to make a leak-proof joint. This bellows is a commercial product and has other possibilities as a hole covering.

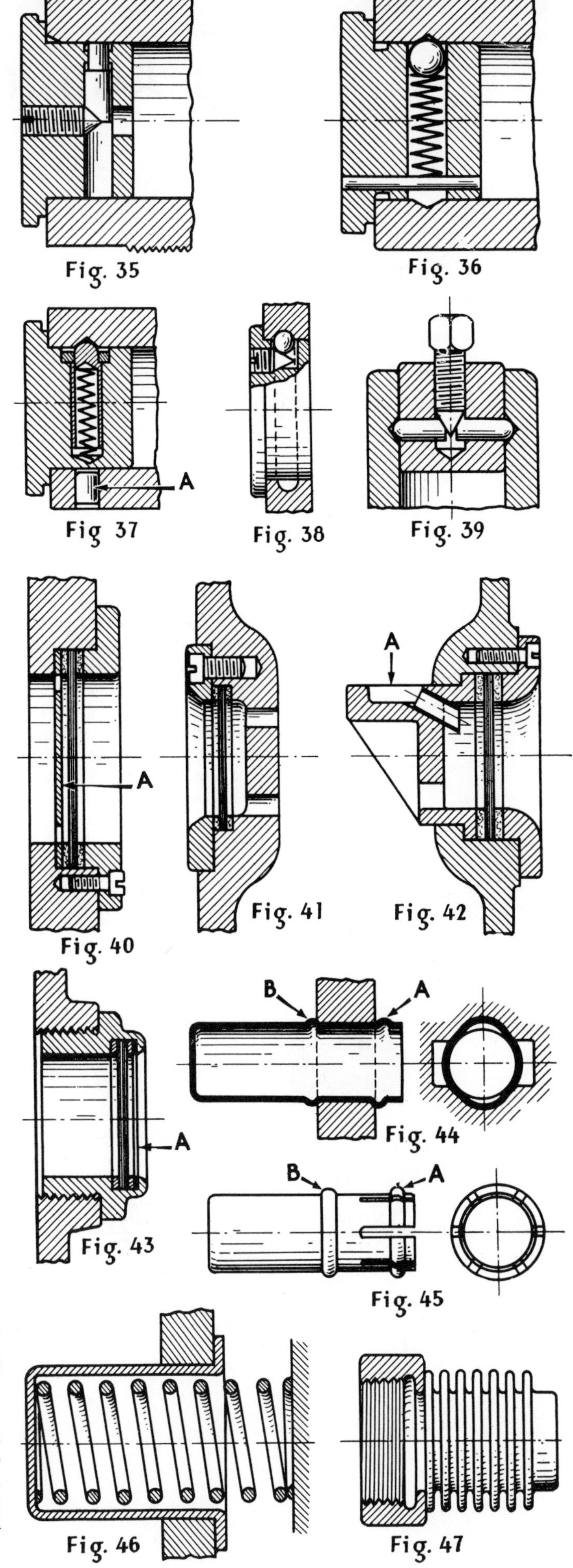

# 6 Applications of Ultrasonic

Douglas C. Greenwood

**They will:**

- **clean**
- **drill**
- **solder**
- **weld**
- **test and measure**

Certain materials undergo a reversible change when subjected to magnetic or electric fields, and thus can serve as transducers for high-frequency vibrations. Cobalt and nickel are magnetostrictive · barium-titanate ceramic is electrostrictive.

Power is supplied to the transducer from an electronic

## 1 Ultrasonic cleaning . . .

of objects immersed in a treatment vessel results from violent cavitation throughout the liquid. Frequency is about 30,000 cps for the type of equipment illustrated. Optimum relationship exists between surface tension, vapor pressure and temperature of the liquid. Water plus a mild detergent is often used, but other liquids such as cyclohexane or trichlor-ethylene with lower viscosity are sometimes more suitable, especially for greasy articles. The liquid is usually heated to about 120 to 140 F for more effective cleaning. Small units such as this require about 450-w input to the ultrasonic generator. Water and oil cooling prevent excessive heating of the transducer.

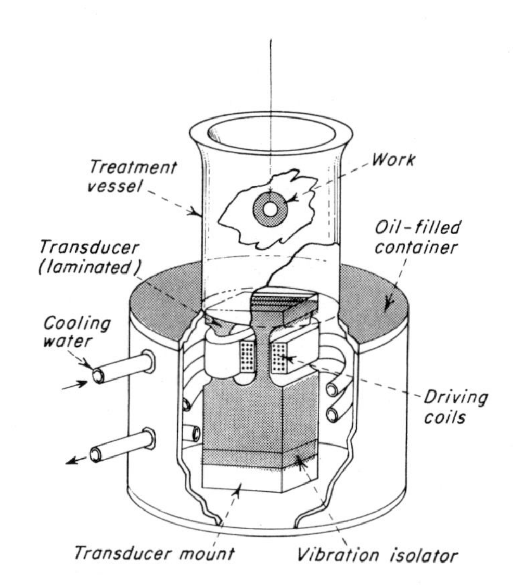


## 4 Ultrasonic welding . . .

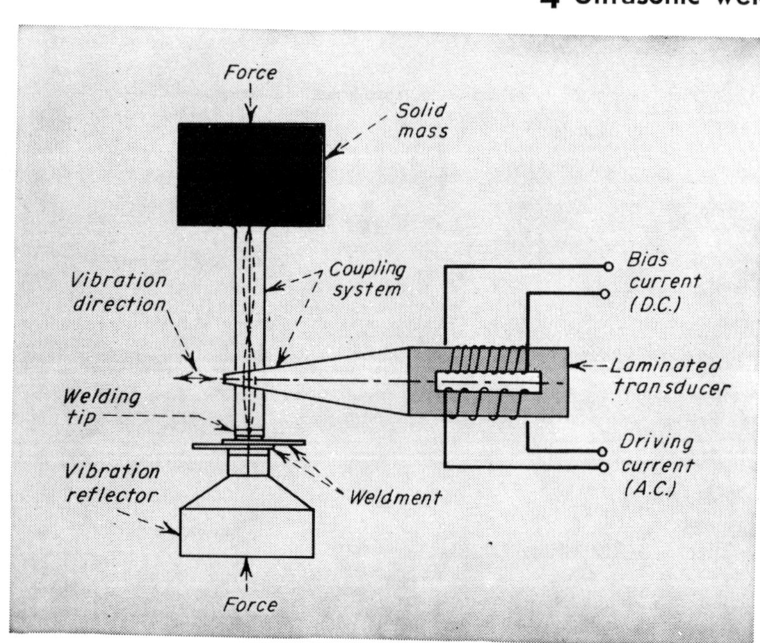


of similar or dissimilar metals gives metallurgical bond with low deformation, low clamping load, and no fusion. Pieces to be joined are clamped between two welding members. Vibratory energy briefly introduced produces a solid-state bond; the phenomenon occurs without high pressures or temperatures at the weld zone. Thicknesses can be 0.00015 to 0.040 in.

## 5 In thickness gage . . .

transducer is applied to test piece and excited at varying frequencies, and an oscilloscope indicates amplitude of vibrations. Resonant frequency of test piece varies with thickness, and when the exciting frequency passes through resonance, the oscilloscope trace shows a peak which corresponds to a previously calibrated thickness. For curved test pieces such as pipe or bearing sleeves, curved crystals are used in searching unit. Accuracy of method is within 1%.

oscillator, usually at frequencies from 20,000 to 60,000 cps for magnetostrictive transducers, and up to 500,000 cps for the electrostrictive type. The transducer is designed to be in resonance with the oscillator at such frequencies—which are in the ultrasonic range (above the frequencies of audible sound).

**2 Vibrating tool . . .** can drill various materials such as hardened steel and glass. Hole shape will be the same as tool section and end face, hence intricate holes or cavities can be sunk quickly. Velocity transformer and detachable toolholder are specially shaped to increase amplitude of transducer vibrations. Cutting tool vibrates at about 20,000 cps on the axis of cone. Abrasive fed between tool and work grinds corresponding outline of tool in work area at about 0.01-0.30 in. per min depending on particle size, material, tool area and pressure. Power for ultrasonic drills varies from 50 w for small units to 2 kw for larger machines.

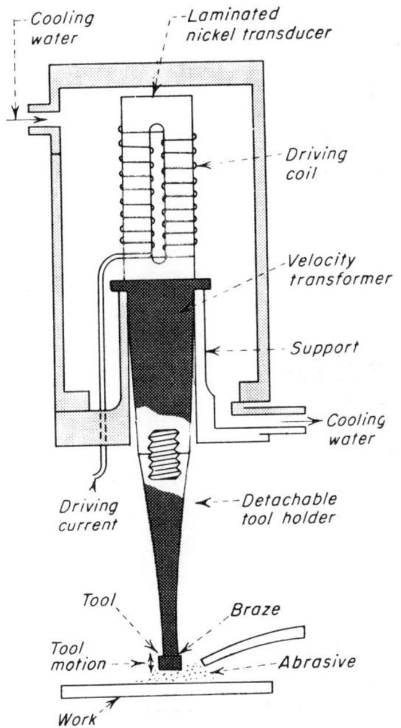
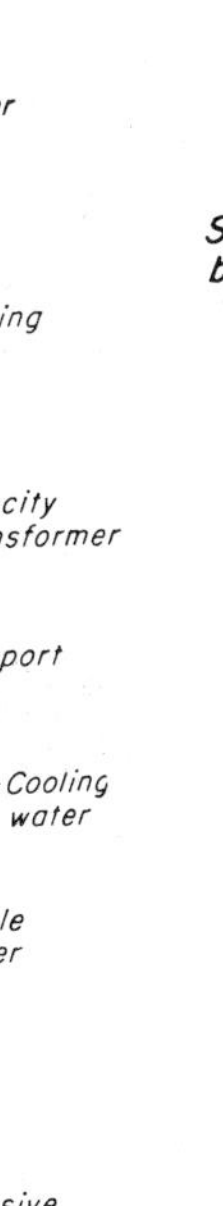


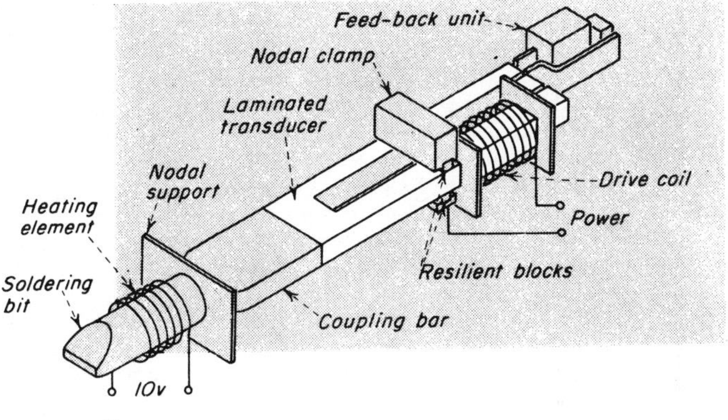


**3 Aluminum soldering . . .** with ultrasonic vibrator can be done without flux. Oxide scale is loosened by cavitation. Soldering bit may be heated directly, or heat may be supplied to work by a hot plate or other external source. Resonance can be maintained by capacitance feedback at end of transducer, but this method, although inexpensive, is not considered efficient by some authorities. The system oscillates at 20,000 to 60,000 cps with an input of 50 w or more depending on bit size. Bits are interchangeable.

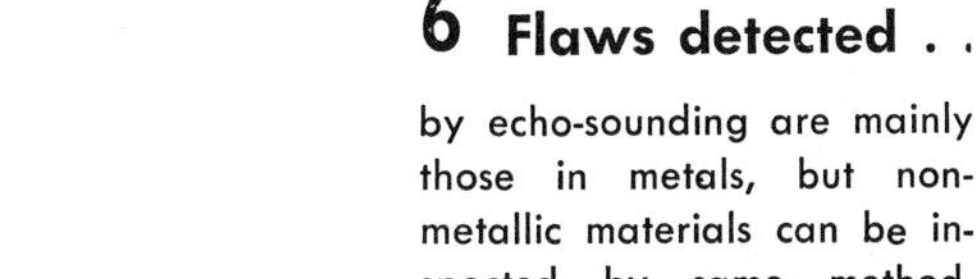


**6 Flaws detected . .** by echo-sounding are mainly those in metals, but nonmetallic materials can be inspected by same method. One vibrating crystal transmits the wave; the other converts the reflected vibrations into electric impulses. Good acoustic contact is essential between transducers and test-piece surface. Film of oil or glycerine between surfaces is simple method for providing good contact. Surface roughness should be less than 125 micro-inches. As only a small volume can be inspected at one time, examination tends to be lengthy for large billets or castings.

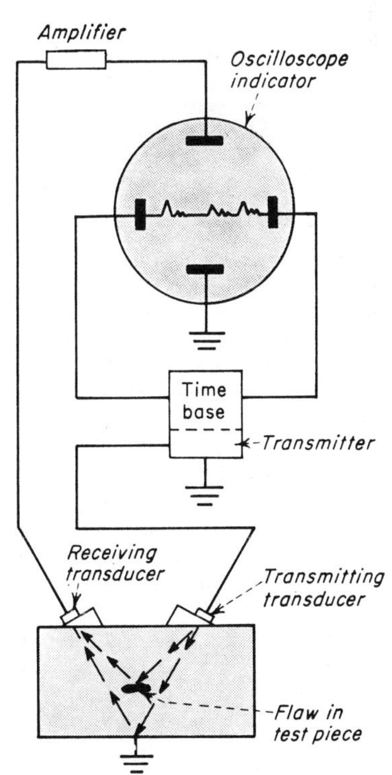

# Selecting Vibration Isolators

RUSSELL LOWE, ASS'T CHIEF ENGR.,
BARRY CONTROLS INCORPORATED

## DESIGN PROCEDURE

1. Locate approximate center of gravity of equipment to be isolated.
2. Determine location of the isolators on the equipment (see illustrations). For optimum isolation and prevention of coupled resonance, isolators should be located in a plane passing through the center of gravity of the equipment. A further refinement is to make distances $x$ and $y$ equal to the radius of gyration about their respective axes.
3. Measure or compute static load $w$ on each isolator position.
4. Measure the disturbing frequency $f$ (usually equal to operating speed of machine).
5. Determine the desired vertical natural frequency $f_n$ of isolated system. The ratio $f/f_n$ should be as large as possible with a minimum value of 2.5. Since disturbing frequencies are usually 1,800 rpm (30 cps) or above most isolators are designed to produce a natural frequency of 8-10 cps at rated loads.

The natural frequency can be computed from the equation:

$$f_n = \frac{1}{2\pi}\sqrt{\frac{Kg}{w}}$$

where $K$ is the spring rate of isolator at load $w$, and $g$ is the acceleration of gravity.

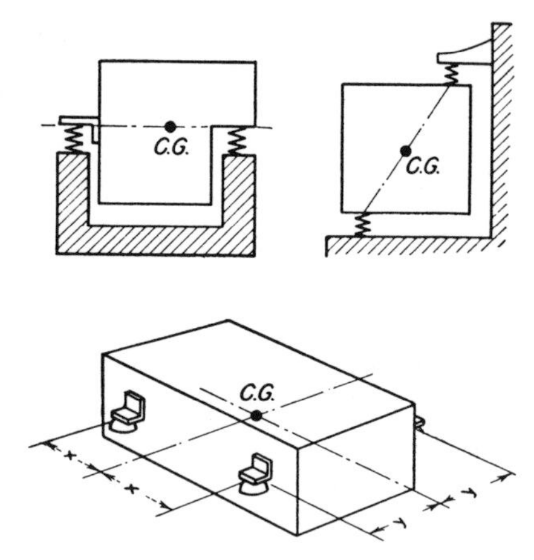


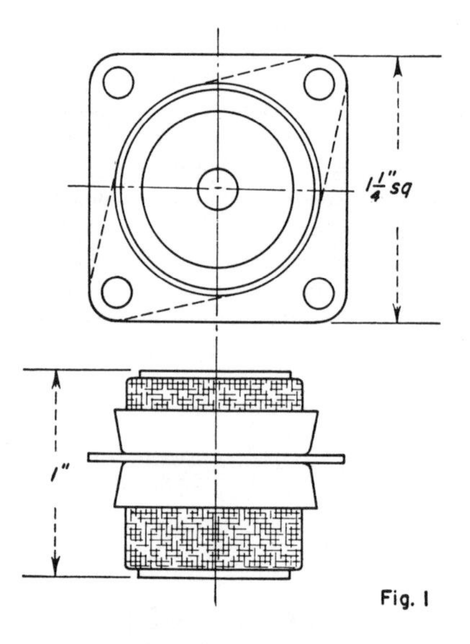


Fig. 1

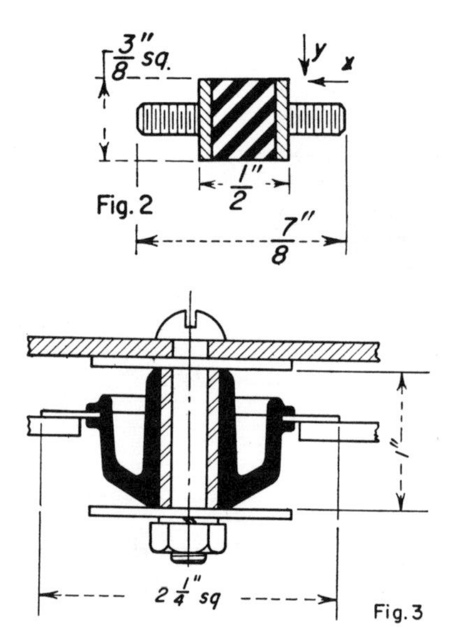


Fig. 3

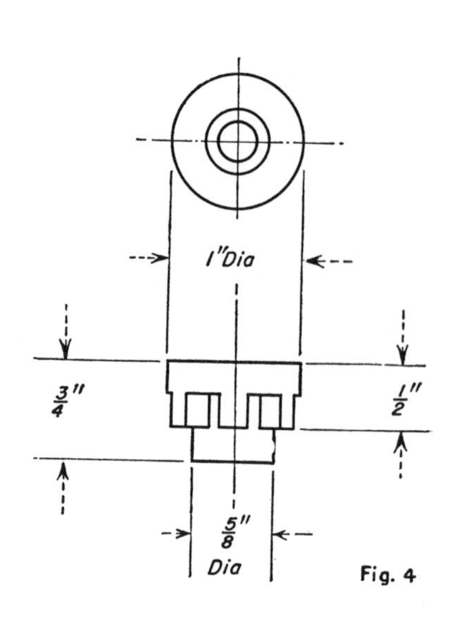


Fig. 4

FIG. 1—Cushion of woven wire acts as a resilient load-carrying spring and has good damping characteristics. Particularly adaptable for the mounting of sensitive electronic equipment in cramped spaces. Load range: ¾ to 2½ lb; effective isolation at disturbing frequencies of 40 cps or higher. *Robinson Aviation Inc.*

FIG. 2—Lightweight motors, instruments and controls can be easily installed with isolators composed of stubs bonded to rubber. Rubber is in compression in $x$-direction and in shear in $y$-direction. Load range in $y$-direction: up to 4½ lb; $f_n$=8 cps at maximum load. *Barry Controls Incorporated.*

FIG. 3—Equal stiffness in all directions is obtained by rubber mount strained in shear when loaded in any direction. Used for isolation of instruments subjected to significant disturbing frequencies in more than one direction. Load range: up to 45 lb; $f_n$ = 10 cps (approx.). *Lord Manufacturing Co.*

FIG. 4—Grommet-type small isolators used on fans, phonograph turntables, instruments. Inexpensive and easy to install. Material: neoprene; durometer hardness influences its spring constant. Available in several degrees of hardness. Load range: up to 120 lb. *The MB Manufacturing Co., Inc.*

FIG. 5—For isolation of lightweight instruments and electronic equipment especially in aircraft. All-metal construction permits operation under high temperatures. Damping is obtained by a knitted stainless steel wire core. Load range: up to 3 lb; $f_n$ = 8 cps at maximum load. *Barry Controls Incorporated.*

FIG. 6—Outer cushions of knitted monel wire restrict de-

**Five-point design procedure for the general case of disturbing amplitudes in the vertical direction. Eleven typical isolators covering loads from ounces to tons and employing woven wire, rubber in shear and compression, conical springs with air damping, and combinations of wire mesh and springs.**

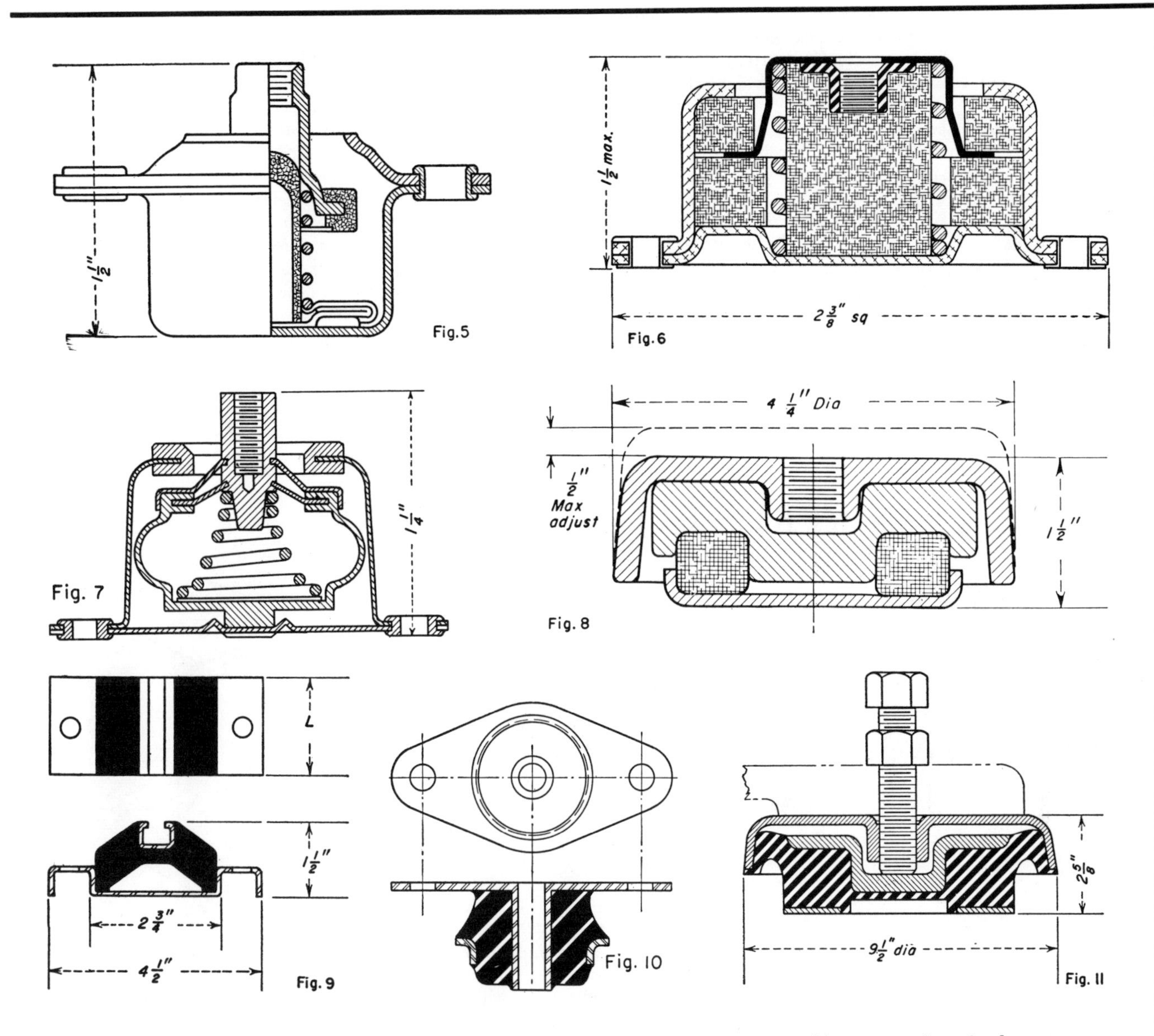


flection while stainless steel spring and inner cushion give non-linear spring characteristics. Used on lightweight machine tools and laboratory equipment. Load range: up to 130 lb. *Robinson Aviation Inc.*

FIG. 7—Conical spring provides necessary resilience while air damping is obtained by air forced through small orifices in bellows during vibration of isolator. For light airborne electronic instruments. Load range: up to 35 lb; $f_n = 8$ cps (approx.). *Barry Controls Incorporated.*

FIG. 8—Industrial machinery mounting has adjustment for leveling and alignment. Uses a stainless steel wire mesh without springs. Applications are in machine tools the operating speed of which is 1,750 rpm or higher or 350 strokes per minute or less. Lagging to the floor is unnecessary. Load capacity: up to 4,000 lb. *Robinson Aviation Inc.*

FIG. 9—Channel-type rubber mounts have load range up to 400 lb depending upon length of channel which varies from 1 to 8 in. Typical applications are light stationary engines and machine tools, compressors and commercial refrigerators. *Lord Manufacturing Co.*

FIG. 10—Vibration mount, designed to stress rubber in both shear and compression when loaded, has almost equal spring rate in all directions. Load range: up to 1,000 lb. Used on combustion engines and heavy electronic equipment. *The MB Manufacturing Co., Inc.*

FIG. 11—Heavy-duty mount for machine tools has load range up to 13,000 lb and can be adjusted for leveling. Does not require lagging to floor even when used under heavy presses. Machines can be easily relocated for optimum work flow; $f_n = 10$ cps at max load. *Barry Controls Inc.*

# Types of Mounts for

TRANSMISSION OF VIBRATION from a machine or motor to the supporting structure can be reduced by using special mountings. Vibrations may be caused by an unbalanced rotor or reciprocating elements of a machine. By flexibly supporting the vibrating member, the disturbance can be greatly absorbed by the resilient mounting.

For most effective vibration isolation, the mounting should be very soft so that its natural frequency is low in comparison to the frequency of the disturbance. As a rule, the disturbing frequency should always be greater than 2.5 times the natural frequency of the mounted system. For *simple linear vibrations,* the natural frequency of the supported mass $f_n$ can be determined by the static deflection of the springs or mount. This is expressed by:

$$f_n = 188 \div \sqrt{d} \text{ (in cpm)}$$

There are many commercial mounts available. Most of these devices utilize rubber for resiliency; this rubber is most often stressed in shear to obtain larger static deflection for a given thickness. Where greater load capacity (per unit volume) is required, rubber in compression is used.

An unrestrained body mounted on four resilient mounts, has six modes of vibration and consequently six natural frequencies.

For single-degree translational motion of a body which is supported on more than one mount, the stiffness of each unit should be in proportion to the weight supported. Besides motion in the vertical plane, the natural frequency in the horizontal plane should be also calculated. For the lowest natural frequency of the system, mountings should be loaded close to the maximum rating given.

In applying vibration mounts, three factors are important: (1) keep mounts far apart for better stability; (2) mounts should be in a plane which passes through the center of gravity of the suspended mass; (3) use of hold-down screws or snubbers to limit large motions.

## STRIP-TYPE ISOLATORS

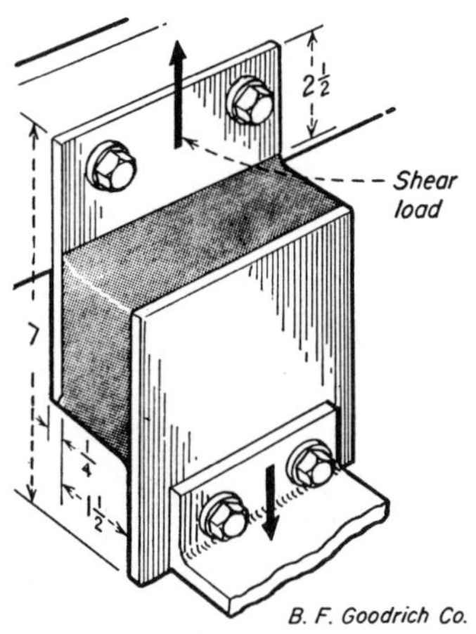


Fig. 1—Rated at 80 lb/linear in. with a shear deflection of 1 in., this type has a compression rating of 250 lb/linear in. at 5/16 in. deflection. Requires drilling and cutting to proper length.

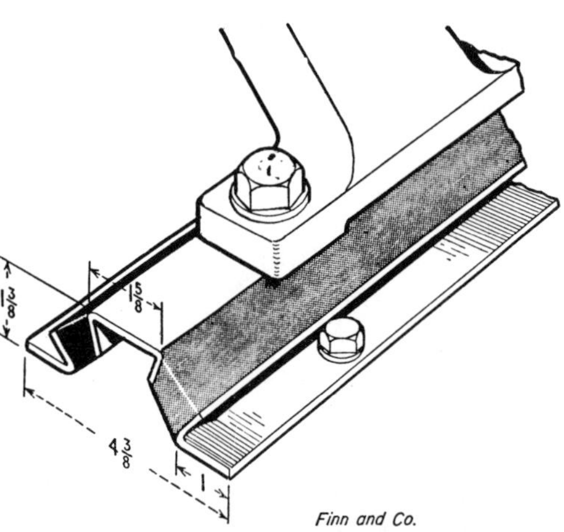


Fig. 2—This mount has ratings from 40 to 105 lb/linear in. and deflects ¼ in. Rubber is stressed in shear and compression. As a result, the load deflection curve is non-linear, which prevents resonance.

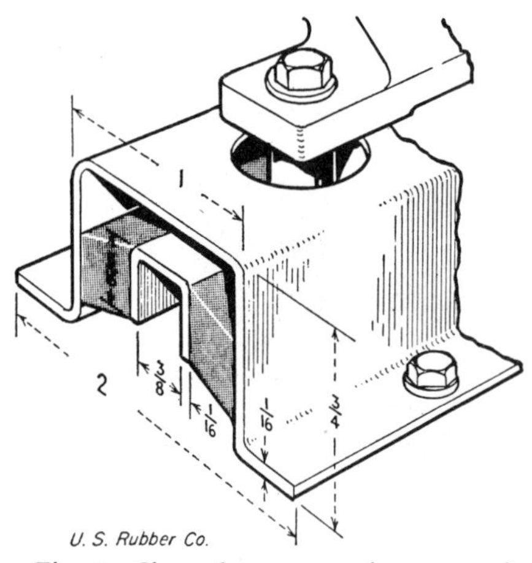


Fig. 3—Channel-type mounting comes in two basic sizes: small, for vertical loads up to 95 lb (illustrated); large, for loads to 420 lb. Lengths are 1 to 7 in. Can be inverted for overhead suspensions.

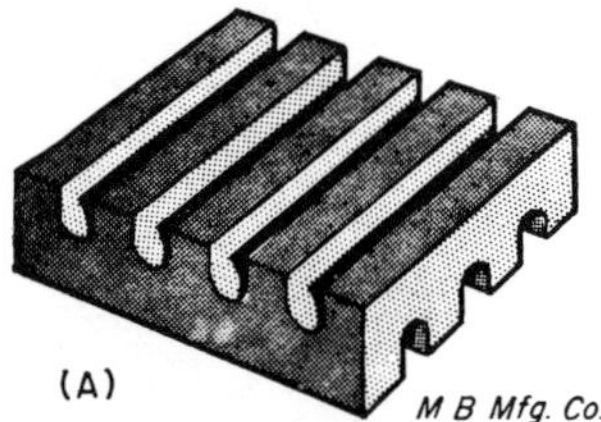


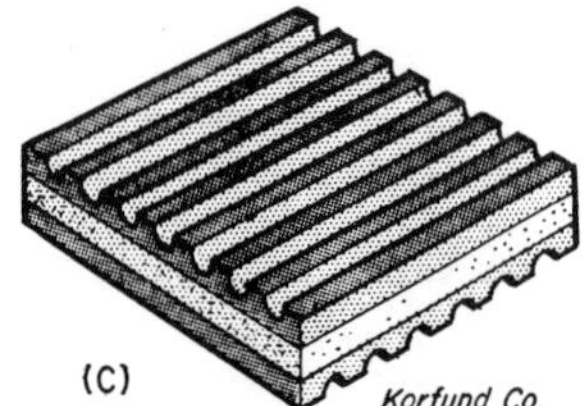


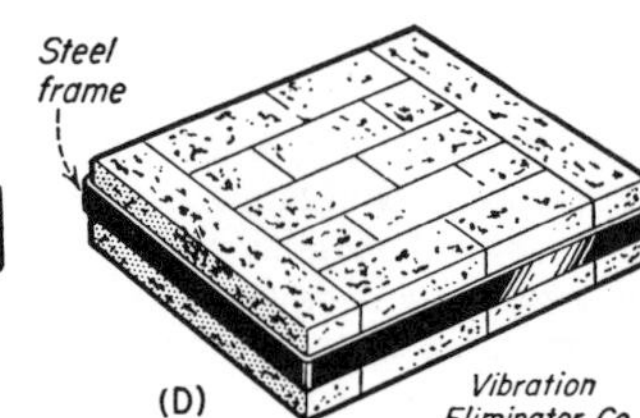




Fig. 4—For heavy, flat base, equipment several materials are available for vibration isolation; these are most effective where very high frequencies and disturbing sound are to be reduced. Sheets of felt, cork, and rubber are most commonly used. (A) *Isomode* pads of neoprene are 5/16 in. thick and are used at loadings of from 10 to 60 psi. Maximum deflected height is ¼ in.; several layers can be used. (B) *Vibrapad* consisting of alternate layers of cotton duck and rubber, ⅝ in. thick, is used for loads up to 200 psi. (C) *Elasto-rib* is a combination of cork and ribbed-rubber which is rated up to 35 psi. Minimum thickness is 1 in. (D) Natural cork plates are held by a steel frame. Thickness ranges from 1 to 4 in. and widths up to 24 inches

# Vibration Isolation

THOMAS R. FINN
Finn and Company

## INDIVIDUAL MOUNTS

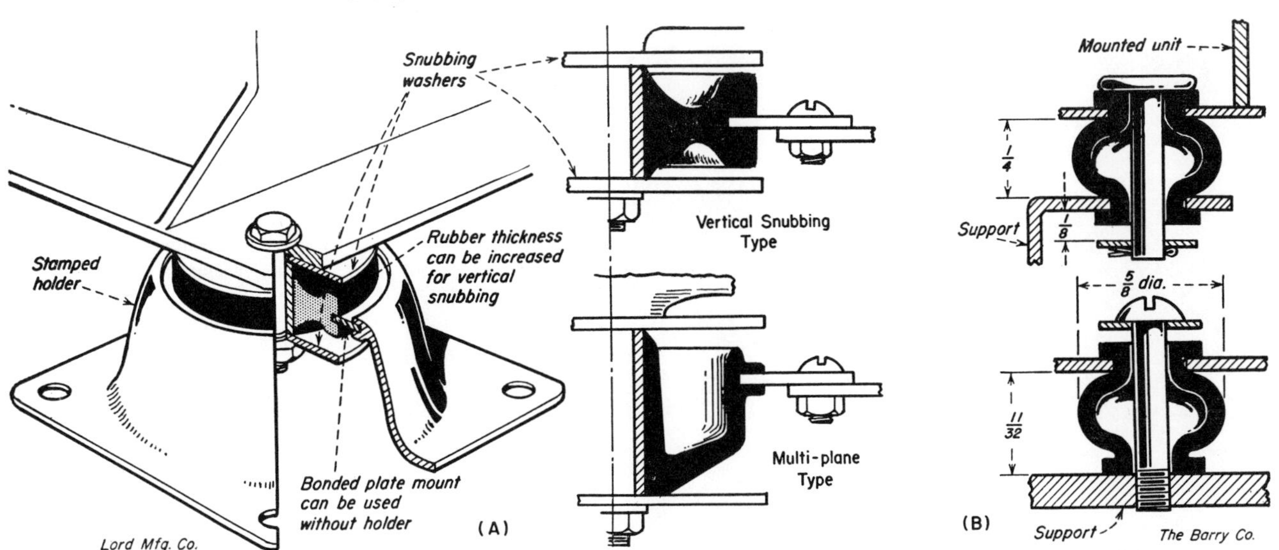


Fig. 5—(A) Axial ratings of this standard series are: 1 to 90 lb at deflections of 1/16 or ⅛ in. Mounting is about twice as stiff radially as compared to the axial direction. For loads up to 310 lb, the rubber is solid. (B) Small mount is rated from 1/3 to 3 lb per unit. Isolates vibrations as low as 900 cpm in all directions. Attachment is simplified by punched or tapped holes.

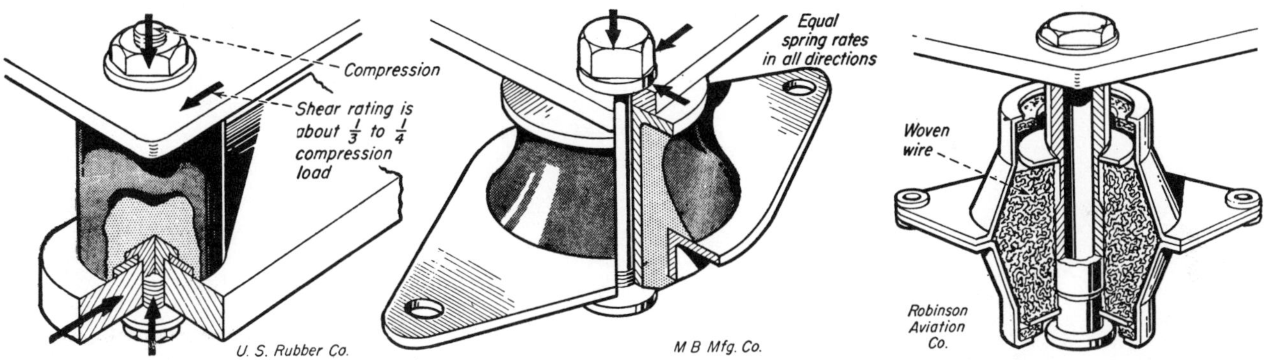


Fig. 6—Stud-type pads are simple to attach and rated in compression from 2 to 270 lb, but limited to applications of shear and compression forces. Pads with tapped holes are made for cap screws.

Fig. 7—Mounts are rated at 150 to 2,670 lb at 0.15 in. deflection. The natural frequency is about 490 cpm. Sleeve ID varies from 0.437 to 1.00 in. and the overall height is from 1.75 to 4.00 in.

Fig. 8—All metal mount for extreme temperature conditions. Woven wire is used for resilient element. With a varying spring rate, it has good damping effects. Unit load ratings are from 2 to 25 lb.

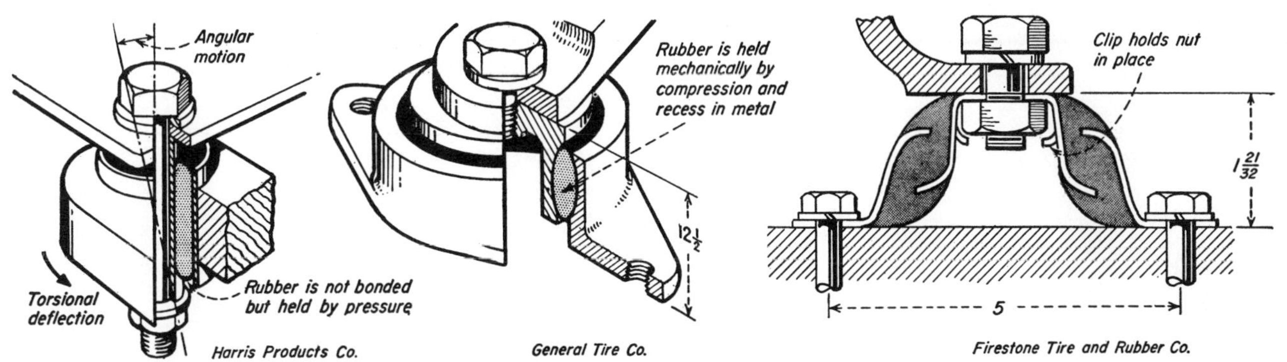


Fig. 9—Specially designed, this mount is widely used in automotive industry. Besides axial and radial resiliency, it is used for torsional and angular motion.

Fig. 10—For heavy loads, rolling joint is useful in applications where alignment is important. Normal ratings range from 125 to 1,500 lb at 0.24 in. deflection.

Fig. 11—Metal reinforced rubber pads are designed for loads of 250 to 1,700 lb. All pads have the same overall dimensions. Attaching bolts are: ½, ⅝ or ¾ inch.

# 2

# ASSEMBLY


| | |
|---|---|
| Properties and Uses of 43 Adhesives | 28 |
| Which Adhesives for What | 32 |
| How to Calculate Stresses in Adhesive Joints | 34 |
| 16 Ways to Align Sheets and Plates with One Screw | 38 |
| Joining Circular Parts without Fasteners | 40 |
| 23 Ways to Attach Small Die-cast Parts | 42 |
| Fundamentals of Selecting Locknuts | 47 |
| Fastening Methods for Flexible Sheet Materials | 50 |
| Methods of Attaching Glass to Metal Structures | 52 |
| Applications of Helical Wire Inserts | 54 |
| Various Methods of Locking Threaded Members | 56 |
| Quick-release Fasteners | 58 |
| Designs for Sheet-metal Boxes with Square and Round Corners | 60 |
| Top and Bottom Attachments for Sheet-metal Containers | 62 |
| Fastening Sheet-metal Parts by Tonques, Snaps and Clinching | 64 |
| Joining Sheet-metal Parts without Fasteners | 66 |
| Liquid-tight Riveted Joints | 68 |
| Simplifying Assemblies with Spring-steel Fasteners | 70 |
| 20 Tamper-proof Fasteners | 72 |
| How to Design Parts That Assemble and Lock by Twisting | 74 |

# Properties and Uses of 43 Adhesives Classified According to Chemical Type

N. J. DeLOLLIS, National Bureau of Standards

| CHEMICAL TYPE | COMPOSITION AND CURING CONDITIONS | STRENGTH | APPLICATIONS | REMARKS |
|---|---|---|---|---|
| **SYNTHETIC THERMOPLASTIC** | | | | |
| 1. VINYL ACETATE | | | Bonds thermosetting plastics, metals, and glass. Used in packaging industry for bonding nonfibrous films, such as cellophane. | Noted for heat sealing properties. Has fair moisture resistance. |
| 2. POLYVINYL ACETATE | Vinyls can be bonded by hot set or cold set.<br><br>***Cold set:*** Permits fabrication of articles quickly and economically before solvents evaporate. However, requires longer to reach ultimate bond strength (when solvent is dispelled). Not adaptable to nonporous surfaces which tend to trap solvent and retard its evaporation.<br><br>***Hot set:*** Baked for a few minutes at 225 to 250 F. May be applied as hot melt or as solution which is dried with or without heat before assembly. The treated surface can be bonded immediately or at later date. Bonding is achieved quickly with heat and low pressure. | Adherent — Tensile psi — Shear psi*<br>Stainless steel — 3600 — 2960<br>Aluminum alloy — 3270 — 3560<br>Paper laminate phenolic — 1060 — 2480<br>Glass — 2430 — 2310<br>Birch wood — 960 — 1990<br>Hard rubber — 400 — 630 | Used for bonding glass and metals. Gives strong, elastic joint. | Resists discoloration from aging. Has about same refractive index as glass. |
| 3. POLYVINYL ALCOHOL | | | Bonds practically any material, porous or nonporous. Used as primary glue for food industry packaging, and as blending agent for improving vegetable glues. | Tough and clear. Soluble in water at 160 F. Resists fats and many organic solvents, and is stable and impermeable to gases. |
| 4. POLYVINYL CHLORIDE | | | | Excellent moisture resistance, but forms a hard, brittle film with less adhesive action than the acetate. Good electrical properties and high chemical resistance. Plasticizers can be used to modify brittleness, but they will impair the adhesive properties. |
| 5. POLYVINYL CHLORIDE-ACETATE | | | | Combines moisture and chemical resistance of the chloride with elastic toughness of the acetate. |
| 6. POLYVINYL BUTYRAL | | | Used as bond for safety glass (stronger than previous adhesives and will not haze). Bonds tough, flexible sheets of mica. Compounded with other resins to produce special adhesives. | Tougher than acetate; more resistant to water but less resistant to weathering. |
| 7. ACRYLATES | | | Bonds wide variety of materials. Especially good for nonporous materials, such as metals or glass. | Moderately soluble with good aging properties. Permanently tacky, pliable, and tough. |
| 8. METHACRYLATES | | | Similar to acrylates. | Moderately soluble with good aging properties. Tougher than acrylates and have higher softening points. Film exhibits water clear transparency and good weathering properties. |
| 9. POLYSTYRENE | | | Electrical installations. | High dielectric strength, and good resistance to water absorption. |
| **SYNTHETIC THERMOSETTING** | | | | |
| 10. FURANE RESINS<br>a. Furfuryl alcohol<br>b. Furfuraldehyde | Can be cured at room temperature or under application of heat. Low pressures required. | | Will bond wide variety of materials. Especially recommended for thermosetting plastics. | Acidity makes it unsuitable for bonding wood and other acid-vulnerable materials. |

| | | | | |
|---|---|---|---|---|
| 11. PHENOL-FORMALDEHYDE | Available as film, powder, or liquid. Film form is spread and dried on paper, which is placed between veneers and heat and pressure applied. Powder form is mixed with water, alcohol, acetone, or combination of the three.<br>Curing temperatures range from 300 F for neutral state to 70 F for varying concentrations of alkali or acid. | | Used as plywood bond where heat can be applied easily. Not used in secondary glueing of large assemblies where heat cannot be conveniently applied. | Extreme resistance to water, weathering, temperature variations, and chemical and bacterial action. |
| 12. RESORCINOL-FORMALDEHYDE | Liquid with separate curing agent. Will cure at room temperature. | Adherent / Tensile psi / Shear psi*<br>Paper laminate phenolic 830 1370<br>Birch wood 1180 1940<br>Hard rubber 1340 590 | High cost limits their use to aircraft and boat building industries. | Chemically related to phenolics with similar good resistance to deteriorating conditions. |
| 13. UREA-FORMALDEHYDE | Available as water soluble powders with curing agent either separate or combined; also syrupy water solutions with separate curing agents. | | Woodworking industry. Replacement for casein in plywood manufacture. | Light in color with good heat resistance up to 160 F. They are less resistant to chemicals, heat, and weathering than phenolics. Film is brittle and tends to craze. This is corrected by using a cellulose filler and thin glue line. |
| 14. MELAMINES | Little pressure required for curing. | | Prefabricated houses, station wagon bodies, furniture, and curved plywood. Can be used to bond imperfectly fitting parts. | More expensive than ureas, but have higher resistance to moisture, boiling water, and acids. |
| 15. ALLYL-MONOMER | | | Used to impregnate fibrous materials to form laminates, and also as optical cement. | Water clear, partially thermoplastic, and insoluble. One form has superior optical qualities, good weathering properties and forms a strong bond with glass. |
| 16. RESORCINOL NYLON | 100 percent solids. Cures at room temperature. | | Used to bond thermosetting and thermoplastic resins. | Reduces strains caused by shrinkage and temperature differentials. Produces a relatively nonporous bond. |
| 17. SILICONES | High temperatures (about 200 C) and pressures required. | Low modulus resins. | Bonding silicone and other types of rubber to nonporous materials. | Most heat resistant adhesive. Producers are: Dow Chamical Co., General Electric Co. |
| 18. ETHOXYLINE RESINS<br>Epoxy<br>Ethoxy | 100% solids available in solid and liquid forms.<br>*Solid form:* available as powder or stick and used like solder. Cures with slight pressure at 390 F for 1 hr. Curing agent may or may not be included.<br>*Liquid form:* Uses separate curing agent. Cures at low temp. (75-120 F) and low pressures. Elevated temp. accelerates cure. | High tensile strength. Typical shear strength with aluminum strips: 4,500 psi. | Will bond wide variety of porous and nonporous materials. Attaching vitreous-coated metal panels. Assembly of metal door and window frames. Cementing carbide tips in reamers.<br>Liquid types are used for bonding glass or ceramics, such as high strength glass cloth laminates. | Low shrinkage and good adhesive qualities Principal Producers: Shell Chemical Co. (EPON series), Ciba Corp. (Aralchite series). |

## CELLULOSE DERIVATIVES

| | | | | |
|---|---|---|---|---|
| 19. CELLULOSE NITRATE | | Adherent / Tensile psi / Shear psi*<br>Stainless steel 2180 1580<br>Aluminum alloy 1500 1360<br>Paper laminate phenolic 860 1680<br>Glass 1080 1680<br>Birch wood 1100 1390<br>Hard rubber 590 1000 | Used to cement glass, metals, leather, cloth, and ceramics. | Flammable and must be handled with care. Discolors on exposure to sunlight. |
| 20. CELLULOSE ACETATE | | | Similar to cellulose nitrate. | Superior to cellulose nitrate in aging properties and fire resistance; inferior in adhesive qualities and moisture resistance, and more expensive. |
| 21. CELLULOSE ACETATE BUTYRATE | | | Similar to cellulose nitrate. | More durable and moisture resistant than cellulose acetate. |

| CHEMICAL TYPE | COMPOSITION AND CURING CONDITIONS | STRENGTH | APPLICATIONS | REMARKS |
|---|---|---|---|---|
| **CELLULOSE DERIVATIVES—Continued** | | | | |
| 22. ETHYL CELLULOSE | | | Bonds cloth, paper, and metal foil. | More resistant to chemical action than other cellulose derivatives. |
| **RUBBER** | | | | |
| 23. NATURAL RUBBER | Best solvents are aliphatic or straight chain hydrocarbons. Some types are air drying at room temperature; others vulcanize at room temperature to produce a more permanent bond. High temp. and pressure curing improves strength. | Best tensile strength, resilience, and elongation of all rubber adhesives.<br>Following strength values are for bonds cured at high temp. and pressure. Solvent benzene.<br>Adherent — Tensile psi — Shear psi*<br>Stainless steel — 260 — 270<br>Aluminum alloy — 390 — 250<br>Paper laminate phenolics — 160 — 130<br>Glass — 34 — 43<br>Birch wood — 170 — 160<br>Hard rubber — 130 — 190<br>Bond strength of air drying adhesives is lower; averages 50 psi. | Automobile industry to bond felt, cloth and rubber to metal; shoe industry; and as household cement. | Poor aliphatic resistance. |
| 24. POLYSULFIDE RUBBER OR " THIOKOL " | Best solvents are chlorinated hydrocarbons. Bonds well under natural aging, and poorly under high temperature aging. | Low tensile strength, resilience, and elongation. | Bonds where oil and solvent resistance are required. | Good resistance to greases, oils, and solvents; excellent resistance to aging; low permeability to gases. |
| 25. NEOPRENE | Solvents are esters or ketones. Bonds well under natural aging, and moderately well under high temperature aging. | Elasticity and tensile strength compare well with natural rubber.<br>Adherent — Tensile psi — Shear psi*<br>Stainless steel — 170 — 90<br>Aluminum alloy — 290 — 130<br>Paper laminate phenolics — 170 — 250<br>Glass — 90 — 100<br>Birch wood — 340 — 180<br>Hard rubber — 240 — 230 | Most popular synthetic rubber adhesive. | Excels in resistance to organic solvents, heat, sunlight, chemical action, and atmospheric oxidation. |
| 26. RECLAIMED RUBBERS | Composed primarily of natural rubbers; recently including some synthetics. | | Similar to natural rubber applications. | Lower cost materials; nearly as good as natural rubber adhesives. |
| 27. GR-S | Best solvent petroleum. Bonds fairly well under both natural and high temperature aging. | Fair tensile strength and elongation; moderately good resilience. | | Comparable to reclaimed rubber adhesives; poor aliphatic resistance. |
| 28. GR-1 (Butyl) | Best solvent petroleum. Best rubber adhesive under natural aging, and good under high temperature aging. | Fair tensile strength, and moderate resiliency and elongation. | | Poor aliphatic resistance. |
| 29. BUTADIENE-STYRENE | High styrene content. | High tensile strength. | | |
| 30. BUTADIENE-ACRYLONITRILE | Must be specially compounded because of tendency to flow under deformation. | | | Good resistance to oils and fuels. |
| 31. NITRILE (Buna N) | Best solvent is Ketone. Bonds fairly well under both natural and high temperature aging. | Moderately good resiliency and tensile strength; fair elongation. | | Good aliphatic resistance, but lack property of welding readily to themselves. Must be bonded with considerable solvent present. |

| | | | | |
|---|---|---|---|---|
| **RUBBER-RESIN** | | | | |
| 32. PHENOLIC ELASTOMERS | Available in three forms: film with liquid primer for metal surfaces; liquid; and two-component powder and liquid combination. High temperature and pressure required for bonding. | Combine strength and adhesive properties of phenolic resins with flexibility and resilience of rubber. | Major application is joining aluminum and stainless steel; can be used for other metals and nonmetals. Not used as wood bond because of curing conditions. | Have characteristic phenolic resistance to corroding conditions. |
| 33. C-3: Phenolic rubber-base resin<br>CB-4: C-3 plus nylon base | Bonds with 200 psi at 325 F for 15 to 20 min. | Typical shear strength with aluminum strips: 4,000 psi. | Bonding aircraft sandwich construction laminating metals. (A similar type is used to bond brake linings to metal drums.) | Good oil resistance. Must be protected from salt water spray. Producer—Chrysler Corp. Tradename—Cycleweld. |
| 34. $M3_c$: Phenolic Synthetic rubber base<br>$MN3_c$: $M3_c$ plus " Nelite " (a thermo-setting plastic) | Requires 50 psi bonding pressure. Cure at 325 F for 30 min. | Typical shear strength with aluminum strips: 3,000 psi. | Attaching reinforcing stringers to sheets. " Waffle " type sandwich construction. | |
| 35. " Reanite " | Cures at 350 F for 10 min. with moderate pressure. | | Metal adhesive. | Producer—U. S. Stoneware Co. Not recommended for shock loading. |
| 36. LIQUID PHENOL-FORMALDEHYDE plus " formvar " powder | Cures with 300–500 psi pressure at 300 F for 15 min. | Typical shear strength with aluminum strips: 5,000 psi. | Aircraft surface reinforcing. Wood to metal structural bonding. | Tradename — Redux. Producer — Resinous Products and Chemical Co. Good adhesive for aluminum alloys, stainless steel, magnesium, and CR steel; not satisfactory for nickel and its alloys, lead, or copper. |
| **VEGETABLE GUM** | | | | |
| 37. GUM ARABIC | Water soluble natural resin. | Adherent / Tensile psi / Shear psi*<br>Stainless steel 110 130<br>Aluminum alloy 110 330<br>Paper laminate phenolic 630 440<br>Glass 0 29<br>Birch wood 400 630<br>Hard rubber 320 240 | Joining paper or wood; packaging for tropical conditions, and in high speed packaging and labeling machines. | Light in color. In high concentrations it is superior to dextrin for difficult packaging operations, despite higher cost. |
| **PROTEIN** | | | | |
| 38. ANIMAL AND FISH GLUE | Produced from hide or bone. | | Woodworking trades. Used as hot melt to give fast bond on cooling. | Higher joint strength and moisture resistance than vegetable glues. Must be applied manually. Fish glues are costly, have bad odor. |
| 39. BLOOD AND EGG ALBUMEN | | | Plywood manufacture. | Become insoluble when heated. Next to synthetic thermosetting adhesives they are most durable and water resistant. |
| 40. CASEIN | Bonds under cold pressure. Made from milk curd. | Adherent / Tensile psi / Shear psi*<br>Stainless steel 510 190<br>Aluminum alloy 110 120<br>Paper laminate phenolics 690 1030<br>Glass 0 29<br>Birch wood 1020 1660<br>Hard rubber 130 150 | Strong water resistant wood joints and durable plywood. | Joints exhibit high strength even with thick glue lines; water resistant for short periods; will withstand low temperatures. |
| 41. SOYBEAN | | | Plywood and wood veneers. Can be used on green veneers with as must as 15% moisture. | Cheapest water resistant glue; lacks tackiness. |
| **STARCH (PASTES)** | | | | |
| 42. STARCH | Acid or alkali modifiers used to make paste thick and tacky. | Low tensile strength. | Used in automatic packaging machines, and in manufacture of low strength, low water resistant plywood. | Low water resistance; cheaper and easier to handle than animal glues. |
| 43. DEXTRIN | Converted starch. | Low strength. | Fast machine packaging. | More expensive than starch, but especially fast setting and tacky. |

# WHICH ADHESIVE FOR WHAT


**R. W. JAMES AND R. W. GORMLY**

*Arthur D. Little, Inc.*
*Cambridge, Mass.*


Ideally, there should be one super-glue to bond anything and everything. But because no one adhesive can do it all, the answer has to be a practical one—especially with the type called structural adhesives.

### Type of Bond

There are two ways the bond is accomplished:

**In mechanical bonding,** the adhesive infiltrates porous materials to give cementing of a physical nature—as in glueing wood.

**In chemical bonding,** the adhesive reacts with the parts, or joins them largely through intermolecular attraction.

The bonding forces act singly or in combination to grip adherends together under shear, tension, peel or cleavage. To resist creep, the adhesives must be quite rigid and also have good shear and tensile strengths. But thermosetting adhesives with these properties have poor resistance to peel and cleavage. For high loads, design the joints so stresses are in shear, and distribute them evenly over large areas.

Nonstructural adhesives, by contrast, being usually thermoplastic, soften and lose strength rapidly with moderately elevated temperature. Since they are somewhat elastic, they resist peel and cleavage well, but have poor shear and tensile strength, and also creep and deform under continuous loading.

### Method of Bonding

Adhesives come in different forms—liquid, paste, films, or powders. The form usually determines method of application: Liquids can be sprayed, roller-coated or brushed; paste, troweled; and powder, dusted. How the adhesive is applied can make a big difference in the labor required. In any case, the film thickness must be accurately controlled—if too thin, the adhesive may give incomplete coverage; if excessively thick, bond strength is decreased.

Thermosetting structural adhesives must react chemically under pressure and heat to become rigid, infusible,

## APPLICATIONS AND PROPERTIES OF

| | BONDING ABILITY | | | | | | | | | | BONDING REQUIREMENTS | | |
|---|---|---|---|---|---|---|---|---|---|---|---|---|---|
| Chemical Type | Aluminum | Other metals | Glass | Plastics, low m.p. | Plastics, heat resistant | Wood | Rubber | Leather | Ceramics | Concrete | Temp., F | Pressure, psi | Time |
| Urea-formaldehyde | P | P | P | P | M | E | P | G | P | P | 70 to 260 | 50 to 200 | 5 min to 4 hr |
| Melmaine-urea formaldehyde | P | P | P | P | G | E | E | E | M | P | 240 to 280 | 100 to 250 | 5 to 20 min |
| Phenol-formaldehyde | F | P | P | P | M | M | E | M | G | P | 70 to 300 | 25 to 250 | 1 min to 4 hr |
| Resorcinol-formaldehyde | F | P | P | G | G | E | E | E | G | P | 70 to 200 | 25 to 250 | 3 min to 2 hr |
| Phenolic-elastomer | E | G | G | P | G | E | G | E | G | G | 350 | 50 to 300 | 30 to 60 min |
| Epoxy | E | E | E | E | E | E | E | E | E | E | 70 to 400 | Contact to 200 | 15 min to 4 hr |
| Epoxy-phenolic | E | E | E | P | G | E | F | E | E | E | 350 | 25 to 200 | About 2 hr |
| Epoxy-polyamide | E | E | E | G | G | E | G | E | E | E | 70 to 300 | Contact to 100 | 10 min to 5 hr |
| Epoxy-thiokol | E | E | E | G | E | E | E | E | E | E | 250 | Contact | 30 min to 2 hr |
| Neporene base (thermoplastic) | F | F | F | G | F | F | E | E | G | G | Solvent release, to 210 | Contact to high pressure; while tacky | Momentary |
| Other curing elastomer systems (thermoplastic) | P | F | F | F | F | F | G | G | P | P | Solvent release, 70 to 325 | 25 to 500 | 5 min to overnight |

Code: E—excellent; G—good; M—moderate; F—fair; P—poor.

Note 1. Soluble in aromatics, esters, chlorinated hydrocarbons.
Note 2. Soluble in ketones and esters.
Note 3. About 50 psi for sustained dead load.

and insoluble. Pressures required range from contact pressure to several hundred psi. Contact is quickest but does not allow adjusting of misaligned joints. High pressures require clamps or presses. Temperatures needed range from room temperature to 400 F. At room temperature adhesives cure slowly and may only develop full bond strength after 2 to 3 days. High bonding temperatures require expensive heated fixtures.

**Trouble-shooting Tips**

An over-all compatibility between adhesive and adherend is required. Here are some warning signals:

- For transparent adherends, avoid an adhesive whose strength would be lowered by the effect of light.
- For plastics, choose adhesives that will not craze or dissolve adherends. If plastic contains plasticizer, it may migrate into the adhesive and soften it.
- For porous materials, an adhesive that soaks through may cause staining.
- On aluminum, alkaline adhesives cause corrosion.
- On a copper-base alloy, natural rubber adhesives will deteriorate.

## STRUCTURAL ADHESIVES

| RESISTANCE OF BOND TO: | | | | ROOM-TEMPERATURE BOND STRENGTH | |
|---|---|---|---|---|---|
| Water | Organic solvents | Heat | Cold | Shear, psi | Peel, lb/in. |
| F | G | G | G | Chief use in plywood. Shears in the wood. Maple, 2800 psi Birch, 450 psi | Not applicable |
| E | E | | | | |
| G | G | | | | |
| E | G | | | | |
| | | | | On Aluminum | |
| E | E | E | E | 3000 to 5000 | 40 to 60 |
| E | E | F to E | E | 1000 to 5000 | 20 to 100 |
| E | E | F to E | E | 2000 to 3000 | 40 to 100 |
| E | G | F to G | E | 2500 to 3500 | 15 to 30 |
| E | E | F | E | 4000 | 20 to 35 |
| E | P Note 1 | P to F | G | 400 on Al Note 3 | 10 to 20 on leather |
| G | P Note 2 | P | G | Up to 1200 on Al | 13 for steel on fir plywood |

## SHOPPING LIST FOR STRUCTURAL ADHESIVES

*(Not necessarily complete, but a good starting point)*

### Urea-formaldehyde

American Cyanamid Co.
Bakelite Co.
Borden Co.
Barrett Div. Allied Chem.
Catalin Corp.
National Casein Co.
Monsanto Chemical Co.
Paisley Products Inc.
Reichhold Chemicals, Inc.
US Plywood Corp.

### Melamime-urea-formaldehiyde

American Cyanamid Co.
Rohm & Haas Co.
Monsanto Chemical Co.

### Phenol-formaldehyde

American-Marietta Co.
Bakelite Co.
H. B. Fuller Co.
B. F. Goodrich Co.
Heresite & Chemical Co.
Imperial Chemical Ind., Ltd
Monsanto Chemical Co.
Minnesota Mining & Mfg. Co.
Varcum Chemical Corp.
Xylos Rubber Co., Div.

### Resorcinol-formaldehyde

B. F. Goodrich Co.
Koppers Co.
Monsanto Chemical Co.
Synco Resins, Inc.
Synvar Corp.
Naugatuck Chemical Div.
Durez Plastics Div.
Schenectady Varnish Co.
Polymer Chemical Co.

### Phenol-elastomer

B. F. Goodrich Co.
B. B. Chemical Co.
Bloomingdale Rubber Co.
Compco Chemical Co.
Cycleweld Cement Products
Pierce & Stevens, Inc.
Rubber & Asbestos Corp.
Borden Co.
Hughes Glue Co.
Narmco Resins & Coatings Co.
Xylos Rubber Co.
Armstrong Cork Co.

### Epoxy

Alkydol Laboratories Inc.
Ciba Co.
Bakelite Co.
Rubber & Asbestos Corp.
Nureco Inc.
Armstrong Products Co.
Furane Platics, Inc.
Loven Chemical of California
Marblette Corp.
American Latex Products
Paisley Products, Inc.
US Stoneware Co.

### Epoxy-phenolic

Adhesive Engineering
Lebec Chemical Corp.
Plastics Engineering Co.
Slomons Laboratories, Inc.
Narmco Resins & Coatings Co.

### Epoxy-polymide

Adhesive Products Corp.
Borden Co.
Rubber & Asbestos Corp.
Chemical Coatings & Engineering Co.
Ohio Adhesive Corp.
Narmco Resins & Coatings
Houghton Laboratories, Inc.
National Starch Products
Lawrence Adhesive & Chemical Co.
General Mills, Inc.

### Epoxy-Thiokol

Rubber & Asbestos Corp.
Polymer Chemical Co.
Borden Co.
Armstrong Cork Co.
Minnesota Mining & Mfg. Co.
Polymer Industries, Inc.
Atlas Mineral Products Co.
Loven Chemical of California

### Neoprene base (thermoplastic)

Goodyear Tire & Rubber Co.
Marbon Chemical Div.
Formica Corp.
Pierce & Stevens, Inc.
US Rubber Co.
Rubber & Asbestos Corp.
Cycleweld Cement Products
Minnesota Mining & Mfg. Co.
Union Paste Co.
Carboline Co.
Dewey & Almy Chemical Co.

### Other curing elastomers (thermoplastic)

US Rubber Co.
Angier Adhesives Div. Interchemical Corp.
Pierce & Stevens, Inc.
Armstrong Cork Co.
US Stoneware Co.
Rubber & Asbestos Corp.
Du Pont

# HOW TO CALCULATE . . .

# STRESSES in ADHESIVE JOINTS

**Structural adhesives are getting stronger and so is the interest in this method of joining load-bearing members. The author analyzes six basic joints and shows how to calculate stress concentrations in each.**

**H. A. PERRY,** *Consultant, Non-Metallic Materials Section, US Naval Ordnance Lab, Silver Spring, Md.*

Rigorous analyses of stress concentrations in adhesive joints are rare. However, two types—lap and scarf joints—are elements in most of today's common joint geometries. And theory on the two types has been checked closely by laboratory tests and service experience. Result is that this theory has been extended to more complex joints. For the six most common of these, here are techniques for calculating stress concentrations.

Answers given by the accompanying equations give conservative results because stress concentrations from three, often interrelated, causes must be accounted for: geometrical assymmetry in the joint, differences in elastic properties between adhesive and adherend, and property differences between adherends. This means that best use of the equations requires information in four main areas: loads and environmental conditions, failure theories for materials used, elastic and plastic properties of adhesive and adherend, and the stress levels both can take without failure. For a critical weight or cost problem, answers obtained from these equations can serve as a starting point —with the final design refined further by testing.

### COLINEAR SCARF JOINT

A pure scarf joint, Fig. 1, is the most efficient of the common structural adhesive joints because it is colinear and introduces no bending. Mathematical analysis finds only an adhesive-to-air shear stress at the ends of the joint, but practice shows a modest stress concentration too. For example, tests on joints with a thickness ratio of 0.1—adhesive to adherend—reveal a shear stress factor of 1.45. For ratios under 0.1, the concentration factor is probably lower, and for ratios under 0.01 it is negligible.

The most important stresses in adhesive joints are the tensile stress at right angles to the adherend faces (tending to pull them apart), and the shear stress parallel to these faces (tending to slide them apart). In a scarf joint, the adhesive line is at an angle to the applied loads—and combined stress theory will give the modifying angle functions for calculating these key adhesive stresses. Modified equations, assuming like and isotropic adherends, are shown in Fig. 1. Adhesive stresses shown by most of these equations are all dependent upon *sin θ*, and can therefore be made small by choosing

## 1 STRESSES IN PURE SCARF JOINTS

| Type | Loading | Nomenclature | Shear stress | Normal stress |
|---|---|---|---|---|
| Flat scarf | Tension or compression |  $F$ = force per unit width | $S_s = \frac{F}{t} \sin\theta \cos\theta$ | $S_n = \frac{F}{t} \sin^2\theta$ |
| | Pure bending |  $M$ = moment per unit width | $S_s = \frac{6M}{t} \sin\theta \cos\theta$ | $S_n = \frac{6M}{t} \sin^2\theta$ |
| Tubular scarf | Tension or compression | 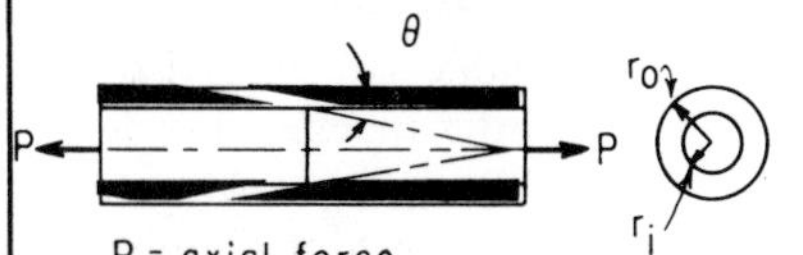 $P$ = axial force | $S_s = \frac{P}{2\pi r_o t} \sin\theta \cos\theta$ | $S = \frac{P}{2\pi r_o t} \sin^2\theta$ |
| | Pure bending | 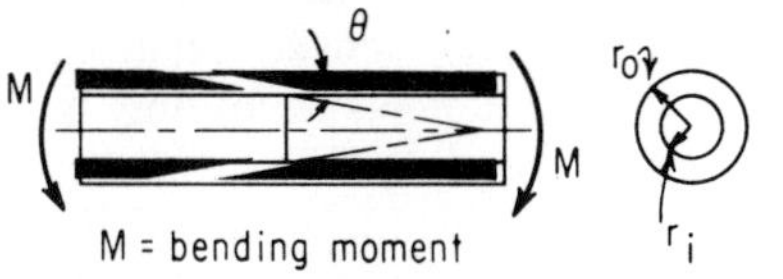 $M$ = bending moment | $S_s = \frac{2M(r_o + r_i)}{\pi(r_o^4 - r_i^4)} \sin\theta \cos\theta$ | $S = \frac{2M(r_o + r_i)}{\pi(r_o^4 - r_i^4)} \sin^2\theta$ |
| | Pure torsion | 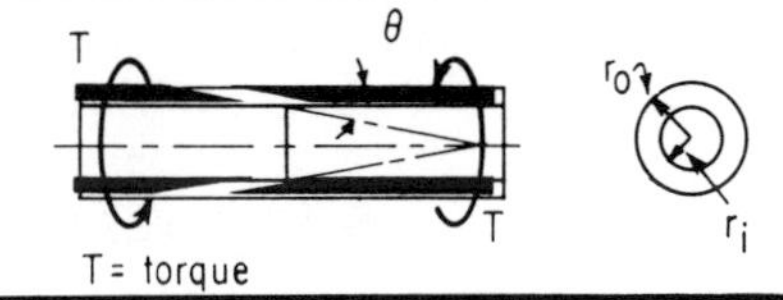 $T$ = torque | $S_s = \frac{2T \sin\theta}{\pi(r_o + r_i)^2}$ | $S_n = 0$ |

a small scarf angle. This makes the joint strength dependent on adherend strength—a desirable design criterion in any adhesive joint.

### BUTT JOINTS

A butt joint is a special case of the scarf joint where the scarf angle $\theta$ is 90° Substitution of this value in the equations of Fig. 1 will give correct adhesive stresses for this configuration under various loading conditions. Under tension, a butt joint does not efficiently exploit the full strength of structural adherends because the bond area is at a minimum. As in other scarf joints, however, stress concentration is low. This means full adherend compressive strength can be developed under compressive loads. Lowest stress concentration occurs if the adhesive film is thin and rigid. Butt joints are often combined, as lands, with scarf and lap joints–a technique discussed on the next page.

### SINGLE FLAT OFFSET LAP JOINT

Flat lap joints in tension are simple to make, so have undergone considerable analysis. Distortion of the joint under load is complex, Fig. 2, but the greatest stresses occur at the lap ends where both adherend and adhesive are deflected sharply as the loads shift from being eccentric to colinear. This is the same problem that occurs with riveted or welded lap joints: as the forces try to readjust, high stresses set up at the ends tend to peel the joint apart.

A useful analysis for like adherends is the Goland-Reissner theory which allows for bending and shear in the adherends. Although such analysis permits calculating stresses throughout the joint, the basic equations in Fig. 2 are for stresses at the lap ends–and give ratios of max shear and max normal stress to the average shear stress in the joint. While cumbersome, these equations can be simplified to cover the majority of cases. For example, the dimensionless factor $B = (3E_a t/Ed)^{1/2}$ is usually greater than 1.5, so $(\coth B)$ $(L/2t)$ approaches $L/2t$ and the shear equation simplifies to

$$S_s/S_{av} = \frac{BL^2}{16\,t^2}(1 + 3K) + (3/4)(1 - K)$$

Similarly, $\lambda$ exceeds 2.0 in most joints, so the normal stress equation reduces to

$$S_n/S_{av} = \lambda^2 K + 2\lambda K'$$

Sizing a lap joint from these equations requires trial and error calculation because the tensile stress $P$ in the adherends appears on both sides of the equation—in $S_{av}$, $K$ and $K'$. To size a joint of a given material ($E$ and $E_a$) for a given load per unit width $F$, successive combinations of overlap $L$ and adherend thickness $t$ are selected until the stresses $S_s$ and $S_n$, calculated from the equations, agree with the allowable stresses for the chosen adhesive. Moment factor $K$ may be taken from the curve in Fig. 2. For quick, ultraconservative results,

**2 SINGLE FLAT OFFSET LAP JOINT (Goland-Reissner Theory)**

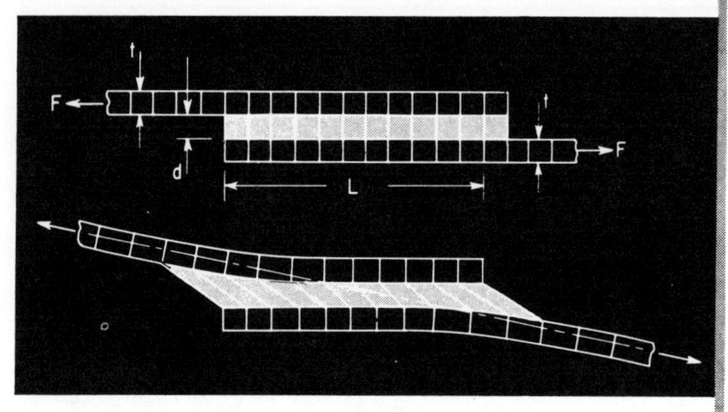


$$S_s/S_{av} = \frac{BL}{8t}(1+3K)(\coth B)\frac{L}{2t}+\frac{3}{4}(1-K)$$

$$S_n/S_{av} = \lambda K\frac{\sinh 2\lambda - \sin 2\lambda}{\sinh 2\lambda + \sin 2\lambda} + 2\lambda K'\frac{\cosh 2\lambda + \cos 2\lambda}{\sinh 2\lambda + \sin 2\lambda}$$

Where:

- $S_s$ = max shear stress at free ends of overlap, psi
- $S_n$ = max normal stress at free ends of overlap, psi
- $S_{av}$ = average shear stress on joint (=Pt/L=F/L), psi
- $F$ = Load in lb per in. of width, psi
- $P$ = unit stress in adherends (=F/t), psi
- $L$ = Length of overlap, in.
- $t$ = thickness of adherend, in.
- $d$ = thickness of adhesive, in.
- $B$ = $\sqrt{3E_at/Ed}$
- $\lambda$ = $\frac{L}{2t}\sqrt[4]{\frac{6E_at}{Ed}}$
- $K$ = dimensionless moment factor (see chart below)
- $K'$ = $\frac{L}{2t}\sqrt{(1-\mu^2)P/E}$
- $E$ = elastic modulus of adherends, psi
- $E_a$ = elastic modulus of adhesive, psi
- $\mu$ = Poisson's ratio of adherend

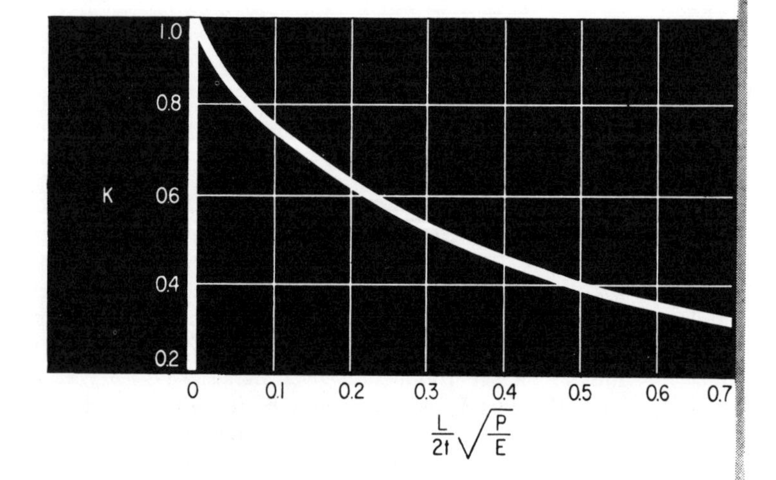


assume $K = 1$. Then

$$S_s/S_{av} = BL^2/4t^2$$

and

$$S_n/S_{av} = \lambda^2 + (\lambda L/t)\sqrt{(1-\mu^2)\,P/E}$$

Stresses obtained this way are not applicable to adherends of unlike elasticity or unequal thickness. Nor are they valid for loading other than simple tension or compression. If the adherends are very stiff or thick, the Volkerson theory for double flat lap joints, discussed next, can be applied to single flat laps where the adherends are dissimilar.

## DOUBLE FLAT LAP JOINTS

Stress concentrations at the free strap ends in a double lap joint, Fig. 3, are set up because the main center member is free of strap restraint at this point—allowing it to elongate more and strain the edge of the adhesive film sharply. When strap and center member are dissimilar, differential elasticity also contributes to stress concentration here. In a properly constructed double lap, loading is symmetrical, eliminating local bending. For symmetrical loading, the straps must be equal thicknesses of like materials, but may differ from the center member in modulus and thickness. The Volkerson theory computes adhesive stresses at the free lap ends by the following equation:

$$S_s/S_{av} = \sqrt{\Delta/W}\bullet\frac{W-1+\cosh\sqrt{\Delta W}}{\sinh\sqrt{\Delta W}}$$

where:

$$\Delta = 3E_aL^2/8E_2t_2d$$

$$W = (E_1t_1 + E_2t_2)/E_2t_2$$

Calculations are simplified by using the plotted form of the Volkerson equation in Fig. 3.

## TUBULAR LAP JOINTS

Tubes in a lap joint are constrained against radial deformation, but beyond the lap there is tendency for the larger tube to neck down and the smaller tube to dilate. This local bending is taken into account, through the thickness-to-dia ratio, by an extension of Goland-Reissner equations for flat lap joints. The tubular lap equations, by Lubkin and Reissner, are cumbersome and have been translated into tabular form in Fig. 4. This table covers tubes of like thickness and material, with modulus ratio $E/G = 8/3$ and Poisson's ratio $\mu = 0.3$ —these are reasonable assumptions for most common metals. If the tube walls are very thick or stiff, the Volkerson theory of Fig. 3 will give the stress concentration factor for dissimilar tubes. Here, the OD of the inner tube replaces $t_1$ in the W-factor calculation.

## LANDED JOINTS

Lands, or stops, perform four functions in lap or scarf joints: control joint length without jigs, control adhesive thickness, avoid hard-to-maintain feathered edges, withstand large compressive forces. Lands may be bonded or left dry. If bonded, the land will help resist tensile loads, although the stress equations must be modified. If left dry, the land will take only com-

### 3 DOUBLE FLAT LAP JOINT (Volkerson Theory)

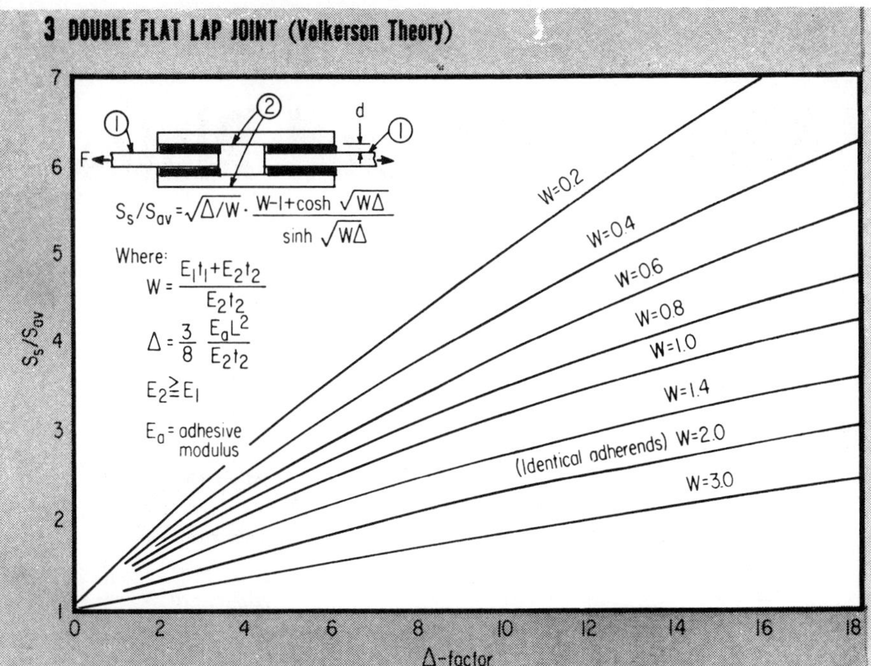


### 5 DIMENSIONS OF LANDED SCARF JOINT

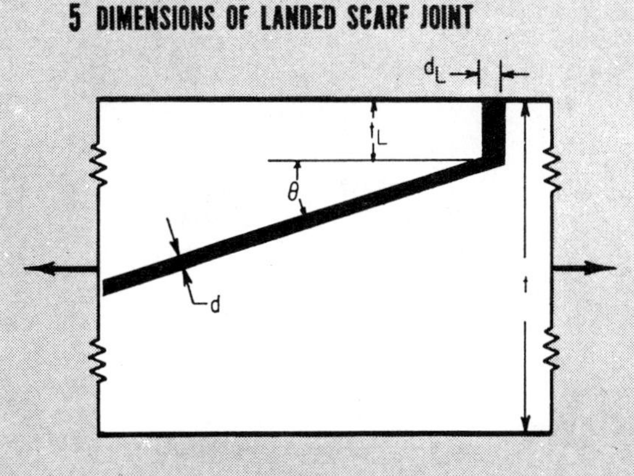


### 4 TUBULAR LAP JOINT (Lubkin-Reissner Theory)

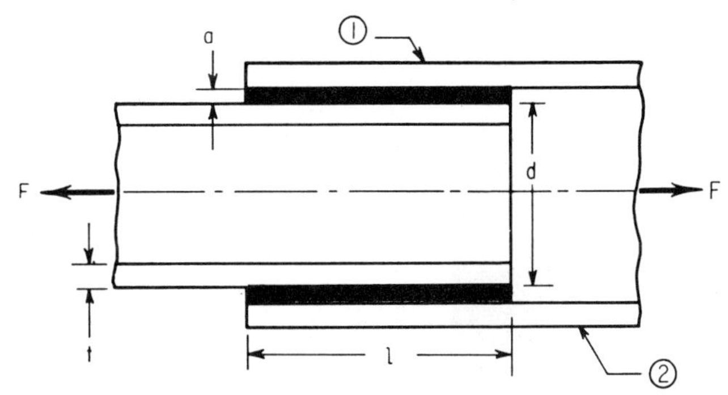


Nomenclature and conditions:

- $t$ = tube thickness ($t_1 = t_2$)
- $E$ = elastic modulus of tube ($E_1 = E_2$)
- $G$ = shear modulus of tube ($G = \frac{3}{8}E$)
- $E_a$ = adhesive elastic modulus
- $l$ = length of overlap
- $d$ = diameter of inner tube
- $\mu$ = Poisson's ratio (=0.3)
- $N = S_n/S_{av}$
- $T = S_s/S_{av}$

| | | t/d | | | | | | | |
|---|---|---|---|---|---|---|---|---|---|
| | | 0.1 | | 0.05 | | 0.025 | | 0.010 | |
| $\frac{aE}{tE_a}$ | $l/t$ | N | T | N | T | N | T | N | T |
| 4 | 1 | 1.22 | 1.06 | 1.20 | 1.05 | 1.17 | 1.05 | 1.15 | 1.05 |
| | 2 | 1.52 | 1.18 | 1.56 | 1.20 | 1.57 | 1.21 | 1.57 | 1.21 |
| | 5 | 1.95 | 1.47 | 2.24 | 1.59 | 2.58 | 1.76 | 2.98 | 1.97 |
| | 10 | 2.97 | 2.17 | 2.85 | 2.12 | 3.12 | 2.56 | 3.89 | 2.68 |
| 20 | 1 | 1.11 | 1.02 | 1.11 | 1.02 | 1.10 | 1.01 | 1.07 | 1.01 |
| | 2 | 1.19 | 1.04 | 1.21 | 1.05 | 1.21 | 1.05 | 1.19 | 1.05 |
| | 5 | 1.37 | 1.13 | 1.52 | 1.18 | 1.68 | 1.24 | 1.87 | 1.32 |
| | 10 | 1.68 | 1.37 | 1.72 | 1.36 | 1.86 | 1.42 | 2.23 | 1.60 |
| 100 | 1 | 1.04 | 1.01 | 1.06 | 1.01 | 1.06 | 1.01 | 1.05 | 1.01 |
| | 2 | 1.06 | 1.01 | 1.08 | 1.01 | 1.09 | 1.01 | 1.08 | 1.01 |
| | 5 | 1.10 | 1.02 | 1.16 | 1.03 | 1.23 | 1.05 | 1.28 | 1.06 |
| | 10 | 1.18 | 1.08 | 1.23 | 1.08 | 1.31 | 1.10 | 1.40 | 1.15 |

pressive loads and the standard equations for lap or scarf joints apply.

Exact analysis of stresses in a bonded land has not yet been made. An approximate solution by Lubkin for normal stresses in the land, Fig. 5, in terms of the normal and shear stresses in the scarf portion is, for like adherends,

$$S_L = \frac{d}{d_L}\left(S_n \sin\theta - S_s \frac{E_a}{G_a}\cos\theta\right)$$

where $G_a$ = shear modulus of adhesive, psi

If the scarf stresses $S_n$ and $S_s$ are approximated by assuming the entire tensile force $F$ is taken by the scarf length $(t - 2t_L)/\sin\theta$, the land stress $S_L$ becomes

$$S_L = \frac{F}{t} \bullet \frac{d}{d_L}\left(1 - \frac{2t_L}{t}\right)^{-1}\left(\sin^3\vartheta - \frac{E_a}{G_a}\sin\theta\cos^2\theta\right)$$

Pressing the lands tightly together during bonding reduces land adhesive thickness compared with scarf adhesive thickness, so that land stresses rise sharply. Cracks may develop under low load and propagate throughout the joint. To minimize these stresses, the lands should not be pressed tightly together; or a rubber adhesive prime coat should be applied to the lands before a tightly pressed joint is made with a more rigid adhesive. Regardless of technique, land stress $S_L$ should not exceed adhesive strength. When length tolerances are critical, however, dry lands are preferable.

# 16 Ways to Align Sheets and Plates

**FEDERICO STRASSER,** *Santiago de Chile*

## Two Flat Parts

**1 Dowels . . .**
accurately align two plates, prevent shear stress in fastening screw. Two pins are necessary because screw can not act as aligning-pin.

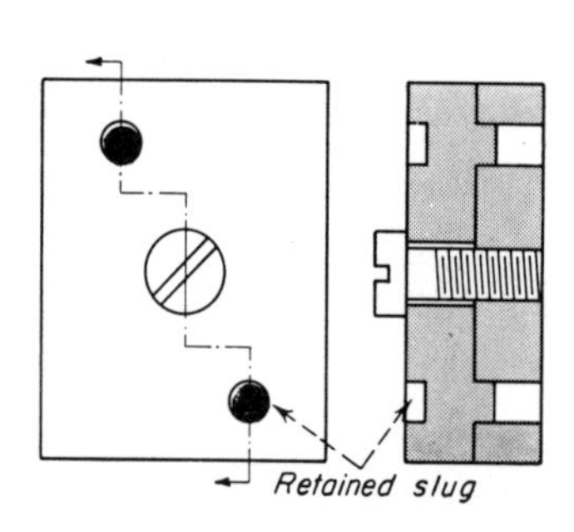


**2 Retained slugs . . .**
act as pins, perform same function as dowels; are cheaper but not as accurate.

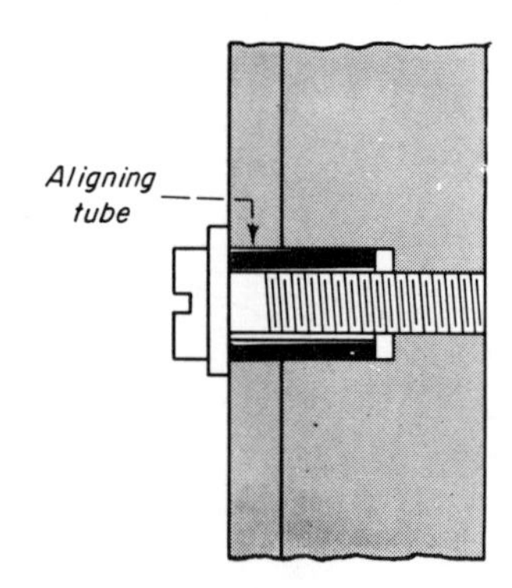


**3 Aligning tube . . .**
fits into counterbored hole through both parts. Screw clearance must be provided in tube.

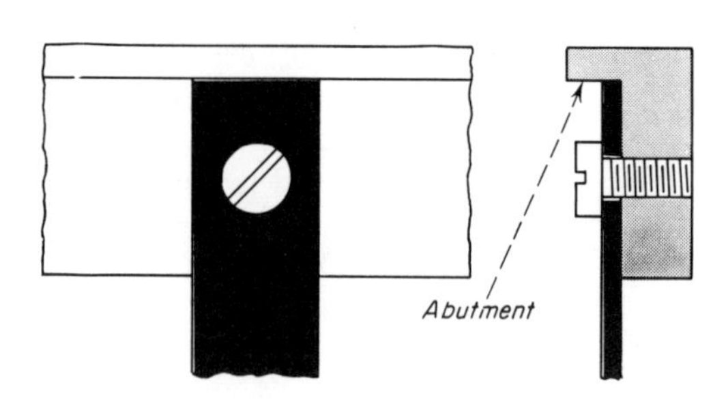


**4 Abutment . . .**
provides positive, cheap alignment of rectangular part.

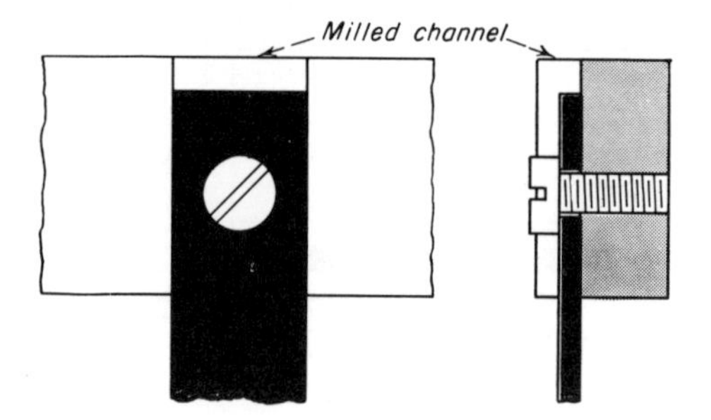


**5 Matching channel . . .**
milled in one part gives more efficient alignment than abutment in preceding method.

## Formed Stampings Assembled with Flat Parts

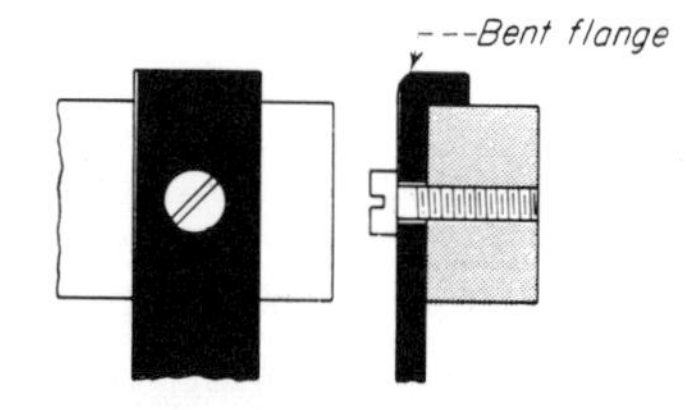


**6 Bent flange . . .**
performs similar function as abutment, but may be more suitable where machining or casting of abutment in large part is not desirable or practical.

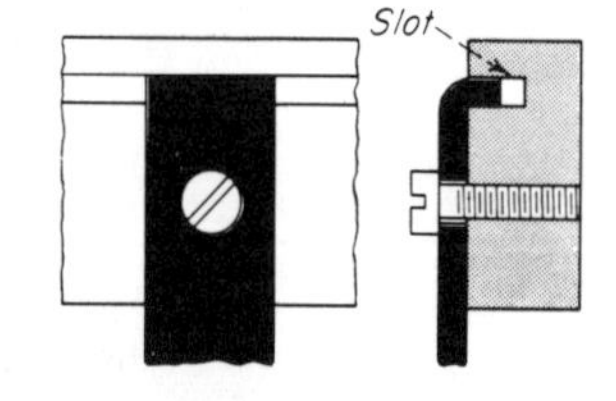


**7 Narrow slot . . .**
receives flange or leg on sheet metal part, allows it to be mounted remote from edge of other part.

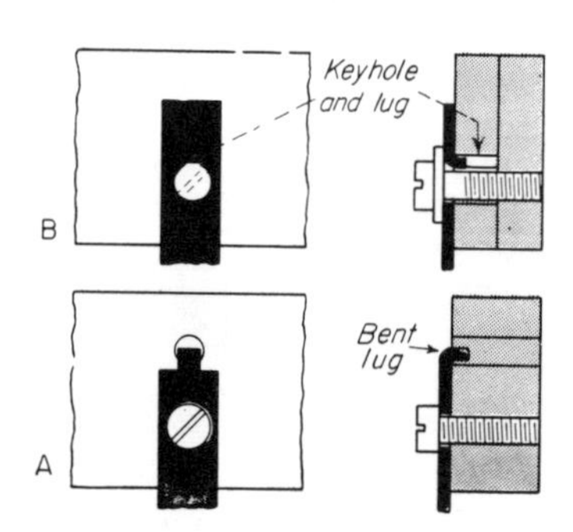


**8 Bent lug . . .**
(A) fits into hole, aligns parts simply and cheaply; or (B) lug formed by slitting clearance hole in sheet metal keys parts together at keyhole.

# with One Screw

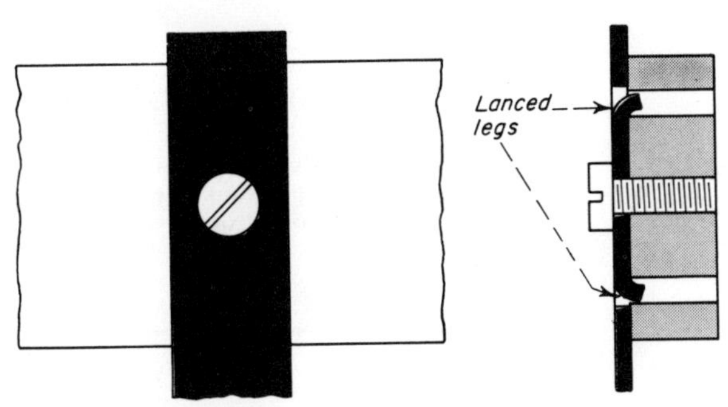


**9 Two legs . . .**
formed by lancing, align parts in manner similar to retained slugs in method 2, but formed legs are only an alternative for sheet too thin to partially extrude slug.

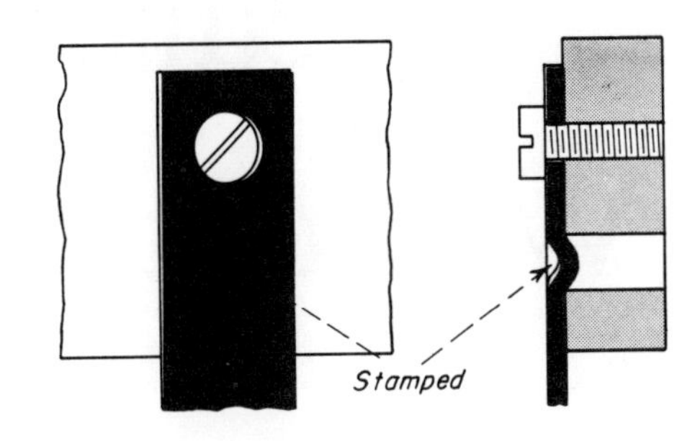


**10 Aligning projection . . .**
formed by slitting and embossing is good locating method, but allows a relatively large amount of play in the assembly.

## Flat Parts and Bars

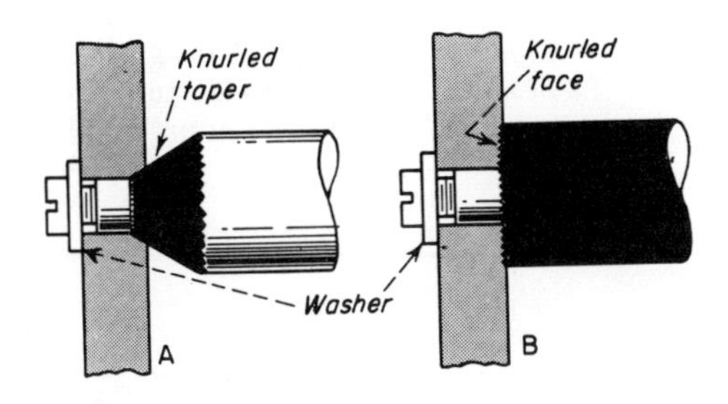


**11 Knurled end . . .**
of round bar (A) has taper which digs into edge of hole when screw is tightened; this gives accurate angular location of bar or sheet. (B) Radial knurling on shoulder is even more positive.

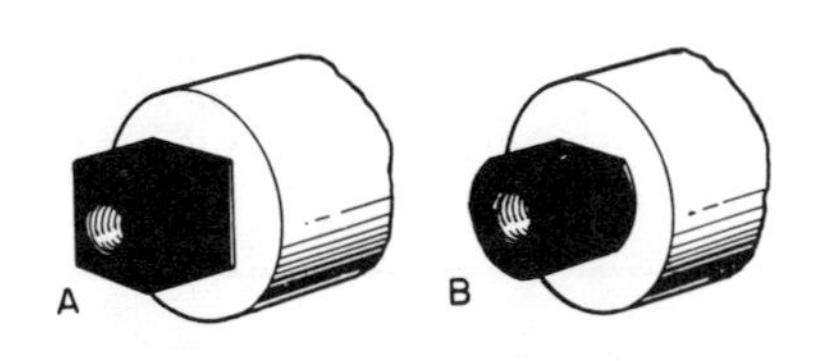


**12 Noncircular end . . .**
on bar may be square (A) or D-shaped (B) and introduced into a similarly-shaped hole. Screw and washer hold parts together as before.

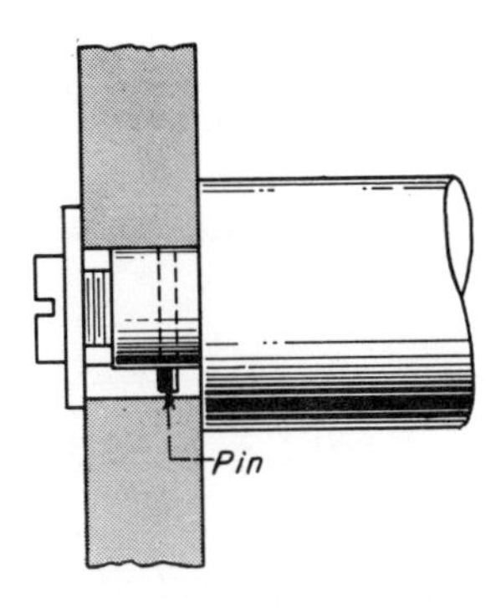


**13 Transverse pin . . .**
in rod end fits into slot, lets rod end be round but nonrotatable.

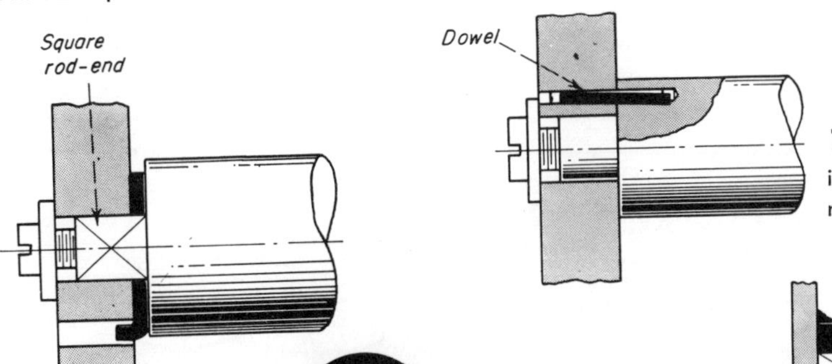


**15 Dowel . . .**
is simple, efficient method of preventing rotation if rod dia is big enough.

**14 Washer over square rod end . . .**
has leg bent to fit in small hole. Washer hole is square, preventing angular movement when all three parts are assembled and fastened with screw and washer.

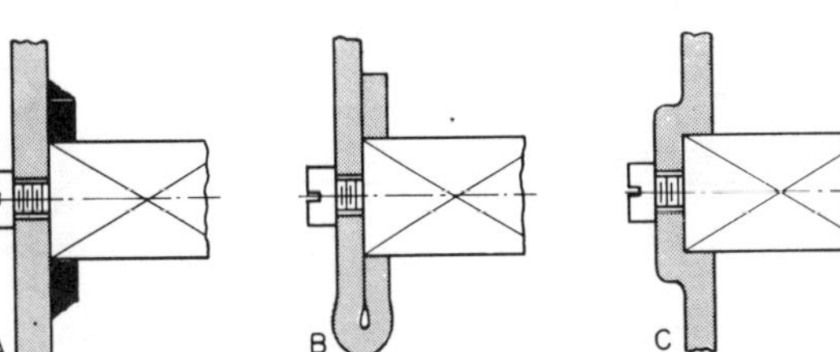


**16 Double sheet thickness . . .**
allows square or hexagonal locating-hole for shaft end to be provided in thin sheet. Extra thickness can be (A) welded (B) folded or (C) embossed.

# Joining Circular Parts Without Fasteners

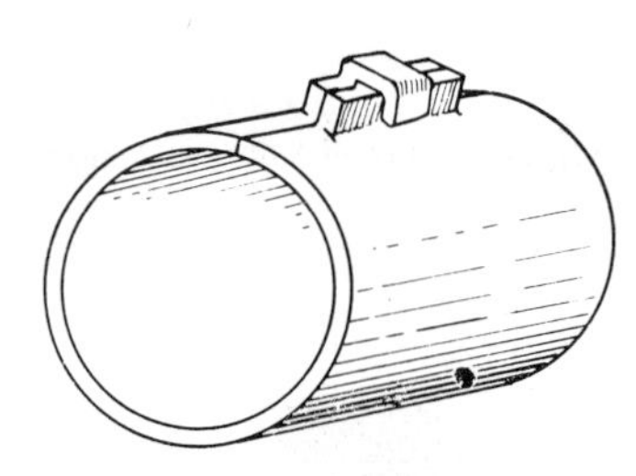

**Fig. 1—Fastening for a rolled circular section. Tabs are integral with sheet; one tab being longer than the other, and bent over on assembly.**

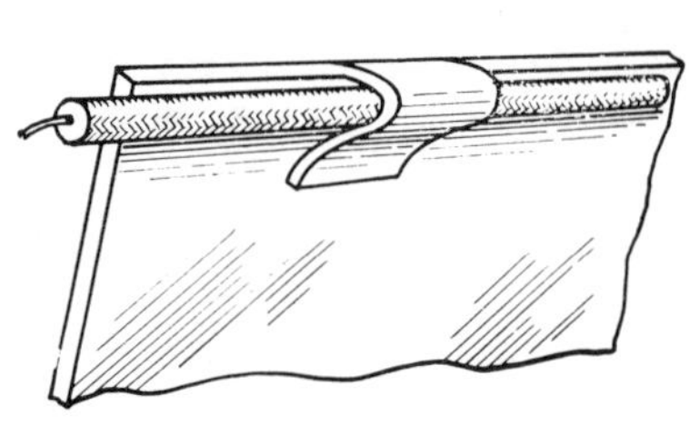

**Fig. 5—Similar to Fig. 4 for supporting electrical wires. Tab is integral with plate and crimped over on assembly.**

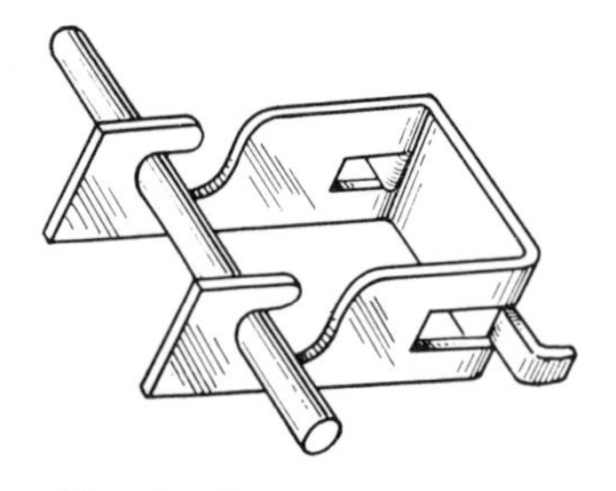

**Fig. 6—For supporting of rods or tubes. Installation can be either permanent or temporary. Sheet metal bracket is held by bent tabs.**

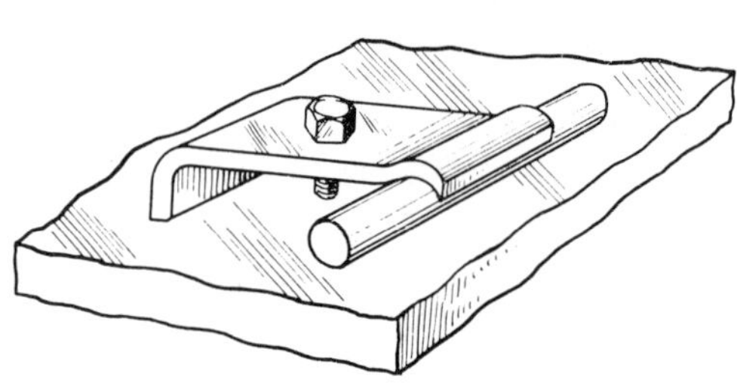

**Fig. 7—Embossed sheet metal bracket to hold rods, tubes or cables. Tension is supplied by screw threaded into lower plate.**

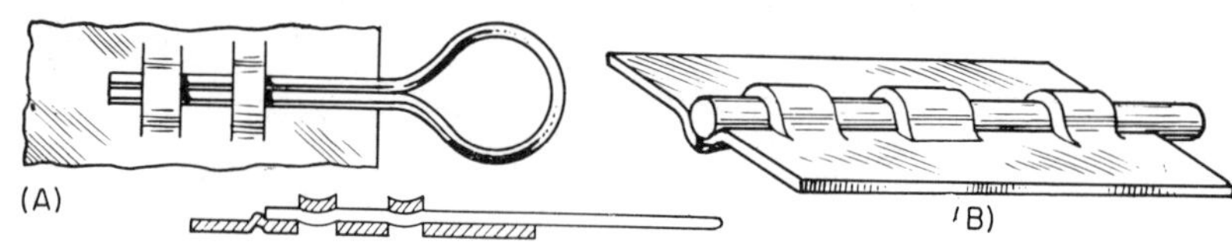


**Fig. 10—Plate is embossed and tabs bent over on assembly. If two plates are used having tab edges (B) a piano-type hinge is formed. (A) and (B) can be combined to form a quick release door mechanism. A cable is passed through the eye of the hinge bolt, and a handle attached to the cable.**

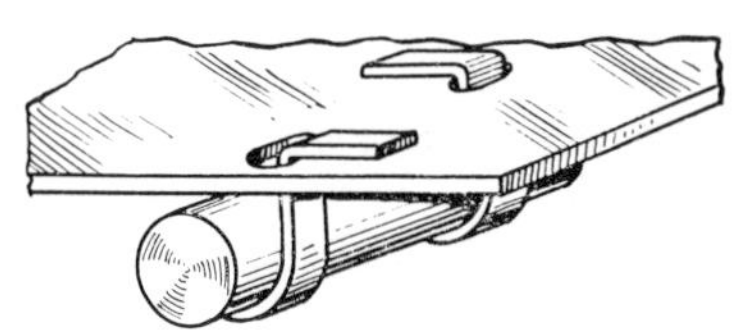

**Fig. 11—Rods and tubes can be supported by sheet metal tabs. Tab is wrapped around circular section and bent through plate.**

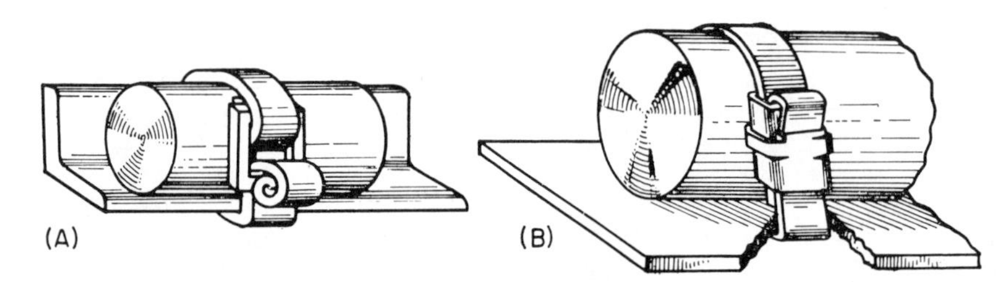


**Fig. 14—Strap fastener to hold a circular section tight against a structural shape. Lock can be made from square bar stock (A) or from sheet metal (B) tabbed as shown. Strap is bent over for additional locking. Slotted holes in sheet should be spaced equal to rod dia to prevent tearing.**

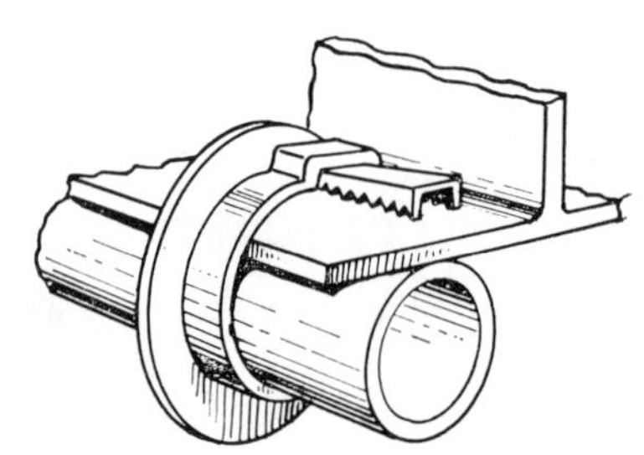

**Fig. 15—*C* clamp support usually used for tubing. Serrated wedge is hammered tight; serrations keep wedge from unlocking.**

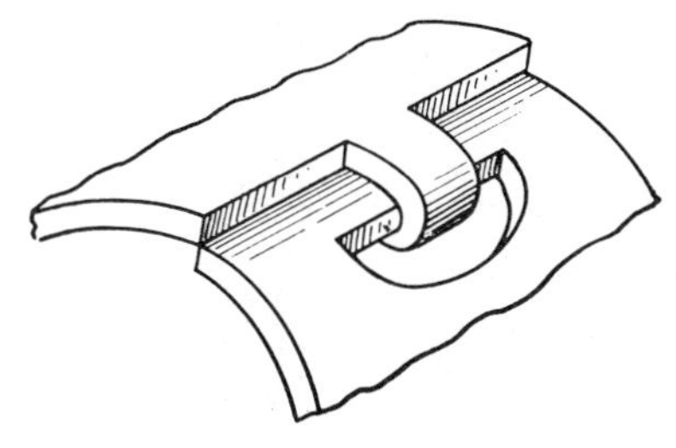

Fig. 2—Similar to Fig. 1 except tube is formed with a lap joint. Tab is bent over and inserted into cut-out on assembly. Joint tension is needed to maintain lock.

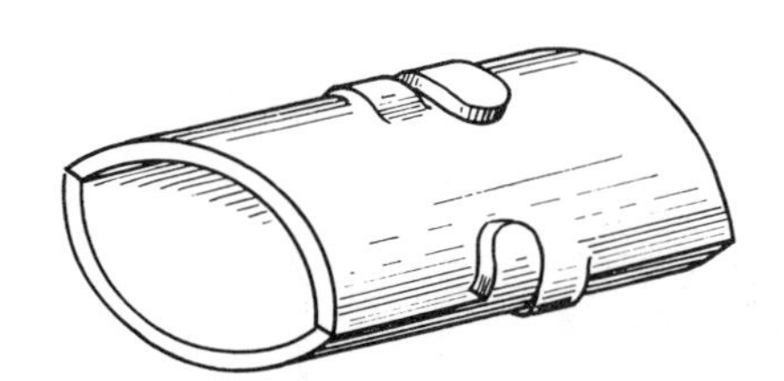

Fig. 3—Tab fastener for elliptical section. Tabs are formed integral with sheet. For best results tabs should be adjacent to each other as shown in sketch above.

Fig. 4—For supporting rod on plate. Tab is formed and bent over rod on assembly. Wedging action holds rod in place. Rod is free to move unless restrained.

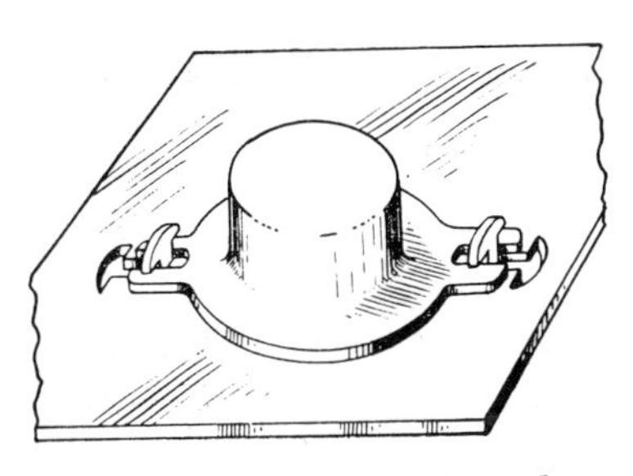

Fig. 8—Fastening of rod to plate. Rod is welded to plate with slotted holes. Tabs in bottom plate are bent on assembly.

Fig. 9—Tabs and bracket (A) used to support rod at right angle to plate. Bracket can be welded to plate. (B) has rod slotted into place. For mass production, the tabs and slots can be stamped into the sheet. For limited production, the tabs and slots can be hand formed.

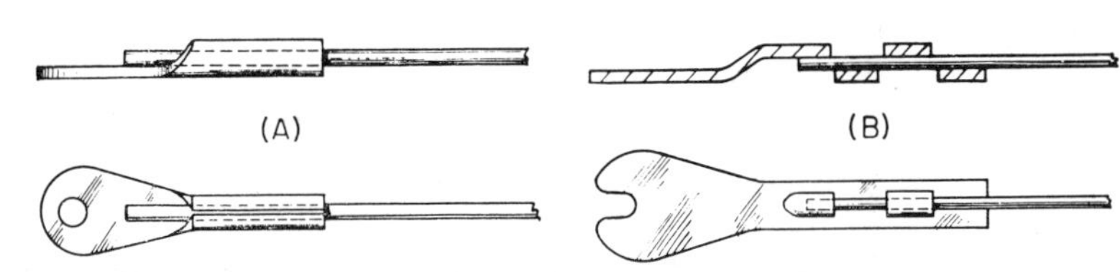


Fig. 12—For connecting wire ends to terminals. Sheet is crimped or tabbed to hold wire in place. Variety of terminal endings can be used. If additional fastening is required, in that parting of the wire and terminal end might create a safety or fire hazard, a drop of solder can be added.

Fig. 13—Spring joins two rods or tubes. Members are not limited in axial motion or rotation except by spring strength.

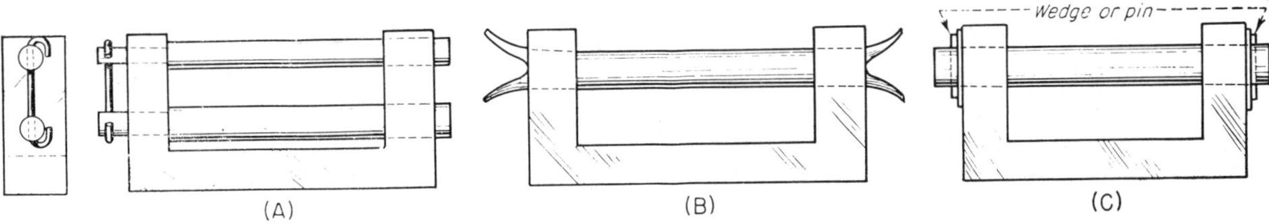


Fig. 16—Methods of locking rods in machine frames. In (A) one end of the rod is machined to a smaller diameter. Shoulder and bent member restrains rod from slipping out of frame. Limited axial and rotational freedom is present.

Split rod in (B) limits axial motion but permits rotation. Rod is split on assembly. Wedge or pin in (C) bear against washers. Axial motion can be restricted but rotation is possible. If rod is to be a roller, bearings can be inserted.

# 23 Ways To Attach Small Die Cast Parts

HIRAM K. BARTON
Barton Die Casting Service
Chinley, England

SMALL DIE CASTINGS are commonly used as non-structural, non-mechanical parts. For example: insignia; trademarks; and decorative trim. One requirement that all such parts have in common is that they must be attached firmly with reasonable cost and effort.

Many of the methods discussed on the following pages are functionally interchangeable; adopting one in preference to another may only be a matter of the degree of accessibility. Where specific advantages for given applications do exist, these are pointed out. Furthermore, most of these techniques take advantage of some particular feature of zinc alloy die castings—ductility, thinness of section, ease of using integral inserts—but some are equally applicable to molded plastic components as well.

**1.** Offset integral lugs, formed either on the outside edge of the die casting or within an internal recess, are so placed that no moving core is required. The screw hole can always be cored. For effective use of offset lugs, thickness of material is held to close a tolerance (plus 0.000, minus 0.010 in.) for rigidity of attachment. Where thickness is likely to vary, cored hole in lug can be tapped for a set-screw.

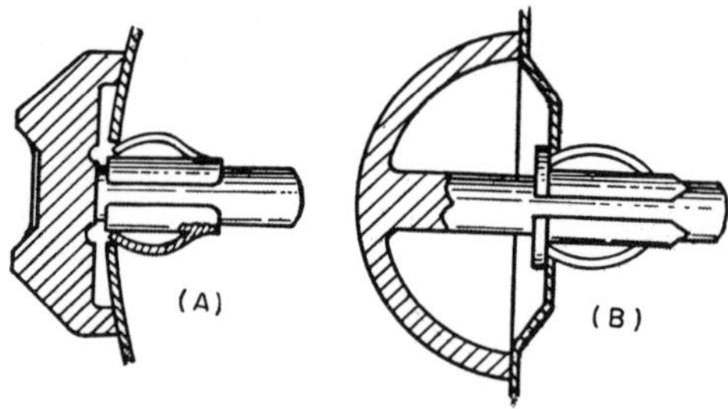


**4.** Integral stud is used with speed-clip which has bowed spring members to hold trim in place. With (A), stud must have shoulder (about 0.010 in.) to locate the clip which is pushed over stud before assembly. In (B), a collar prevents clip from passing through hole. Clip is inserted first; the stud (on trim) is pushed into it. Curved segments at rear of clip bite into the stud and prevent it from drawing out. The final position on the stud is not critical.

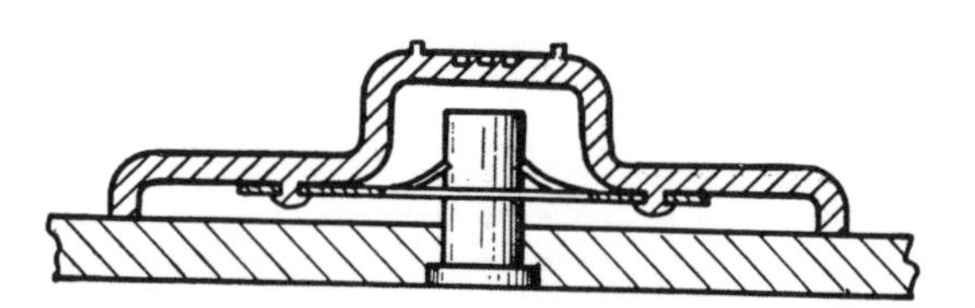

**7.** Extended speed clip with two punched holes is attached by integral rivets to the underside of the trim and is forced over a stud projecting from the main part of the assembly. After being pushed over the stud, trim is given a quarter turn to tighten down the clip. This is advantageous only when rear of assembly is inaccessible, and service is not severe. If the trim is circular, the stud can be threaded.

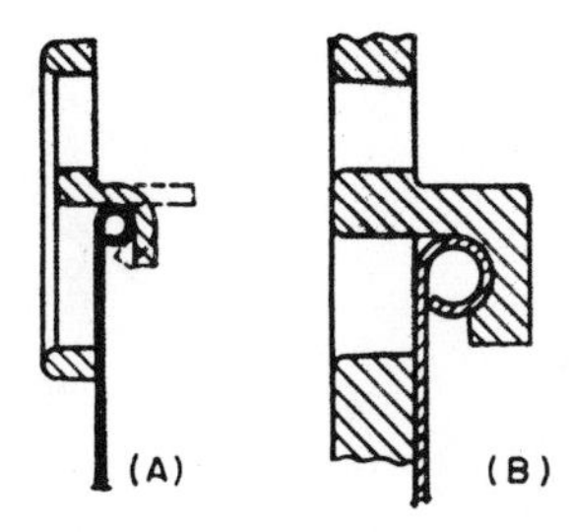


**2.** Small offset lug overhanging the bead can be used for attaching trim to rolled edges. Lug can either be formed to drop over the edge of the bead and then pressed to retain it as in sketch (A), or it can be formed to hold the bead as illustrated in (B). The latter is only practicable when the trim piece can be slid into position from one end of the rolled edge.

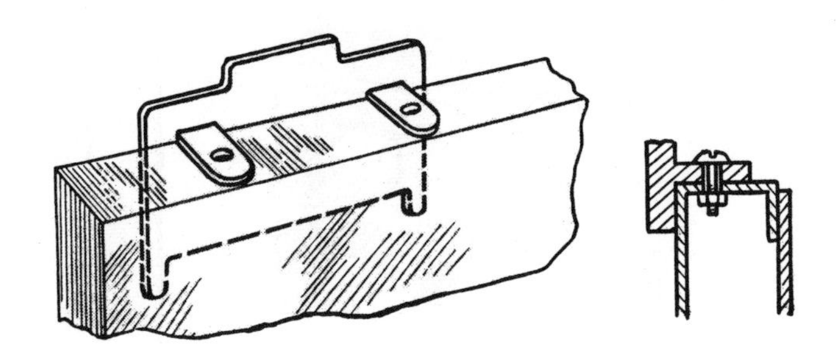

**3.** Flat integral lugs are applicable where the trim extends beyond the edge of the part to which it is attached. It is not usually economical to core the holes in the lugs; they can be drilled or punched if the lug is 0.080 in. or less in thickness. These lugs can be used for attaching trim to wood, fiber, or molded plastic, and sheet metal (with a bolt and nut) if part is formed to a box section as in small sketch. This method is commonly used for display stands and vending machines.

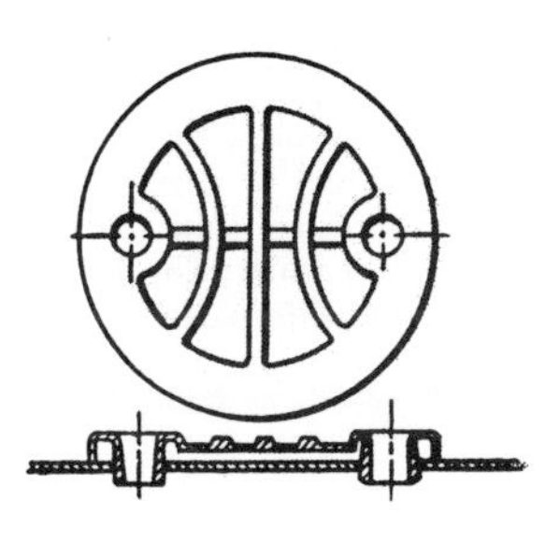

**5.** Hollow expandable studs, cast integral with trim, engage holes punched in the sheet metal as shown. A bullet-nosed punch is forced through the die casting, expanding the cored hole to lock the trim in place. The initial clearance between stud and punched hole diameters should not be greater than 0.008 inches. Applications are restricted to assemblies where the cored holes in the trim are desirable.

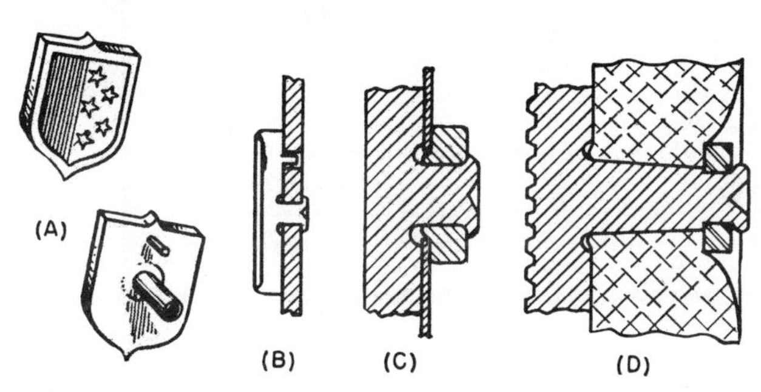


**6.** Integral, riveted stud is one of the most widely applicable fastening methods. When the thickness of the material is not less than the diameter of the stud, it can be peened as shown. If the material is thin, a washer, (C) and (D), is recommended, since short studs are severely stressed by heading. It is always advisable to specify a filleted channel around studs, to reduce stress concentration. Hard fiber washers can be used where vibration might cause chatter. Small pieces of trim held by a single stud need some means of preventing rotation like a stud.

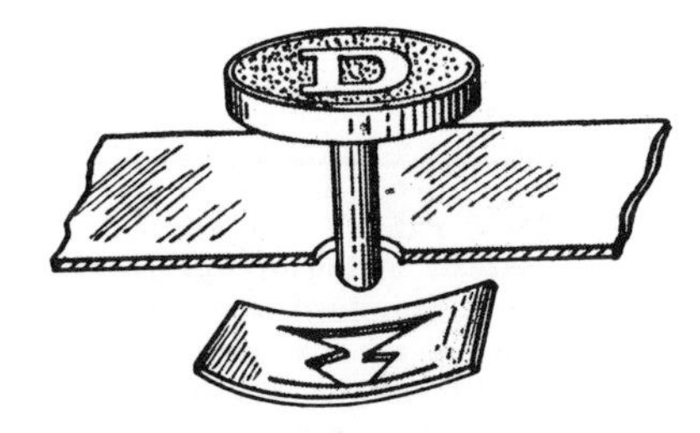

**8.** Integral stud and speed clip is widely used in the automobile and appliance fields. This method permits quick manual assembly. Speed clips can be made, and are also commercially available, in almost any shape to suit specific requirements and applications; the type illustrated here is useful when the rear of the assembly is accessible and clear of other components.

**9.** Tee shaped lugs for locating trim at the junction of two structural elements of an assembly are cast on the underside of the trim. These lugs are located between the two parts and, by twisting through 90 deg, are wedged as shown. Width of lug should be twice the thickness, and length of the portion subject to twisting at least four times the width. This method is only economical when the lugs can be formed in the parting-plane of the die casting die.

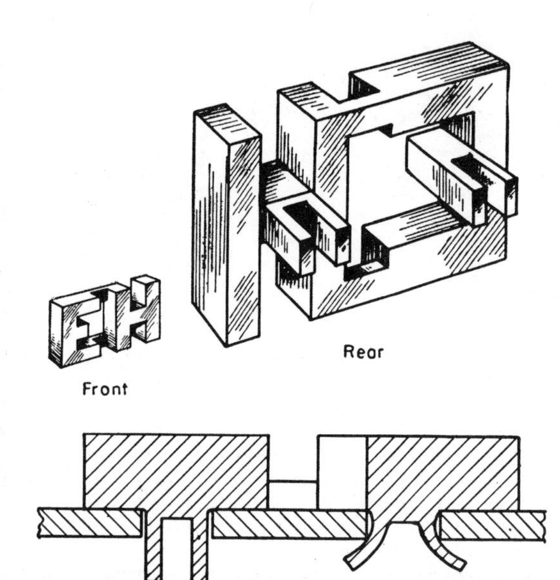


**10.** Rectangular lugs, projecting from the rear of the trim, pass through a square punched hole in the casting and are peened in place. This gives a sturdy assembly and is preferred to riveted round studs that project from the surface and are susceptible to failure or damage by accidental impact. For small pieces, a single pair of lugs usually suffices; the square holes prevent rotation.

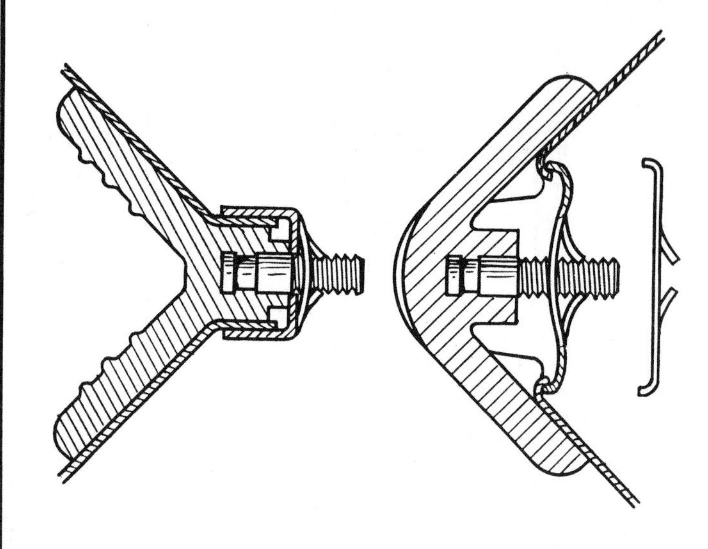

**11.** Trim strips covering joints of sheets at right angles give support in addition to their decorative function. Inserted threaded studs, used with rigid steel clips and/or speed nuts, provide a simple, reliable attachment. Edge of sheet must be turned-in to engage clip. For filler trim, left, speed nut separate from clip is necessary; for bead trim, right, single clip is usually satisfactory.

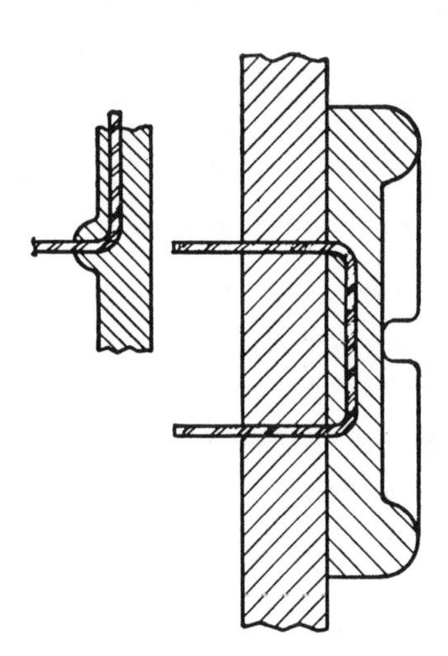

**14.** Round wire staples, cast as inserts, will hold thin-section insignia securely to almost any sheet materials. Holes should be pre-punched. Staples are clinched against filleted anvils. For trim of exceptionally thin sections, points of staple protrusion should be strengthened, left, the pips so formed are forced into the surface of the sheet if it is deformable. If not, dimples can be used if holes are punched.

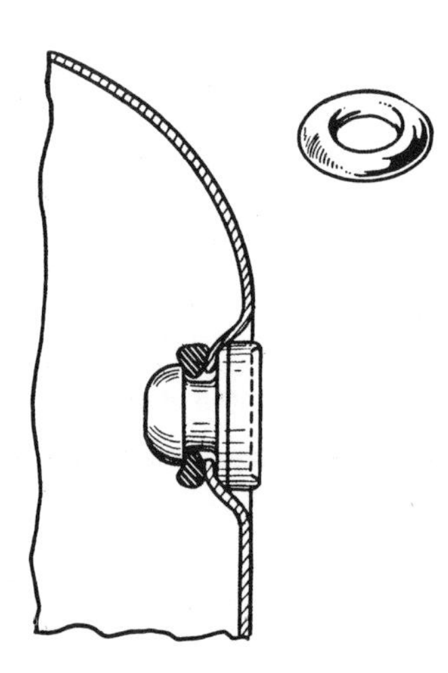

**15.** Rubber grommet stretched over button is a simple, but effective means for fastening small insignia located in recesses on the face of a sheet-metal housing. The rear of the die casting has a head that projects through a punched hole in the sheet; thick rubber ring slips over this between the projecting head and the edge of the sheet. The assembly is vibration proof and will keep insignia from rotating.

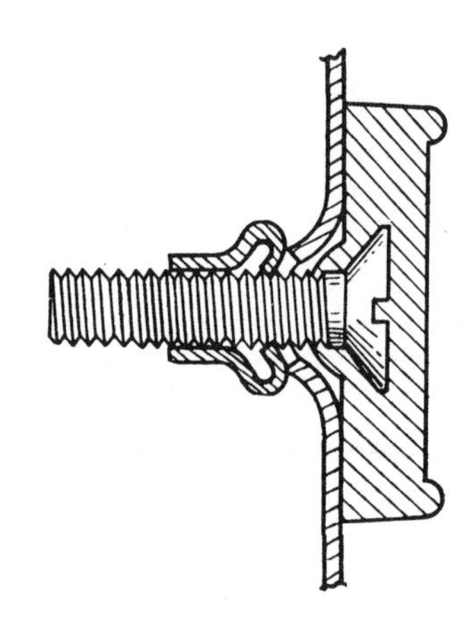

**16.** Screwthread insert and rubber nut can be used instead of speed nut for light duty. Standard machine screws can be used as inserts. These pass through punched holes in the main assembly, and are secured by forcing short lengths of extruded synthetic rubber sheathing over the ends. If the sheath is forced down until it doubles upon itself, as shown in the above sketch, the assembly will be vibration-resistant.

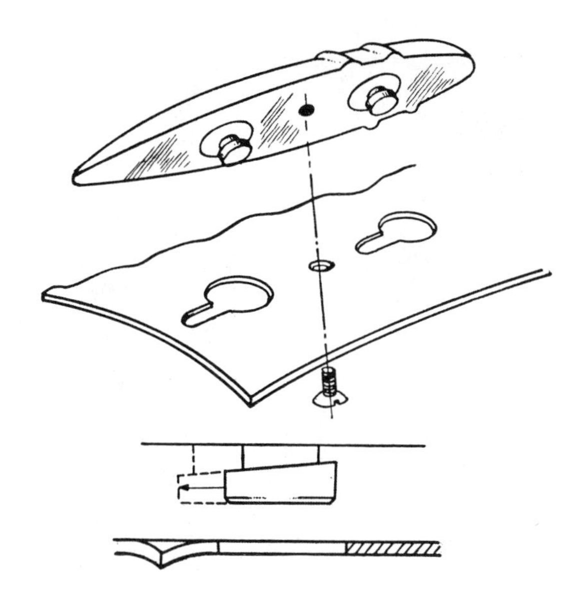

**12.** Long trim strips can be fastened with headed studs which enter key-holes punched in the sheet; after inserting, the whole strip is displaced about 3/16 in. to lock the heads under the narrow portions of the holes. A screw can then be added to prevent any further movement of the trim element. The heads of the studs are skewed to give a wedging action.

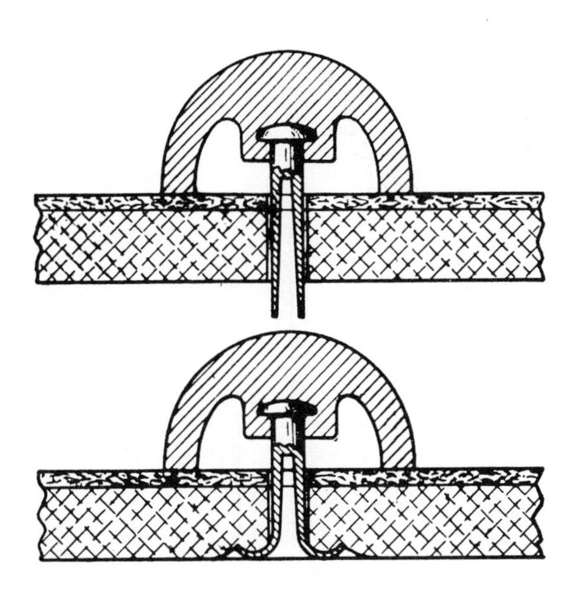

**13.** Bifurcated rivets can be cast in die casting, and in some applications are better than loose rivets for attaching small parts to fiber-board and similar materials. These fasteners are particularly suitable to heavy baggage items, display cases, and the like. Locks, hinges and protective ribs can also be attached. For added security, a washer can be placed over each rivet before clinching.

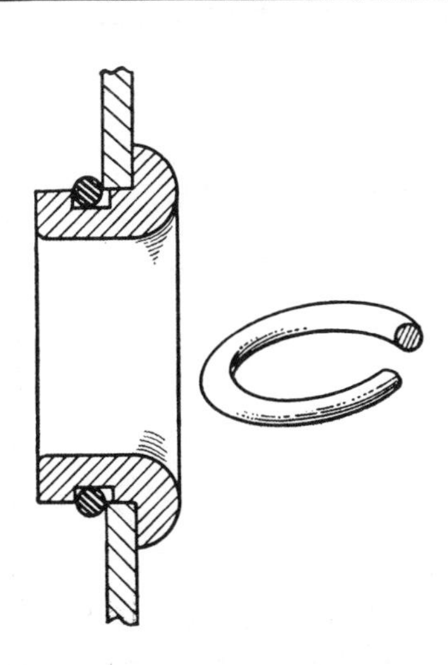

**17.** Die cast ring, which protects electrical leads from damage, is held in place with a snap ring. Groove for ring can be cast but a collet die will be required to avoid flash-lines on the exposed surfaces. Trimming will also be necessary to make certain that the ring will seat securely throughout its length. As a result of these requirements, the manufacturing costs may be prohibitive for many types of applications.

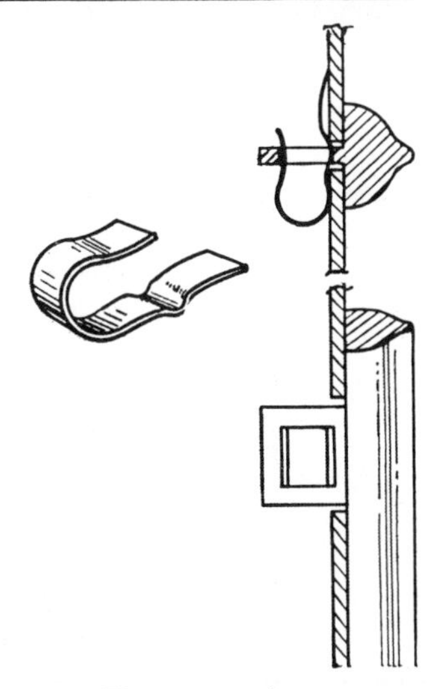

**18.** Flat spring loop engages a cored lug that projects from the back of the trim and passes between adjoining metal sheets or, as shown, through a punched rectangular hole. This method is primarily for applications where the trim strips are located on each side of a removable panel, but with rear access. It is well adapted to machines where front access is essential for servicing, but the rear should be protected.

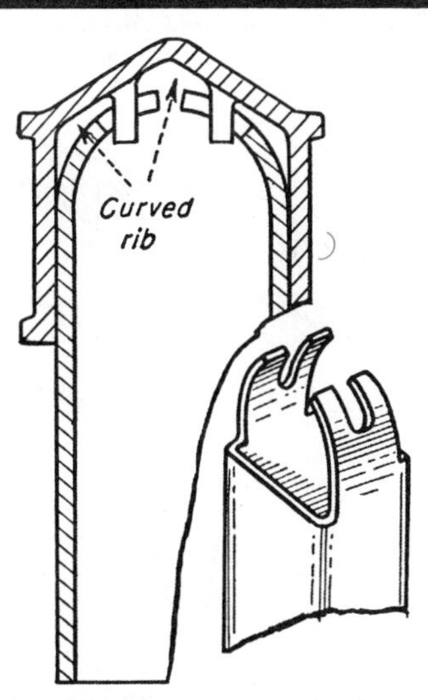


**19.** Cover for rectangular, sheet metal tube. Ends of tube are notched. Inside of cover has a pair of studs cast between two narrow curved ribs. When die casting is pushed over end of tube, the two notched lugs are forced to follow curve of ribs until they engage the two studs; both fittings are assembled simultaneously. Method is suitable for medium and large size tubing.

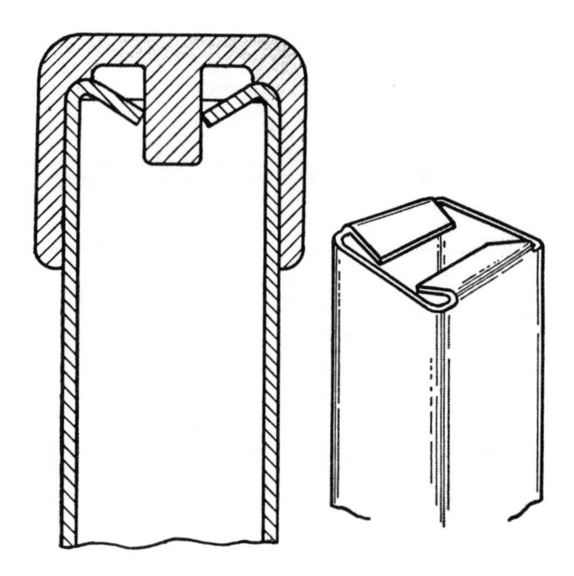

**20.** Simple cover suitable only to elastic materials, such as steel or brass. Ends of the tube are cut, sheared, and turned inward at an angle of 10-12 deg as shown. Die cast cover has a projecting lug of rectangular section; when cover is forced on, the lug passes between the turned-in tangs and presses them inward, resulting in a spring locking force.

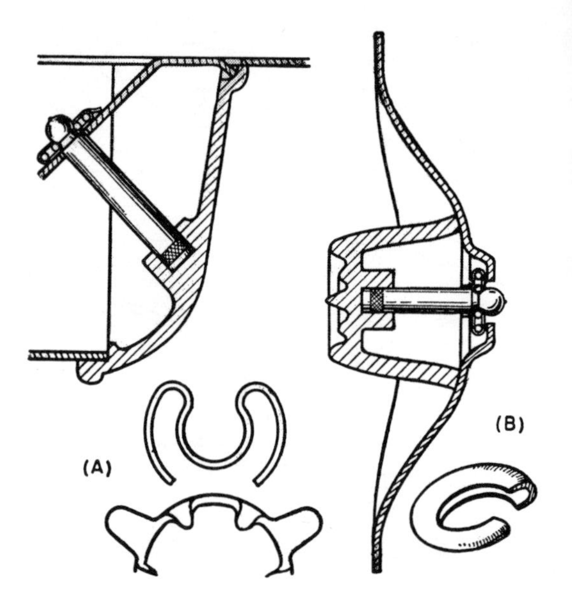


**21.** Diecast trim is provided with rod inserts having spherical ends. Rods engage spring clips of snap-button type located on the main assembly. Clips consist of a spring wire loop with reverse bends held in a stamped retainer. Latter may have projecting lugs (A), or a plain hollow ring (B), that is held by lugs formed on main assembly. (B), trim is readily removed and replaced.

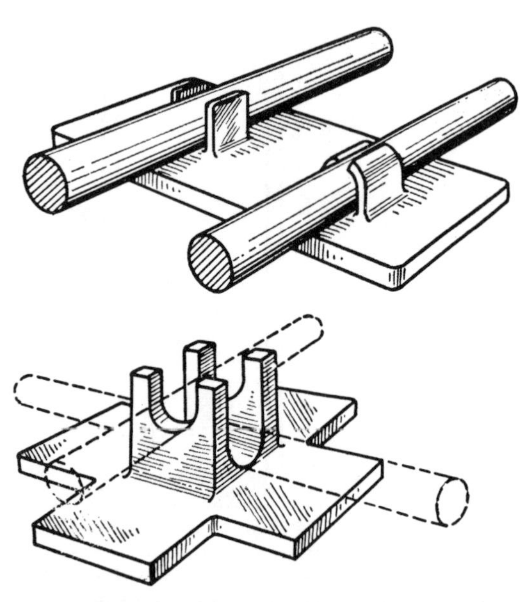

**22.** Deformable lugs can be used to hold die castings that are to be mounted on a grille or part with cylindrical elements. Lugs are bent around the rod as shown. When two lugs are used, the shape of the recess between them should conform to that of the rods (if the latter, as is likely, are non-circular in section). If trim is large, attachment to vertical and horizontal rods should be made by separate pairs of lugs. Small insignia can be located at an intersection using forked lugs.

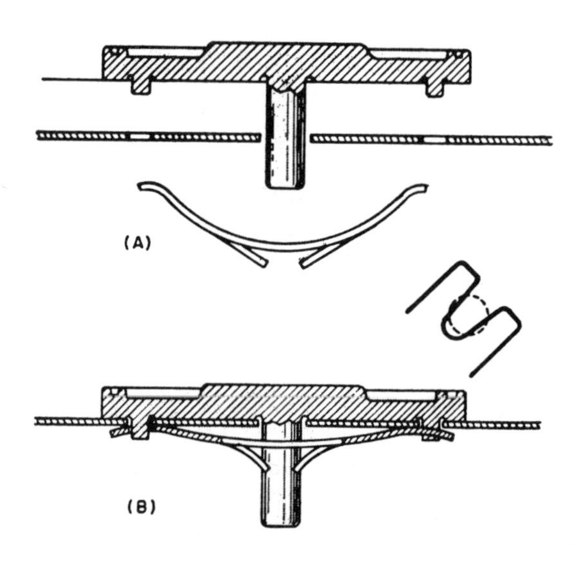


**23.** Integral studs and wedging speed clip pass through punched holes in the bodywork. The speed clip, which is more curved than usual, has extended, fork-shaped end as shown in the auxiliary view, the slots being slightly tapered and, at their outer ends, wide enough to engage the two shorter studs on the trim. Clip can be pushed over the central stud with the forked ends partially engaging the outer studs. When the clip is engaged, as illustrated in (B), the ends are held rigidly.

# The Fundamentals of Selecting Locknuts

C. C. FERONI
Chief Product Engineer, Elastic Stop Nut Corporation

A CONFUSING VARIETY OF LOCKNUTS differing in principle, materials, method of manufacture, and cost is now available. While the primary purpose of all of them is to resist loosening under service conditions, one or more of several other factors may also be important if the best balance between performance and over-all cost is desired. In many applications, fastener cost itself is incidental when compared to the potential cost of a joint failure. In others, it may not be economically justifiable to use the best quality fastener.

Unfortunately, there are no simple rules or formulas that will pinpoint the best nut for a given application. All that can be offered is a list of factors that should be considered, with supplementary data based on years of observing the mistakes, questions, and needs of users in many different fields.

A guide to locknut performance for critical applications is given by aircraft specifications AN-N-5 and AN-N-10. After establishing strength and materials criteria, these specifications go into the question of torque. First, they set a maximum "first-on" torque. If this is too high, assembly will be difficult.

Next, starting torque is specified. This is the torque required to move a nut from any position where all threads are engaged, but the nut is not seated. A minimum value is set, since this is a measure of the tendency of a nut to stay in place.

A minimum prevailing torque is also specified. This is the torque required to keep a nut turning, once it has been started. It, too, is a measure of the nut's tendency to stay in place after assembly.

Regarding reusable life, the specifications set minimum starting and prevailing torques for nuts, to be measured after they have been turned on, and removed from, a bolt a given number of times. They specify that a nut damages neither bolt threads nor mating surfaces.

Finally, a vibration test is outlined. A test rig is described, which will subject a joint to impact and to vibration of different frequencies. After a given period, a nut must not have moved with respect to the bolt, and must not be capable of being turned by hand.

Most important in specifying locknuts is the question of how resistant to loosening and how reliable the joint must be. The different types vary widely in this respect. Several factors affect the answer. Some are pointed out in the accompanying table, but others defy such generalization. For example, in many applications maximum joint reliability is imperative—loss of life or damage to valuable equipment could result from loosening of a joint. In such cases, cost is no factor unless there is a cost difference between two nuts of the same reliability.

On the other hand, there are applications where reliability can be sacrificed to some extent in favor of cost. A fastener that meets the rigid AN specs may not be needed. Consider a household appliance which has a component attached with locknuts to prevent rattling. If an occasional joint loosens, no harm is done except that a service call is required. It might be needlessly expensive to use a top quality fastener to obtain maximum reliability. But it would not be permissible to have several joints loosen; service costs would become excessive, and the manufacturer's reputation would suffer. Somewhere in-between lies the answer. This is the type of decision that the designer must make in the light of his particular requirements. Some of the factors that determine what class of locknut should be used are:

1. *Can the risk be run that occasionally something may accidentally be forgotten in assembly?* Class I nuts depend on a secondary element, such as a lock washer or a cotter pin. Not only might this be overlooked in assembly, but lock wires or cotter pins can break under severe vibration. Without the secondary element, these nuts are as free to spin as plain nuts. There seems to be little justification

for the popularity that still exists for this class. Probably it is partially inertia in resisting a change from practice and partially a reluctance to trust what seems like less positive methods. Actually many types often perform better.

2. *Can the nut be seated tight against the work?* Class V types lock only when they are tightly seated against one of the surfaces being joined. If any relaxation occurs, they are free to turn off. Such relaxation can occur due to creep of the bolt, wear or corrosion of mating surfaces.

3. *Is locking pressure spread evenly, or concentrated on a few threads?* If vibration is severe, the friction which prevents a nut from loosening should be spread over as large an area as possible. This requires expensive, precise manufacture. Some nuts lock with interference between a few threads or load some threads more than others.

4. *Are all nuts of a given type equally reliable?* Some classes of nuts are not uniform in their locking ability. The amount of distortion of shape which causes locking may vary from nut to nut. Also, these types cannot adjust to the normal variations of bolt diameter within normal tolerance limits. Thus, on bolts with diameters near the low limit, some nuts may not lock at all.

5. *Will the joint be exposed to high temperatures?* Some types of Class IV locknuts have plastic or other non-metallic inserts to obtain locking action, and are not recommended for use above 250F.

6. *Will the nut be frequently removed and reused?* Often, bolted joints must be broken periodically for inspection, repairs, access, or maintenance. The types of nuts which jam a few threads together for locking can damage bolt threads to the extent that the joint cannot be remade unless bolts are replaced.

7. *Is speed of assembly important?* It obviously takes more assembly time for nuts which require extra motions to lock, such as insertion of cotter pins or lock wires. With large volume production, this is an important factor.

8. *Is ease of assembly important?* When a locknut is required for a relatively inaccessible location, the free-spinning type, Class V, may be preferred.

9. *Will the nut damage the bolt or the work surfaces?* When a joint design is critical and maximum strength is required, stress raisers which jammed-on nuts can cause should be noted.

10. *Is there relative motion between parts bolted together?* If this is so, a castellated nut with pin or lock wire may be best, since repeated rotary motion might loosen other types.

All of these factors add up to locknut performance. Performance costs money. A precision-made nut, which fits all bolts within normal tolerance range, distributes locking friction evenly over all threads, can be removed and reused dozens of times, and locks anywhere on the bolt, whether seated or not, costs more to make than a nut which locks by means of threads distorted slightly by a hammer blow. A design must be analyzed to determine how much reliability and performance is required. On the other hand, it is poor economy to underestimate performance and reliability requirements.

## BASIC CLASSES

### CLASS I—Pins, Keys, Tabs, Safety Wire

**Assembly is more costly than for other types, and there is the possibility of forgetting the locking member at assembly. Also, the very vibration which the locknuts are intended to withstand may cause fatigue failure of the pin, tab or wire. On the other hand, these types of nuts are most satisfactory where some relative motion exist between the respective part of the joint.**

### CLASS II—Deformed Threads

**Widely used for moderate service conditions because of low original cost. Several factors limit their performance: (1) Locking friction is a result of high pressure in a few localized places; (2) Performance is usually erratic; (3) Cyclic loading and vibration can materially wear the interference points. Often merely loading the nut causes local yielding and a loss of locking torque. Also, deformed threads do not lend themselves to frequent reuse.**

### CLASS III—Secondary Spring Elements

**These often give an attractive balance between cost and performance. Initial cost is usually low and reliability is adequate for many applications. Limitations: (1) Spring member may fail from vibrations; (2) Locking effectiveness is reduced or even lost if bolt stretches or mating surfaces wear; (3) Most nuts of this type tend to score the surfaces on which they bear.**

### CLASS IV—Frictional Interference

**Highest performance, but more expensive than any of the other classes. Locking action comes from plastic deformation of elements of the nut itself. These elements are either non-metallic inserts, or slotted collars. With either, spring-back tendency which produces action is uniform and does not depend solely upon bolt tension. Also these types need not be seated to lock. Reuseability is high. Operating temperature is limited.**

### CLASS V—Free Spinning

**These include types that are free to spin until seated. Advantages: Easy and inexpensive installation; easily removed and replaced. Disadvantages: Any loosening of joint or loss of bolt tension converts them into free running plain nuts. Also, since the clamping force is in addition to bolt stress and is usually confined to the lower threads where load is already high, combined stress in these threads lowers the fatigue and impact resistance.**

# OF COMMERCIAL LOCKNUTS

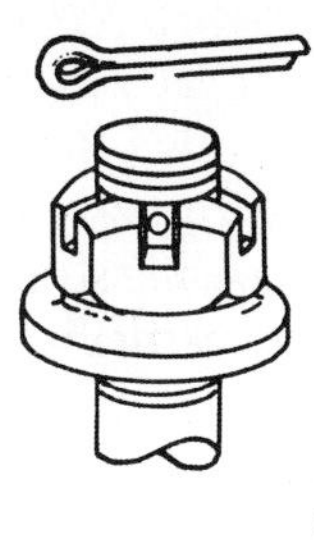

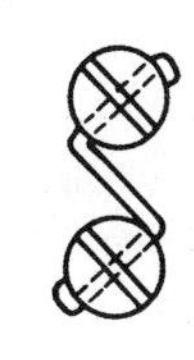

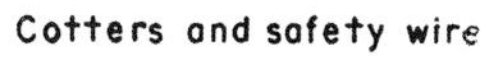

Cotters and safety wire

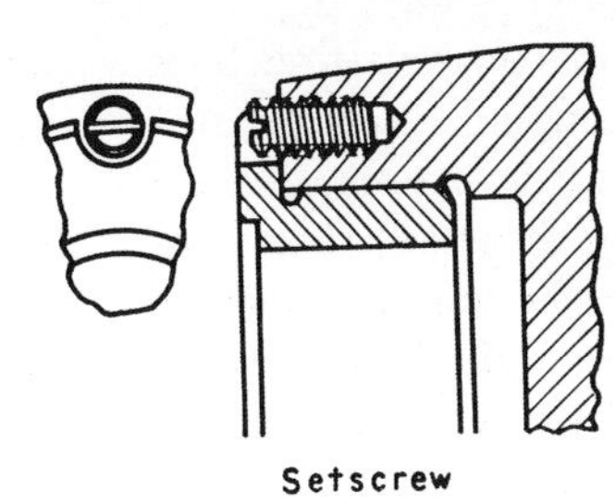

Setscrew

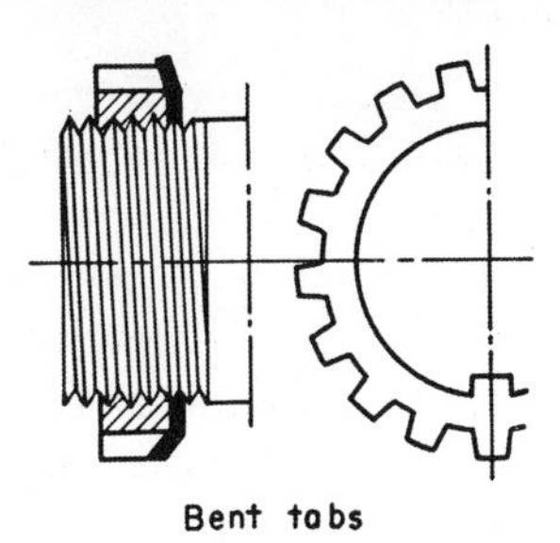

Bent tabs

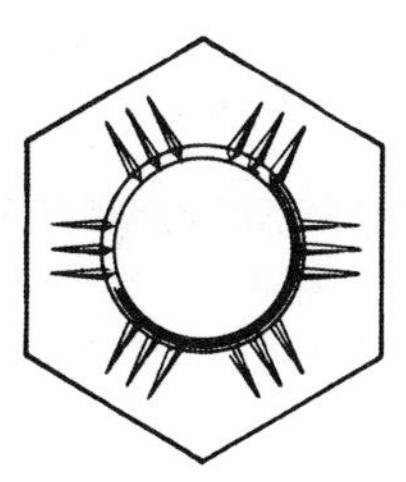

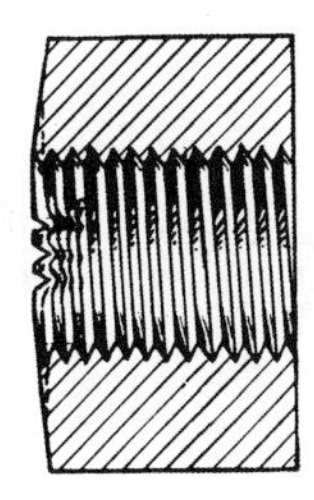

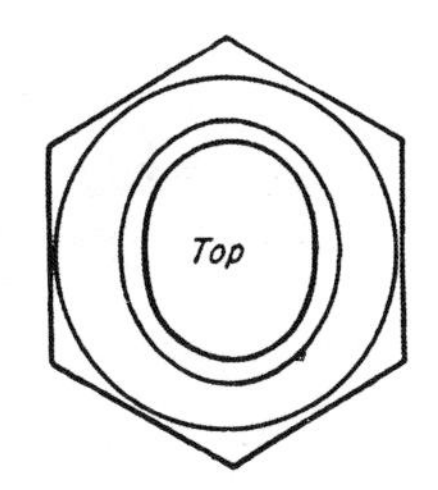


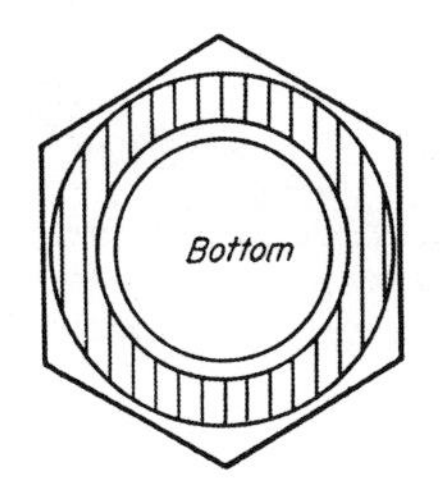


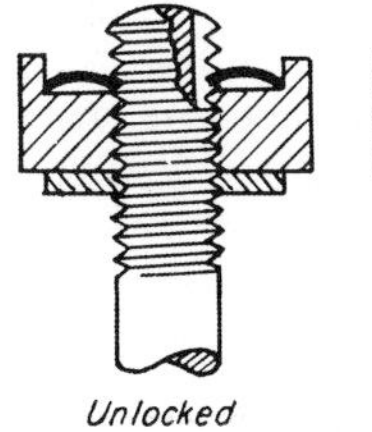

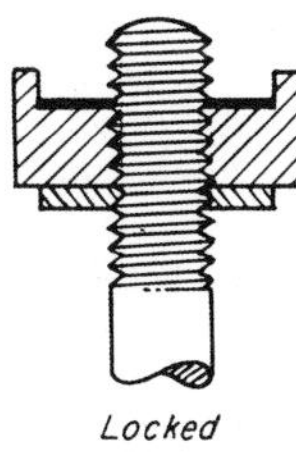

Unlocked Locked

Disk Type Spring

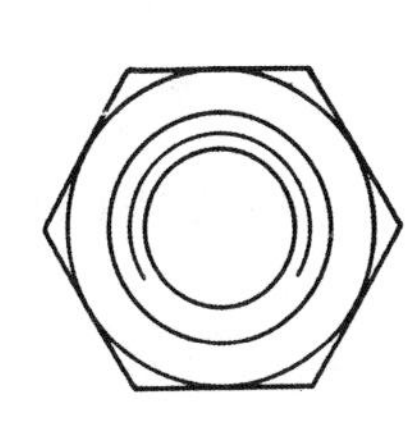

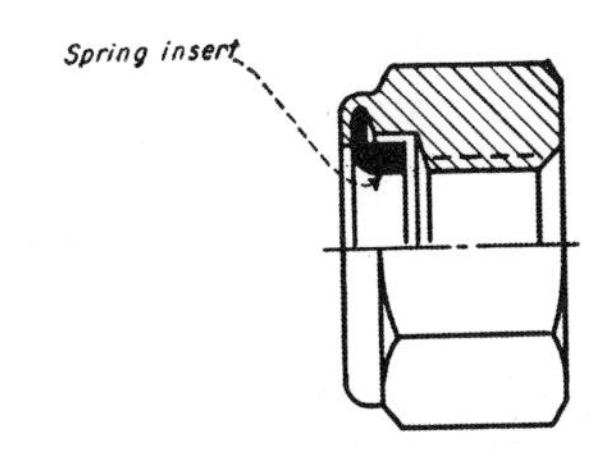


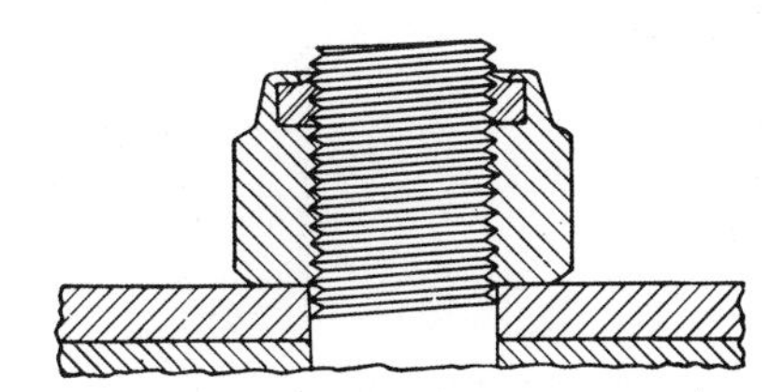

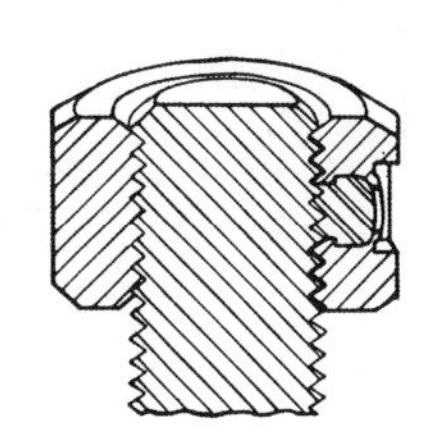

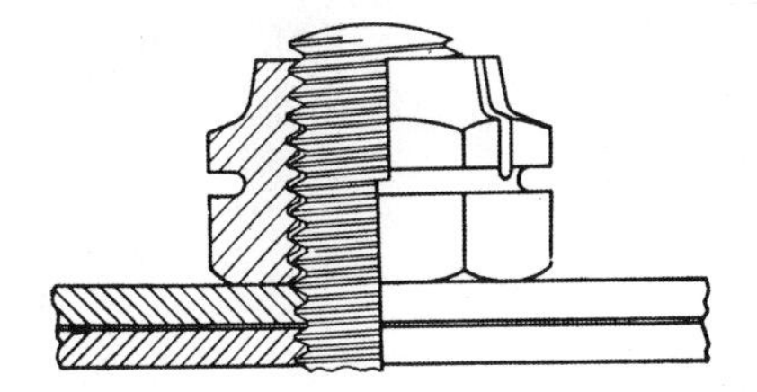

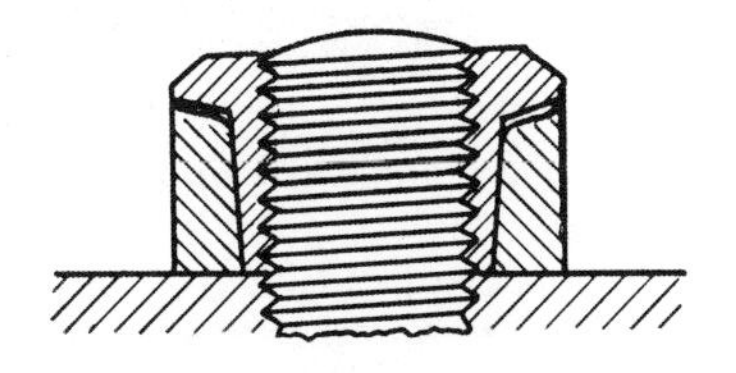

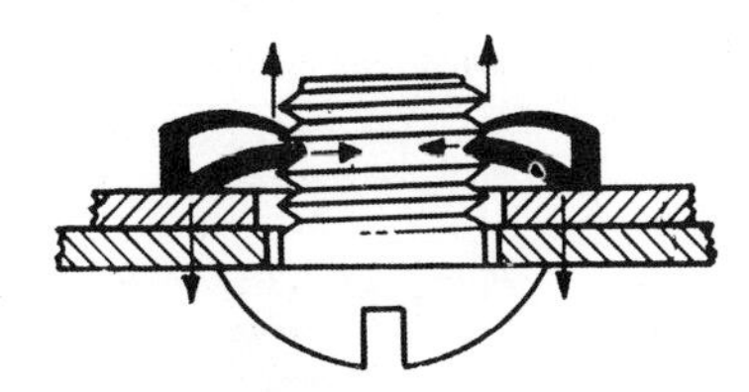

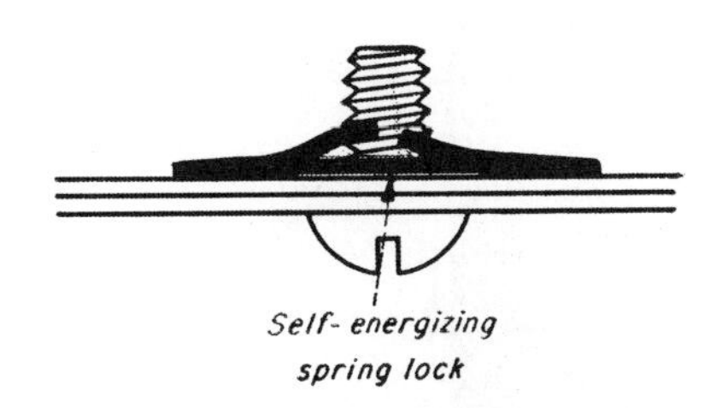

Self-energizing spring lock

# FASTENING METHODS FOR

Methods for fastening flexible materials such as cloth, felt, sheet rubber, leather, and insulating material include sewing, adhesives, vulcanization with rubber, tacking, patented snap fasteners, screw fasteners, riveting, wire stitching, and crimping. Sketches on these pages show typical applications in upholstery, insulation, floor coverings, and for decorative or protective purposes in automobiles, radios, industrial machinery, and other products.

Although adhesives and sewing are by far the most popular methods, they are too commonplace to be given much attention here. When using other methods, clinch the fastener over as large an area as possible by using washers, grommets, eyelets, or molding strips. Also, sharp corners should be avoided, particularly where the material is subject to tension or repeated flexures at the joint. Free edges should be protected to prevent fraying, curling, or catching on pointed objects.

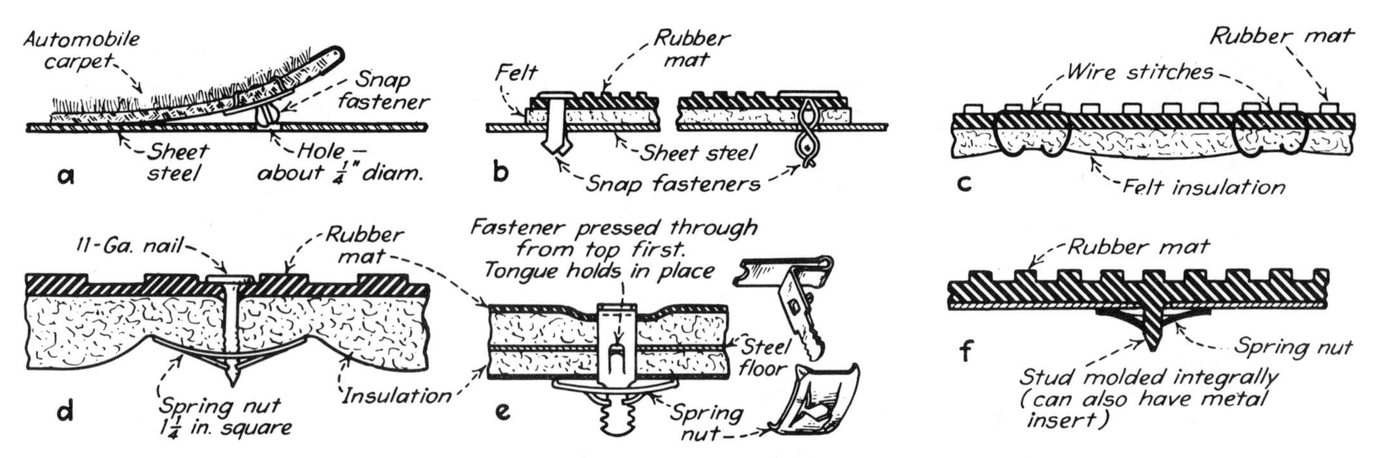


FIG. 1 — FASTENINGS FOR CARPET AND INSULATION

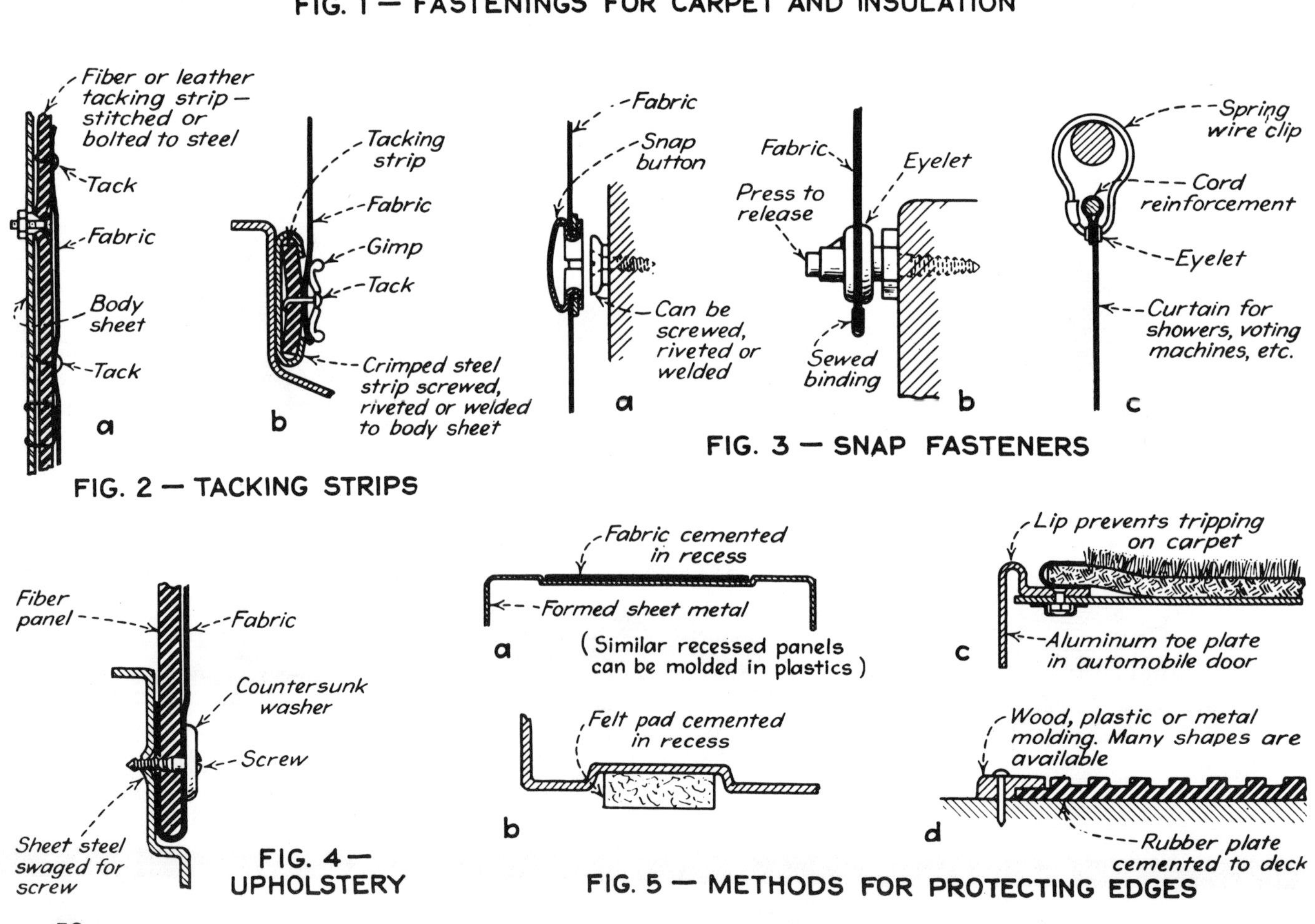


FIG. 2 — TACKING STRIPS

FIG. 3 — SNAP FASTENERS

FIG. 4 — UPHOLSTERY

FIG. 5 — METHODS FOR PROTECTING EDGES

# FLEXIBLE SHEET MATERIALS

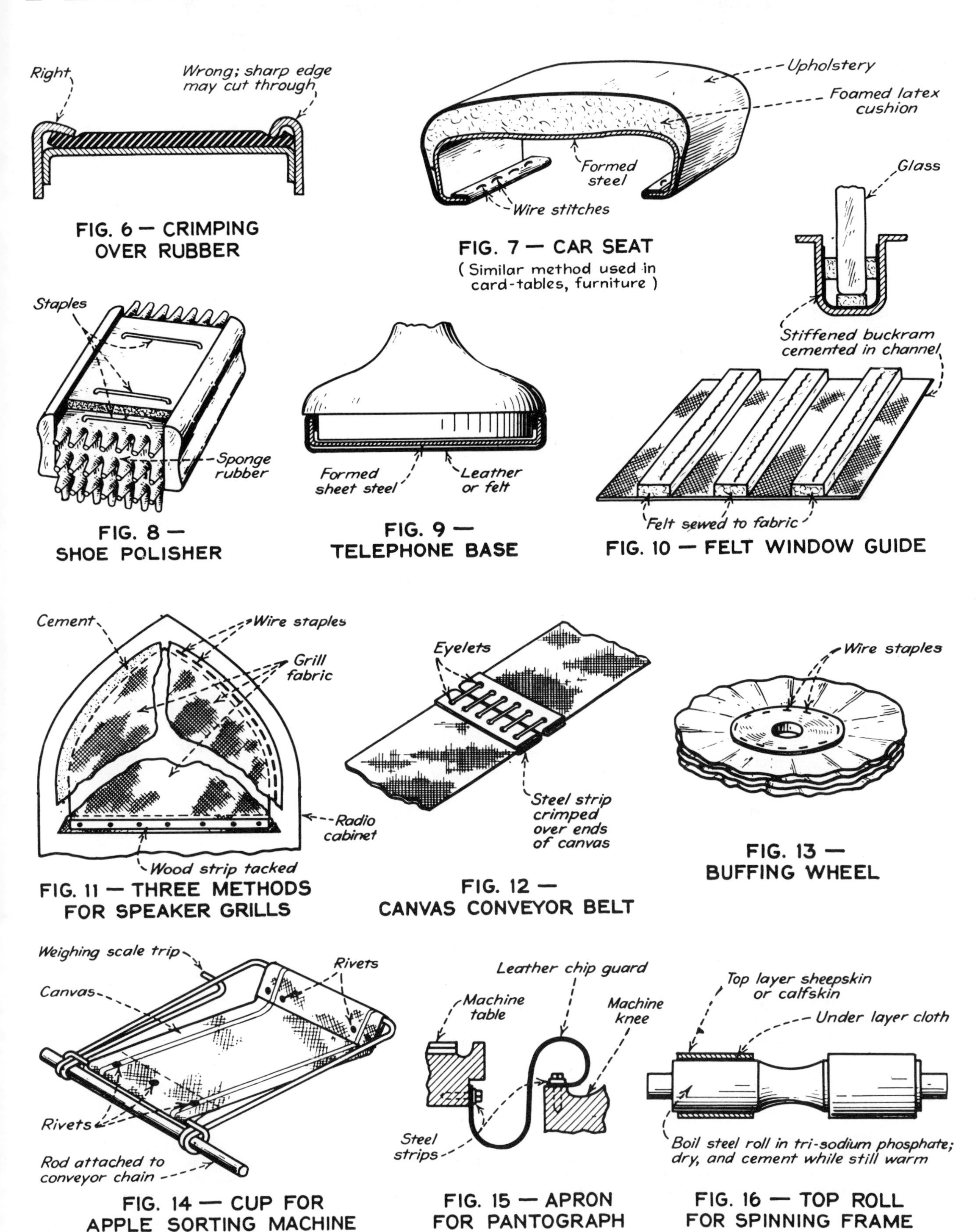


FIG. 6 — CRIMPING OVER RUBBER

FIG. 7 — CAR SEAT (Similar method used in card-tables, furniture)

FIG. 8 — SHOE POLISHER

FIG. 9 — TELEPHONE BASE

FIG. 10 — FELT WINDOW GUIDE

FIG. 11 — THREE METHODS FOR SPEAKER GRILLS

FIG. 12 — CANVAS CONVEYOR BELT

FIG. 13 — BUFFING WHEEL

FIG. 14 — CUP FOR APPLE SORTING MACHINE

FIG. 15 — APRON FOR PANTOGRAPH

FIG. 16 — TOP ROLL FOR SPINNING FRAME

# Methods of Attaching Glass

GENERAL RECOMMENDATIONS

1. Design the glass for compression loads whenever possible.

2. Load the glass uniformly. Avoid concentrated or point loads, by gasketing.

3. Design mounting to compensate for differential expansions. Glass have an expansion range from $1.8-5.0 \times 10^{-6}$ in/deg F.

4. Glass under load should be clamped in a rigid fixture that will not deform when load is applied.

5. Glass not under load can be mounted in a flexible support so that loads on the structure are not transmitted to the glass.

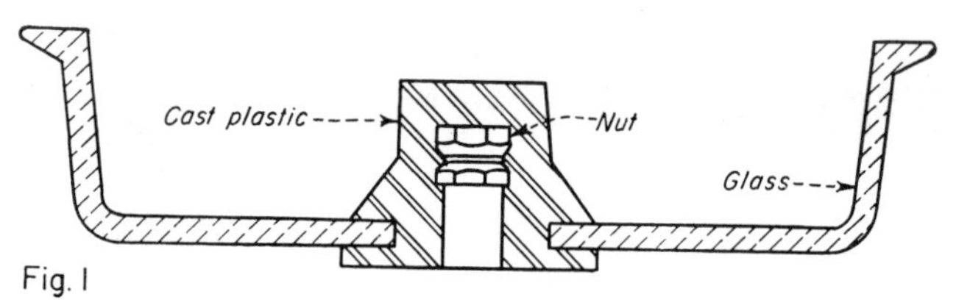


CAST PLASTIC JOINT. Glass part is attached to threaded metal insert by cast plastic hub. For moderate loads.

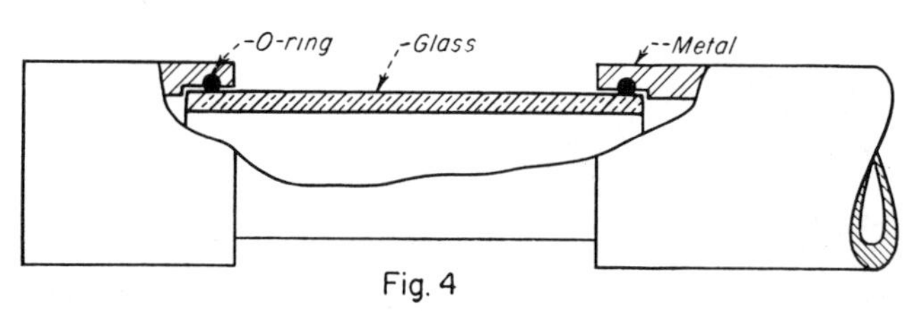


O-RINGS SEAL glass tubing to metal fixture. Shoulders in the metal tubes retain the glass member in its proper place.

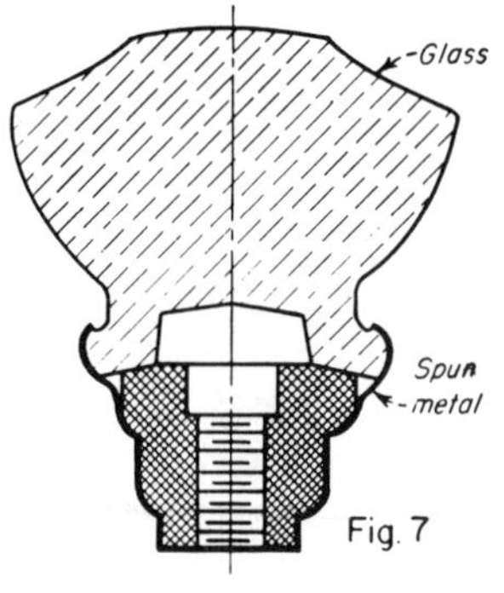


SPUN METAL JOINT. For light loads, a thin metal sheet can be spun around a glass shoulder to form an attachment as illustrated above.

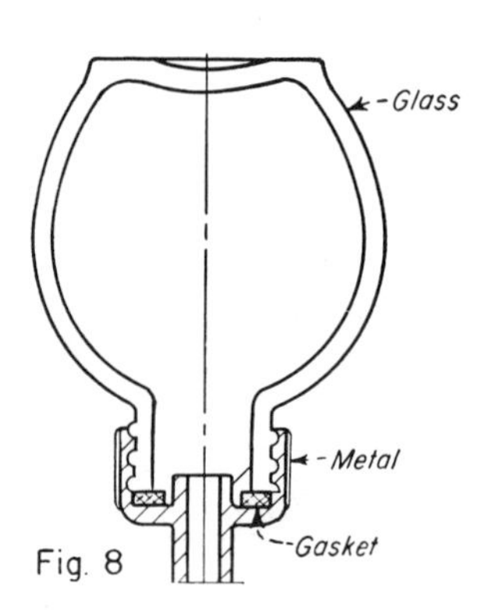


THREADER JOINT. For low pressure applications threaded glass container can be screwed into metal fixture. Cement can also be added.

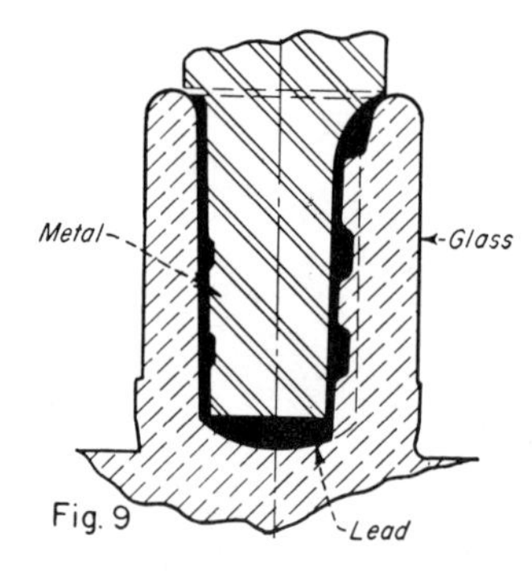


PERMANENT SEAL. Poured lead locks metal arbor to hubbed glass part. Use is restricted, however, to moderate loads and temperatures.

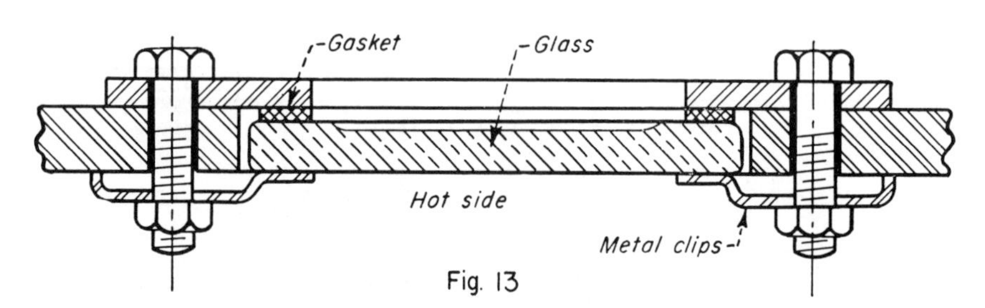


FURNACE SIGHT GLASS. Small metal spring clips hold glass in place and allow for different rates of expansion of glass and metal fixture. This method of suspension also insures that the glass will be heated uniformly, reducing thermal gradients and internal stresses.

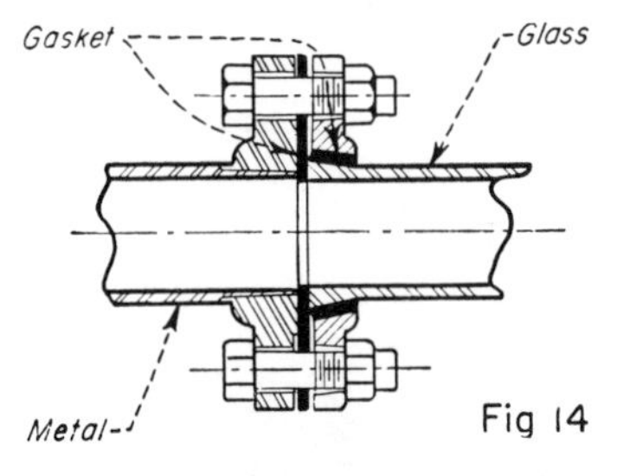


GLASS PIPE TO METAL PIPE. Gasketed conical flange reduces bending stresses. Rubber gasket seals flat end of glass pipe.

# to Metal Structures

M. H. HUNT
Corning Glass Works

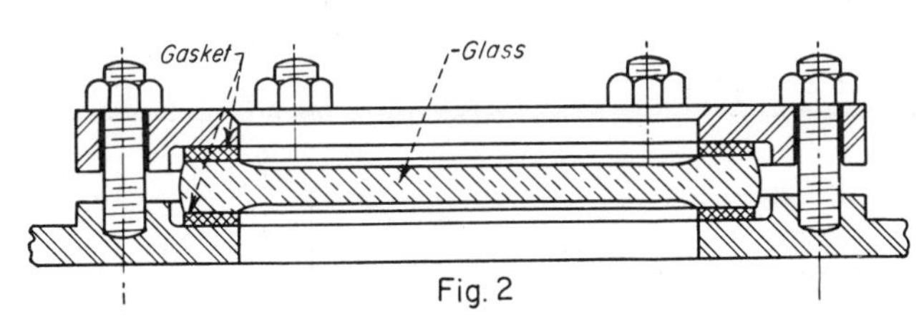


PRESSURE-TIGHT JOINT. Glass sheet is clamped uniformly in compression using rubber or asbestos gaskets.

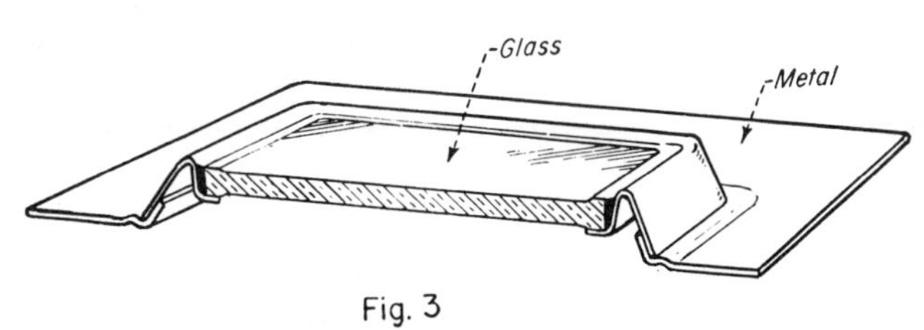


SOLDERED JOINT. Glass has metallized edge for soldering. Frame is semi-rigid to allow for differential expansion.

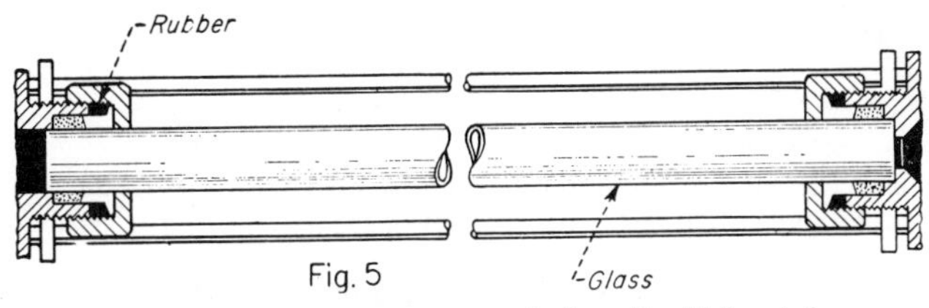


TUBULAR GAGE GLASS is attached to liquid level fixture. Rubber gasket is compressed to effect a leak-proof seal.

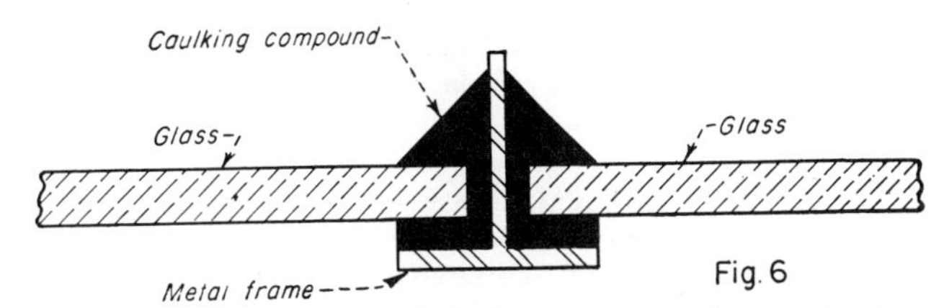


WINDOW JOINT. Resilient caulking compounds and rubber cement make liquid-tight seals with metal frames.

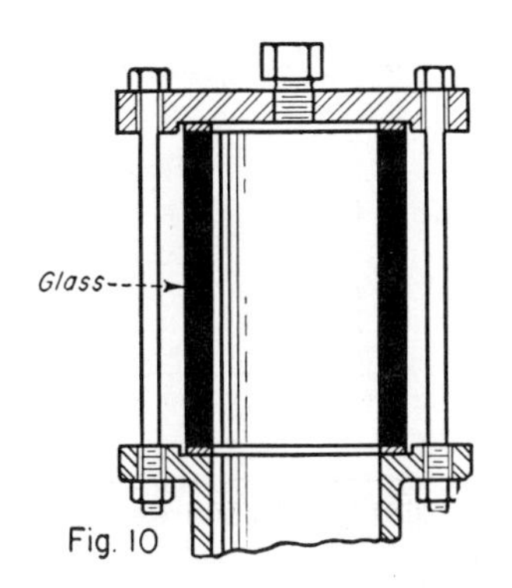


GLASS TUBE IN PRESSURIZED OIL SYSTEM. Glass is clamped in compression; resilient gaskets are added to help distribute load uniformly.

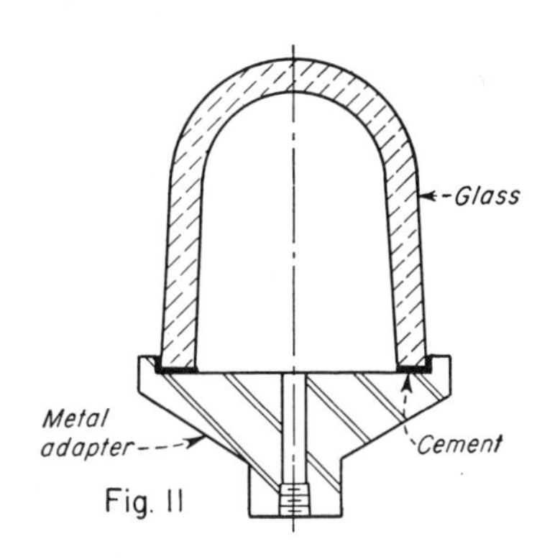


CEMENTED JOINT. Container shown carries explosive in oil wells. High external pressure keeps cemented joint in compression to maintain the seal.

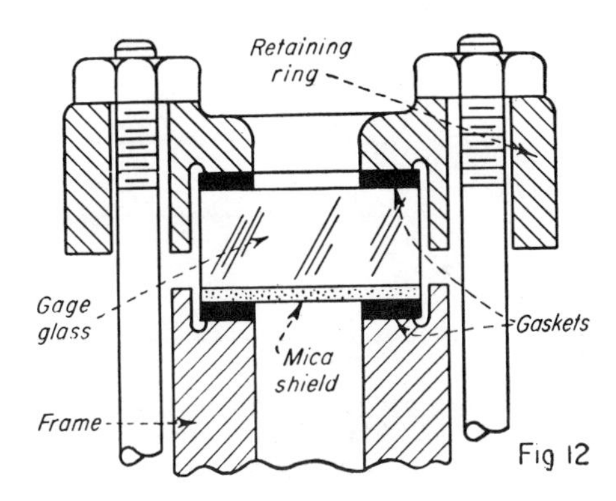


GAGE GLASS attached to water level column on a steam boiler. The mica shield protects the glass from corrosion by the alkaline boiler water.

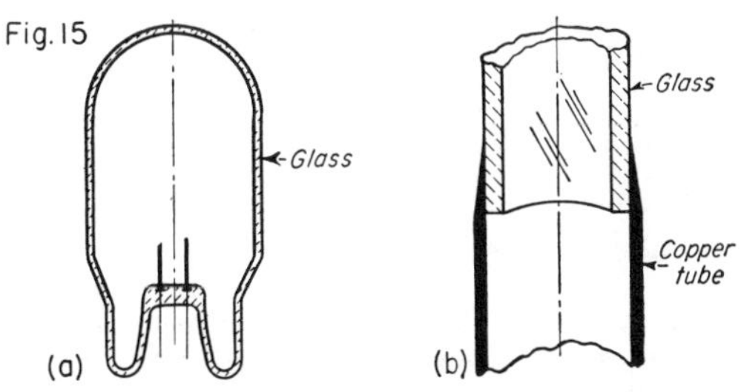


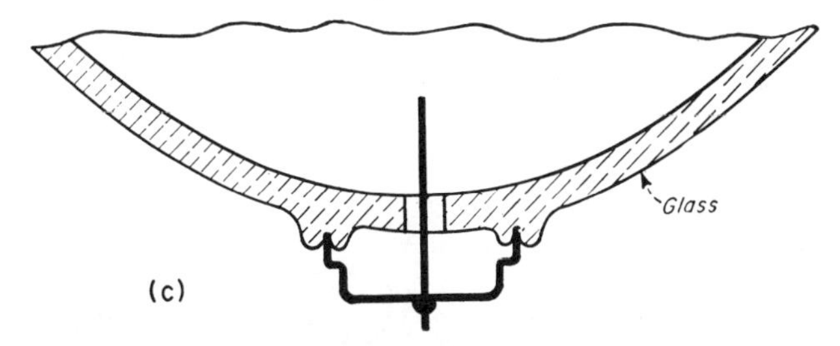


ELECTRICAL CONNECTIONS. (a) Inside of vacuum tubes. First, wire is sealed to tubular glass flare by pressing the hot glass around the wire. The flare and wire assembly is then sealed to the tube body. Dumet (42 per-cent nickel iron coated with copper) is commonly sealed to high expansion glasses. Tungsten and Kovar are sealed to low expansion glasses. (b) Inside of air and water cooled transmitting tubes. Thin copper tube sealed to glass tubing is thin enough to compensate for different expansion coefficients (c) Inside of glass sealed-beam headlight. Thin 42% nickel iron metal cup is sealed directly to glass. Kovar or similar metals can also be used. Flexible cup does not stress the glass.

# Applications of Helical

**Originally devised to reduce thread wear and stripping between steel fasteners and soft materials, helical wire inserts are now being used in plastic and wood for similar pur-**

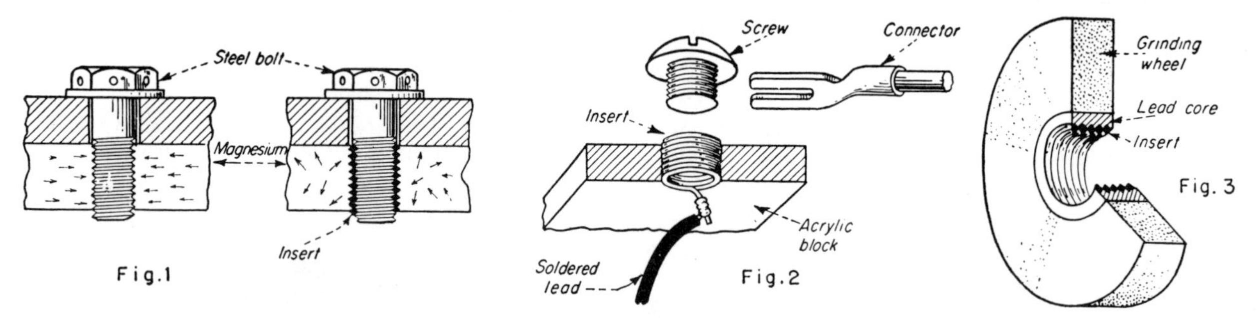


**Fig. 1—Galvanic action between steel bolt and magnesium part, (Left), attacks thread causing part failure. Stainless steel insert, (Right), reduces galvanic action to a negligible amount while strengthening threads.**

**Fig. 2—Insert used as an electrical connection as well as a thread reinforcement. Unit is threaded into plastic and tang is bent to form soldering lug.**

**Fig. 3—Direct connection of grinding wheel onto a threaded shaft by using a wire insert. Washers and nut are not required thus simplifying assembly.**

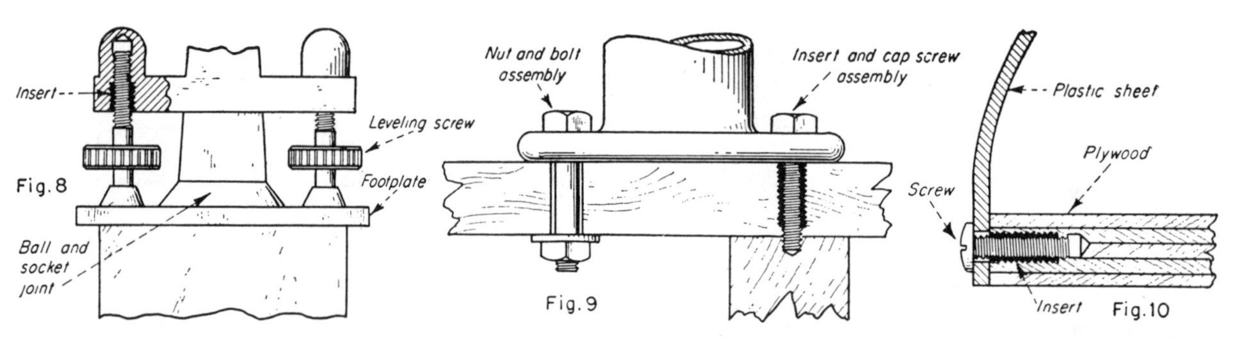


**Fig. 8—Wear and backlash can be reduced on adjusting threads. Clamping strength is not needed but intermittent thread travel makes reinforcement desirable.**

**Fig. 9—Combination of inserts and capscrews permits installation of machinery and other equipment on wood floors and walls. Access to opposite face of the wood is not a factor nor are joists and other obstructions.**

**Fig. 10—Plastic-to-wood connector. Repeated assembly and disassembly does not affect protected threads in wood or plastic parts.**

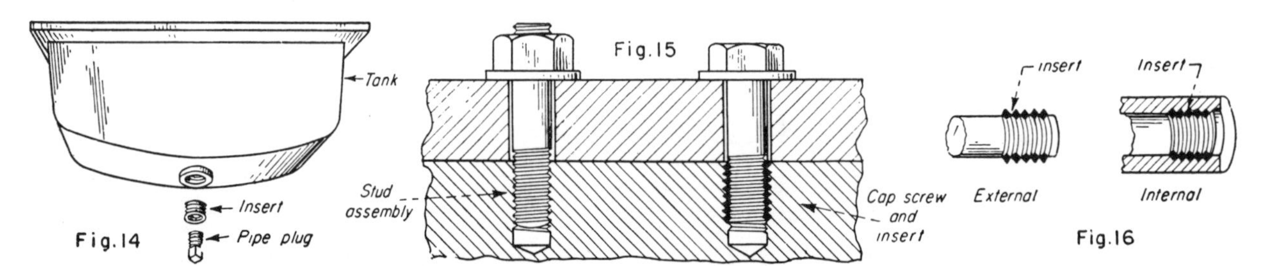


**Fig. 14—Seizure and corrosion of pipe threads on compression, fuel and lubricant tanks, pipe lines, fittings, pumps and boilers are prevented by using an insert.**

**Fig. 15—Stud (Left) transfers thread wear from the tapped hole and into the expendable threads on the stud and nut. Interference fit in the tapped hole is mandatory. Insert (Right) prevents wear and makes stud unnecessary. Lower cost cap screw can be used.**

**Fig. 16—Thread series can be changed from special to standard, from fine to coarse or vice versa, or corrected in case of a production error by redrilling and retapping. Inserts giving desired thread are then used.**

# Wire Inserts

PAUL E. WOLFE
Director of Project Engineering,
Heli-Coil Corporation, Danbury, Conn.

poses. Other applications are to prevent galvanic action, act as electrical connectors, and to reduce weight, cost and number of parts needed in an assembly.

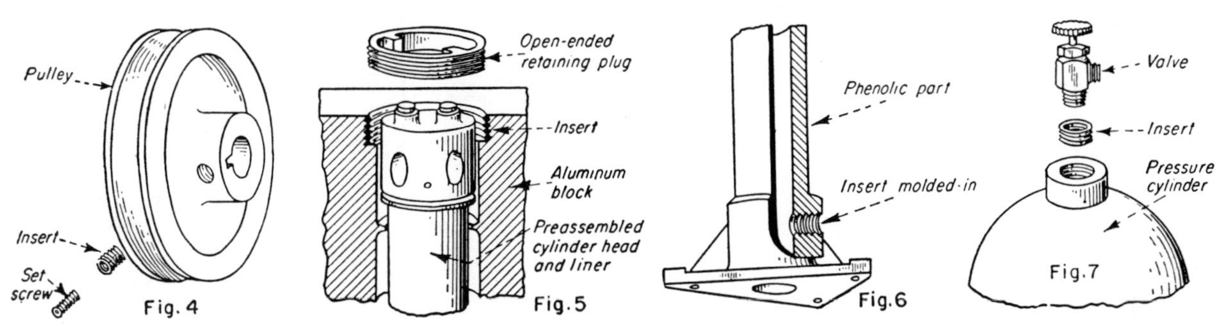


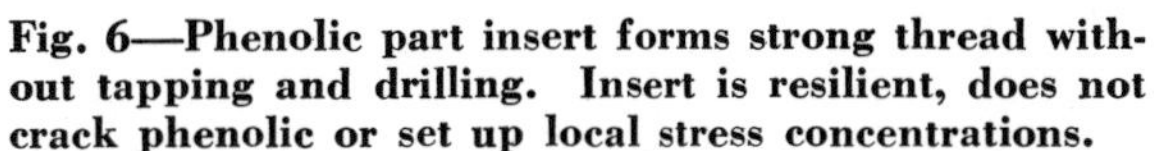

**Fig. 4—Loosening of the set screw by vibration is reduced by using an insert. As pulley is a soft metal die casting set screw tightening often stripped threads.**

**Fig. 5—Insert withstands combustion thrust of diesel cylinder. It prevents heat seizure and scale on threaded plug making cylinder replacement and servicing easy.**

**Fig. 6—Phenolic part insert forms strong thread without tapping and drilling. Insert is resilient, does not crack phenolic or set up local stress concentrations.**

**Fig. 7—Enlargement of taper pipe threads in necks of pressure vessels, caused by frequent inspection and interchange of fittings, can be minimized.**

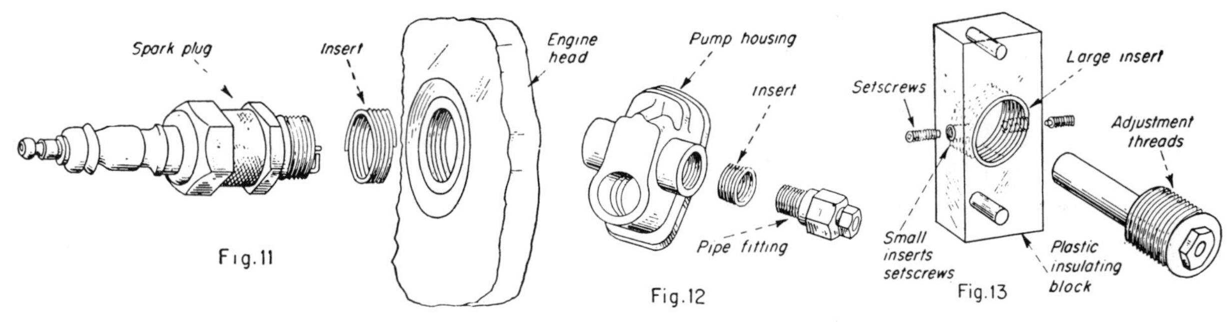


**Fig. 11—Threads are protected from stripping in new aluminum engine heads by inserts. Also can be used to repair stripped spark plug holes in engine heads.**

**Fig. 12—Insert prevents pipe fittings from peeling chips out of tapped aluminum threads. Introduction of chips into the lines could cause a malfunction.**

**Fig. 13—Center insert serves as brakeband to lock adjustable bushing. Small inserts keep set screws from stripping plastic when tightened. Also, adjustment threads can not be marred by end of set screw.**

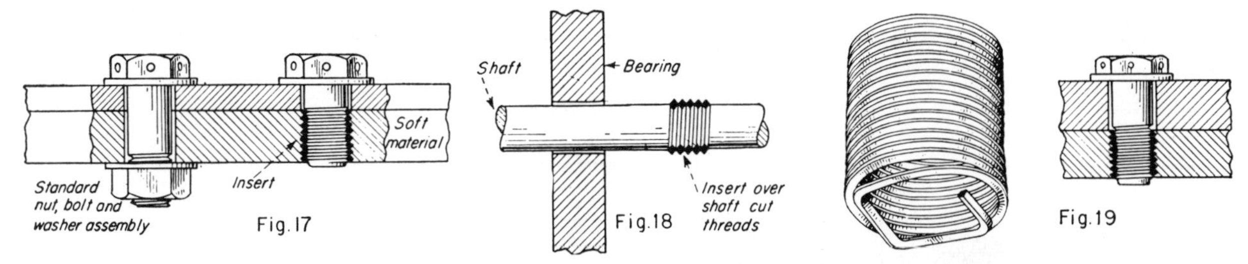


**Fig. 17—Assembly weight may be reduced. Left view shows the standard method of attaching front and mid frame of a compressor. Lockwire is used after assembly is completed. Right view is new method resulting in weight and space economies.**

**Fig. 18—Assembly of a shaft through a bearing is simplified by adding external insert over cut shaft threads. Machining the full shaft length is also unnecessary.**

**Fig. 19—Square tang insert—called screw lock—automatically locks the screw so that lock washers, nuts or wires are unnecessary. Insert locks itself into parent material without need for pins, rings, or staking.**

# Various Methods of Locking

Locking devices can generally be classified as either form or jam locking. Form locking units utilize mechanical interference of parts whereas the jam type depends on friction developed between the threaded elements. Thus their performance is a function of the torque required to tighten them. Both types are illustrated below.

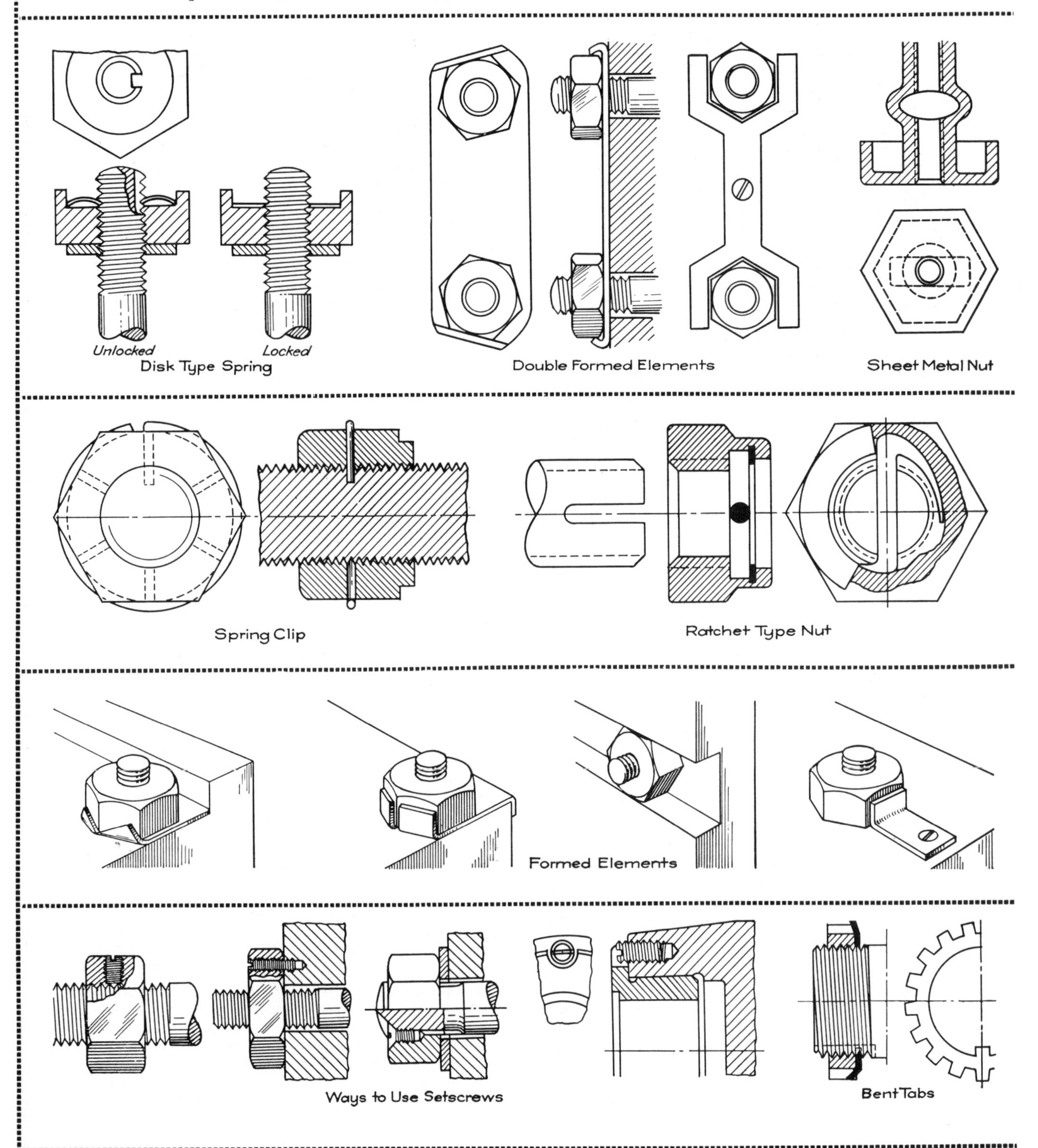

# Threaded Members

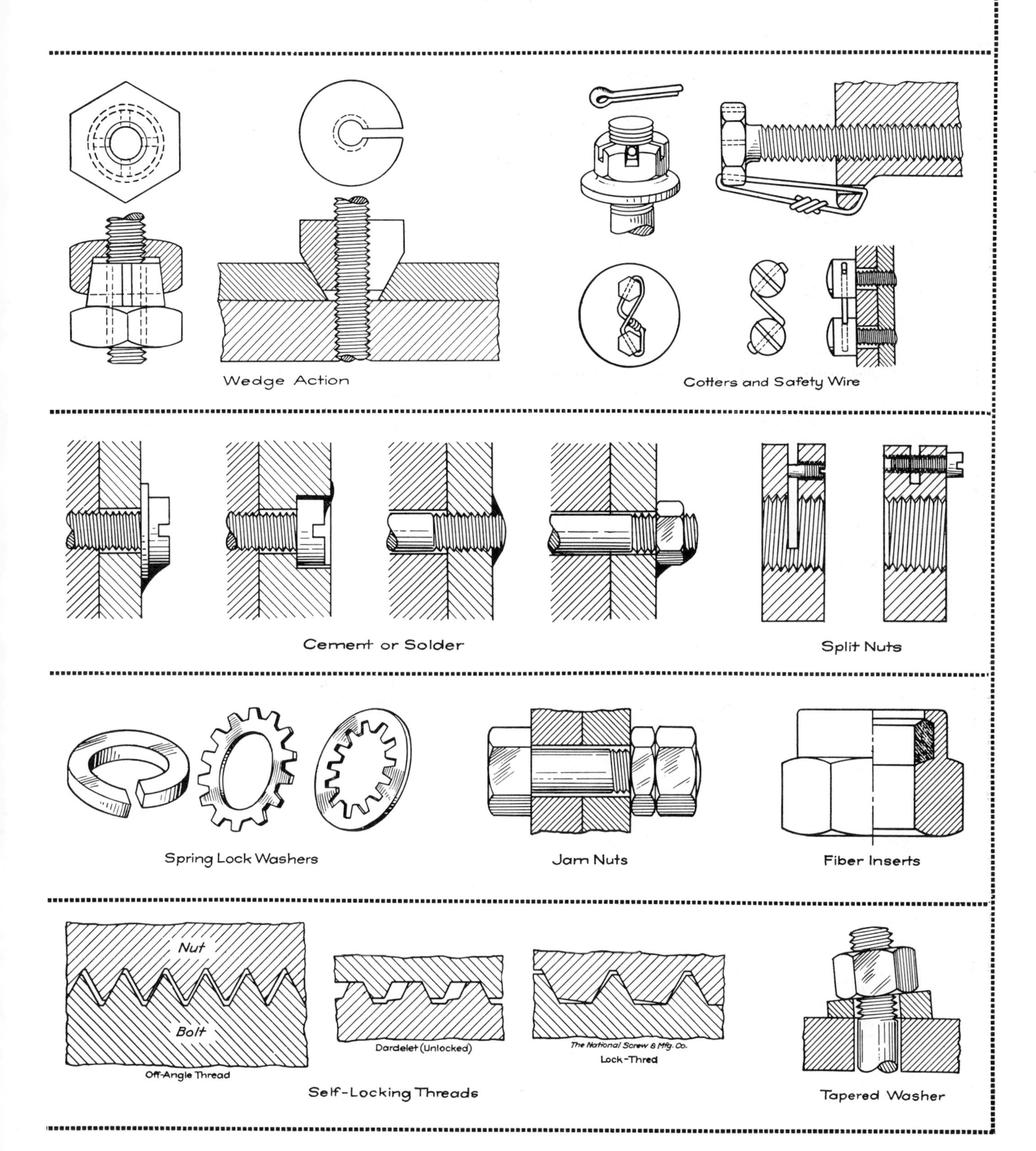

# Quick-release fasteners

MARK LEVINE, project engineer, Grant Pulley & Hardware Corp.

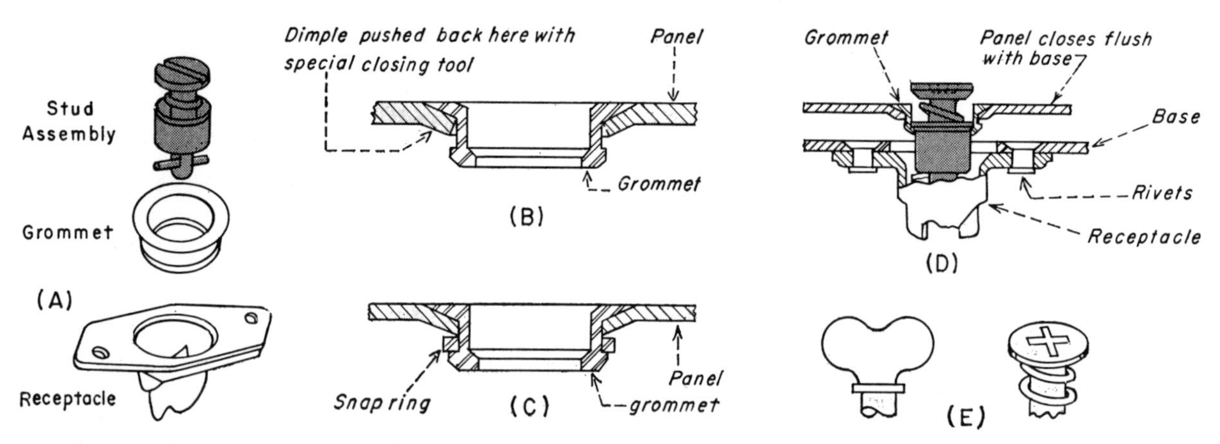


**Fig. 1—GENERAL-PURPOSE FASTENER composed of (A) stud assembly, grommet and receptacle. Grommet secured to removable panel by dimpling or countersinking panel and then (B) pushing dimple back with special closing tool or (C) using snap ring. Receptacle is fastened to underside of base of cabinet (D) with two rivets. Locks or unlocks with quarter turn. Fabrication required: one large hole and two countersunk holes in base; one dimpled or countersunk hole in panel. Studs (E) also available with winged or cross-recessed head. Tension and shear strength: 700 to 1200 lb depending on size of fastener.** *Camlock Fastener Corporation.*

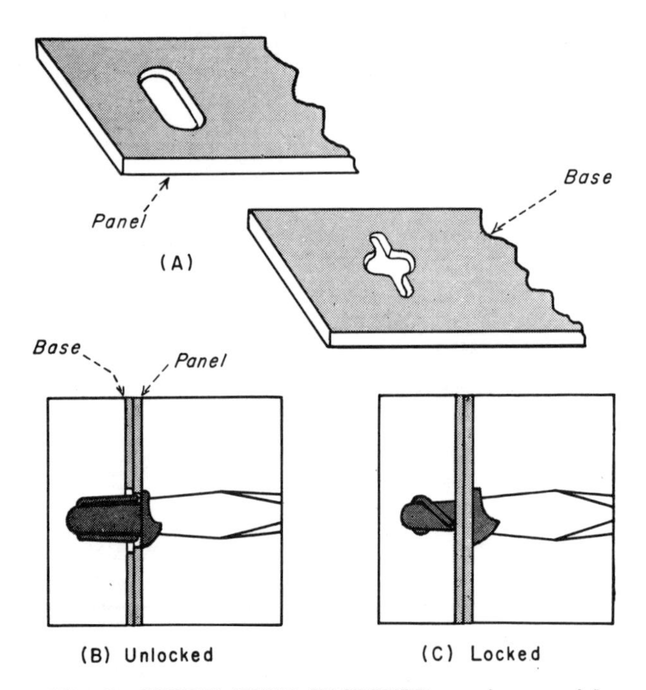


**Fig. 2—SCREW TYPE FASTENER requires special-shaped holes (A) in panel and base. Dies for piercing holes available from manufacturer. Inserted through panel and base (B) and locked (C) with quarter turn. Inexpensive but not self-retained to panel when unlocked. Special-shaped heads for supporting shelves.**
*Simmons Fastener Corporation. (Licensed by Illinois Tool Works)*

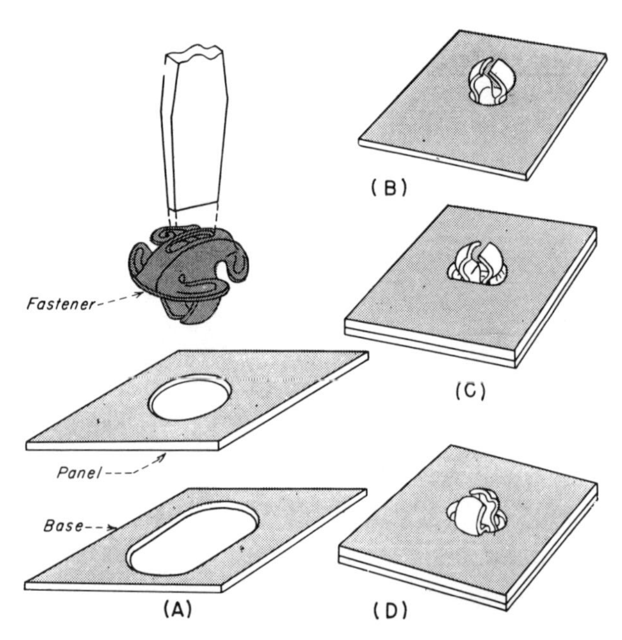


**Fig. 18—ONE-PIECE FASTENER very low in cost but requires special hole (A) in base. Snaps into panel hole (B) where it is retained in place, then passes through elongated hole (C) of base. Quarter turn (D) locks panel to base but fastener can be turned past quarter turn position unless panel is dimpled for a stop.**
*Fastex, Division of Illinois Tool Works*

**Seven types of fasteners that engage or disengage with a partial twist—usually a quarter turn—as contrasted to fasteners that require several turns.**

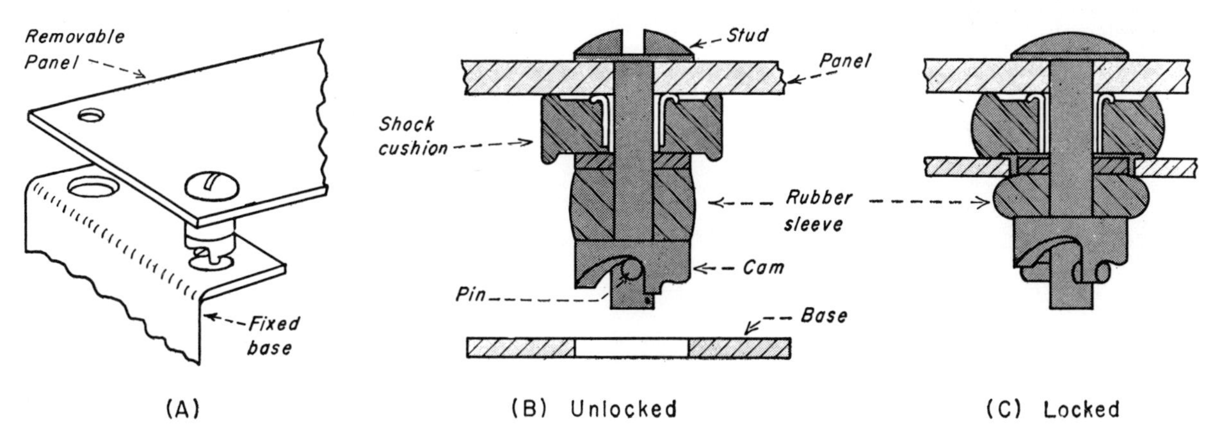


**Fig. 4—SHOCK-MOUNTED FASTENER reduces shock, vibration and noise between panel and fixed base (A). One hole drilled or pierced in both panel and base. Fastener permanently attached to panel (B) by assembling components and inserting pin with special pliers. Half turn of stud locks panel to base (C) by compressing rubber sleeve against base. Base "floats" between rubber sleeve and shock cushion.** *The General Tire & Rubber Co.*

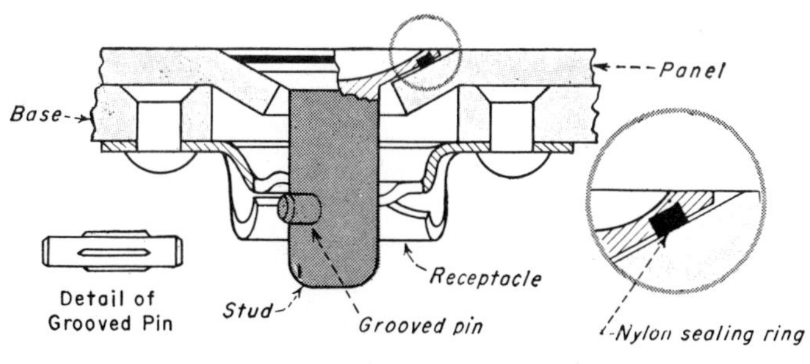


**Fig. 5—SEALING FASTENER for moisture and pressure sealing has nylon sealing ring held captive in groove on underside of stud head. Fabrication requirements similar to fastener in Fig. 1—two counter sunk holes and large hole in base; one dimpled or countersunk hole in panel. Grooved pin supplied separately and driven into stud during installation.** *Manadnock Mills.*

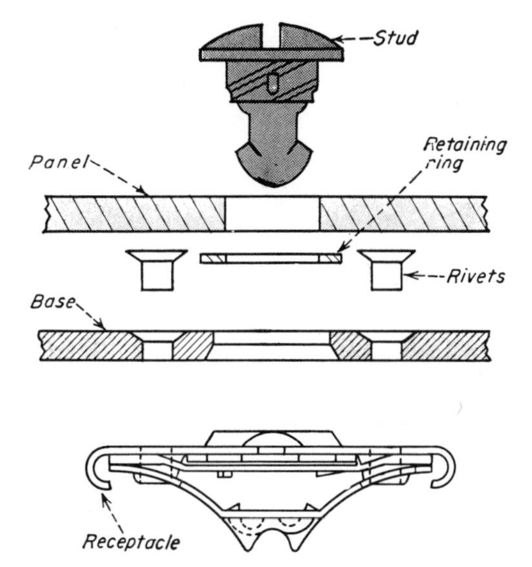


**Fig. 6—HIGH-STRENGTH FASTENER has load capacity up to 3,000 lb in tension—4,750 lb in shear. Stud captivated to panel by retaining ring. Base drilled and countersunk for two rivets and to accurate dimensions in receptacle. Special countersink tool available from manufacturer for countersinking receptacle. Locks or unlocks with quarter turn. Up to 0.125 in. misalignment and 0.030 take-up permissible.** *Southco Div., South Chester Corporation.*

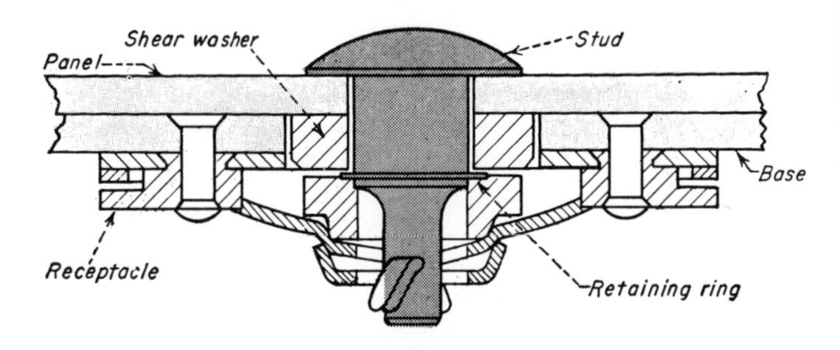


**Fig. 7—HIGH-SHEAR FASTENER prevents shift between panel and base when carrying shear loads. Shear washer held captive to stud with retaining ring installed during panel assembly. Hole in base pierced or drilled to outside dimension of shear washer using close tolerances. Close fit between washer and hole necessary to reduce shifting of panel under shear load. Base also requires two countersunk holes for receptacle. Locks or unlocks with quarter turn.** *Scovill Manufacturing Company.*

# Designs for Sheet Metal Boxes

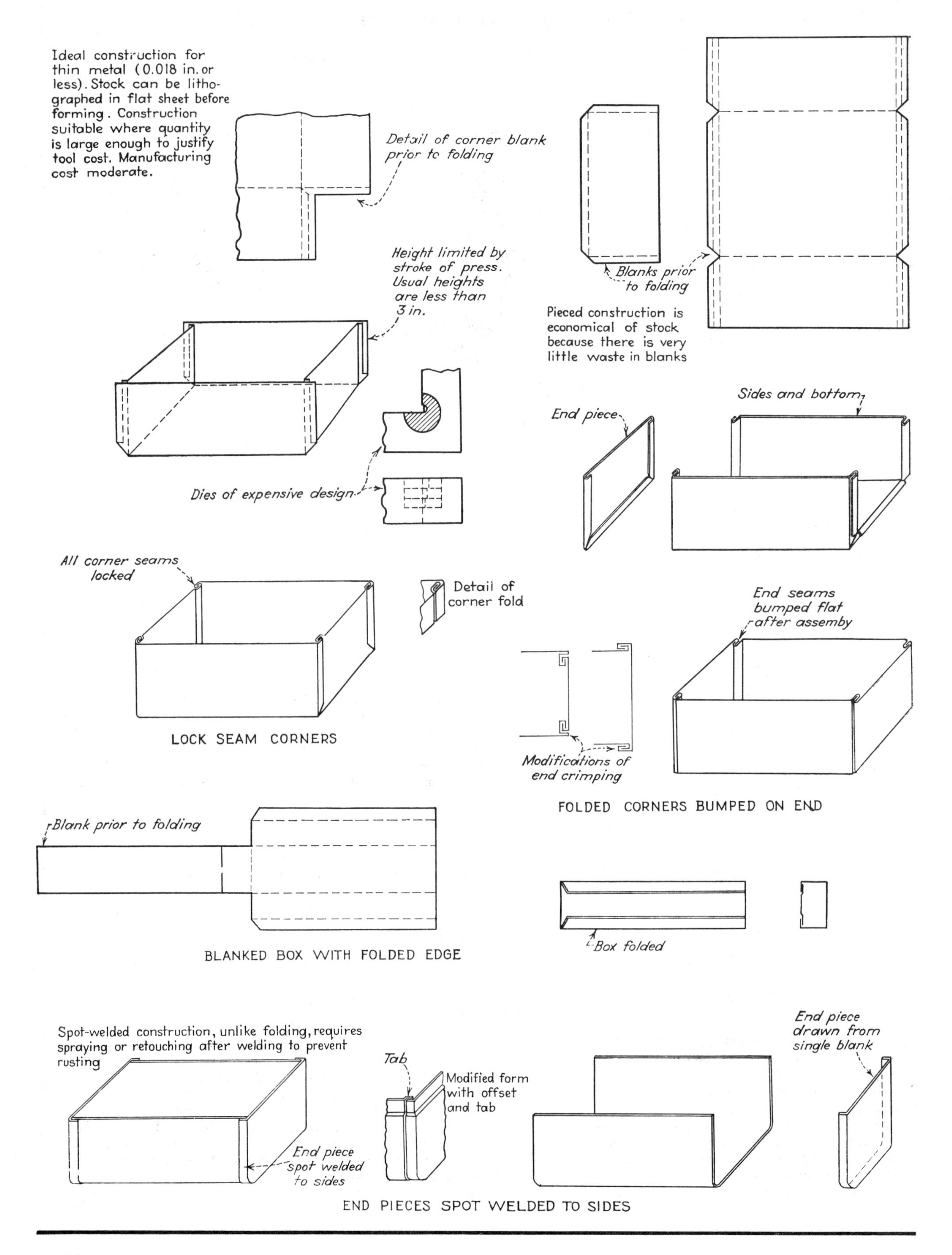

# With Square and Round Corners

WALLACE C. MILLS

The problem of designing a sheet metal box or box-like structure occurs frequently. For some purposes, ordinary can and box structures are satisfactory. For others, special structures for utility, appearance, or low cost may be required. Designs are shown for seamed, welded, and folded constructions. Blank sizes and methods of manufacture are included.

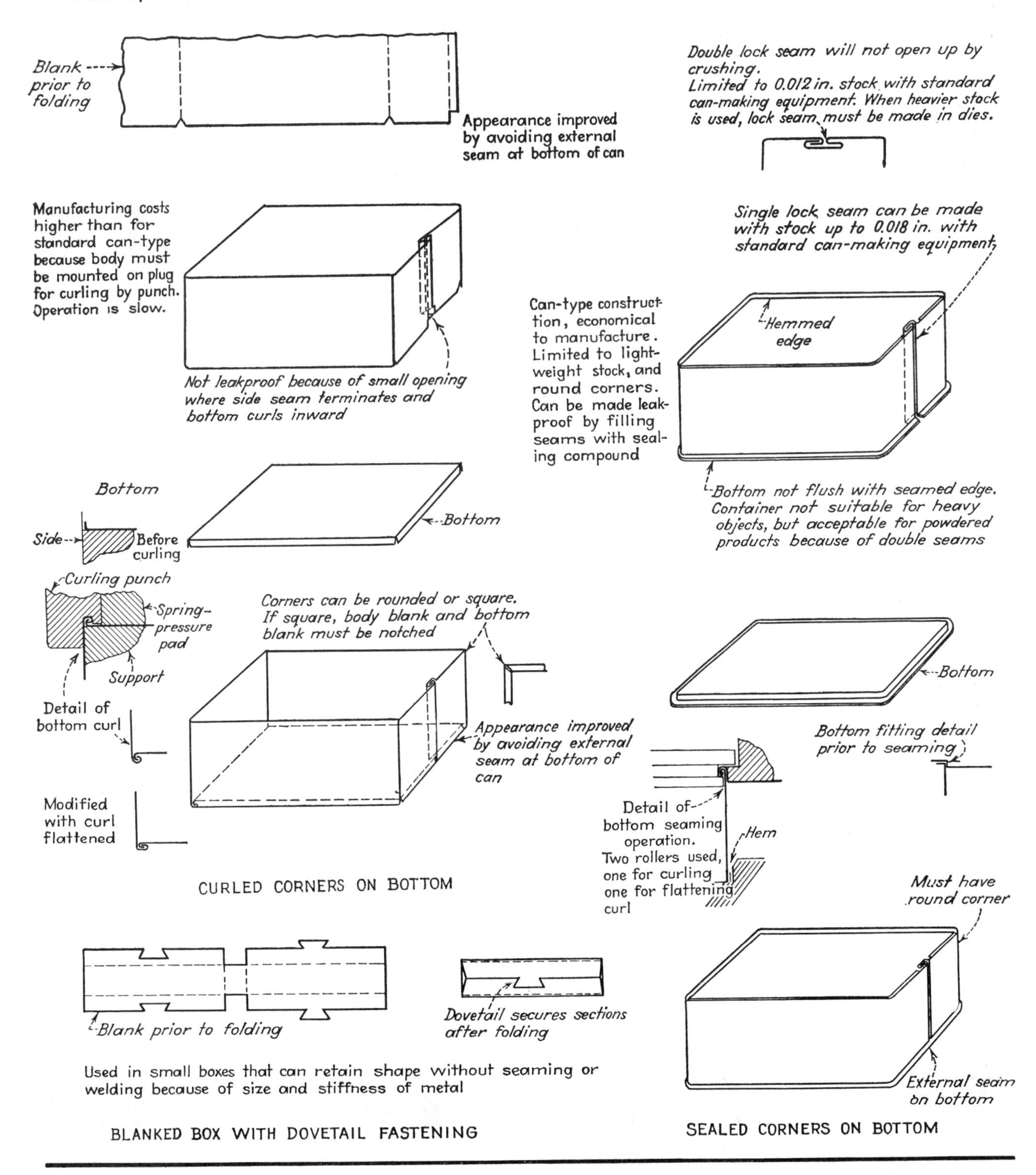

# TOP AND BOTTOM ATTACHMENT

WALLACE C. MILLS

**Numerous methods of attaching bottoms and tops on sheet metal containers such as cans, buckets, drums, tanks, wash tubs, and tea kettles, have been developed to meet different requirements in service and in economy of production. These methods usually make an interlocking seam although for some applications press fits or snap fits are satisfactory. In many of these applications sealing compounds, or no sealing material at all, are as satisfactory as solder. The accompanying illustrations include commonly used seam designs produced by specially developed automatic machines.**

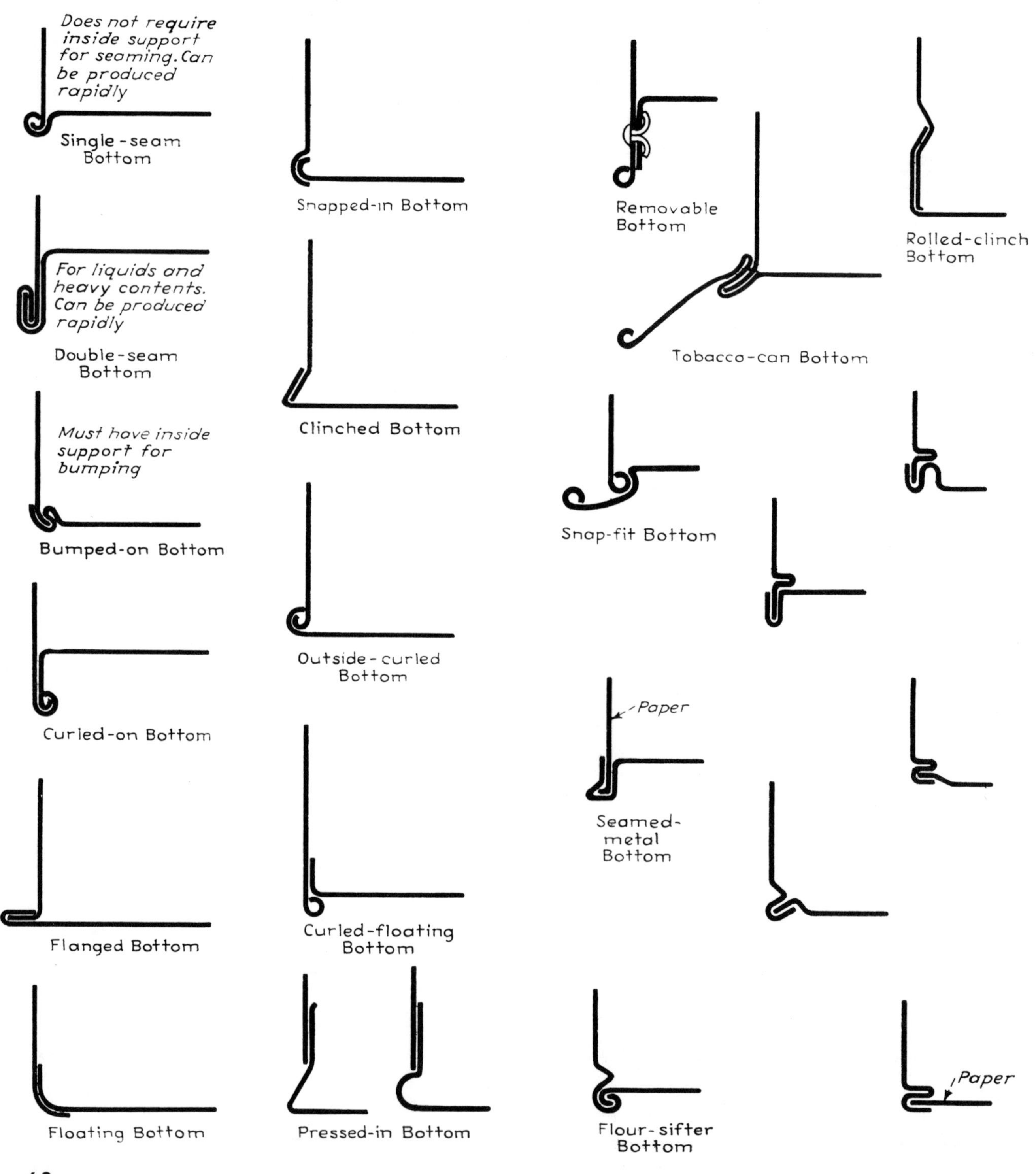

# FOR SHEET METAL CONTAINERS

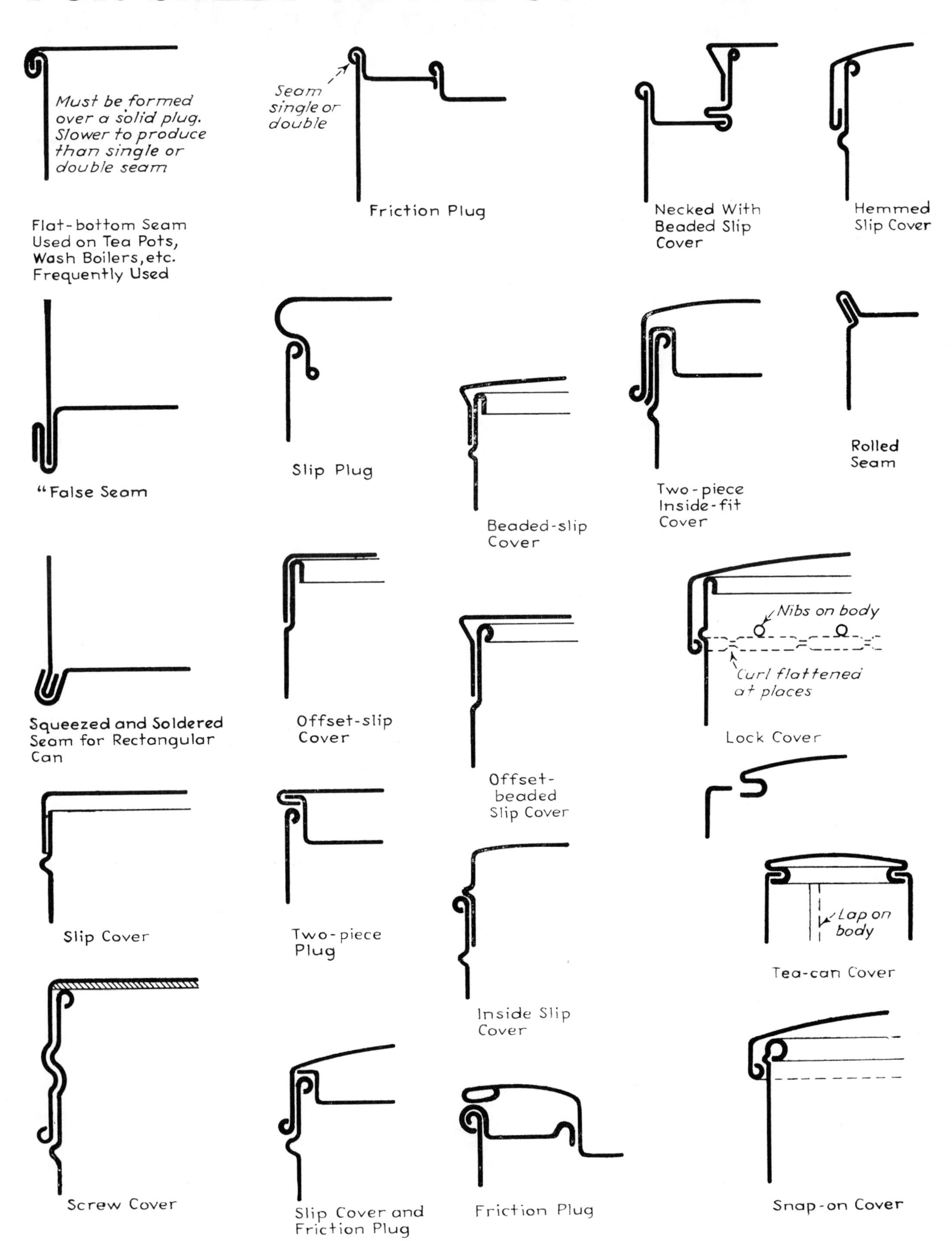

# Fastening Sheet-Metal Parts by

## Detachable and permanent assembly of sheet metal

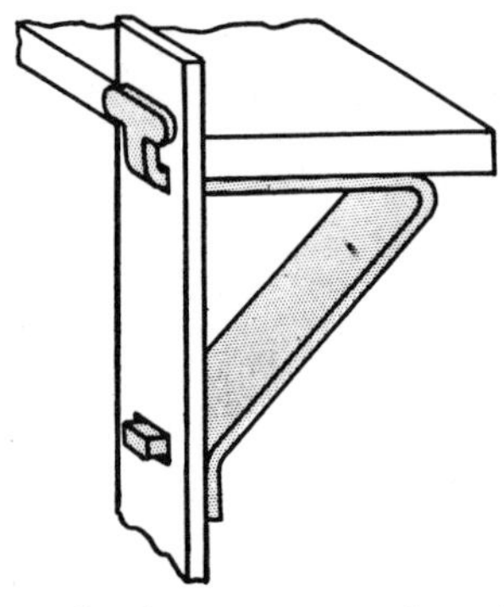

**Fig. 1—Supporting bracket formed from sheet metal and having integral tabs. Upper tab is inserted into structure and bent. Ledge weight holds lower tab.**

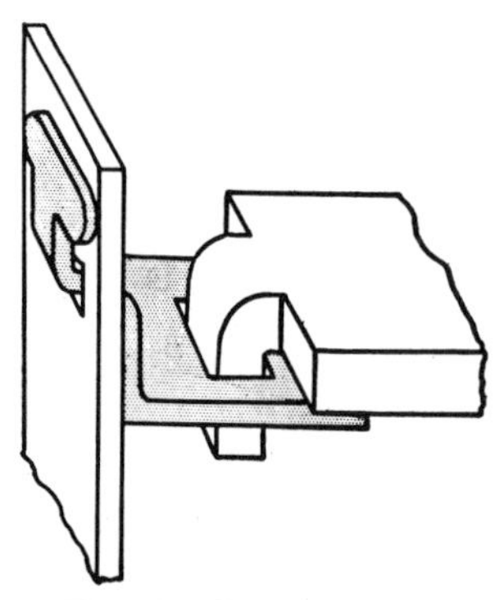

**Fig. 2—Supporting bracket similar to Fig. 1 but offering restraint to shelf or ledge. Tabs are integral with sheet metal part and are bent on assembly.**

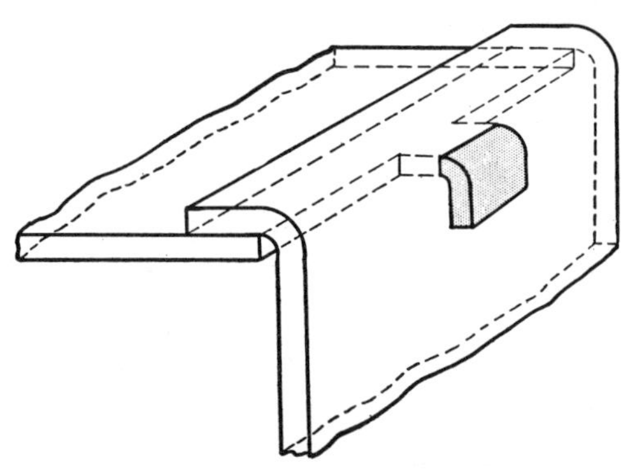

**Fig. 3—Supporting ledge or shelf by direct attachment. Tab is integral and bent on assembly. Additional support is possible if sheet is placed on flange and tabbed.**

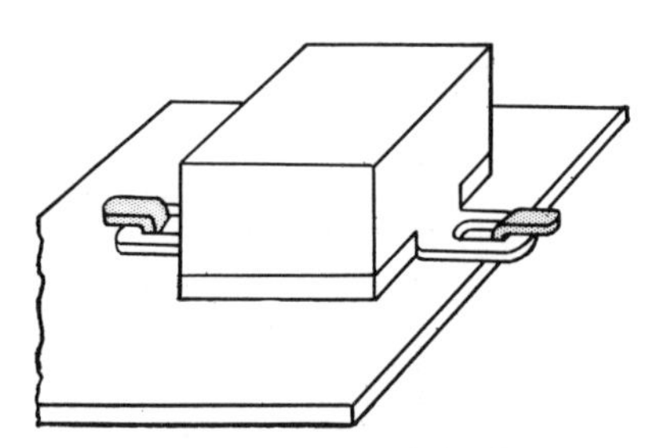

**Fig. 7—Box section joined to a flat sheet or plate. Elongated holes are integral with box section and tabs are integral with plate. Design is not limited to edge location.**

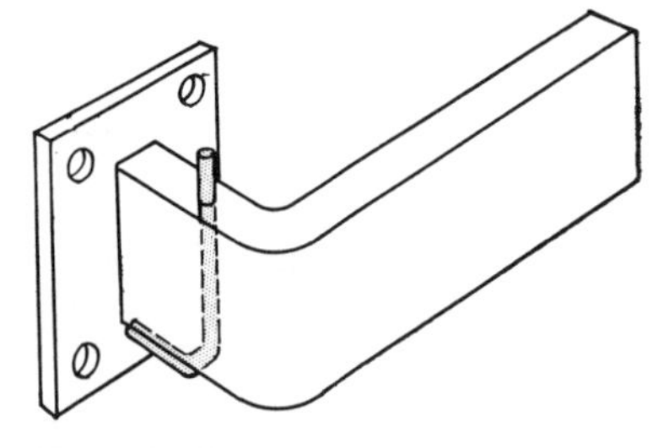

**Fig. 8—Bar is joined to sheet metal bracket by a pin or rod. Right angle bends in pin restrict sidewise or rocking motion or bar. Bracket end of pin is peened.**

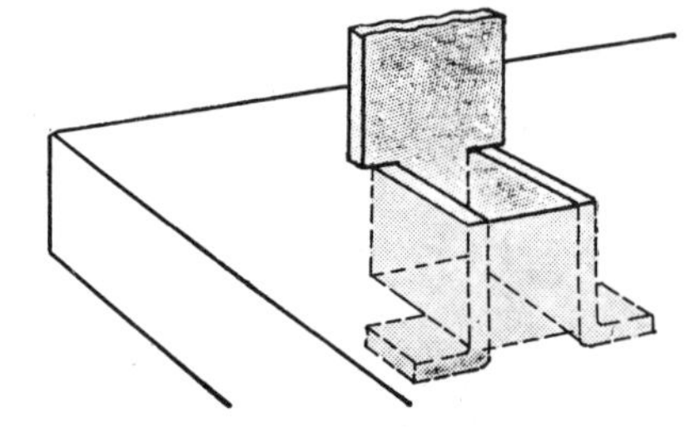

**Fig. 9—To support and join sheet metal support at right angle to plate. Motion is restricted in all directions. Bottom surface can be grooved for tabs.**

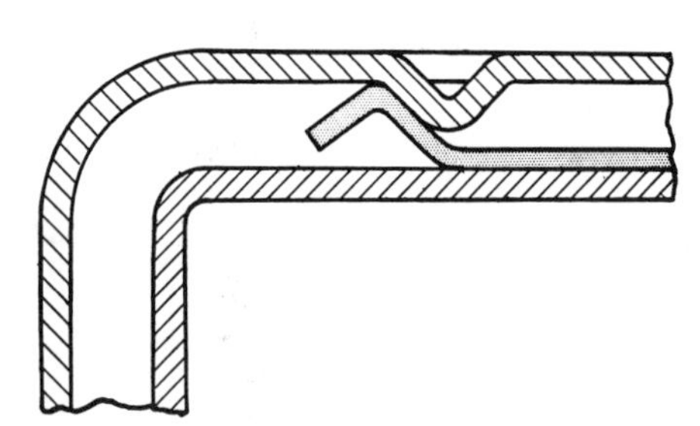

**Fig. 13—A spacing method that can be used for circular sections. Formed sheet metal member support outer structure at set distance. Bead centers structure.**

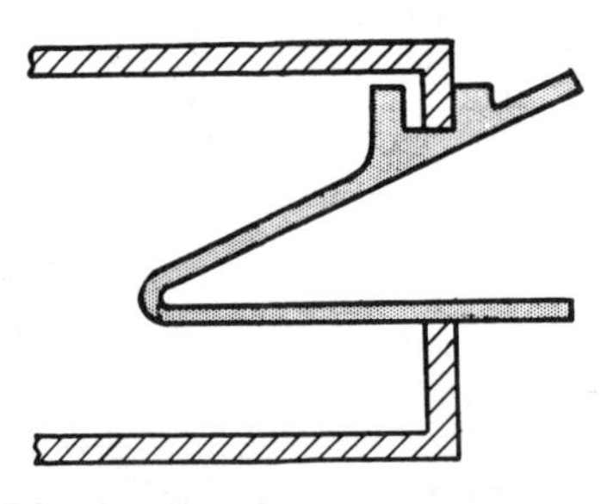

**Fig. 14—A removable section held in place by elasticity of material. Design shown is a temporary or a removable cover for an elongated slotted hole in a sheet metal part.**

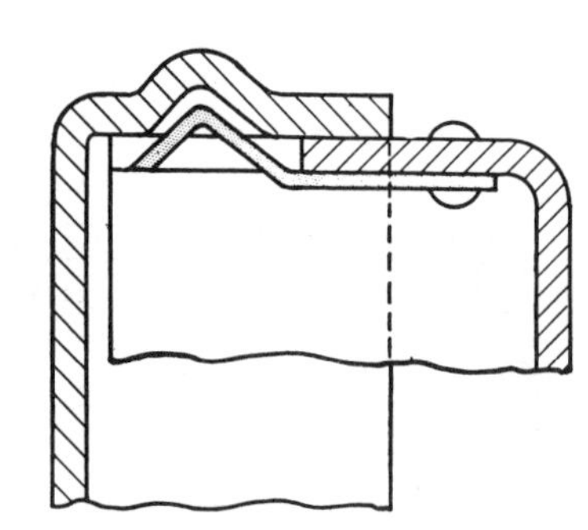

**Fig. 15—A cover held in position by bead and formed sheet. Cover is restrained from motion but can be rotated. Used for covers that must be removable.**

# Tongues, Snaps or Clinching

## parts without using rivets, bolts or screws

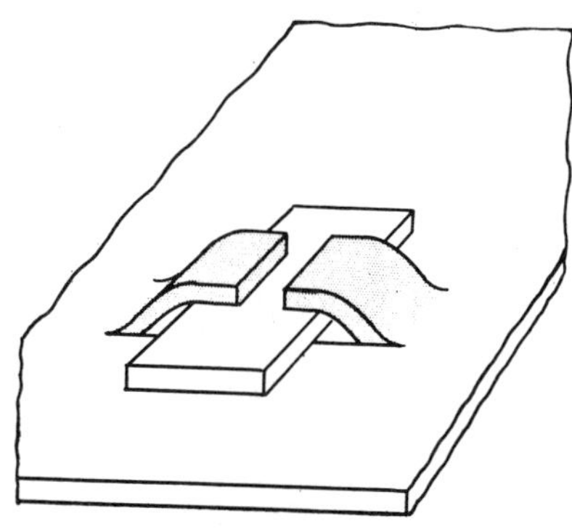

Fig. 4—To support or join a flat sheet metal form on a large plate. Tabs are integral with plate and bent over on assembly. Only sidewise motion is restricted.

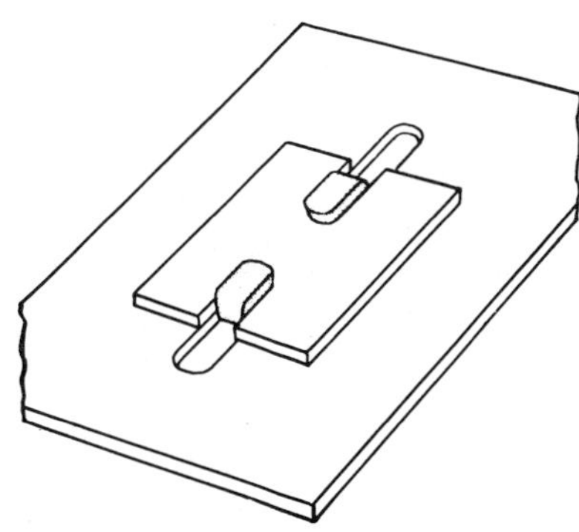

Fig. 5—Similar to Fig. 4 but motion is restricted in all directions. Upper sheet is slotted, and tabs are bent over and into slots on assembly.

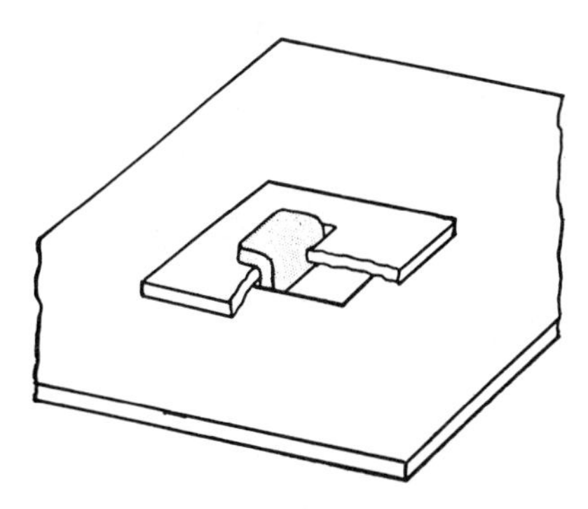

Fig. 6—Single tab design for complete restriction of motion. Upper plate has an elongated hole that matches width and thickness of integral lower plate tab.

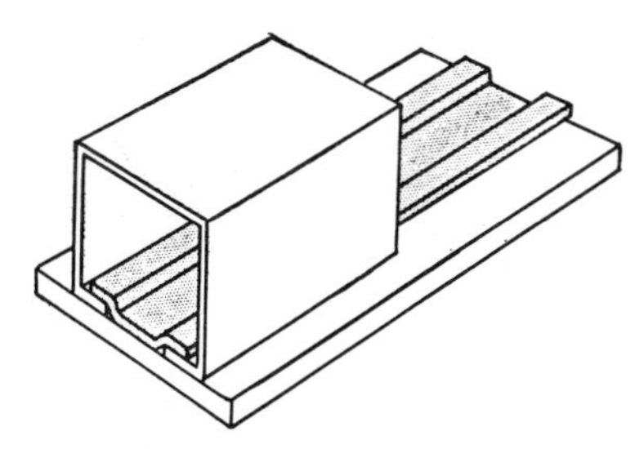

Fig. 10—Channel section spot welded to plate forms bottom surface and joins box section to plate. Channel edges can be crimped or spot welded to restrict motion.

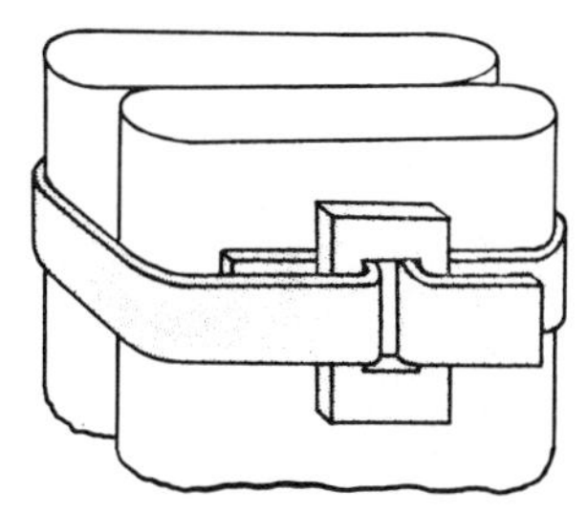

Fig. 11—Sheet metal strap used to join two flat surfaces. Edges of plate are rounded to allow strap to follow contour and prevent cutting of plate by the metal strap.

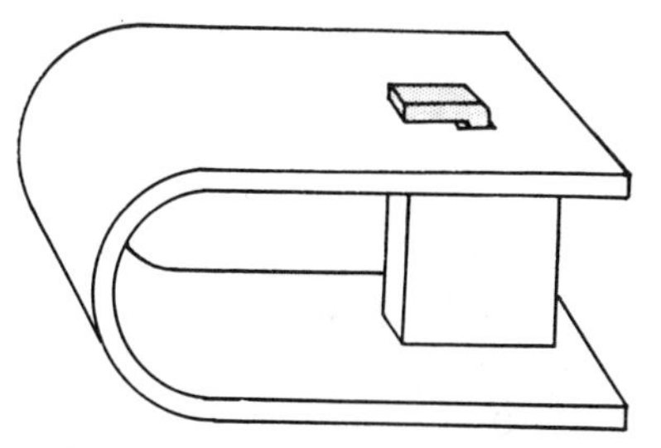

Fig. 12—Sheet metal structures can be spaced and joined by use of a tabbed block. Formed sheet metal U section is held to form by the block as shown.

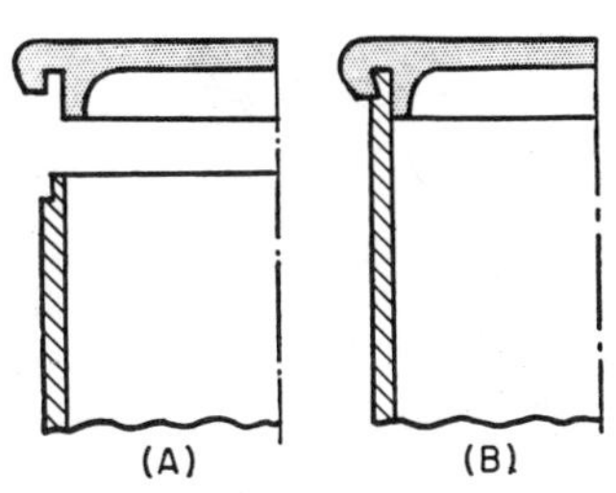


Fig. 16—A non-removable cover design. The vessel is notched as shown in A, and the cover crimped over, B, on assembly. This is a permanent cover assembly.

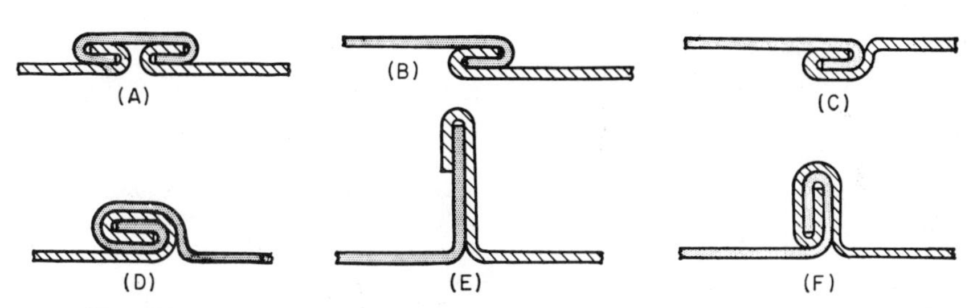


Fig. 17—Six methods of joining two sheet metal parts. These can be temporary or permanent joints. If necessary, joints can be riveted, bolted, screwed or welded for added strength and support. Such joints can also be used to make right angle corner joints on sheet metal boxes, or for attaching top and bottom covers on sheet metal containers.

# Joining Sheet Metal Parts

## Methods of attaching rods, tubing and sheet metal parts in permanent

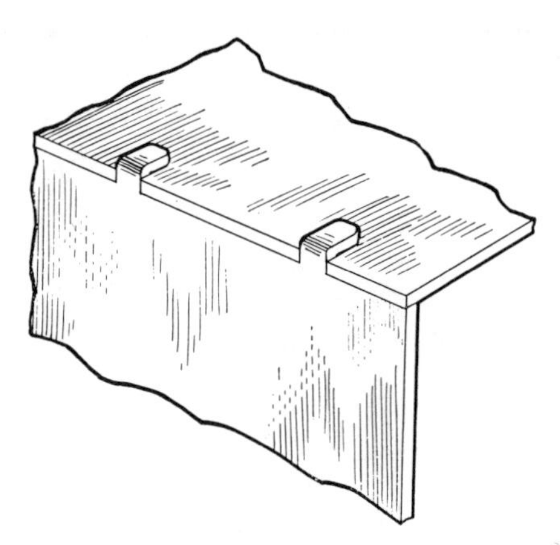

FIG. 1—For butting sheet metal parts at or near a right angle. Two or more tabs are required.

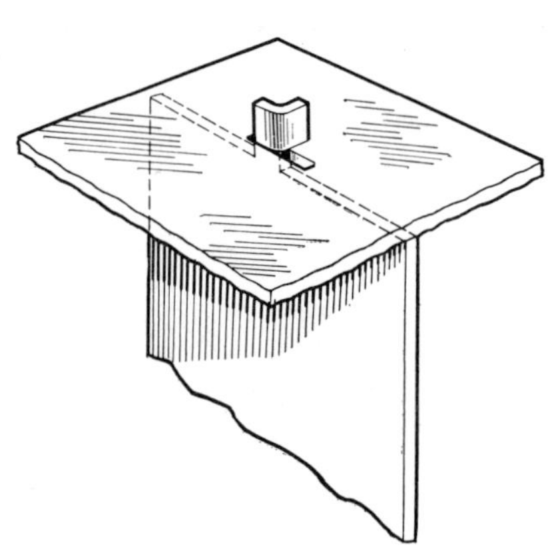

FIG. 2—For applications similar to Fig. 1, but where inaccessibility forbids use of bulky tools.

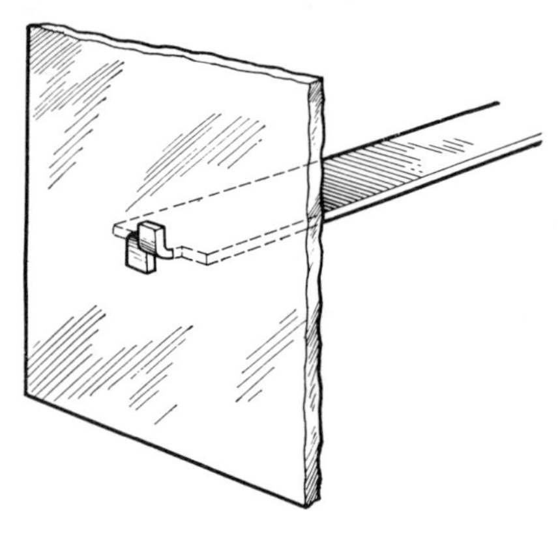

FIG. 3—For use as in Fig. 2, but where greater rigidity is needed.

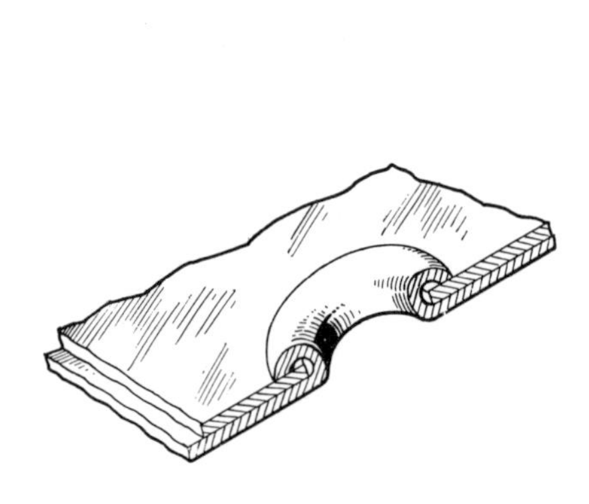

FIG. 4—For permanent attachment of parallel sections. Material must be quarter-hard or softer.

FIG. 5—For positioning shouldered rods in sheet metal. Center-punch as well as "nicking" punch can be used.

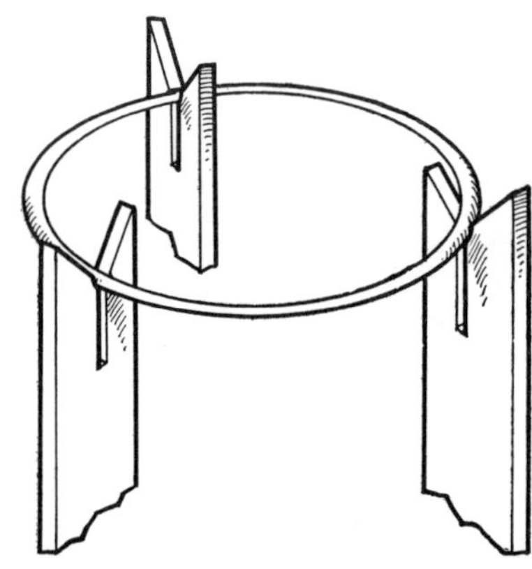

FIG. 6—For temporary support of rods or tubes. Jaws are sprung open during installation and removal.

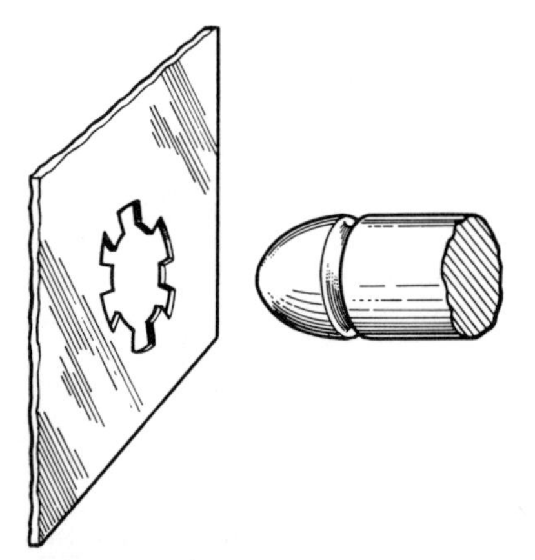

FIG. 7—For semi-rigid attachment. Installation can be permanent or temporary, depending on roundness of rod end and "springiness" of teeth.

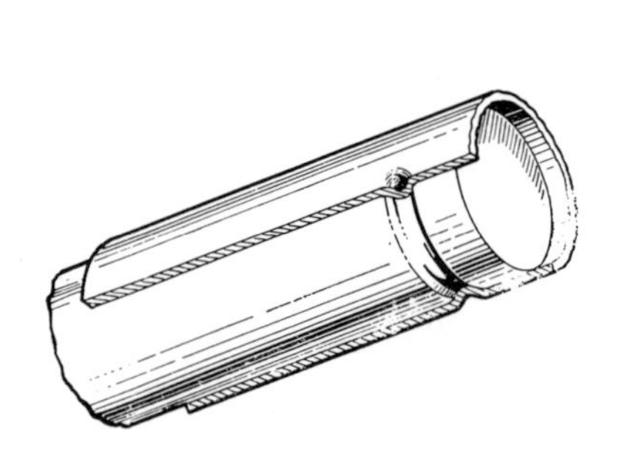

FIG. 8—For permanent location of rod in tube. Dents are made after rod is in place. Axial movement can be obtained by lengthening recess.

FIG. 9—For temporary positioning of rods in sheet metal. Slots in rod prevent axial movement.

# Without Fasteners

ERWIN RAUSCH
Chief Engineer,
General Automotive Specialty Co., Inc.

or temporary assemblies without using bolts, screws or rivets.

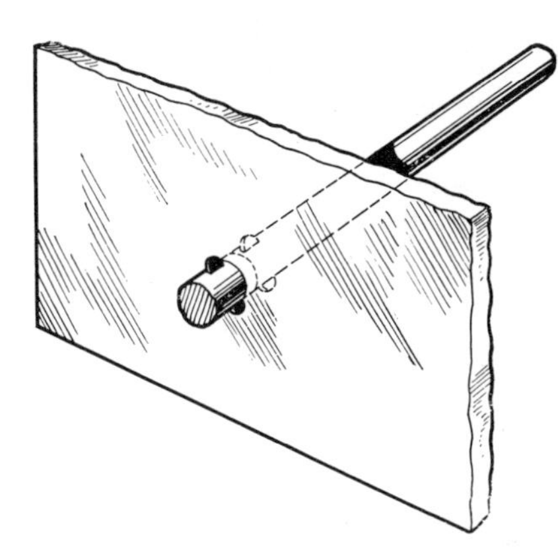

FIG. 10—For permanent assemblies where split forming die can be applied on one or both sides of sheet.

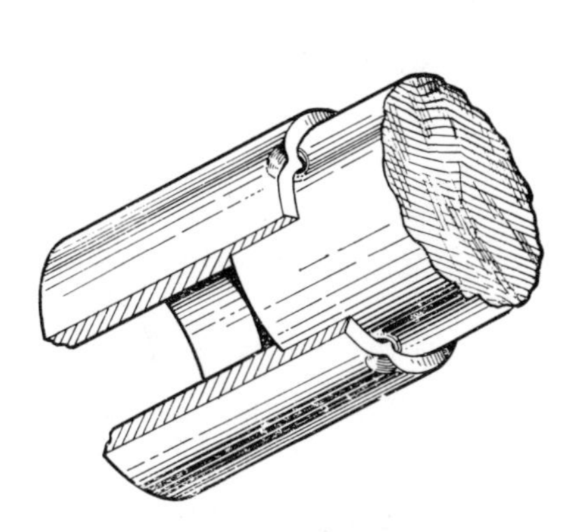

FIG. 11—For attachment of recessed, solid metallic or non-metallic parts in thick-sectioned cylindrical parts.

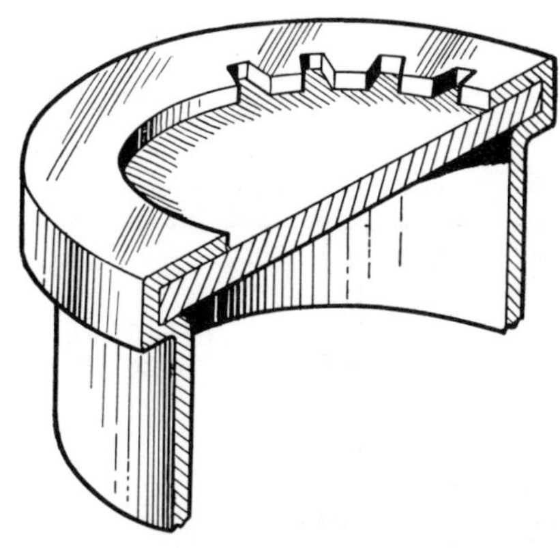

FIG. 12—For closing the end(s) of cylindrical parts. "Toothed" section permits removal of cap, "solid" section does not.

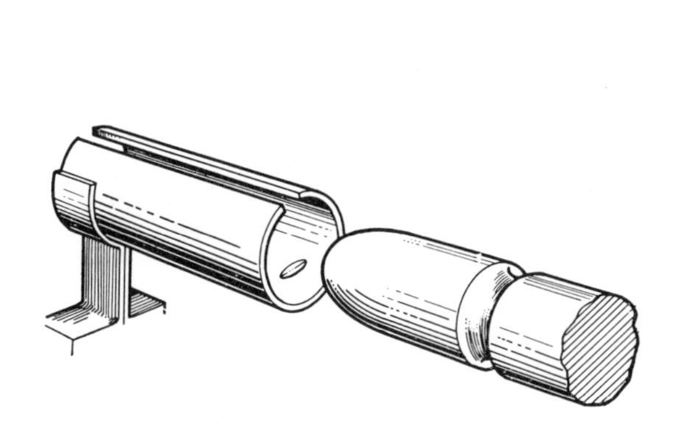

FIG. 13—For bayonet-type electrical fittings. Sleeve and mounting bracket can be modified to suit requirements.

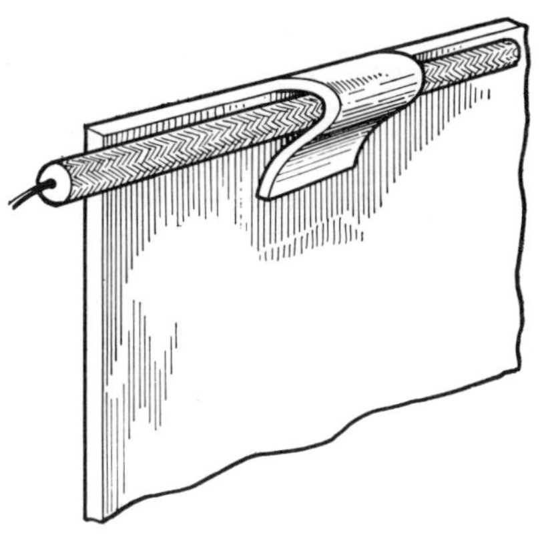

FIG. 14—For supporting electrical conductors.

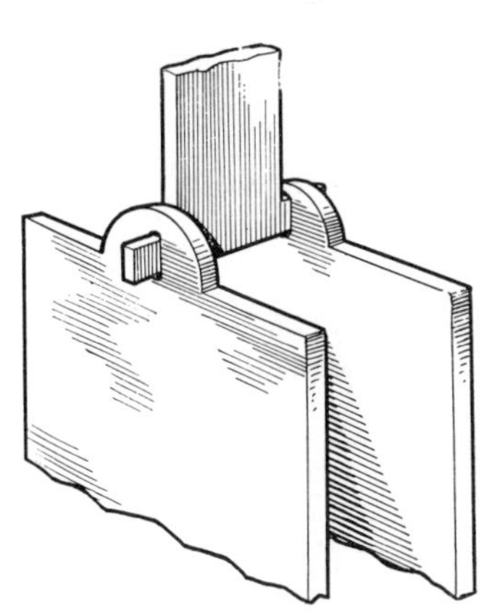

FIG. 15—For permanent assemblies requiring relative movement, but minimum rigidity.

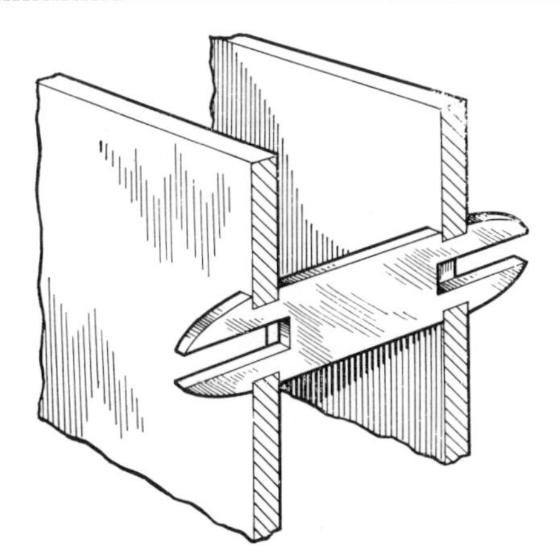

FIG. 16—For temporary use as spacer. Sheet holes can be round or rectangular.

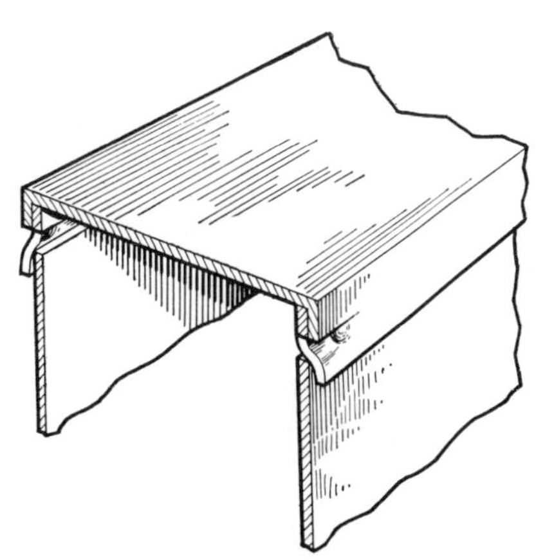

FIG. 17—For attachment of cover to channel or similar assemblies.

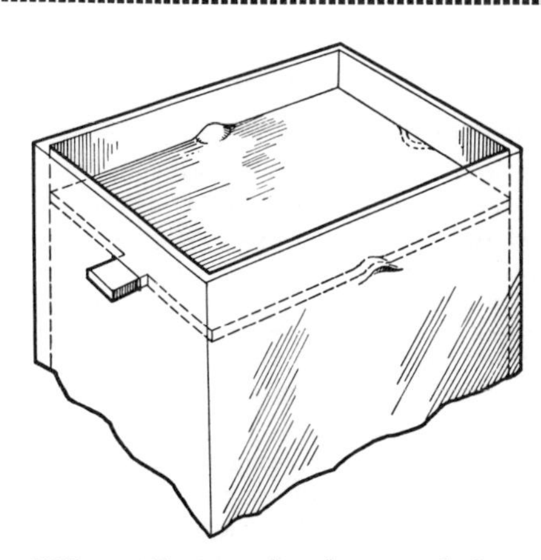

FIG. 18—For recessing sheet metal plates in rectangular housings.

# LIQUID-TIGHT

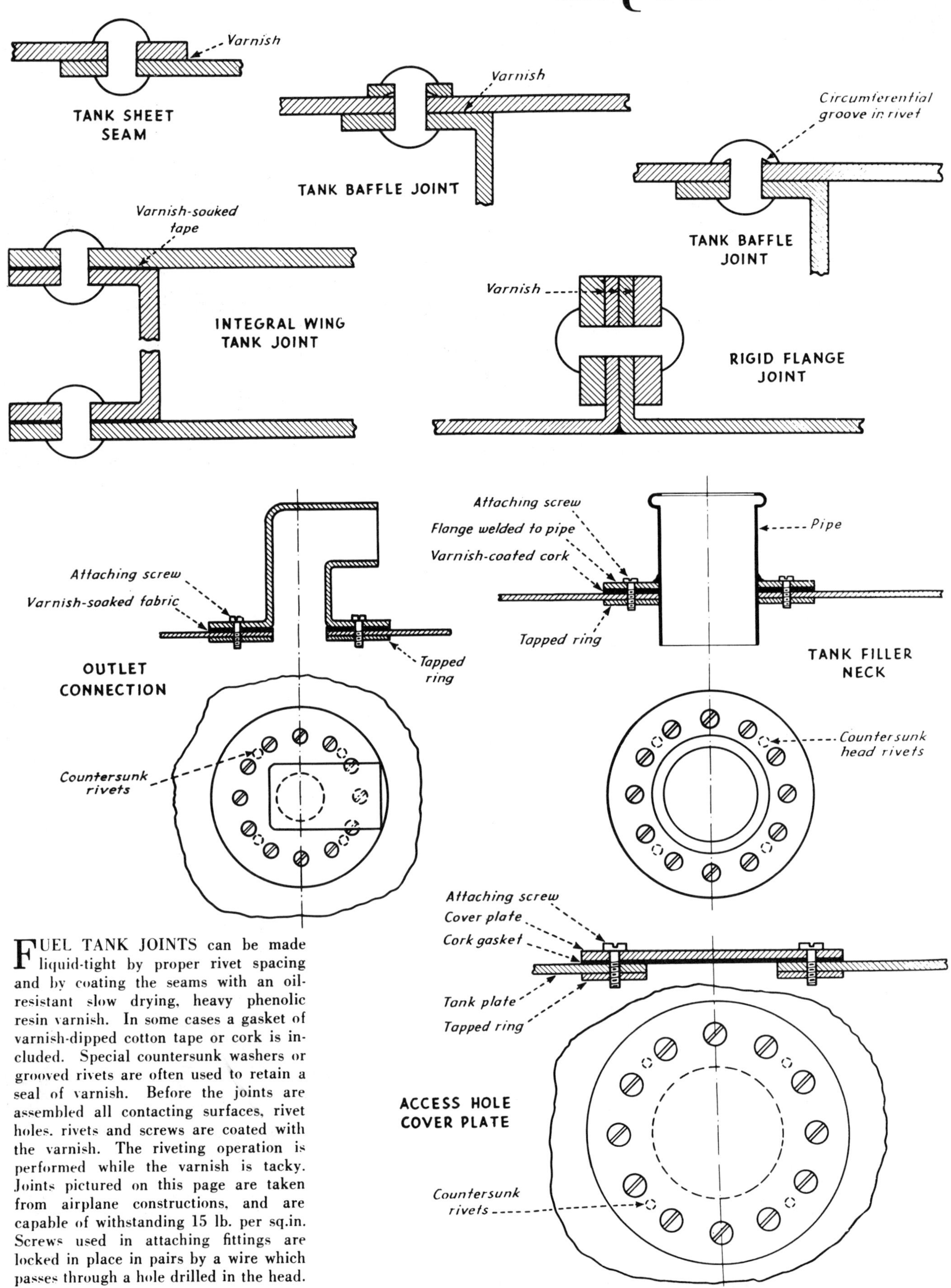


FUEL TANK JOINTS can be made liquid-tight by proper rivet spacing and by coating the seams with an oil-resistant slow drying, heavy phenolic resin varnish. In some cases a gasket of varnish-dipped cotton tape or cork is included. Special countersunk washers or grooved rivets are often used to retain a seal of varnish. Before the joints are assembled all contacting surfaces, rivet holes, rivets and screws are coated with the varnish. The riveting operation is performed while the varnish is tacky. Joints pictured on this page are taken from airplane constructions, and are capable of withstanding 15 lb. per sq.in. Screws used in attaching fittings are locked in place in pairs by a wire which passes through a hole drilled in the head.

# RIVETED JOINTS

FRED M. MORRIS

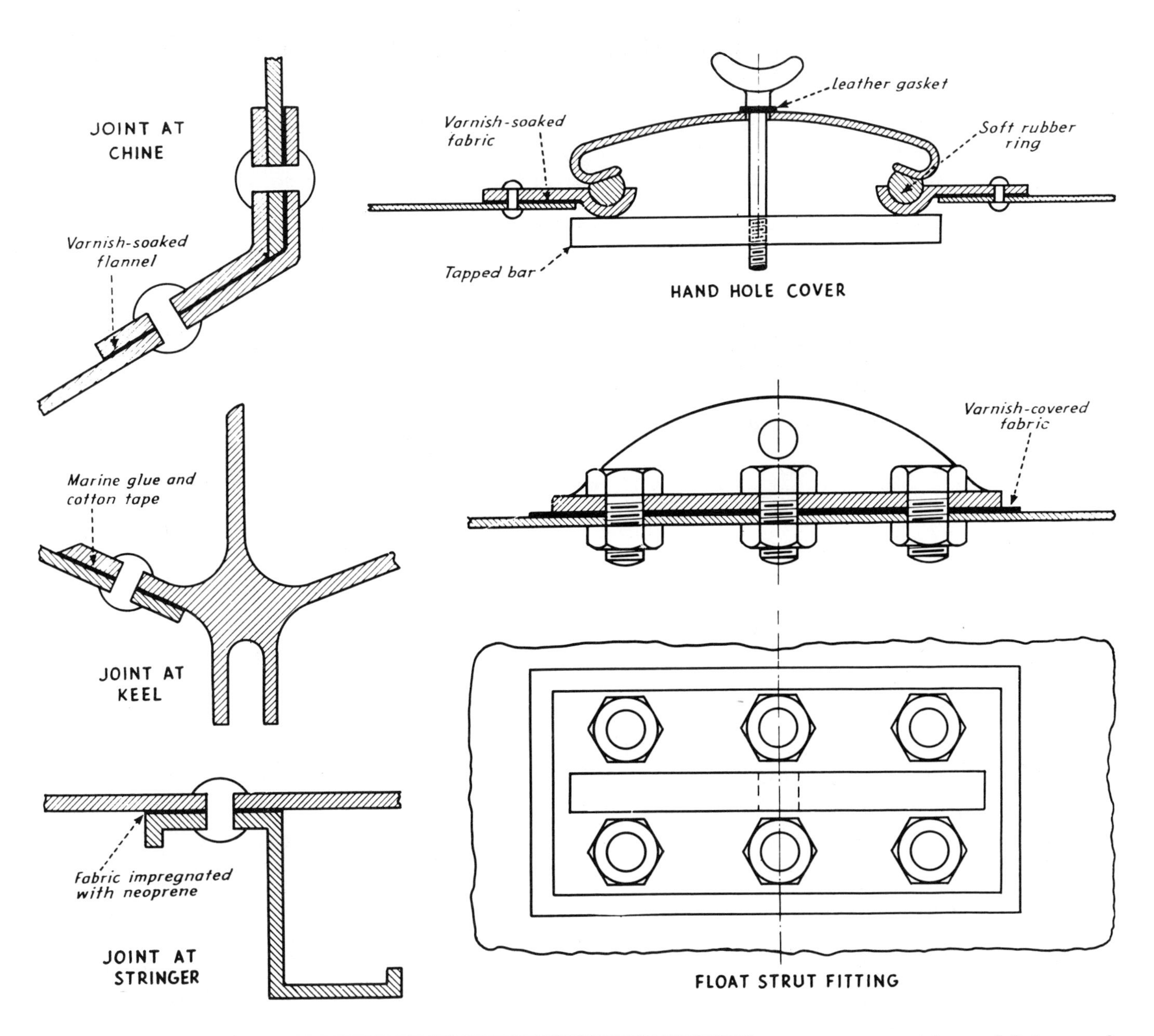


## LIQUID-TIGHT RIVET SPACING

For 17ST Aluminum Alloy Round Head Rivets and Plate

| SMALLEST PLATE THICKNESS | RIVET DIAMETER | EDGE DISTANCE | PITCH SINGLE ROW | PITCH DOUBLE ROW | DISTANCE BETWEEN RIVET ROWS |
|---|---|---|---|---|---|
| 0.023 in. | 3/32 in. | 1/4 in. | 3/8 in. | 13/32 in. | 5/16 in. |
| 0.029 | 3/32 | 1/4 | 3/8 | 13/32 | 5/16 |
| 0.036 | 1/8 | 5/16 | 1/2 | 9/16 | 13/32 |
| 0.045 | 5/32 | 3/8 | 5/8 | 23/32 | 1/2 |
| 0.051 | 5/32 | 3/8 | 5/8 | 23/32 | 1/2 |
| 0.064 | 3/16 | 3/8 | 3/4 | 1 3/32 | 1/2 |
| 0.081 | 1/4 | 15/32 | 1 | 1 17/32 | 5/8 |
| 0.100 | 5/16 | 19/32 | 1 1/4 | 1 15/16 | 25/32 |
| 0.125 | 3/8 | 11/16 | 1 1/2 | 2 5/16 | 15/16 |
| 0.187 | 1/2 | 27/32 | 2 | 3 3/32 | 1 1/4 |
| 0.250 | 5/8 | 1 | 2 1/2 | 3 7/8 | 1 9/16 |

WATER-TIGHT joints and fittings used, for example, in amphibious aircraft construction may be sealed in the same manner as fuel tank joints. Neoprene-impregnated tape, wet with kerosene to make it sticky, is used for straight seams. Domet flannel or cotton tape soaked in a water-resistant soy bean oil varnish or in marine glue is used at corners and complicated joints. All contacting surfaces, rivet holes, rivets, and screws in these joints are varnished before the joints are assembled. Before it has dried, excess varnish is removed with a solvent to save time in the painting operation. Seams made in this manner are capable of withstanding 15 lb. per sq. in. Practically all metallic parts of these joints are constructed of aluminum alloy for lightness.

# Simplifying Assemblies With

Since annealed spring steel can be stamped, twisted, or bent into any desirable shape and then heat-treated to develop spring characteristics, it can be designed for multiple functions in addition to that of fastening. Shown below are 16 ways

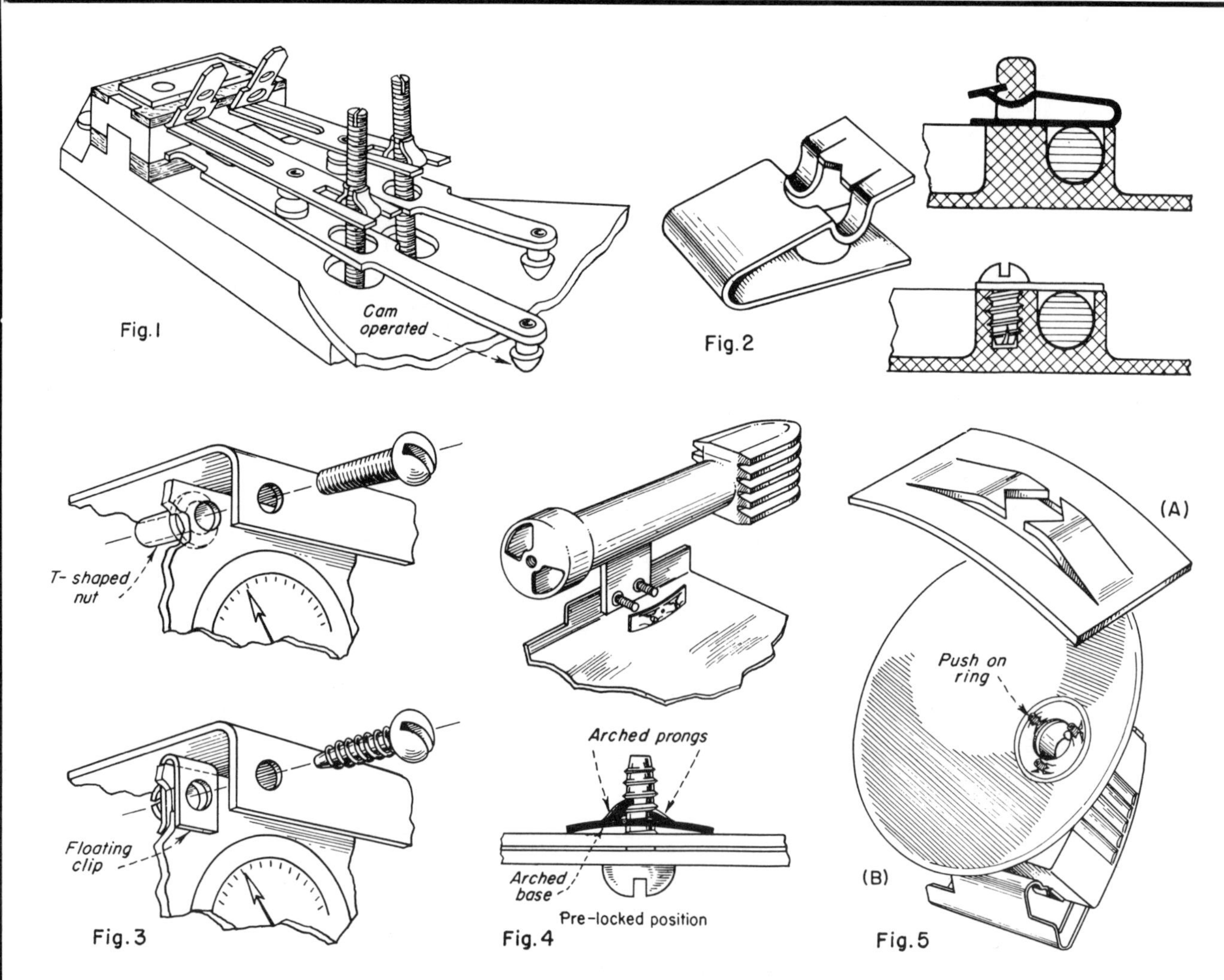


FIG. 1—Multiple-purpose flat spring has stamped hole helically formed to accept adjusting screw. Replaces a locknut, bushing with internal threads and spring blade. Screw adjusts gap of contacts thus changing duration of current flow. Application: thermostatic timing control unit of an automatic beverage percolator.

FIG. 2—Spring-steel clip firmly holds stud of control knob while allowing it to turn freely on its bearing surface. Clip is removed by merely compressing spring arms and pulling off stud. Replaces screw and machined plate.

FIG. 3—Floating clip, which snaps in place by hand, reduces hole misalignment problems by permitting sufficient shift of mounting holes to offset normal manufacturing tolerances of main parts. Replaces welded T-shape nuts.

FIG. 4—Twin-hole nut removes need for hand wrench in hard-to-reach location and replaces two nuts and lockwashers. Combined force of arched prongs and base when nut is compressed creates high resistance to vibration loosening. Application: gas burner assembly of an automatic household clothes dryer heating unit.

FIG. 5—Push-on nuts (A) easily press over studs, rivets, tubing and other unthreaded parts. Their steel prongs securely bite into smooth surfaces under load. Application: (B) triple-hole push-on ring on a flash bulb reflector.

FIG. 6—Previous method of assembling desk calendar (A) required seven parts: wire guide, spring clip, plate, two bolts and two nuts. Multi-purpose spring clip (B) replaces above although using same retaining principle.

# Spring-Steel Fasteners

Tinnerman Products, Inc.

in which spring-steel fasteners can be used to: Compensate for hole or part misalignment; prevent vibration loosening; eliminate parts; speed assembly; permit fastener removals; lock on unthreaded studs; and permit blind installations.

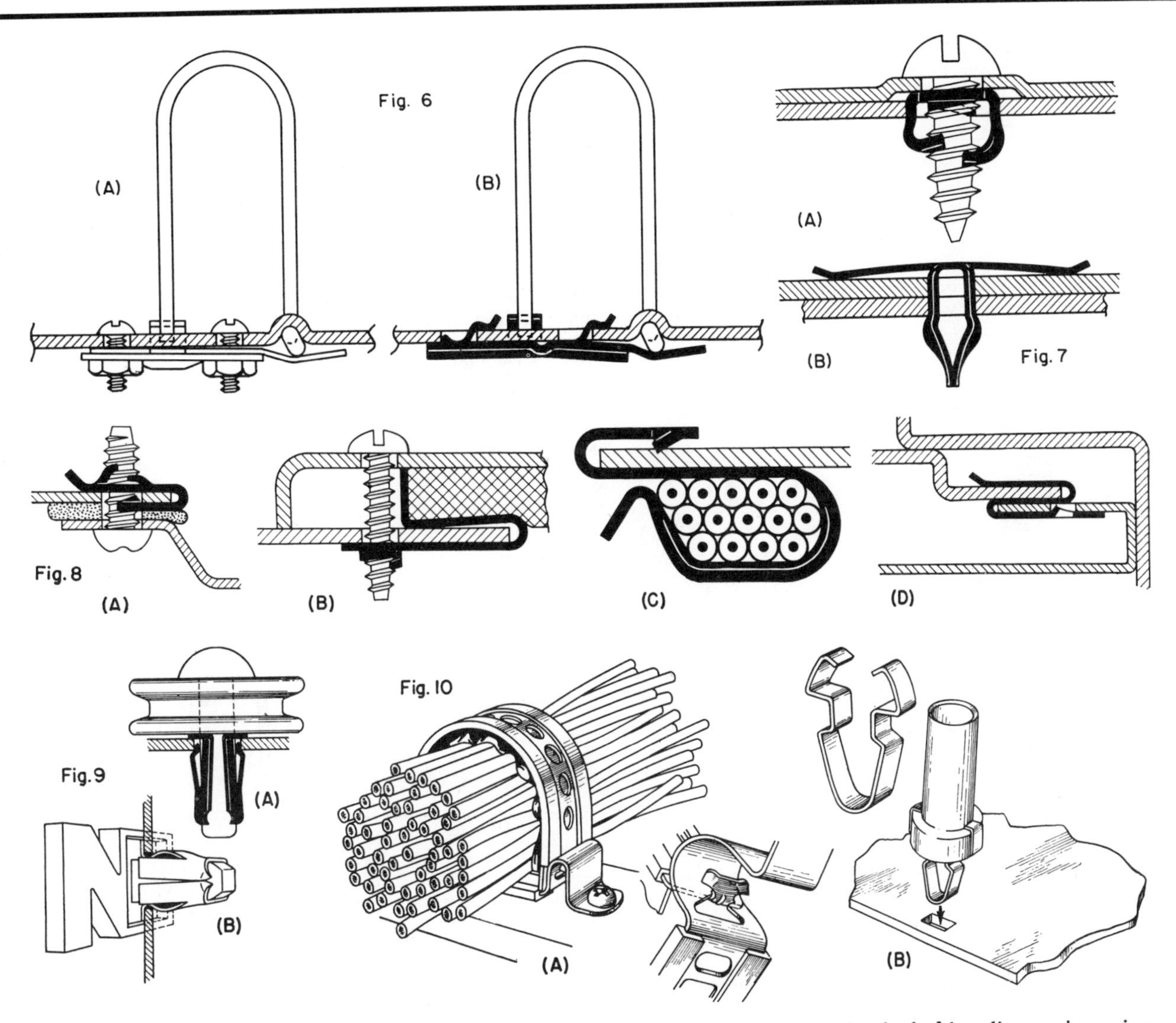


FIG. 7—Expansion-type fastener permits blind installation where access is from one side of an assembly only. Insertion of screw (A) spreads fastener arms apart thus producing a wedging effect in the hole. Dart-type fastener (B) can be quickly snapped in place; one common application is attaching molding trim strips to auto panels.

FIG. 8—End clips pressed by hand on panel edges have barbed retaining leg which either bites into the metal or snaps into a mounting hole. Applications: (A) sheet metal screw and J-clip with arched prongs compresses insulating material between panels; (B) bent leg of clip acts as spacer between two panels to support sheet of insulating material; (C) barbed-leg-clip retains wires without need for mounting hole; (D) S-clip spring-steel fastener secures removable panel in inaccessible position.

FIG. 9—Tubular-type fastener has cam-like prongs which spring out after insertion to hold fastener in position. Applications: (A) radio dial pulley; (B) attaching automotive name plate to panel.

FIG. 10—Special-function fasteners for quick assembly and disassembly of components. (A) Wire harness clamp using torque and slot principle; (B) dart-shaped clip for attaching coils and other parts to electronic chassis.

# 20 Tamper-proof

FEDERICO STRASSER *Santiago de Chile*

Ways to prevent or indicate unauthorized removal of fasteners in vending machines, instruments, radios, TV sets and other units. Included are positively retained fasteners to prevent loss where retrieval would be difficult.

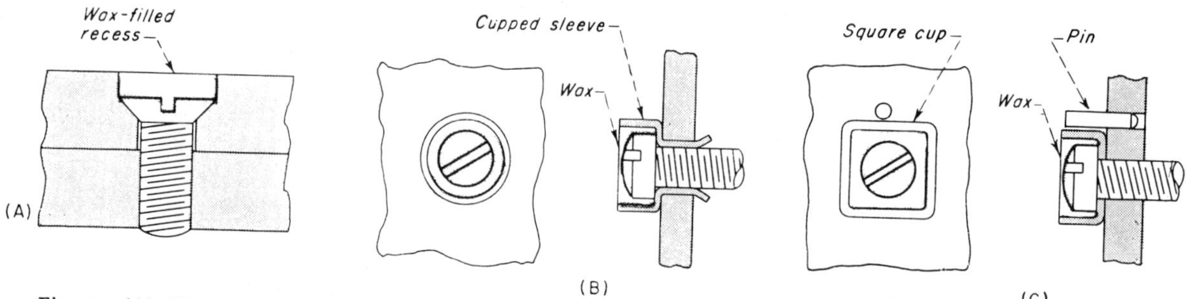


Fig. 1—(A) Wax or other suitable material fills recess above screw. Wax flush with plate hides screw position if surface is painted. (B) Cupped sleeve riveted in screw hole provides cavity for wax when plate is too thin for recessing. (C) Pin prevents rotation of square cup which would allow screw to be removed without disturbing wax.

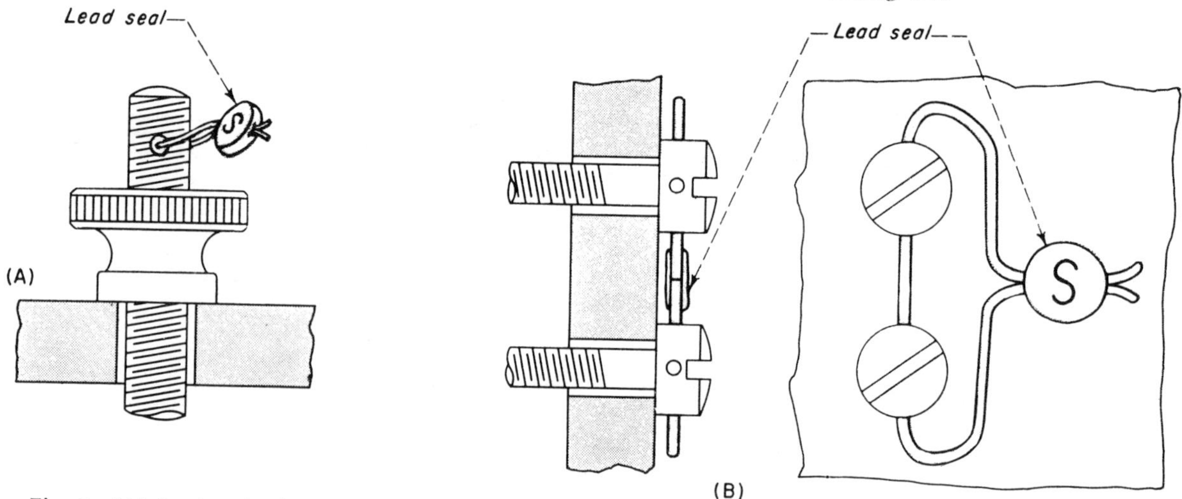


Fig. 2—(A) Lead seal crimped over twisted ends of wire passing through screw allows only limited slackening of nut. (B) Two or more screws strung through heads with wire are protected against unauthorized removal by only one seal. Code or other signet can be embossed on seals during crimping.

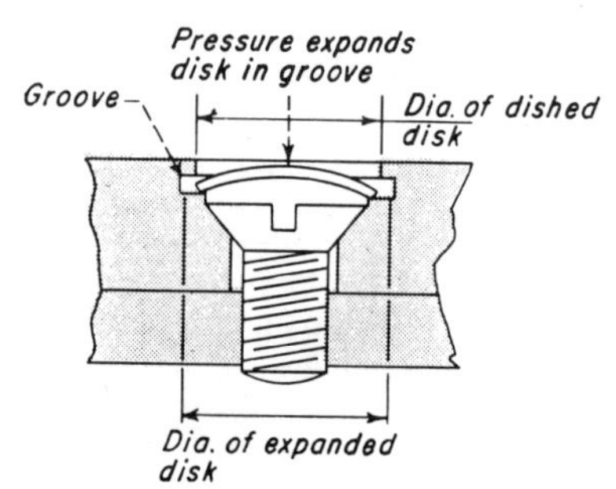


Fig. 3—Sheet-metal disk pressed into groove can only be removed with difficulty and discourages tampering.

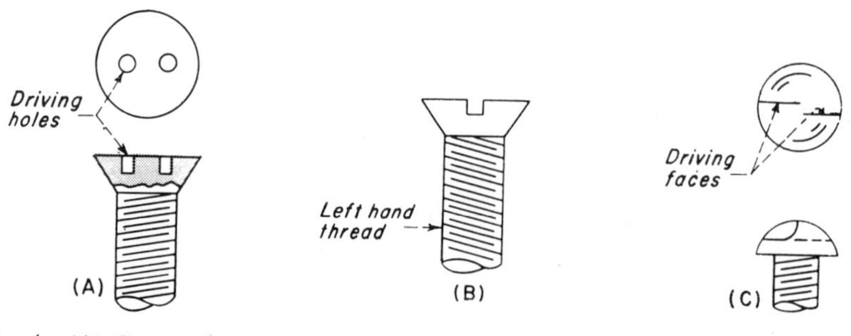


Fig. 4—(A) Spanner-head screws are available in all standard heads and sizes from U.S. manufacturers. Special driver is required for each screw size except ¼-in. dia and above. (B) Left-hand screw thread is sometimes sufficient to prevent unauthorized loosening, or (C) special head lets screw be driven but not unscrewed.

# Fasteners

## POSITIVELY RETAINED FASTENERS

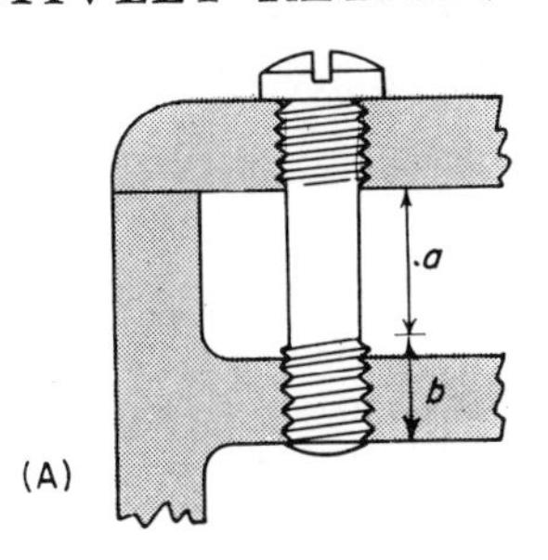


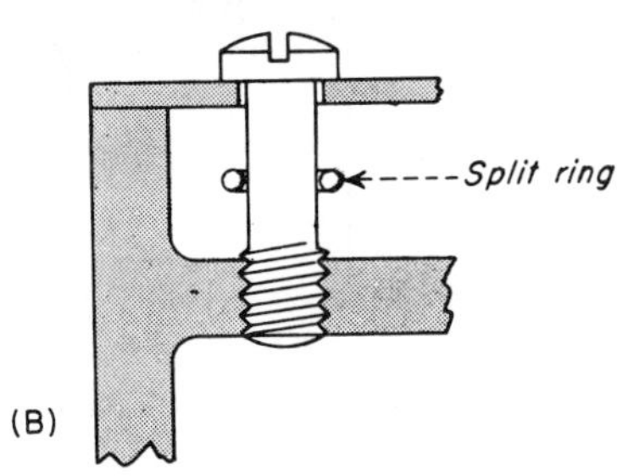


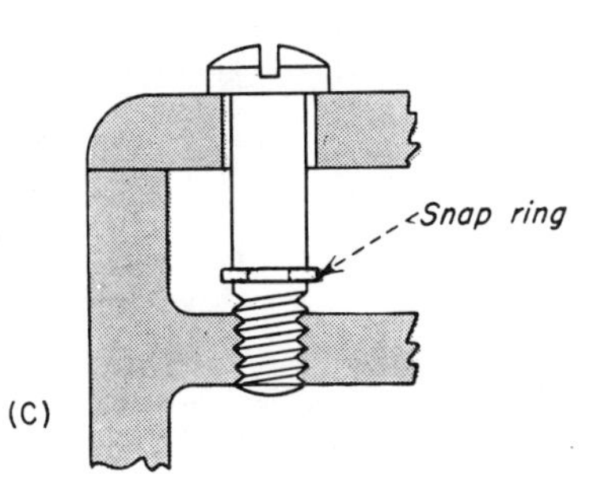


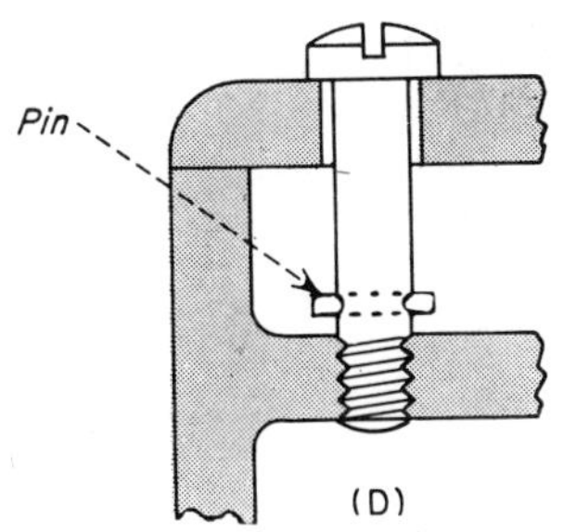


Fig. 5—(A) Tapped cover and casing allows screw (a > b) with reduced shank diameter to be completely unscrewed from casing yet retained positively in cover. For thin sheet-metal covers, split ring on reduced shank (B) is preferable. Snap ring in groove (C) or transverse pin (D) are effective on unreduced shank. Simple and cheap method (E) is fiber washer pushed over thread.

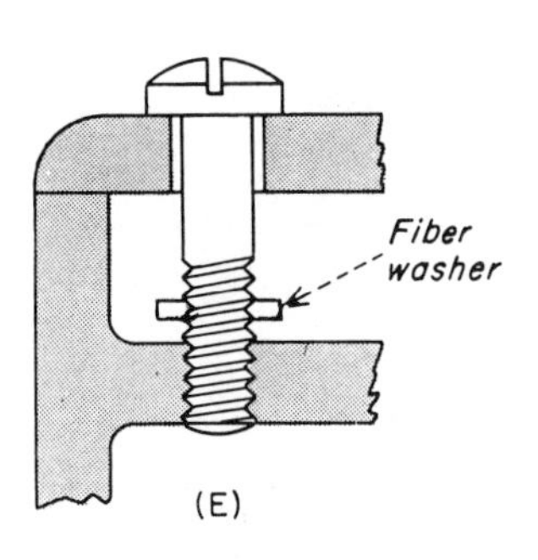


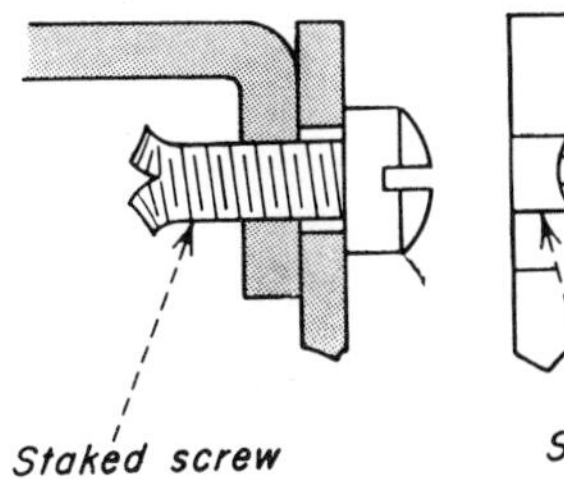


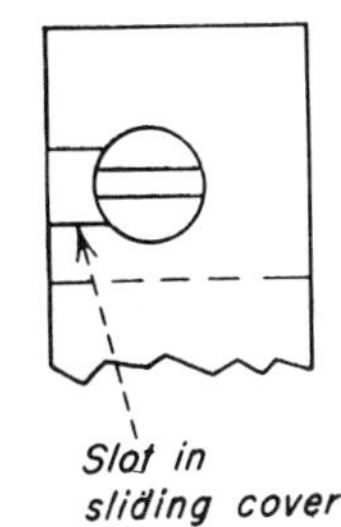


Fig. 6—Open-ended slot in sliding cover allows screw end to be staked or burred so screw cannot be removed, once assembled.

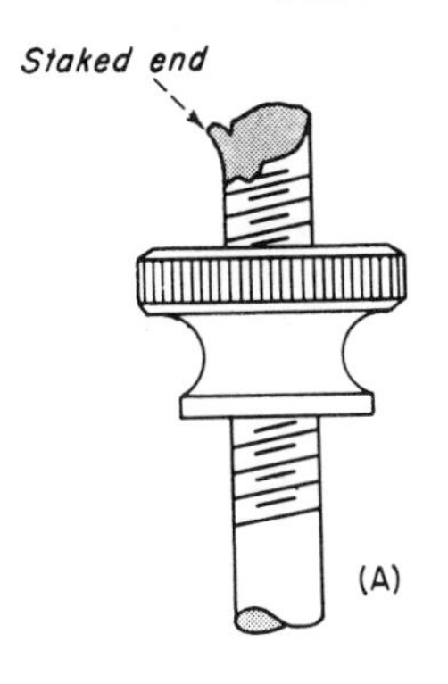


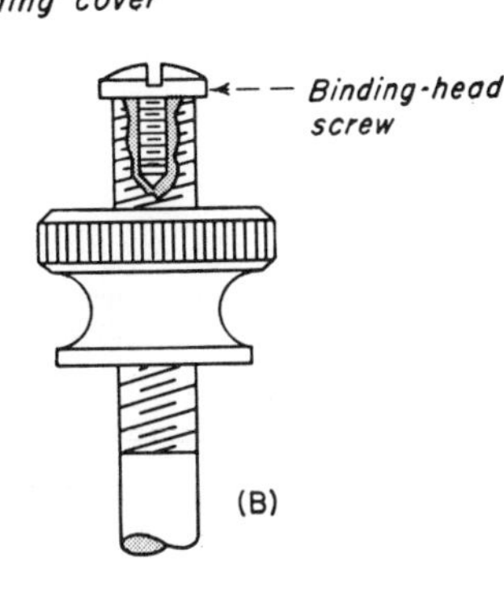


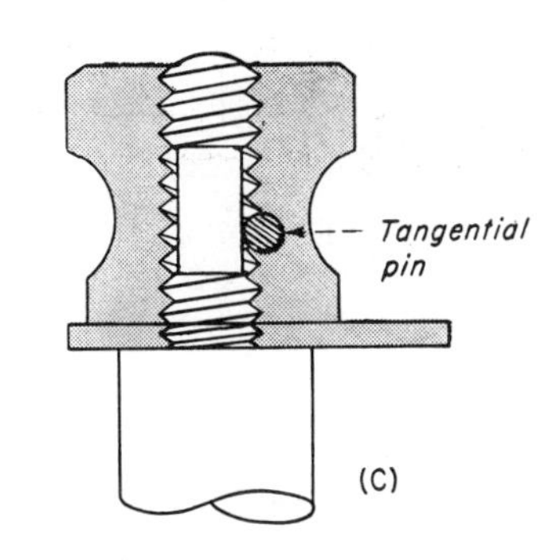


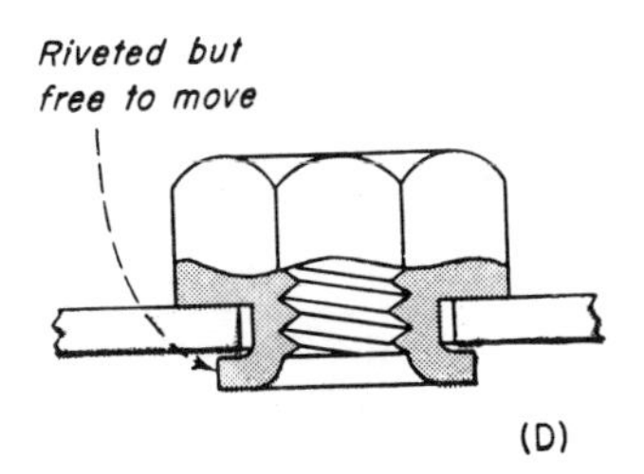


Fig. 7—(A) Nut is retained on screw by staking or similar method but, if removal of nut is occasionally necessary, coaxial binding-head screw (B) can be used. Where screw end must be flush with nut, pin through nut tangential to undercut screw (C) limits nut movement. Rotatable nut (D) or screw (E) should have sufficient lateral freedom to accommodate slight differences in location when two or more screws are used.

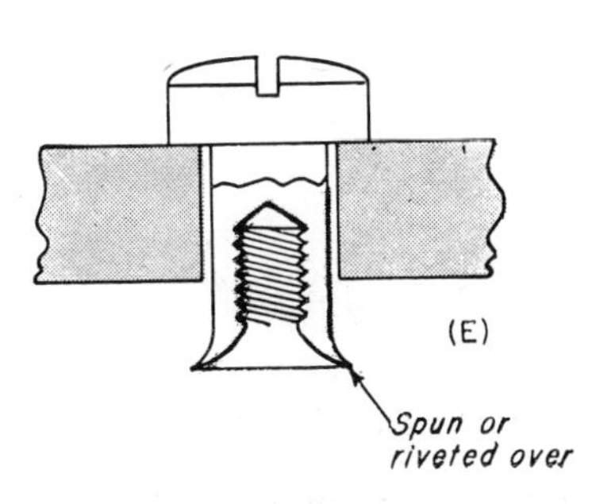

# How to Design Parts that Assemble

Methods of joining tubes, sheet metal and machine parts in

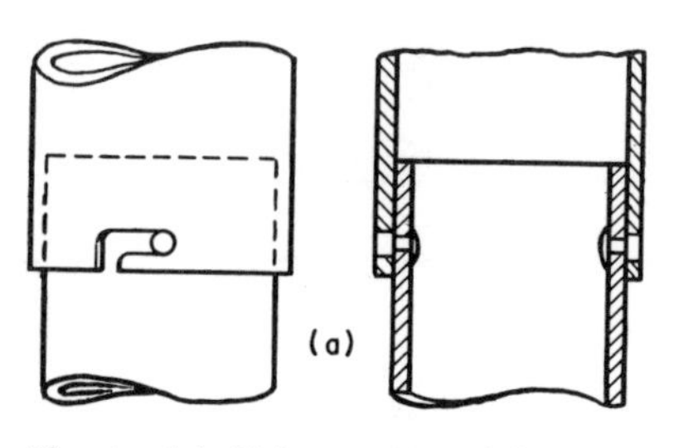


Fig. 1—(*a*) Telescope-type joint uses headless bayonet pin to assemble tubes. Method can also be used to fasten cover on cylindrical member. (*b*) Pins such as those in (*a*) can be eliminated by using formed tabs.

Fig. 2—Rotating cylindrical member 90 deg forces spring elements apart and results in a clamping force on milled section.

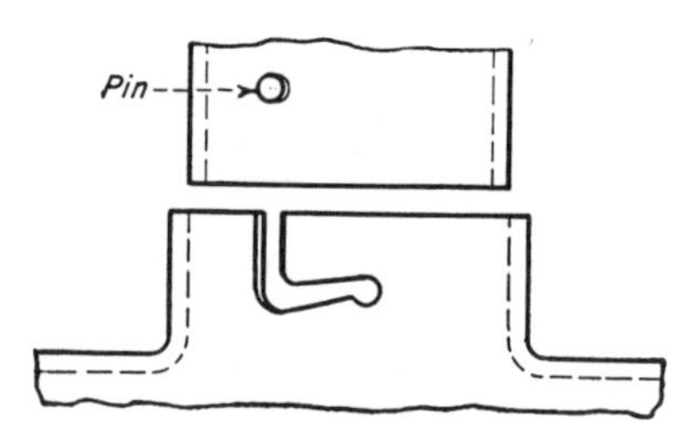


Fig. 6—Two diametrically opposite pins engage slots in lower member. Slot has slight incline with recess to provide a locking tendency. This method can provide leak proof sealing.

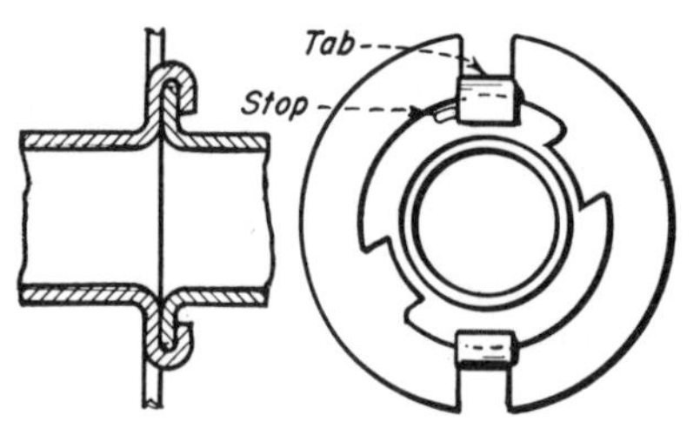


Fig. 7—Tubes of equal diameter are joined by engaging tabs in slots and twisting. Provides means of quick assembling and disassembling without bending tabs each time.

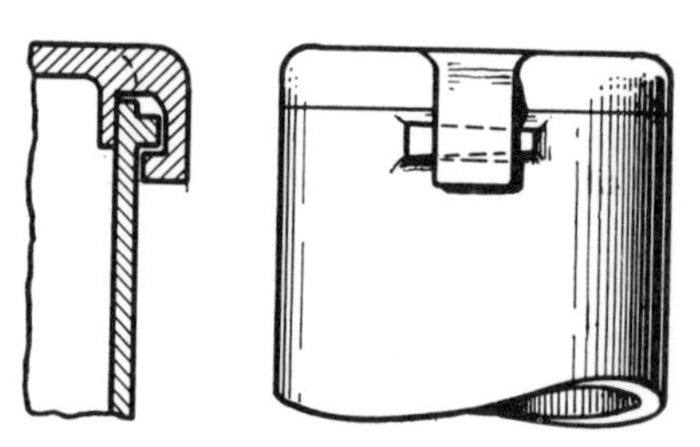

Fig. 8—Tapered projections on diametrically opposite sides of cast structure engage tabs on cast cover to provide for clamping cover tightly. Machined pins can replace the tabs.

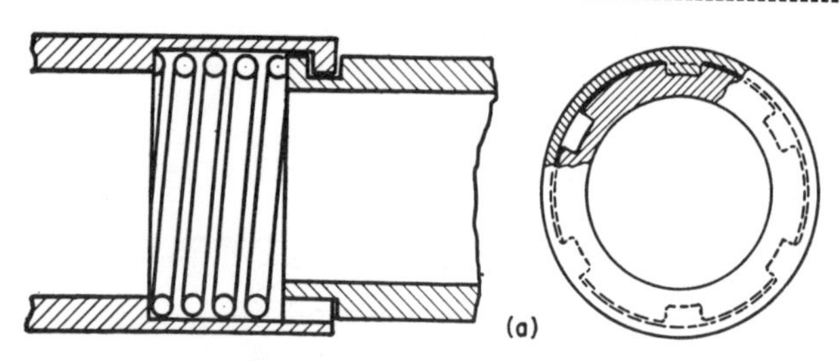


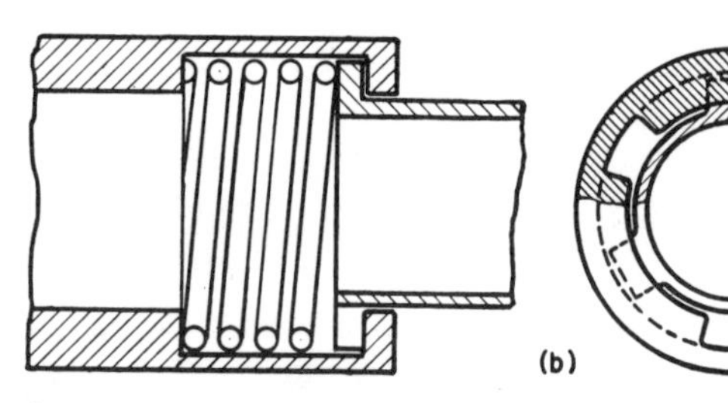


Fig. 12—Two different versions of a spring-loaded type coupling for metal or plastic tubing. In both cases, the two parts can be assembled or disassembled by twisting one of the elements. In sketch (a), the holding tabs are located on the larger member and the receiving slots are on the smaller one. In (b), the relative positions of tabs and slots are reversed.

QUICK-RELEASE FASTENERS that engage or disengage with a partial twist as contrasted to fasteners that require several turns for complete engagement or disengagement.

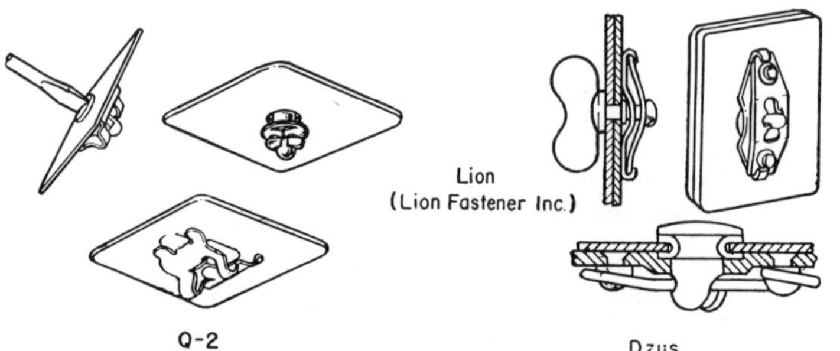

# and Lock by Twisting

assemblies that engage and disengage with a twist of one element.

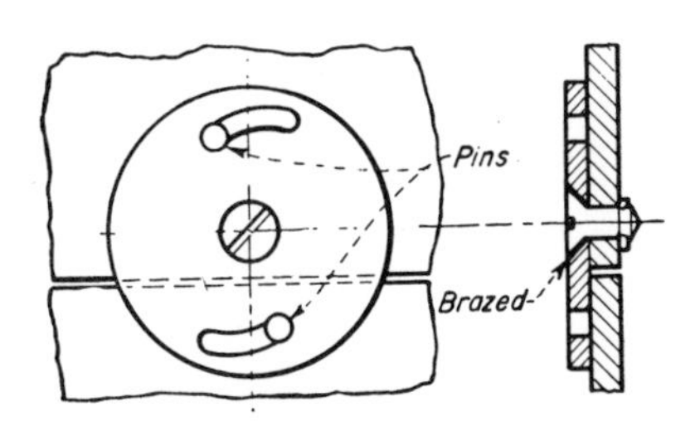


Fig. 3—Spiral type slots force pins to pull two elements together. Recesses at ends of slots serve for locking.

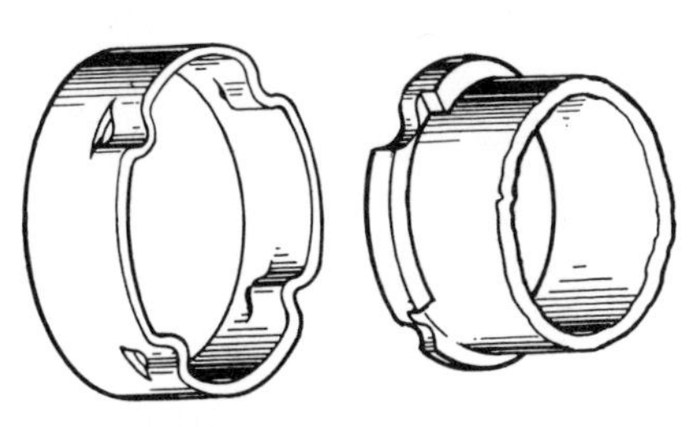

Fig. 4—Formed projections on cover engage recesses in flange of cylindrical vessel. Taper on flange provides locking.

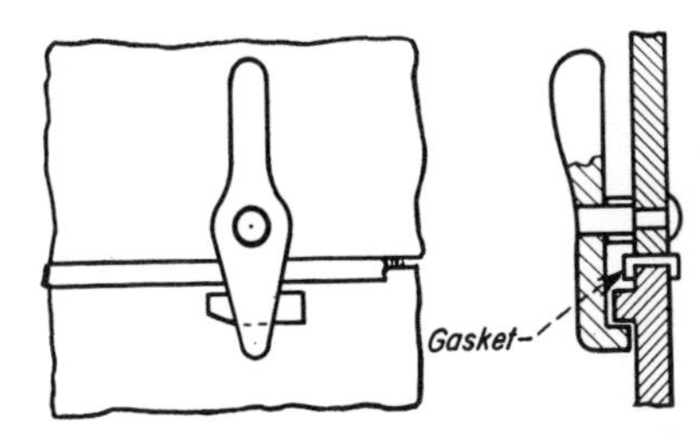


Fig. 5—Wedge principle can also be used for locking vent windows on cars or fastening two metal sheets.

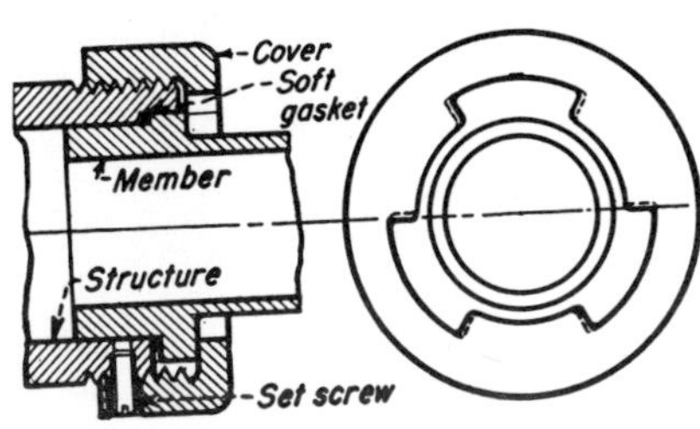


Fig. 9—Structure and cover are cut away in three places to receive projections on member. Rotating cover 60 deg permits member to be inserted or removed. Setscrew provides means for locking cover.

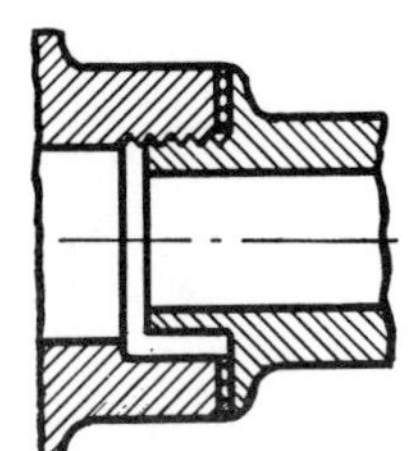

Fig. 10—Intermittent threads on both members require only a partial revolution for complete engagement. Tightness of joint can be controlled within limits by varying the thread lead.

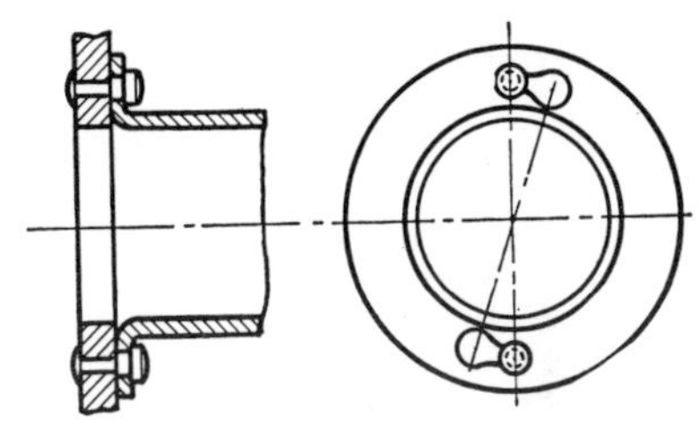

Fig. 11—Flanged tube is fastened to plate by twisting. Bayonet-type pin with head engages slot in tube. Seal can be made leak-proof by spring-loading the pin and adding a soft gasket.

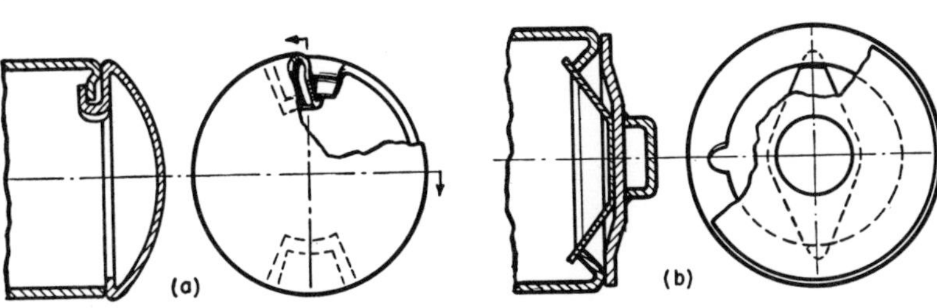


Fig. 13—Two other methods of attaching covers to cylindrical vessels. In *(a)* flanges on structure hold the cover tightly after it has been engaged and rotated. Elongated tabs are used in *(b)* to mate with notches in the structure.

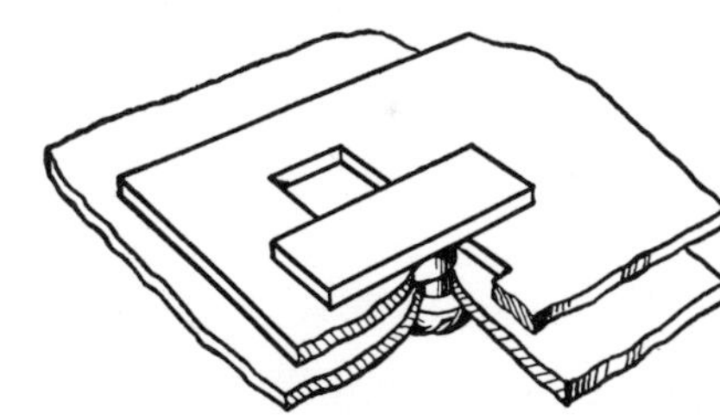

Fig. 14—Method of fastening a member to a structure by twisting a third element through a 90 degree arc.

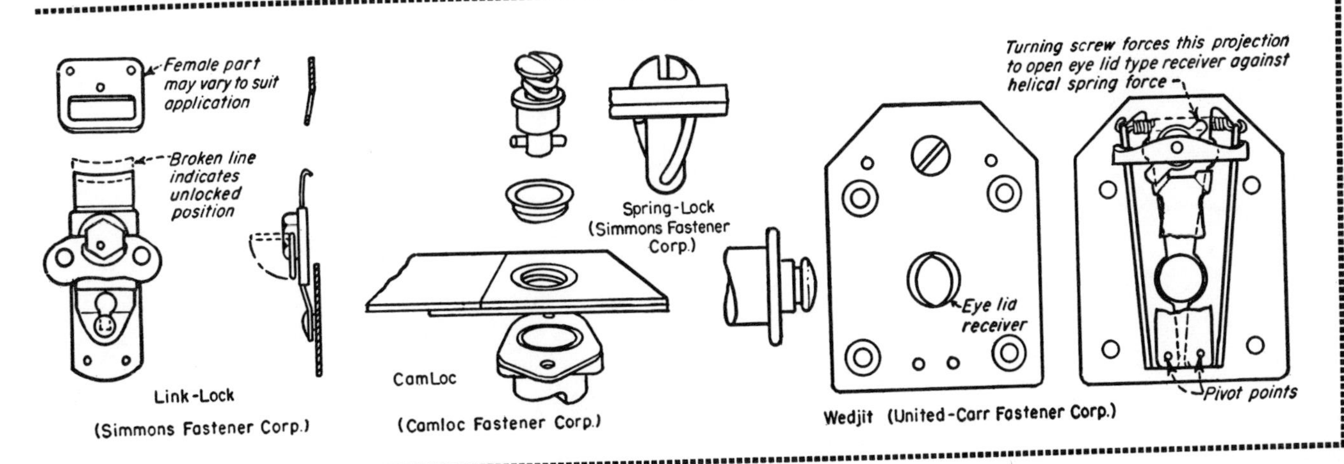

# 3

# *CLUTCHES*


Basic Types of Mechanical Clutches 78
Construction Details of Over-riding Clutches 80
10 Ways to Apply Overrunning Clutches 82
Small Mechanical Clutches for Precise Service 84
Serrated Clutches and Detents 86

# Basic Types of Mechanical

Sketches include both friction and positive types. Figs. 1-7 are classified as externally controlled; Figs. 8-12 are internally controlled. The latter are further divided into

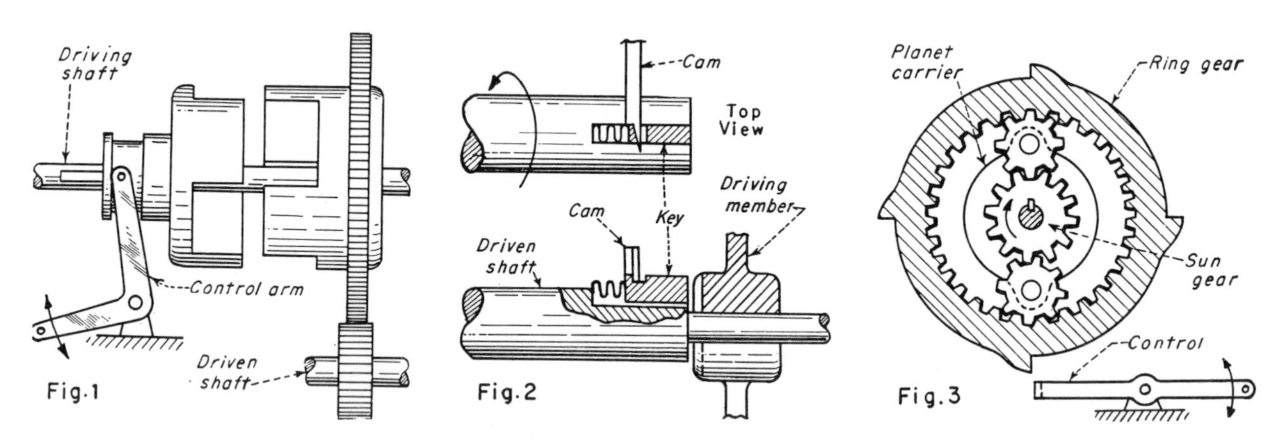


**1. JAW CLUTCH.** Left sliding half is feathered to the driving shaft while right half rotates freely. Control arm activates the sliding half to engage or disengage the drive. This clutch, though strong and simple, suffers from disadvantages of high shock during engagement, high inertia of the sliding half, and considerable axial motion required for engagement.

**2. SLIDING KEY CLUTCH.** Driven shaft with a keyway carries freely-rotating member which has radial slots along its hub; sliding key is spring loaded but is normally restrained from engaging slots by the control cam. To engage the clutch, control cam is raised and key enters one of the slots. To disengage, cam is lowered into the path of the key; rotation of driven shaft forces key out of slot in driving member. Step on control cam limits axial movement of the key.

**3. PLANETARY TRANSMISSION CLUTCH.** In disengaged position shown, driving sun gear will merely cause the free-wheeling ring gear to idle counter-clockwise, while the driven member, the planet carrier, remains motionless. If motion of the ring gear is blocked by the control arm, a positive clockwise drive is established to the driven planet carrier.

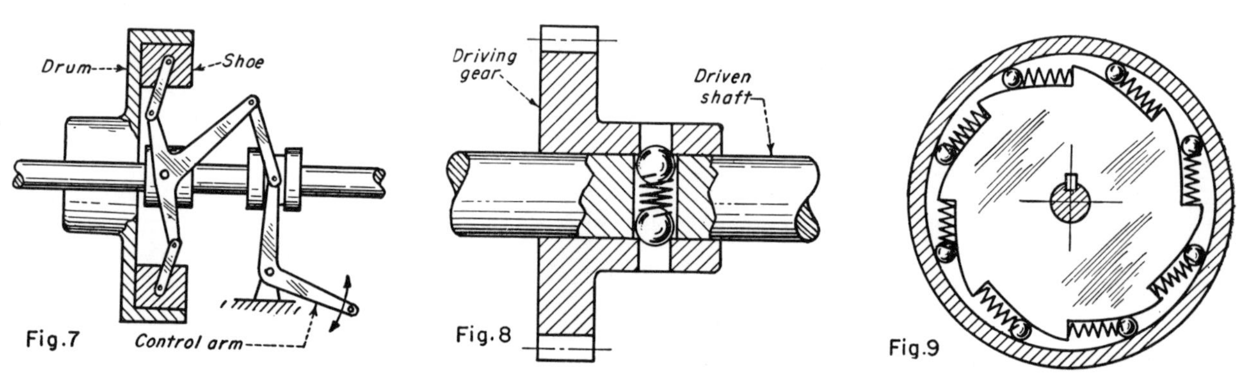


**7. EXPANDING SHOE CLUTCH.** In sketch above, engagement is obtained by motion of control arm which operates linkages to force friction shoes radially outward into contact with inside surface of drum.

**8. SPRING AND BALL RADIAL DETENT CLUTCH.** This design will positively hold the driving gear and driven shaft in a given timing relationship until the torque becomes excessive. At this point the balls will be forced inward against their spring pressure and out of engagement with the holes in the hub, thus permitting the driving gear to continue rotating while the driven shaft is stationary.

**9. CAM AND ROLLER CLUTCH.** This over-running clutch is suited for higher speed free-wheeling than the pawl and ratchet types. The inner driving member has camming surfaces at its outer rim and carries light springs that force rollers to wedge between these surfaces and the inner cylindrical face of the driven member. During driving, self-energizing friction rather than the springs forces the roller to tightly wedge between the members and give essentially positive drive in a clockwise direction. The springs insure fast clutching action. If the driven member should attempt to run ahead of the driver, friction will force the rollers out of a tight wedging position and break the connection.

# Clutches

MARVIN TAYLOR
Mechanical Research & Development Department.
Monroe Calculating Machine Company

overload relief, over-riding, and centrifugal types.

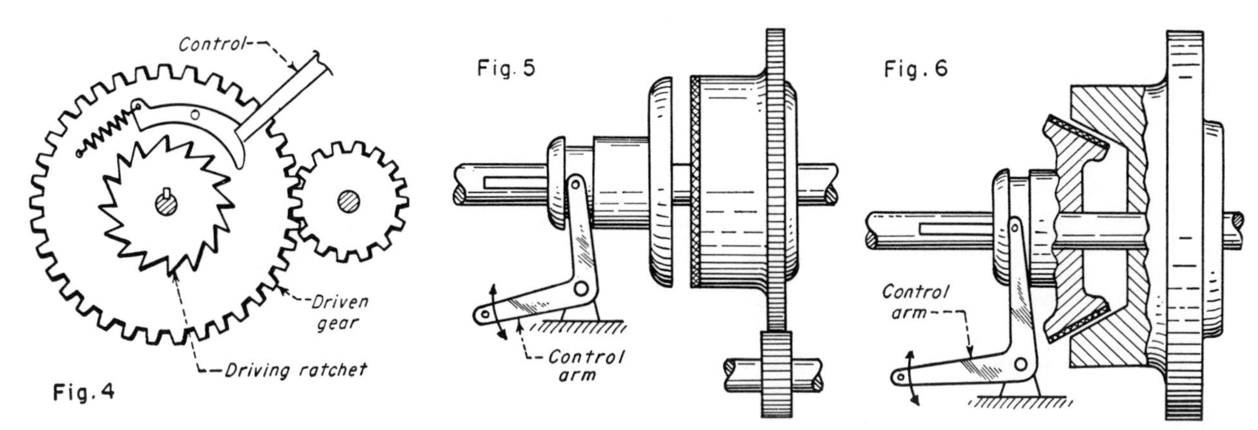


**4.** PAWL AND RATCHET CLUTCH. (External Control). Ratchet is keyed to the driving shaft; pawl is carried by driven gear which rotates freely on the driving shaft. Raising the control member permits the spring to pull the pawl into engagement with the ratchet and drive the gear. Engagement continues until control member is lowered into the path of a camming surface on the pawl. The motion of the driven gear will then force the pawl out of engagement and bring the driven assembly to a solid stop against the control member. This clutch can be converted into an internally controlled type of unit by removing the external control arm and replacing it with a slideable member on the driving shaft.

**5.** PLATE CLUTCH. Available in many variations, with single and multiple plates, this unit transmits power through friction force developed between the faces of the left sliding half which is fitted with a feather key and the right half which is free to rotate on the shaft. Torque capacity depends upon the axial force exerted by the control member when it activates the sliding half.

**6.** CONE CLUTCH. This type also requires axial movement for engagement, but the axial force required is less than that required with plate clutches. Friction material is usually applied to only one of the mating surfaces. Free member is mounted to resist axial thrust.

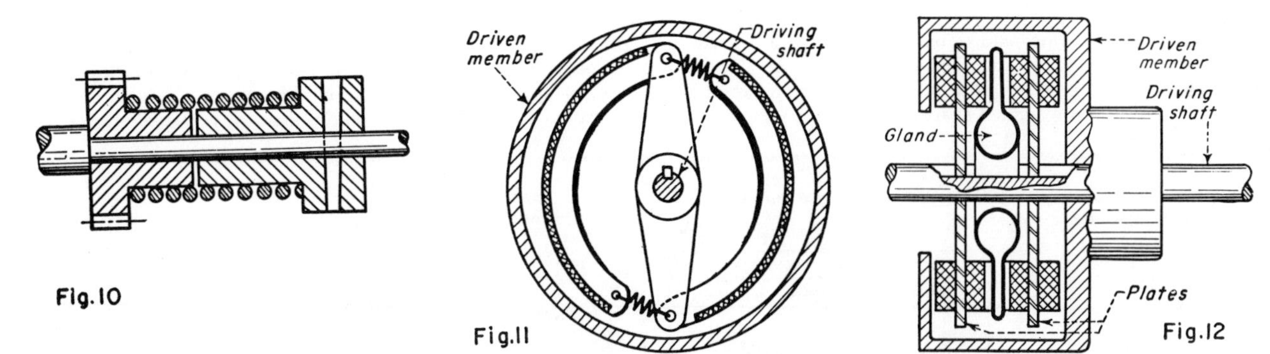


**10.** WRAPPED SPRING CLUTCH. Makes a simple and inexpensive uni-directional clutch consisting of two rotating members connected by a coil spring which fits snugly over both hubs. In the driving direction the spring tightens about the hubs producing a self energizing friction grip; in the opposite direction it unwinds and will slip.

**11.** EXPANDING SHOE CENTRIFUGAL CLUTCH. Similar in action to the unit shown in Fig. 7 with the exception that no external control is used. Two friction shoes, attached to the driving member, are held inward by springs until they reach the "clutch-in" speed, at which centrifugal force energizes the shoes outward into contact with the drum. As the driver rotates faster the pressure between the shoes and the drum increases thereby providing greater torque capacity.

**12.** MERCURY GLAND CLUTCH. Contains two friction plates and a mercury filled rubber gland, all keyed to the driving shaft. At rest, mercury fills a ring shaped cavity near the shaft; when revolved at sufficient speed, the mercury is forced outward by centrifugal force spreading the rubber gland axially and forcing the friction plates into driving contact with the faces of the driven housing. Axial thrust on driven member is negligible.

# Construction Details of

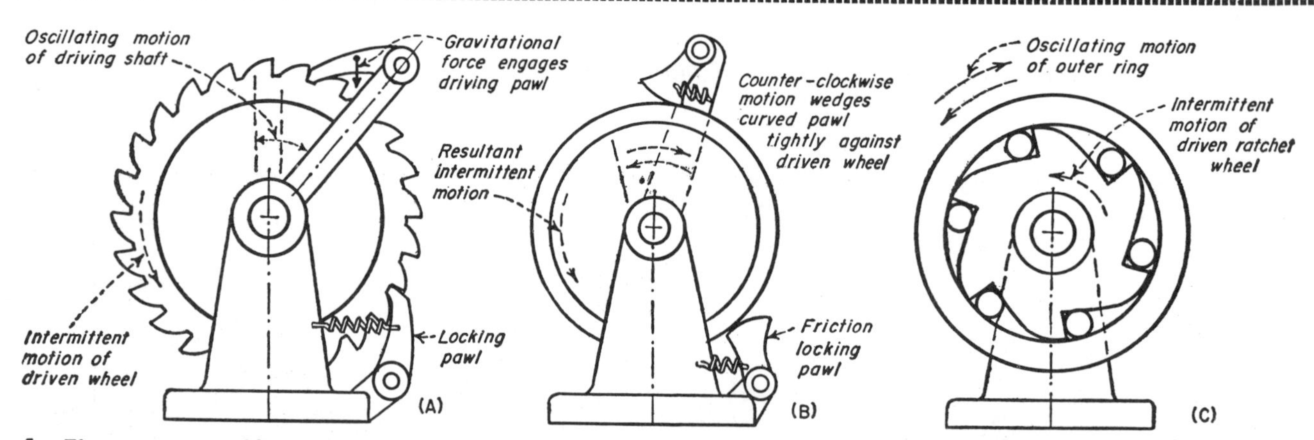


**1** Elementary over-riding clutches: (A) Ratchet and Pawl mechanism is used to convert reciprocating or oscillating movement to intermittent rotary motion. This motion is positive but limited to a multiple of the tooth pitch. (B) Friction-type is quieter but requires a spring device to keep eccentric pawl in constant engagement. (C) Balls or rollers replace the pawls in this device. Motion of the outer race wedges rollers against the inclined surfaces of the ratchet wheel.

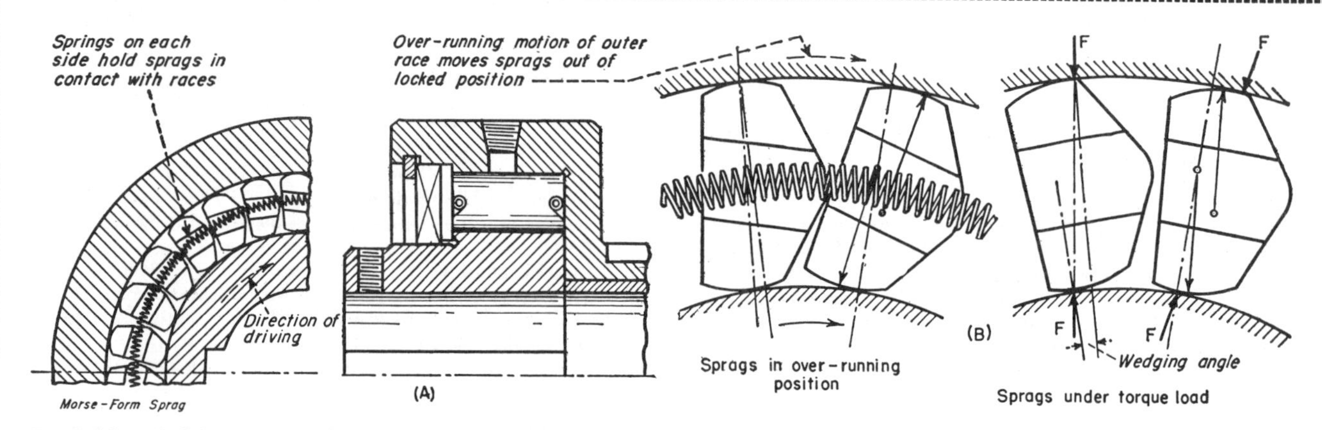


**4** With cylindrical inner and outer races, sprags are used to transmit torque. Energizing springs serves as a cage to hold the sprags. (A) Compared to rollers, shape of sprag permits a greater number within a limited space; thus higher torque loads are possible. Not requiring special cam surfaces, this type can be installed inside gear or wheel hubs. (B) Rolling action wedges sprags tightly between driving and driven members. Relatively large wedging angle insures positive engagement.

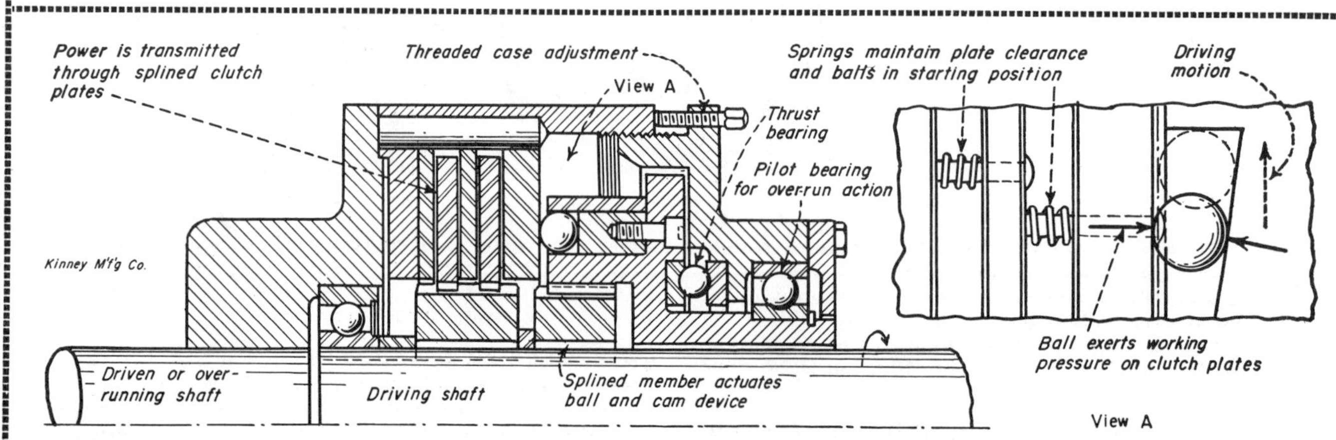


**6** Multi-disk clutch is driven by means of several sintered-bronze friction surfaces. Pressure is exerted by a cam actuating device which forces a series of balls against a disk plate. Since a small part of the transmitted torque is carried by the actuating member, capacity is not limited by the localized deformation of the contacting balls. Slip of the friction surfaces determine the capacity and prevent rapid, shock loads. Slight pressure of disk springs insure uniform engagement.

# Over-Riding Clutches

A. DeFEO
Design Engineer
Wright Aeronautical Corporation

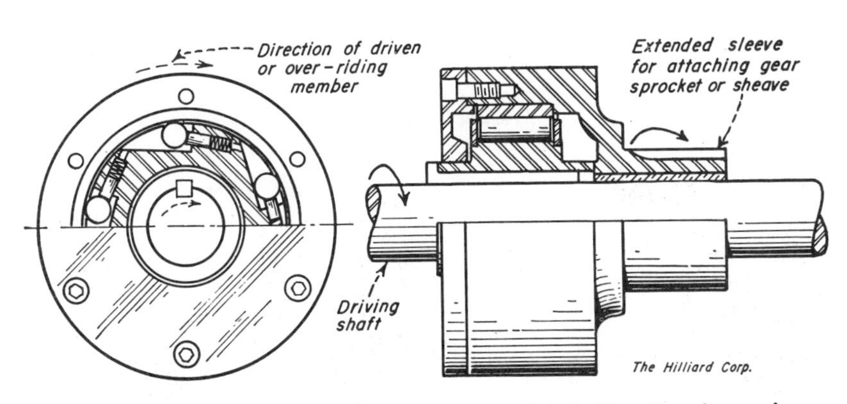


**2** Commercial over-riding clutch has springs which hold rollers in continuous contact between cam surfaces and outer race; thus there is no backlash or lost motion. This simple design is positive and quiet. For operation in the opposite direction, the roller mechanism can easily be reversed in the housing.

**3** Centrifugal force can be used to hold rollers in contact with cam and outer race. Force is exerted on lugs of the cage which controls the position of the rollers.

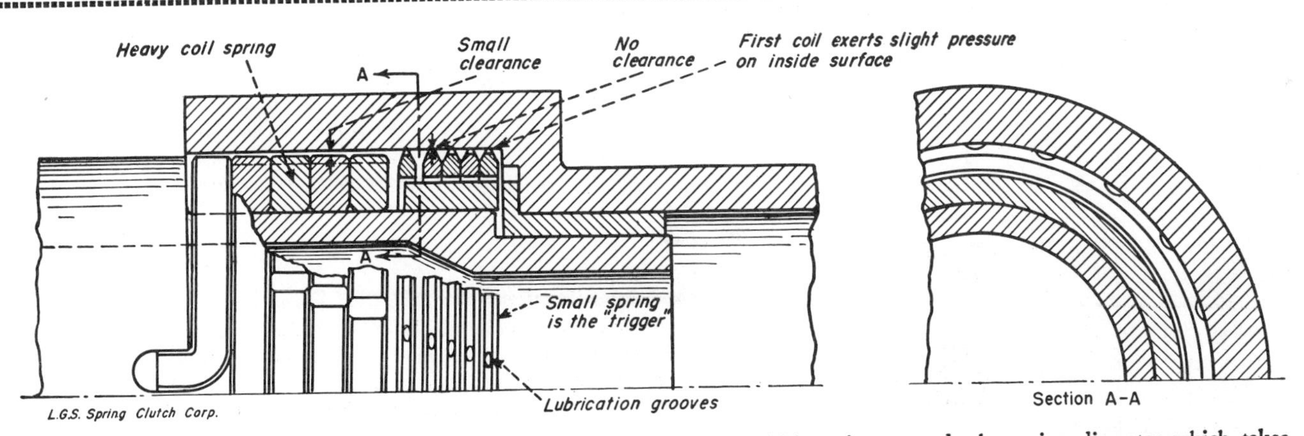


**5** Engaging device consists of a helical spring which is made up of two sections: a light trigger spring and a heavy coil spring. It is attached to and driven by the inner shaft. Relative motion of outer member rubbing on trigger causes this spring to wind-up. This action expands the spring diameter which takes up the small clearance and exerts pressure against the inside surface until the entire spring is tightly engaged. Helix angle of spring can be changed to reverse the over-riding direction.

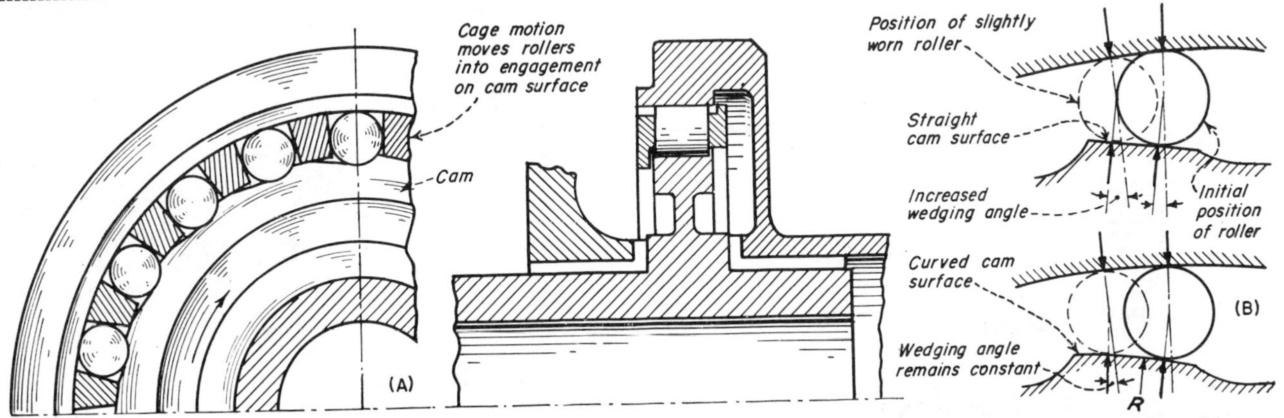


**7** Free-wheeling clutch widely used in power transmission has a series of straight-sided cam surfaces. An engaging angle of about 3 deg is used; smaller angles tend to become locked and are difficult to disengage while larger ones are not as effective. (A) Inertia of floating cage wedges rollers between cam and outer race. (B) Continual operation causes wear of surfaces; 0.001 in. wear alters angle to 8.5 deg. on straight-sided cams. Curved cam surfaces maintain constant angle.

# 10 Ways to Apply OVERRUNNING

**These clutches allow freewheeling, indexing and backstopping applicable to many design**

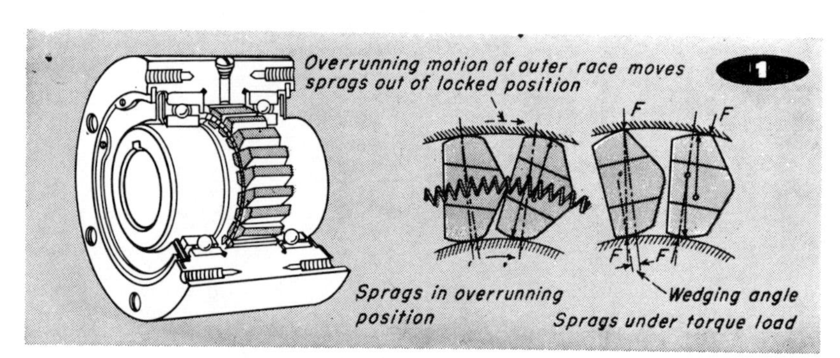


**Precision Sprags . . .**
act as wedges and are made of hardened alloy steel. In the Formsprag clutch, torque is transmitted from one race to another by wedging action of sprags between the races in one direction; in other direction the clutch freewheels.

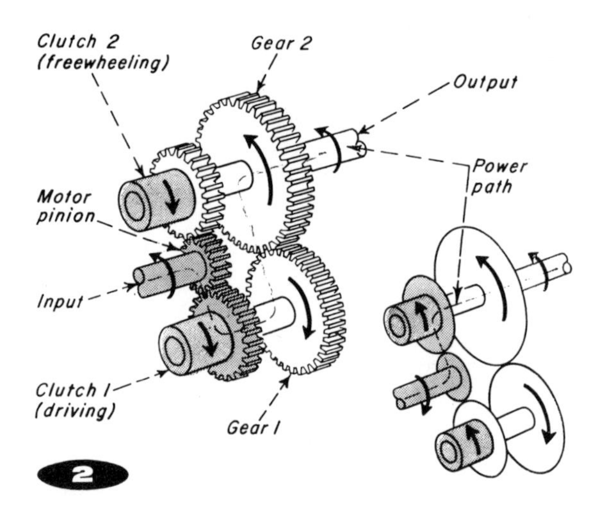


**2-Speed Drive—I . . .**
requires input rotation to be reversible. Counterclockwise input as shown in the diagram drives gear 1 through clutch 1; output is counterclockwise; clutch 2 over-runs. Clockwise input (schematic) drives gear 2 through clutch 2; output is still counterclockwise; clutch 1 over-runs.

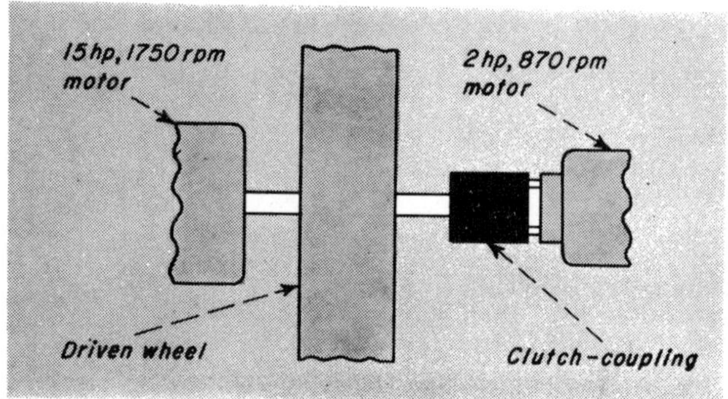


**2-Speed Drive—II . . .**
for grinding wheel can be simple, in-line design if over-running clutch couples two motors. Outer race of clutch is driven by gearmotor; inner race is keyed to grinding-wheel shaft. When gearmotor drives, clutch is engaged; when larger motor drives, inner race over-runs.

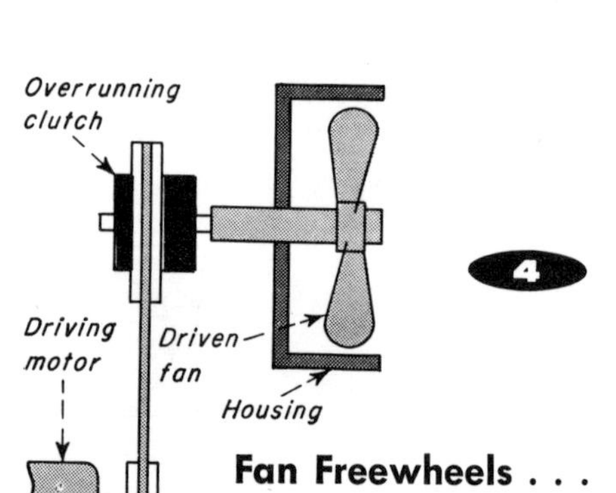


**Fan Freewheels . . .**
when driving power is shut off. Without over-running clutch, fan momentum can cause belt breakage. If driving source is a gearmotor, excessive gear stress may also occur by feedback of kinetic energy from fan.

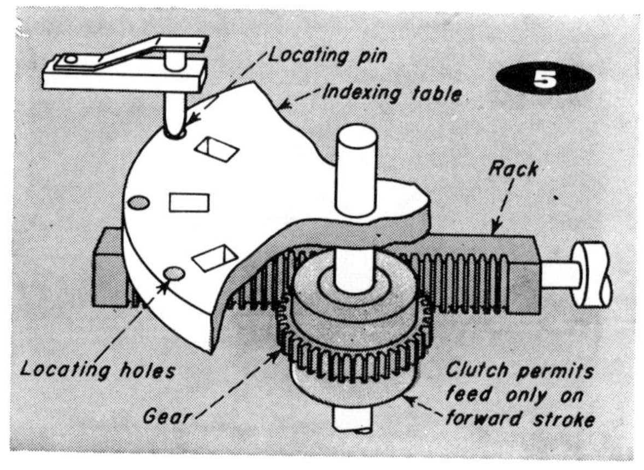


**Indexing Table . . .**
is keyed to clutch shaft. Table is rotated by forward stroke of rack, power being transmitted through clutch by its outer-ring gear only during this forward stroke. Indexing is slightly short of position required. Exact position is then located by spring-loaded pin, which draws table forward to final positioning. Pin now holds table until next power stroke of hydraulic cylinder

problems. Here are some clutch setups.

W. EDGAR MULHOLLAND, Executive Sales Engineer
JOHN L. KING, JR., Project Engineer
Formsprag Company, Warren, Mich.

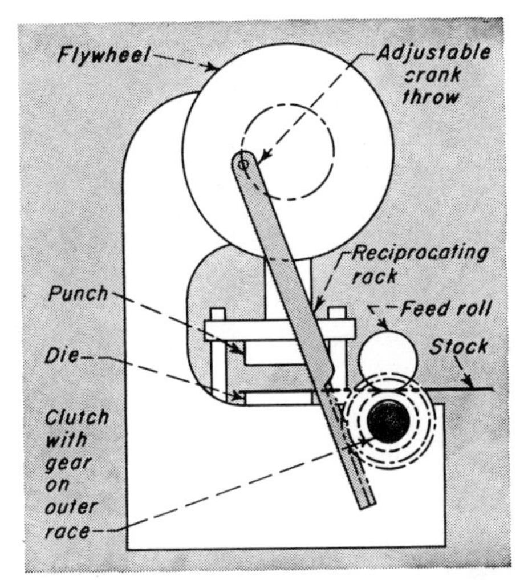

### Punch Press Feed . .

is so arranged that strip is stationary on downstroke of punch (clutch freewheels); feed occurs during upstroke when clutch transmits torque. Feed mechanism can easily be adjusted to vary feed amount.

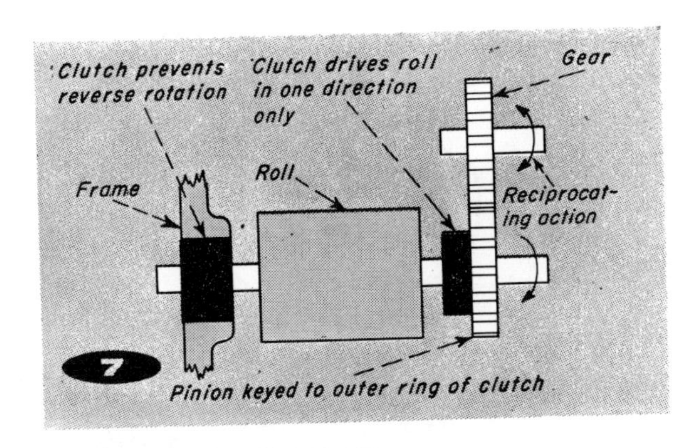


### Indexing and Backstopping . . .

is done with two clutches so arranged that one drives while the other freewheels. Application here is for capsuling machine; gelatin is fed by the roll and stopped intermittently so blade can precisely shear material to form capsules.

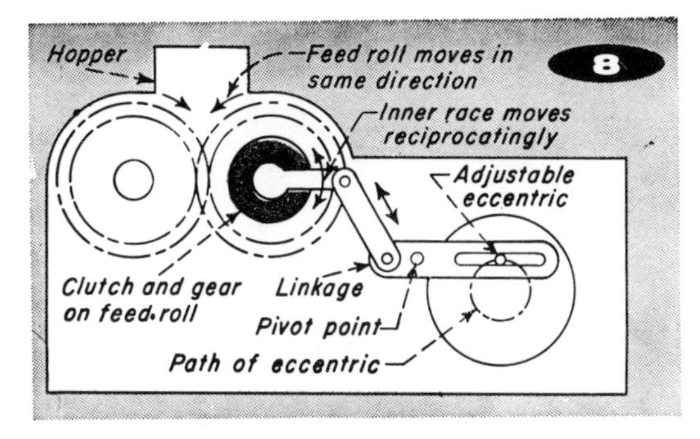


### Intermittent Motion . . .

of candy machine is adjustable; function of clutch is to ratchet the feed rolls around. This keeps the material in the hopper agitated.

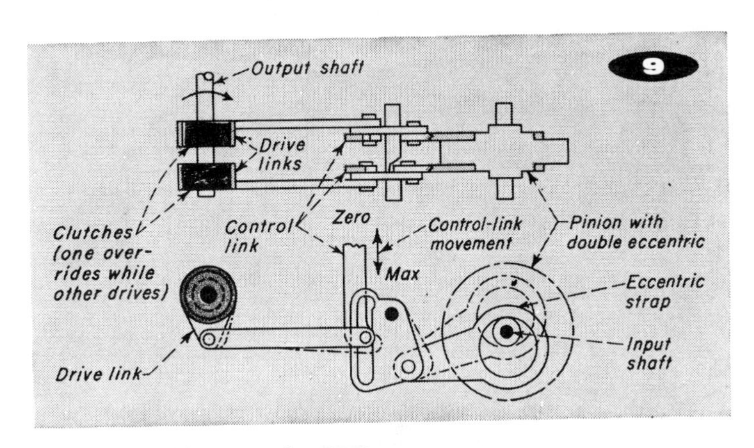


### Double-impulse Drive . . .

employs double eccentrics and drive clutches. Each clutch is indexed 180° out of phase with the other. One revolution of eccentric produces two drive strokes. Stroke length, and thus the output rotation, can be adjusted from zero to max by the control link.

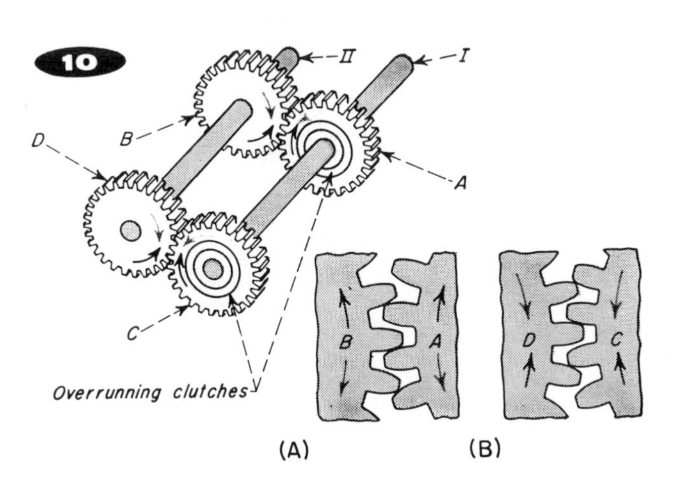


### Anti-backlash Device . . .

uses over-running clutches to insure that no backlash is left in the unit. Gear *A* drives *B* and shaft II with the gear mesh and backlash as shown in (A). The over-running clutch in gear *C* permits gear *D* (driven by shaft II) to drive gear *C* and results in the mesh and backlash shown in (B). The over-running clutches never actually over-run. They provide flexible connections (something like split and sprung gears) between shaft I and gears *A*, *C* to allow absorption of all backlash.

# Small Mechanical Clutches for

Clutches used in calculating machines must have: (1) Quick response—lightweight moving parts; (2) Flexibility—permit multiple members to

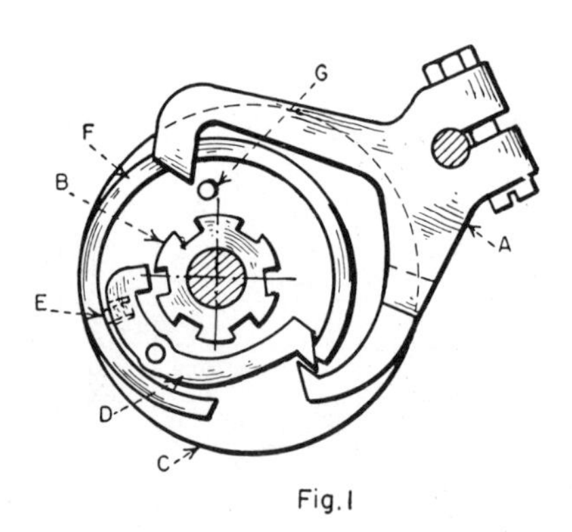

Fig. 1

**PAWL AND RATCHET SINGLE CYCLE CLUTCH (Fig. 1).** Known as Dennis Clutch, parts *B, C* and *D*, are primary components, *B*, being the driving ratchet, *C*, the driven cam plate and, *D*, the connecting pawl carryied by the cam plate. Normally the pawl is held disengaged by the lower portion of clutch arm *A*. When activated, arm *A* rocks counter-clockwise until it is out of the path of rim *F* on cam plate *C* and permits pawl *D* under the effect of spring *E* to engage with ratchet *B*. Cam plate *C* then turns clockwise until, near the end of one cycle, pin *G* on the plate strikes the upper part of arm *A* camming it clockwise back to its normal position. The lower part of *A* then performs two functions: (1) cams pawl *D* out of engagement with the driving ratchet *B* and (2) blocks further motion of rim *F* and the cam plate.

**PAWL AND RATCHET SINGLE CYCLE DUAL CONTROL CLUTCH—(Fig. 2).** Principal parts are: driving ratchet, *B*, directly connected to the motor and rotating freely on rod *A*; driven crank, *C*, directly connected to the main shaft of the machine and also free on A; and spring loaded ratchet pawl, *D*, which is carried by crank, *C*, and is normally held disengaged by latch *E*. To activate the clutch, arm *F* is raised, permitting latch *E* to trip and pawl *D* to engage with ratchet *B*. The left arm of clutch latch *G*, which is in the path of the lug on pawl *D*, is normally permitted to move out of interference by the rotation of the camming edge of crank *C*. For certain operations block *H* is temporarily lowered, preventing motion of latch *G*, resulting in disengagement of the clutch after part of the cycle until subsequent raising of block *H* permits motion of latch *G* and resumption of the cycle.

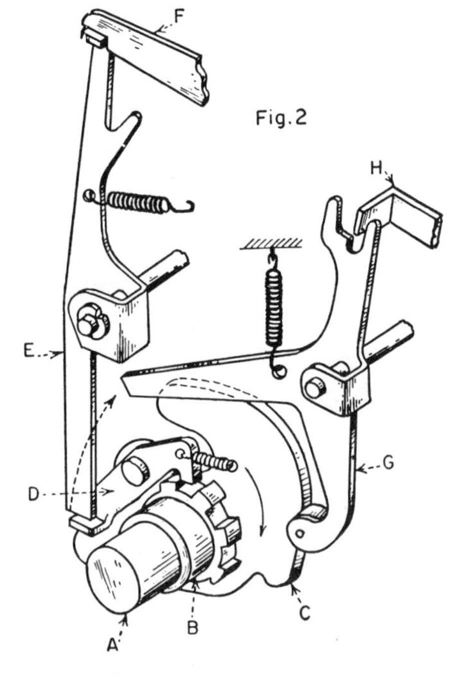


**PLANETARY TRANSMISSION CLUTCH (Fig. 3).** A positive clutch with external control, two gear trains to provide bi-directional drive to a calculator for cycling the machine and shifting the carriage. Gear *A* is the driver, gear *L* the driven member is directly connected to planet carrier *F*. The planet consists of integral gears *B* and *C*; *B* meshing with sun gear *A* and free-wheeling ring gear *G*, and *C* meshing with free-wheeling gear *D*. Gears *D* and *G* carry projecting lugs, *E* and *H* respectively, which can contact formings on arms *J* and *K* of the control yoke. When the machine is at rest, the yoke is centrally positioned so that the arms *J* and *K* are out of the path of the projecting lugs permitting both *D* and *G* to free-wheel. To engage the drive, the yoke rocks clockwise as shown, until the forming on arm *K* engages lug *H* blocking further motion of ring gear *G*. A solid gear train is thereby established driving *F* and *L* in the same direction as the drive *A* and at the same time altering the speed of *D* as it continues counter-clockwise. A reversing signal rotates the yoke counter-clockwise until arm *J* encounters lug *E* blocking further motion of *D*. This actuates the other gear train of the same ratio.

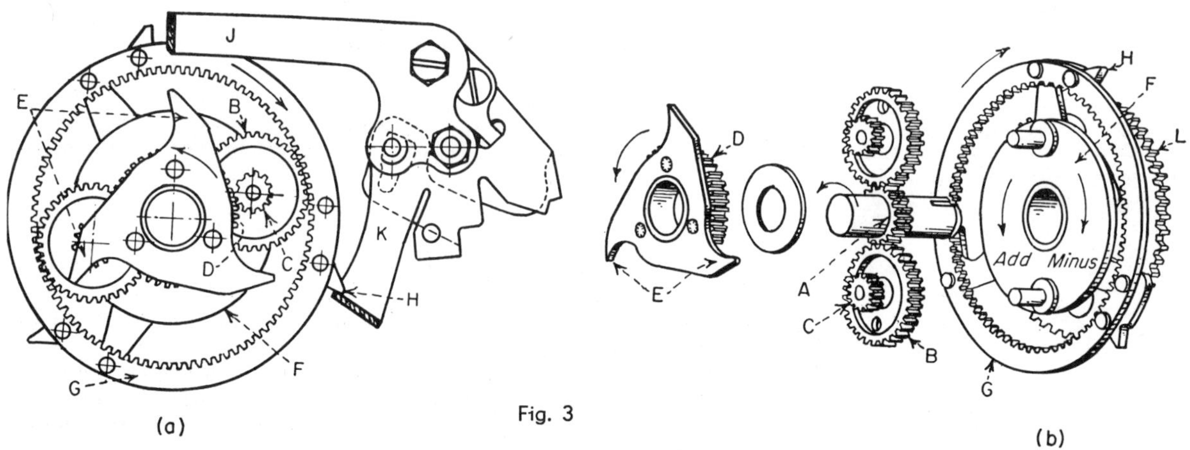

Fig. 3

# Precise Service

MARVIN TAYLOR
Monroe Calculating Machine Company

control operation; (3) Compactness—for equivalent capacity positive clutches are smaller than friction; (4) Dependability; and (5) Durability.

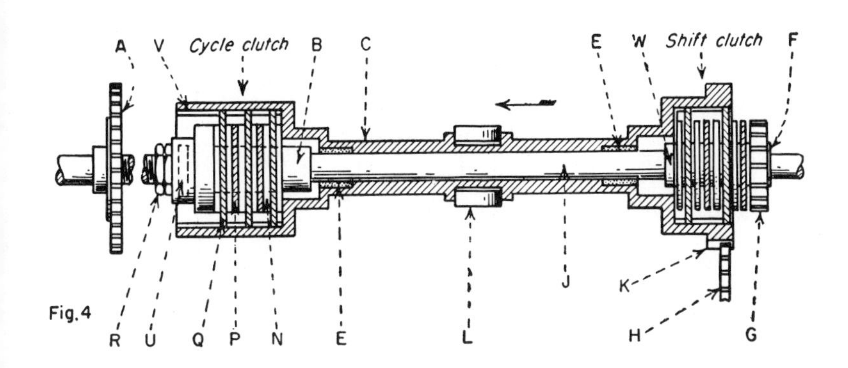

Fig. 4

**MULTIPLE DISK FRICTION CLUTCH (Fig. 4).** Two multiple disk friction clutches are combined in a single two-position unit which is shown shifted to the left. A stepped cylindrical housing *C* enclosing both clutches is carried by self-lubricated bearing *E* on shaft *J* and is driven by the transmission gear *H* meshing with the housing gear teeth *K*. At either end, the housing carries multiple metal disks *Q* that engage keyways *V* and can make frictional contact with formica disks *N* which, in turn, can contact a set of metal disks *P* which have slotted openings for coupling with flats on sleeves *B* and *W*. In the position shown, pressure is exerted through rollers *L* forcing the housing to the left making the left clutch compact against adjusting nuts *R*, thereby driving gear *A* via sleeve *B* which is connected to jack shaft *J* by pin *U*. When the carriage is to be shifted, rollers *L* force the housing to the right, first relieving the pressure between the adjoining disks on the left clutch then passing through a neutral position in which both clutches are disengaged and finally making the right clutch compact against thrust bearing *F*, thereby driving gear *G* through sleeve *W* which rotates freely on the jack shaft.

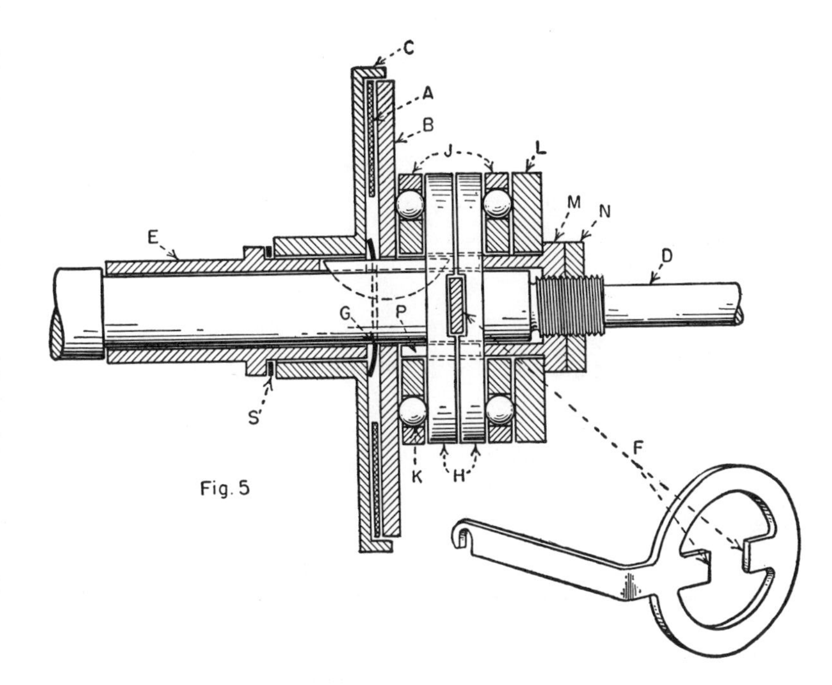

Fig. 5

**SINGLE PLATE FRICTION CLUTCH (Fig. 5).** The basic clutch elements, formica disk *A*, steel plate *B* and drum *C*, are normally kept separated by spring washer *G*. To engage the drive, the left end of a control arm is raised, causing ears *F*, which sit in slots in plates *H*, to rock clockwise spreading the plates axially along sleeve *P*. Sleeves *E* and *P* and plate *B* are keyed to the drive shaft; all other members can rotate freely. The axial motion loads the assembly to the right through the thrust ball bearings *K* against plate *L* and adjusting nut *M*, and to the left through friction surfaces on *A*, *B* and *C* to thrust washer *S*, sleeve *E* and against a shoulder on shaft *D*, thus enabling plate *A* to drive the drum *C*.

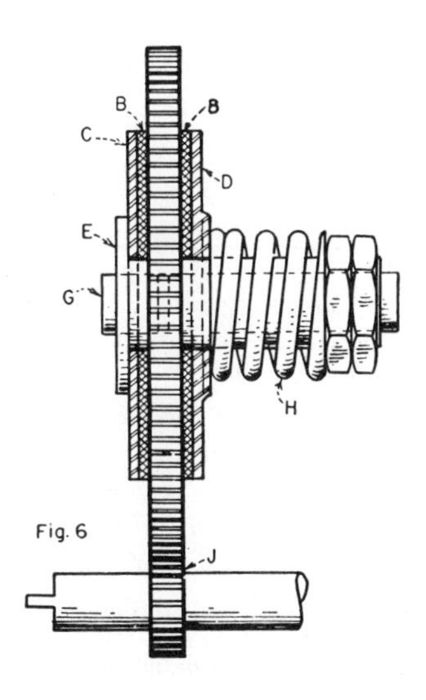

Fig. 6

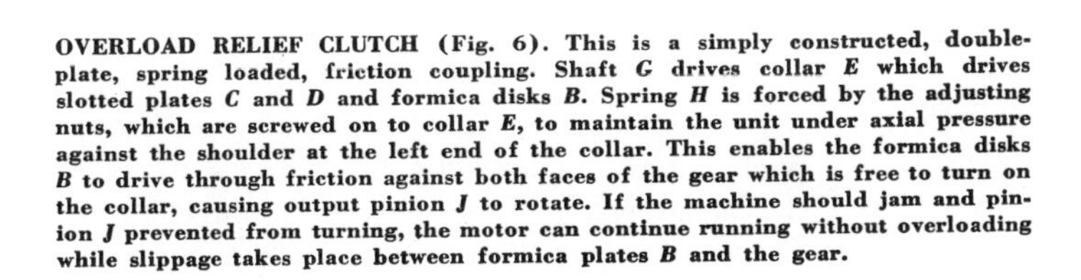

**OVERLOAD RELIEF CLUTCH (Fig. 6).** This is a simply constructed, double-plate, spring loaded, friction coupling. Shaft *G* drives collar *E* which drives slotted plates *C* and *D* and formica disks *B*. Spring *H* is forced by the adjusting nuts, which are screwed on to collar *E*, to maintain the unit under axial pressure against the shoulder at the left end of the collar. This enables the formica disks *B* to drive through friction against both faces of the gear which is free to turn on the collar, causing output pinion *J* to rotate. If the machine should jam and pinion *J* prevented from turning, the motor can continue running without overloading while slippage takes place between formica plates *B* and the gear.

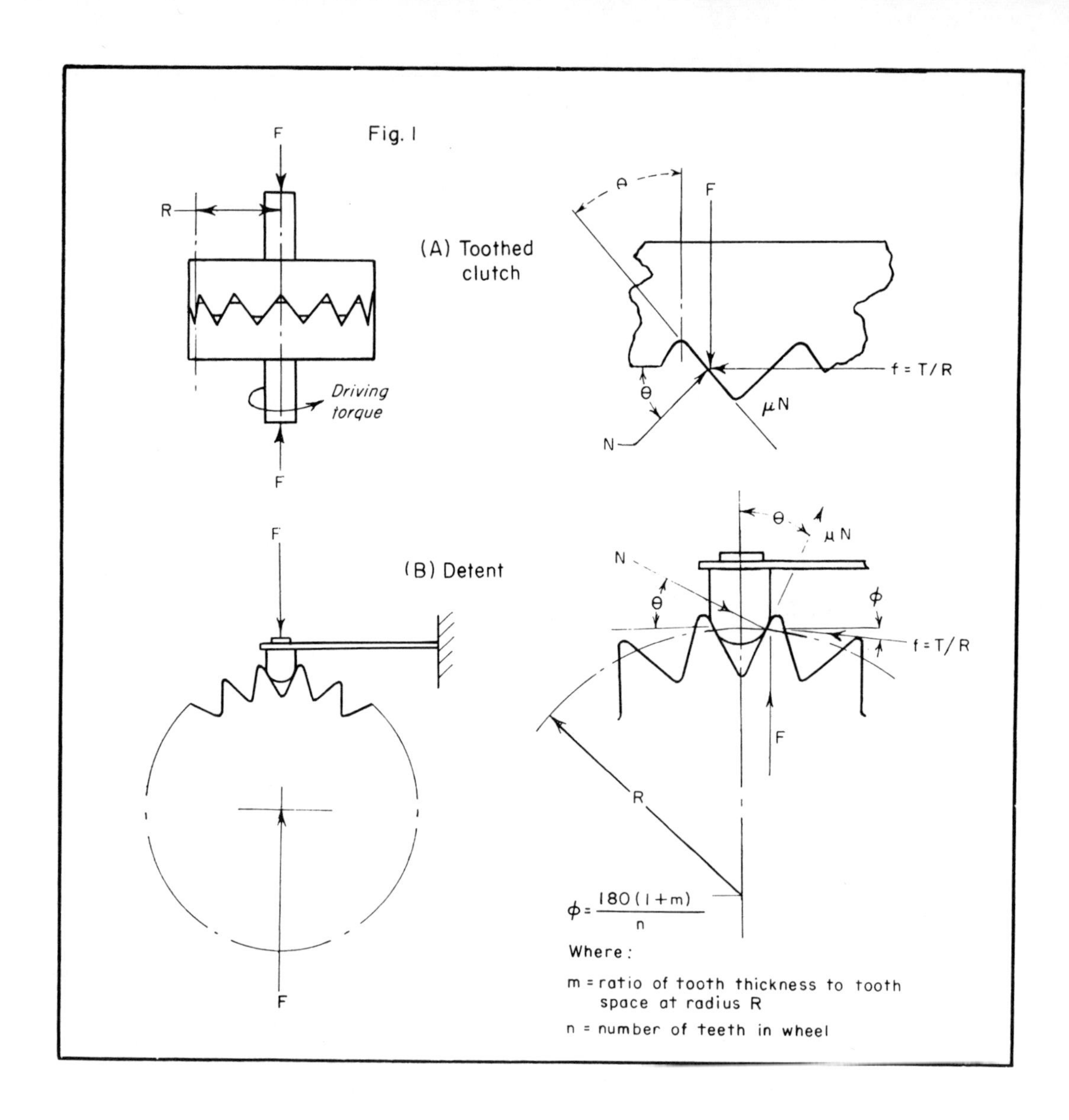


# Serrated Clutches and Detents

**L. N. CANICK,** Servomechanisms, Inc.

IN THE DESIGN OF straight toothed components such as serrated clutches, Fig. 1(A), and detent wheels, Fig. 1(B), the effective pitch radius is usually set by size considerations. The torque transmitting capacity of the clutch, or the torque resisting capacity of the detent wheel, is then obtained by assigning suitable values to the engaging force, tooth angle, and coefficient of friction.

The nomogram, Fig. 2, is designed to be a convenient means for considering the effect of variations in the values of tooth angle and coefficient of friction. For a given coefficient of friction, there is a tooth angle below which the clutch or detent is self-locking and will transmit torque limited only by its structural strength. Where

$T$ = torque transmitted without clutch slip, or torque resisted by detent wheel, lb in.
$R$ = effective clutch, or detent wheel, radius, in.
$F$ = axial, or radial, force, lb
$f$ = tangential force acting at radius $R$, lb
$N$ = reaction force of driven tooth, or detent, acting normal to tooth face, lb
$\mu$ = coefficient of friction of tooth material
$\theta$ = see Fig. 1(B)
= angle of tooth face, deg
$K = (1 + \mu \tan \theta)/(\tan \theta - \mu)$

a statement of the conditions of equilibrium for the forces acting on a clutch tooth will lead to the following equation

$$T = RFK \quad (1$$

A similar statement of the conditions of equilibrium for the forces acting on a tooth of the detent wheel shown in Fig. 1(B) will lead to the following equation:

$$T = \frac{RF}{[(\cos \varphi)/K] - \sin \varphi} \quad (2)$$

From Eqs (1) and (2), when all other terms have constant values, it is obvious that the required axial force, or the radial force, diminishes as the value of $K$ increases. Dependent upon the values of $\theta$ and $\mu$, the value of $K$ can vary from zero to infinity.

The circular nomogram shown in Fig. 2 relates the values of the parameters $K$, $\theta$, and $\mu$ that satisfy the basic equation

$$K = (1 + \mu \tan \theta)/(\tan \theta - \mu)$$

EXAMPLE I. Find the maximum tooth angle for a self-locking clutch, or for which $K$ is infinity, taking the coefficient of friction as 0.4 minimum.

SOLUTION I. Line I through these values for $K$ and $\mu$ on the nomogram gives a maximum tooth angle slightly less than 22 deg for the self-locking condition.

EXAMPLE II. Find the minimum value of $K$ to be expected for a clutch having a tooth angle of 30 deg and a coefficient of friction of 0.2 minimum.

SOLUTION II. Line II through these values for $\theta$ and $\mu$ on the nomogram gives a value for $K$ of 3 approximately.

EXAMPLE III. Find the value of $K$ for a flat-face ($\theta$ equals 90 deg) friction clutch, the face material of which has a coefficient of friction of 0.2. Compare its torque transmitting capacity with that of the toothed clutch of Example II.

SOLUTION III. Line III through these values for $\theta$ and $\mu$ on the nomogram gives a value for $K$ of 0.2.

Torque transmitting capacity of flat-face clutch:

$$T = 0.2\,R\,F$$

Torque transmitting capacity of toothed-clutch:

$$T = 3\,R\,F$$

Thus for equal effective radii and engaging forces, the torque capacity of the toothed-clutch is 3/0.2, or 15, times greater than that of the flat face clutch.

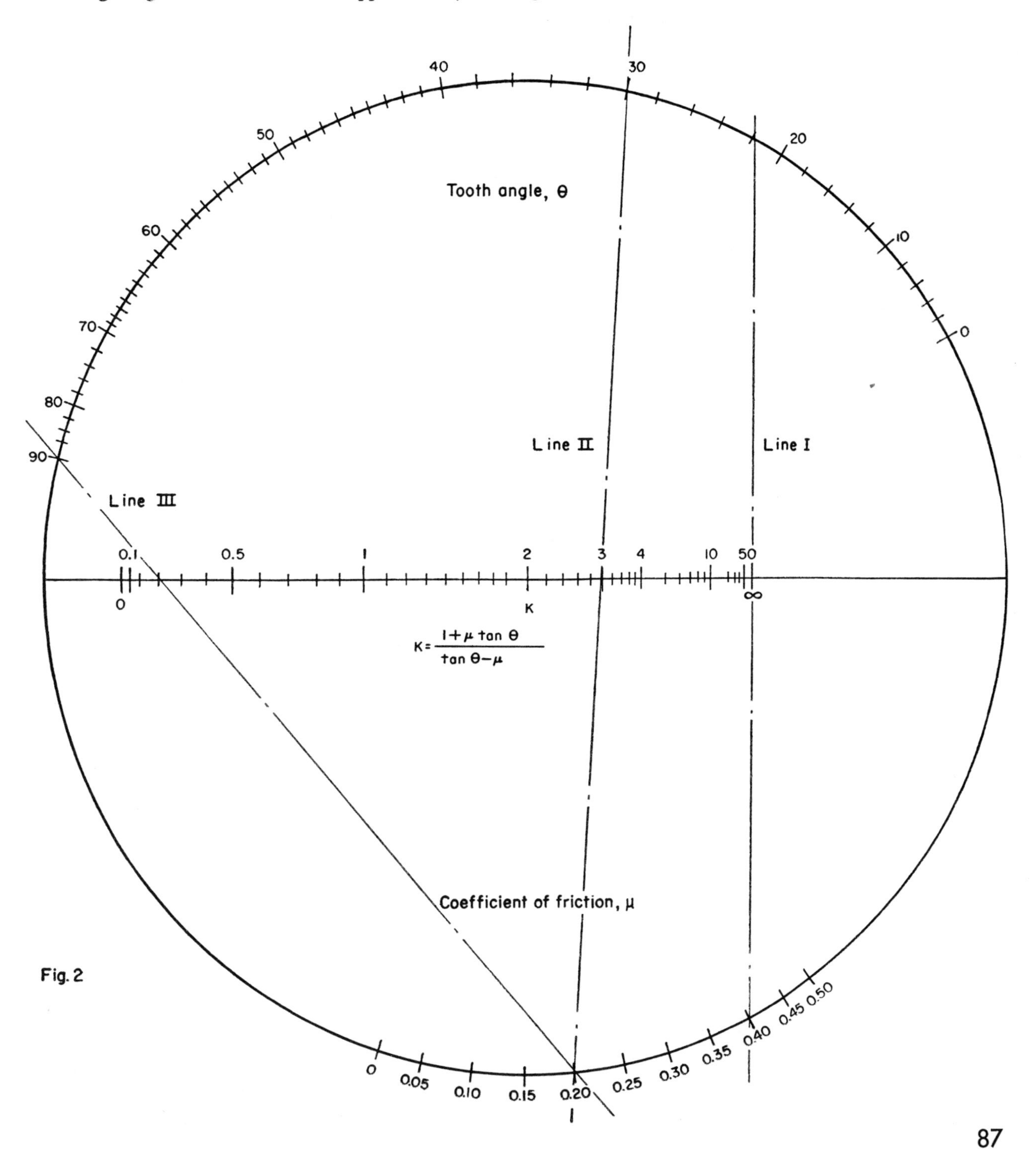


Fig. 2

# 4

# COUPLINGS


Typical Designs of Flexible Couplings – I 90
Typical Designs of Flexible Couplings – II 92
Typical Designs of Flexible Couplings – III 94
Low-cost Methods of Coupling Small Diameter Shafts 96
Coupling of Parallel Shafts 98
Ten Different Types of Splined Connections 100
How to Connect Tubing – Cross and Tee Joints 102
Different Mechanical Methods for Attaching Tubing 104

# Typical Designs of

CYRIL DONALDSON
*Rochester Mechanics Institute*

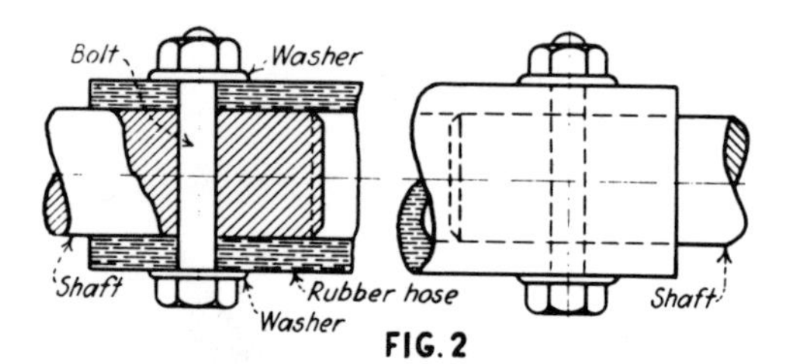


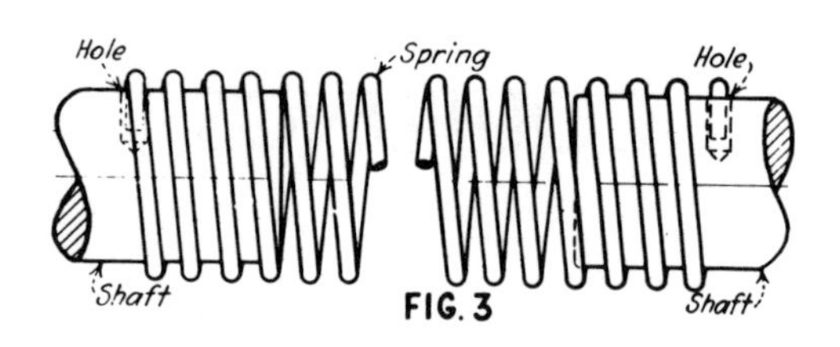


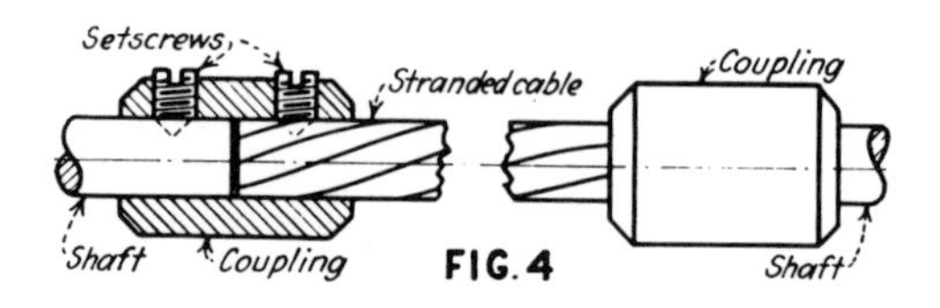


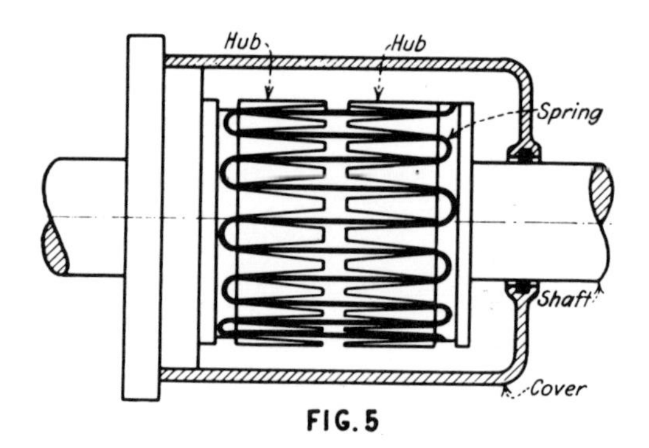


**Fig. 1**—A rubber hose clamped to two shafts. For applications where the torque is low and slippage unimportant. It is easily assembled and disconnected without disturbing either machine element. Adaptable to changes in longitudinal distance between machines. This coupling absorbs shocks, is not damaged by overloads, does not set up end thrusts, requires no lubrication and compensates for both angular and offset misalignment.

**Fig. 2**—Similar to Fig. 1, but positive drive is assured by bolting hose to shafts. Has same advantages as type in Fig. 1, except there is no overload protection other than the rupture of the hose.

**Fig. 3**—The use of a coiled spring fastened to shafts gives the same action as a hose. Has excellent shock absorbing qualities, but torsional vibrations are possible. Will allow end play in shafts, but sets up end thrust in so doing. Other advantages are same as in types shown in Figs. 1 and 2. Compensates for misalignment in any direction.

**Fig. 4**—A simple and effective coupling for low torques and unidirectional rotation. Stranded cable provides a positive drive with desirable elasticity. Inertia of rotating parts is low. Easily assembled and disconnected without disturbing either shaft. Cable can be encased and length extended to allow for right angle bends such as used on dental drills and speedometer drives. Ends of cable are soldered or bound with wire to prevent unraveling.

**Fig. 5**—A type of Falk coupling that operates on the same principle as design shown in Fig. 6, but has a single flat spring in place of a series of coiled springs. High degree of flexibility obtained by use of tapered slots in hubs. Smooth operation is maintained by inclosing the working parts and packing with grease.

**Fig. 6**—Two flanges and a series of coiled springs give a high degree of flexibility. Used only where the shafts have no free end play. Needs no lubrication, absorbs shocks and provides protec-

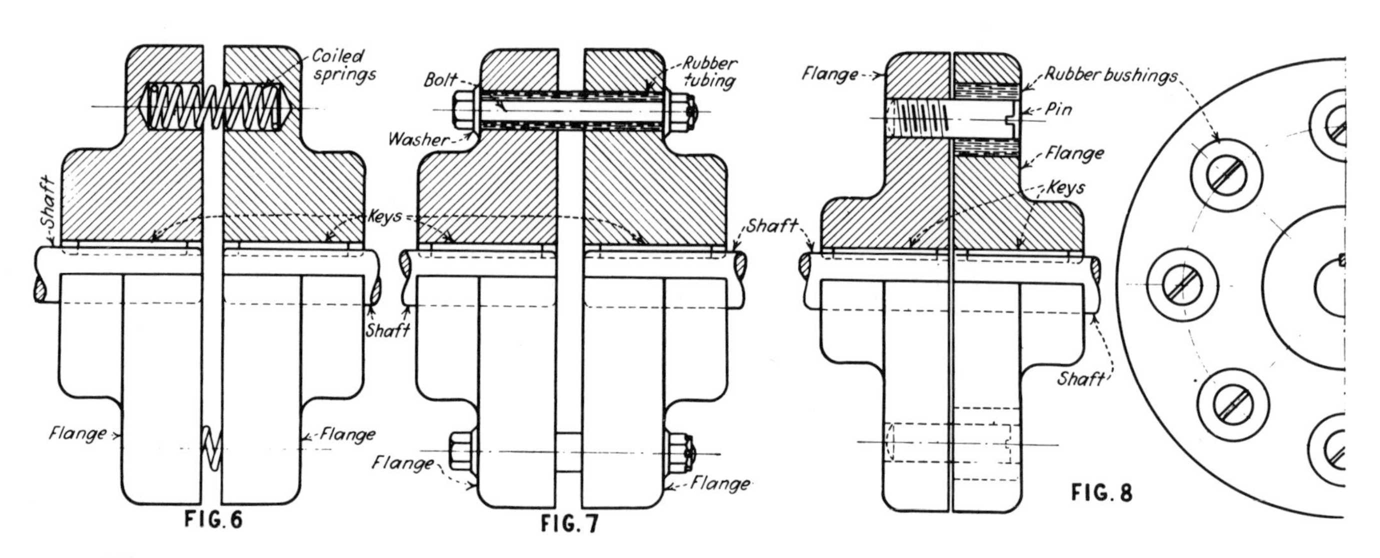

# *Flexible Couplings—I*

tion against overloads, but will set up torsional vibrations. Springs can be of round or square wire with varying sizes and pitches to allow for any degree of flexibility.

**Fig. 7**—Is similar to Fig. 6, except that rubber tubing, reinforced by bolts, is used instead of coiled springs. Is of sturdier construction but more limited in flexibility. Has no overload protection other than shearing of the bolts. Good anti-vibration properties if thick rubber tubing is used. Can absorb minor shocks. Connection can be quickly disassembled.

**Fig. 8**—A series of pins engage rubber bushings cemented into flange. Coupling is easy to install. Flanges being accurately machined and of identical size makes accurate lining-up with spirit level possible. Will allow minor end play in shafts, and provides a positive drive with good flexibility in all direction.

**Fig. 9**—A Foote Gear Works flexible coupling which has shear pins in a separate set of bushings to provide overload protection. Construction of studs, rubber bushings and self-lubricating bronze bearings is in principle similar to that shown in Fig. 10. Replaceable shear pins are made of softer material than the shear pin bushings.

**Fig. 10**—A design made by the Ajax Flexible Coupling Company. Studs are firmly anchored with nuts and lock washers and bear in self-lubricating bronze bushings spaced alternately in both flanges. Thick rubber bushings cemented in flanges are forced over the bronze bushings. Life of coupling said to be considerably increased because of self-lubricated bushings.

**Fig. 11**—Another Foote Gear Works coupling. Flexibility is obtained by solid conically-shaped pins of metal or fiber. This type of pin is said to provide a positive drive of sturdy construction with flexibility in all directions.

**Fig. 12**—In this Smith & Serrell coupling a high degree of flexibility is obtained by laminated pins built-up of tempered spring steel leaves. Spring leaves secured to holder by keeper pin. Phosphor bronze bearing strips are welded to outer spring leaves and bear in rectangular holes of hardened steel bushings fastened in flange. Pins are free to slide endwise in one flange, but are locked in the other flange by a spring retaining ring. This type is used for severe duty in both marine and land service.

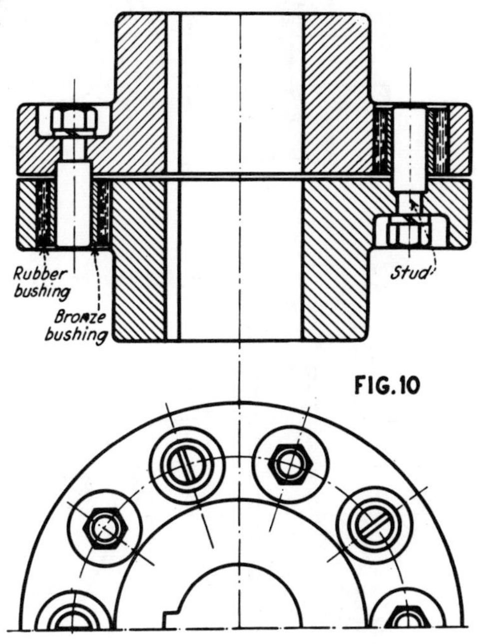


FIG.10

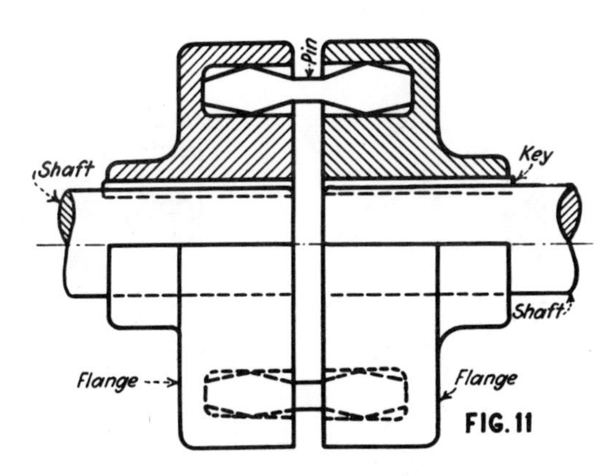


FIG.11

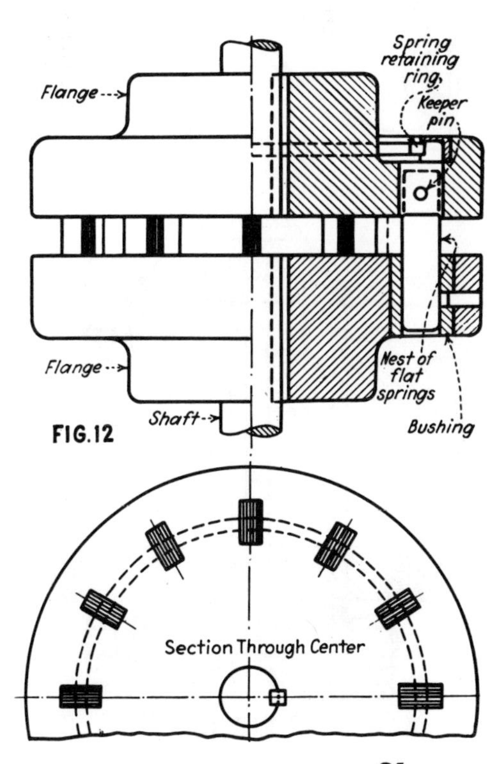


FIG.12

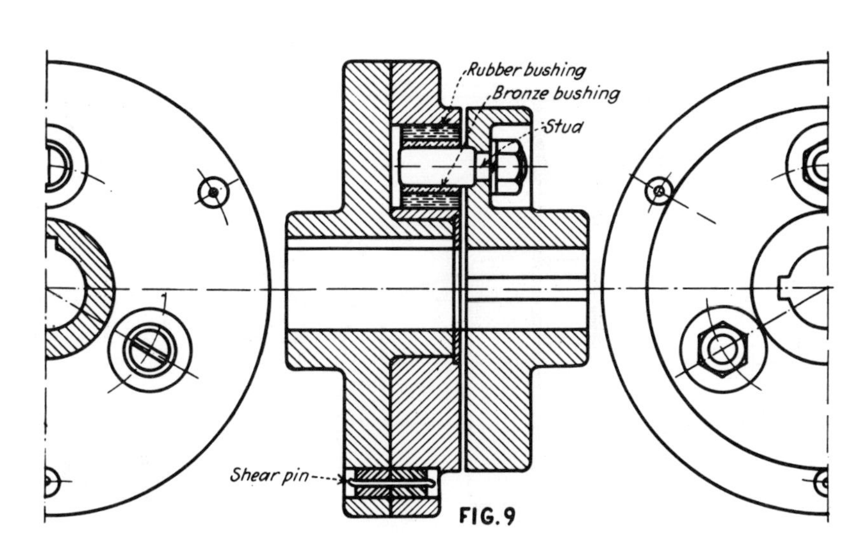


FIG.9

# Typical Designs of

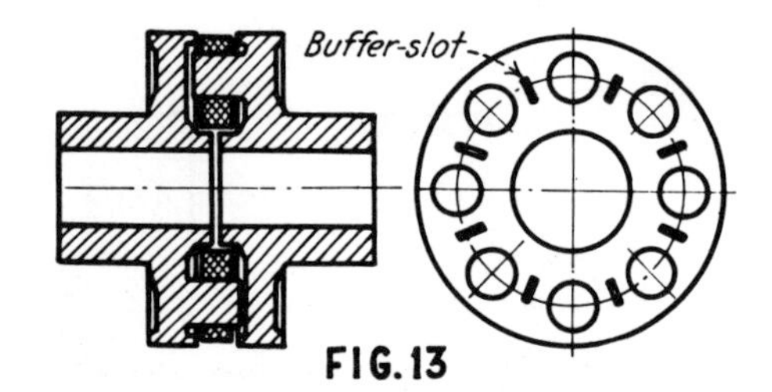


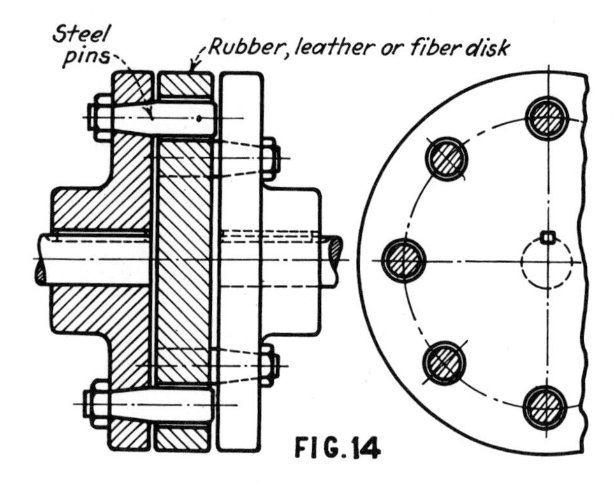


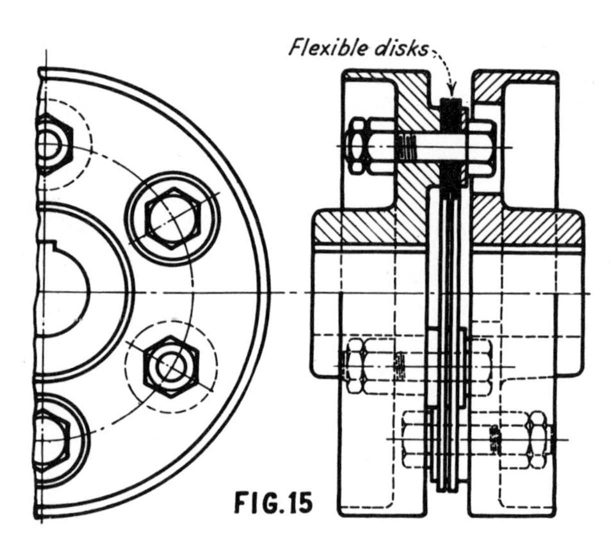


**F g. 13**—In this Brown Engineering Company coupling flexibility is increased by addition of buffer-slots in the laminated leather. These slots also aid in the absorption of shock loads and torsional vibration. Under parallel misalignment or shock loads, buffer slots will close over their entire width, but under angular misalignment buffer slots will close only on one side.

**Fig. 14**—Flexibility is provided by resilience of a rubber, leather, or fiber disk in this W. A. Jones Foundry & Machine Company coupling. Degree of flexibility is limited to clearance between pins and holes in the disk plus the resilience of the disk. Has good shock absorbing properties, allows for end play and needs no lubrication.

**Fig. 15**—A coupling made by Aldrich Pump Company, similar to Fig. 14, except bolts are used instead of pins. This coupling permits only slight endwise movement of the shaft and allows machines to be temporarily disconnected without disturbing the flanges. Driving and driven members are flanged for protection against projecting bolts.

**Fig. 16**—Laminated metal disks are used in this coupling made by Thomas Flexible Coupling Company. The disks are bolted to each flange and connected to each other by means of pins supported by a steel center disk. The spring action of the center ring allows torsional flexibility and the two side rings compensate for angular and offset misalignment. This type of coupling provides a positive drive in either direction without setting up backlash. No lubrication is required.

**Fig. 17**—A design made by Palmer-Bee Company for heavy torques. Each flange carries two studs upon which are mounted square metal blocks. The blocks slide in the slots of the center metal disk.

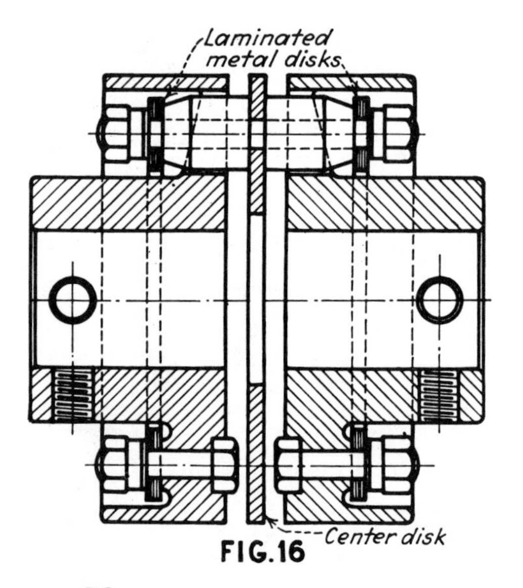


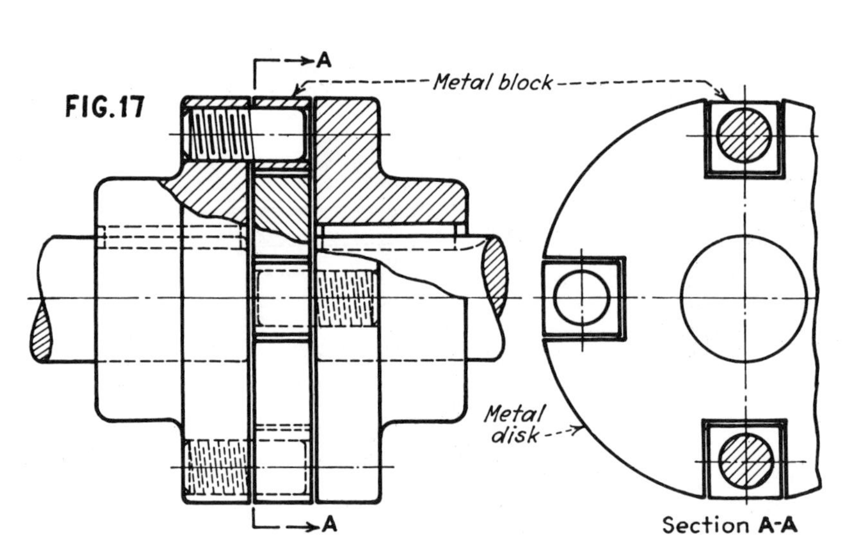

# *Flexible Couplings—II*

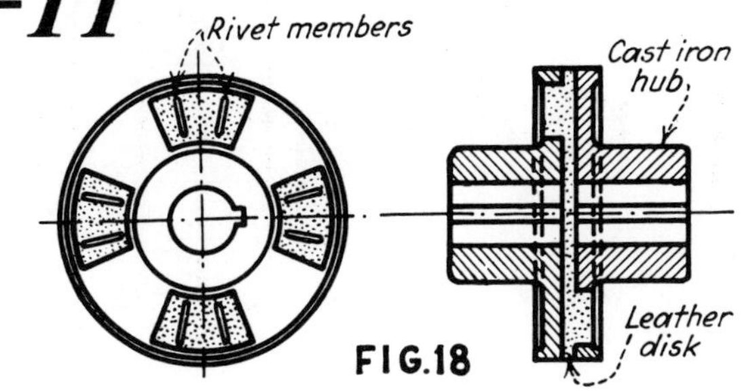


**Fig. 18**—In this Charles Bond Company coupling a leather disk floats between two identical flanges. Drive is through four laminated leather lugs cemented and riveted to the leather disk. Compensates for misalignment in all directions and sets up no end thrusts. The flanges are made of cast iron and the driving lug slots are cored.

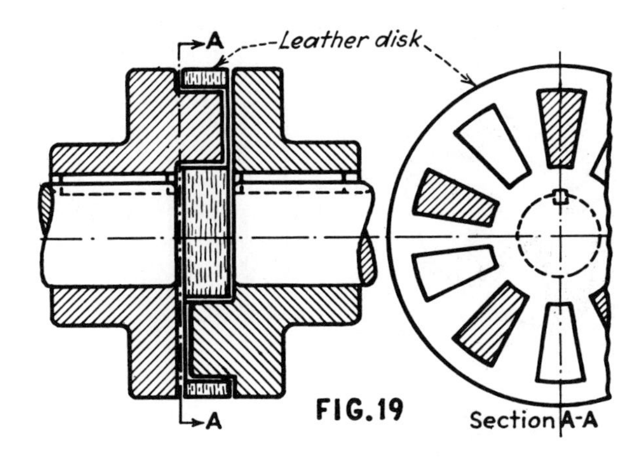


**Fig. 19**—The principle of the T. B. Wood & Sons Company coupling is the same as Fig. 18, but the driving lugs are cast integrally with the metal flanges. The laminated leather disk is punched out to accommodate the metal driving lugs of each flange. This coupling has flexibility in all directions and does not require lubrication.

**Fig. 20**—Another design made by Charles Bond Company. The flanges have square recesses into which a built-up leather cube fits. Endwise movement is prevented by through bolts set at right angles. The coupling operates quietly and is used where low torque loads are to be transmitted. Die-castings can be used for the flanges.

**Fig. 21**—Similar to Fig. 20, being quiet in operation and used for low torques. This is also a design of Charles Bond Company. The floating member is made of laminated leather and is shaped like a cross. The ends of the intermediate member engage the two cored slots of each flange. The coupling will withstand a limited amount of end play.

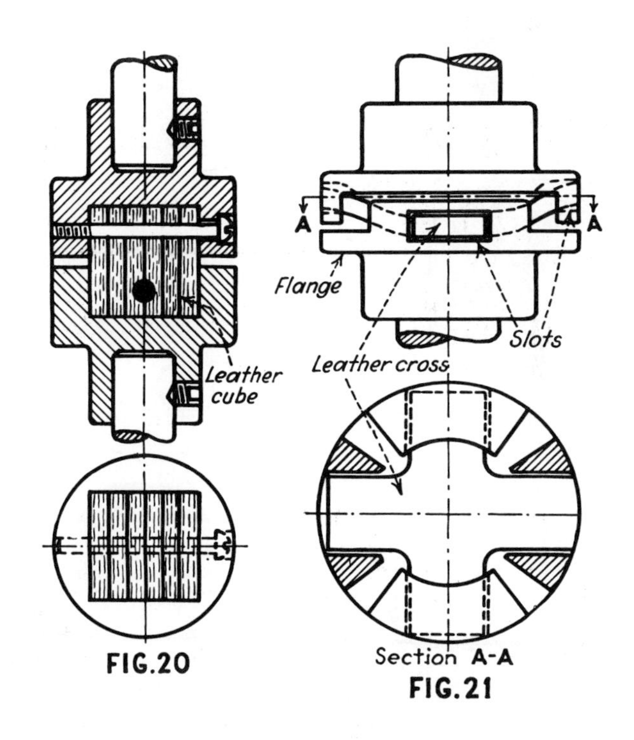


**Fig. 22**—Pins mounted in flanges are connected by leather, canvas, or rubber bands. Coupling is used for temporary connections where large torques are transmitted, such as the driving of dynamometers by test engines. Allows for a large amount of flexibility in all directions, absorbs shocks but requires frequent inspection. Machines can be quickly disconnected, especially when belt fasteners are used on the bands. Driven member lags behind driver when under load.

**Fig. 23**—This Bruce-Macbeth Engine Company coupling is similar to that of Fig. 22, except that six endless wire cable links are used, made of plow-steel wire rope. The links engage small metal spools mounted on eccentric bushings. By turning these bushings the links are adjusted to the proper tension. The load is transmitted from one flange to the other by direct pull on the cable links. This type of coupling is used for severe service.

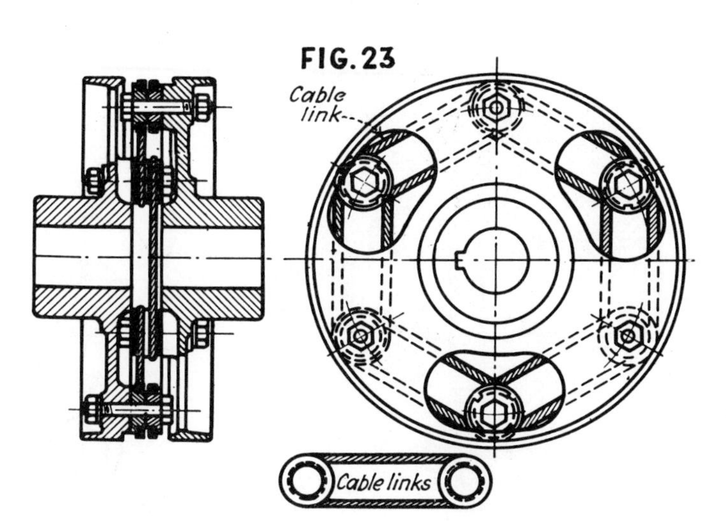


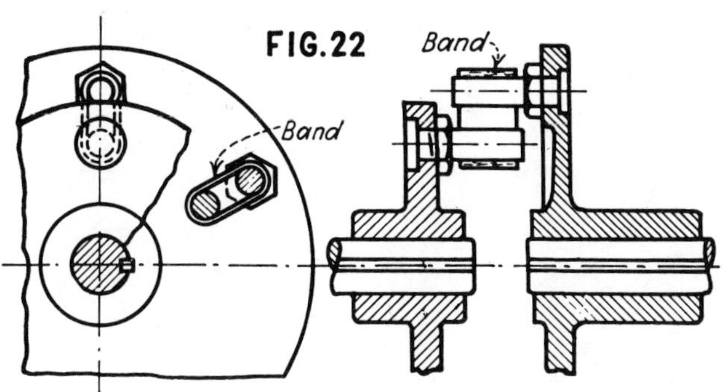

# *Typical Designs of*

FIG. 24

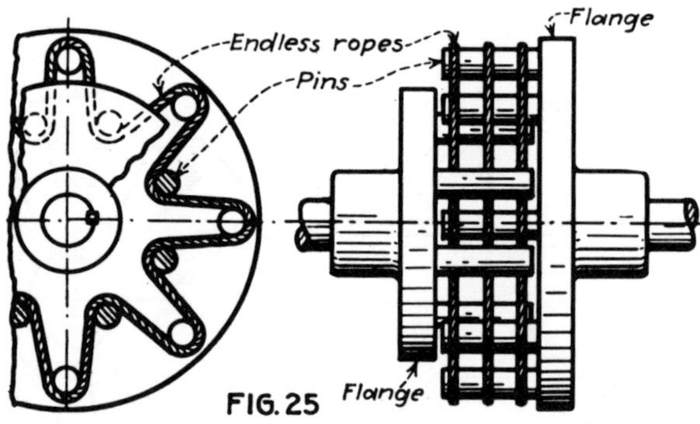

FIG. 25

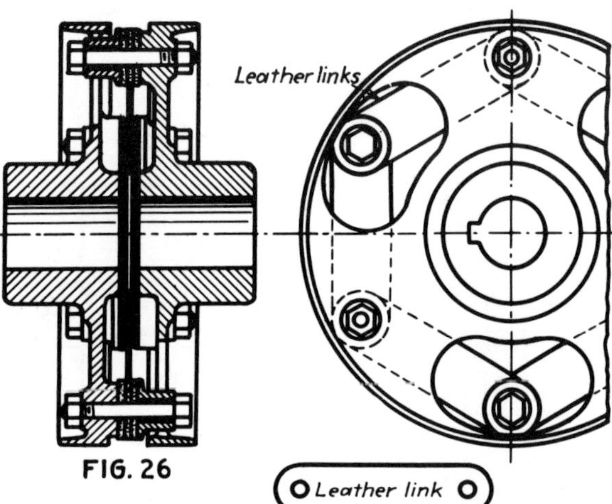


FIG. 26

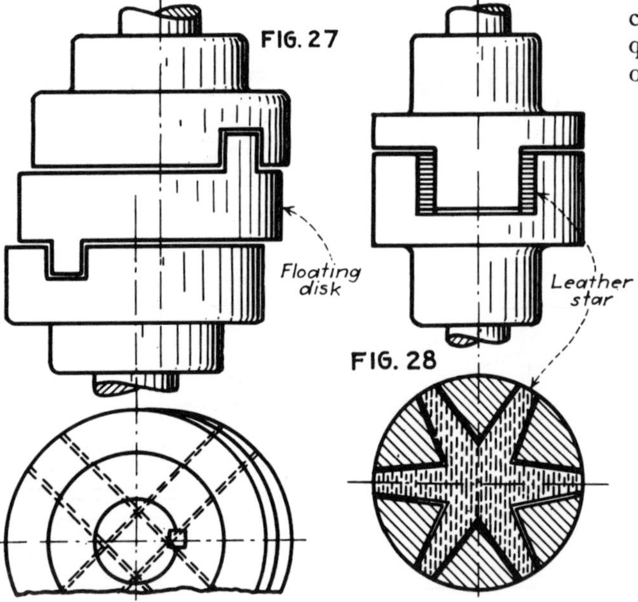


FIG. 27

FIG. 28

**Fig. 24**—This Webster Manufacturing Company coupling uses a single endless leather belt instead of a series of bands, as in Fig. 22. The belt is looped over alternate pins in both flanges. Has good shock resisting properties because of belt stretch and the tendency of the pins to settle back into the loops of the belt.

**Fig. 25**—This coupling made by the Weller Manufacturing Company is similar to the design in Fig. 24, but instead of a leather belt uses hemp rope, made endless by splicing. The action under load is the same as in the endless belt type.

**Fig. 26**—This Bruce-Macbeth design uses leather links instead of endless wire cables, as shown in Fig. 23. The load is transmitted from one flange to the other by direct pull of the links, which at the same time allows for the proper flexibility. Intended for permanent installations requiring a minimum of supervision.

**Fig. 27**—The Oldham form of coupling made by W. A. Jones Foundry and Machine Company is of the two-jaw type with a metal disk. Is used for transmitting heavy loads at low speed.

**Fig. 28**—The Charles Bond Company star coupling is similar to the cross type shown in Fig. 21. The star-shaped floating member is made of laminated leather. Has three jaws in each flange. Torque capacity is thus increased over the two-jaw or cross type. Coupling takes care of limited end play.

**Fig. 29**—Combination rubber and canvas disk is bolted to two metal spiders. Extensively used for low torques where compensation for only slight angular misalignment is required. Is quiet in operation and needs no lubrication or other attention. Offset misalignment shortens disk life.

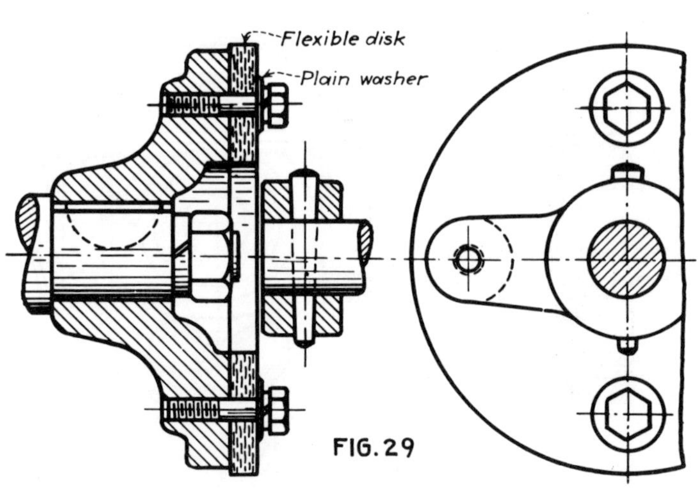


FIG. 29

# Flexible Couplings–III

**Fig. 30**—A metal block as a floating center is used in this American Flexible Coupling Company design. Quiet operation is secured by facing the block with removable fiber strips and packing the center with grease. The coupling sets up no end thrusts, is easy to assemble and does not depend on flexible material for the driving action. Can be built in small sizes by using hardwood block without facings, for the floating member.

**Fig. 31**—This Westinghouse Nuttall Company coupling is an all-metal type having excellent torsional flexibility. The eight compression springs compensate for angular and offset misalignment. Allows for some free endwise float of the shafts. Will transmit high torques in either direction. No lubrication is needed.

**Fig. 32**—Similar to Fig. 29, but will withstand offset misalignment by addition of the extra disk. In this instance the center spider is free to float. By use of two rubber-canvas disks, as shown, coupling will withstand a considerable angular misalignment.

**Fig. 33**—In this Smith & Serrell coupling a flexible cross made of laminated steel strips floats between two spiders. The laminated spokes, retained by four segmental shoes, engage lugs integral with the flanges. Coupling is intended for the transmission of light loads only.

**Fig. 34**—This coupling made by Brown Engineering Company is useful for improvising connections between apparatus in laboratories and similar temporary installations. Compensates for misalignment in all directions. Will absorb varying degrees of torsional shocks by changing the size of the springs. Springs are retained by threaded pins engaging the coils. Overload protection is possible by the slippage or breakage of replacable springs.

**Fig. 35**—In another design by Brown Engineering Company, a series of laminated spokes transmit power between the two flanges without setting up end thrusts. This type allows free end play. Among other advantages are absorption of torsional shocks, has no exposed moving parts, and is well balanced at all speeds. Wearing parts are replacable and working parts are protected from dust.

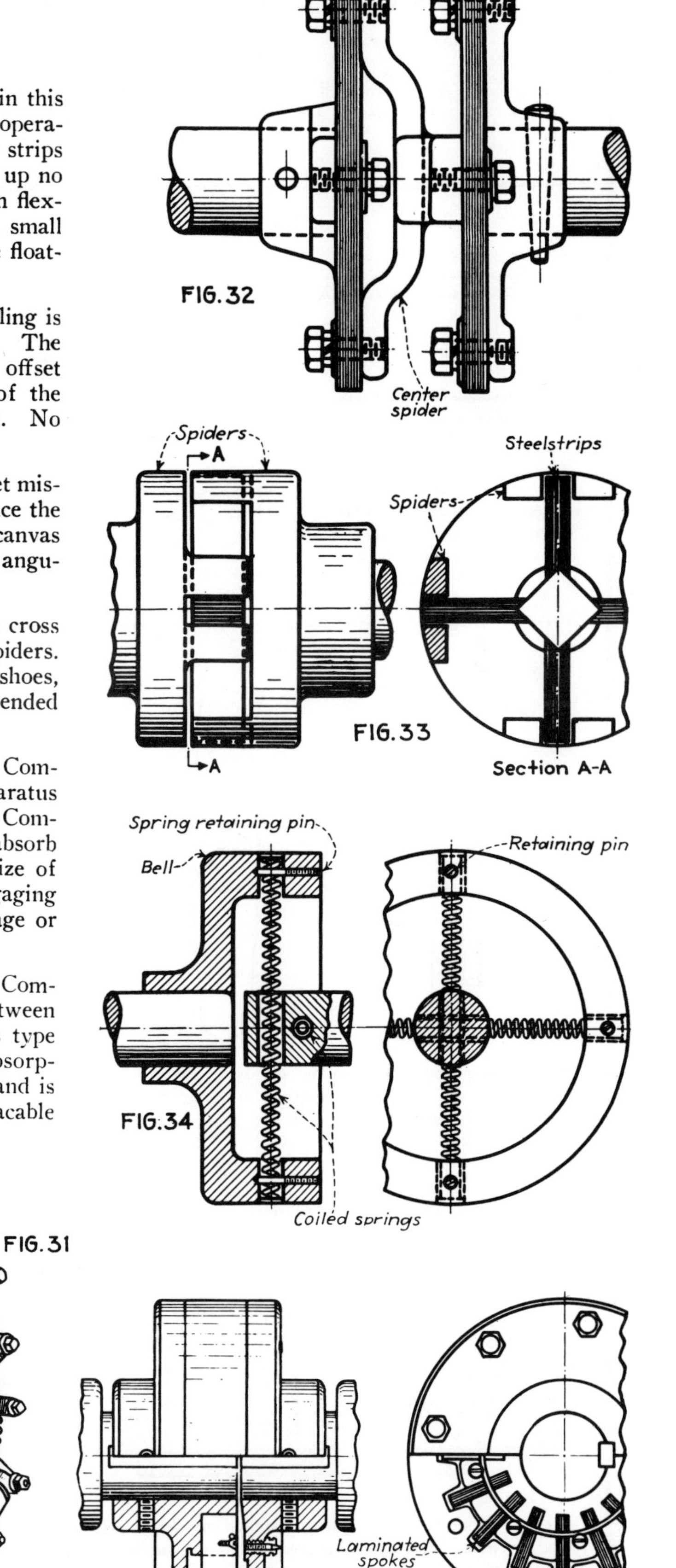

# Low Costs Methods of Coupling

Sixteen types of low cost couplings, including flexible and non-flexible types. Most are for small diameter, lightly loaded shafts, but a few of them can also be

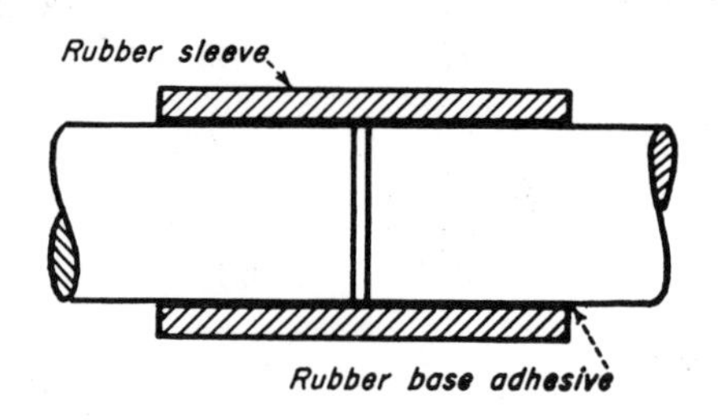


Fig 1—Rubber sleeve has inside diameter smaller than shaft diameters. Using rubber-base adhesive will increase the torque capacity.

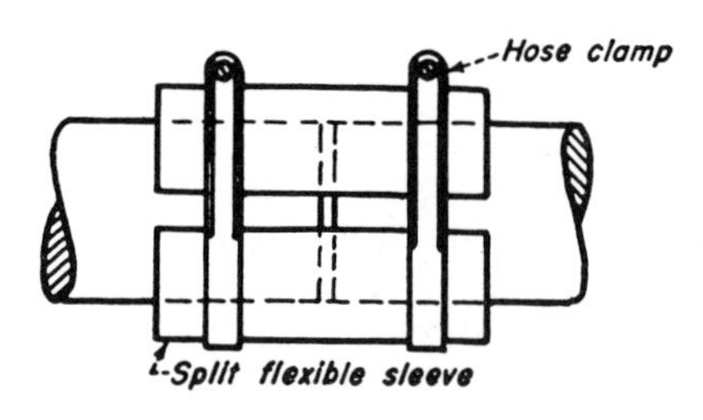


Fig 2—Slit sleeve of rubber or other flexible material is held by hose clamps. Easy to install and remove. Absorbs vibration and shock loads.

Fig 3—Ends of spring extend through holes in shafts to form coupling. Dia of spring determined by shaft dia, wire dia determined by loads.

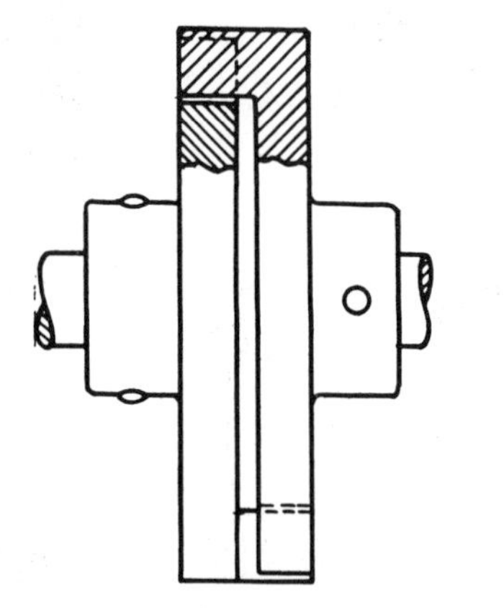

Fig 7—Jaw-type coupling is secured to shafts with straight pins. Commercially available; some have flexible insulators between jaws.

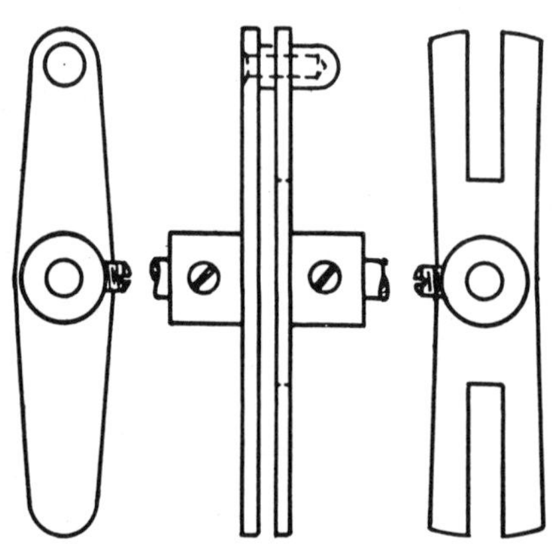

Fig 8—Removable type coupling with insulated coupling pin. A set screw in the collar of each stamped member is used to fasten it to the shaft.

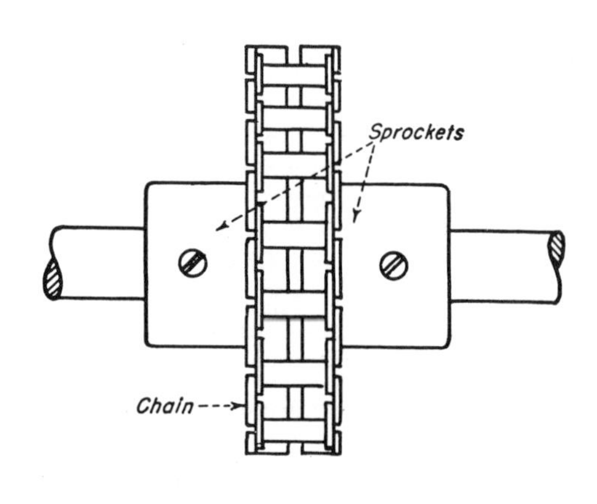


Fig 9—Sprockets mounted on each shaft are linked together with roller chain. Wide range of torque capacity. Commercially available.

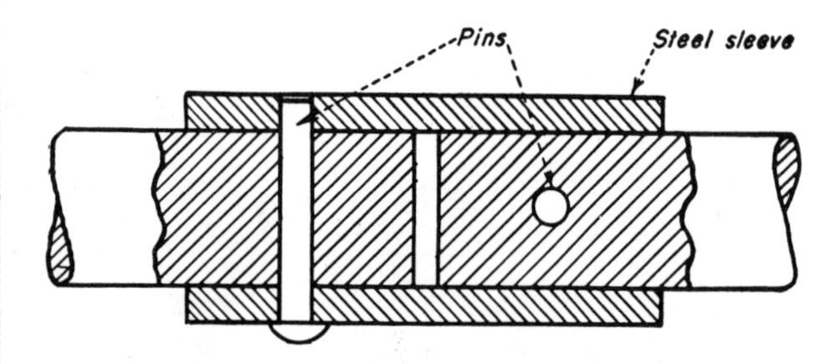


Fig 13—Steel sleeve coupling fastened to shafts with two straight pins. Pins are staggered at 90 degree intervals to reduce the stress concentration.

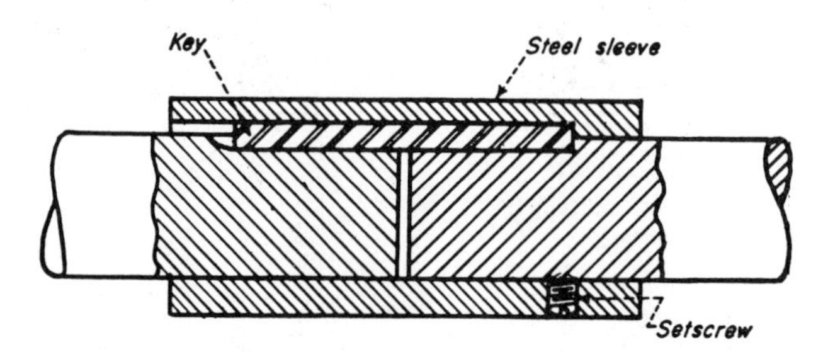


Fig 14—Single key engages both shafts and metal sleeve which is attached to one shaft with setscrew. Shoulder on sleeve can be omitted to reduce costs.

# Small Diameter Shafts

adapted to heavy duty shafts. Some of them are currently available as standard commercial parts.

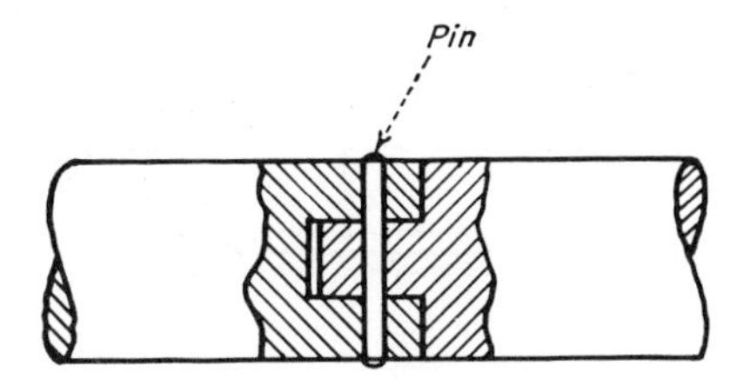


Fig 4—Tongue-and-groove coupling made from shaft ends is used to transmit torque. Pin or set screw keeps shafts in proper alignment.

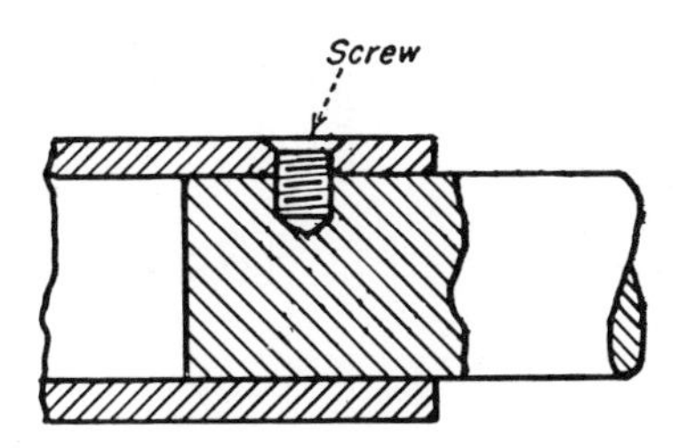


Fig 5—Screw fastens hollow shaft to inner shaft. Set screw can be used for small shafts and low torque by milling a flat on the inner shaft.

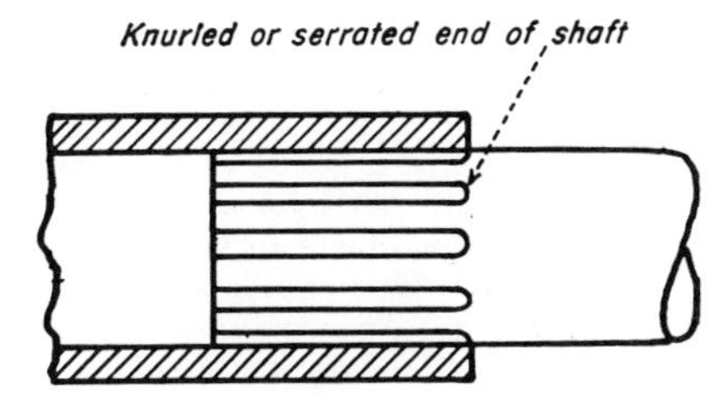


Fig 6—Knurled or serrated shaft is pressed into hollow shaft. Effects of misalignment must be checked to prevent overloading the bearings.

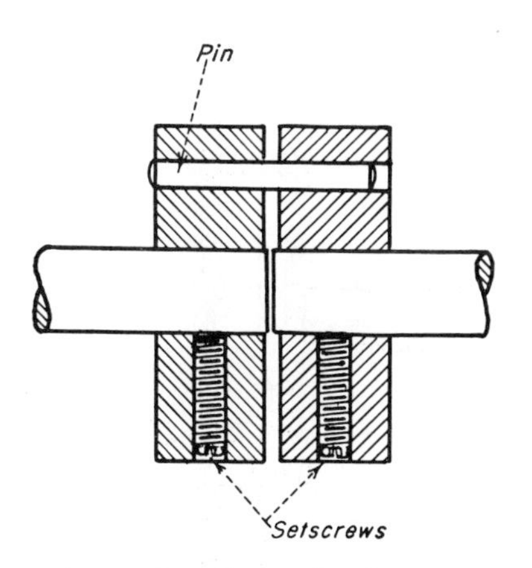


Fig 10—Coupling made of two collars fastened to shafts with set screws. Pin in one collar engages hole in other. Soft spacer can be used as cushion.

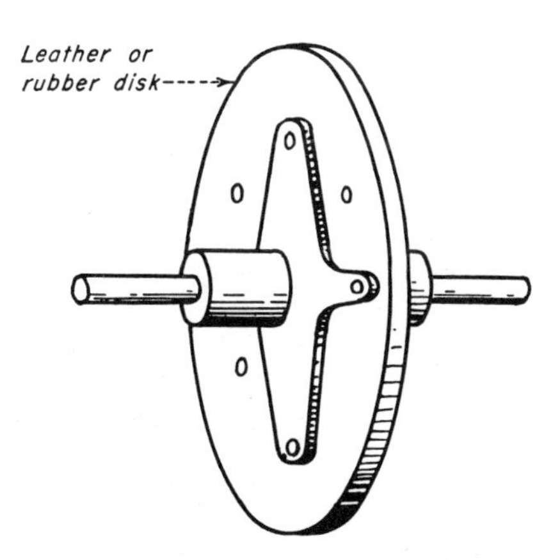


Fig 11—Coupling is made from two flanges rivited to leather or rubber center disk. Flanges are fastened to the shafts by means of setscrews.

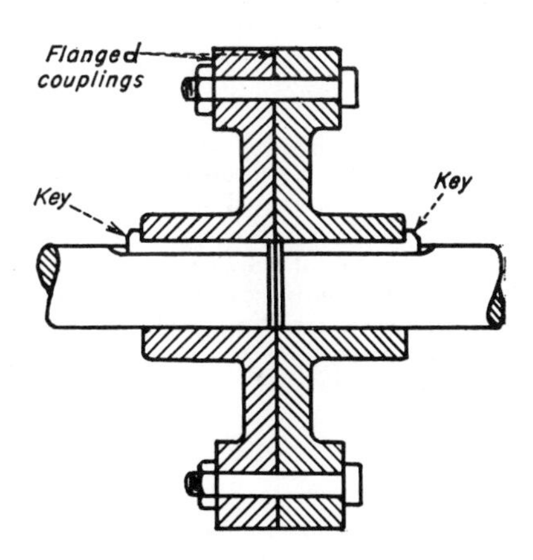


Fig 12—Bolted flange couplings are used on shafts from one to twelve inches in diameter. Flanges are joined by four bolts and are keyed to shafts.

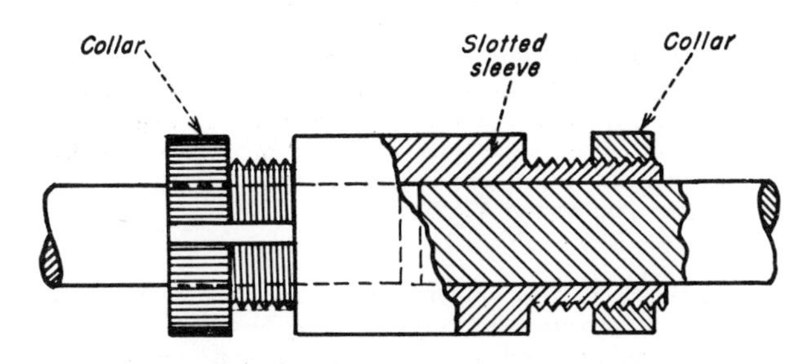


Fig 15—Screwing split collars on tapered threads of slotted sleeve tightens coupling. For light loads and small shafts, sleeve can be made of plastic material.

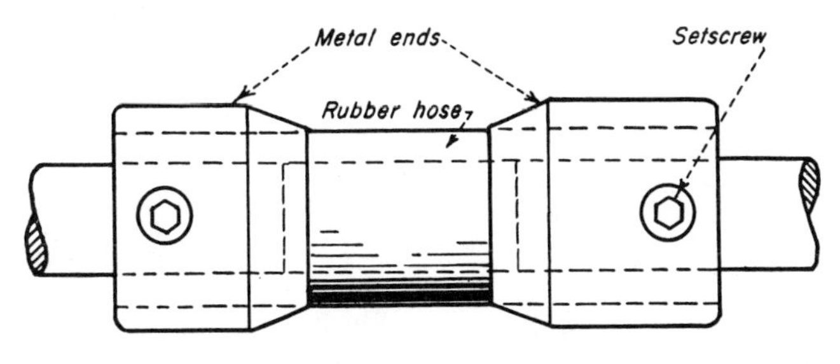


Fig 16—One-piece flexible coupling has rubber hose with metallic ends that are fastened to shafts with set screws. Commercially available in several sizes and lengths.

# Coupling of Parallel Shafts

H. G. CONWAY
Cheltenham, England

**FIG. 1**—A common method of coupling shafts is with two gears; for gears may be substituted chains, pulleys, friction drives and others. Major limitation is need for adequate center distance; however, an idler can be used for close centers as shown. This can be a plain pinion or an internal gear. Transmission is at constant velocity and axial freedom is present.

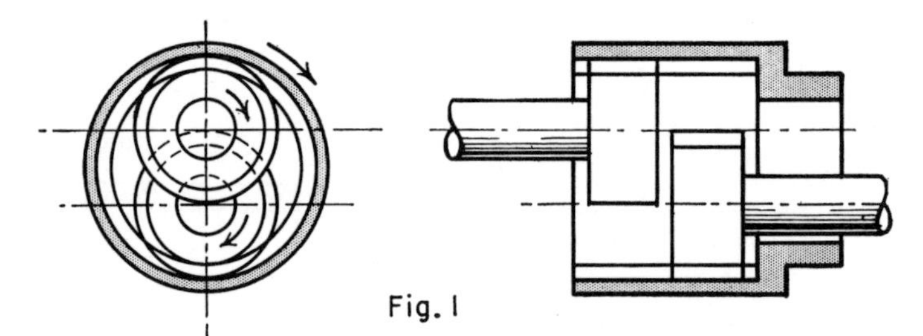

Fig. 1

**FIG. 2**—Two universal joints and a short shaft can be used. Velocity transmission is constant between input and output shafts if the shafts remain parallel and if the end yokes are disposed symmetrically. Velocity of the central shaft fluctuates during rotation and at high speed and angles may cause vibration. The shaft offset may be varied but axial freedom requires a splined mounting of one shaft.

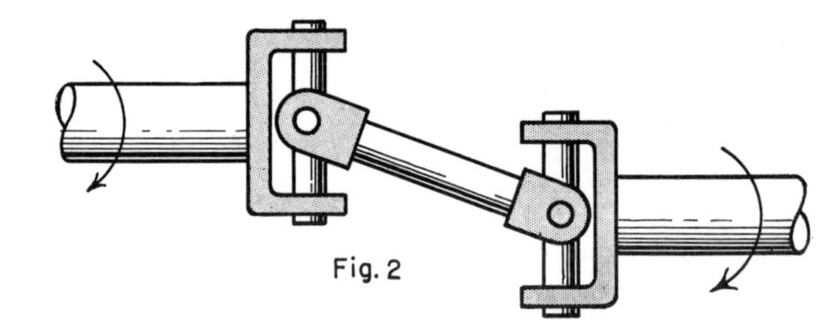

Fig. 2

**FIG. 3**—Crossed axis yoke coupling is a variation of the mechanism in Fig. 2. Each shaft has a yoke connected so that it can slide along the arms of a rigid cross member. Transmission is at a constant velocity but the shafts must remain parallel, although the offset may vary. There is no axial freedom. The central cross member describes a circle and is thus subjected to centrifugal loads.

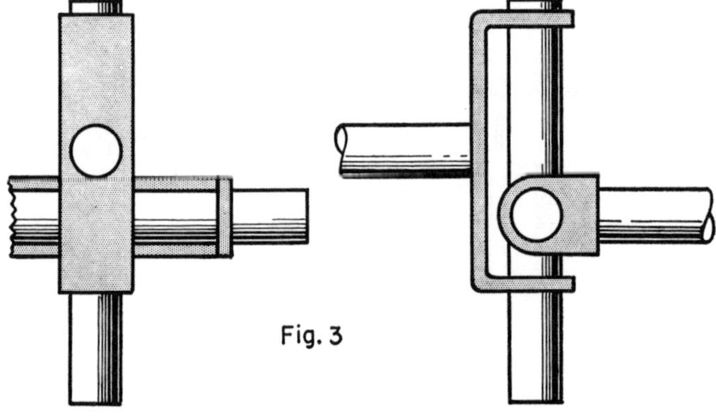

Fig. 3

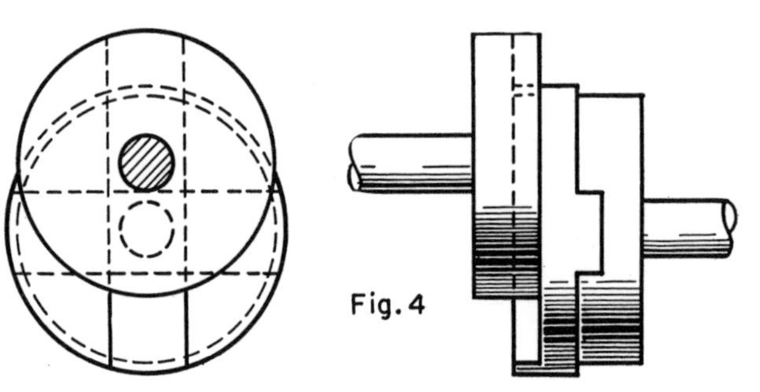

Fig. 4

**FIG. 4**—Another often used method is the Oldham coupling. The motion is at constant velocity, the central member describing a circle. The shaft offset may vary but the shafts must remain parallel. A small amount of axial freedom is possible. A tilting action of the central member can occur caused by the offset of the slots. This can be eliminated by enlarging the diameter and milling the slots in the same transverse plane.

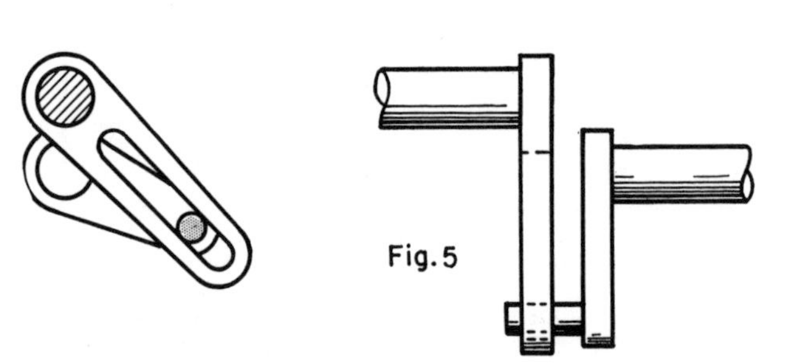

Fig. 5

**FIG. 5**—If the velocity does not have to be constant a pin and slot coupling can be used. Velocity transmission is irregular as the effective radius of operation is continually changing, the shafts must

The coupling of parallel shafts so that they rotate together is a common machine design problem. Illustrated are several methods where a constant 1:1 velocity ratio is possible and others where the velocity ratio may fluctuate during rotation. Some of the couplings have particular value for joining two shafts that may deflect or move relative to each other.

remain parallel unless a ball joint is used between the slot and pin. Axial freedom is possible but any change in the shaft offset will further affect the fluctuation of velocity transmission.

**FIG. 6**—The parallel-crank mechanism is sometimes used to drive the overhead camshaft on engines. Each shaft has at least two cranks connected by links and with full symmetry for constant velocity action and to avoid dead points. By using ball joints at the ends of the links, displacement between the crank assemblies is possible.

**FIG. 7**—A mechanism kinematically equivalent to Fig. 6, can be made by substituting two circular and contacting pins for each link. Each shaft has a disk carrying three or more projecting pins, the sum of the radii of the pins being equal to the eccentricity of offset of the shafts. The lines of center between each pair of pins remain parallel as the coupling rotates. Pins do not need to be of equal diameter. Transmission is at constant velocity and axial freedom is possible.

**FIG. 8**—Similar to the mechanism in Fig. 7, but with one set of pins being holes. The difference of radii is equal to the eccentricity or offset. Velocity transmission is constant; axial freedom is possible, but as in Fig. 7, the shaft axes must remain fixed. This type of mechanism is sometimes used in epicyclic reduction gear boxes.

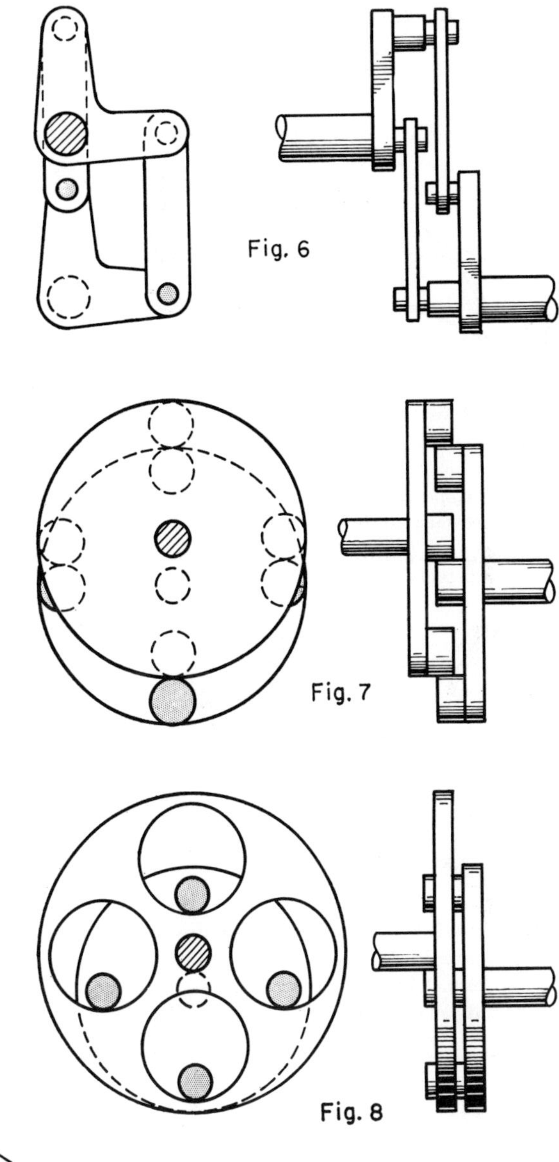


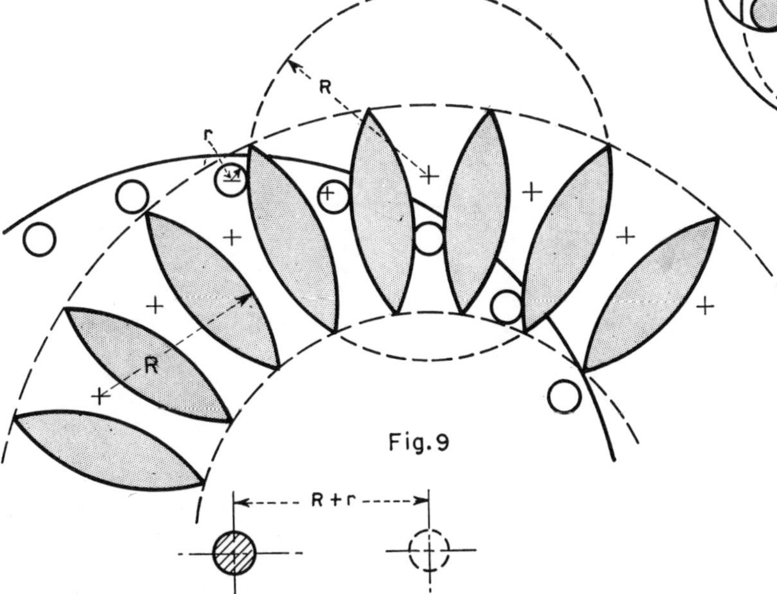


**FIG. 9**—An unusual development of the pin coupling is shown left. A large number of pins engage lenticular or shield shaped sections formed from segments of theoretical large pins. The axes forming the lenticular sections are struck from the pitch points of the coupling and the distance $R + r$ is equal to the eccentricity between shaft centers. Velocity transmission is constant; axial freedom is possible but the shafts must remain parallel.

# Ten Different Types of

## CYLINDRICAL TYPES

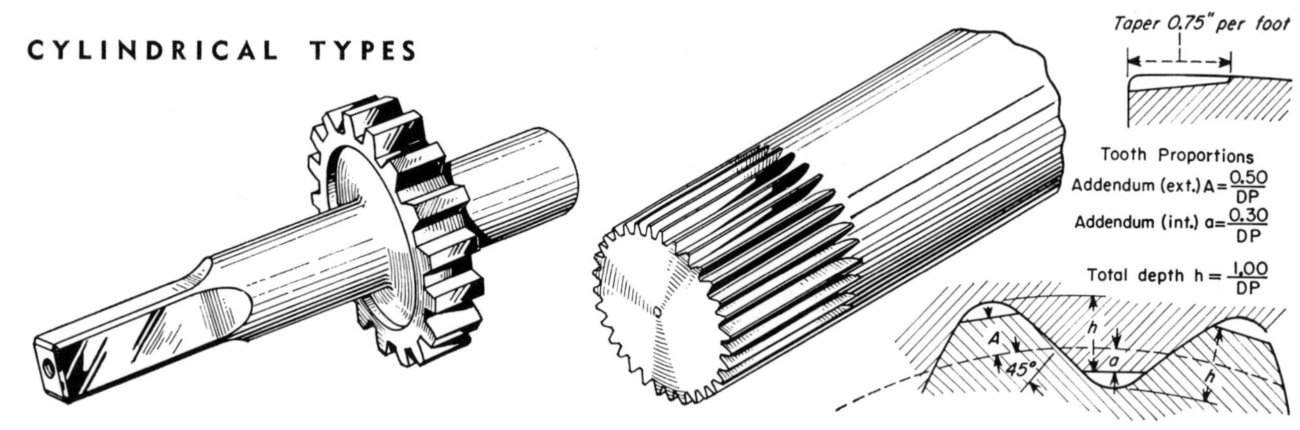


**1** SQUARE SPLINES make a simple connection and are used mainly for applications of light loads, where accurate positioning is not important. This type is commonly used on machine tools; a cap screw is necessary to hold the enveloping member.

**2** SERRATIONS of small size are used mostly for applications of light loads. Forcing this shaft into a hole of softer material makes an inexpensive connection. Originally straight-sided and limited to small pitches, 45 deg serrations have been standardized (SAE) with large pitches up to 10 in. dia. For tight fits, serrations are tapered.

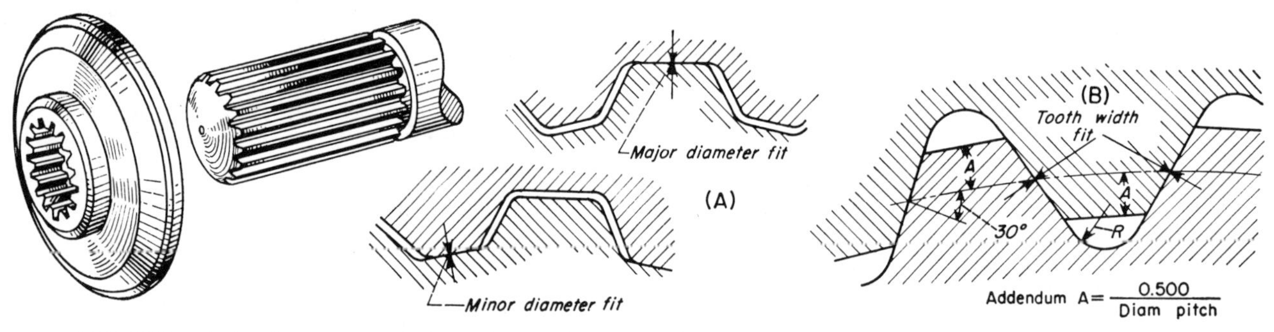


**5** INVOLUTE-FORM splines are used where high loads are to be transmitted. Tooth proportions are based on a 30 deg stub tooth form. (A) Splined members may be positioned either by close fitting major or minor diameters. (B) Use of the tooth width or side positioning has the advantage of a full fillet radius at the roots. Splines may be parallel or helical. Contact stresses of 4,000 psi are used for accurate, hardened splines. Diametral pitch above is the ratio of teeth to the pitch diameter.

## FACE TYPES

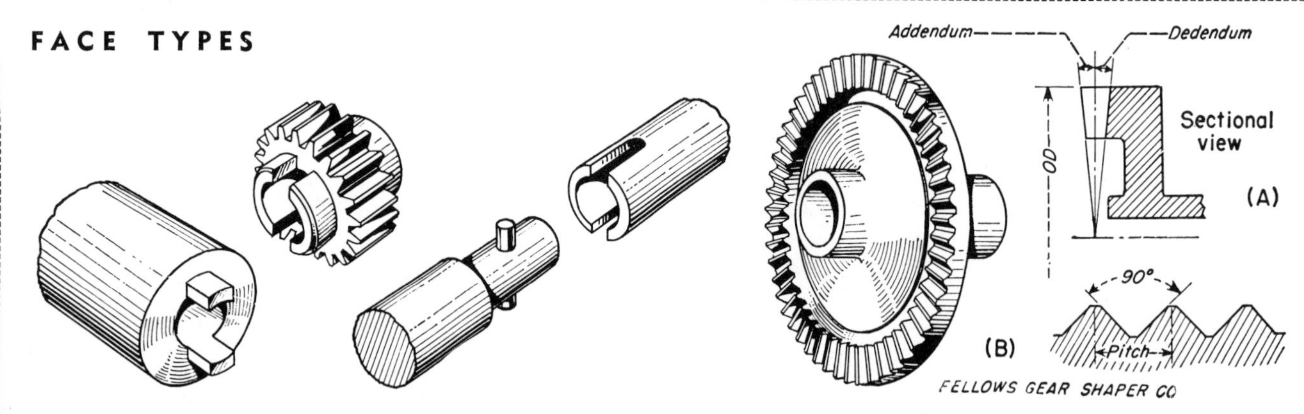


**8** MILLED SLOTS in hubs or shafts make an inexpensive connection. This type is limited to moderate loads and requires a locking device to maintain positive engagement. Pin and sleeve method is used for light torques and where accurate positioning is not required.

**9** RADIAL SERRATIONS by milling or shaping the teeth make a simple connection. (A) Tooth proportions decrease radially. (B) Teeth may be straight-sided (castellated) or inclined; a 90 deg angle is common.

# Splined Connections

W. W. HEATH

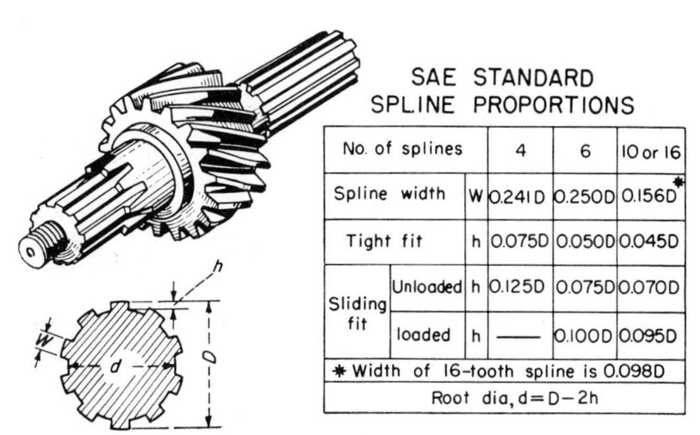


SAE STANDARD SPLINE PROPORTIONS

| No. of splines | | | 4 | 6 | 10 or 16 |
|---|---|---|---|---|---|
| Spline width | | W | 0.241D | 0.250D | 0.156D* |
| Tight fit | | h | 0.075D | 0.050D | 0.045D |
| Sliding fit | Unloaded | h | 0.125D | 0.075D | 0.070D |
| | loaded | h | — | 0.100D | 0.095D |
| * Width of 16-tooth spline is 0.098D | | | | | |
| Root dia, d = D − 2h | | | | | |

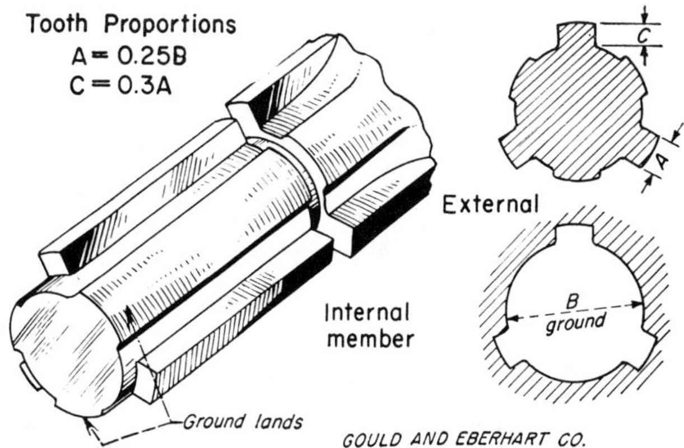


**3** STRAIGHT-SIDED splines have been widely used in the automotive field. Such splines are often used for sliding members. The sharp corner at the root limits the torque capacity to pressures of approximately 1,000 psi on the spline projected area. For different applications, tooth height is altered as shown in the table above.

**4** MACHINE-TOOL spline has a wide gap between splines to permit accurate cylindrical grinding of the lands—for precise positioning. Internal parts can be ground readily so that they will fit closely with the lands of the external member.

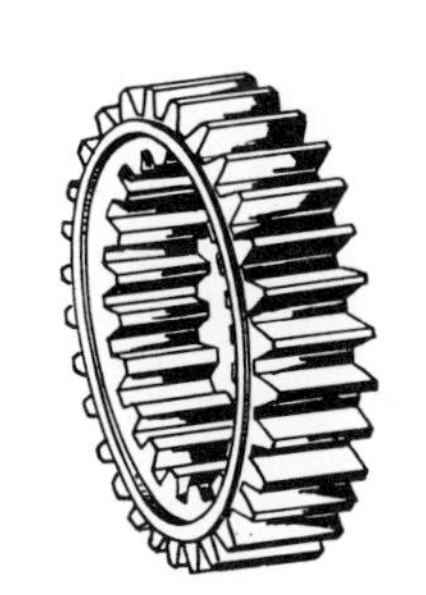

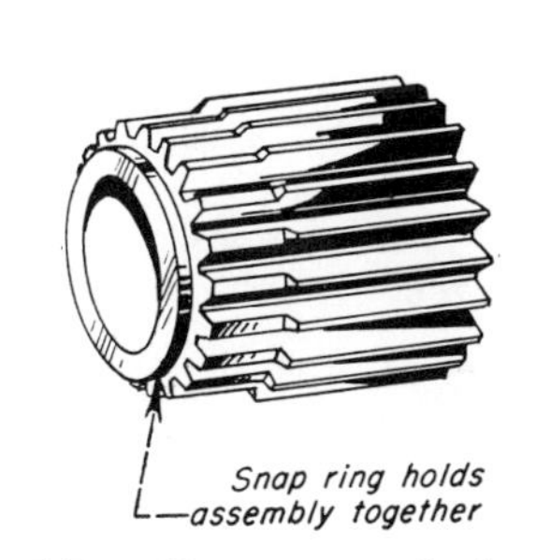


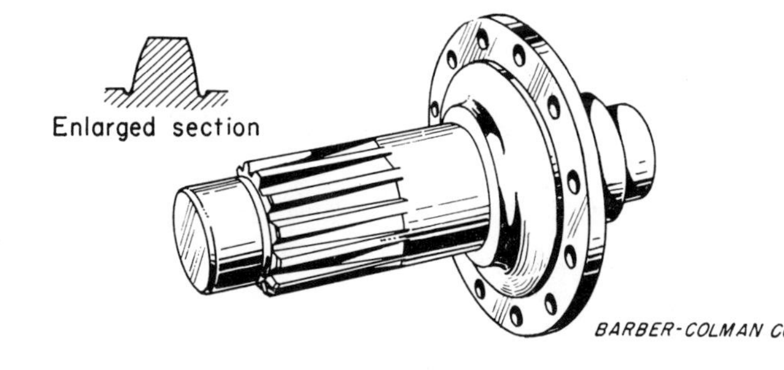


**6** SPECIAL INVOLUTE splines are made by using gear tooth proportions. With full depth teeth, greater contact area is possible. A compound pinion is shown made by cropping the smaller pinion teeth and internally splining the larger pinion.

**7** TAPER-ROOT splines are for drives which require positive positioning. This method holds mating parts securely. With a 30 deg involute stub tooth, this type is stronger than parallel root splines and can be hobbed with a range of tapers.

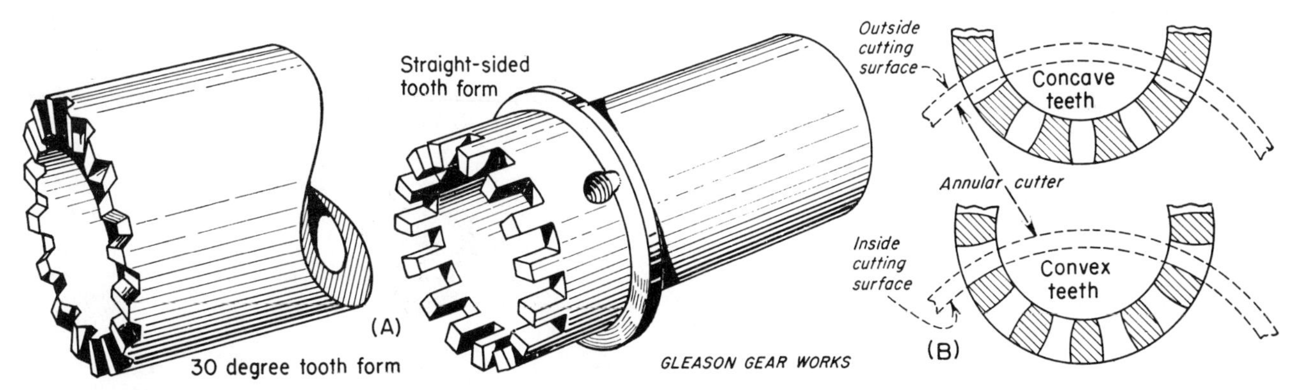


**10** CURVIC COUPLING teeth are machined by a face-mill type of cutter. When hardened parts are used which require accurate positioning, the teeth can be ground. (A) This process produces teeth with uniform depth and can be cut at any pressure angle, although 30 deg is most common. (B) Due to the cutting action, the shape of the teeth will be concave (hour-glass) on one member and convex on the other—the member with which it will be assembled.

# How to Connect Tubing—

## TEE JOINTS

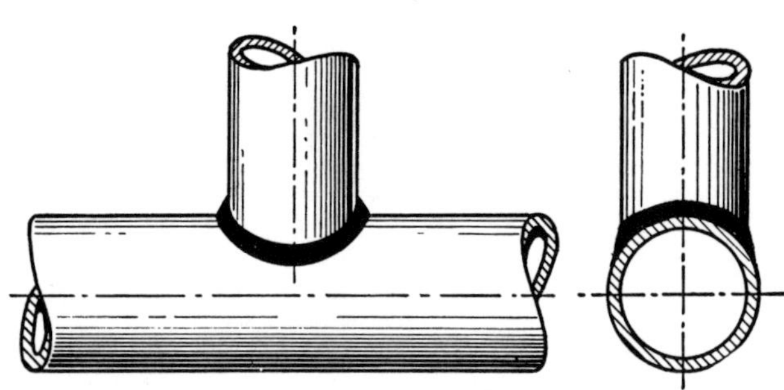

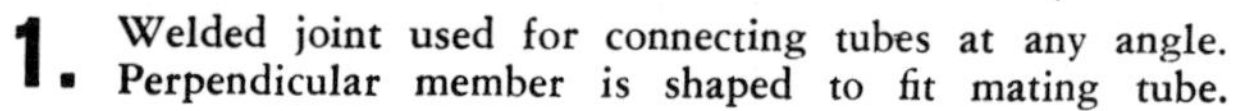

**1.** Welded joint used for connecting tubes at any angle. Perpendicular member is shaped to fit mating tube.

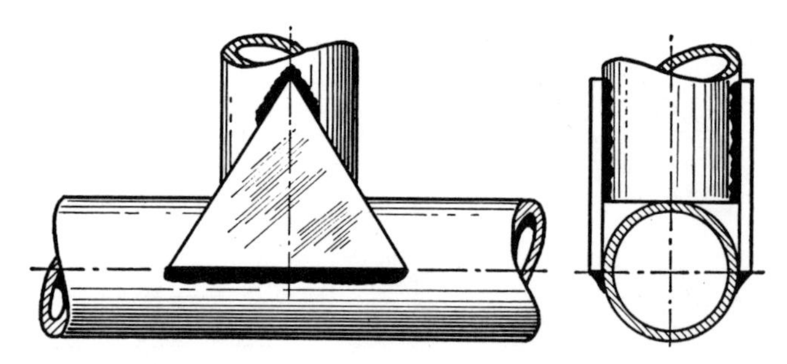

**2.** Gusset plate makes a reinforced tubing connection. Attachment can be made by means of a weld, rivets or bolts.

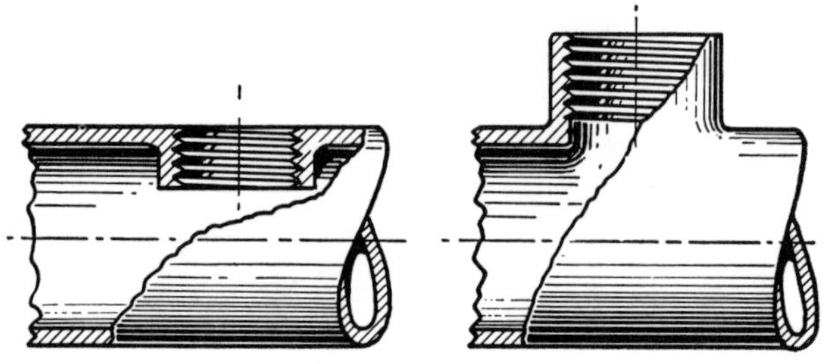

**5.** Formed flanges simplifies tubing connection. Attachment can be made by using threads or welding operation.

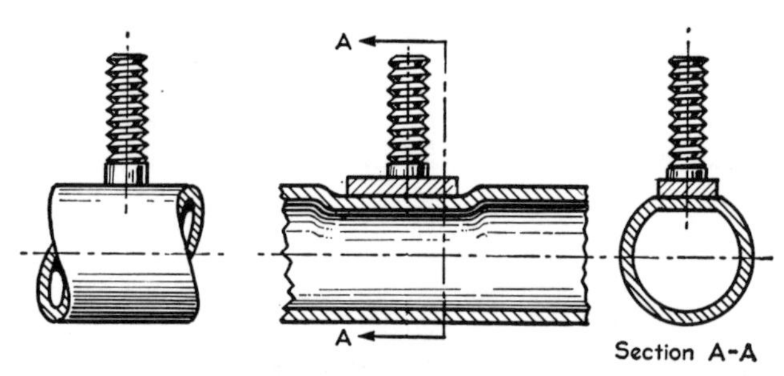


**6.** Threaded welding studs make a convenient connection. Welding stud to a pad on tubing reinforces joint.

## CROSS JOINTS

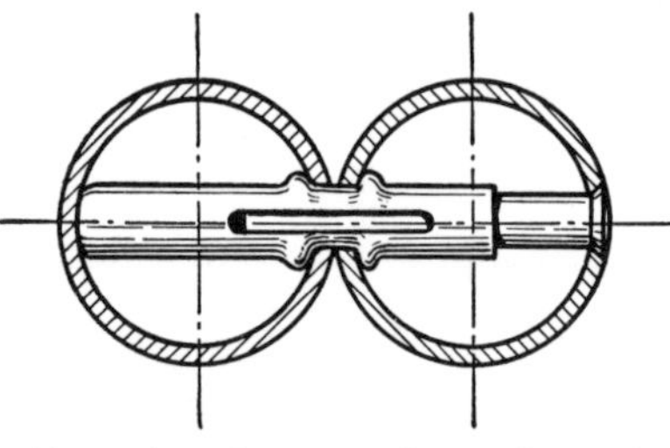

**1.** Expansion sleeve can be used to join tubes. Split-sleeve is expanded in tubing holes by means of steel drive plug.

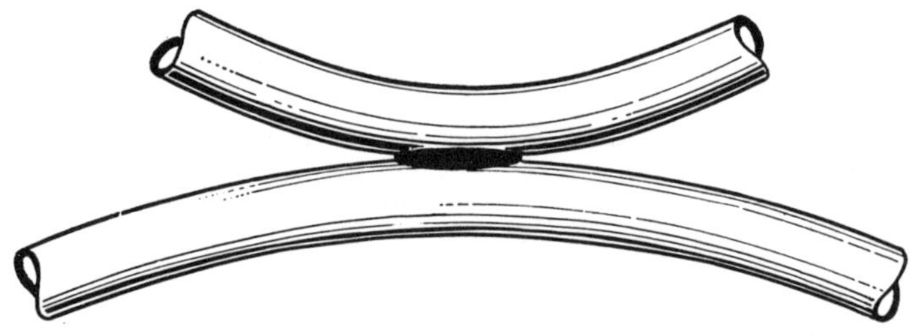

**2.** Curved tubing used as a cross joint; this can easily be connected by projection welding.

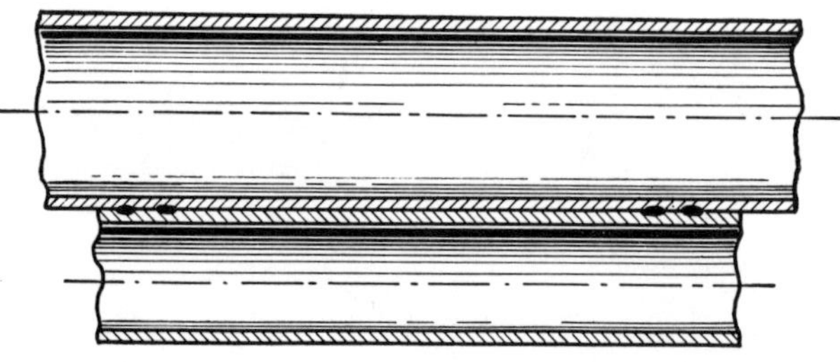

**3.** Projection welding parallel tubes makes more rigid construction. Air vent tubes can be attached in this way.

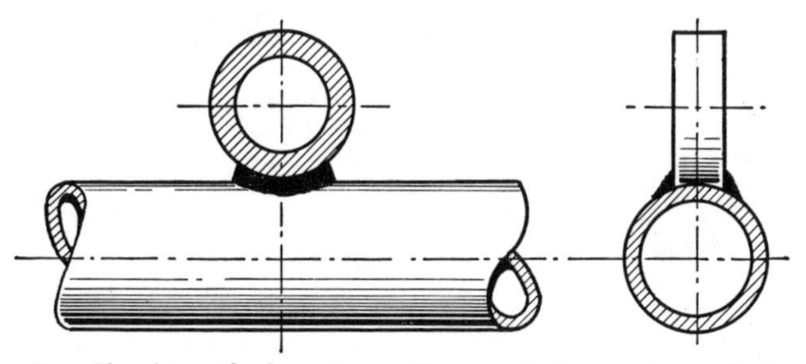

**4.** Simple guide for other tubing or shafts can be made by projection welding ring to tubing.

# Cross and Tee Joints

Prepared by the Product Development Committee, Formed Steel Tube Institute, Cleveland, Ohio.

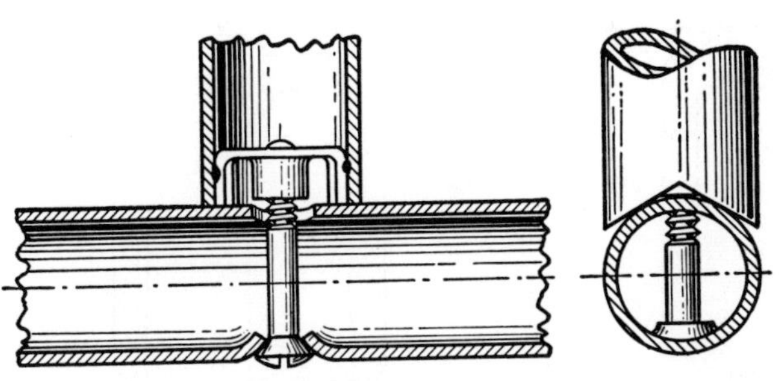

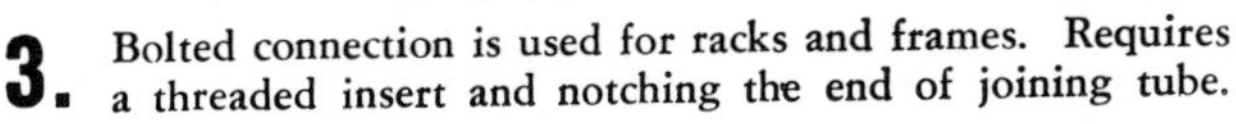

**3.** Bolted connection is used for racks and frames. Requires a threaded insert and notching the end of joining tube.

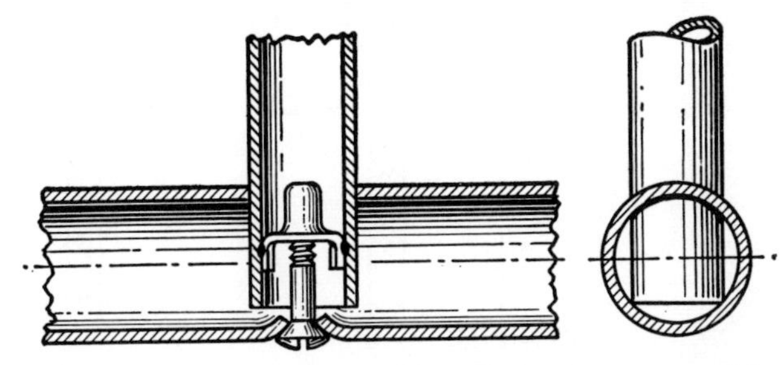

**4.** Tubes of different sizes can be more rigidly connected by inserting the smaller tube inside the larger one.

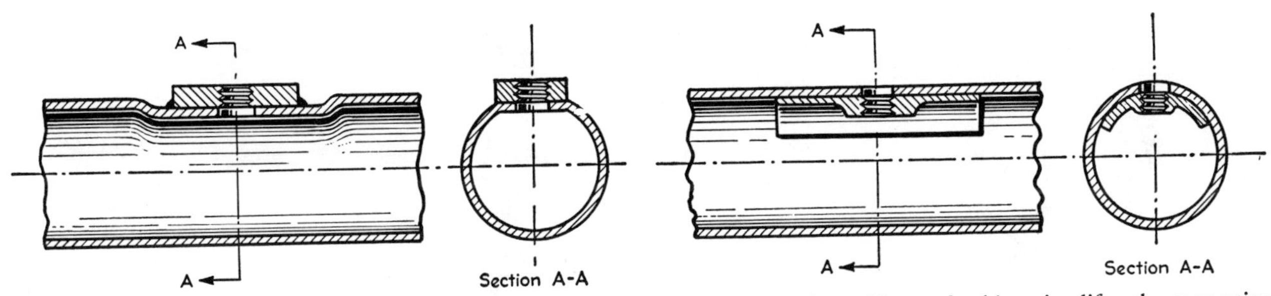


**7.** Reinforced connection can be made by attaching pads to outside or inside surface. Flattened tubing simplifies the connection for brazing or projection welding. Pads may be threaded either for pipe or screw fitting.

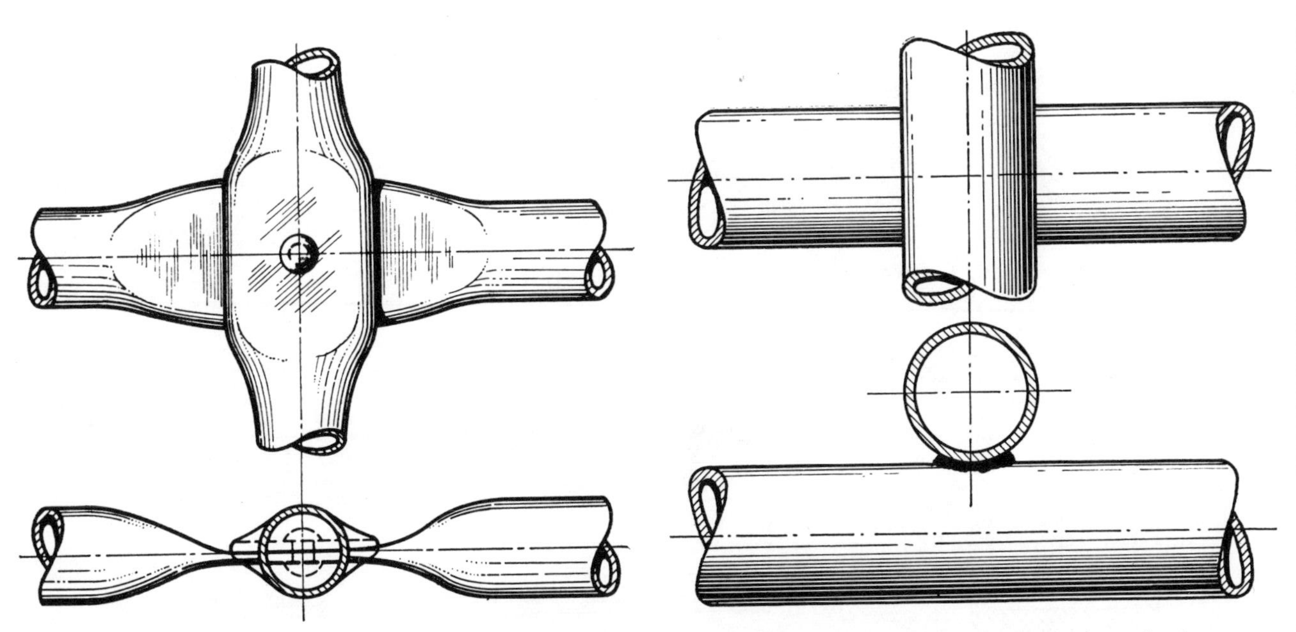

**5.** Flattened tubing makes joint when tubes intersect in same plane. Connection is welded, riveted or bolted together.

**6.** Welded connection can be made between tubes in contact. Tubes may cross at any angle.

# Different Mechanical Methods

## ATTACHMENTS

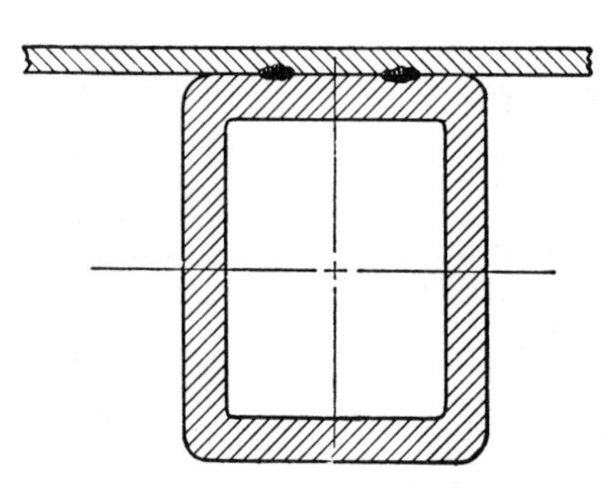

Fig. 1—Welded attachment of plates to tubing by spot or projection welding.

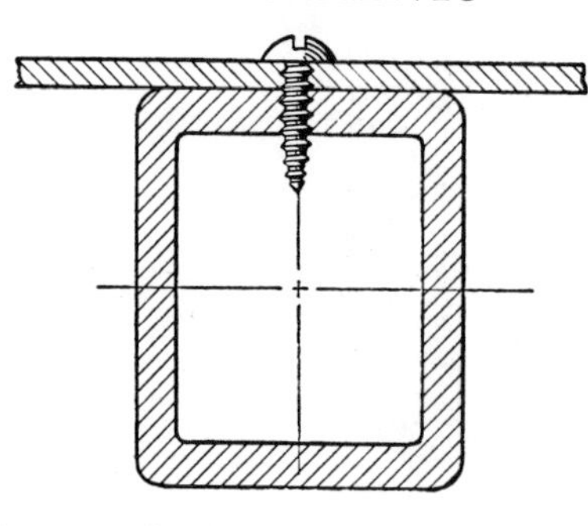

Fig. 2—Self-tapping or drive screws can be used to attach devices to tubing.

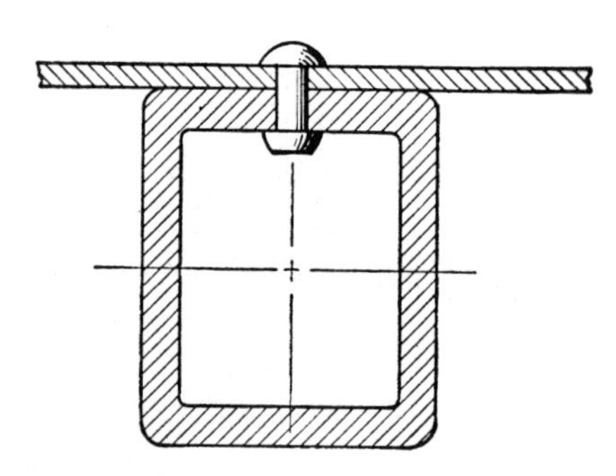

Fig. 3—Blind rivets, mechanical or explosive, can also be used for connection.

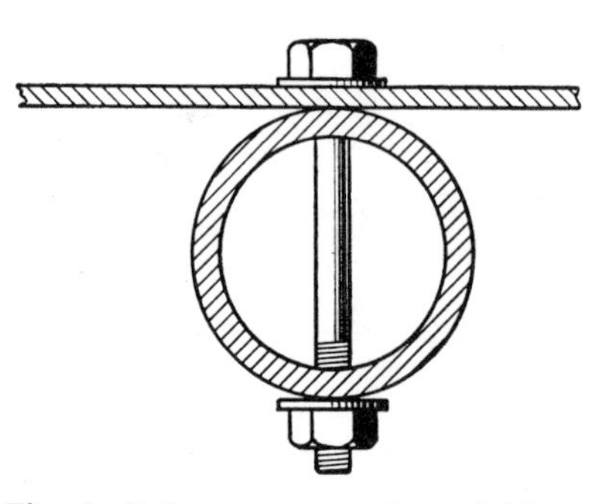

Fig. 4—Bolts or rivets make a rigid connection of plates to tubing.

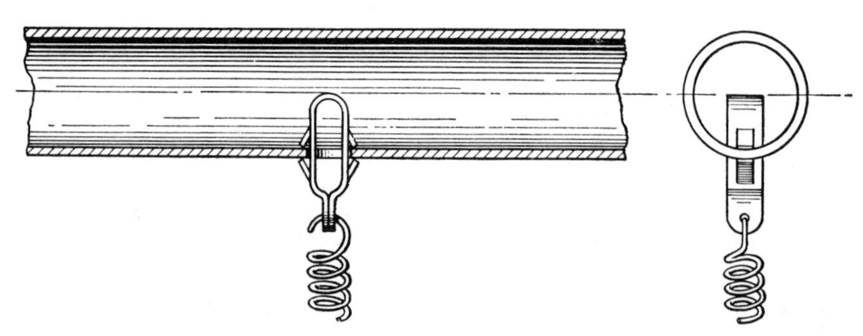

Fig. 5—Clip device is a simple means for holding springs or hangars to tubing. Requires only single hole in tubing and metal prongs lock clip in position.

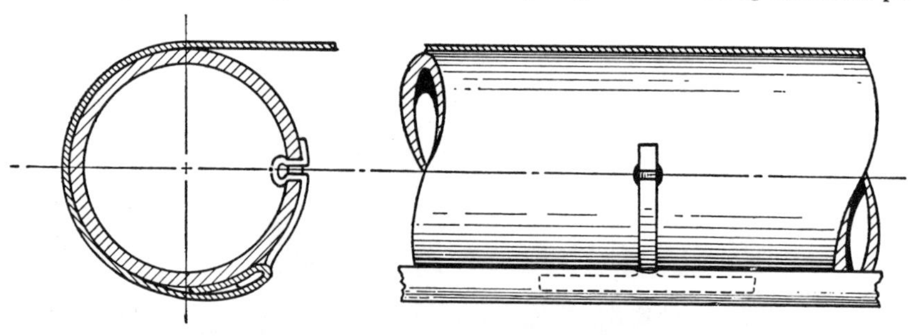

Fig. 6—Fabrics are conveniently fastened to tubing using a metal clip. This clip is inserted into fold of fabric and snapped into hole in tube.

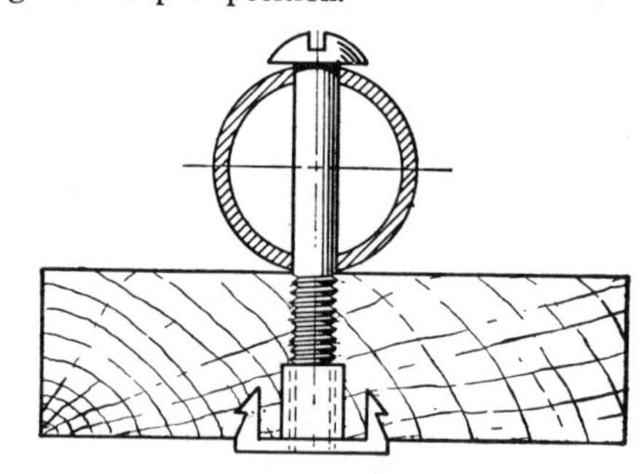

Fig. 7—Wood attachment uses a threaded "T" nut. Prongs hold nut in position.

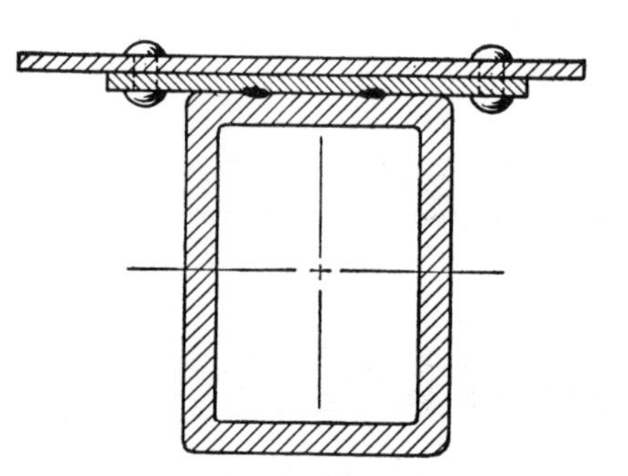

Fig. 8—Fabricated attachment consists of welded strip and riveted joint.

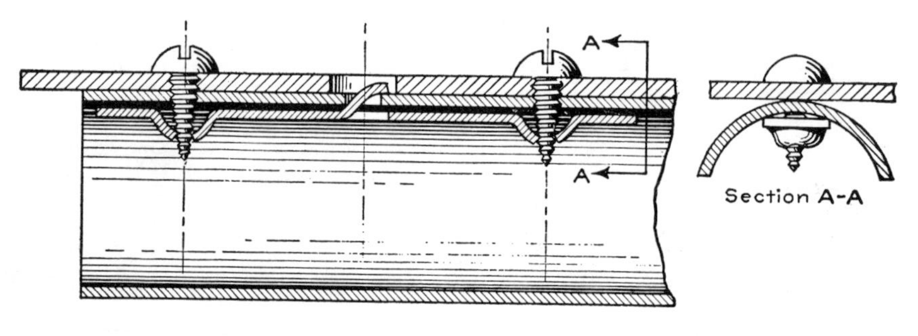


Fig. 9—Threaded clip is held inside tubing by lug and insures a positive connecting device. Clip is used when tubing wall is too thin for screw attachment.

# for Attaching Tubing

This construction data has been prepared by the Product Development Committee of the Formed Steel Tube Institute, Cleveland, Ohio.

## END FITTINGS

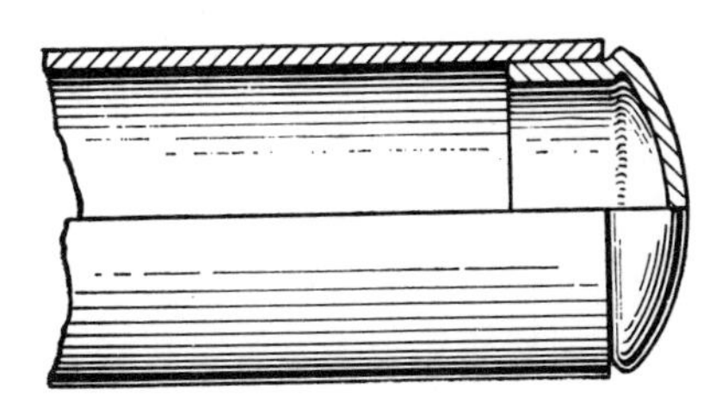

Fig. 10—End plug may be threaded or press-fitted on the OD or ID of tubing.

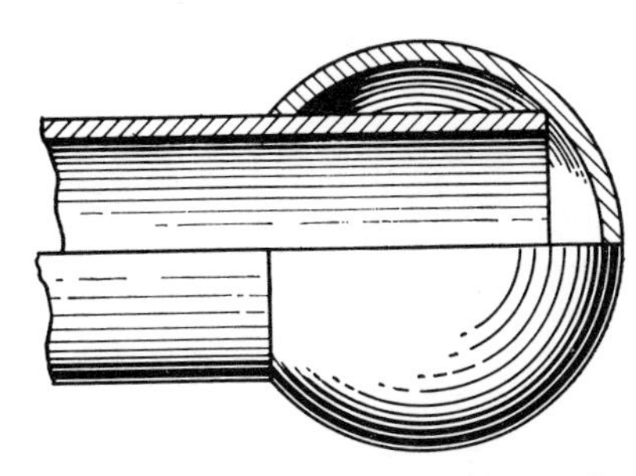

Fig. 11—Ornamental cap may be used to give tubing finished appearance.

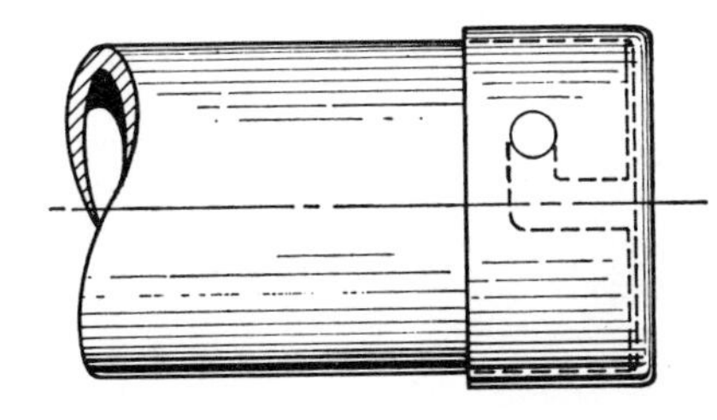

Fig. 12—Detachable cap uses bayonet-type connection for convenient removal.

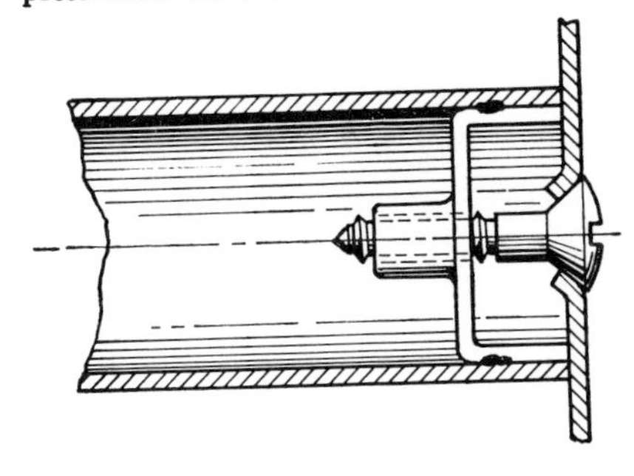

Fig. 13—Threaded insert, welded inside, facilitates the assembly of plates.

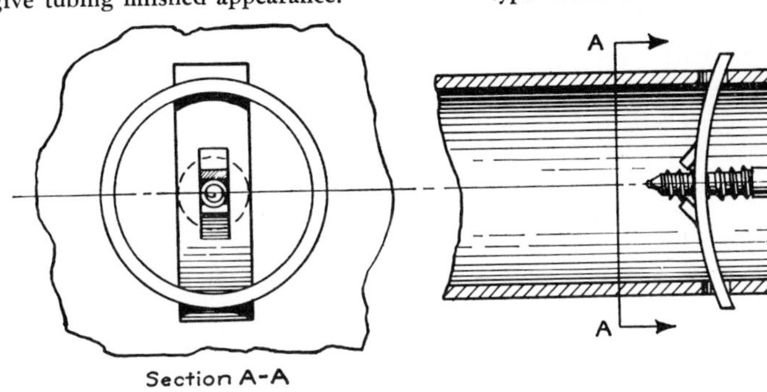


Fig. 14—Spring-steel nut is used for attaching objects to the ends of tubing. Nut requires two slots in the tubing and is held in position by screw.

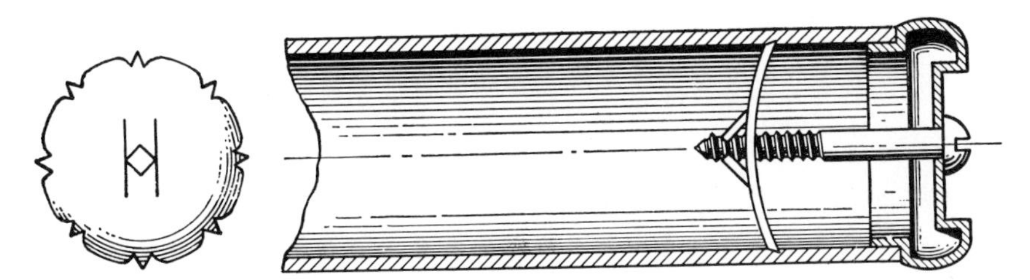

Fig. 15—Arched-nut attachment requires no machining. Nut is first rammed into tube; tightening of screw forces barbs into metal for positive anchorage.

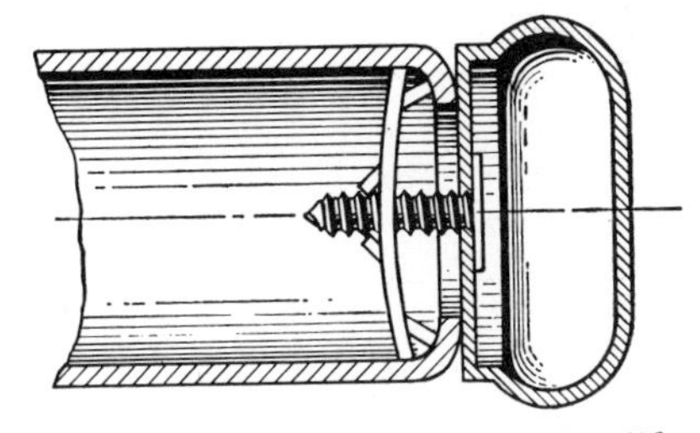

Fig. 16—Rolled-end of tubing simplifies attachment of end fitting.

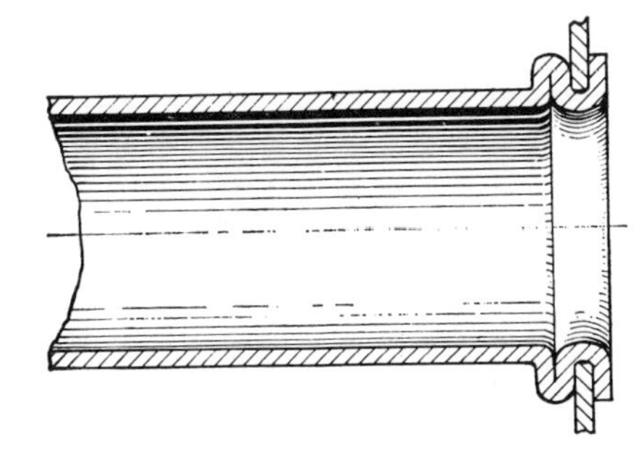

Fig. 17—Roll and press operation can be used to firmly hold tubing in plate.

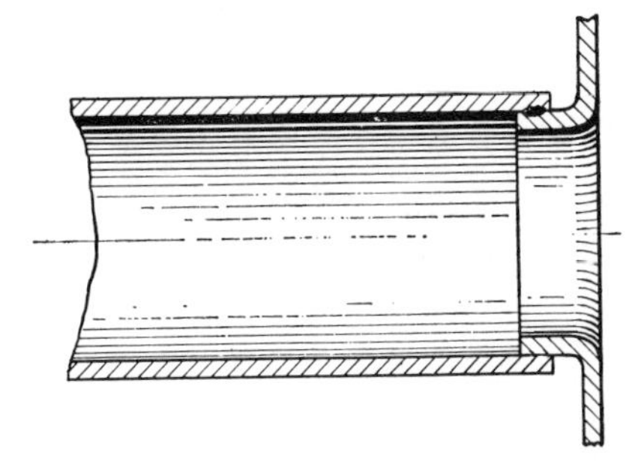

Fig. 18—Formed plate facilitates attachment of tubing by welding or brazing.

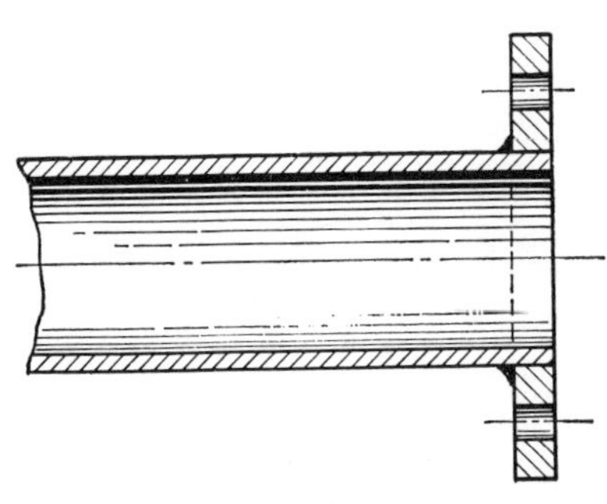

Fig. 19—Flange fitting for piping is made by welding flange to tubing.

# 5

# BEARINGS AND MOUNTS


Eleven Ways to Oil Lubricate Ball Bearings 108
Lubrication of Small Bearings 110
Nine Types of Ball Slides for Linear Motion 112
Unusual Applications of Miniature Bearings 114
Rolling Contact Bearing Mounting Units 116

# Eleven Ways to Oil Lubricate

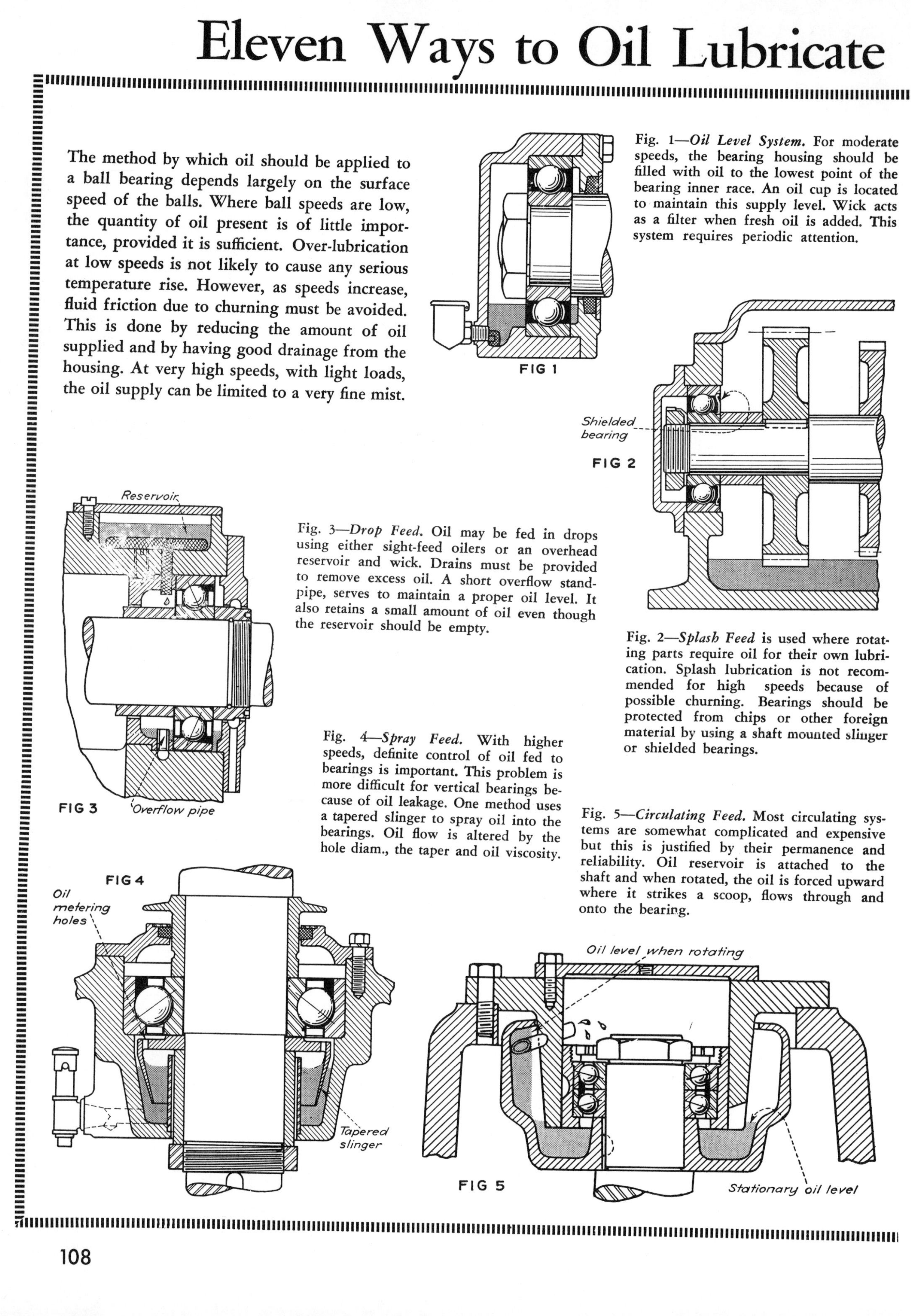

The method by which oil should be applied to a ball bearing depends largely on the surface speed of the balls. Where ball speeds are low, the quantity of oil present is of little importance, provided it is sufficient. Over-lubrication at low speeds is not likely to cause any serious temperature rise. However, as speeds increase, fluid friction due to churning must be avoided. This is done by reducing the amount of oil supplied and by having good drainage from the housing. At very high speeds, with light loads, the oil supply can be limited to a very fine mist.

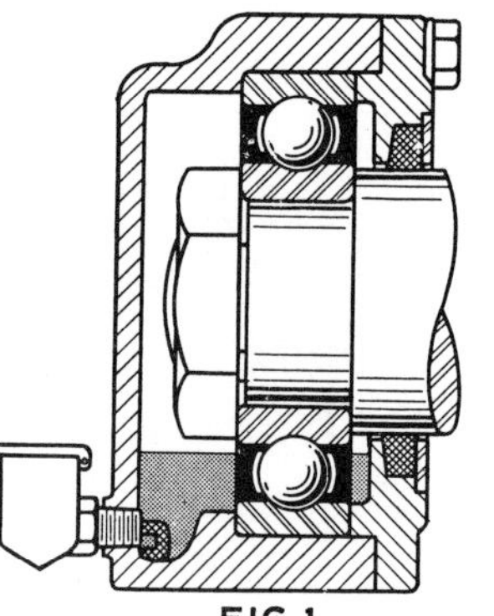


Fig. 1—*Oil Level System.* For moderate speeds, the bearing housing should be filled with oil to the lowest point of the bearing inner race. An oil cup is located to maintain this supply level. Wick acts as a filter when fresh oil is added. This system requires periodic attention.

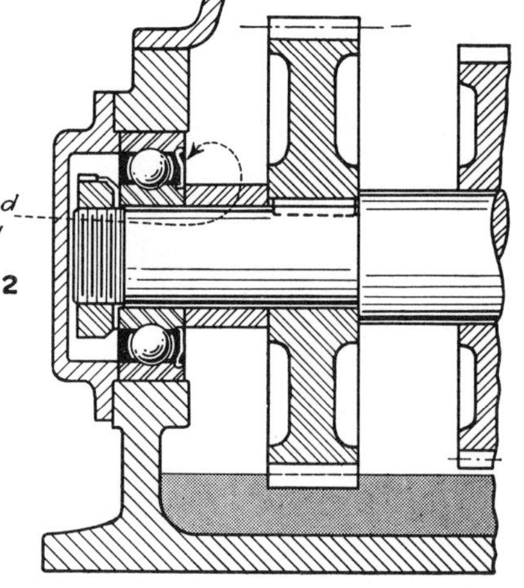


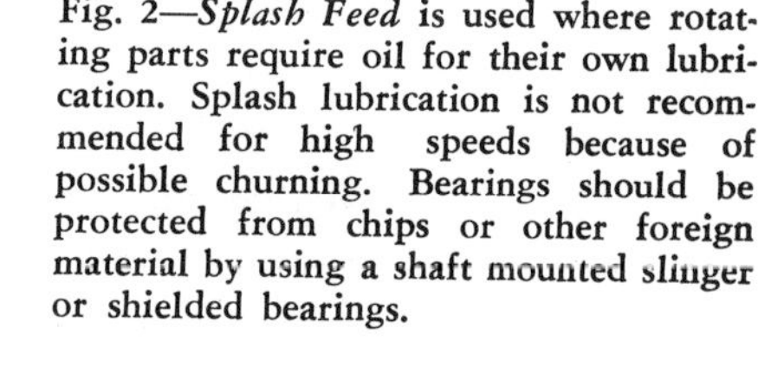

Fig. 2—*Splash Feed* is used where rotating parts require oil for their own lubrication. Splash lubrication is not recommended for high speeds because of possible churning. Bearings should be protected from chips or other foreign material by using a shaft mounted slinger or shielded bearings.

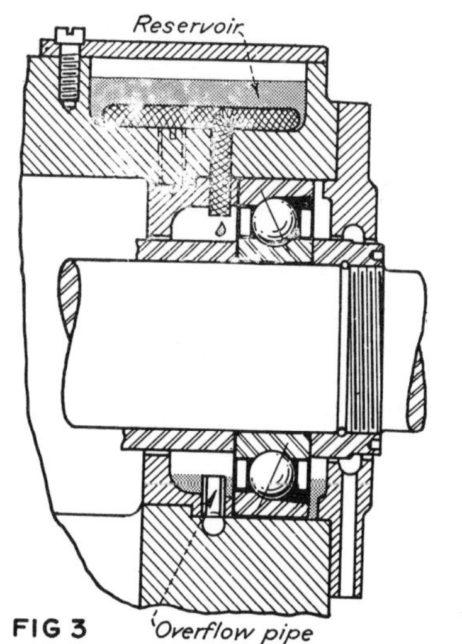


Fig. 3—*Drop Feed.* Oil may be fed in drops using either sight-feed oilers or an overhead reservoir and wick. Drains must be provided to remove excess oil. A short overflow standpipe, serves to maintain a proper oil level. It also retains a small amount of oil even though the reservoir should be empty.



Fig. 4—*Spray Feed.* With higher speeds, definite control of oil fed to bearings is important. This problem is more difficult for vertical bearings because of oil leakage. One method uses a tapered slinger to spray oil into the bearings. Oil flow is altered by the hole diam., the taper and oil viscosity.



Fig. 5—*Circulating Feed.* Most circulating systems are somewhat complicated and expensive but this is justified by their permanence and reliability. Oil reservoir is attached to the shaft and when rotated, the oil is forced upward where it strikes a scoop, flows through and onto the bearing.

# Ball Bearings

D. L. WILLIAMS
New Departure Div., General Motors Corporation

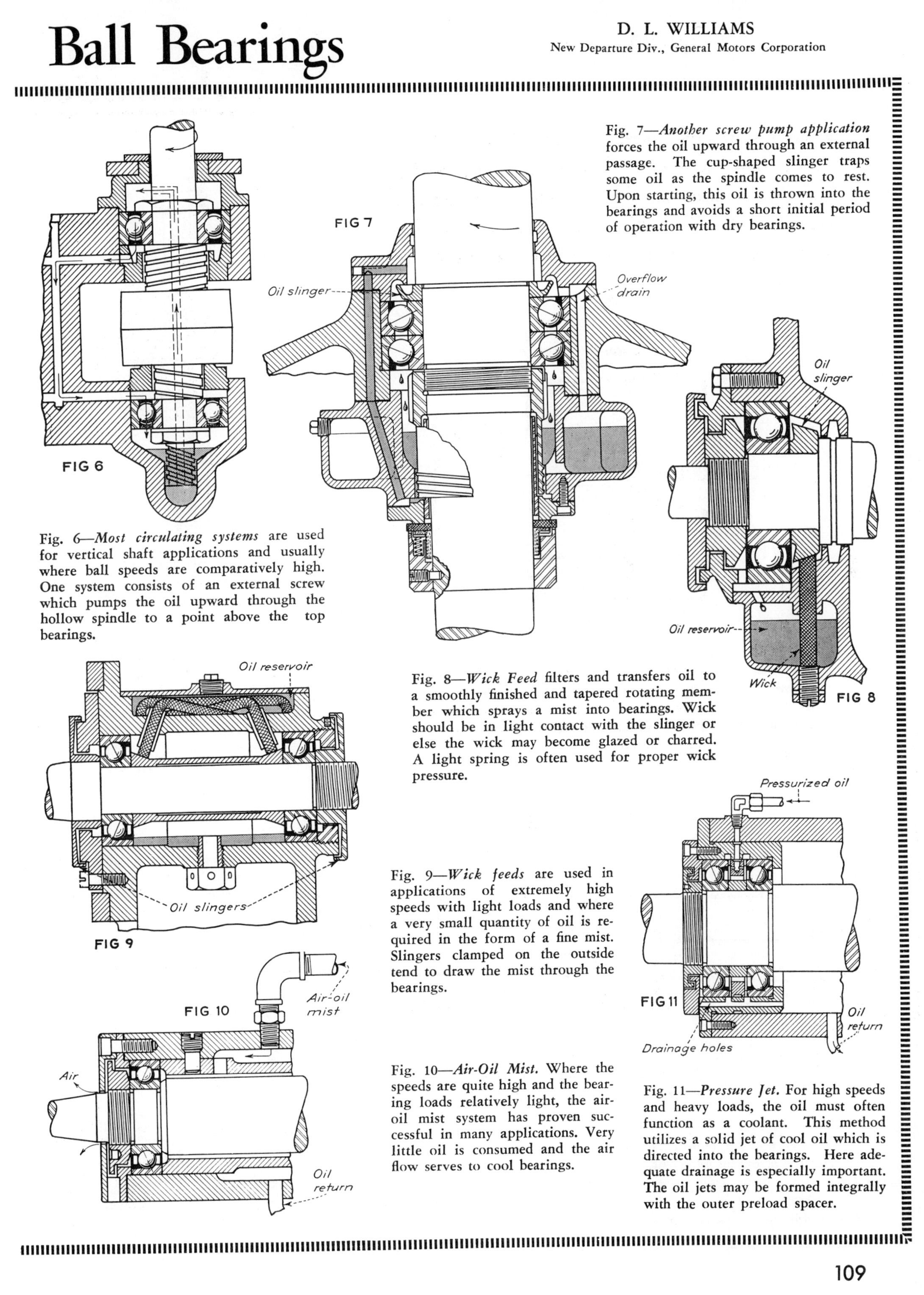


Fig. 7—*Another screw pump application* forces the oil upward through an external passage. The cup-shaped slinger traps some oil as the spindle comes to rest. Upon starting, this oil is thrown into the bearings and avoids a short initial period of operation with dry bearings.

Fig. 6—*Most circulating systems* are used for vertical shaft applications and usually where ball speeds are comparatively high. One system consists of an external screw which pumps the oil upward through the hollow spindle to a point above the top bearings.

Fig. 8—*Wick Feed* filters and transfers oil to a smoothly finished and tapered rotating member which sprays a mist into bearings. Wick should be in light contact with the slinger or else the wick may become glazed or charred. A light spring is often used for proper wick pressure.

Fig. 9—*Wick feeds* are used in applications of extremely high speeds with light loads and where a very small quantity of oil is required in the form of a fine mist. Slingers clamped on the outside tend to draw the mist through the bearings.

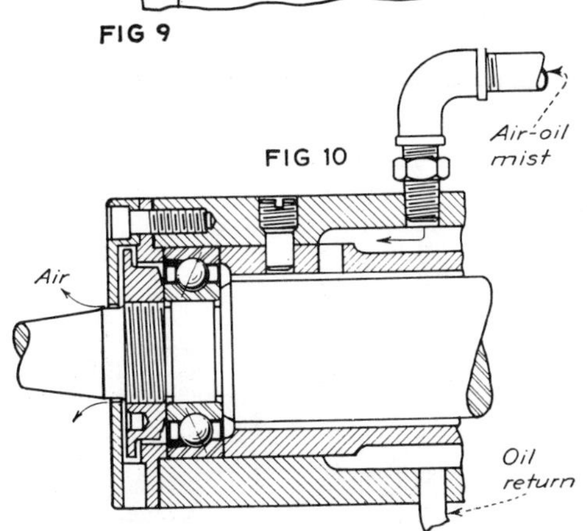


Fig. 10—*Air-Oil Mist.* Where the speeds are quite high and the bearing loads relatively light, the air-oil mist system has proven successful in many applications. Very little oil is consumed and the air flow serves to cool bearings.

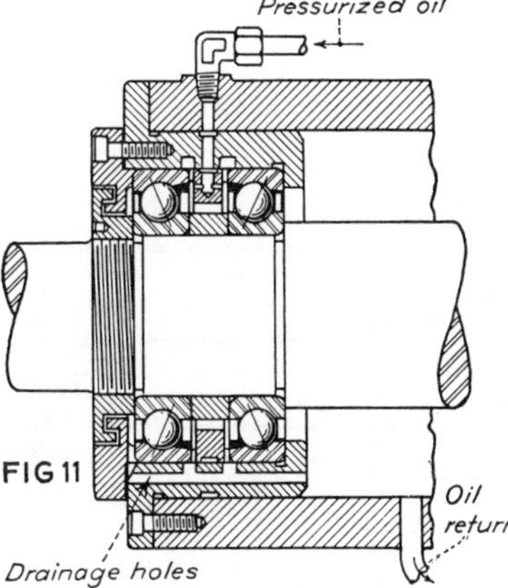


Fig. 11—*Pressure Jet.* For high speeds and heavy loads, the oil must often function as a coolant. This method utilizes a solid jet of cool oil which is directed into the bearings. Here adequate drainage is especially important. The oil jets may be formed integrally with the outer preload spacer.

# Lubrication

Examples of good modern practice in the lubrication of small bearings. These designs have the feature that no attention to lubrication is required over long periods. Several of them show applications of porous bronze bushings for long-life lubrication

▼

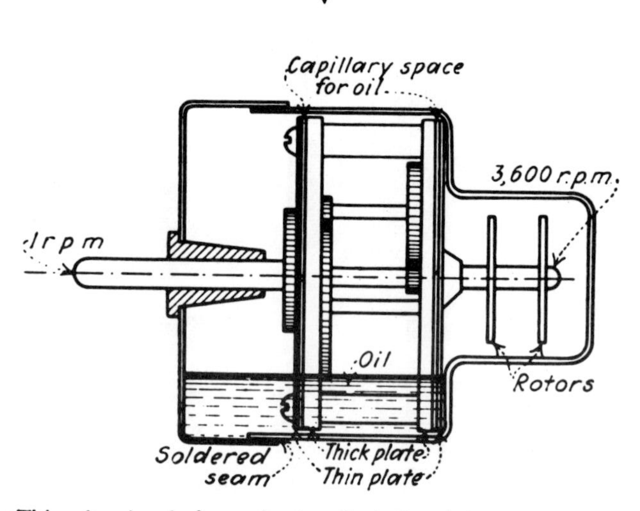


This electric clock mechanism is inclosed in an oil-tight case with only a single opening for the 1-r.p.m. shaft. The bushing for this shaft projects sufficiently far into the case so that the oil level is below its inner end regardless of how the case is tilted. Oil feeds by capillary action between the plates as indicated and works out along the shaft.

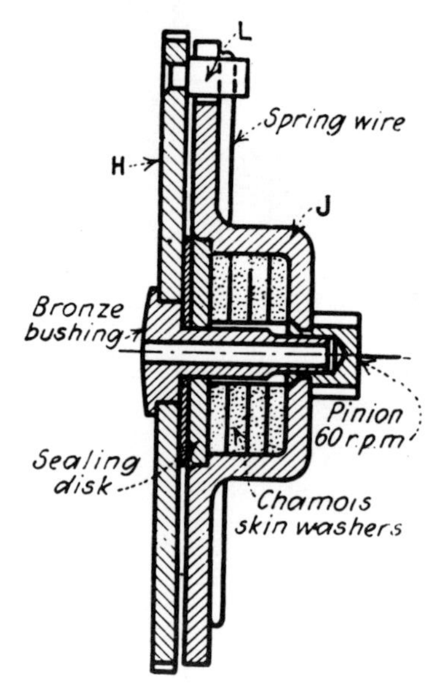


In this design, the main plate *H* of the rotor tends, because of magnetic attraction, to stay in the same place. The central bronze bushing is a press fit in the main rotor plate and turns with it. On the outside of the bushing is a floating flywheel assembly made of the cupped stamping *J*, the pinion, and the sealing disk, the latter inclosing a series of chamois-skin washers that have been soaked in oil. Oil seeps slowly through holes in the bushing, and thus to the bearing surfaces.

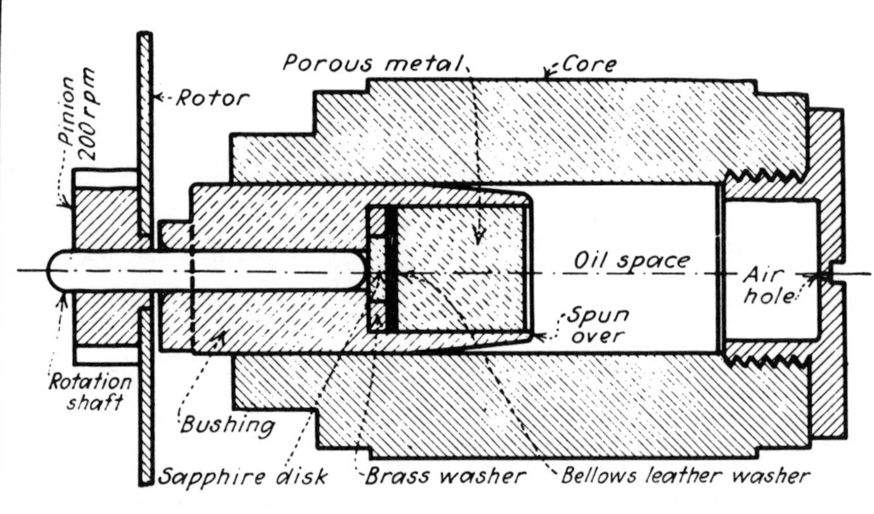


The rotor shaft turns in a phosphor bronze bushing pressed into the stainless steel core of the motor. The oil seeps in minute quantities through the cylindrical porous plug, through the bellows leather washer, around the floating thrust jewel or sapphire disk, and finally into the bore of the bushing in which the rotor shaft turns. The bearing clearance is held within the limits of 0.0008 to 0.0003 in.

Below, to the left is an oil resrevoir cast integrally; to the right, a drawn metal shell is screwed into the hub for an oil reservoir. In both designs, lubricant is fed through the porous metal bushing. Increased bearing temperature brings additional oil to the bearing surface because of expansion and increased oil fluidity.

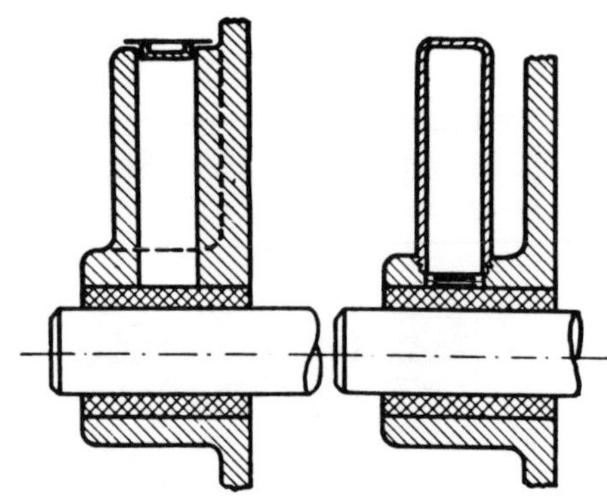

# of Small Bearings

HERBERT CHASE

Below are shown three different designs of bearings for extremely light duty, as in clocks and meters. To the left is a self-aligning bearing having a porous bushing seated in a two-piece cadmium-plated steel shell which also holds a felt washer saturated with sufficient oil to last a year or more in ordinary service. In the center is shown a design wherein a pressed steel frame forms a spherical seat for the porous bearing. A light stamping incloses an oil-soaked felt washer that contains sufficient lubrication for the life of the motor. To the right is shown the bearing for an electric clock, the light cupped stamping that contains the oil-soaked felt being pressed over the bearing flange. In both of these designs, the lubricant in the felt is sufficient for the life of the motor.

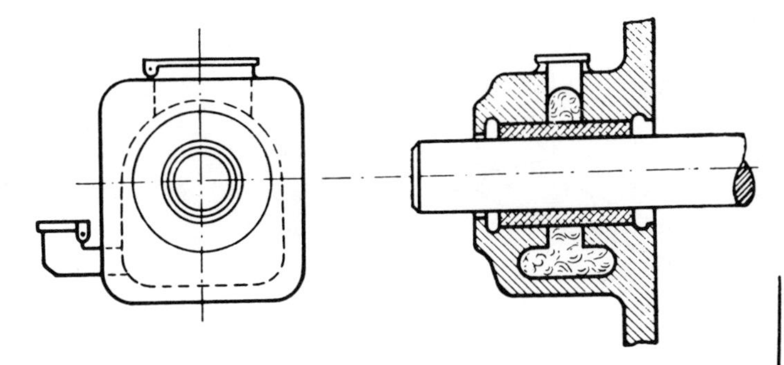

In this design, an annular groove for oil or light grease is cored in the housing of the bearing. As indicated, oil-soaked wool waste can be packed in the cored recess. The concentric grooves at each end of the porous bushing are for the purpose of catching any end leakage. A refinement would be the addition of drilled holes to lead the oil back to the cellar. The additional oiler at the bottom is optional.

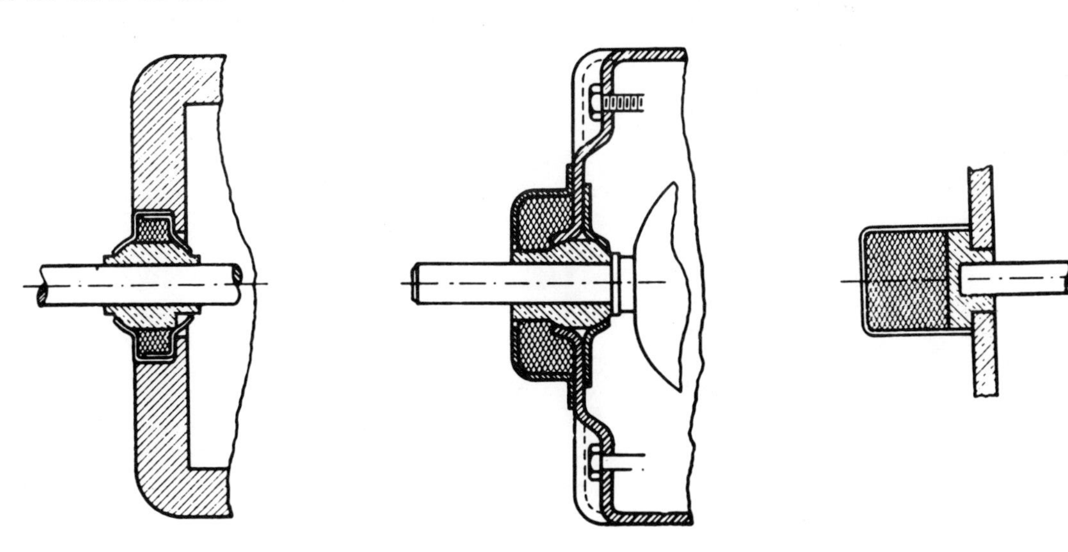

A cup-shaped stamping (below) holds the oil-soaked felt in the housing, and also acts as a dust shield. Lubrication of the bearing is through the porous bushing which is die-pressed from powdered metal. The lubricant is always fed to the outer wall of the bushing, which acts as a wick. To the right is shown an optional construction wherein an annular groove in the bearing housing forms the oil reservoir.

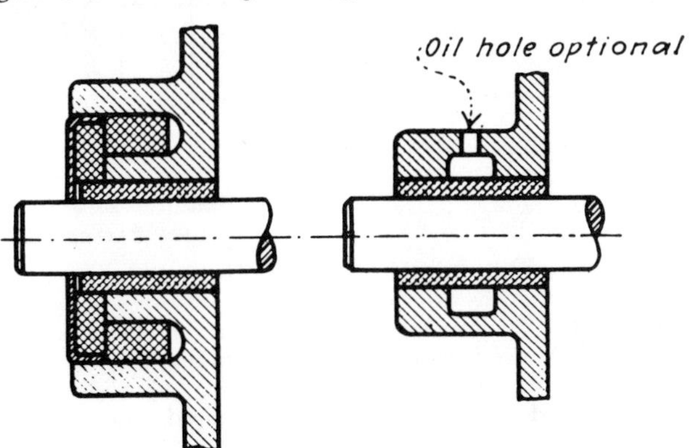


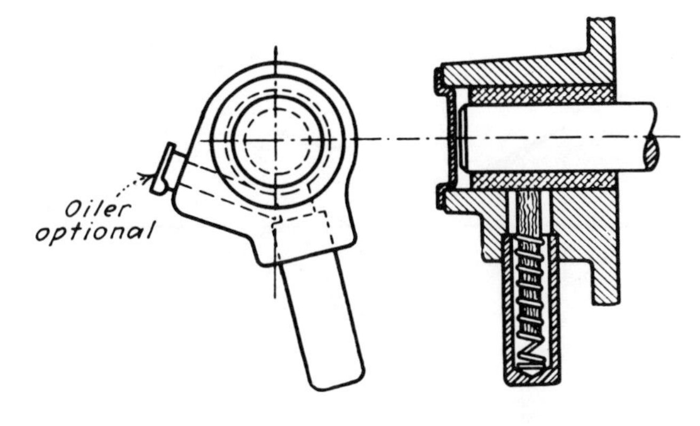


In this wick-feed arrangement for oil or grease, capillary action feeds the lubricant to the surface of the porous bushing. For light service, sufficient lubricant is contained in the cup to last over a long period of time. The addition of the oiler shown in the end view is for the purpose of convenience so that the machine user will be more apt to give some attention to lubrication. Note the metal dust shield.

# Nine Types of Ball Slides

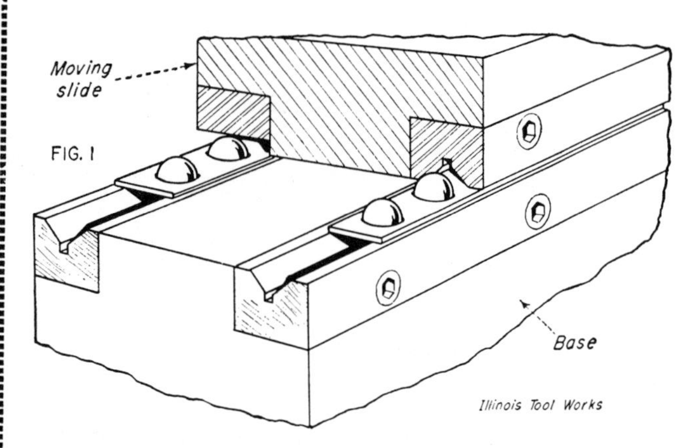


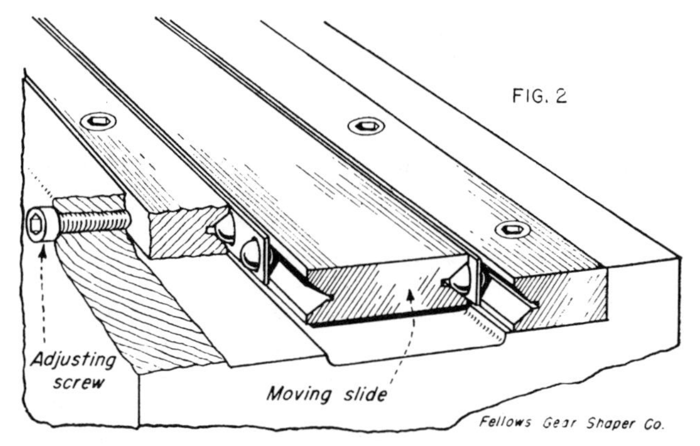


**1** V grooves and flat surface make simple horizontal ball slide for reciprocating motion where no side forces are present and a heavy slide is required to keep balls in continuous contact. Ball cage insures proper spacing of balls; contacting surfaces are hardened and lapped.

**2** Double V grooves are necessary where slide is in vertical position or when transverse loads are present. Screw adjustment or spring force is required to minimize looseness in the slide. Metal-to-metal contact between the balls and grooves insure accurate motion.

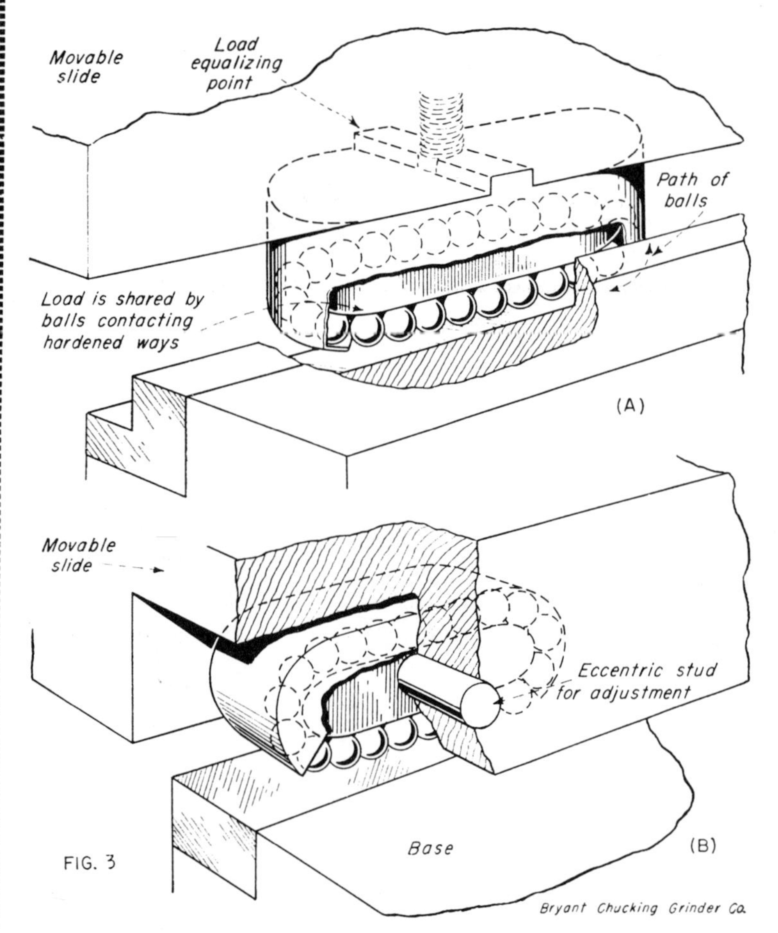


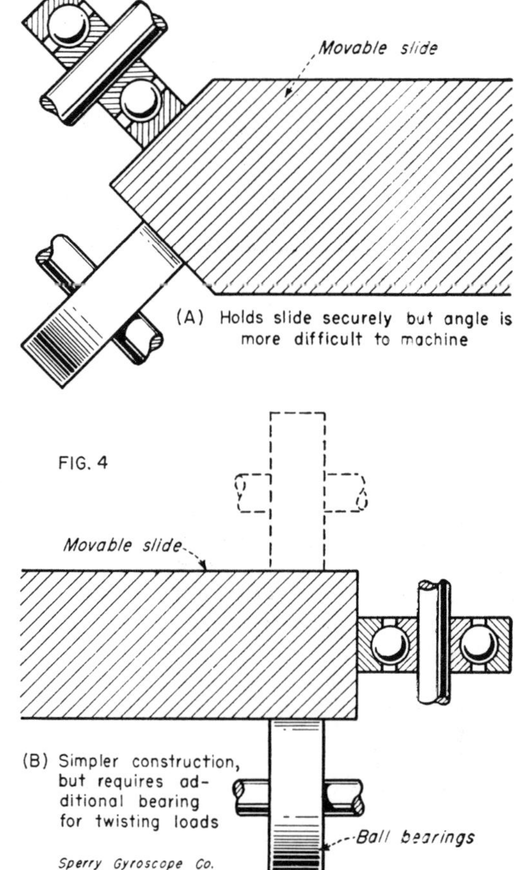


**3** Ball cartridge has advantage of unlimited travel since balls are free to recirculate. Cartridges are best suited for vertical loads. (A) Where lateral restraint is also required, this type is used with a side preload. (B) For flat surfaces cartridge is easily adjusted.

**4** Commercial ball bearings can be used to make a reciprocating slide. Adjustments are necessary to prevent looseness of the slide. (A) Slide with beveled ends, (B) Rectangular-shaped slide.

# For Linear Motion

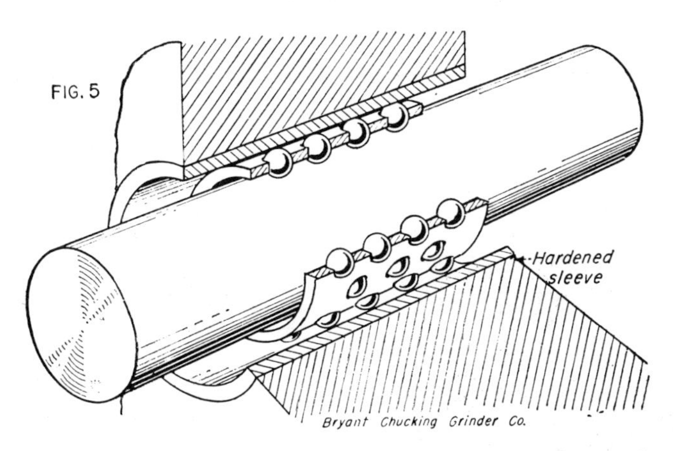


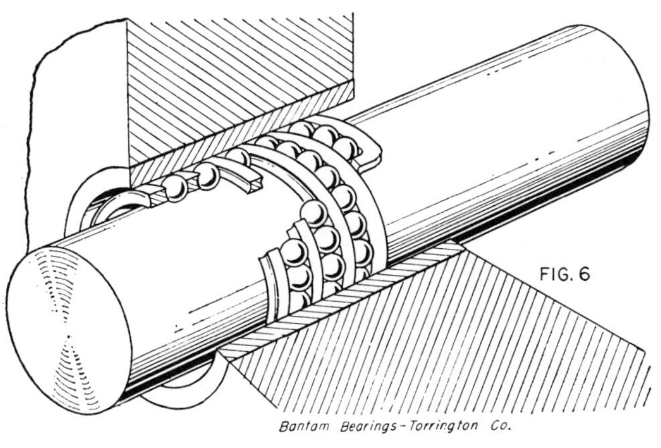


**5** Sleeve bearing consisting of a hardened sleeve, balls and retainer, can be used for reciprocating as well as oscillating motion. Travel is limited similar to that of Fig. 6. This type can withstand transverse loads in any direction.

**6** Ball reciprocating bearing is designed for rotating, reciprocating or oscillating motion. Formed-wire retainer holds balls in a helical path. Stroke is about equal to twice the difference between outer sleeve and retainer length.

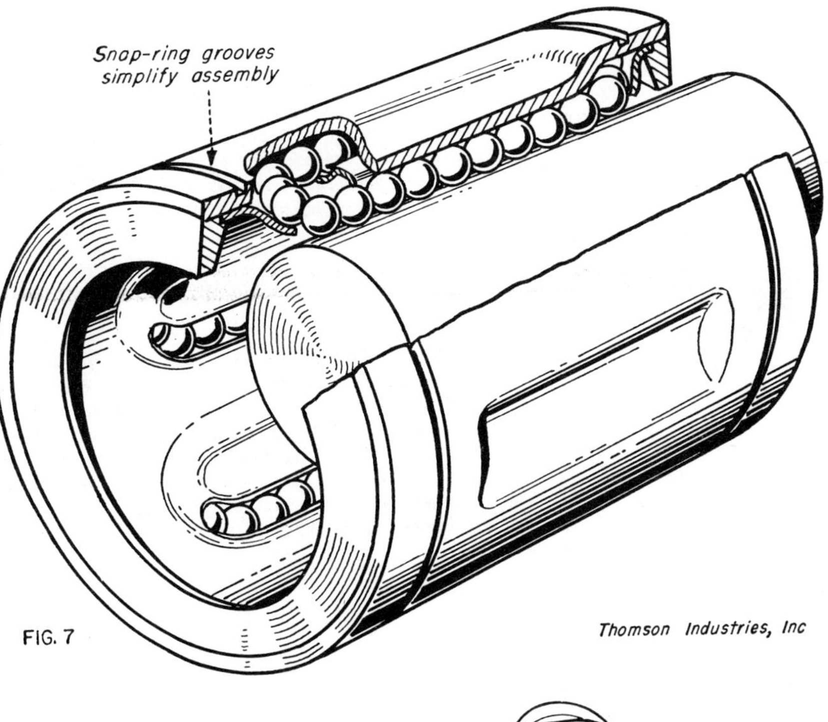


**7** Ball bushing with several recirculating systems of balls permit unlimited linear travel. Very compact, this bushing simply requires a bored hole for installation. For maximum load capacity a hardened shaft should be used.

**8** Cylindrical shafts can be held by commercial ball bearings which are assembled to make a guide. These bearings must be held tightly against shaft to prevent looseness.

**9** Curvilinear motion in a plane is possible with this device when the radius of curvature is large. However, uniform spacing between grooves is important. Circular - sectioned grooves decrease contact stresses.

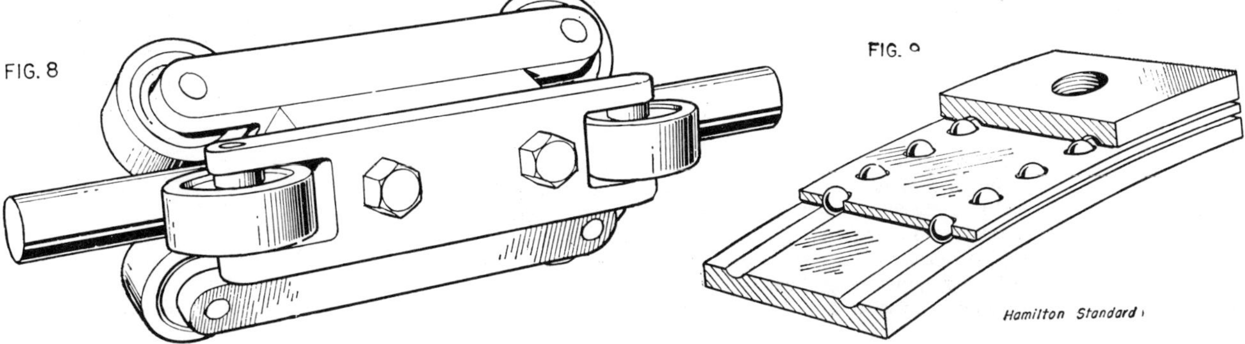

# Unusual applications of

Fig. 1—BALL-BEARING SLIDES. Six miniature bearings accurately support a potentiometer shaft to give low-friction straight line motion. In each end housing, three bearings are located 120 deg radially apart to assure alignment and freedom from binding of the potentiometer shaft.

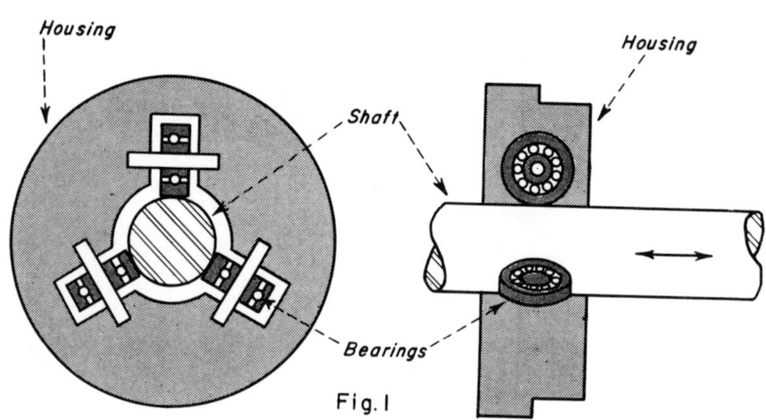


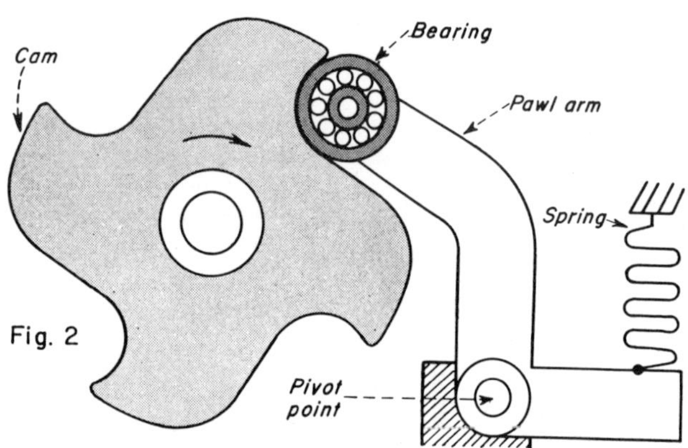


Fig. 2—CAM-FOLLOWER ROLLER. Index pawl on a frequency selector switch uses bearing for a roller. Bearing is spring loaded against cam and extends life of unit by reducing cam wear. This also retains original accuracy in stroke of swing of the pawl arm.

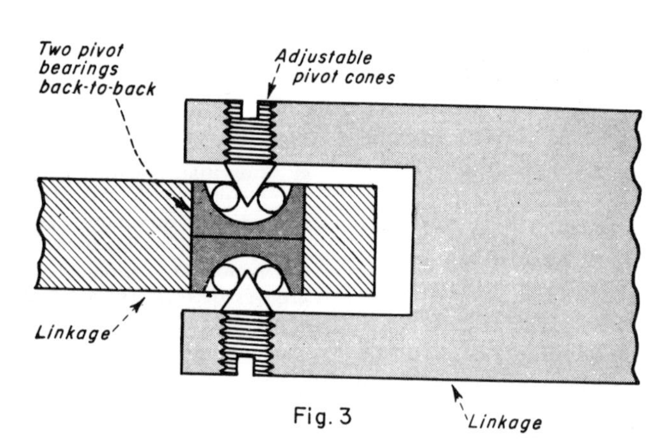


Fig. 3—SEAT FOR PIVOTS. Pivot-type bearings reduce friction in linkages especially when manually operated such as in pantographing mechanisms. Minimum backlash and maximum accuracy are obtained by adjusting the threaded pivot cones. Mechanism is used to support diamond stylus that scribes sight lines on the lenses of gunnery telescopes.

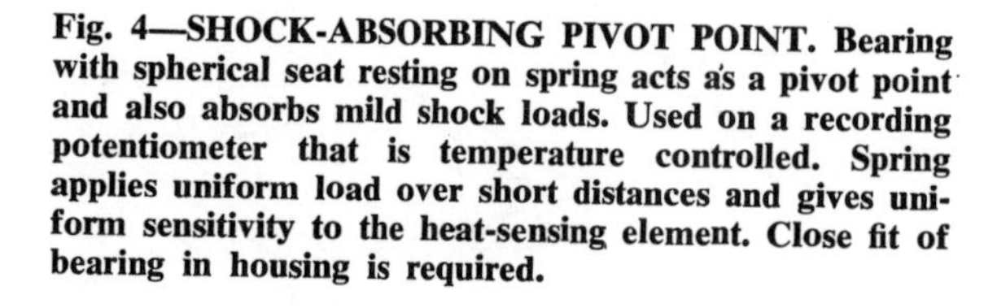

Fig. 4—SHOCK-ABSORBING PIVOT POINT. Bearing with spherical seat resting on spring acts as a pivot point and also absorbs mild shock loads. Used on a recording potentiometer that is temperature controlled. Spring applies uniform load over short distances and gives uniform sensitivity to the heat-sensing element. Close fit of bearing in housing is required.

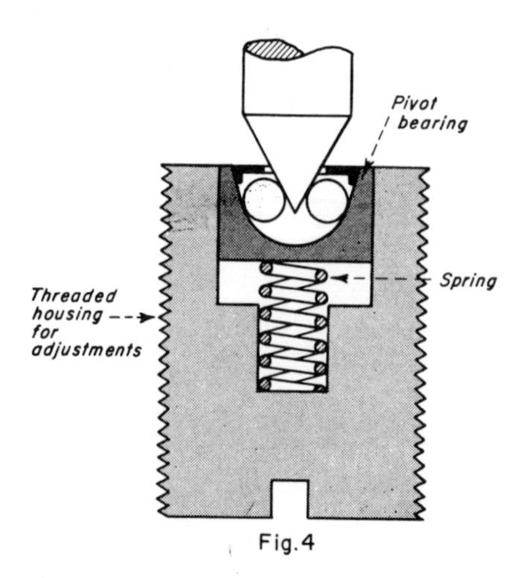

# miniature bearings

R. H. CARTER,
chief engineer
Miniature Precision Bearings, Inc.

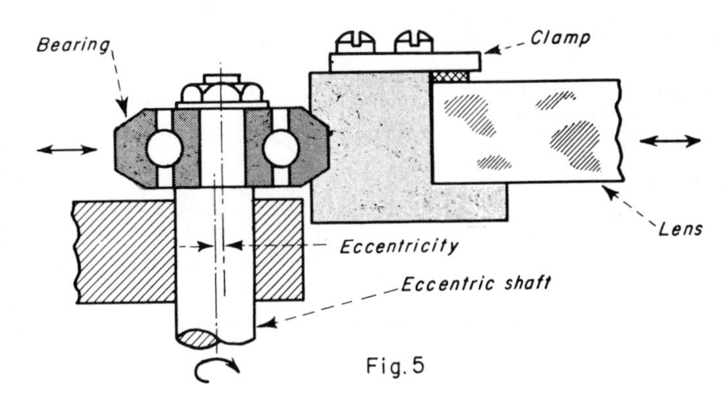


**Fig. 5—PRECISE RADIAL ADJUSTMENTS obtained by rotating the eccentric shaft thus shifting location of bearing. Bearing has special-contoured outer race with standard inner race. Application is to adjust a lens with grids for an aerial survey camera.**

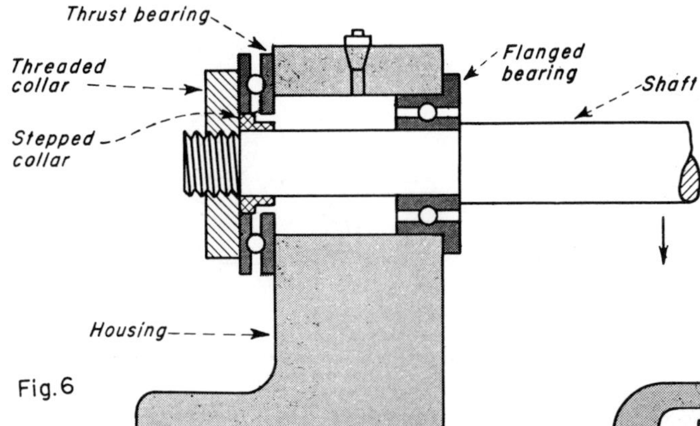


**Fig. 6—SUPPORT FOR CANTILEVERED SHAFT obtained with combination of thrust and flanged bearings. Stepped collar provides seat for thrust bearing on the shaft but does not interfere with stationary race of thrust bearing when shaft is rotating.**

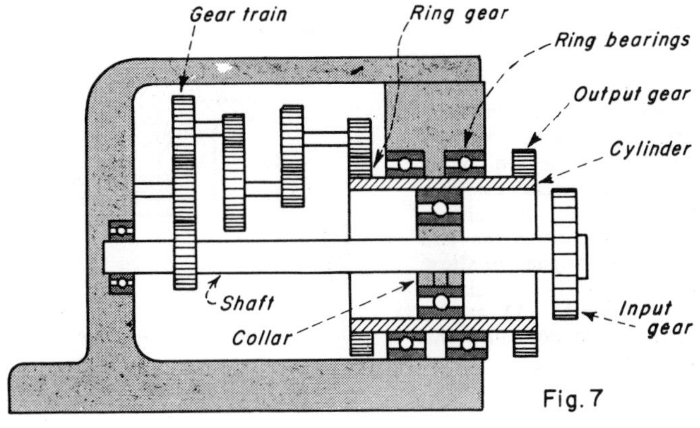


**Fig. 7—GEAR-REDUCTION UNIT. Space requirements reduced by having both input and output shafts at same end of unit. Output shaft is a cylinder with ring gears at each end. Cylinder rides in miniature ring bearings that have relative large inside diameters in comparison to the outside diameter.**

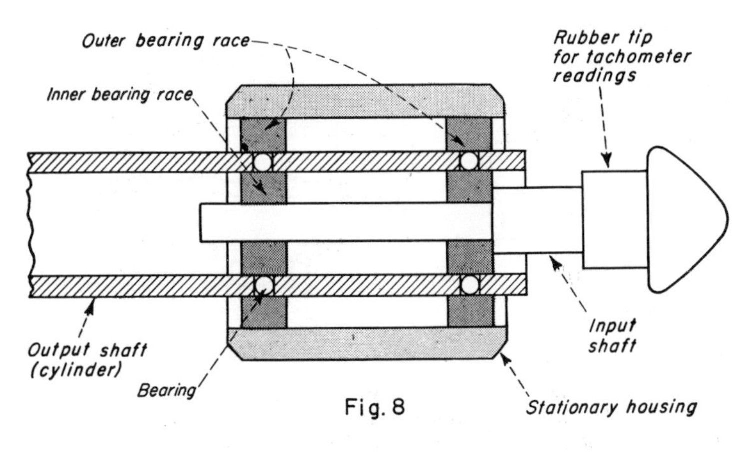


**Fig. 8—BEARINGS USED AS GEARS. Manually operated tachometer must take readings up to 6000 rpm. A 10-to-1 speed reduction was obtained by having two bearings function both as bearings and as a planetary gear system. Input shaft rotates the inner race of the inner bearings, causing the output shaft to rotate at the peripheral speed of the balls. Bearings are preloaded to prevent slippage between races and balls. Outer housing is held stationary. Pitch diameters and ball sizes must be carefully calculated to get correct speed reduction.**

# Rolling Contact Bearing

FIG. 1—Pillow blocks are for supporting shafts running parallel to the surface on which they are mounted. Provision for lubrication and sealing are incorporated in the pillow block unit. Assembly and disassembly are easily accomplished. For extremely precise installations, mounting units are inadvisable.

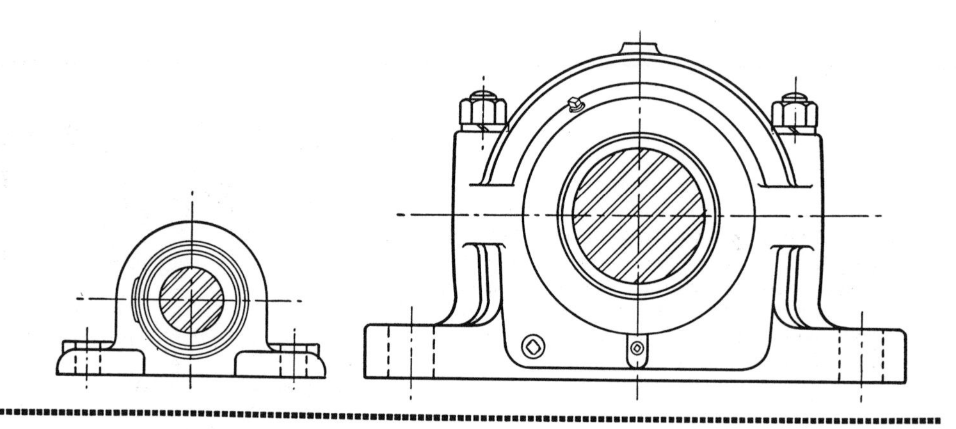

FIG. 2—Pillow blocks can be designed to prevent the transmission of noise to the support. One design (A) consists of a bearing mounted in rubber. The rubber in turn is firmly supported by a steel casing. Another design (B) is made of synthetic rubber. Where extra rigidity is required the synthetic rubber mount can be reinforced by a steel strap bolted around it.

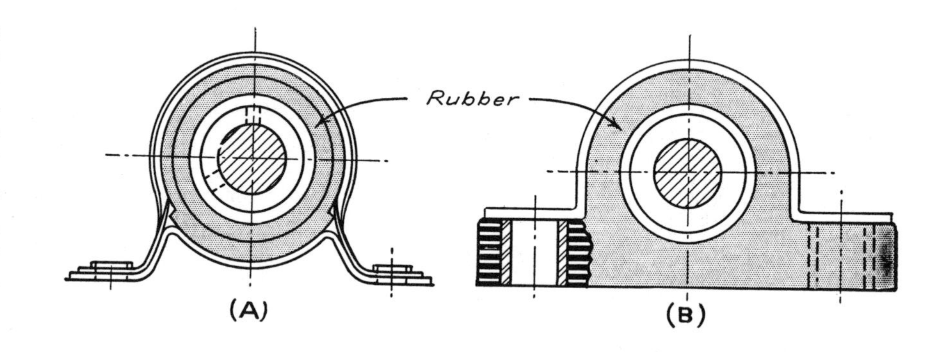


FIG. 3—Changes in the temperature are accompanied by changes in the length of a shaft. To compensate for this change in length, the pillow block (B) supporting one end of the shaft is designed to allow the bearing to shift its position. The pillow block (A) at the other end should not allow for longitudinal motion.

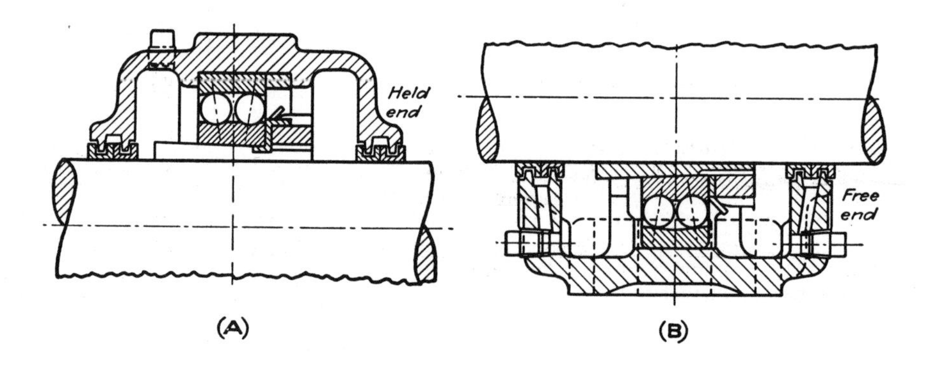


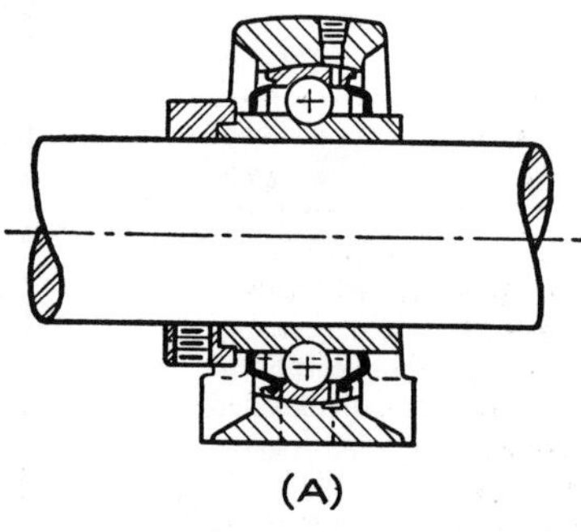


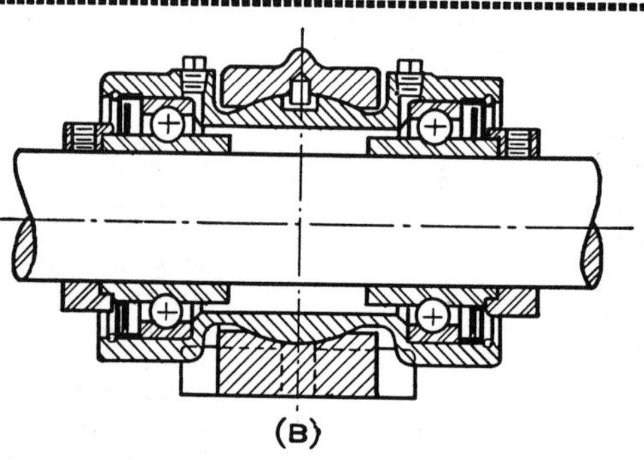


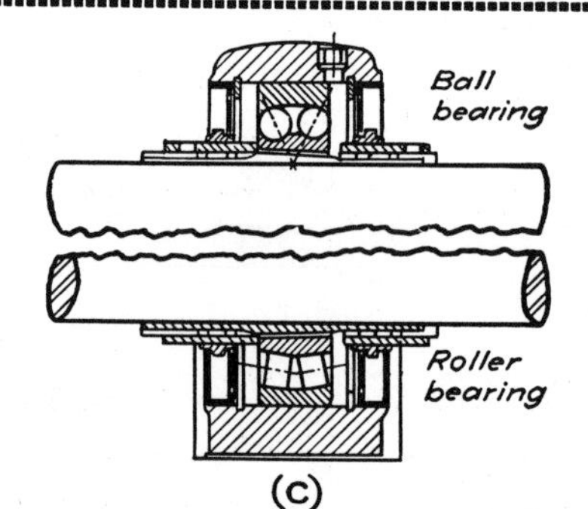


Fig. 4—Compensation for misalignment can be incorporated into pillow blocks in various ways. One design (A) uses a spherical outer surface of the outer race. Design (B) uses a two-part housing. The spherical joint compensates for misalignment. Another design (C) uses a spherical inner surface of the outer race.

# Mounting Units

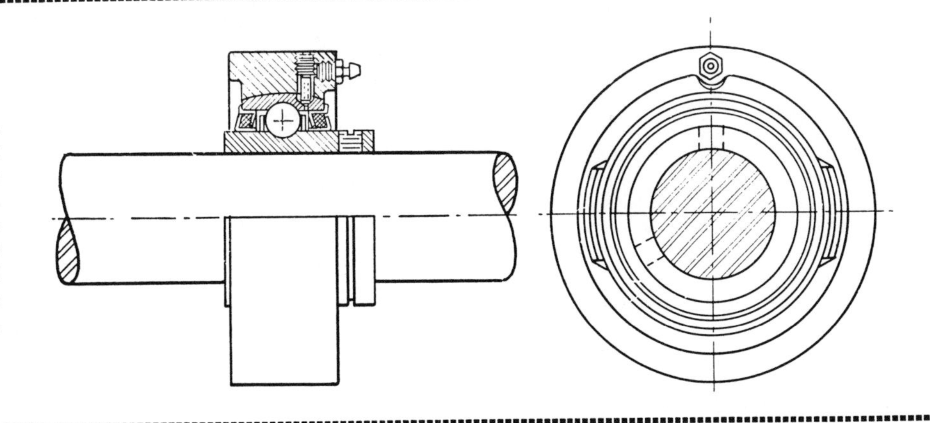

Fig. 5—The cylindrical cartridge is readily adaptable to various types of machinery. It is fitted as a unit into a straight bored housing with a push fit. A shoulder in the housing is desirable but not essential. The advantages of a predesigned and preassembled unit found in pillow blocks also apply here.

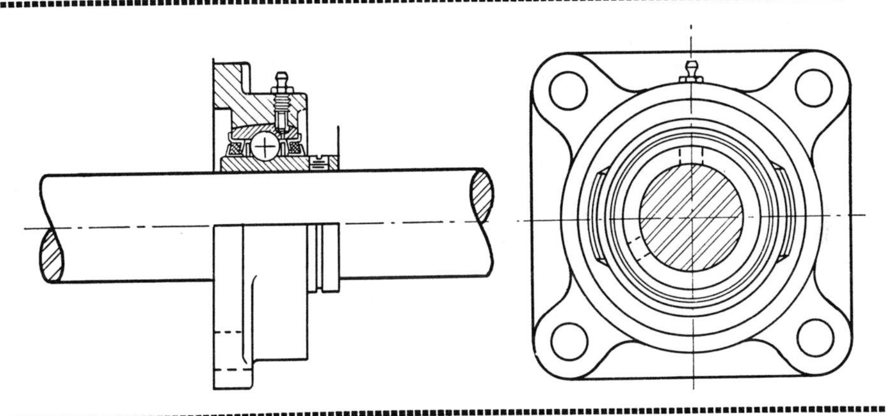

FIG. 6—The flange mounting unit is normally used when the machine frame is perpendicular to the shaft. The flange mounting unit can be assembled without performing the special boring operations required in the case of the cartridge. The unit is simply bolted into the housing when it is being installed.

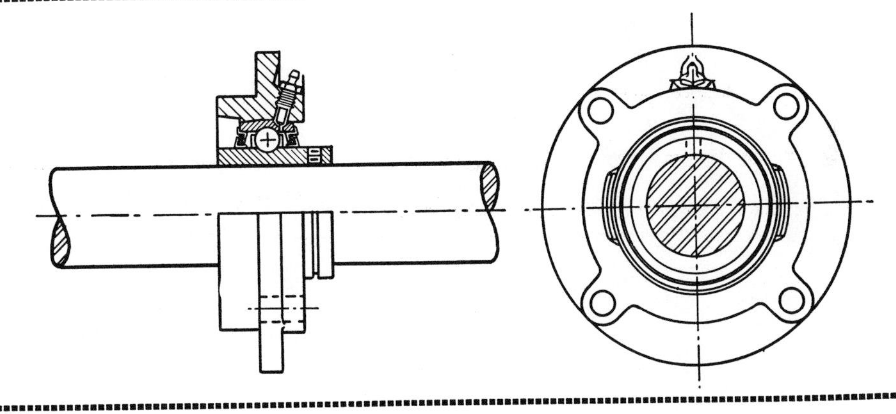

FIG. 7—The flange cartridge unit projects into the housing and is bolted in place through the flange. The projection into the housing absorbs a large part of the bearing loads. A further use of the cylindrical surface is the location of the mounting unit relative to the housing.

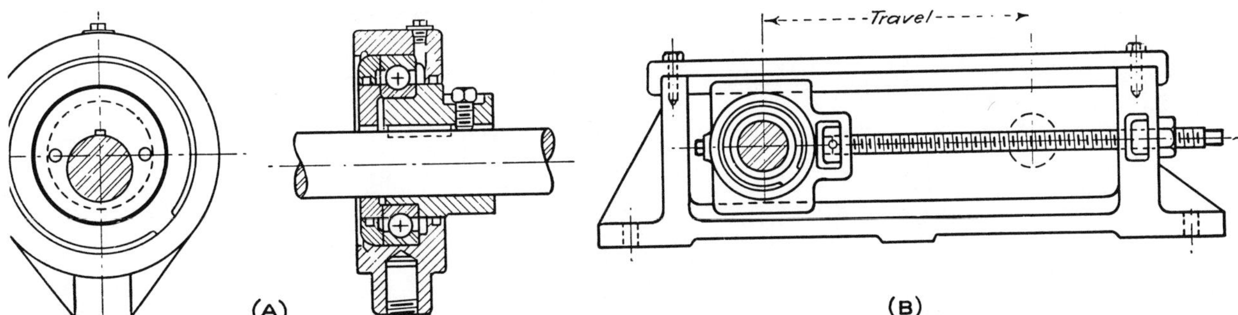


FIG. 8—Among specialized types of mounting units are (A) Eccentrics used particularly for cottonseed oil machinery and mechanical shakers and (B) Take-up units which make possible an adjustment in the position of the shaft for conveyor units. Many other types of special rolling contact bearing mounting units are made.

# 6

# CONTROL AND MEASUREMENT DEVICES


| | |
|---|---|
| Methods of Electric Control | 120 |
| Limit Switching Mechanisms – Methods of Mechanical Actuation | 122 |
| Liquid Level Indicators and Controllers | 124 |
| Mechanisms of Indicators and Controllers | 126 |
| Photoelectric Controls | 128 |
| Automatic Stop Mechanisms Protect Machines and Work | 130 |
| Electric Automatic Stop Mechanisms | 132 |
| Mechanical Automatic Stop Mechanisms | 134 |
| How to Obtain Constant Speed Motion Below 10 RPM | 136 |
| 12 Ways of Measuring Speed | 140 |
| Temperature Regulators | 142 |
| Thermocouple Details for Temperature Measurement | 144 |
| Types of Automatic Timers | 146 |

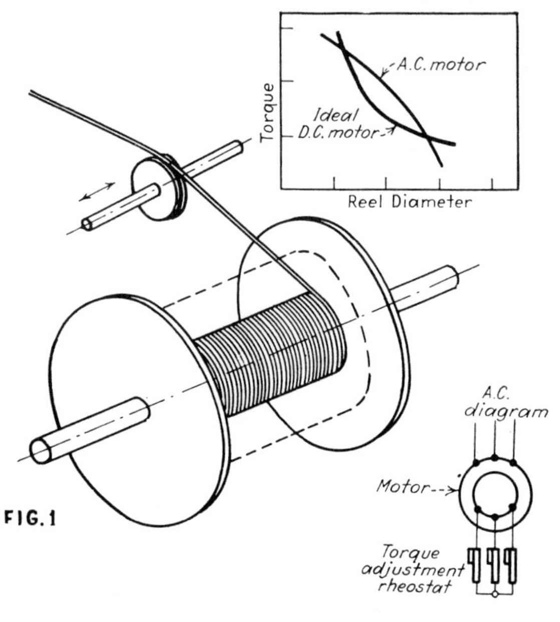


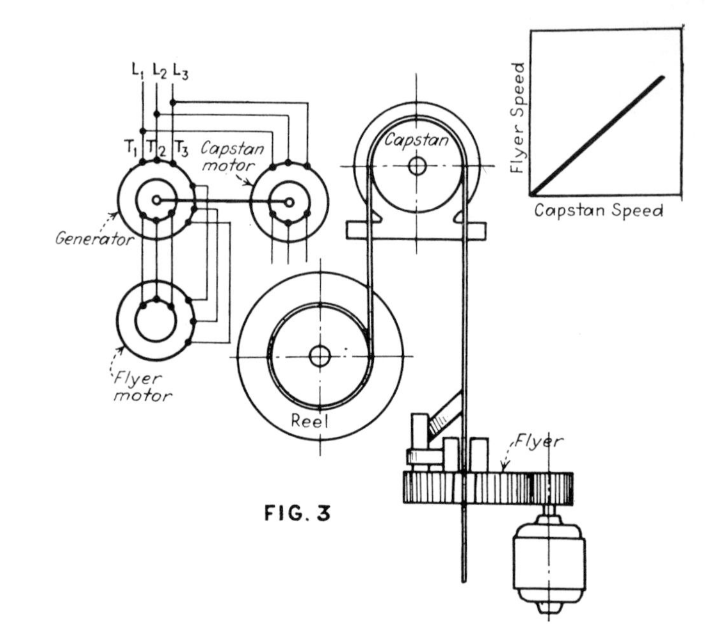


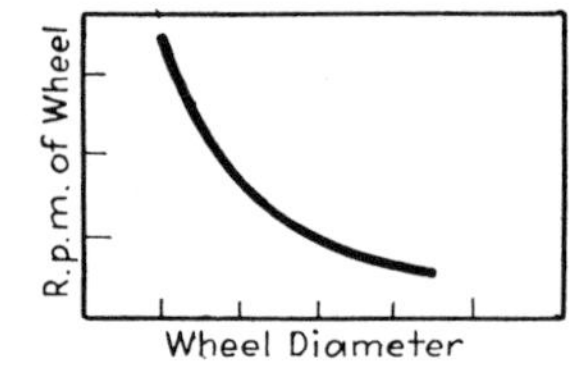


# Methods of Electric Control

R. S. ELBERTY, JR.
*Machinery Electrification Division*
*Westinghouse Electric & Manufacturing Company*

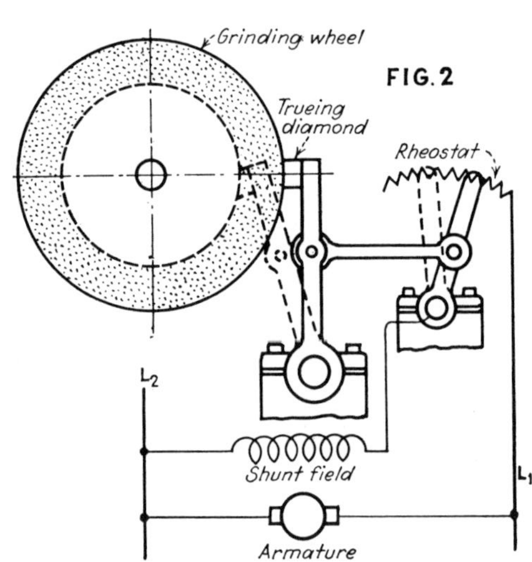


How electrical controls can be applied to machinery for regulating peripheral speeds, rates of feed, controlling speed ratios, maintaining constant loads and for machine reversal

**Fig. 1**—Constant tension with constant peripheral speed is required in this wire reel application. The application can be used on wire drawing machines, insulating machines or any other reeling operation. As the reeling diameter increases, the reel speed decreases, and at the same time the reeling torque is increased. The required constant hp. characteristic is obtained accurately with a d.c. motor and regulator type of control on shunt field. An a.c. wound rotor motor with secondary resistance control approximates ideal conditions

**Fig. 2**—For automatically limiting the peripheral speed of a grinding wheel, the truing diamond is mechanically interlocked with the wheel motor field rheostat. The wheel r.p.m. is increased as the wheel diameter decreases.

**Fig. 3**—A wire insulating machine requires a constant speed ratio between capstan motor and flyer for starting and running. The capstan motor drives a frequency changer or transmitter electrically connected to the synchronous motor of the flyer. The speed ratio between flyer and capstan is constant at all times.

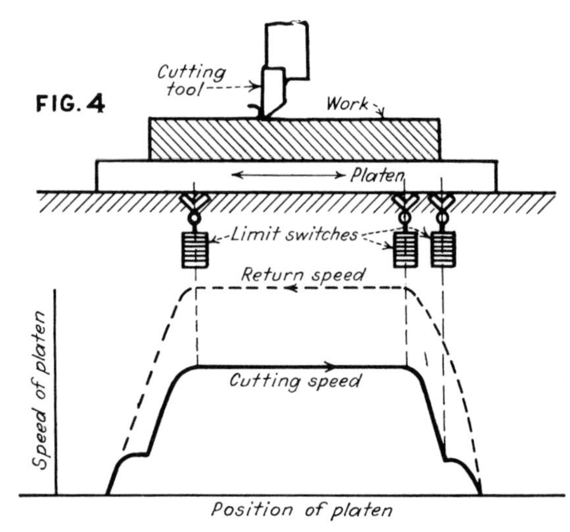


**Fig. 4**—For high-speed cutting on a metal planer, the tool enters the work at a slow speed to prevent tool breakage, cutting speed is then increased and near the end of the cut the cutting speed is reduced to prevent breaking out of edge of work. This speed control is accomplished by limit switches which put full field on the motor before the tool leaves work. After the return stroke de-

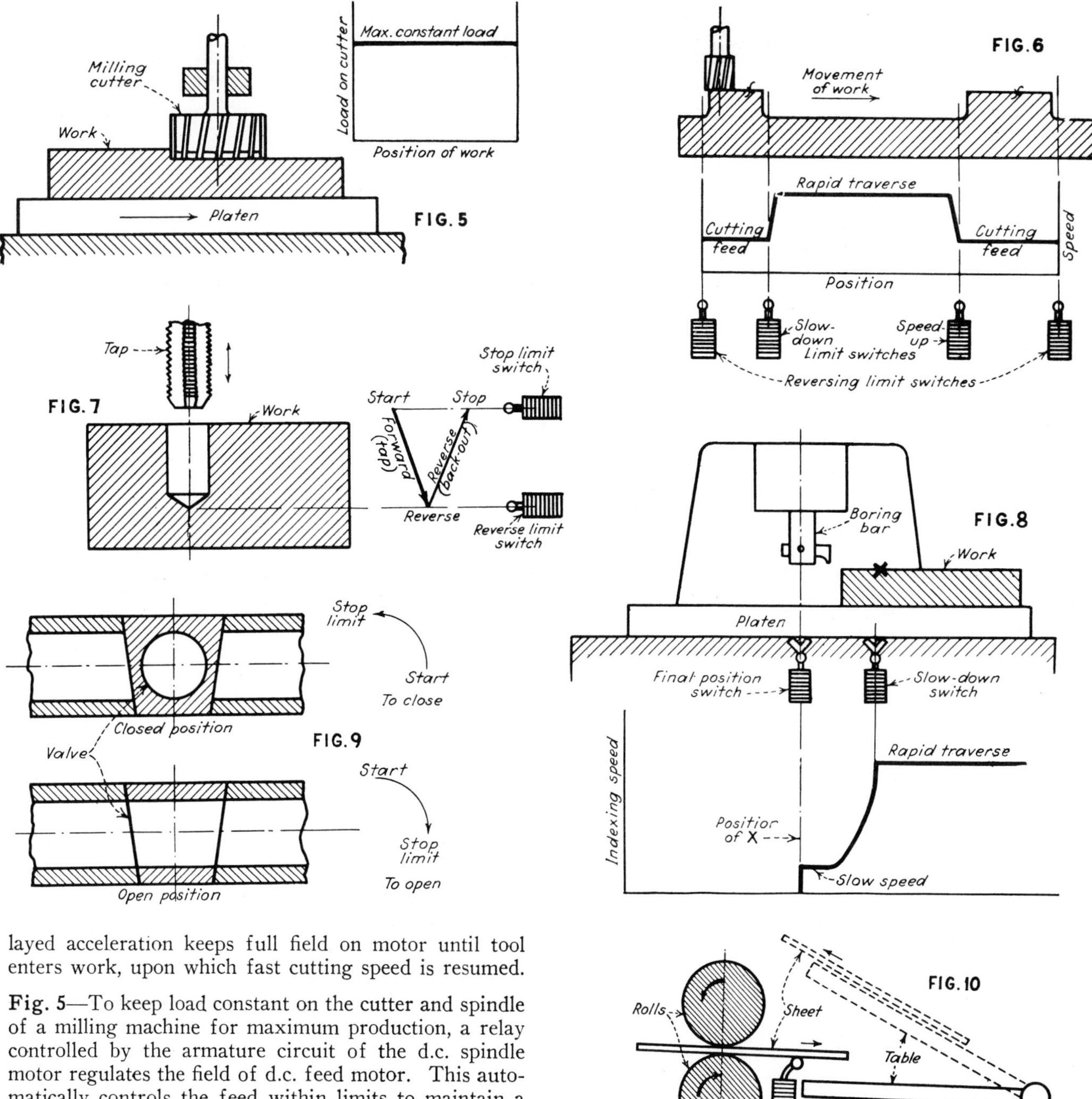


layed acceleration keeps full field on motor until tool enters work, upon which fast cutting speed is resumed.

**Fig. 5**—To keep load constant on the cutter and spindle of a milling machine for maximum production, a relay controlled by the armature circuit of the d.c. spindle motor regulates the field of d.c. feed motor. This automatically controls the feed within limits to maintain a maximum constant load on the spindle motor.

**Fig. 6**—When milling work having a gap between machined surfaces, production is increased by rapid traverse between machining positions. Jump feed control is accomplished by means of adjustable limit switches, multi-speed motors, and suitable magnetic controls.

**Fig. 7**—Accurate positioning of reversing and stop limits is necessary on tapping machines, especially when tapping blind holes. Special a.c. reversing motors for tapping service permit as many as 60 reversals per min. The use of two or four-speed motors reduces the number of gear changes required. Accurate limit switches, quick-acting contactors, and high torque motors are used. A plug stop is used for braking at the "out" position.

**Fig. 8**—Accurate location of boring tool for indexing requires extremely slow speed of work table to prevent over-travel when stop limit is reached. A d.c. motor and control is used; heavy armature series resistance and armature parallel resistance provide for creep speeds for final positioning.

**Fig. 9**—For accurately stopping a slowly moving body, such as a rotary type valve, a geared switch is mounted on the motor, operating at motor speed. The motor can be stopped within one-fourth revolution and with a gear reduction of 100 to 1 between the motor and valve, the valve can be accurately located to within less than 1 degree.

**Fig. 10**—On a sheet catcher, the table must reverse and return the sheet as soon as it passes through the rolls. Since the length of the sheet varies, the sheet itself is used to operate the limit switch which reverses the table. This application requires specially designed motors and exceptional ruggedness in the control equipment.

# Limit Switching Mechanisms—

Limit switches, used to confine or restrain the travel or rotation of moving parts within certain predetermined points, are actuated by varying methods. Some of these, such as cams, rollers, push-rods, and traveling nuts, are described and illustrated below.

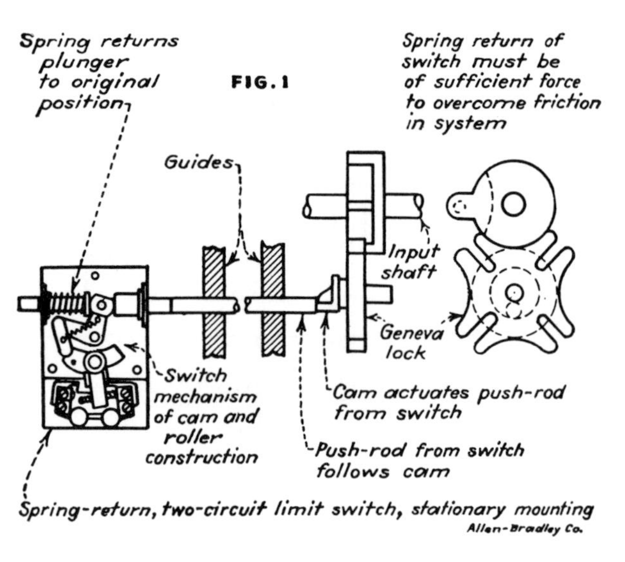


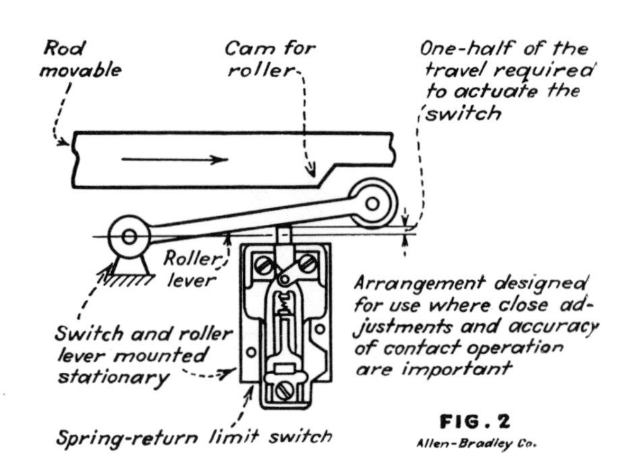


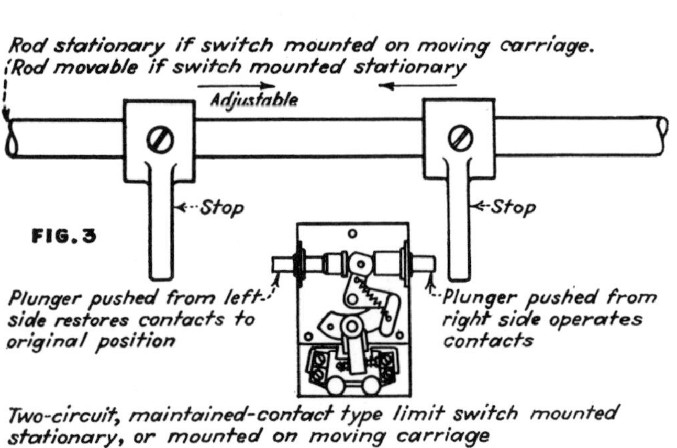


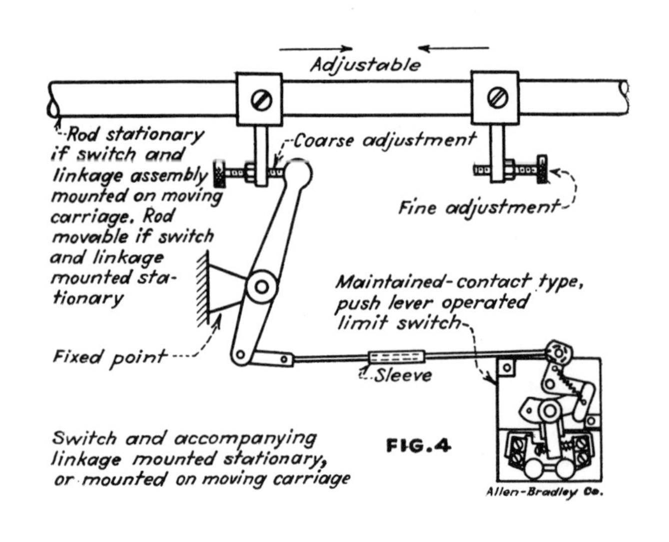


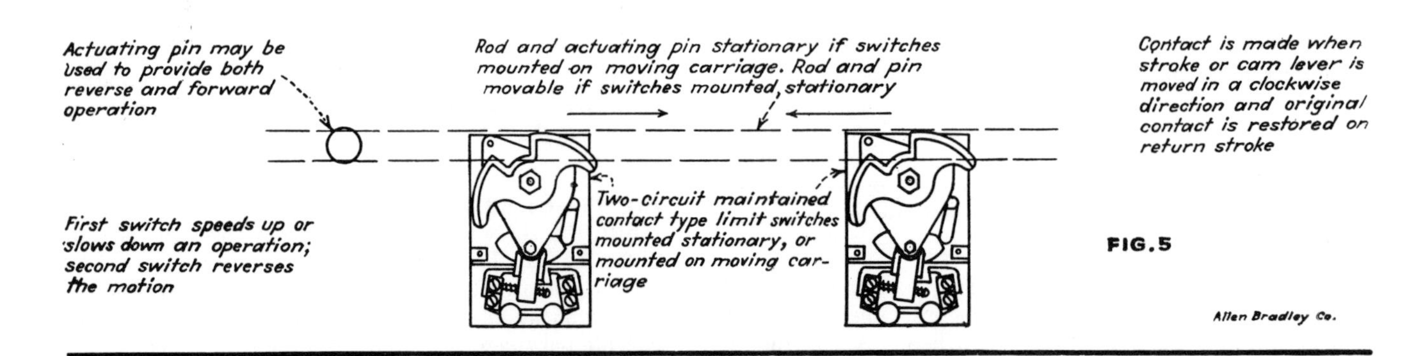

# Methods of Mechanical Actuation

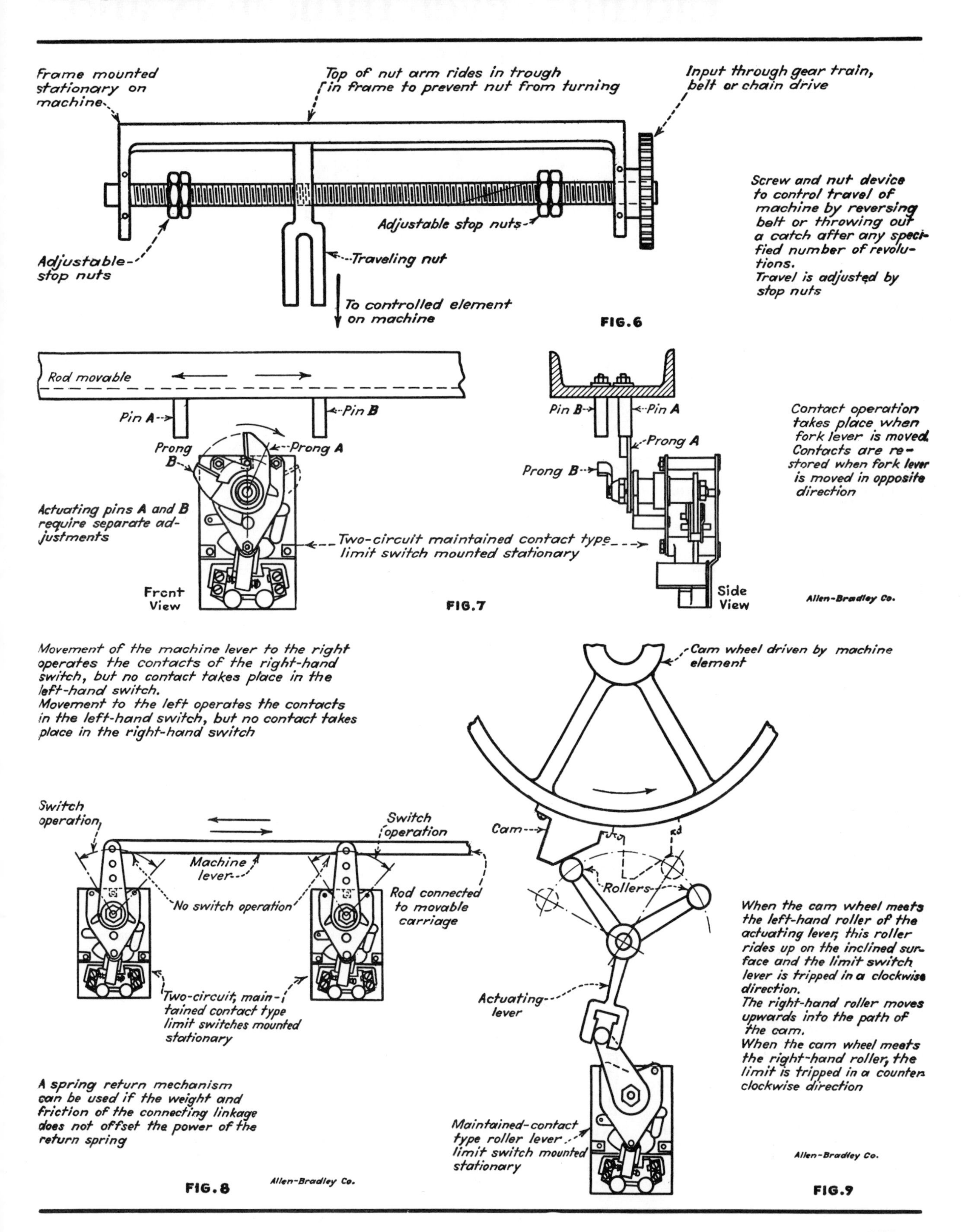

# Liquid Level Indicators and

Thirteen different systems of operation are shown. Each one represents at least

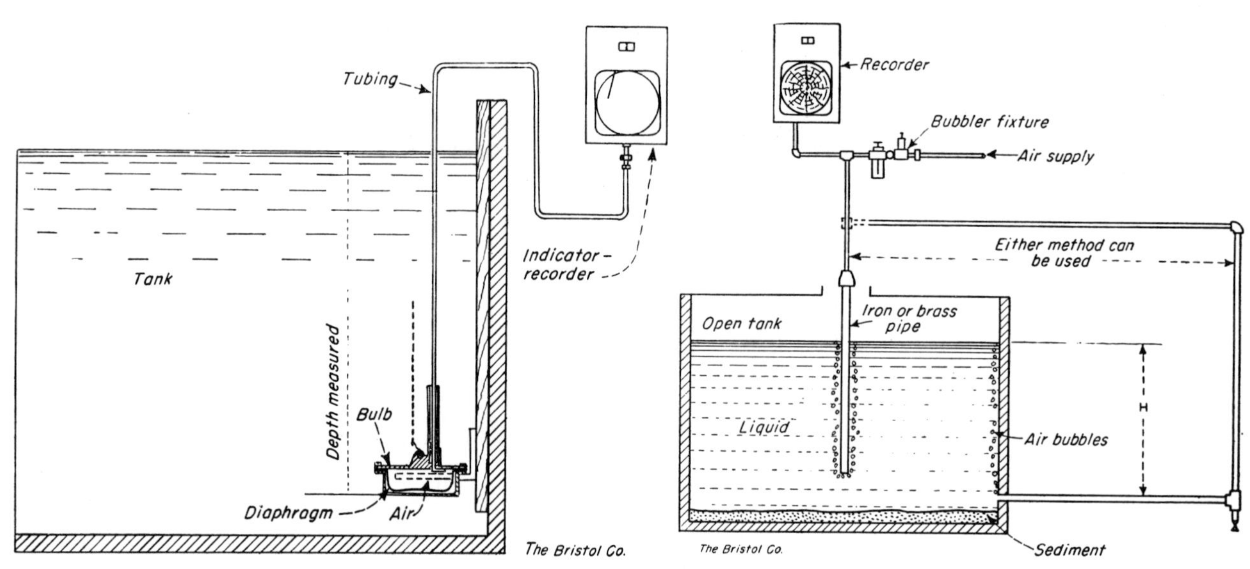


**DIAPHRAGM ACTUATED INDICATOR.** Can be used with any kind of liquid, whether it be flowing, turbulent, or carrying solid matter. Recorder can be mounted above or below the level of the tank or reservoir.

**BUBBLER TYPE RECORDER** measures height $H$. Can be used with all kinds of liquids, including those carrying solids. Small amount of air is bled into submerged pipe. Gage measures pressure of air that displaces fluid.

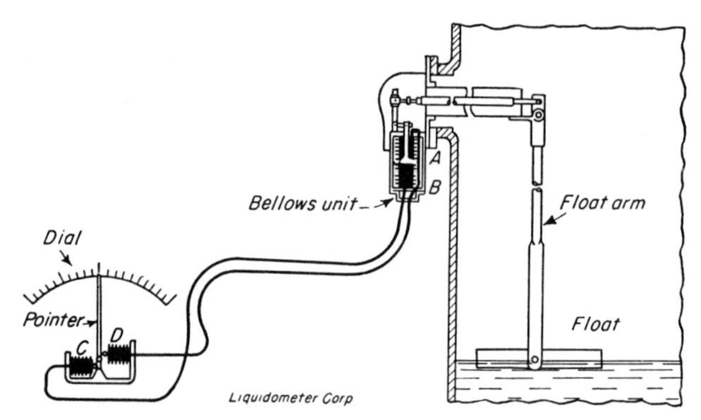


**BELLOWS ACTUATED INDICATOR.** Two bellows and connecting tubing are filled with incompressible fluid. Change in liquid level displaces transmitting bellows and pointer.

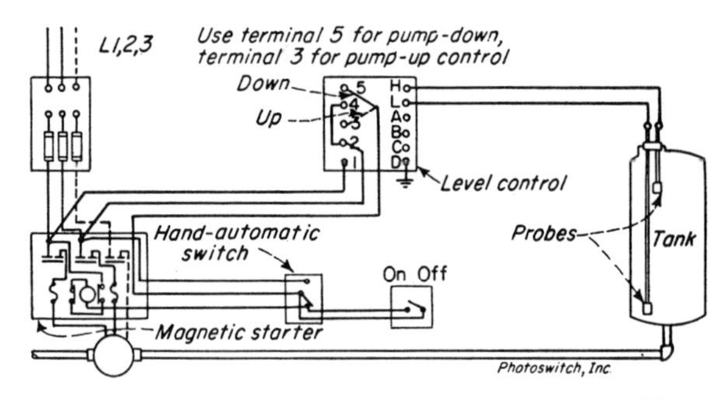


**ELECTRICAL TYPE LEVEL CONTROLLER.** Positions of probes determine duration of pump operation. When liquid touches upper probe, relay operates and pump stops. Through auxiliary contacts, lower probe provides relay holding current until liquid drops below it.

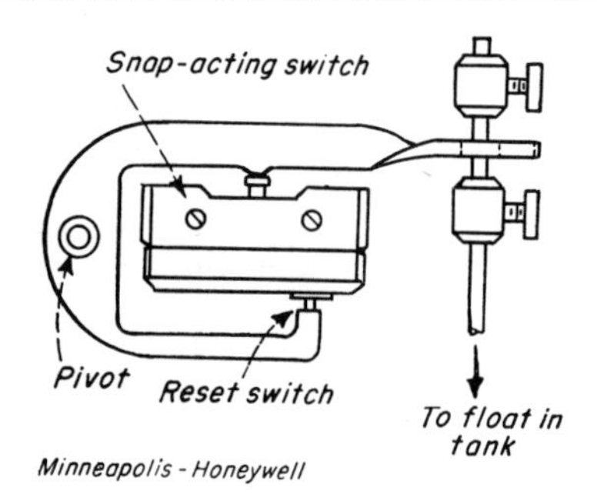


**FLOAT-SWITCH TYPE CONTROLLER.** When liquid reaches predetermined level, float actuates switch through horseshoe-shape arm. Switch can operate valve or pump, as required.

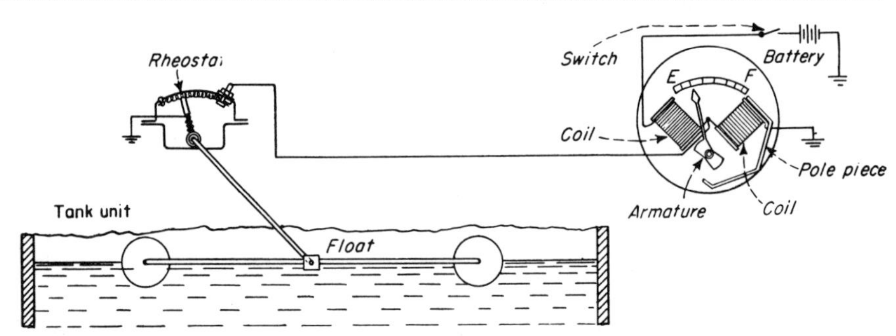


**AUTOMOTIVE TYPE LIQUID LEVEL INDICATOR.** Indicator and tank unit are connected by a single wire. As liquid level in tank increases, brush contact on tank rheostat moves to the right, introducing an increasing amount of resistance into circuit that grounds the "F" coil. Displacement of needle from empty mark is proportional to the amount of resistance introduced into this circuit.

# Controllers

H. W. HAMM

one commercial instrument. Some of them are available in several modified forms.

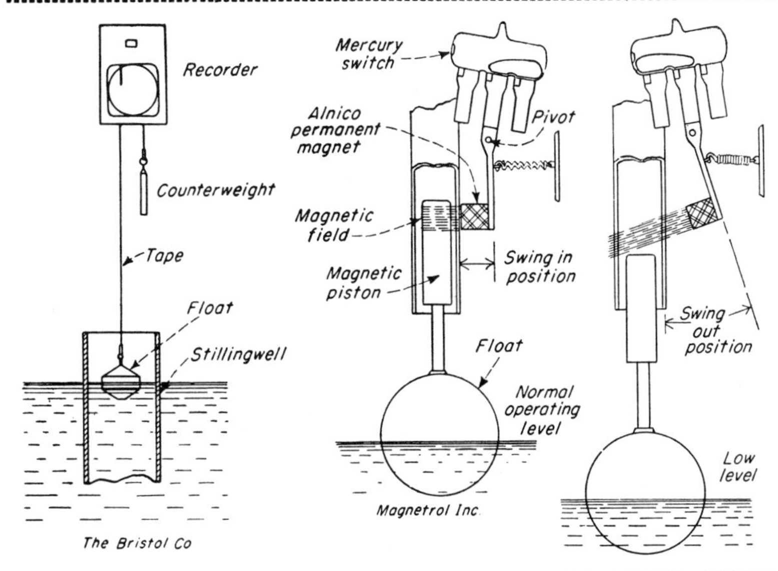


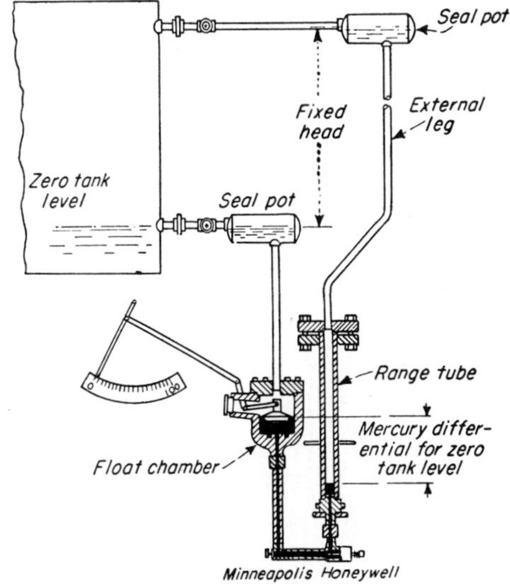


**FLOAT TYPE RECORDER.** Pointer can be attached to a calibrated float tape to give an approximate instantaneous indication of fluid level.

**MAGNETIC LIQUID LEVEL CONTROLLER.** When liquid level is normal, common-to-right leg circuit of mercury switch is closed. When level drops to predetermined level, magnetic piston is drawn below the magnetic field.

**DIFFERENTIAL PRESSURE SYSTEM.** Applicable to liquids under pressure. Measuring element is mercury manometer. Mechanical or electric meter body can be used. Seal pots protect meter body.

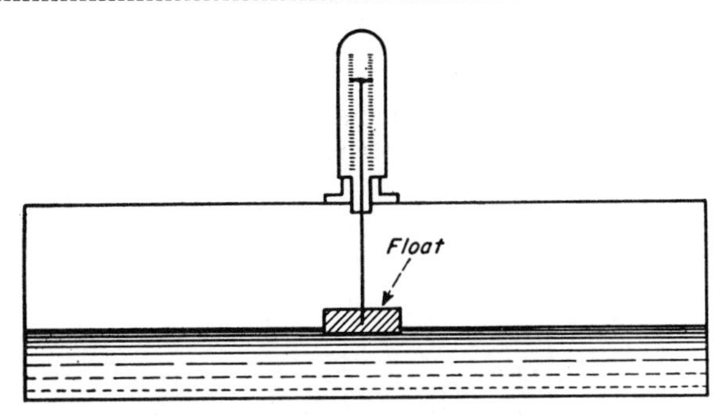


**DIRECT READING FLOAT TYPE GAGE.** Inexpensive, direct-reading gage has dial calibrated to tank volume. Comparable type as far as simplicity is concerned has needle connected through a right-angle arm to float. As liquid level drops, float rotates the arm and the needle.

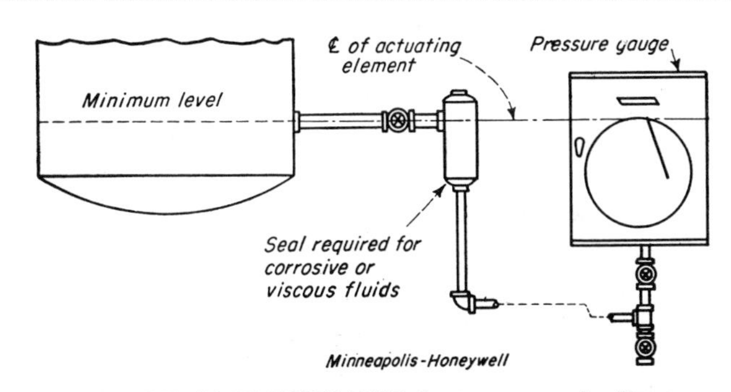


**PRESSURE GAGE INDICATOR** for open vessels. Pressure of liquid head is imposed directly upon actuating element of pressure gage. Center line of the actuating element must conincide with the minimum level line, if the gage is to read zero when the liquid reaches the minimum level.

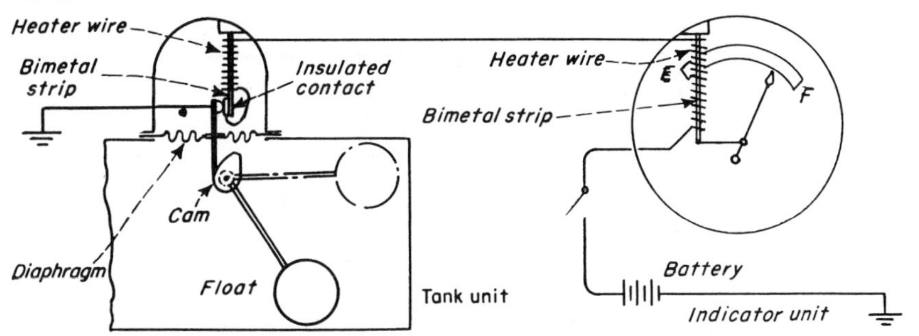


**BIMETALLIC TYPE INDICATOR.** When tank is empty, contacts in tank unit just touch. With switch closed, heaters cause both bimetallic strips to bend. This opens contacts in tank and bimetals cool, closing circuit again. Cycle repeats about once per sec. As liquid level increases, float forces cam to bend tank bimetal. Action is similar to previous case, but current and needle displacement are increased.

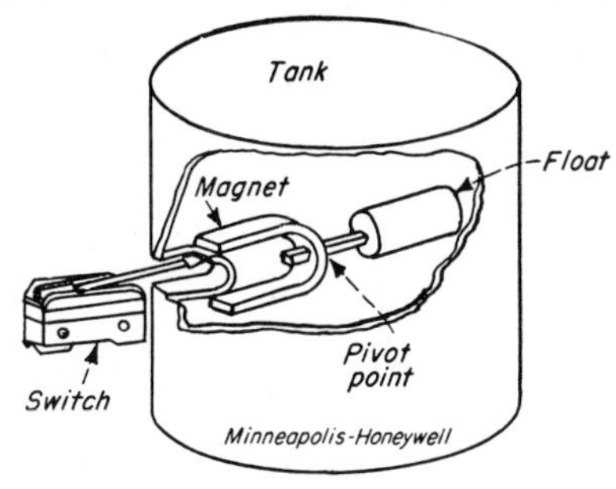


**SWITCH ACTUATED LEVEL CONTROLLER.** Pump is actuated by switch. Float pivots magnet so that upper pole attracts switch contact. Tank wall serves as other contact.

# Liquid Level Mechanisms—

Means of determining liquid level, detection of changes in liquid level, transmission of indicated levels, or warnings of changes beyond set limits; and means of using level changes for level control, or control of other conditions such as temperature and pressure, have been accomplished by numerous mechanisms. The most popular devices employ floats or pressure measurement with instruments such as the U-tube manometer, bourdon tube, and bellows.

The methods shown here are largely indicating methods or simple devices for automatic control of liquid level although they can conceivably be applied to control other conditions such as temperature and pressure. Methods using electric resistance of a column of liquid and measurement of pressure changes by means of piezo-electric crystals are not shown. Patent No. 2,162,180 describes a method involving determination of change in air pressure when a measured volume of air is introduced into a tank.

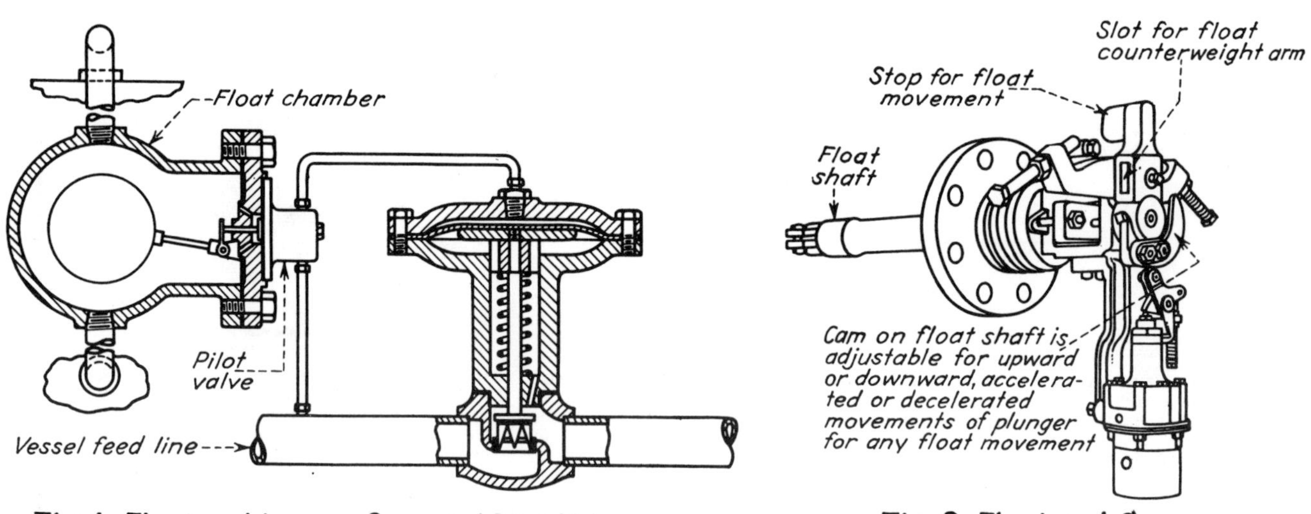


**Fig. 1**-Float and Lever- Operated Pilot Valve

**Fig. 2**-Float and Cam-Operated Pilot Valve

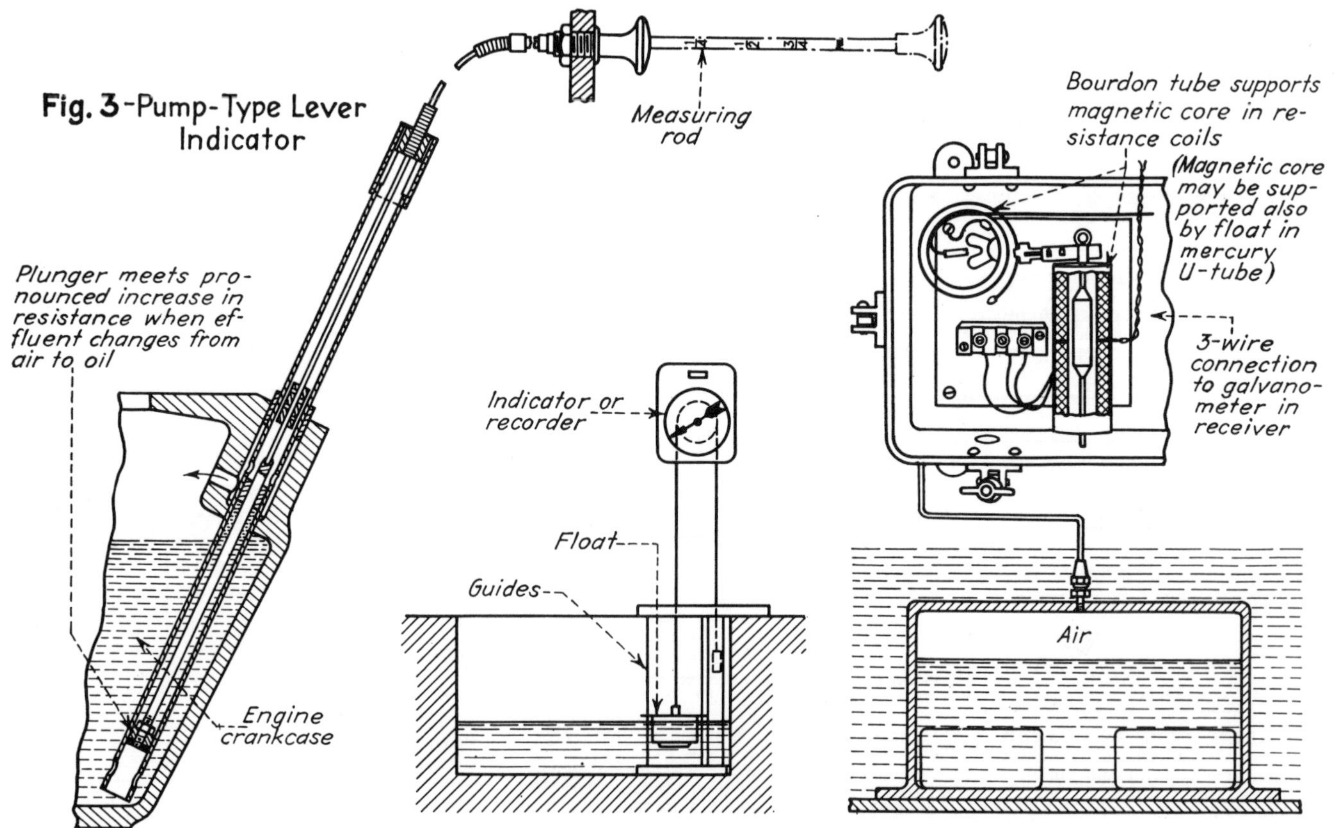


**Fig. 3**-Pump-Type Lever Indicator

**Fig. 4**-Float and Pulley Indicator

**Fig. 5**-Pressure Dome Indicator

# Indicators and Controllers

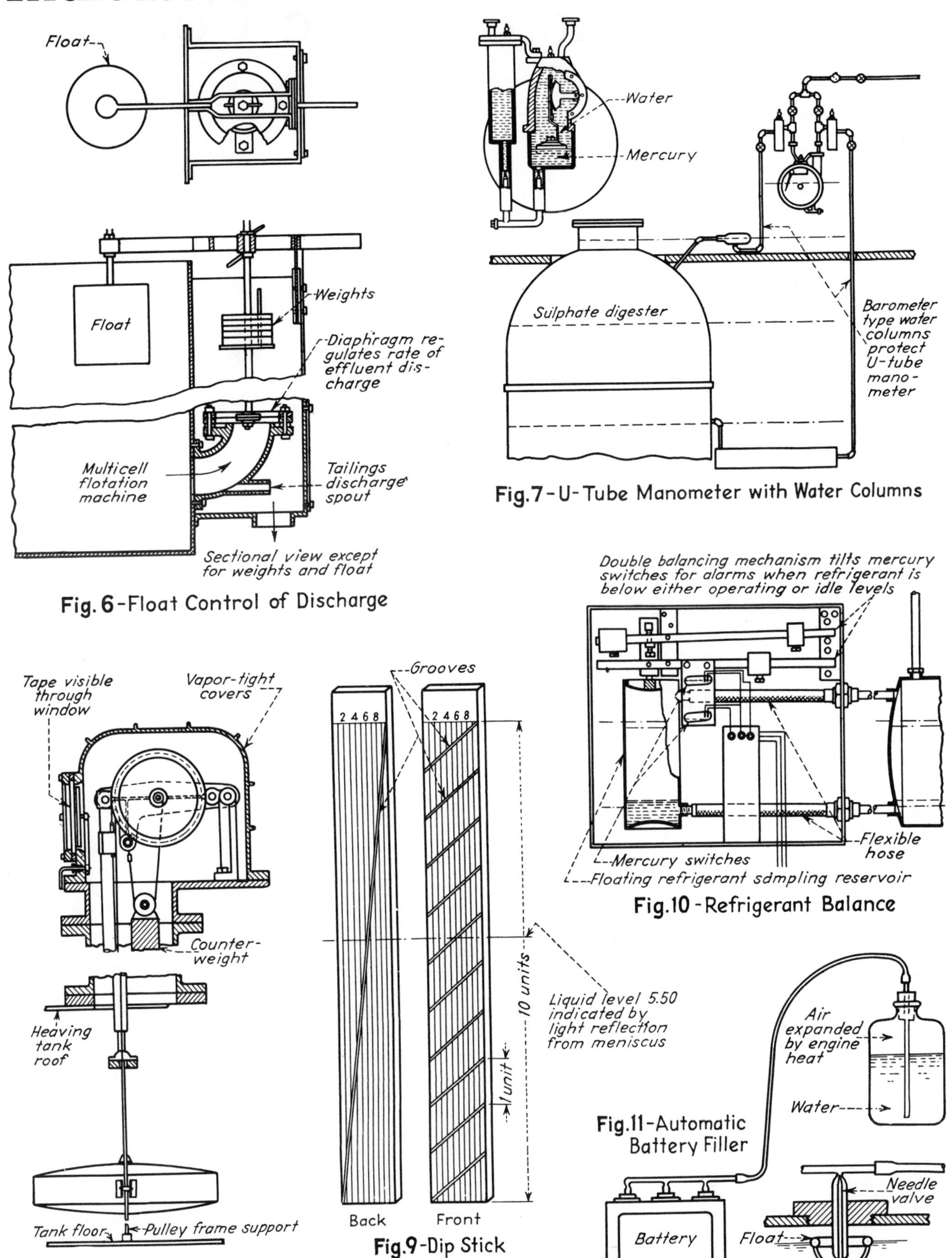


Fig. 6-Float Control of Discharge

Fig. 7-U-Tube Manometer with Water Columns

Fig. 10-Refrigerant Balance

Fig. 8-Tank Roof Indicator

Fig. 9-Dip Stick Indicator

Fig. 11-Automatic Battery Filler

# Photoelectric Controls

PHOTOSWITCH DIVISION, ELECTRONICS CORPORATION OF AMERICA

FIG. 1—Automatic weighing and filling. Problem is to fill each box with exact quantity of products such as screws. Electric feeder vibrates parts through chute and into box on small balance. Photoelectric control is mounted at rear of scale. Light beam is restricted to very small dimensions by optical slit. Control is positioned so that light is interrupted by balanced cantilever arm attached to scale when proper box weight is reached. Photoelectric control then stops flow of parts by de-energizing feeder. Simultaneously, indexing mechanism is activated to remove filled box and replace with empty one. Completion of indexing re-energizes feeder which starts flow of screws.

FIG. 2—Operator safeguard. Most presses operate by foot pedal leaving hands free for loading and unloading operations. This creates a safety hazard. Use of mechanical gate systems reduce production speeds. With photoelectric controls, curtain of light is set up with multiple series of photoelectric scanners and light sources. When light is broken at any point by operator's hand, control energizes a locking mechanism which prevents punch press drive from being energized. Wiring is such that power or tube failure causes control to function as though light beam was broken. In addition, the light is frequently used as the actuating control since clutch is thrown as soon as operator removes his hand from the die on the press table.

FIG. 3—Sorting cartons of three types of electronic tubes. Since cartons containing one type differ widely as to size, it is not feasible to sort by carton size and shape. Solution: small strip of reflecting tape is placed on cartons by a packer during assembly. For one type of tube, strip is placed along one edge of bottom and extends almost to the middle. For second type, strip is located along same edge but from middle to opposite side. No tape is used for third type. Cartons are placed on conveyor so that tape is at right angle to direction of travel. Photoelectric controls, shown in *A*, "see" the reflecting tape pass and operate pusher bar mechanism, shown in *B*, which pushes carton on to proper distribution conveyor. Cartons without tape pass through.

FIG. 4—Cut-off machine uses photoelectric control for strip material which does not have sufficient mass to operate a mechanical limit switch satisfac-

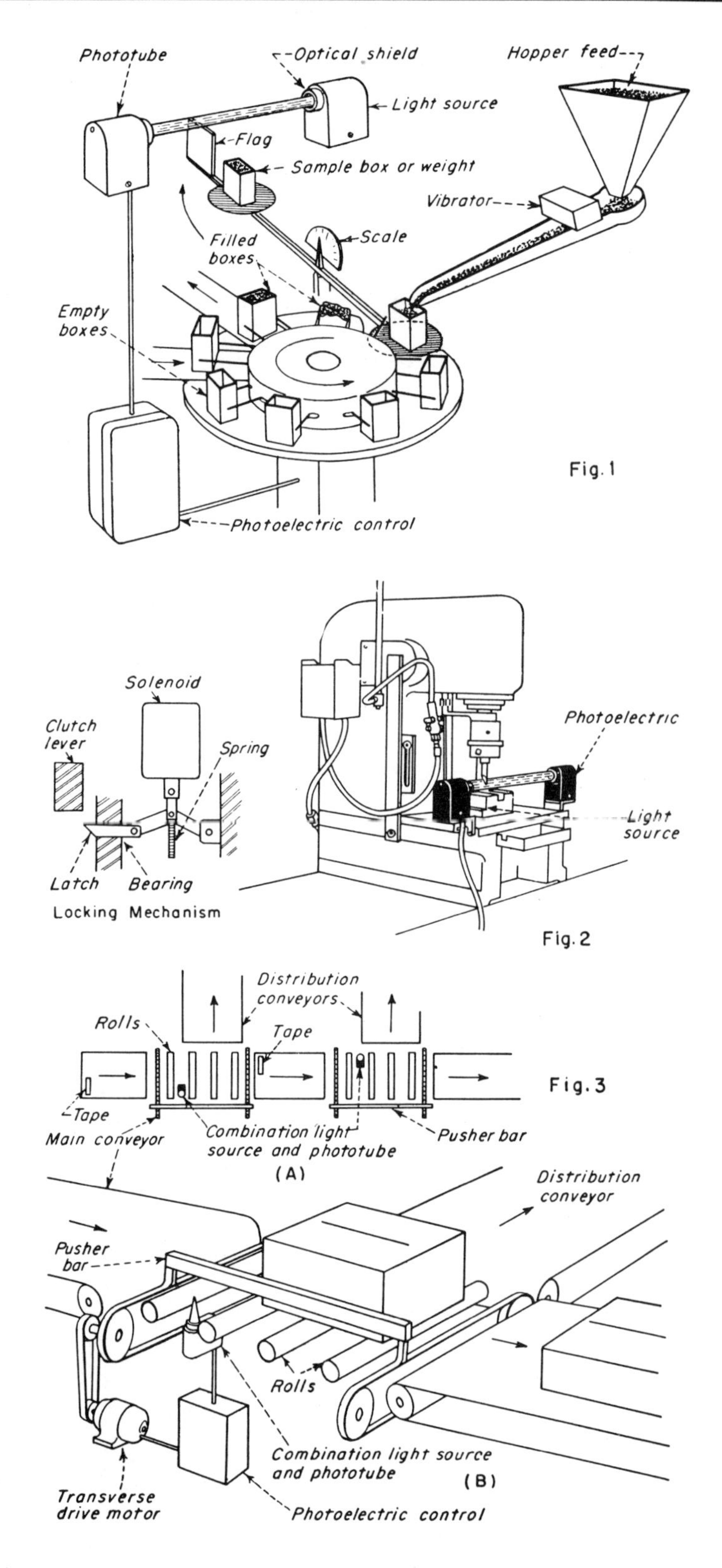

Some typical applications for reducing production costs and increasing operator safeguards by precisely and automatically controlling the feed, transfer or inspection of products from one process stage to another.

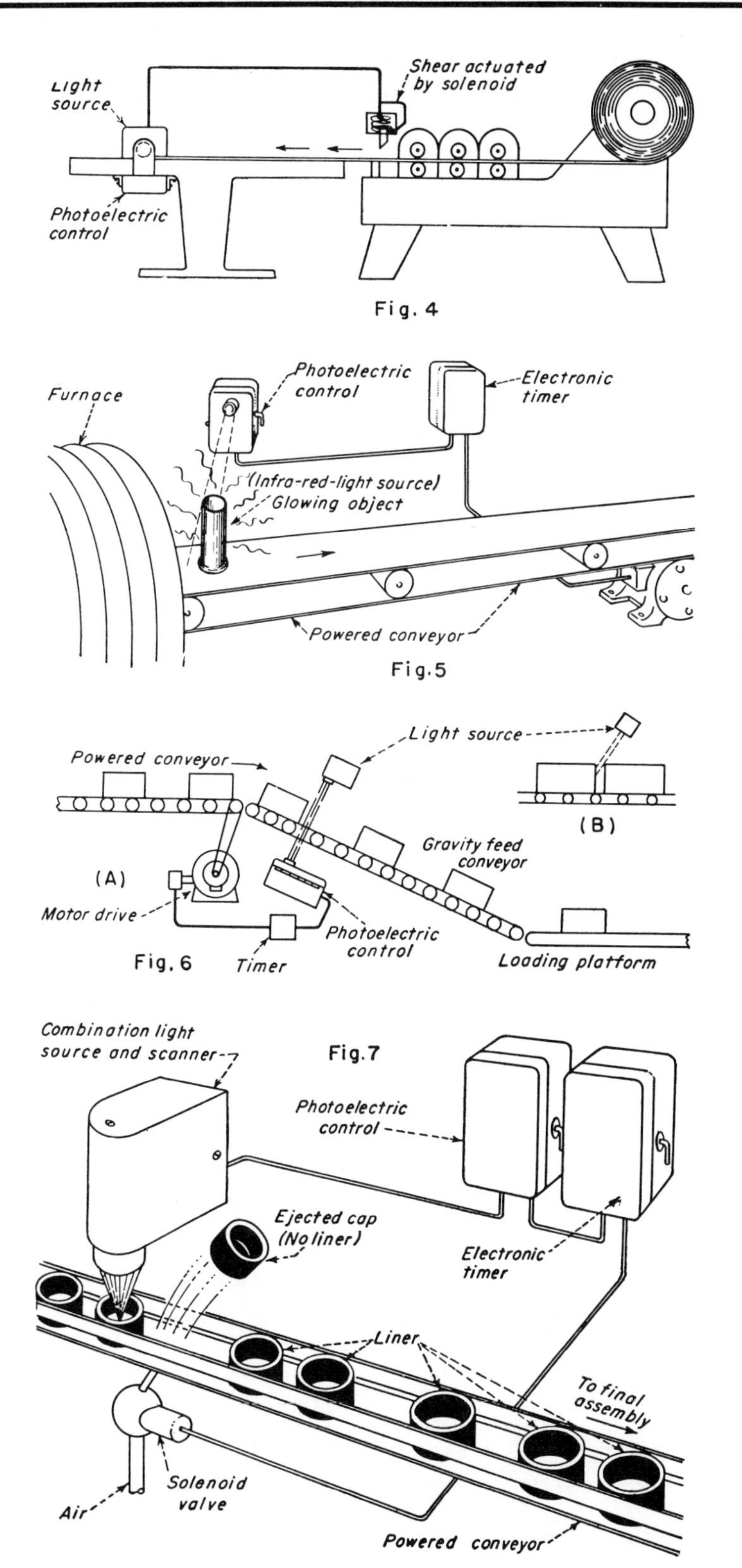


torily. Forward end of strip breaks light beam thus actuating the cut-off operation. Light source and control is mounted on adjustable stand at end of machine to vary length of finished stock.

FIG. 5—Heat-treating conveyor uses electronic timer in conjunction with photoelectric control to carry parts emerging from furnace at 2300 F. Problem is to operate conveyor only when a part is placed on it and only for distance required to reach next process stage. Parts are ejected on to conveyor at varying rates. High temperatures caused failures when mechanical switches were used. Glowing white-hot part radiates infra-red rays which actuates photoelectric control as soon as part comes in view. Control operates conveyor which carries part away from furnace and simultaneously starts timer. Conveyor is kept running by timer for pre-determined length of time required to position part for next operation.

FIG. 6—Jam detector. Cartons jamming on conveyor cause loss of production time and damage to cartons, products and conveyors. Detection is accomplished with a photoelectric control using a timer as shown in (A). Each time a carton passes light source, control beam is broken which starts timing interval in the timer. Timing circuit is reset without relay action each time beam is restored before preset timing interval has elapsed. If jam occurs causing cartons to butt one against the other, light beam cannot reach control. Timing circuit will then time out, opening load circuit which stops conveyor motor. By locating light source at an angle to conveyor, as shown in (B), power conveyor can be delayed if cartons are not butting but are too close to each other.

FIG. 7—Automatic inspection. As steel caps are conveyed to final assembly, they pass intermediate stage where an assembler inserts insulation liner into cap. Inspection point for missing liners has reflection-type photoelectric scanner which incorporates both light source and phototube with common lens system to instantly recognize difference in reflection between dark liner and light steel cap. When it detects a cap without a liner, a relay operates ejector device composed of air blast controlled by solenoid valve. Start and duration of air blast is accurately controlled by timer so that no other caps are displaced.

# AUTOMATIC STOP MECHANISMS

**This group of stop mechanisms includes types and applications for machines in different fields. Such devices, which prevent automatic machines from damaging themselves or the work passing through them, make use of mechanical, electrical, hydraulic, and pneumatic principles. Typical mechanisms illustrated prevent excess speed, misweaving, jamming of toggle press and food-canning machines, operation of printing presses when the paper web breaks, improper feeding of wrapping paper, and uncoordinated operation of glass-making machines.**

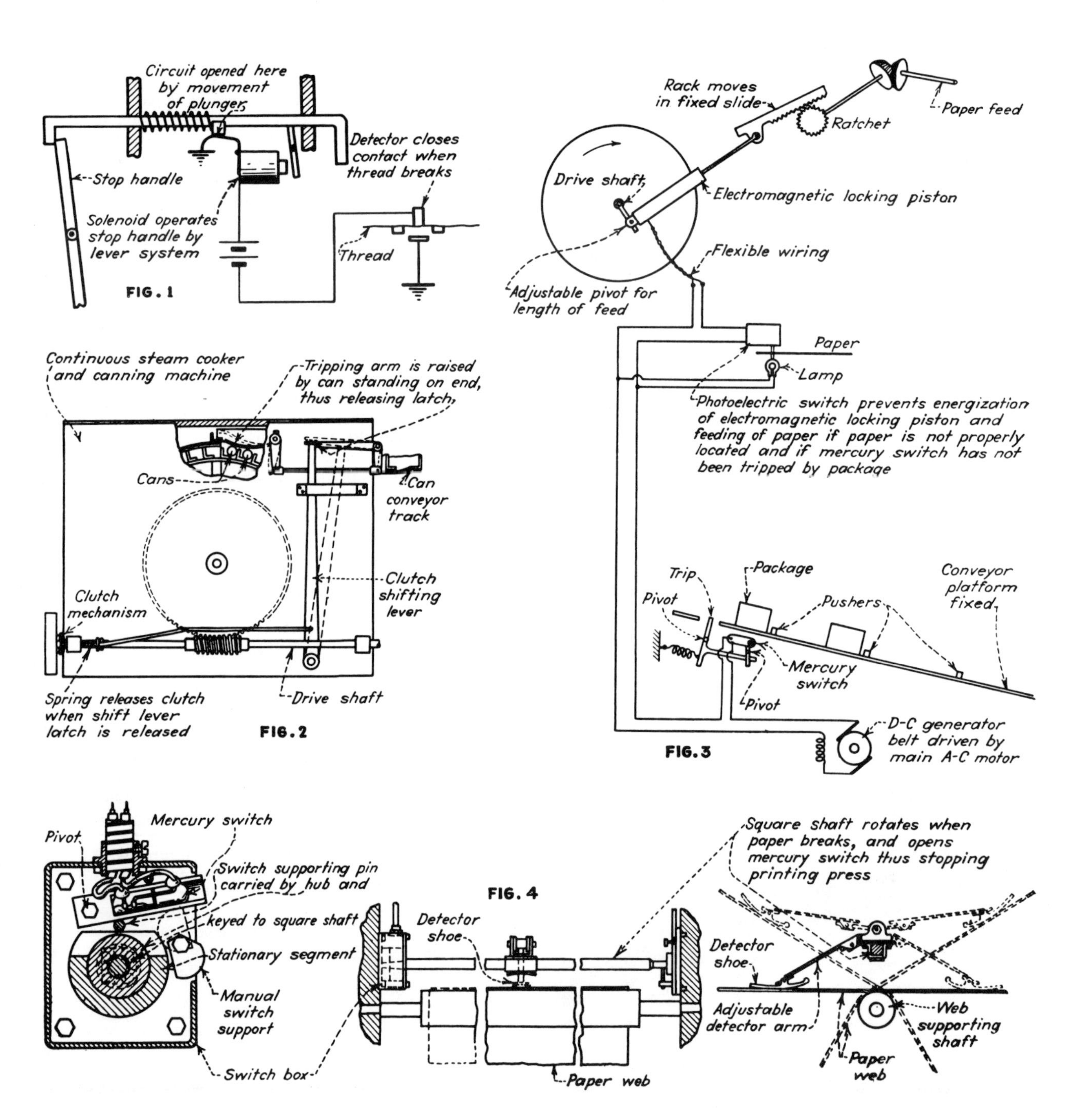

# PROTECT MACHINES AND WORK

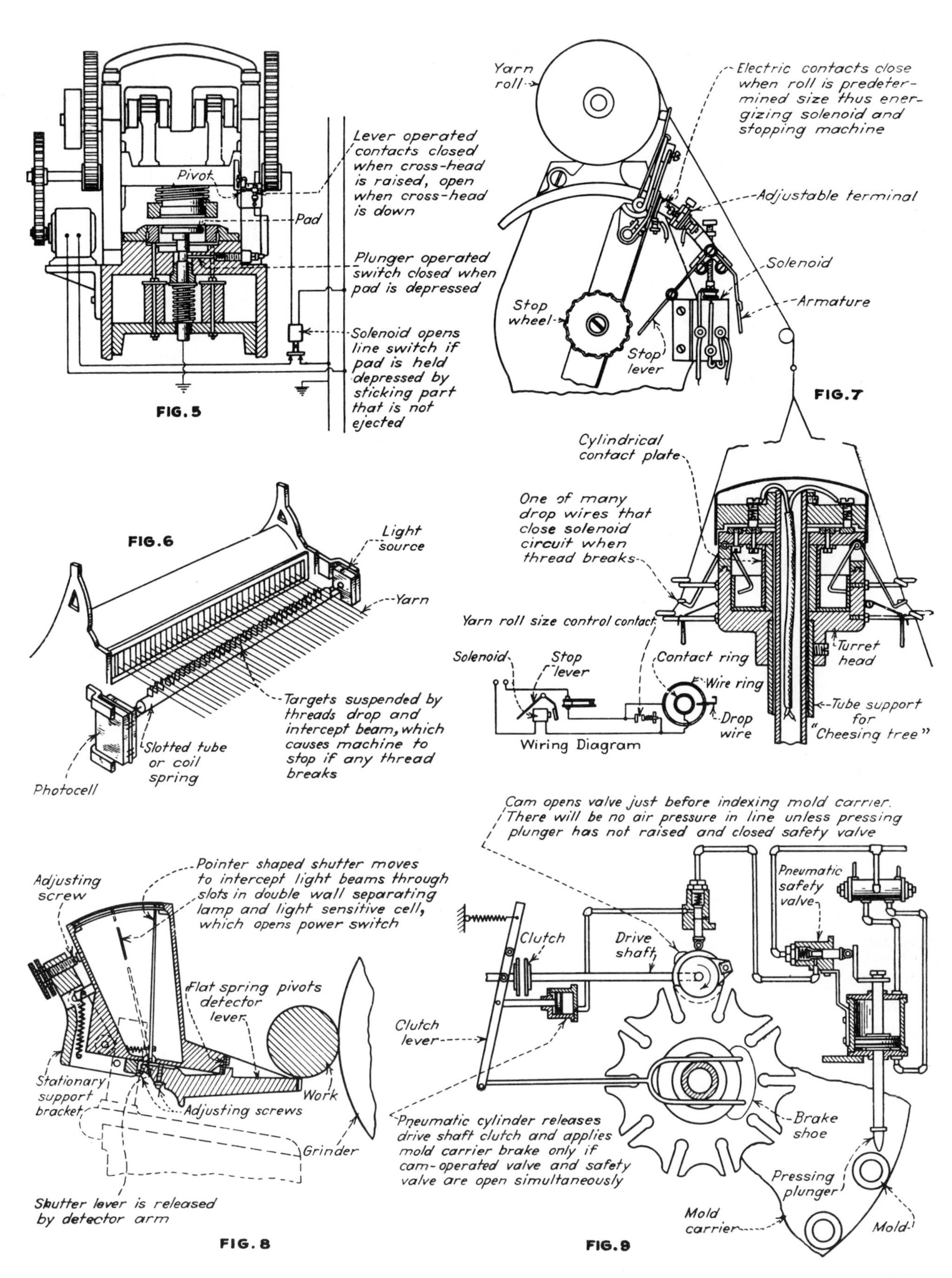

# ELECTRIC AUTOMATIC

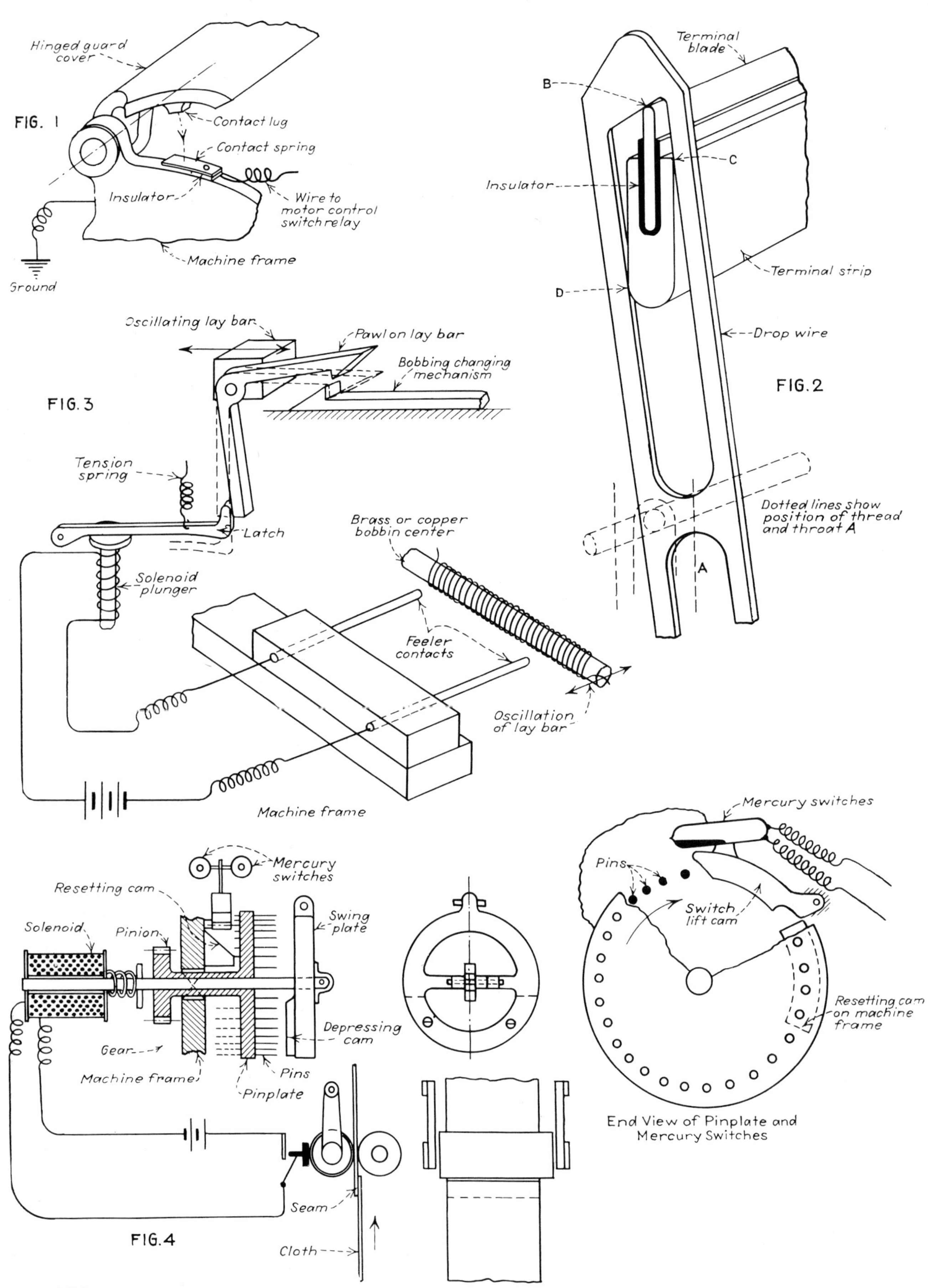

# STOP MECHANISMS

**Arrangements and designs diagrammatically illustrated here were taken from packaging machines and textile machines. Modifications of them to perform other operations can be easily devised**

Fig. 1—Safety arrangement used on some machines to stop motor when guard cover is lifted. Circuit is complete only when cover is down, in which position contact lug has metal to metal connection with contact spring, completing circuit to relay.

Fig. 2—Electrical 3-point wedging type "warp stop" shown after thread has broken and drop wire has fallen and tilted to close circuit. Dotted lines indicate normal position of drop wire when riding on thread. When thread breaks the drop wire falls and strikes the top of terminal blade at *B*, the inclined top of the slot causing a wedging effect which tilts drop wire against the terminal strip at *C* and *D* intensifying the circuit closing action.

Fig. 3—Bobbin changer. When bobbin is empty the feeders contact the metal bobbin center, completing the circuit through a solenoid which pulls a latch that causes bobbin changing mechanism to operate and put a new bobbin in the shuttle. As long as the solenoid remains deenergized, the pawl on the lay bar will be raised clear of the hook on the bobbin changing mechanism.

Fig. 4—Control for automatic shear. When a seam of two thicknesses of cloth passes between the rolls the swing roller is moved outward and closes a sensitive switch which energizes a solenoid. Action of solenoid pulls in an armature the outer end of which is attached to a hinged ring to which a cam plate is fastened. The cam plate depresses a number of pins in a rotating plate. As the plate rotates the depressed pins lift a hinged cam arm on which are mounted two mercury switches which, when tilted, complete circuits in two motor controls. Fastened on the frame of the machine is a resetting cam for pushing the depressed pins back to their original position. In this arrangement two motors are stopped and reversed until seam has passed through rollers, then stopped and reversed again.

Fig. 5—Electric stop for loom. When thread breaks or slackens the drop wire falls and contact *A* rides on contact *C*. The drop wire being supported off center swings so that contact *B* is pulled against inner terminal strip *D* completing solenoid circuit.

Fig. 6—Automatic stop for folder or yarder to stop machine always in the same position when a seam in the cloth passes between the rolls. A seam passing between the rolls causes swivel mounted roll to lift slightly. This motion closes contacts in sensitive switch which throws relay in control box, so that the next time the cam closes the limit switch the power of motor with integral magnetic brake is shut off. The brake stops the machine always in the same place.

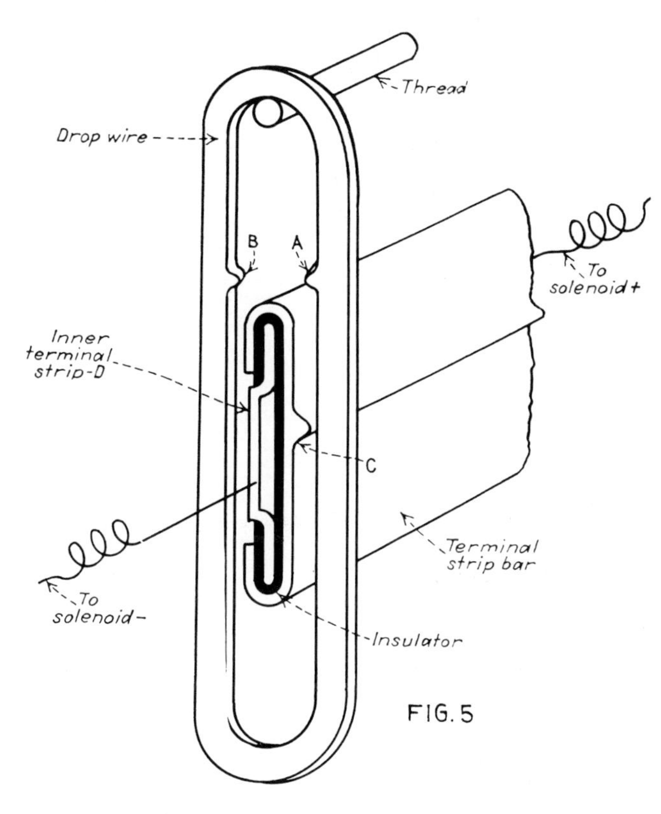


FIG. 5

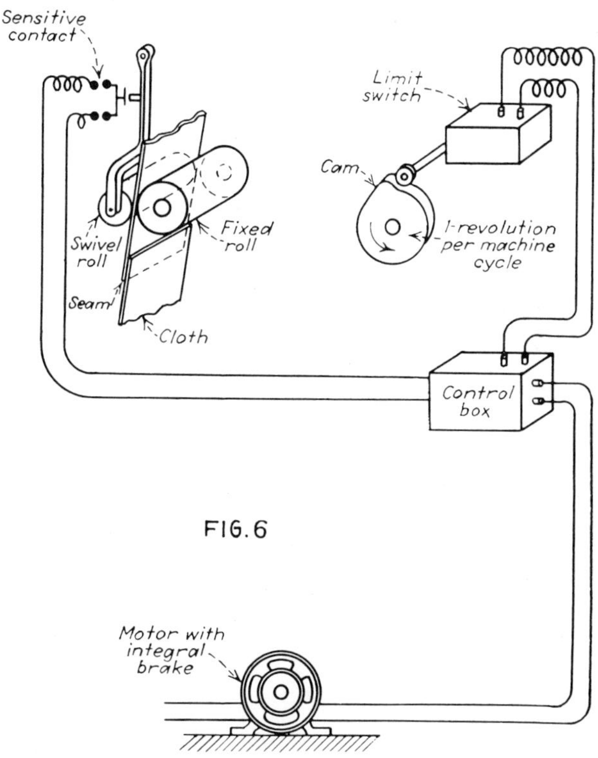


FIG. 6

# MECHANICAL AUTOMATIC

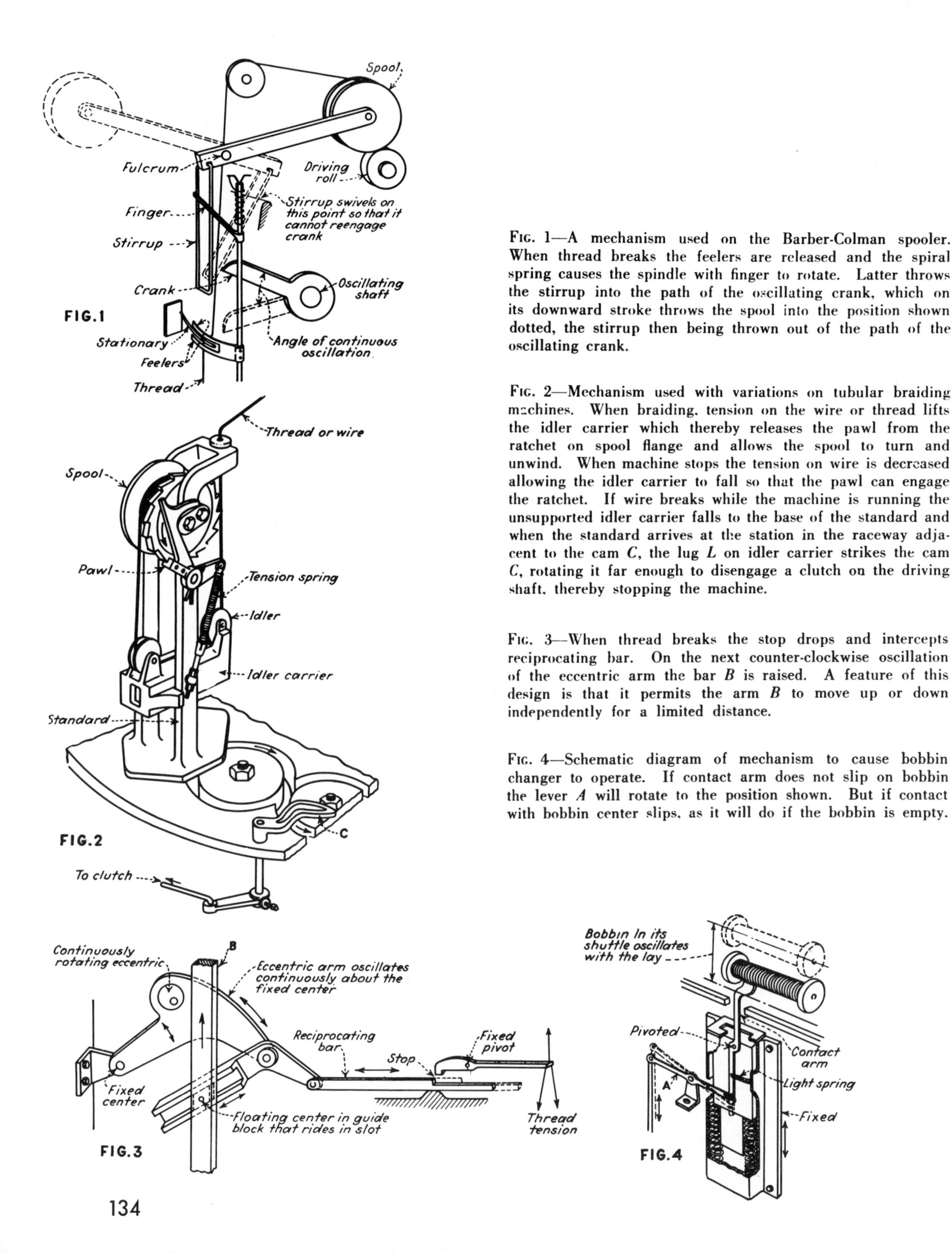


Fig. 1—A mechanism used on the Barber-Colman spooler. When thread breaks the feelers are released and the spiral spring causes the spindle with finger to rotate. Latter throws the stirrup into the path of the oscillating crank, which on its downward stroke throws the spool into the position shown dotted, the stirrup then being thrown out of the path of the oscillating crank.

Fig. 2—Mechanism used with variations on tubular braiding machines. When braiding, tension on the wire or thread lifts the idler carrier which thereby releases the pawl from the ratchet on spool flange and allows the spool to turn and unwind. When machine stops the tension on wire is decreased allowing the idler carrier to fall so that the pawl can engage the ratchet. If wire breaks while the machine is running the unsupported idler carrier falls to the base of the standard and when the standard arrives at the station in the raceway adjacent to the cam *C*, the lug *L* on idler carrier strikes the cam *C*, rotating it far enough to disengage a clutch on the driving shaft, thereby stopping the machine.

Fig. 3—When thread breaks the stop drops and intercepts reciprocating bar. On the next counter-clockwise oscillation of the eccentric arm the bar *B* is raised. A feature of this design is that it permits the arm *B* to move up or down independently for a limited distance.

Fig. 4—Schematic diagram of mechanism to cause bobbin changer to operate. If contact arm does not slip on bobbin the lever *A* will rotate to the position shown. But if contact with bobbin center slips, as it will do if the bobbin is empty.

**Designs shown here diagrammatically were taken from textile machines, braiding machines and packaging machines. Possible modifications of them to suit other applications will be apparent**

# STOP MECHANISMS

lever *A* will not rotate to position indicated by dashed line, thereby causing bobbin changer to come into action.

FIG. 5—Simple type of stop mechanism for limiting the stroke of a reciprocating machine member. Arrows indicate the direction of movements.

FIG. 6—When the predetermined weight of material has been poured on the pan, the movement of the scale beam pushes the latch out of engagement, allowing the paddle wheel to rotate and thus dump the load. The scale beam drops, thereby returning the latch to the holding position and stopping the wheel when the next vane hits the latch.

FIG. 7—In this textile machine any movement that will rotate the stop lever counter-clockwise will bring it in the path of the continuously reciprocating shaft. This will cause the catch lever to be pushed counter-clockwise and the hardened steel stop on the clutch control shaft will be freed. A spiral spring then impels the clutch control shaft to rotate clockwise, which movement throws out the clutch and applies the brake. Initial movement of the stop lever may be caused by the breaking of a thread, a moving dog, or any other means.

FIG. 8—Arrangement used on some package loading machines to stop machine if a package should pass loading station without receiving an insert. Pawl finger *F* has a rocking motion obtained from crankshaft, timed so that it enters the unsealed packages and is stopped against the contents. If the box is not filled the finger enters a considerable distance and the pawl end at bottom engages and holds a ratchet wheel on driving clutch which disengages the machine driving shaft.

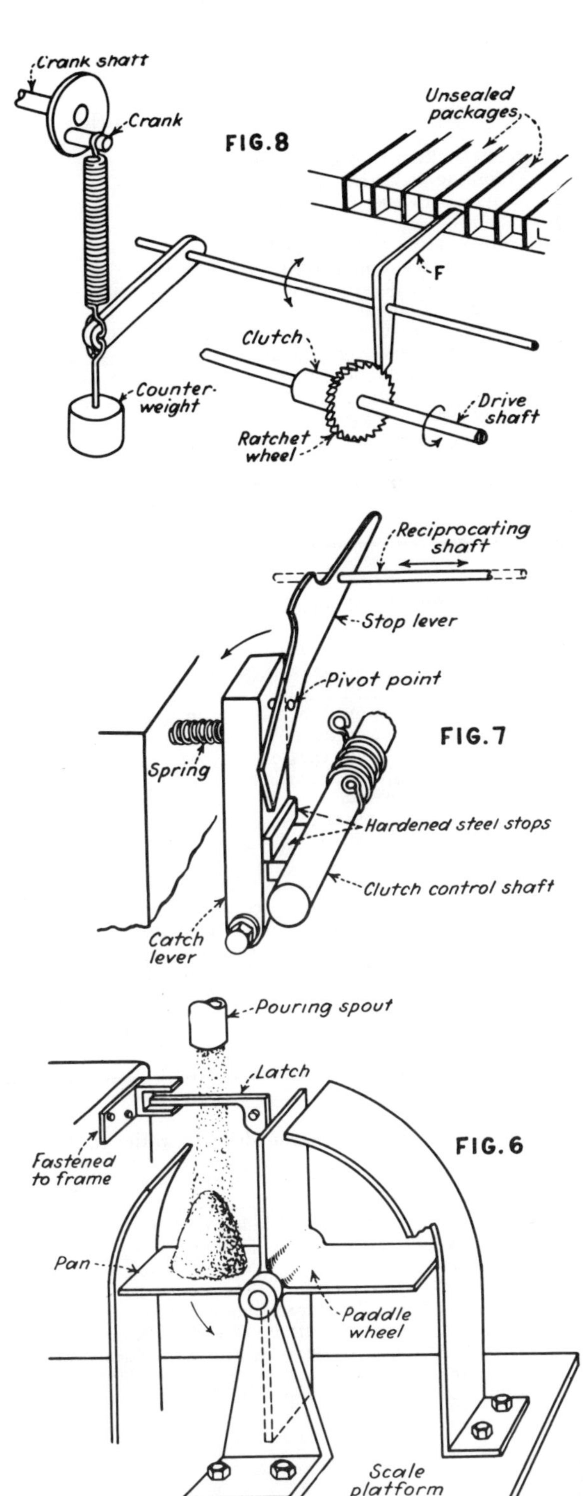


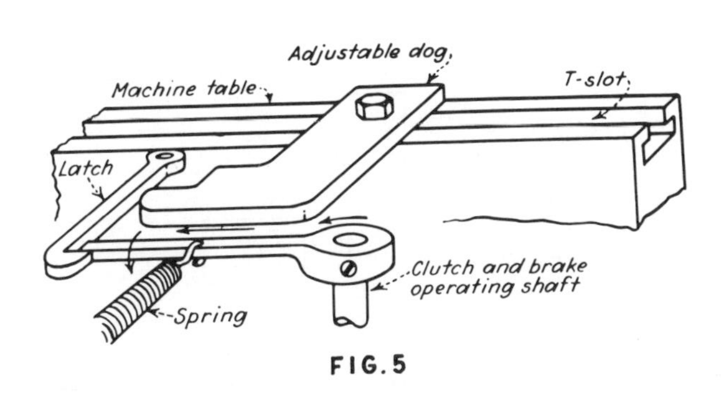

# How to Obtain Constant Speed

ACCURATE METHODS for controlling constant speed devices operating at extremely low linear speeds (less than 1 in. per sec) are the basis of the development of a slow speed photographic recording machine. Attainment of constant linear velocity has presented a problem in the recording art since Edison invented the phonograph. The rigorous analysis of constant speed regulation—presented on the following pages—was made necessary by the requirements of geophysical exploration, in which time of arrival of seismic waves, excited by explosives, must be accurately recorded. The accuracy of the recorded information depends on the linear velocity of the medium, which is a function of the speed regulation of the drive mechanism.

At angular velocities below 10 rpm (1 in. per sec linear

METHOD 1—Flywheel with capstan string—drive from synchronous motor.

SPEED REGULATION: Excellent—most constant speed device in recording industry. Accuracy, 0.02 percent.
SYNCHRONISM: Not exact, since there is some slippage in string belt.
REFERENCE STANDARDS:
Primary—tuning fork; accuracy, 1 part in 500,000.
Secondary—synchronous motor; accuracy, 1 part in 50,000.
EFFECTS CONTRIBUTING TO ERROR:
Primary—flywheel bearing rumble; flywheel oscillation.
Secondary—synchronous motor bearings; standard frequency source; synchronous motor angular error; bearing eccentricities.
SIZE AND WEIGHT: Flywheel effect demands prohibitive mass.
NECESSARY SERVICING IN FIELD: Replacing string belt; electronic components; bearings; lubrication.
MANUFACTURING DIFFICULTIES: Few.
REMARKS: Used in recording to generate constant frequency tone of 0.05 percent accuracy.

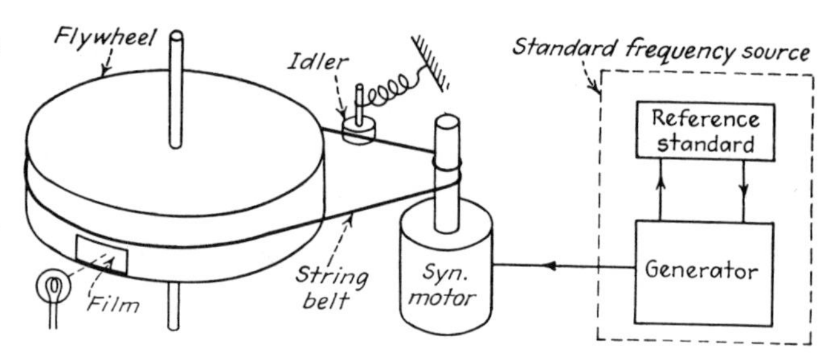


METHOD 2—Magnetic low pass torsional filter with worm gear reduction from synchronous motor.

SPEED REGULATION: Excellent—accuracy 0.01 percent.
SYNCHRONISM: Exact with secondary standard.
REFERENCE STANDARDS:
Primary—tuning fork; 1 part in 500,000.
Secondary—synchronous motor; accuracy, 1 part in 50,000.
EFFECTS CONTRIBUTING TO ERROR:
Primary—flywheel bearing rumble; flywheel oscillation; low pass filter characteristic.
Secondary—standard frequency source; reduction gear eccentricity; bearing eccentricities; magnetic filter eccentricities; synchronous motor angular error.
SIZE AND WEIGHT: Appreciable flywheel weight and size—about 10 in. dia and 35 lb for 1 cycle low frequency cutoff.
NECESSARY SERVICING IN FIELD: Electronic components; lubrication.
MANUFACTURING DIFFICULTIES: Precision cone type gear; magnetic filter symmetry; precision bearings.
REMARKS: Used in optical dial division to divide circle accurate to 0.01 percent. Eddy current effect damping can be used in combination with the hystereisis effect compliance of the magnetic low pass filter to provide superior filter characteristic. Loaded cone type worm gear is expensive.

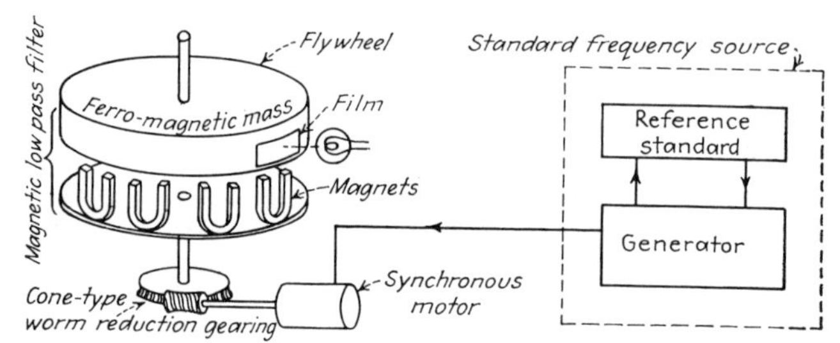


Method 3—Flywheel—rim drive by roller in magnetic circuit from gear reduction and synchronous motor.

SPEED REGULATION: Fair—accuracy 0.2 percent.
SYNCHRONISM: Not synchronous because of slippage.
REFERENCE STANDARDS: Timing track necessary.
EFFECTS CONTRIBUTING TO ERROR:
Primary—wear of roller; flexure of shaft bearings.
Secondary—gear eccentricities.
SIZE AND WEIGHT: Appreciable size and weight.
NECESSARY SERVICING IN FIELD: Lubrication of drive components.
MANUFACTURING DIFFICULTIES: Precision gears; precision bearings.
REMARKS: Simple and effective low pass filter action of rim drive irons out gear ripple and eccentricities.

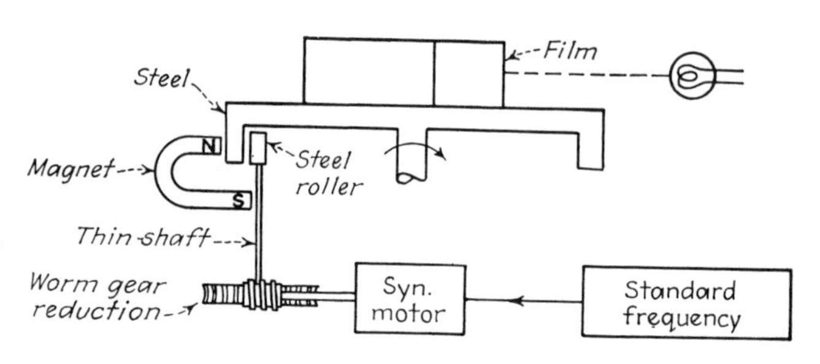

# Motion Below 10 Rpm

WILLIAM HOTINE
Potter Instrument Company

speeds), a desired percentage accuracy of speed regulation is much more difficult to maintain than at comparatively high speeds. One reason for the difficulty can be attributed to the decreasing flywheel effect.

Another limitation in obtaining speed control at low velocities is created by the accuracy of frequency or time standards. The most accurate practical reference standard is the constant frequency signal broadcast by the National Bureau of Standards. This signal has an accuracy in the order of one part in 50 million. All of the techniques listed are based, directly or indirectly, on this standard.

In the following list of constant speed control techniques, it should be noted that the most inaccurate method is the one utilizing a mechanical drive.

**METHOD 4—Synchronous friction drive. Servomechanism adjusts variable ratio friction drive to be in synchronism with reference gears.**

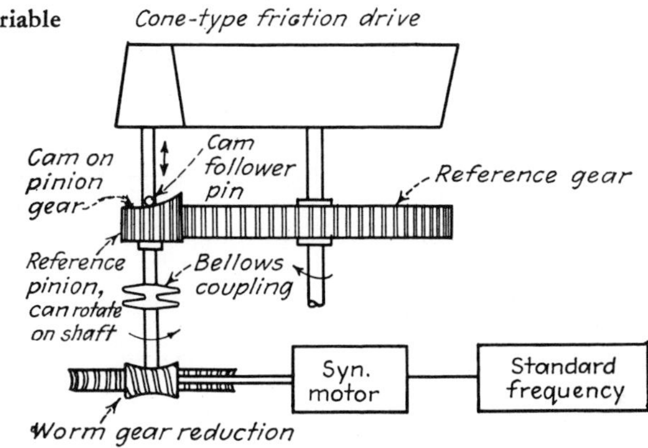


SPEED REGULATION: Good—accuracy 0.1 percent.
SYNCHRONISM: In synchronism with secondary standard within 1 part in 1,000.
REFERENCE STANDARDS:
Primary—synchronous motor.
Secondary—reference gears.
EFFECTS CONTRIBUTING TO ERROR:
Primary—bearings; gear eccentricities.
Secondary—friction drive eccentricity; synchronous motor.
SIZE AND WEIGHT: Fairly bulky.
NECESSARY SERVICING IN FIELD: Lubrication of drive components.
MANUFACTURING DIFFICULTIES: Precision machining required; precision gears; precision bearings.
REMARKS: Quick starting. High torque.

**METHOD 5—Electronically controlled d-c motor. Optical pickup of mechanically generated rotating standard frequency record compared with reference standard frequency regulates d-c motor current.**

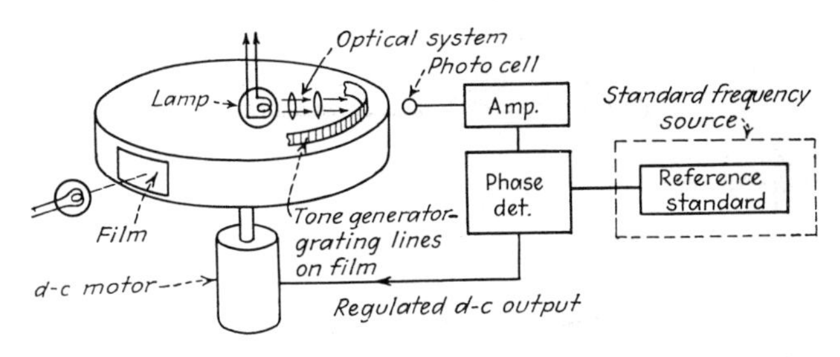


SPEED REGULATION: Excellent—accuracy 0.01 percent.
SYNCHRONISM: In synchronism with standard to mechanical accuracy of optical tone generator.
REFERENCE STANDARDS:
Primary—tuning fork; accuracy 1 part in 500,000.
Secondary—tone generator; mechanical accuracy 1 part in 10,000.
EFFECTS CONTRIBUTING TO ERROR:
Primary—tone generator segment spacing accuracy and uniformity; temperature effects.
Secondary—d-c motor bearings; d-c motor commutation; resolution of optical system; time constants of masses and regulating circuit.
SIZE AND WEIGHT: Light and compact.
NECESSARY SERVICING IN FIELD: Motor Brushes; electronic components.
MANUFACTURING DIFFICULTIES: Manufacture of mechanical tone generator to required tolerance; low inertia d-c motor.

**METHOD 6—Magnetic recording controlled motor. Standard frequency is magnetically recorded on ferro-magnetic material. Picked up 360 deg out of phase and applied to phase comparator, d-c output of which regulates motor speed.**

SPEED REGULATION: Excellent—accuracy, 0.01 percent.
SYNCHRONISM: Synchronous.
REFERENCE STANDARDS:
Primary—tuning fork; accuracy 1 part in 500,000.
Secondary—tone frequency recorded.
EFFECTS CONTRIBUTING TO ERROR:
Primary—definition of magnetic aperture and of ferro-magnetic material; main bearing eccentricity.
Secondary—temperature.
SIZE AND WEIGHT: Fairly bulky.
NECESSARY SERVICING IN FIELD: Mainly electronic.
MANUFACTURING DIFFICULTIES: High precision machining.
REMARKS: Very elaborate electronic equipment—high power consumption.

# How to Obtain Constant Speed Motion Below 10 Rpm (continued)

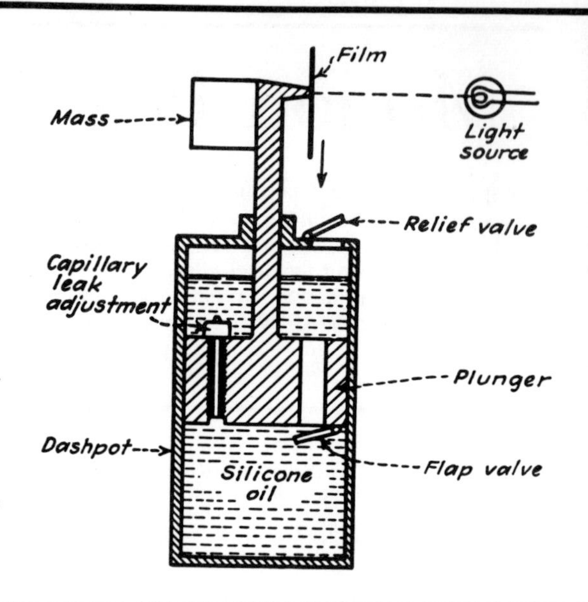


**METHOD 7—Oil dashpot—falling mass controlled by oil viscosity.**

SPEED REGULATION: Poor—accuracy 2 percent.
SYNCHRONISM: Not synchronous.
REFERENCE STANDARDS: Timing track necessary.
EFFECTS CONTRIBUTING TO ERROR:
Primary—Variation in oil viscosity; temperature; friction, vibration; turbulence.
Secondary—wear.
SIZE AND WEIGHT: Small and compact.
NECESSARY SERVICING IN FIELD: Little.
MANUFACTURING DIFFICULTIES: High precision fits and fine finishes on some components.
REMARKS: Viscosimeter art applies. Not used at present in high precision timing.

**METHOD 8—Magnetic dashpot—falling mass controlled by hysteresis and eddy current effects.**

SPEED REGULATION: Fair—accuracy 0.2 percent.
SYNCHRONISM: Not synchronous.
REFERENCE STANDARDS: Timing track necessary.
EFFECTS CONTRIBUTING TO ERROR:
Primary—control voltage; vibration; friction.
Secondary—flexure; windage.
SIZE AND WEIGHT: Small and compact.
NECESSARY SERVICING IN FIELD: Little.
MANUFACTURING DIFFICULTIES: High precision machining; uniformity of cylinder material important.
REMARKS: Considerably superior to oil dashpot. Inverse feedback used to improve linearity.

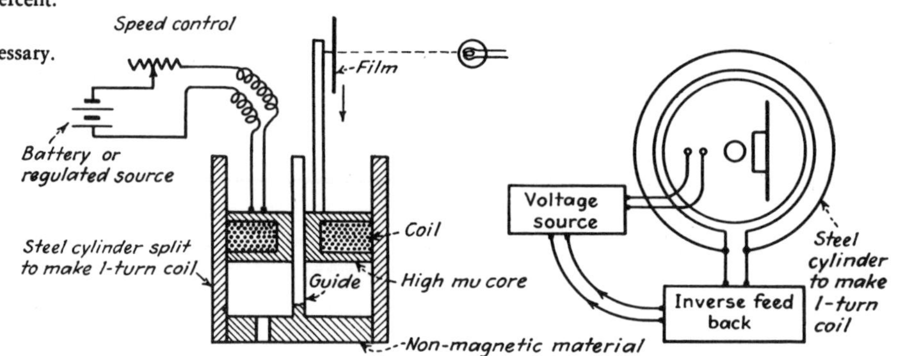


**METHOD 9—Lathe—moving element driven by screw from synchronous motor.**

SPEED REGULATION: Poor—accuracy 2 percent.
SYNCHRONISM: Not exact, affected by mechanical non-linearity.
REFERENCE STANDARDS:
Primary—tuning fork; accuracy 1 part in 500,000.
Secondary—lead screw; accuracy 1 part in 50,000.
EFFECTS CONTRIBUTING TO ERROR:
Primary—backlash and wear in guides; nuts; thrust bearing; play in bearing; non-linearity of screw; wear on screw; variable friction; temperature effects.
Secondary—synchronous motor bearings; standard frequency source; synchronous motor error.

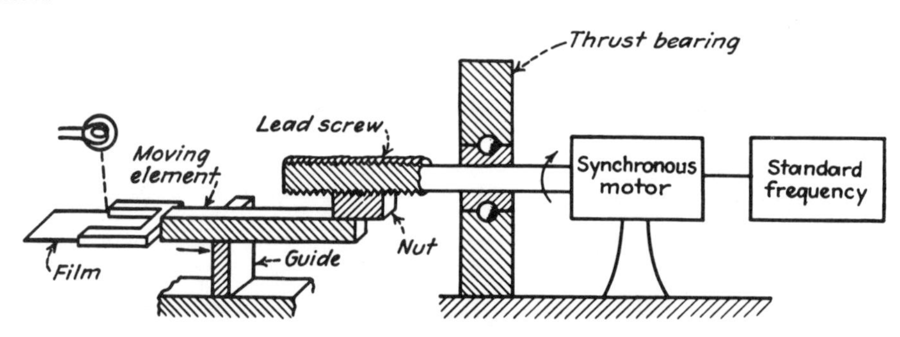


SIZE AND WEIGHT: Small—compact drive mechanism. Synchronous motor should be large enough to override variable load.
NECESSARY SERVICING IN FIELD: Lubrication of drive components.
MANUFACTURING DIFFICULTIES: Very high precision fits and fine finishes on drive components; precision bearings, temperature compensation construction.
REMARKS: Best lead screw manufacturing practice introduces errors up to 0.01 percent.

**METHOD 10—Galvanometer—mirror driven by moving coil.**

SPEED REGULATION: Poor—accuracy 0.5 percent.
SYNCHRONISM: In synchronism with secondary standard, within 1 part in 5,000.
REFERENCE STANDARDS: Timing track necessary.
EFFECTS CONTRIBUTING TO ERROR:
Primary—non-linear magneto-motive torque; non-linear shaper amplifier; variable friction in bearing; play in bearings; flexure; windage, optical amplification.
Secondary—Standard frequency source.
SIZE AND WEIGHT: Small and compact; may require much electronic equipment and service.
NECESSARY SERVICING IN FIELD: Electronic components.
MANUFACTURING DIFFICULTIES: High precision required; delicate assembly.
REMARKS: Small mechanical errors in moving system multiplied by optical system. Optical difficulties. Electronic difficulties.

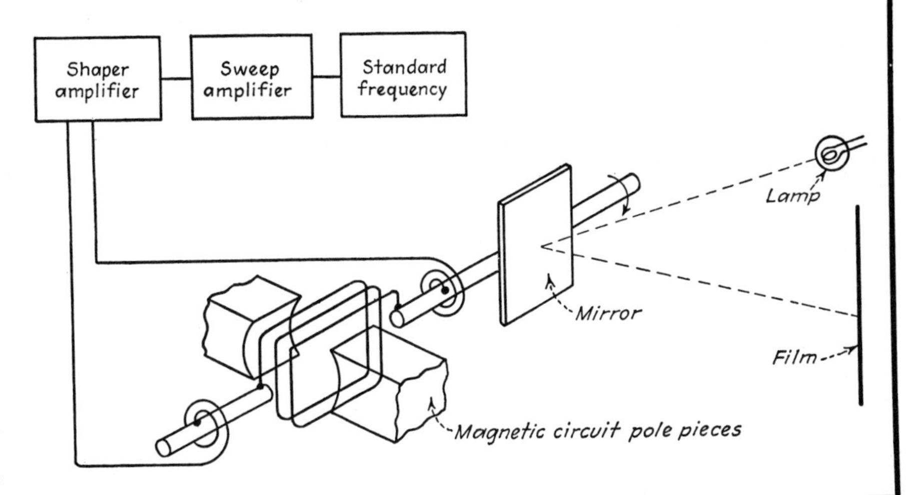


**METHOD 11—Electrostatic voltmeter—mirror driven by moving vane.**

SPEED REGULATION: Poor—accuracy 0.5 percent.

SYNCHRONISM: In synchronism with secondary standard within 1 part in 100.

REFERENCE STANDARDS: Timing track necessary.

EFFECTS CONTRIBUTING TO ERROR:
Primary—Vibration; flexure; windage.
Secondary—standard frequency source; linearity of variable voltage.

SIZE AND WEIGHT: Fairly bulky.

NECESSARY SERVICING IN FIELD: Electronic only.

MANUFACTURING DIFFICULTIES: High precision and delicate workmanship required.

REMARKS: Would need to be carefully handled. Small errors in moving system multiplied by optical system. Optical difficulties.

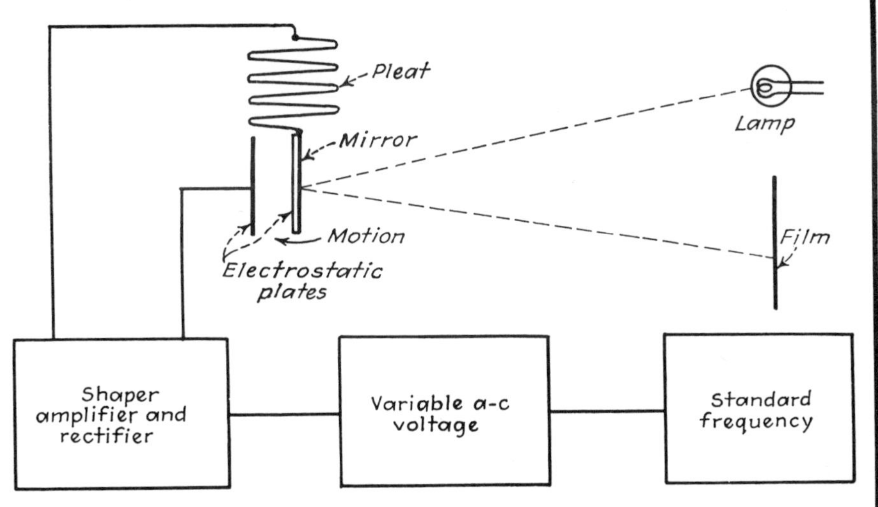


**METHOD 12—Cathode ray recorder, flying spot type—raster (scan pattern) embodying desired line pattern intensity modulated by control grid of cathode ray tube.**

SPEED REGULATION: Good—accuracy 0.1 percent.

SYNCHRONISM: Exact with secondary standard.

REFERENCE STANDARDS:
Primary—tuning fork.
Secondary—sweep circuit.

EFFECTS CONTRIBUTING TO ERROR:
Primary—non-linearity of sweep; stability of sweep components; thermal stability of cathode ray tube.
Secondary—halation on cathode ray tube face; spot size; optical system.

SIZE AND WEIGHT: Bulky and heavy.

NECESSARY SERVICING IN FIELD: Electronic only.

MANUFACTURING DIFFICULTIES: Linear sweep; cathode ray tube spot size; halation.

REMARKS: Unaffected by vibration; non-linearity may cancel out if same sweep is used for playback.

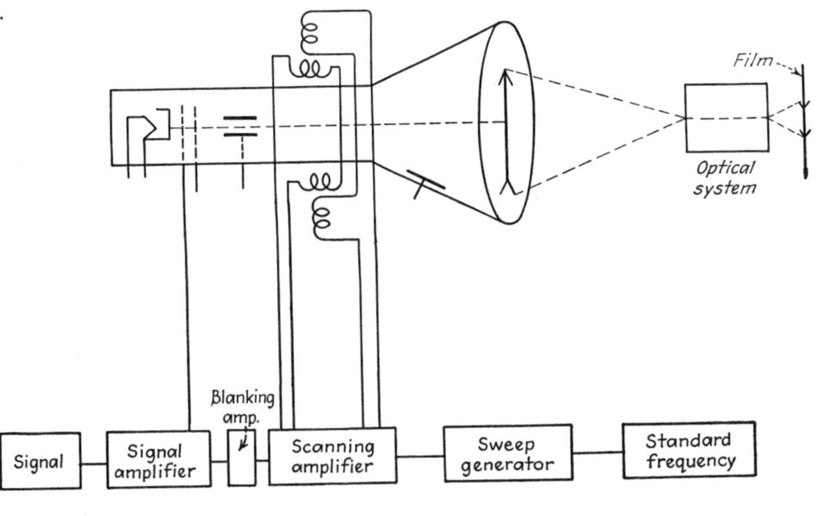

KIRK CARLSTEN, Chief Engineer
and KARSTEN HELLEBUST
Metron Instrument Company

# 12 Ways of

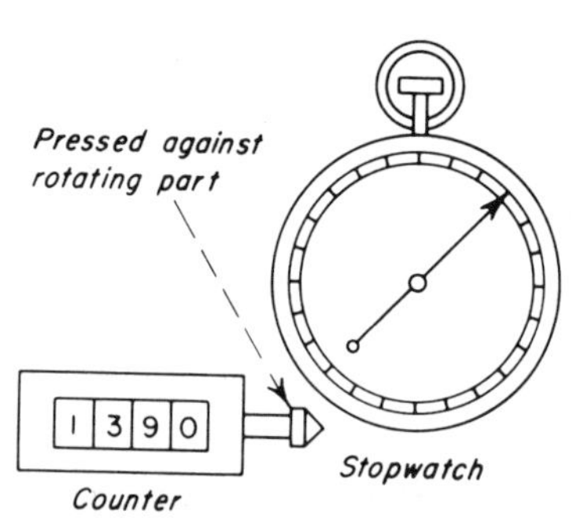


Fig. 1—Simple counter and separate stop watch: Requires dexterity since counter is held with one hand—stop watch with other hand. Average speed measured during one minute interval by noting the counter readings at beginning and end of the time interval.

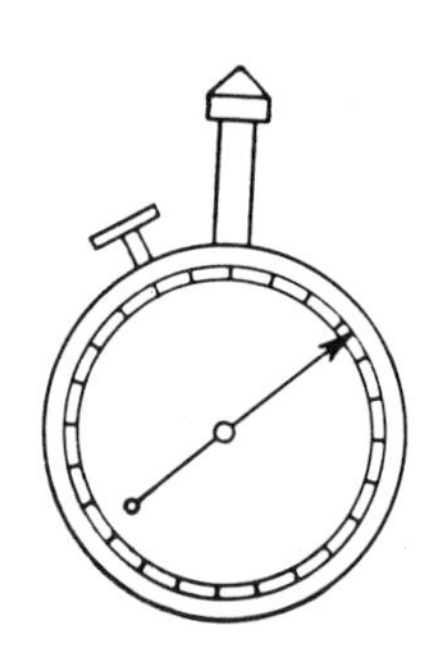

Fig. 2—Combination counter and stop watch: Counter starts when button is pressed and then watch automatically stops counter hand at end of a given time (say six seconds). Dial is calibrated to read in rpm. Measures average speed.

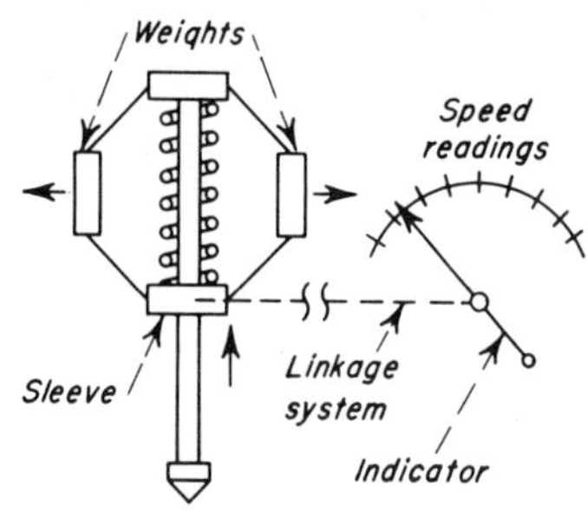


Fig. 3—Centrifugal tachometers: Centrifugal governor type mechanisms with weights which move outwardly from shaft as speed increases causing sleeve to move up. Sleeve through linkage moves indicator pointer in proportion to speed. Care must be taken to avoid overspeeding.

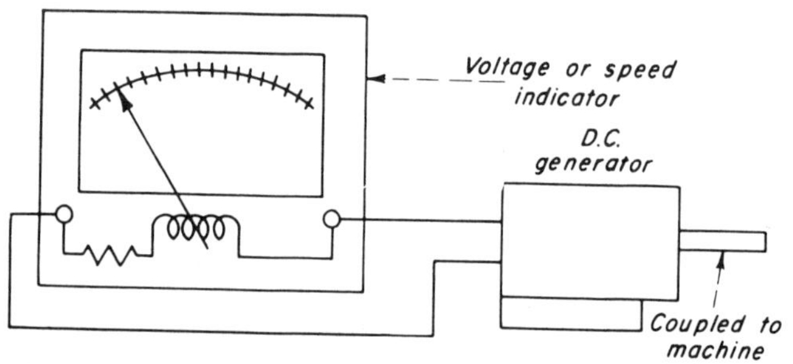


Fig. 4—D.C. Generator tachometers: D.C. voltage generated in proportion to speed and displayed on D.C. indicator or recorder. System does not use gearing and is usually limited to 150-2000 rpm at full scale. Generator matched with a particular indicator because voltage not exactly linear with speed.

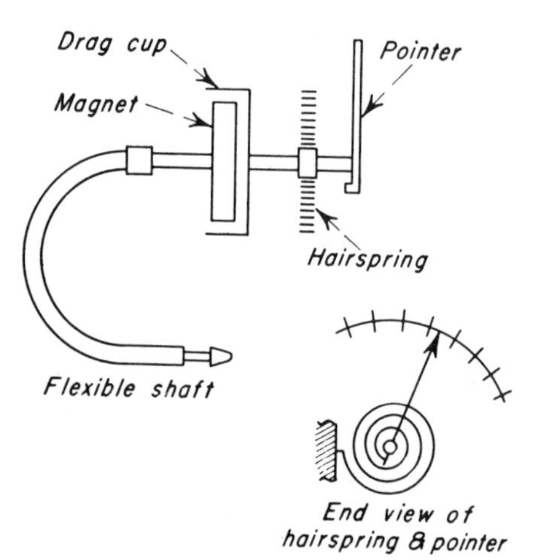


Fig. 6—Drag cup tachometers: Flexible shaft drives small permanent magnet in drag cup. Motion of magnet sets up eddy currents in wall of drag cup. Resulting rotary force causes drag cup to follow. Hair spring balanced to produce a pointer deflection proportional to speed. System simple and inexpensive, but with limited accuracy and speed range.

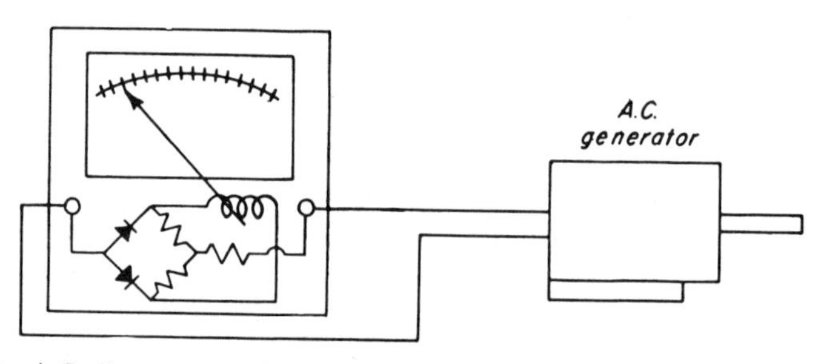


Fig. 5—A.C. Generator tachometers: A.C. voltage generated in proportion to speed and displayed on an A.C. indicator. System does not use gearing. Usually limited to full scale readings of 500 rpm minimum and 5000 rpm maximum. Not as accurate as D.C. systems, but covers wider speed range and does not require brushes.

# Measuring Speed

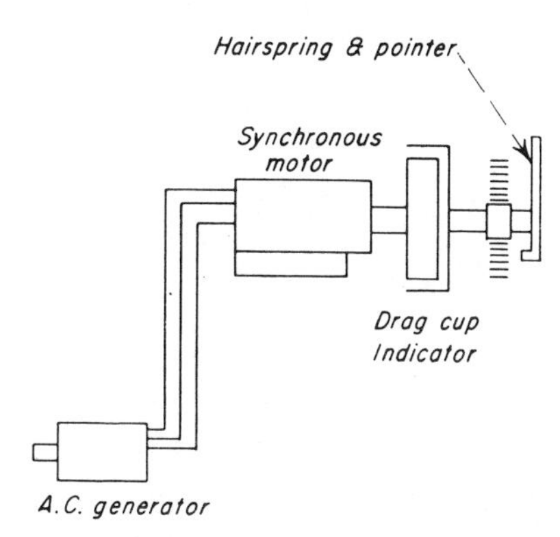


Fig. 7—Drag cup indicator driven by synchronous motor whose speed in controlled by remote A.C. generator. Synchronous motor drives small permanent magnet in drag cup similar to arrangement shown in Fig. 6. System expensive but without need for long and cumbersome flexible shafting.

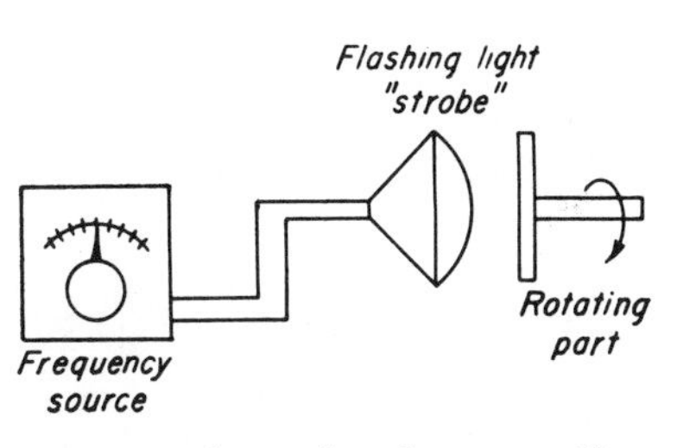


Fig. 8—Stroboscopic tachometers: Very fast flashing light used to view rotating part. Frequency of flashing adjusted until rotating part appears to stand still. Frequency reading corresponds to rpm of rotating part. Requires no accessible shaft and absorbs no power from rotating part. However, system requires constant adjustment if speed changes.

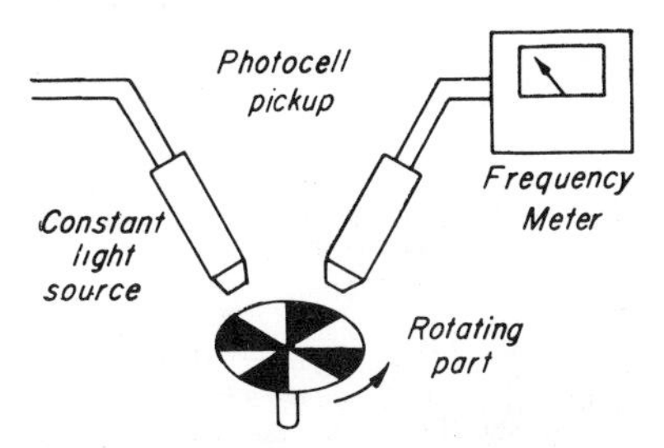


Fig. 9—Photo electric tachometers: Constant light source focused to shine on rotating disc; light reflects into photocell pickup. Intensity of light modulated by dark and light spots on rotating part. Frequency of modulations proportional to rotating speed and measured by frequency meter. Speeds up to 3,000,000 rpm can be measured but expensive equipment required.

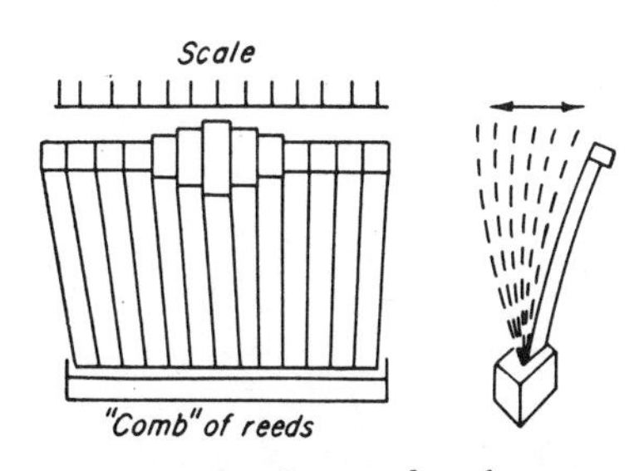


Fig. 10—Vibrating reed tachometers: "Comb" made up of a set of accurately tuned steel reeds: Comb held against any part of rotating machine, such as against case of motor. Vibration of machine vibrates only reeds "in tune" with machine's frequency. This speed, or frequency, read directly on a scale above the comb. Requires no access to a rotating shaft. Range is from 800 to 12,500 rpm with an accuracy within 50 rpm for individual reeds when new.

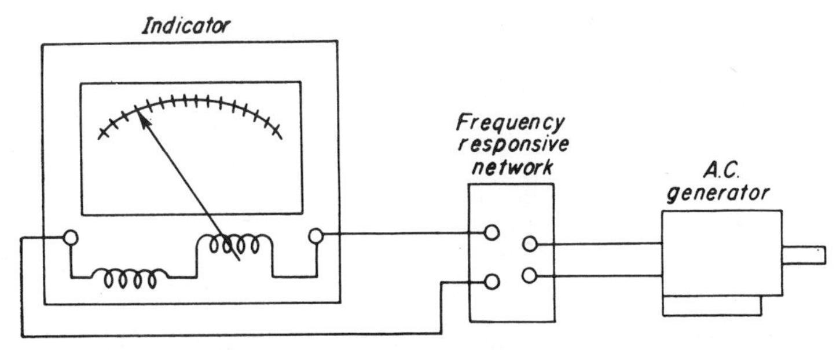


Fig. 11—A.C. frequency responsive tachometers: Similar to A.C. generator tachometers but measures generated frequency (not voltage). Range: 500 to 80,000 rpm with good accuracy. Expensive.

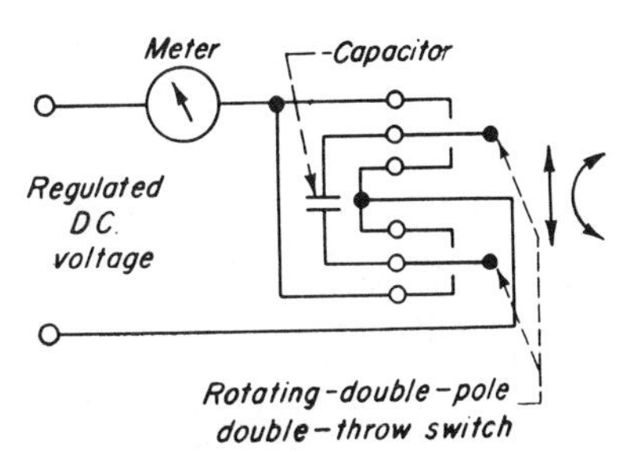


Fig. 12—Commutated capacitor tachometers: Rotating double pole, double throw switch charges and discharges capacitor. Resulting average D.C. current linear with speed. Current indicated on a D.C. microammeter as rpm. System very accurate. Linearity allows multiple range indicators and rotating switches are interchangeable.

# Temperature Regulators

L. C. BLAUVELT

Hoffman LaRoche

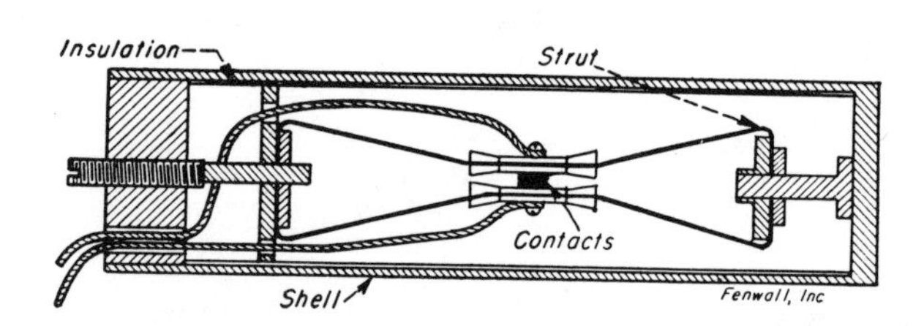


**1.** Bimetallic device is simple, compact and precise. Contacts mounted on low-expansion struts determine slow make-and-break action. Shell contracts or expands with temperature changes, opening or closing the electrical circuit that controls a heating or cooling unit. Adjustable and resistant to shock and virbration. Range: —100 to 1,500 F. Accuracy: Operates on less than 0.5 deg temperature change.

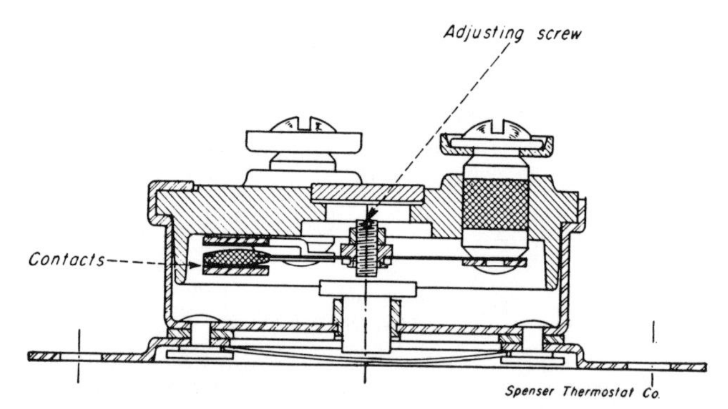


**2.** Typical inclosed disk-type, snap-action control has fixed operating temperature. Suitable for unit and space heaters, small hot water heaters, clothes dryers and other applications requiring non-adjustable temperature control. Useful where dirt, dust, oil or corrosive atmosphere is involved. Available with various temperature differentials and with manual reset. Depending on model, temperature setting range is from —10 to 550 F and minimum differential may be 10, 20, 30, 40 or 50 deg F.

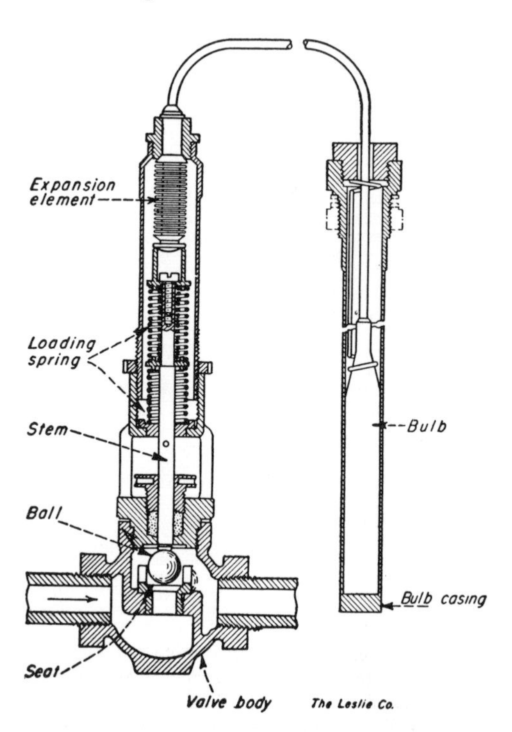


**5.** Self-contained regulator is actuated by expansion or contraction of liquid or gas in temperature sensitive bulb that is immersed in medium being controlled. Signal is transmitted from bulb to sealed expansion element that opens or closes the ball-valve. Range: 20 to 270 F. Accuracy: ±1 deg. Max. press. 100 psi for dead end service, 200 psi continuous flow.

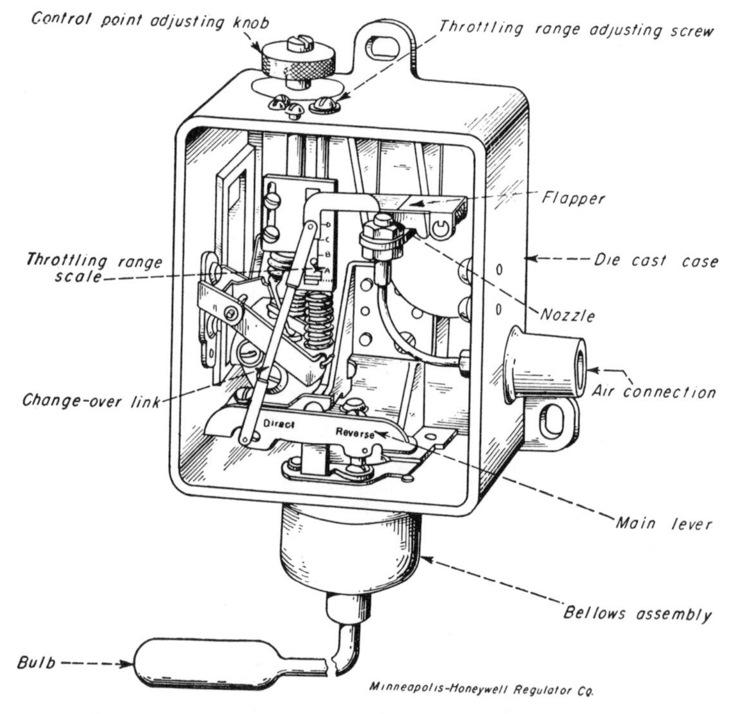


**6.** Remote bulb, non-indicating regulator uses a bellows assembly to operate a flapper. This allows air pressure in the control system to build up or bleed depending upon the position of the change-over link. Unit can be direct or reverse acting. Control knob adjusts the setting and the throttling range adjustment determines the percentage of the control range in which full output pressure (3-15 psi) is obtained. Range: 0-700 F. Accuracy: About ±0.5 percent of full scale range depending upon installation factors and associated variables.

Temperature regulators are either of the on-off or throttling type. The characteristics of the process determine which should be used. Within each group, selection of a device is governed by the accuracy required, space limitations, simplicity and cost.

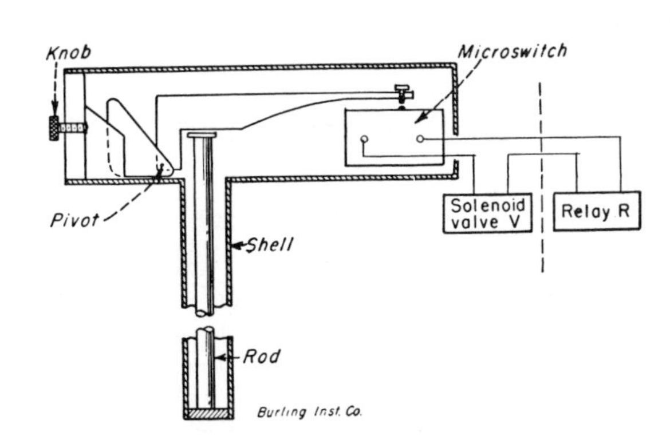


**3.** Bimetallic unit has rod with a low coefficient of expansion and a shell with a high coefficient. Microswitch gives a snap action to the electrical control circuit. Current can be large enough to operate a solenoid valve or relay directly. Set point is adjusted by knob that moves the pivot point of the lever. Range: —20 to 1,750 F. Accuracy: 0.25 to 0.50 degrees.

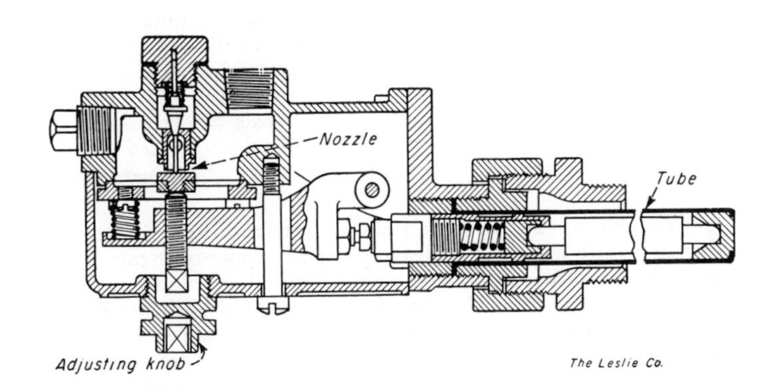


**4.** Bimetallic actuated, air piloted control. Expansion of rod causes air signal (3-15 psi) to be transmitted to a heating or cooling pneumatic valve. Position of the pneumatic valve depends upon the amount of air bled through the pilot valve of the control. This produces a throttling type of temperature control as contrasted to the on-off characteristic that is obtained with the three units described previously. Range: 32 to 600 F. Accuracy: ± 1 to ±3 F depending upon the range.

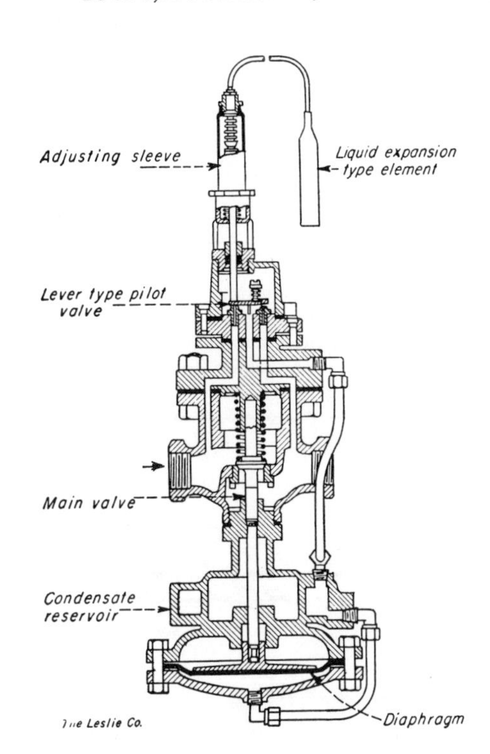


**7.** Lever-type pilot valve is actuated by temperature sensitive bulb. Motion of lever causes water or steam being controlled to exert pressure on a diaphragm that opens or closes the main valve. Range: 20 to 270 F. Accuracy: ±1 to 4 degrees. Pressure: 5-125 psi, steam; 5-175 psi, water.

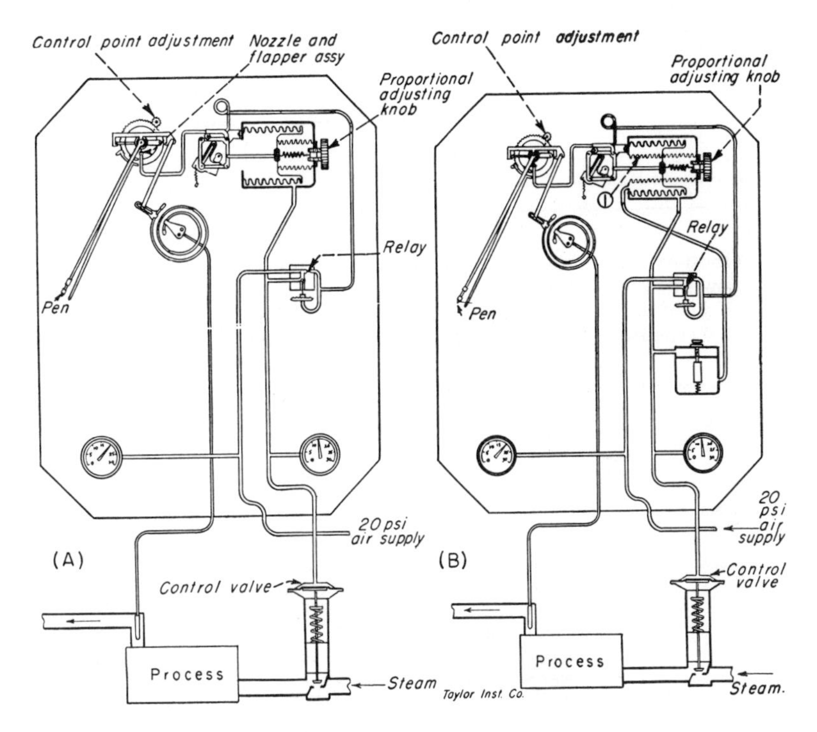


**8.** Two recording and controlling instruments with adjustable proportional ranges. In both, air supply is divided by a relay valve. A small portion of the flow goes through nozzle and flapper assembly. The main flow goes to the control valve. Unit *B* has an extra bellows for automatic resetting. It is designed for systems with continuously changing control points and can be used where both heating and cooling are required for one process. Both *A* and *B* are easily changed from direct to reverse acting. Accuracy: One percent of range of —40 to 800 F.

# Some Thermocouple Details for

Howard W. Cole, Jr.
Consultant
Mountain Lakes, N. J.

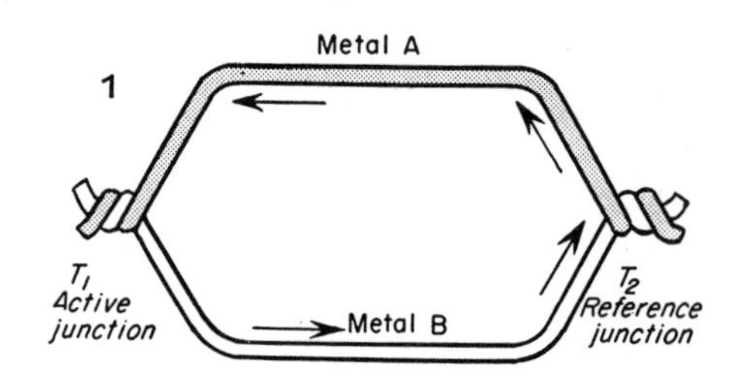


**1** Connecting two dissimilar conductors together to form a closed loop makes a simple thermocouple. With the two junctions at different temperatures, an electromotive force (emf) will be generated in the loop. The magnitude of this emf is dependent on the chemical composition of the dissimilar metals and is proportional to the temperature difference ($T_1$—$T_2$) of the junctions. In practice, the junction $T_2$ is usually the terminals of the measuring instrument and is maintained at room temperature. Thus the indicated temperature is the difference between the room temperature and the actual measured temperature.

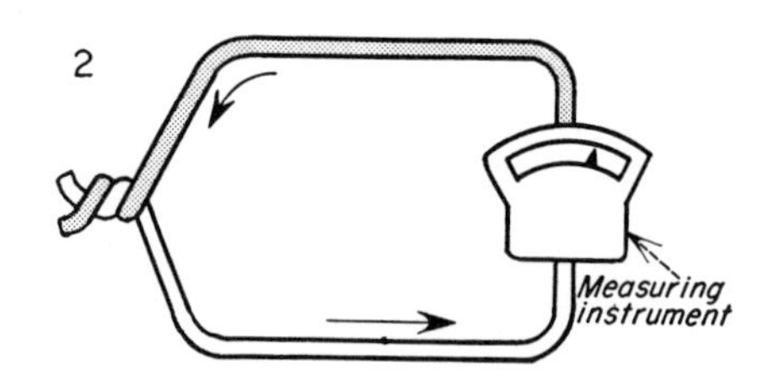


**2** Most modern measuring instruments have a built-in cold junction compensation which is usually a resistance element whose resistance varies as its temperature. Such devices provide a temperature indication that is referenced to the zero temperature of the scale being used. Although the electrical portion of the instrument may be constructed of a material different from either of the thermocouple metals, the possible thermoelectric effect is canceled since the terminals are closely spaced and there is no temperature difference.

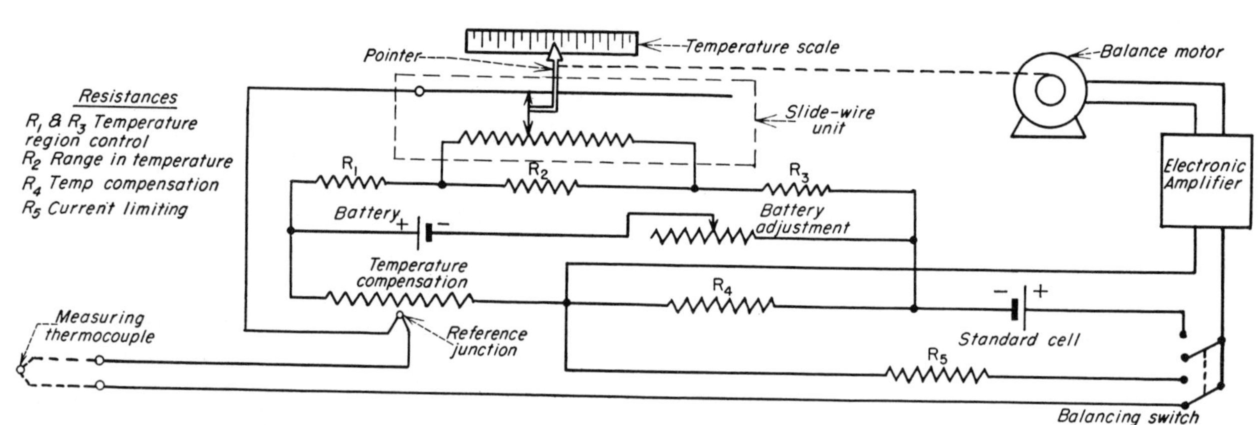


**3** A null-balance potentiometer is an indicating device to show when the emf in the instrument exactly balances the generated emf in the thermocouple. With this type, the resistance of the measuring thermocouple, the lead wires and the instrument is unimportant; this condition allows almost unlimited use of all types of thermocouples, lead wire size and length. Internal resistances of the instrument permit altering both the temperature range and the region being observed.

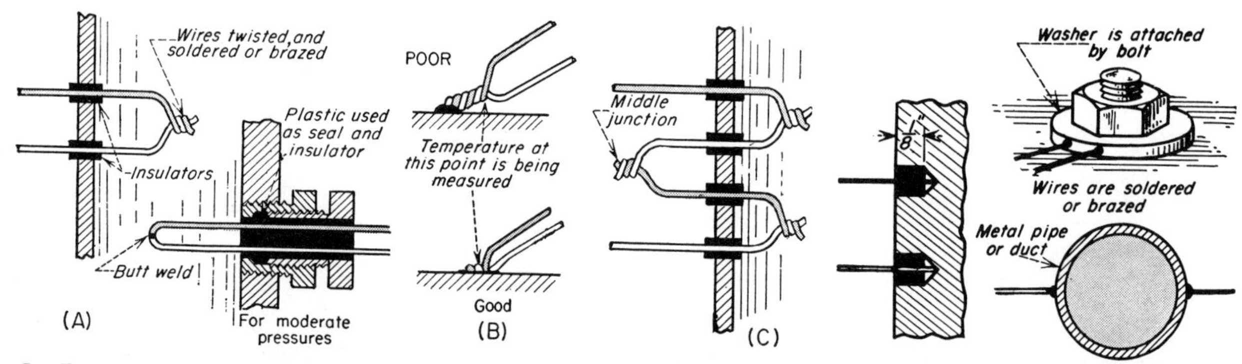


**6** For point-source measurement, a simple but positive connection of the wires is required. (A) Twisted connection should be short and welded to insure a good junction. (B) To properly obtain the temperature of a body, a point nearest to the wire junction should be in contact. (C) Connecting thermocouples in series doubles the emf; however middle junction should be at cold junction temperature.

**7** For averaging the temperature, staking each of the wires with small copper ferrules can be done. One method uses a copper washer to which thermocouples are attached. Tubes can serve as junctions.

# Temperature Measurement

| Thermocouple* Material | Recommended Operating Range | | Millivolts at 100 C Reference Junction at 0 C | Accuracy Using Thermocouple Wire as Leads | | Accuracy Using Low cost Lead Wires | | | Suitable Surrounding Atmosphere | Application | Remarks |
|---|---|---|---|---|---|---|---|---|---|---|---|
| | Deg. C | Deg. F | | Temperature Range, F | Error, ± | Material | Temperature Range F** | Error in ± F | | | |
| copper and constantan | −200 to 400 | −328 to 752 | 4.276 | −200 to −75<br>− 76 to +200<br>201 to 700 | 1.0%<br>1.0 F<br>0.5% | copper and constantan | −75 to +200 | 1 | oxidizing or reducing | precision low temperature measurements | Closest tolerance and most stable of base metal thermocouples |
| iron and constantan | −100 to 300 | −148 to 1652 | 5.40 | 0 to 530<br>531 to 1400 | 3.0 F<br>0.32% | iron and constantan | 0 to 400 | 3 | oxidizing or reducing | general purpose use | Relatively high millivolt output—close tolerance |
| chromel and alumel | 0 to 1200 | + 32 to 2192 | 4.10 | 0 to 530<br>531 to 2300 | 5.0 F<br>0.75% | chromel and alumel | 0 to 400 | 4 | oxidizing | general purpose high temperature measurements | Best of base metal Thermocouples for higher temperatures |
| chromel and alumel | 0 to 1200 | + 32 to 2192 | 4.10 | | | copper and constantan | 75 to 200 | 6 | oxidizing | | |
| platinum and platinum rhodium | 500 to 1400 | 932 to 2552 | 0.643 | 0 to 1000<br>1000 to 2700 | 2.5 F<br>0.25% | copper alloy | 75 to 400 | 12 | oxidizing | precision high temperature measurements—standardization | Close tolerance, low millivolt output, highest temperature range |

* Also see Instrument Society of America Standards. ** Temperature range of lead-in wires.

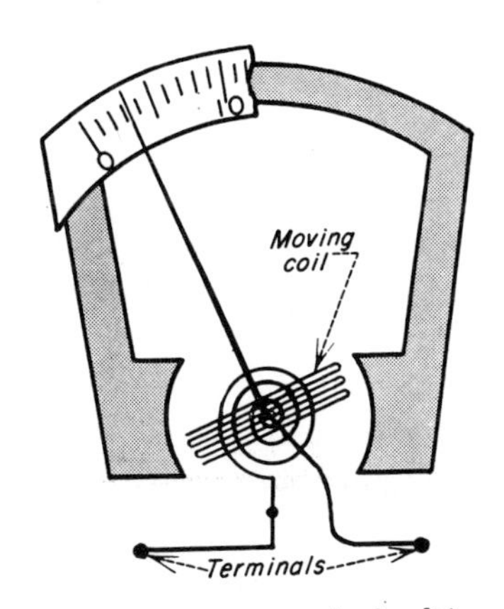


**4** A simple instrument is the d'Arsonval type. It measures the current and thus all resistances, including the thermocouple wire, must be considered.

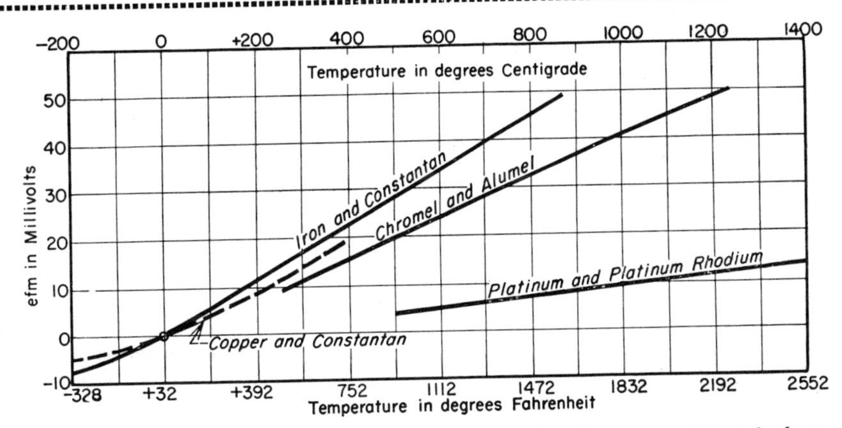


**5** Although any two dissimilar metals can be used as thermocouples, only four types are commonly used. The alloy content of these materials is closely maintained to standards of close tolerances. Curves show the electromotive characteristics of each thermocouple combination over the full operating range.

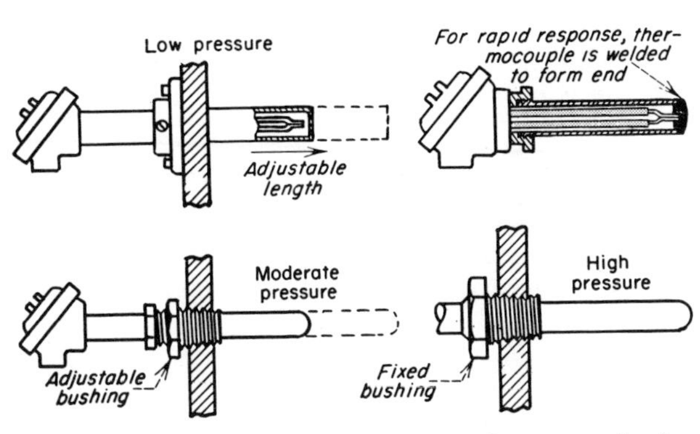


**8** To protect thermocouples from corrosion, contamination, and mechanical damage, there are many styles and types of protecting tubes available. Such tubes increase the time lag due to a longer heat path. These protecting tubes may also act as a conductor of heat away from the junction.

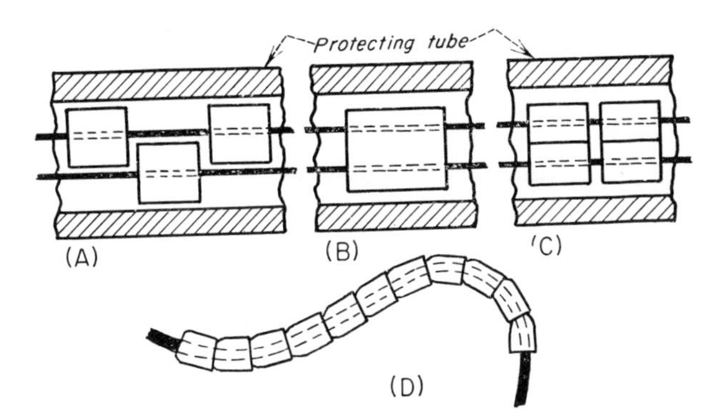


**9** For insulating bare wires inside of the protecting tubes, asbestos covering is used up to 400F and Fiberglas to 1000F. Also ceramic beads are available: (A) large, single wire bead, (B) double wire bead, (C) small single wire bead and (D) fish-spine beads permit the wire to form a bend.

# TYPES OF AUTOMATIC TIMERS

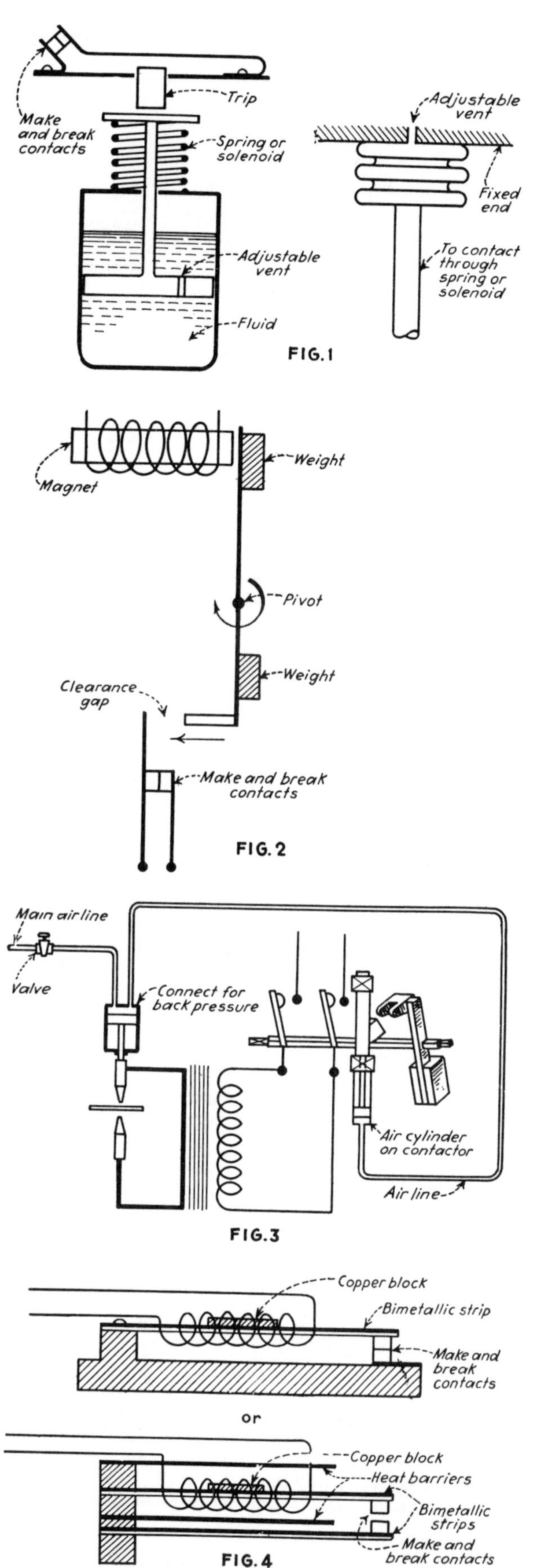


**Thermal, Mechanical and Electronic Methods for Automatic Timing of Industrial Machinery, Welding Operations and Motor Control Devices**


GILBERT SMILEY

*Chief Engineer, General Control Co.*


Fig. 1—Dashpot principle. Simplest form consists of a piston or plunger operating in oil, mercury, or air. Adjustable small orifices or bleeders provide time adjustment. A by-pass may be provided near the end of the piston travel for snap action closing of the contact. Widely used because of its simplicity and low cost. When air is used, changing clearances due to dust, gumming of lubricant, and leakage affect the timing. If oil is used, the temperature will change oil viscosity and affect the timing. Also subject to error because of clearance changes from wear.

Fig. 2—Inertia mechanism. Time-delay is by virtue of the inertia of two weights mounted on a pivoted arm, and the length of arc to be traversed before mechanical contact is made. Tilted by gravity, this device gives a relatively short interval and becomes clumsy for long time intervals.

Fig. 3—Contactor works on back pressure from the main cylinder on the welder, assuring pressure between the welding points before the welding contactor closes. When the back pressure has built up to a predetermined value, the plunger moves upward at a definite rate of speed and the hardened cam closes the main contacts. After a predetermined time, the cam moves by the roller that it engages and the main contacts open. One adjustment sets the back pressure at which the contactor plunger starts to move, and therefore determines the lag in applying the current after pressure has been applied. A second adjustment changes the needle valve opening to the contactor air cylinder and thus times the upstroke. This determines the welding time. A third adjustment varies the time of the down stroke and is of importance only when used with a repeater.

Fig. 4—Thermal relays. Inexpensive time-delay utilizing the effect of a heating coil around a bimetallic strip. Least accurate device. Has a slow make and break action. For longer time intervals, a copper block may be mounted to absorb some of the heat; the larger the block of copper the longer the time interval. Time intervals ranging from ½ sec. to 5-10 min. are possible with this device.

Fig. 5—Magnetic time-delay, used on direct current only. Relatively inexpensive, effects time-delays up to 10 sec. by means of residual magnetism. Magnet may be copper jacketed, or may have copper rings, or may have short-circuited turns around the magnet. Variation in the amount of copper, or in the resistance of short-circuited turns will affect the time-delay.

Fig. 6—Magnetic drag time-delay. A small electro-magnet is used, and the motion of the relay plunger is made to revolve a metal disk in the field of the magnet. The rotation of the disk is retarded by magnetic induction. Reliable device, trouble free, but relatively expensive.

Fig. 7—Vacuum tube. Condenser charged or discharged through a resistor closes a relay after definite time, using direct current. When switch is open, the condenser discharges slowly through shunt resistor. This lowers the negative potential on the grid, and at the critical value the plate current will rise enough to operate the relay. Full line voltage may be applied to the condenser to obtain longer time delay.

Fig. 8—In this circuit, operation is maintained for a pre-determined time after the starting impulse has stopped. When the button has been pressed, the filament gets current in series with relay winding No. 1, and the relay pulls up, locking in the circuit. The second contact charges the condenser negative, and no plate current flows. When button is released, the relay stays closed until condenser discharges. Then the plate current flows through the second relay winding in opposition to the first, releasing the armature. Applicable to direct current or rectified alternating current only.

Fig. 9—In the Westinghouse electronic relay there is no temperature error, reset is instantaneous, adjustment is easy and first cost is low. When the switch is closed, the tube passes current. As the current increases, the increasing IR drop from the potentiometer causes a charging current through condenser. The IR drop across the resistor due to this current applies the negative bias to the grid. Plate current cannot build up very rapidly, because the faster it increases the more negative the grid becomes. After a time period, adjustable through potentiometer, the plate current will operate relay. The time-delay is proportional to the product of resistance and capacitance. Long delays require large resistors, and short delays correspondingly small resistors. Maximum time-delay with this device is about 3 min. About 0.05 sec. is the minimum.

Various mechanical devices have been adapted to replace the dashpot: one uses an escapement pendulum similar to those for clocks. Time is varied by changing the length of the pendulum. Mechanical clocks may be used, driving a cam or traveller, or tilting a mercury switch, or any type of make and break mechanism.

Motor-driven timers are used where long time intervals are required and where starting is infrequent. Generally a small pilot motor drives a contact cylinder through reduction gearing. For motor-driven timing devices a synchronous motor is generally used, driving through gearing. These motors offer timing intervals from 1 sec. to infinity. A non-synchronous direct current shunt or an a.c. induction motor of the capacitor type may also be used to drive a timer, since these motors maintain approximately constant speed. However, the accuracy of this type of timing is variable.

Electronic time-delay devices for long time intervals are expensive. Circuits employing thermionic tubes will give a time-delay of more than 30 min. by the use of carbon resistors. Condensers and resistors vary with use, humidity and temperature, and adjustable resistors do not always hold their characteristics.

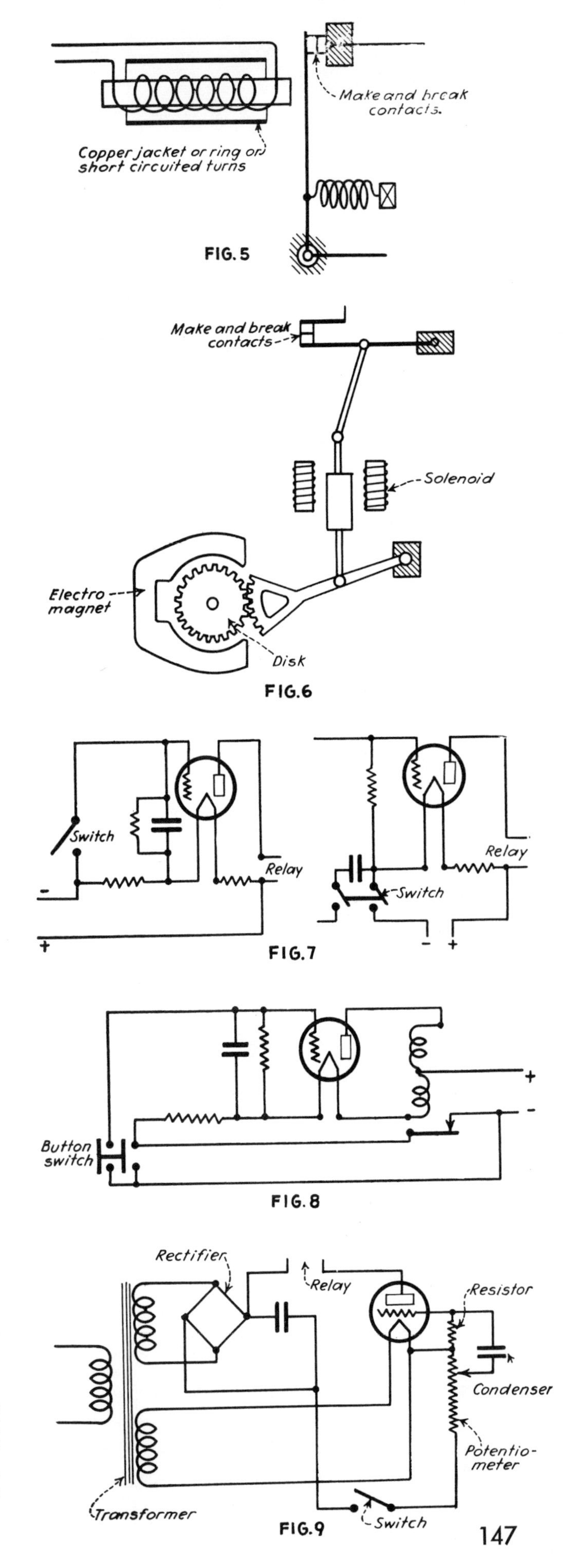

FIG. 5

FIG. 6

FIG. 7

FIG. 8

FIG. 9

# 7

# DIMENSIONS AND DESIGN


Check-chart for Single-part Complexity 150
Formulas and Curves Give Quick Answers for Partially Loaded Continuous Beams 152
Nomogram to Determine Parallel Axis Moment of Inertia 154
The 6 Types of Corrosion, A Preventive Guide 156
Designing and Fabricating with Clad Steel Plate 158
Universal Tables for Circular Segments and Sectors 160
How to Specify Precision Knurls 164
How Much Horsepower to Pipe Liquids 166
Design Fundamentals for Molded Plastic Parts 168
Recommended Practices for Designing Porcelain Enamel Products 172
How to Calculate Stresses in Press-fit Bushings 174
Increasing the Holding Power of Press Fits 176
Ratchet Layout Analyzed 178
Tips for Designing Rubber Parts and Assemblies 180
Shrink-fit Nomograph 182
The Right Finishes for Stainless Steel 184
How to Find Final Dimensions of Stretch-formed Parts 186
Design of Toothed Mechanical Components 188
Design Data for Metal Wire Stitching 190

# Automatic Assembly:

# *CHECK-CHART for SINGLE-PART COMPLEXITY...*

- ***can guide design***
- ***reveal total complexity***

Human hands can grasp a complicated part easily and position it for assembly; no simple mechanism can do this. Engineers should, therefore, design parts shaped for easy machine assembly. This sometimes requires a viewpoint change. The design engineer usually has the underlying thought that the parts he designs will be assembled manually. His emphasis is on the design that serves its functional purpose best, with the least material.

For automatic assembly, he should be alert to the difference between a part that is easily handled by machinery and one that is not. For example, a cylinder with one domed end has to be selected, then oriented for assembly. More machinery is required to do this than would be needed for a cylinder with symmetrical ends. A slight difference in shape, insignificant to the designer, may prevent automatic assembly and increase costs by more than the slight material saving involved.

This chart shows some basic shapes in steps of increasing complexity. Each step is significant to a production engineer. In each column to the right an inherent property of the basic shape has been lost.

Multiplied column-numbers (vertical by horizontal) give a rough measure of comparative complexity for the shape located by the numbers. The "A" numbers will represent a somewhat higher score than the others.

For a complete parts list, the average "complexity number" would be a good guide to how easily the whole unit could be assembled automatically.

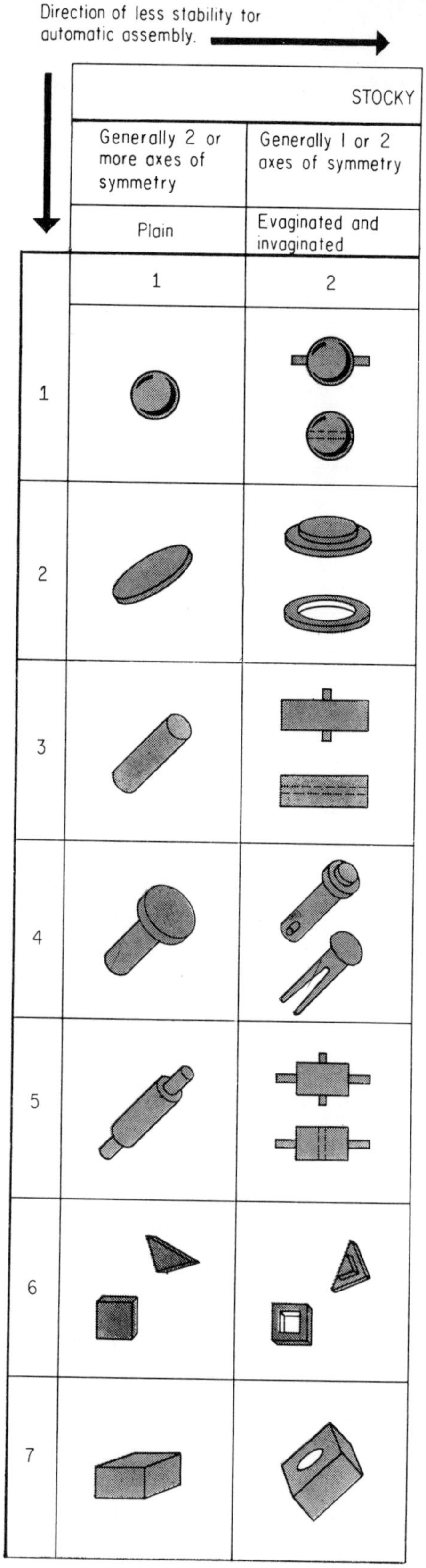

Direction of less stability for automatic assembly. →

| | STOCKY | |
|---|---|---|
| | Generally 2 or more axes of symmetry | Generally 1 or 2 axes of symmetry |
| | Plain | Evaginated and invaginated |
| | 1 | 2 |
| 1 | | |
| 2 | | |
| 3 | | |
| 4 | | |
| 5 | | |
| 6 | | |
| 7 | | |

| SHAPES | ELONGATED SHAPES | | |
|---|---|---|---|
| Generally 1 or no axis of symmetry | Generally 2 or more axes of symmetry | Generally 1 or 2 axes of symmetry | Generally 1 or no axis of symmetry |
| Off center | Plain | Evaginated and invaginated | Off center |
| 3 | 1A | 2A | 3A |


**PETER C. NOY**
*Manufacturing Engineer*
*Canadian General Electric Co., Ltd.*
*Barrie, Ont.*


**Stocky and elongated shapes** are graded according to complexity. The last three vertical columns show shapes that are similar to the first set of shapes but are not stocky. Rigidity is important although it cannot be depicted. For example, a needle and a piece of soft wire may be similar in shape but they could not be handled the same way. The stiffer needle is more stable; the wire would bend easily. Parts for automatic assembly should best be: symmetrical, simple, stocky, stable.

*Different basic shapes* may be chosen by other engineers. A sphere, however, would always head the list; it has an infinite number of axes of symmetry. The disk has two axes of symmetry, the evaginated disk one, and the disk with the off-center evagination has none.

*formulas and curves give quick answers for*

# PARTIALLY LOADED CONTINUOUS BEAMS

ALEXANDER KRIVETSKY, *Bell Aircraft Corp, Buffalo*

Problems of partially loaded, continuous beams are usually tough to analyze and tedious to calculate. The formulas and curves given here remove the headaches and hard work from many such problems.

The 3-moment equation for a constant-section, continuous beam on level, rigid supports—the most usual case—can become:

$$\frac{M_1L_1}{I_1} + \frac{2\,M_2L_1}{I_1} + \frac{2\,M_2L_2}{I_2} + \frac{M_3L_2}{I_2} =$$

$$f\,(w_1, L_1, I_1) + f\,(w_2, L_2, I_2)$$

The right-hand side of the equation differs with each different loading or loading combination. Formulas for 10 different loadings are tabulated below; curves on the continuing page speed answers to these formulas. To solve any 3-moment equation for a continuous beam with spans loaded as illustrated in table:

1. Select appropriate formula
2. Find value of loading coefficient from curves
3. Substitute numerical values $L$, $w$, and loading coefficients in the above equation. (There will be as many equations as there are indeterminate moments, e.g. a 2-span continuous beam will require a simultaneous equation with two unknowns.)

Sign conventions for the equations are: Positive moments cause compression in upper fiber; positive loads act upwards.

### Symbols

- $I_1$ = moment of inertia of left-hand span
- $I_2$ = moment of inertia of right-hand span
- $L$ = span length in general, in.
- $L_1$ = Left-hand span length, in.
- $L_2$ = right-hand span length, in.
- $M_1$, $M_2$, $M_3$ = moments at three consecutive supports
- $m$, $k$ = load-location/span-length ratios
- $w$ = distributed loading, lb/in. (cases 2 to 10) or concentrated load, lb (case 1 only)

**Loading Formulas** *(Left-hand span only is shown; right-hand span is a mirror image. Loading coefficients are in square brackets.)*

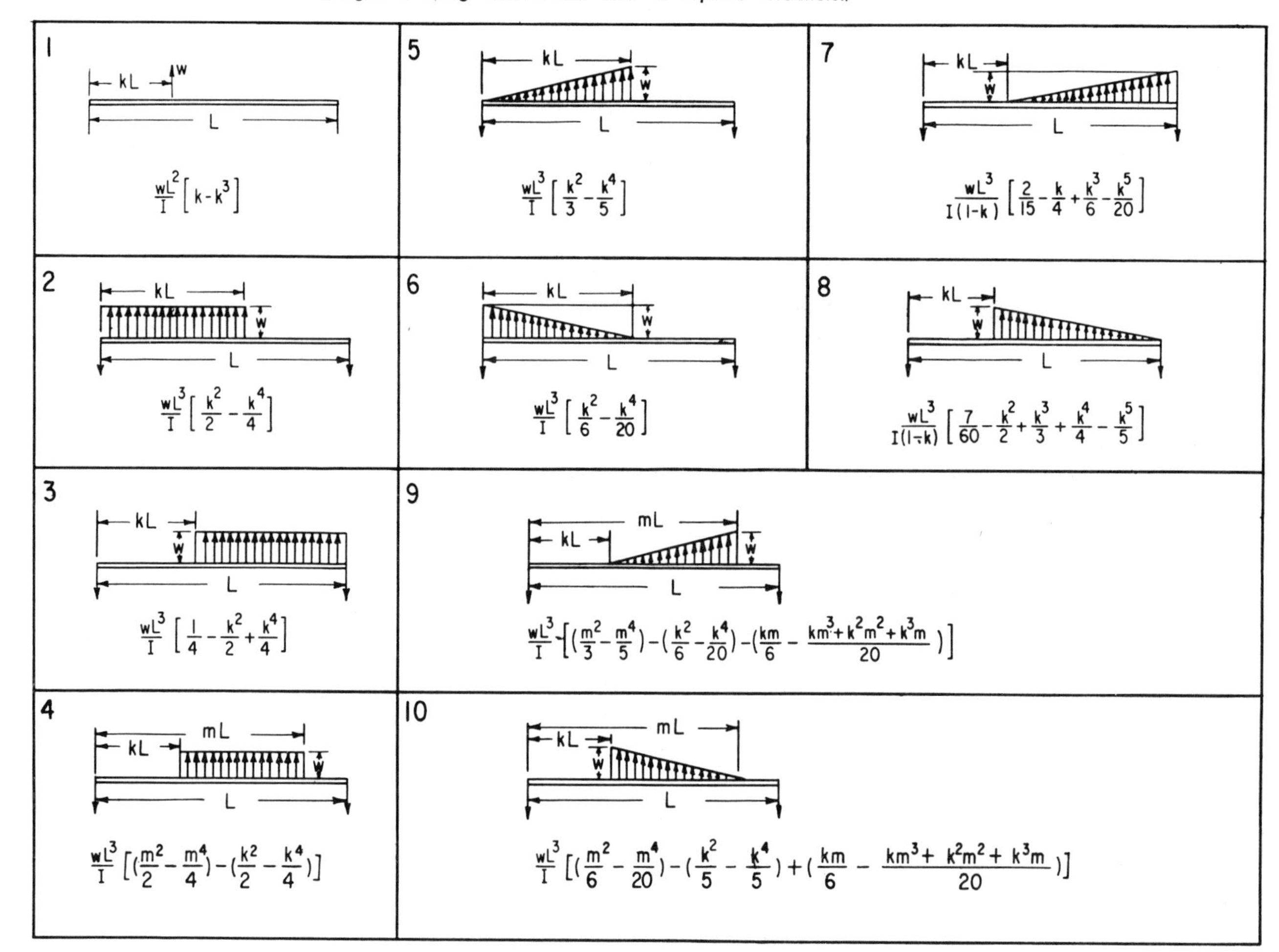

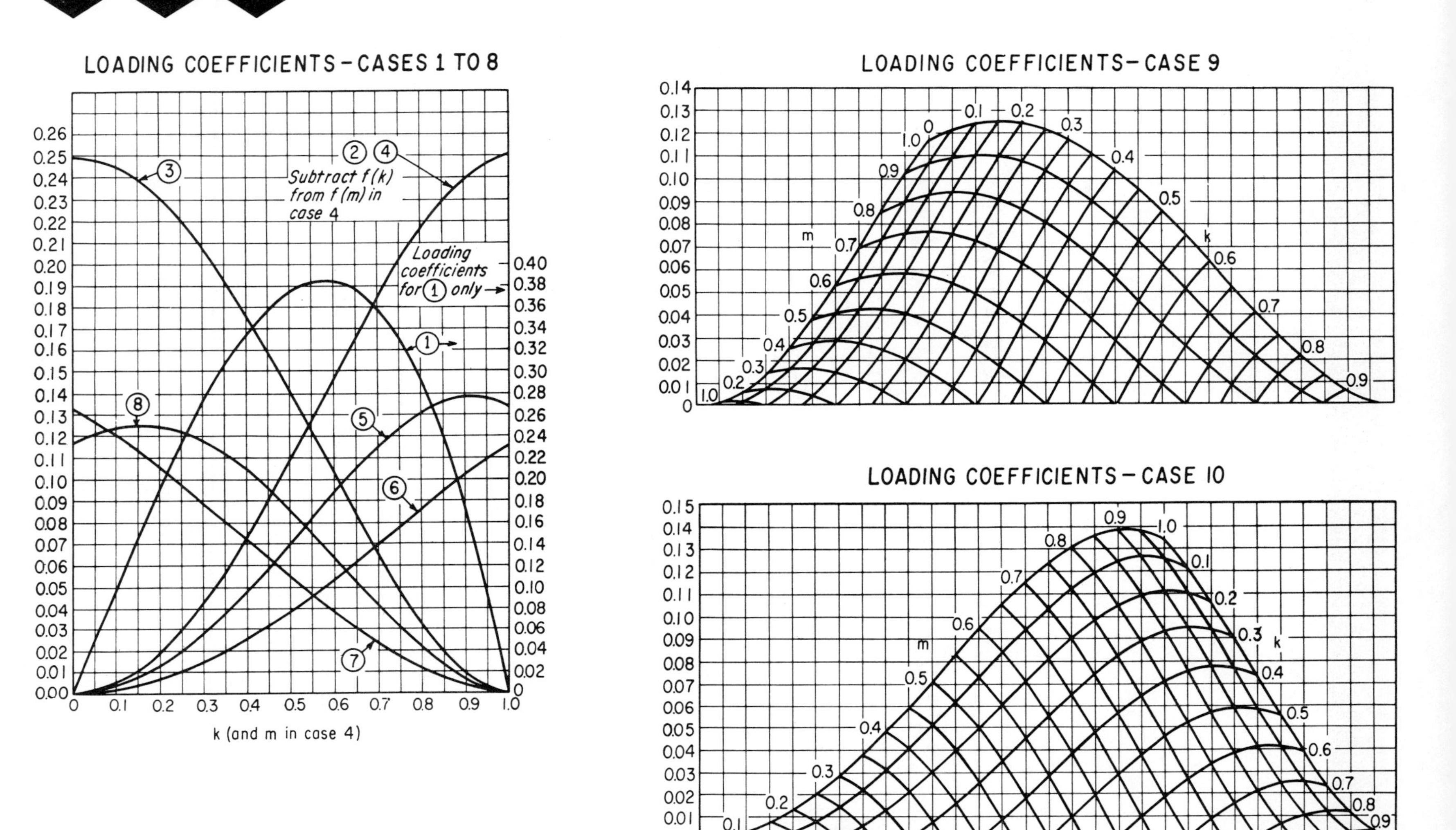
LOADING COEFFICIENTS - CASES 1 TO 8
Subtract f(k) from f(m) in case 4
Loading coefficients for ① only
k (and m in case 4)
LOADING COEFFICIENTS - CASE 9
m
k
LOADING COEFFICIENTS - CASE 10
m
k

# Nomogram to Determine Parallel Axis Moment of Inertia

HERBERT F. BARIFFI
Consulting Engineer

ONE METHOD of determining the moment of inertia of a mass through its center of gravity is to suspend the mass from a knife-edge. Such an arrangement, using a piece of keystock for a knife-edge, is shown in Fig. 1. If this compound gravity pendulum, which is in effect what the setup amounts to, is set swinging and the arc of swing is limited to about 6 deg double amplitude, the moment of inertia about the support axis is equal to

$$I_o = \frac{W}{4\pi^2} t^2 L \qquad (1)$$

where

$I_o$ = moment of inertia about support axis
$W$ = weight of mass, lb
$t$ = oscillation period, sec
$L$ = distance between center of gravity and support axis, in.

To find the moment of inertia about center of gravity the following equation can be used.

$$I_{cg} = I_o - L^2 \frac{W}{g} \qquad (2)$$

where

$g$ = acceleration of gravity, ft per sec²

Substituting $I_o$ value from Eq (1) into Eq (2)

$$I_{cg} = \frac{W}{4\pi^2} t^2 L - L^2 \frac{W}{32.2 \times 12}$$

Simplifying

$$I_{cg} = WL\,[0.02533\, t^2 - 0.002588L] \qquad (3)$$

This Eq (3) will give values of $I_{cg}$ in in. lb sec².

The nomogram illustrated can be used to solve for $I_{cg}$ with values of $t$, $L$ and $W$ known. To use this nomogram start with a value $t$ on the inner left-hand A scale, pass through a value of $L$ on the A curve to the right-hand, or turning axis. Then, follow a straight line from the turning point through a value of $W$ on the A diagonal, and return to the left-hand axis, reading $I_{cg}$ on the outer scale. By using the right-hand $I$ and $t$ scales, and working to the left over the B curve for $L$, the same problem can be solved with larger values of $t$.

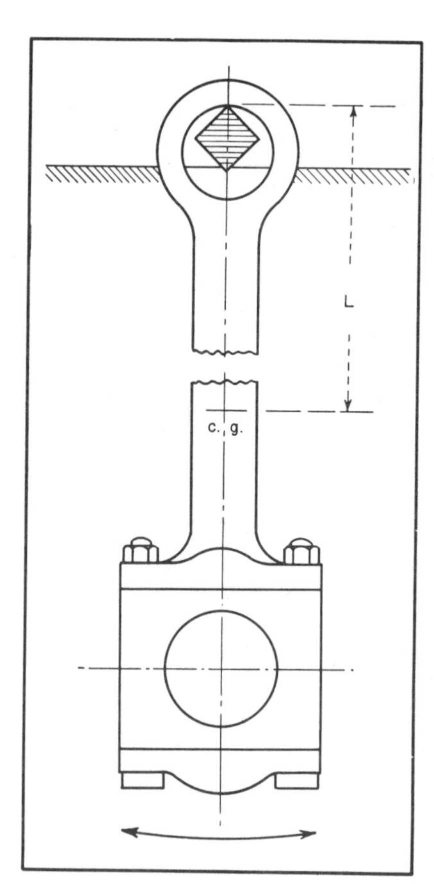


Fig. 1—Connecting rod suspended from keystock.

EXAMPLE: Given a pulley of 8 lb weight, $L$ equals 10 in., $t$ equals 1.5 sec. Find $I_{cg}$.

SOLUTION: Starting at $t$ equals $1\frac{1}{2}$ on inner A scale of left-hand axis, follow the dotted line through curve A at $L$ equals 10 to the turning line; go back through $W$ equals 8 on diagonal A to $I_{cg}$ equals 2.5 in. lb sec ².

It should be noted that the effects of $L$ and $t$ are screened in the equation. However, $I_{cg}$ is directly proportional to $W$, and this fact permits the unlimited extension of the nomogram's use.

Suppose, in the above example, $W$ had been 880 lb. It would appear that this value is too large for the $W$ scale, but Eq (3) can be written:

$$\frac{I_{cg}}{k} = \frac{WL}{k} [0.02533t^2 - 0.02588\, L]$$

So, in this case, assume $k$ equals 110, and use the nomogram as above with $W$ equaling 8, and read $I_{cg}/k$ equals 2.5. Now merely multiply this value by 110 to find 275 in. lb sec². This is the true value of $I_{cg}$.

An unusual feature of this nomogram is the relation between the values of $t$ and $L$; these quantities cannot be assumed at random for trial of nomogram accuracy. For example, a scant passing through $t = 1.81$ and $L = 30$ also passes through $L = 2$, thus suggesting that two different suspension lengths will satisfy the same result. One of these lengths is an extraneous value as the previously outlined experiment will indicate. Furthermore, $t$ is always larger than $0.31936\sqrt{L}$ when inch and second units are used.

# Nomogram to Determine Parallel Axis Moment of Inertia

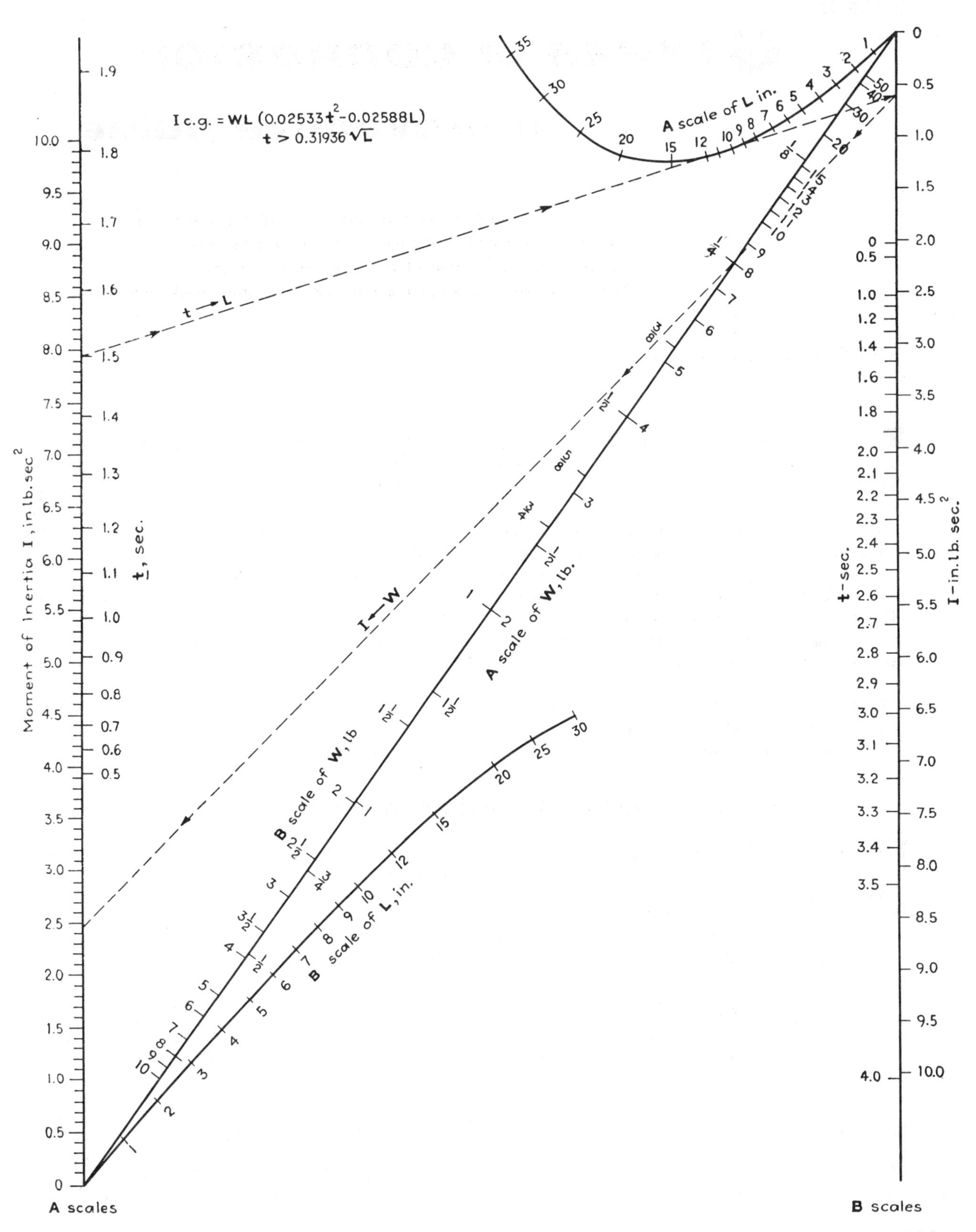

# THE 6 TYPES of CORROSION

## . . . a preventive guide

**To prevent trouble—know the cause. Though there still are puzzles concerning just how corrosion works, the destruction can be traced to six basic beginnings. The author describes each, along with examples and preventives.**

PETER C NOY, *Production Engineer*
*Canadian General Electric Co Ltd, Barrie, Ontario*

The problem of corrosion is simplified here by breaking it down into six basic causes and showing what design can do to minimize or overcome each kind of trouble.

Such a listing is necessarily artificial because corrosion is often puzzling—even to experts. All the physical and chemical phenomena involved are not yet fully understood. Moreover, the corrosive attack is often the end result of a combination of causes. However, focusing on individual causes can help combat even the more complicated challenges that face design in this field.

Corrosion resistance usually depends on some form of protective film (oxide coating, electroplating, paint, etc). But mechanical action—such as air bubbles, scraping, flexing, cavitation and moving liquid—can destroy this film and expose the base metal to the corrosive environment. When this happens corrosion sometimes occurs at a dramatically swift rate, e.g. aluminum scratched under mercury will oxidize visibly at the scratch when exposed to the air.

**Three basic recommendations** are:

1. Have the design examined by a corrosion expert.

2. Test the product exhaustively. Here, keep as close to service conditions as possible, and include the most severe conditions.

3. Watch out for environmental surprises (for example, hot coffee dissolves nickel plate a thousand times faster than will drinking water).

## THE MECHANISMS OF CORROSION

**1 Direct Attack.** The most important example here is formation of oxide films when metals are exposed to air These films are usually protective, as with aluminum and lead. Some metals, however, act differently—for example,

Aluminum — *Invisible aluminum oxide film: impermeable, protects metal*

Lead — *Visible film of lead oxide: impermeable, protects metal but destroys luster*

Calcium — *$CA(OH)_2$ forms a loose unprotective film*

those that reluctantly oxidize in dry air but readily form unprotective hydroxides in moist air. (Calcium, melting at 1436 F, and harder than tin, would be a useful manufacturing metal if not fpr this destructive affinity for water.) So:

**Select materials that form stable films when exposed to service conditions.**

**Foodstuffs (vinegar, salt, etc.) can also cause direct attack.**

**2 Complex Attack.** In some media, metals that ordinarily form protective films are prevented from doing so because of an intervening reaction at the surface.

Thus iron immersed in water that contains no impurities but a very small amount of oxygen will first form ferrous hydroxide at the surface; only then will insoluble ferric oxide start forming—but at a distance from the surface, say, 1 mm.

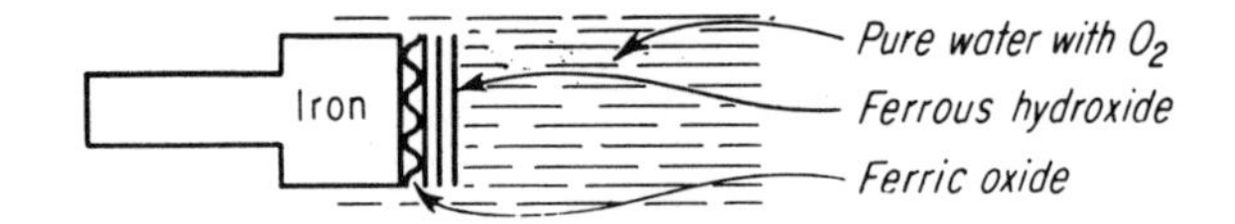

If, however, the supply of oxygen is increased, the protective oxide will be formed at the surface, thus slowing the attack.

**Remember that natural waters vary greatly in content of such dissolved gases as oxygen and carbon dioxide. Life-testing is the safest and often the only answer.**

**3 Galvanic Action.** Dissimilar metals, when in contact in a medium capable of carrying a current, will form a cell in which one metal becomes a cathode and the other an anode. The anode will dissolve (the rate depending largely on where the metals stand in the electromotive series) and the cathode will usually be protected.

The best example here is in electroplating, where metals come in intimate contact with each other.

Copper-plated steel, for instance, is poor protection against

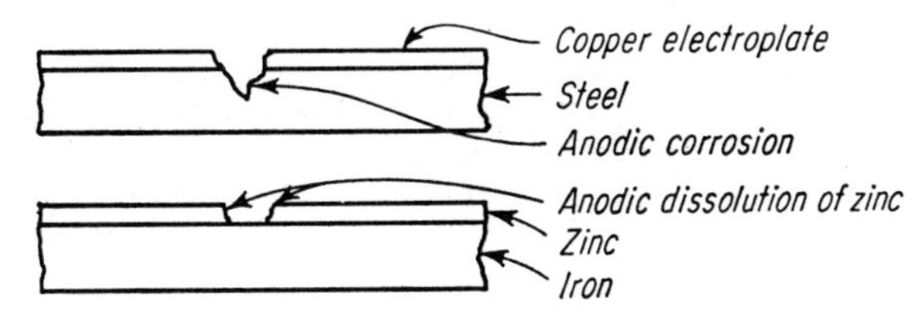


corrosion because steel is anodic to copper. Pores in the electroplate become cells that promote the dissolution of the steel.

On the other hand, zinc is anodic to iron, and zinc electroplate or galvanized coating will dissolve while the base metal is protected. Cathodic action is at base of the pores. This is a small area in comparison with the anodic area and insures long protection of the steel.

**Consult the electromotive series to select those metals for contact that are close to each other in potential. When specifying electroplate, the coating should be anodic; if this is impractical, it should be specified as nonporous.**

**4 Concentration Cell.** If for any reason the metal-ion concentration of the corroding medium drops when in contact with a metal, then ions of metal will be forced into solution to maintain a balance. This condition exists at the surface

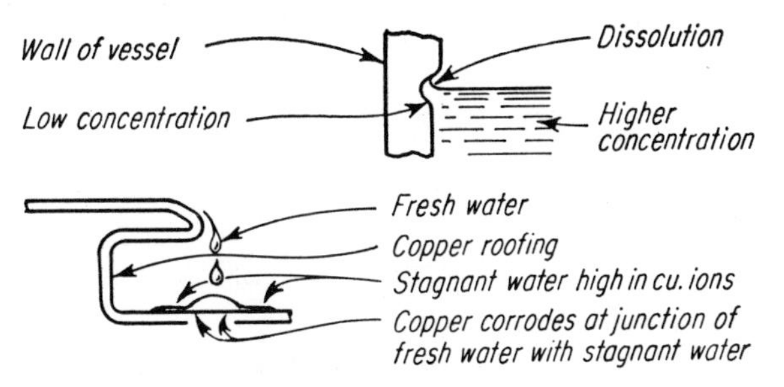


of a liquid, accounting for corrosion at solution lines. Circulation of the liquid, and cracks or crevices, lodgement of organic matter such as slime, or even stagnation beneath loose films of corrosion products, would lead to this type of corrosion. Illustration shows corrosion of copper roofing due to dripping of relatively low-concentration electrolyte (rain water in industrial area).

**Where the medium is a moving one, avoid causes of irregular circulation (lap joints, crevices, etc.)**

**Condensation streams should be watched where condensate runs back into a vessel, causing differences in concentration.**

**5 Dezincification.** Brass is subject to this kind of corrosion. In very soft waters with high concentrations of carbon dioxide the zinc constituent of brass dissolves out, leaving a spongy mass of copper.

Brass water pipes can be deceptive. Sometimes a small

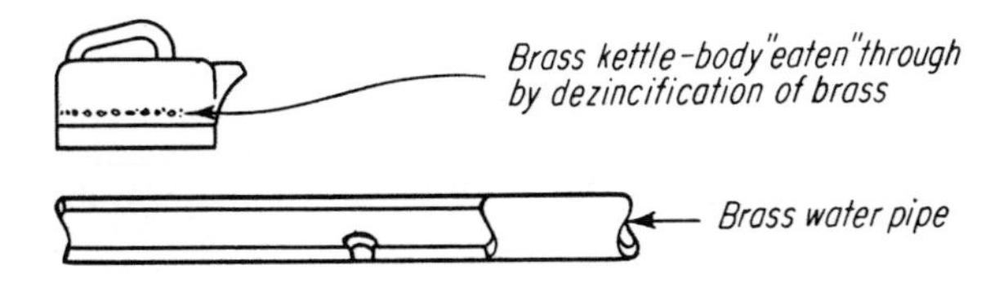


area or plug of porous copper (dezincified brass) in the wall of the pipe causes a slow leak, which subsequent liming may stop. Any disturbance or increase in pressure, however, may cause sudden failure.

**Where soft water conditions occur, keep brass low in zinc—less than 15%.**

**6 Fatigue and Stress Corrosion.** Stressed metal, when in a corrosive medium, is subject to harm greater than from the sum of the corrosion and stress effects. If the stress is cyclic, the mechanism is called *corrosion fatigue*; if static, it is *stress corrosion*. Machine parts operating in corrosive places, engine blocks, boiler parts are all prone to such corrosion. The best example of stress corrosion is shown in

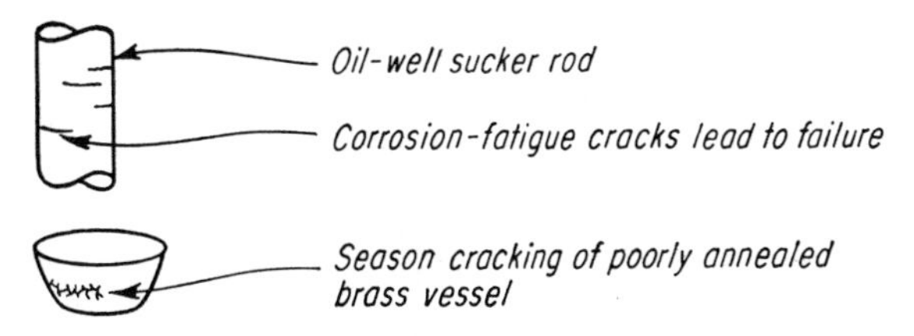


brass having internal stress. In corrosive media, including most natural waters, deep-drawn brass vessels will crack right through in a pattern called season cracking—it resembles the cracks in seasoned wood.

**Anneal all machine parts to be used in corrosive media.**

**Fatigue corrosion can be avoided by proper coatings on the part—if a corrosion-resistant electroplate will remain bonded under cyclic stress, you need only design to resist the fatigue effect itself.**

# Designing and Fabricating With

Clad steel is a composite plate—a carbon or low alloy steel backing plate inseparably bonded to cladding of a corrosion-resistant metal. Cladding thickness ranges from 5 to 50 per

## BUTT JOINTS

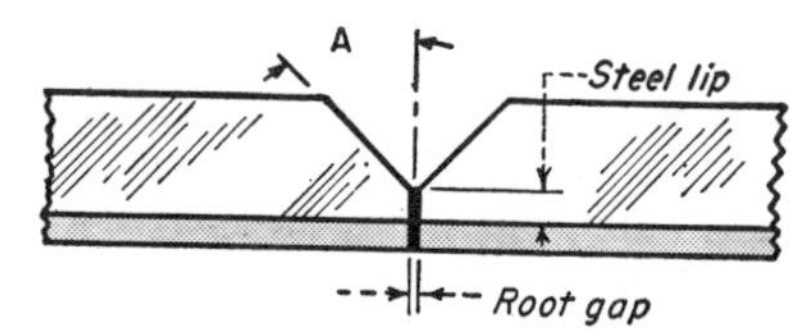


**Fig. 1—Single "V" butt joint for material ¼ to 1¼ in. thick. Angle A is 45 deg for ¼ to ½ in. thick; 35 deg for ¾ in.; and 30 deg for 1 to 1¼ in. sections. The lip of steel prevents pickup of the high alloy cladding in mild steel weld.**

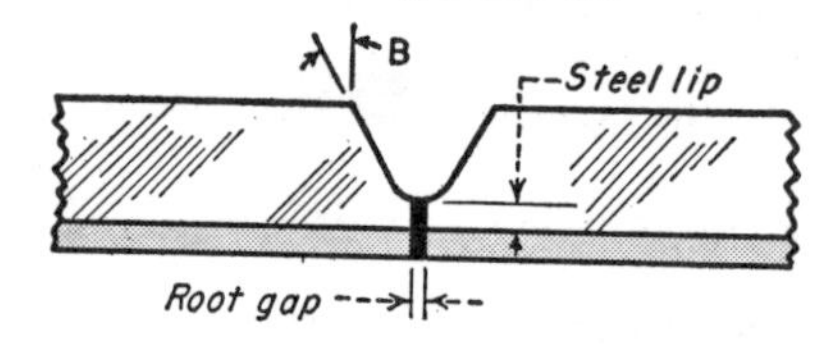


**Fig. 2—Single "U" butt joint for material ¾ to 2 in. thick. Angle B is 10 deg for all thicknesses. The lip of steel should be 1/16 to 3/32 in. for single "V" and "U" butt joints. Root gap for single "V" and single "U" is zero to 1/16 inch.**

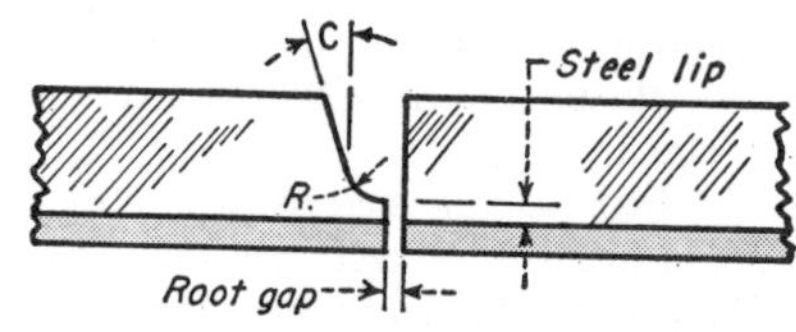


**Fig. 3—Single "J" butt joint for material ½ to 2 in. thick. A 30 deg angle is used for all thicknesses. The lip of steel should be 1/16 to 3/32 in. Radius for "U" and "J" butt joints is ¼ in. Root gap can vary between zero and 1/16 inch.**

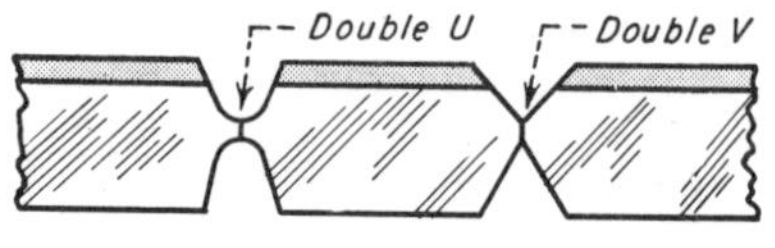


**Fig. 4—Double "U" and "V" butt joints are used on the heavier gages of material. The center lines of the bevels should be offset to favor the clad side. This will result in the use of less alloy weld metal.**

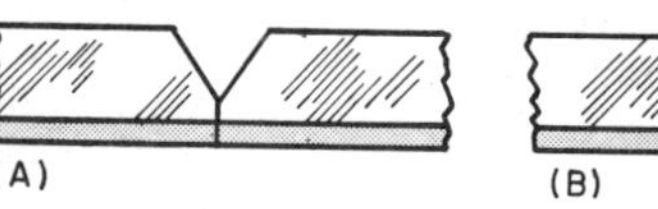


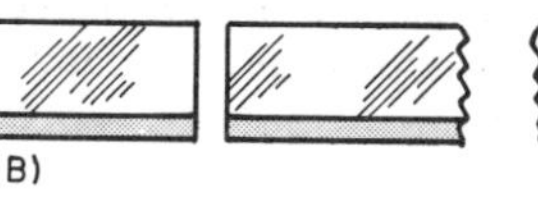


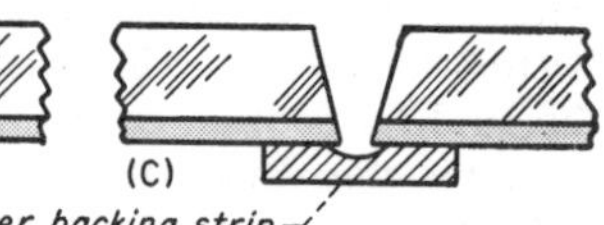


**Fig. 5—For lighter gages, 3/16 in. and less, the minimum clad should be 20 per cent to allow for sufficient cladding and adequate corrosion-resistant welds. For a butt weld, (A), the plate is beveled approximately 70 deg on the steel side. Where unfused areas will not be detrimental a butt joint similar to (B) can be used. If backing strip is used, (C), groove should be 5/32 in. wide by 1/16 in. deep. A ⅛ in. root gap is used.**

## TEE AND LAP JOINTS

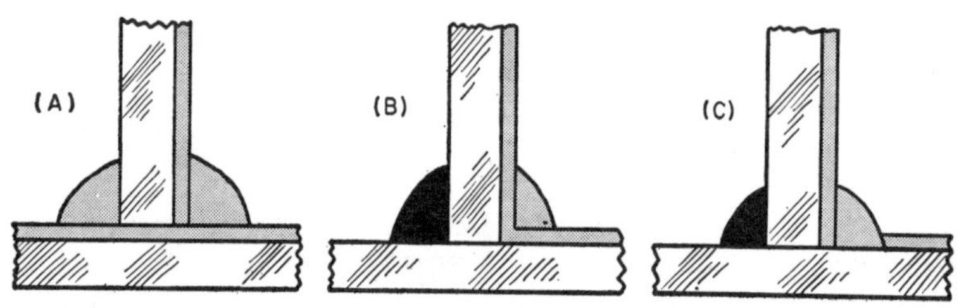


**Fig. 1—Whenever possible the steel side should be welded first. When the vertical member is butted to the vertical clad surface, (A), alloy welds must be used. If the clad surface is stripped back, (B) and (C), a steel weld can be used on the stripped side as shown.**

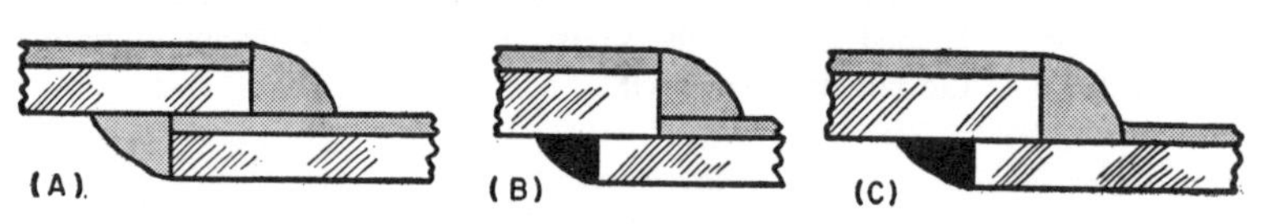


**Fig. 2—Lap joints are similar to Tee joints. Alloy welds must be used whenever clad is present. Full alloy welds are needed for joint (A). In joints (B) and (C), steel welds can be used on bottom surfaces.**

## DOUBLE CLAD STEEL

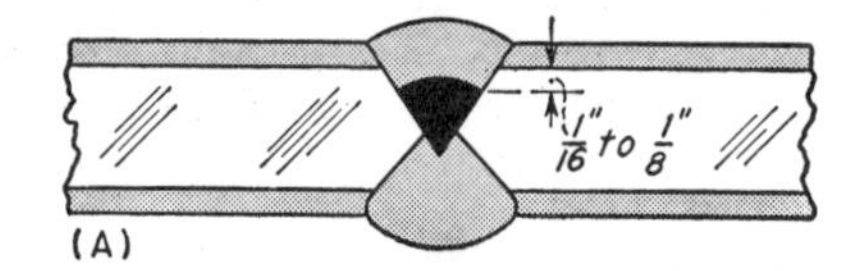


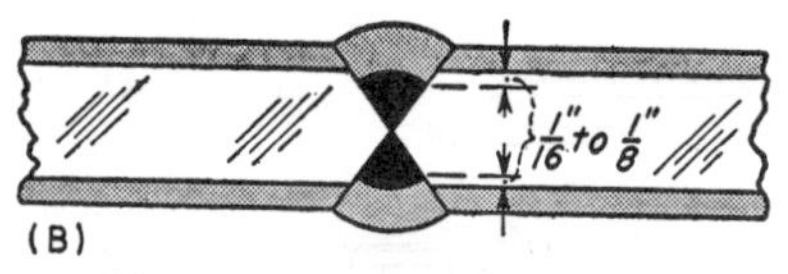


**Fig. 1—Two methods are in general use for welding double clad steel; in both double "V" or "U" edges are used. For light gages, (A), a steel weld is made on one side to about 1/16 to ⅛ in. of the clad line; then the other side is back chipped and alloy welded. For heavy gages, (B), steel welds are first made on both sides and then followed by alloy welds.**

# Clad Steel Plate

H. F. PETERS
Manager, Technical Service Department
Lukens Steel Company

cent of the total plate thickness. The backing plate material is selected to contribute strength properties; the cladding to supply the required corrosion or abrasion resistance.

## CORNER JOINTS

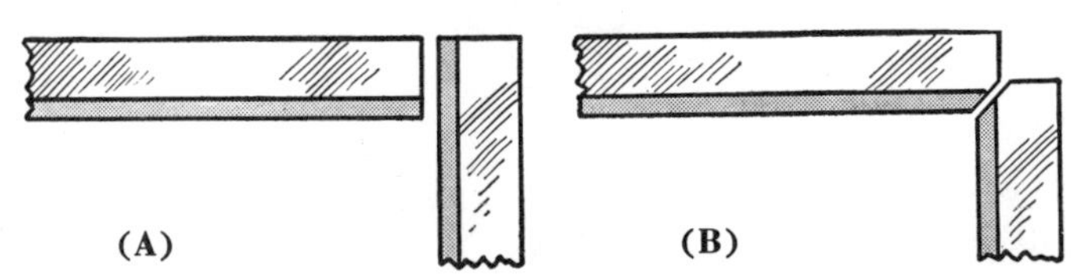


**Fig. 1—Used on unstressed or heavily supported structures.—(A) The top surface and the inside of the joint are welded with an alloy electrode. This type of joint has an undesirable unfused area.—(B) Two methods can be used. Clad side can be welded with an alloy electrode. The outside is then back chipped to the diagonal cut and the remainder of the joint alloy welded. Second method is to weld steel side, back chip from inside, and finish weld with an alloy electrode.**

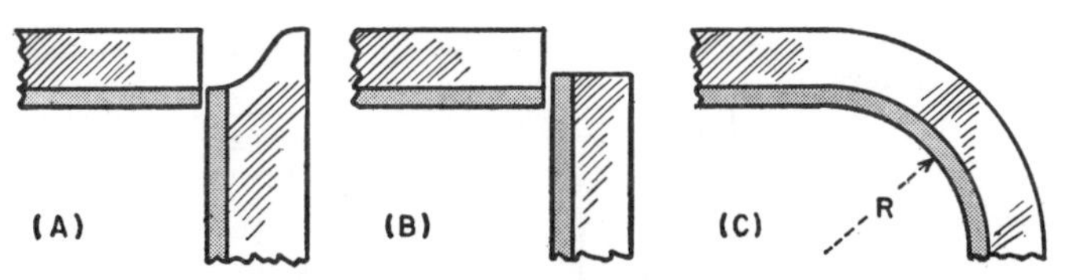


**Fig. 2—Joints (A) and (B) should be welded throughout with an alloy electrode. After welding the clad side, the steel side should be back chipped or ground, and welded with an alloy rod. Whenever possible, corner joints should be avoided and a formed section (C) used. This design is more desirable from a fabrication standpoint.**

## BLIND JOINTS

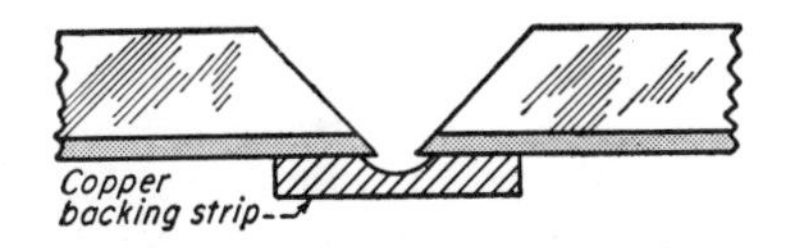


**Fig. 1—When clearance is available, a copper backing strip is used. When cladding is on blind side, entire cross section of joint must be welded with an alloy electrode. If steel section is on the blind side, steel electrodes are used up to 1/16 to ⅛ in. of clad line, and then finished with alloy.**

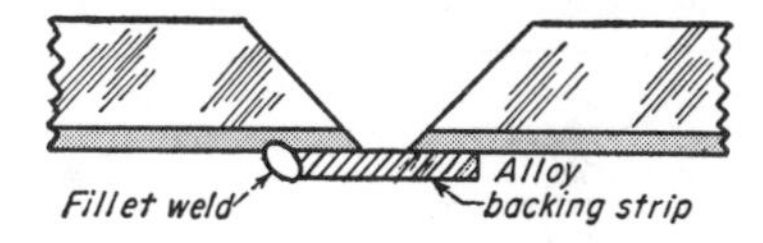


**Fig. 2—An alloy backing strip can also be used. When the clad material is on the blind side, a backing strip of the same analysis as the cladding must be used. When the clad surface is exposed, a steel backing strip can be used.**

## SOME TYPICAL CLAD STEEL ASSEMBLIES

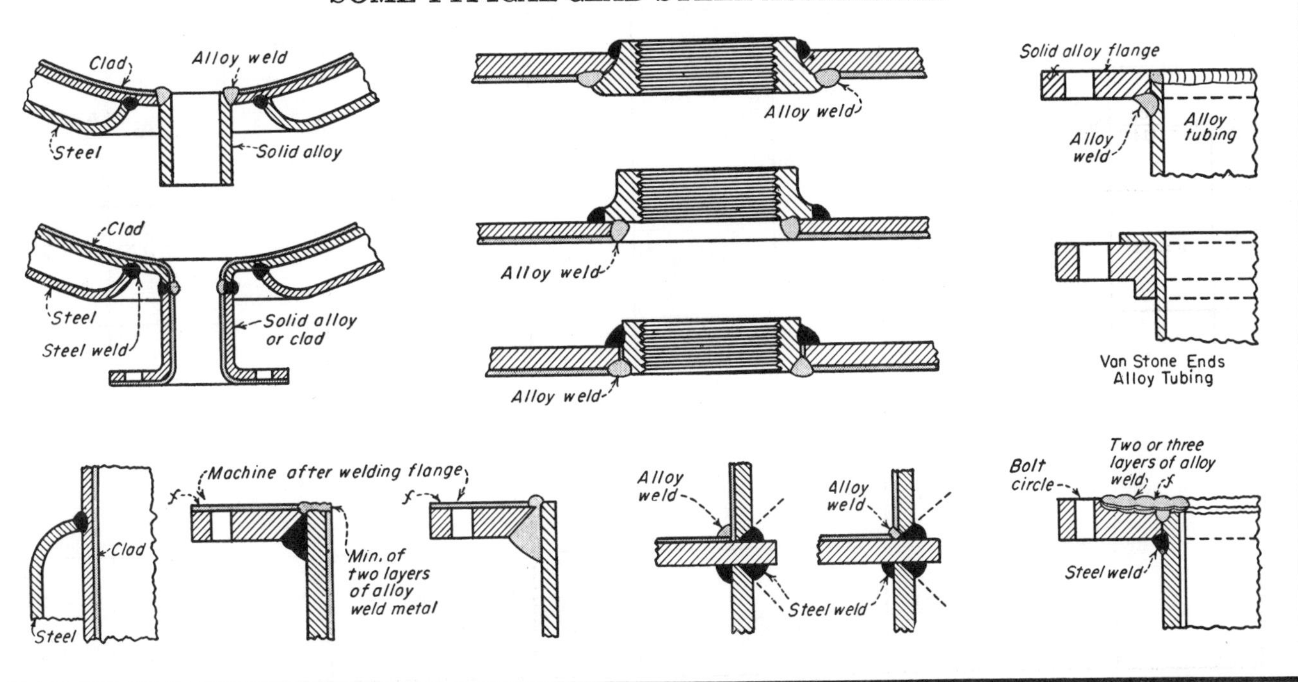

# UNIVERSAL TABLES

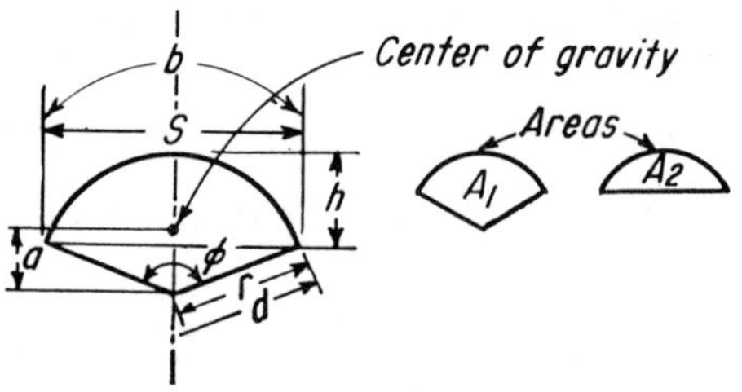

Every dimension of a circular segment or sector can be found from the following tables. Usual tables, by contrast, are based on a given radius or diameter, which may not be known to begin with. Mr. Huber's tables will work with other starting points just as well.

Table I provides equations from which unknown dimensions may be calculated. Table 2 lists numerical values of various factors used in the equations.

| Examples | 1 | 2 |
|---|---|---|
| GIVEN: | $\phi = 53°$, $s = 71.4$ in. | $s = 19.77$ in., $b = 20$ in. |
| REQUIRED: | diameter, $d$ | $h$ |
| FACTOR (from Table 2) | $= i = 0.446$ | $p = b/s = 1.0116$ |
| SOLUTION: | $d = s/i = 160$ in. dia. | $h = 57.3bk/\phi = 13.02$ in. |

In Ex. 2 it is first necessary to find $p$, from which $\phi$ and $k$ can then be found from Table 2.

## TABLE 1—FORMULAS

| Required | Known factors | | | | | | | | | | |
|---|---|---|---|---|---|---|---|---|---|---|---|
| | $\phi$ and . . . | | | | | | d or r and . . . | | | | |
| | d or r | s | b | h | $A_1$ | $A_2$ | s | b | h | $A_1$ | $A_2$ |
| $\phi$ | | | ("$\phi \rightarrow$ T2" means "Find equivalent $\phi$ in T2") | | | | $i = \frac{s}{d}$<br>$\phi \rightarrow$ T2 | $\phi = \frac{57.3\,b}{r}$ | $k = \frac{h}{r}$<br>$\phi \rightarrow$ T2 | $\phi = \frac{114.6\,A_1}{r^2}$ | $n = \frac{A_2}{d^2}$<br>$\phi \rightarrow$ T2 |
| d<br>r | | $d = \frac{s}{i}$ | $d = \frac{114.6\,b}{\phi}$ | $d = \frac{2\,h}{k}$ | $d = 21.4\sqrt{\frac{A_1}{\phi}}$ | $d = \sqrt{\frac{A_2}{n}}$ | | | | | |
| s | $s = di$ | | $s = \frac{114.6\,bi}{\phi}$ | $s = hu$ | $s = \frac{\sqrt{A_1}}{v}$ | $s = \frac{\sqrt{A_2}}{w}$ | | $\phi = \frac{57.3\,b}{r}$<br>$s = di$ | $k = \frac{h}{r}$<br>$s = di$ | $\phi = \frac{114.6\,A_1}{r^2}$<br>$s = di$ | $n = \frac{A_2}{d^2}$<br>$s = di$ |
| b | $b = \frac{d\phi}{114.6}$ | $b = \frac{s\phi}{114.6i}$ | | $b = hq$ | $b = 0.1868\sqrt{A_1\phi}$ | $b = \frac{\sqrt{A_2}}{t}$ | $i = \frac{s}{d}$<br>$b = \frac{r\phi}{57.3}$ | | $k = \frac{h}{r}$<br>$b = \frac{r\phi}{57.3}$ | $b = \frac{2\,A_1}{r}$ | $n = \frac{A_2}{d^2}$<br>$b = \frac{r\phi}{57.3}$ |
| h | $h = rk$ | $h = \frac{sk}{2i}$ | $h = \frac{57.3\,bk}{\phi}$ | | $h = 10.7k\sqrt{\frac{A_1}{\phi}}$ | $h = \frac{k}{2}\sqrt{\frac{A_2}{n}}$ | $i = \frac{s}{d}$<br>$h = rk$ | $\phi = \frac{57.3\,b}{r}$<br>$h = rk$ | | $\phi = \frac{114.6\,A_1}{r^2}$<br>$h = rk$ | $n = \frac{A_2}{d^2}$<br>$h = rk$ |
| $A_1$ | $A_1 = \frac{r^2\phi}{114.6}$ | $A_1 = (sv)^2$ | $A_1 = 28\,65\,\frac{b^2}{\phi}$ | $A_1 = (hx)^2$ | | $A_1 = A_2z$ | $i = \frac{s}{d}$<br>$A_1 = \frac{r^2\phi}{114.6}$ | $A_1 = \frac{rb}{2}$ | $k = \frac{h}{r}$<br>$A_1 = \frac{r^2\phi}{114.6}$ | | $n = \frac{A_2}{d^2}$<br>$A_1 = \frac{r^2\phi}{114.6}$ |
| $A_2$ | $A_2 = d^2n$ | $A_2 = (sw)^2$ | $A_2 = (bt)^2$ | $A_2 = (hy)^2$ | $A_2 = \frac{A_1}{z}$ | | $i = \frac{s}{d}$<br>$A_2 = d^2n$ | $\phi = \frac{57.3\,b}{r}$<br>$A_2 = d^2n$ | $k = \frac{h}{r}$<br>$A_2 = d^2n$ | $\phi = \frac{114.6\,A_1}{r^2}$<br>$A_2 = d^2n$ | |
| a | $a = \frac{d}{3p}$ | $a = \frac{38.2\,s}{\phi}$ | $a = \frac{38.2\,b}{p\phi}$ | $a = \frac{hqu}{1.5\,k}$ | $a = \frac{38.2\sqrt{A_1}}{\phi v}$ | $a = \frac{\sqrt{A^2}}{12w}$ | $i = \frac{s}{d}$<br>$a = \frac{38.2\,s}{\phi}$ | $\phi = \frac{57.3\,b}{r}$<br>$a = \frac{70.4\,ri}{\phi}$ | $k = \frac{h}{r}$<br>$a = \frac{ir}{6}$ | $\phi = \frac{114.6\,A_1}{r^2}$<br>$a = \frac{ir}{6}$ | $n = \frac{A_2}{d^2}$<br>$a = \frac{ir}{6}$ |

# for Circular Segments and Sectors

## TABLE 2—FACTORS

| $\phi°$ | i | k | n | p | q | t | u | v | w | x | y | z |
|---|---|---|---|---|---|---|---|---|---|---|---|---|
| 0 | 0 | 0 | 0 | 1 | ∞ | 0 | — | — | 0 | — | ∞ | ∞ |
| 0.5 | 0.0044 | 0.000009 | 0.00000001 | 1 | 970 | 0.0278 | 970 | 7.45 | 0.0278 | 7230 | 49.7 | 30000 |
| 1 | 87 | 0.000038 | 0.00000016 | 1 | 459 | 405 | 459 | 5.35 | 405 | 2460 | 18.6 | 17450 |
| 1.5 | 0.0137 | 86 | 36 | 1.00005 | 304 | 436 | 304 | 4.37 | 436 | 1330 | 13.3 | 10070 |
| 2 | 175 | 0.000152 | 82 | 7 | 230 | 520 | 238 | 3.78 | 520 | 869 | 11.9 | 5300 |
| 2.5 | 218 | 238 | 0.00000117 | 9 | 183 | 598 | 183 | 39 | 598 | 621 | 11.0 | 3210 |
| 3 | 262 | 343 | 298 | 1.00012 | 153 | 648 | 153 | 09 | 648 | 472 | 9.89 | 2275 |
| 3.5 | 305 | 466 | 468 | 14 | 131 | 701 | 131 | 2.86 | 701 | 375 | 18 | 1669 |
| 4 | 349 | 609 | 701 | 19 | 115 | 751 | 115 | 68 | 752 | 307 | 8.61 | 1269 |
| 4.5 | 393 | 771 | 0.0000101 | 25 | 102 | 804 | 102 | 52 | 803 | 257 | 18 | 987 |
| 5 | 0.0436 | 992 | 138 | 33 | 91.7 | 0.0846 | 91.6 | 2.39 | 0.0846 | 219 | 7.75 | 800 |
| 5.5 | 480 | 0.00115 | 184 | 39 | 83.3 | 889 | 83.3 | 28 | 889 | 190 | 41 | 659 |
| 6 | 523 | 137 | 238 | 46 | 76.4 | 928 | 76.4 | 19 | 928 | 167 | 10 | 554 |
| 6.5 | 567 | 161 | 304 | 53 | 70.6 | 969 | 70.5 | 10 | 969 | 148 | 6.83 | 470 |
| 7 | 611 | 187 | 380 | 62 | 65.5 | 0.101 | 65.5 | 02 | 0.101 | 132 | 59 | 405 |
| 7.5 | 654 | 214 | 467 | 72 | 61.1 | 04 | 61.1 | 1.96 | 04 | 119 | 37 | 352 |
| 8 | 698 | 244 | 567 | 81 | 57.3 | 08 | 57.3 | 89 | 08 | 108 | 17 | 310 |
| 8.5 | 741 | 275 | 680 | 91 | 53.9 | 11 | 53.9 | 84 | 11 | 99 | 5.99 | 274 |
| 9 | 785 | 308 | 806 | 1.0010 | 50.9 | 14 | 50.9 | 79 | 14 | 90.9 | 81 | 245 |
| 9.5 | 828 | 343 | 948 | 12 | 48.3 | 17 | 48.2 | 74 | 17 | 83.8 | 66 | 219 |

(TABLE 2 CONTINUED ON PAGE 162)

| Known factors | | | | | | | | | |
|---|---|---|---|---|---|---|---|---|---|
| s and... | | | | b and... | | | h and... | | $A_1$ and... |
| b | h | $A_1$ | $A_2$ | h | $A_1$ | $A_2$ | $A_1$ | $A_2$ | $A_2$ |
| $p=\frac{b}{s}$<br>$\phi \rightarrow T2$ | $u=\frac{s}{h}$<br>$\phi \rightarrow T2$ | $v=\frac{\sqrt{A_1}}{s}$<br>$\phi \rightarrow T2$ | $w=\frac{\sqrt{A_2}}{s}$<br>$\phi \rightarrow T2$ | $q=\frac{b}{h}$<br>$\phi \rightarrow T2$ | $\phi=28.65\frac{b^2}{A_1}$ | $t=\frac{\sqrt{A_2}}{b}$<br>$\phi \rightarrow T2$ | $x=\frac{\sqrt{A_1}}{h}$<br>$\phi \rightarrow T2$ | $y=\frac{\sqrt{A_2}}{h}$<br>$\phi \rightarrow T2$ | $z=\frac{A_1}{A_2}$<br>$\phi \rightarrow T2$ |
| $p=\frac{b}{s}$<br>$d=\frac{114.6\,b}{\phi}$ | $u=\frac{s}{h}$<br>$d=\frac{s}{i}$ | $v=\frac{\sqrt{A_1}}{s}$<br>$d=\frac{s}{i}$ | $w=\frac{\sqrt{A_2}}{s}$<br>$d=\sqrt{\frac{A_2}{n}}$ | $q=\frac{b}{h}$<br>$d=\frac{114.6\,b}{\phi}$ | $r=\frac{2\,A_1}{b}$ | $t=\frac{\sqrt{A_2}}{b}$<br>$d=\sqrt{\frac{A_2}{n}}$ | $x=\frac{\sqrt{A_1}}{h}$<br>$r=10.7\sqrt{\frac{A_1}{\phi}}$ | $y=\frac{\sqrt{A_2}}{h}$<br>$d=\sqrt{\frac{A_2}{n}}$ | $z=\frac{A_1}{A_2}$<br>$r=10.7\sqrt{\frac{A_1}{\phi}}$ |
| | | | | $q=\frac{b}{h}$<br>$s=\frac{b}{p}$ | $\phi=28.65\frac{b^2}{A_1}$<br>$s=\frac{b}{p}$ | $t=\frac{\sqrt{A_2}}{b}$<br>$s=\frac{\sqrt{A_2}}{w}$ | $x=\frac{\sqrt{A_1}}{h}$<br>$s=\frac{\sqrt{A_1}}{v}$ | $y=\frac{\sqrt{A_2}}{h}$<br>$s=\frac{\sqrt{A_2}}{w}$ | $z=\frac{A_1}{A_2}$<br>$s=\frac{\sqrt{A_1}}{v}$ |
| | $u=\frac{s}{h}$<br>$b=sp$ | $v=\frac{\sqrt{A_1}}{s}$<br>$b=sp$ | $w=\frac{\sqrt{A_2}}{s}$<br>$b=\frac{\sqrt{A_2}}{t}$ | | | | $x=\frac{\sqrt{A_1}}{h}$<br>$b=0.1868\sqrt{A_1\phi}$ | $y=\frac{\sqrt{A_2}}{h}$<br>$b=\frac{\sqrt{A_2}}{t}$ | $z=\frac{A_1}{A_2}$<br>$b=0.1868\sqrt{A_1\phi}$ |
| $p=\frac{b}{s}$<br>$h=\frac{57.3\,bk}{\phi}$ | | $v=\frac{\sqrt{A_1}}{s}$<br>$h=\frac{sk}{2i}$ | $w=\frac{\sqrt{A_2}}{s}$<br>$h=\frac{\sqrt{A_2}}{y}$ | | $\phi=28.65\frac{b^2}{A_1}$<br>$h=\frac{2\,A_1k}{b}$ | $t=\frac{\sqrt{A_2}}{b}$<br>$h=\frac{\sqrt{A_2}}{y}$ | | | $z=\frac{A_1}{A_2}$<br>$h=\frac{\sqrt{A_1}}{x}$ |
| $p=\frac{b}{s}$<br>$A_1=(sv)^2$ | $u=\frac{s}{h}$<br>$A_1=(sv)^2$ | | $w=\frac{\sqrt{A_2}}{s}$<br>$A_1=A_2z$ | $q=\frac{b}{h}$<br>$A_1=28.65\frac{b^2}{\phi}$ | | $t=\frac{\sqrt{A_2}}{b}$<br>$A_1=A_2z$ | | $y=\frac{\sqrt{A_2}}{h}$<br>$A_1=A_2z$ | |
| $p=\frac{b}{s}$<br>$A_2=(sw)^2$ | $u=\frac{s}{h}$<br>$A_2=(sw)^2$ | $v=\frac{\sqrt{A_1}}{s}$<br>$A_2=(sw)^2$ | | $q=\frac{b}{h}$<br>$A_2=(bt)^2$ | $\phi=28.65\frac{b^2}{A_1}$<br>$A_2=\frac{A_1}{z}$ | | $x=\frac{\sqrt{A_1}}{h}$<br>$A_2=\frac{A_1}{z}$ | | |
| $p=\frac{b}{s}$<br>$a=\frac{38.2\,s}{\phi}$ | $u=\frac{s}{h}$<br>$a=\frac{38.2\,s}{\phi}$ | $v=\frac{\sqrt{A_1}}{s}$<br>$a=\frac{38.2\,s}{\phi}$ | $w=\frac{\sqrt{A_2}}{s}$<br>$a=\frac{38.2\,s}{\phi}$ | $q=\frac{b}{h}$<br>$a=\frac{38.2\,b}{p\phi}$ | $\phi=28.65\frac{b^2}{A_1}$<br>$a=\frac{38.2\,b}{\phi p}$ | $t=\frac{\sqrt{A_2}}{b}$<br>$a=\frac{38.2\,b}{\phi p}$ | $x=\frac{\sqrt{A_1}}{h}$<br>$a=817\frac{i}{\phi}\sqrt{\frac{A_1}{\phi}}$ | $y=\frac{\sqrt{A_2}}{h}$<br>$a=38.2\frac{i}{\phi}\sqrt{\frac{A_2}{n}}$ | $z=\frac{A_1}{A_2}$<br>$a=\frac{38.2\sqrt{A_1}}{\phi v}$ |

**TABLE 2 (cont.)**

| φ° | | i | k | n | p | q | t | u | v | w | x | y | z |
|---|---|---|---|---|---|---|---|---|---|---|---|---|---|
| 1 | 0 | 0.0872 | 0.00381 | 0.000110 | 1.0013 | 45.9 | 0.120 | 45.8 | 1.69 | 0.120 | 77.6 | 52 | 198 |
| | 1 | 960 | 460 | 15 | 15 | 41.7 | 27 | 41.7 | 62 | 27 | 67.3 | 28 | 163 |
| | 2 | 0.105 | 548 | 19 | 18 | 38.1 | 33 | 38.1 | 55 | 33 | 59 | 06 | 136 |
| | 3 | 13 | 643 | 25 | 22 | 35.4 | 38 | 35.2 | 49 | 38 | 52.4 | 4.88 | 116 |
| | 4 | 22 | 745 | 30 | 25 | 33.0 | 42 | 32.8 | 43 | 43 | 46.9 | 67 | 101 |
| | 5 | 31 | 856 | 37 | 29 | 30.4 | 47 | 30.5 | 39 | 48 | 42.3 | 51 | 87.9 |
| | 6 | 39 | 973 | 45 | 33 | 28.8 | 52 | 28.6 | 34 | 52 | 38.4 | 36 | 77.6 |
| | 7 | 48 | 0.0110 | 54 | 37 | 27.0 | 57 | 26.9 | 30 | 58 | 35.1 | 24 | 68.4 |
| | 8 | 56 | 23 | 64 | 41 | 25.5 | 61 | 25.4 | 27 | 62 | 32.2 | 12 | 61.1 |
| | 9 | 65 | 37 | 76 | 46 | 24.2 | 66 | 24.1 | 23 | 67 | 29.7 | 01 | 54.9 |
| 2 | 0 | 0.174 | 52 | 88 | 1.0051 | 23 | 0.170 | 22.9 | 1.20 | 0.171 | 27.5 | 3.91 | 49.6 |
| | 1 | 82 | 68 | 0.00102 | 56 | 22 | 74 | 21.8 | 17 | 75 | 25.6 | 82 | 44.8 |
| | 2 | 91 | 84 | 117 | 62 | 20.9 | 78 | 20.8 | 15 | 79 | 23.9 | 72 | 41 |
| | 3 | 99 | 0.0201 | 134 | 67 | 20 | 82 | 19.9 | 12 | 83 | 22.2 | 64 | 37.5 |
| | 4 | 0.208 | 19 | 152 | 74 | 19.2 | 86 | 19 | 10 | 87 | 20.9 | 57 | 34.5 |
| | 5 | 16 | 37 | 171 | 80 | 18.4 | 90 | 18.3 | 08 | 91 | 19.7 | 49 | 31.8 |
| | 6 | 25 | 56 | 193 | 86 | 17.7 | 94 | 17.6 | 06 | 95 | 18.6 | 43 | 29.4 |
| | 7 | 34 | 76 | 0.00216 | 93 | 17.1 | 97 | 16.9 | 04 | 99 | 17.6 | 36 | 27.3 |
| | 8 | 42 | 97 | 240 | 1.0100 | 16.5 | 0.201 | 16.3 | 02 | 0.203 | 16.6 | 30 | 25.4 |
| | 9 | 50 | 0.0319 | 267 | 108 | 15.9 | 04 | 15.7 | 00 | 06 | 15.8 | 24 | 23.7 |
| 3 | 0 | 0.259 | 41 | 295 | 115 | 15.4 | 08 | 15.2 | 0.989 | 10 | 15 | 3.19 | 22.2 |
| | 1 | 67 | 64 | 0.00325 | 123 | 14.9 | 11 | 14.7 | 73 | 13 | 14.3 | 14 | 20.8 |
| | 2 | 76 | 87 | 357 | 13 | 14.4 | 14 | 14.2 | 59 | 17 | 13.6 | 09 | 19.5 |
| | 3 | 84 | 0.0412 | 392 | 14 | 14 | 17 | 13.8 | 45 | 20 | 13 | 04 | 18.4 |
| | 4 | 92 | 37 | 428 | 15 | 13.6 | 20 | 13.4 | 32 | 24 | 12.5 | 2.99 | 17.3 |
| | 5 | 0.301 | 63 | 466 | 16 | 13.2 | 24 | 13 | 19 | 27 | 11.9 | 95 | 16.4 |
| | 6 | 09 | 89 | 0.00507 | 17 | 12.9 | 27 | 12.6 | 07 | 30 | 11.4 | 91 | 15.5 |
| | 7 | 17 | 0.0517 | 549 | 18 | 5 | 30 | 12.3 | 0.895 | 34 | 11 | 87 | 14.7 |
| | 8 | 26 | 45 | 595 | 19 | 2 | 33 | 12 | 84 | 37 | 10.6 | 83 | 13.9 |
| | 9 | 34 | 74 | 0.00642 | 20 | 11.9 | 35 | 11.6 | 74 | 40 | 10.2 | 79 | 13.3 |
| 4 | 0 | 0.342 | 0.0603 | 692 | 1.021 | 6 | 0.238 | 11.3 | 0.864 | 0.243 | 9.80 | 2.76 | 12.6 |
| | 1 | 50 | 33 | 0.00744 | 22 | 3 | 41 | 11.1 | 54 | 46 | 45 | 72 | 12 |
| | 2 | 58 | 64 | 799 | 23 | 0 | 44 | 10.8 | 45 | 49 | 11 | 69 | 11.5 |
| | 3 | 67 | 96 | 0.00856 | 24 | 10.8 | 47 | 10.5 | 36 | 52 | 8.80 | 66 | 11 |
| | 4 | 75 | 0.0728 | 916 | 25 | 6 | 49 | 10.3 | 27 | 55 | 51 | 63 | 10.5 |
| | 5 | 83 | 61 | 979 | 26 | 3 | 52 | 10.1 | 19 | 58 | 23 | 60 | 10 |
| | 6 | 91 | 95 | 0.0104 | 27 | 1 | 55 | 9.83 | 11 | 61 | 7.97 | 57 | 9.61 |
| | 7 | 99 | 0.0829 | 11 | 28 | 9.89 | 57 | 62 | 03 | 64 | 72 | 54 | 22 |
| | 8 | 0.407 | 65 | 18 | 30 | 70 | 60 | 41 | 0.796 | 67 | 49 | 52 | 8.86 |
| | 9 | 15 | 0.0900 | 26 | 31 | 61 | 62 | 21 | 89 | 70 | 26 | 49 | 51 |
| 5 | 0 | 23 | 37 | 0.0133 | 1.032 | 32 | 0.265 | 02 | 0.781 | 0.273 | 05 | 2.46 | 18 |
| | 1 | 31 | 74 | 41 | 34 | 14 | 67 | 8.84 | 75 | 76 | 6.85 | 44 | 7.88 |
| | 2 | 38 | 0.101 | 50 | 35 | 8.97 | 70 | 66 | 68 | 79 | 66 | 42 | 59 |
| | 3 | 46 | 05 | 58 | 37 | 80 | 72 | 49 | 62 | 82 | 47 | 39 | 32 |
| | 4 | 54 | 09 | 67 | 38 | 65 | 74 | 33 | 56 | 85 | 30 | 37 | 06 |
| | 5 | 62 | 13 | 76 | 39 | 50 | 76 | 17 | 50 | 87 | 13 | 35 | 6.82 |
| | 6 | 70 | 17 | 85 | 41 | 35 | 79 | 02 | 45 | 90 | 5.98 | 33 | 60 |
| | 7 | 77 | 21 | 95 | 42 | 21 | 81 | 7.87 | 39 | 93 | 82 | 30 | 38 |
| | 8 | 85 | 25 | 0.0205 | 44 | 07 | 83 | 73 | 34 | 95 | 67 | 28 | 18 |
| | 9 | 92 | 30 | 16 | 46 | 7.95 | 85 | 60 | 29 | 98 | 54 | 27 | 5.9 |
| 6 | 0 | 0.500 | 0.134 | 0.0227 | 1.047 | 7.81 | 0.288 | 7.46 | 0.724 | 0.301 | 5.40 | 2.25 | 5.78 |
| | 1 | 08 | 38 | 38 | 49 | 69 | 90 | 33 | 19 | 04 | 27 | 23 | 60 |
| | 2 | 15 | 43 | 49 | 50 | 58 | 92 | 21 | 14 | 07 | 15 | 21 | 44 |
| | 3 | 23 | 47 | 62 | 52 | 46 | 94 | 09 | 10 | 10 | 03 | 20 | 25 |
| | 4 | 30 | 52 | 73 | 54 | 35 | 96 | 6.98 | 05 | 12 | 4.92 | 18 | 12 |
| | 5 | 37 | 57 | 85 | 56 | 24 | 98 | 86 | 01 | 14 | 81 | 16 | 4.98 |
| | 6 | 45 | 61 | 98 | 57 | 14 | 0.300 | 75 | 0.697 | 17 | 70 | 14 | 84 |
| | 7 | 52 | 66 | 0.0311 | 59 | 04 | 02 | 65 | 93 | 20 | 60 | 12 | 70 |
| | 8 | 60 | 71 | 25 | 61 | 6.94 | 04 | 54 | 89 | 22 | 50 | 11 | 57 |
| | 9 | 66 | 76 | 38 | 63 | 85 | 06 | 44 | 85 | 25 | 41 | 09 | 45 |
| 7 | 0 | 0.574 | 0.181 | 53 | 1.065 | 6.76 | 0.308 | 35 | 0.681 | 0.328 | 4.32 | 2.08 | 4.33 |
| | 1 | 81 | 86 | 67 | 67 | 67 | 09 | 25 | 78 | 30 | 23 | 06 | 22 |
| | 2 | 88 | 91 | 82 | 69 | 58 | 11 | 15 | 74 | 33 | 15 | 05 | 11 |
| | 3 | 95 | 96 | 97 | 71 | 50 | 13 | 07 | 71 | 35 | 07 | 03 | 01 |
| | 4 | 0.602 | 0.201 | 0.0413 | 73 | 41 | 15 | 5.98 | 68 | 38 | 3.99 | 02 | 3.91 |
| | 5 | 09 | 07 | 29 | 75 | 34 | 16 | 89 | 64 | 40 | 92 | 01 | 82 |
| | 6 | 16 | 12 | 45 | 77 | 26 | 18 | 81 | 61 | 43 | 84 | 1.99 | 73 |
| | 7 | 23 | 17 | 62 | 79 | 18 | 20 | 73 | 58 | 45 | 77 | 98 | 64 |
| | 8 | 29 | 23 | 79 | 82 | 11 | 22 | 65 | 55 | 48 | 70 | 96 | 55 |
| | 9 | 36 | 28 | 97 | 84 | 04 | 23 | 57 | 53 | 50 | 64 | 95 | 47 |
| 8 | 0 | 0.643 | 0.234 | 0.0514 | 1.086 | 5.97 | 0.325 | 5.49 | 0.650 | 0.353 | 3.57 | 1.94 | 3.40 |
| | 1 | 50 | 40 | 32 | 88 | 90 | 26 | 42 | 47 | 55 | 51 | 93 | 32 |
| | 2 | 56 | 45 | 51 | 91 | 83 | 28 | 35 | 45 | 58 | 45 | 91 | 25 |
| | 3 | 63 | 51 | 70 | 93 | 77 | 30 | 28 | 42 | 60 | 39 | 90 | 18 |
| | 4 | 69 | 57 | 89 | 96 | 71 | 31 | 21 | 40 | 63 | 33 | 89 | 11 |
| | 5 | 76 | 63 | 0.0609 | 98 | 65 | 33 | 14 | 37 | 65 | 28 | 88 | 3.04 |
| | 6 | 82 | 69 | 29 | 1.100 | 59 | 34 | 08 | 35 | 68 | 22 | 87 | 2.98 |
| | 7 | 88 | 75 | 50 | 03 | 53 | 36 | 01 | 33 | 70 | 17 | 86 | 92 |
| | 8 | 95 | 81 | 71 | 05 | 47 | 37 | 4.95 | 31 | 73 | 12 | 85 | 86 |
| | 9 | 0.701 | 87 | 92 | 08 | 42 | 39 | 89 | 29 | 75 | 07 | 84 | 81 |
| 9 | 0 | 0.707 | 0.293 | 0.0714 | 1.111 | 5.36 | 0.340 | 4.83 | 0.627 | 0.378 | 3.02 | 1.825 | 2.75 |
| | 1 | 13 | 99 | 35 | 13 | 31 | 41 | 77 | 25 | 80 | 2.98 | 14 | 70 |
| | 2 | 19 | 0.305 | 58 | 16 | 26 | 43 | 71 | 23 | 83 | 93 | 04 | 65 |
| | 3 | 25 | 12 | 81 | 19 | 21 | 44 | 66 | 21 | 85 | 89 | 1.793 | 60 |
| | 4 | 31 | 18 | 0.0804 | 22 | 16 | 46 | 60 | 19 | 88 | 85 | 83 | 55 |
| | 5 | 37 | 24 | 27 | 24 | 11 | 47 | 55 | 17 | 90 | 81 | 73 | 51 |
| | 6 | 43 | 31 | 51 | 27 | 06 | 48 | 49 | 16 | 93 | 77 | 63 | 46 |
| | 7 | 49 | 37 | 76 | 30 | 02 | 50 | 44 | 14 | 95 | 73 | 54 | 42 |
| | 8 | 55 | 44 | 0.0900 | 33 | 4.99 | 51 | 39 | 13 | 98 | 69 | 45 | 38 |
| | 9 | 60 | 51 | 28 | 36 | 93 | 53 | 34 | 11 | 0.401 | 65 | 37 | 34 |

**TABLE 2 (cont.)**

| φ° | | i | k | n | p | q | t | u | v | w | x | y | z |
|---|---|---|---|---|---|---|---|---|---|---|---|---|---|
| 10 | 0 | 0.766 | 0.357 | 51 | 1.139 | 4.89 | 0.354 | 4.29 | 0.610 | 03 | 2.62 | 1.727 | 2.30 |
| | 1 | 72 | 64 | 77 | 42 | 84 | 55 | 24 | 08 | 05 | 58 | 18 | 26 |
| | 2 | 77 | 71 | 0.100 | 45 | 80 | 56 | 19 | 07 | 07 | 55 | 09 | 22 |
| | 3 | 83 | 78 | 02 | 49 | 76 | 57 | 15 | 06 | 10 | 51 | 00 | 19 |
| | 4 | 88 | 84 | 06 | 52 | 72 | 58 | 10 | 05 | 12 | 48 | 1.691 | 15 |
| | 5 | 93 | 91 | 08 | 55 | 68 | 59 | 4.06 | 03 | 15 | 45 | 83 | 12 |
| | 6 | 99 | 98 | 11 | 58 | 65 | 60 | 01 | 02 | 17 | 42 | 74 | 08 |
| | 7 | 0.804 | 0.405 | 14 | 61 | 51 | 61 | 3.97 | 01 | 20 | 39 | 66 | 05 |
| | 8 | 09 | 12 | 17 | 65 | 57 | 62 | 93 | 00 | 22 | 36 | 58 | 02 |
| | 9 | 14 | 19 | 20 | 68 | 54 | 64 | 88 | 0.599 | 25 | 33 | 50 | 1.99 |
| 11 | 0 | 0.819 | 0.426 | 0.123 | 1.172 | 4.50 | 0.365 | 3.84 | 98 | 0.427 | 2.30 | 1.642 | 1.96 |
| | 1 | 24 | 34 | 26 | 75 | 47 | 66 | 80 | 97 | 30 | 27 | 34 | 93 |
| | 2 | 29 | 41 | 29 | 79 | 43 | 67 | 76 | 96 | 32 | 24 | 27 | 90 |
| | 3 | 34 | 48 | 32 | 83 | 40 | 68 | 72 | 95 | 35 | 22 | 19 | 87 |
| | 4 | 39 | 55 | 35 | 86 | 37 | 69 | 68 | 95 | 37 | 19 | 11 | 85 |
| | 5 | 43 | 63 | 38 | 90 | 34 | 70 | 65 | 94 | 40 | 16 | 03 | 82 |
| | 6 | 48 | 70 | 41 | 94 | 31 | 71 | 61 | 93 | 42 | 14 | 1 596 | 80 |
| | 7 | 53 | 78 | 44 | 98 | 28 | 71 | 57 | 93 | 45 | 12 | 89 | 77 |
| | 8 | 57 | 85 | 47 | 1.201 | 25 | 72 | 53 | 92 | 47 | 09 | 82 | 75 |
| | 9 | 62 | 93 | 50 | 50 | 05 | 22 | 73 | 91 | 50 | 07 | 74 | 73 |
| 12 | 0 | 0.866 | 0.500 | 0.154 | 1.209 | 4.19 | 0.3743 | 3.46 | 0.5908 | 0.453 | 2.05 | 1.568 | 1.70 |
| | 1 | 70 | 08 | 57 | 13 | 16 | 50 | 43 | 03 | 55 | 02 | 60 | 68 |
| | 2 | 75 | 15 | 60 | 17 | 13 | 58 | 40 | 0.5899 | 57 | 00 | 53 | 66 |
| | 3 | 79 | 23 | 63 | 21 | 11 | 67 | 36 | 95 | 60 | 1.982 | 47 | 64 |
| | 4 | 83 | 31 | 67 | 25 | 08 | 75 | 33 | 90 | 63 | 61 | 40 | 62 |
| | 5 | 87 | 38 | 70 | 30 | 05 | 83 | 29 | 87 | 65 | 41 | 34 | 60 |
| | 6 | 91 | 46 | 74 | 34 | 03 | 92 | 26 | 85 | 68 | 21 | 27 | 58 |
| | 7 | 95 | 54 | 77 | 38 | 4.00 | 99 | 23 | 82 | 71 | 01 | 21 | 56 |
| | 8 | 99 | 62 | 81 | 43 | 3.98 | 0.3807 | 20 | 80 | 73 | 1.882 | 14 | 54 |
| | 9 | 0.903 | 70 | 84 | 47 | 95 | 14 | 17 | 77 | 76 | 63 | 08 | 53 |
| 13 | 0 | 06 | 0.577 | 89 | 1.252 | 3.93 | 0.3821 | 3.14 | 0.5876 | 0.478 | 45 | 02 | 1.51 |
| | 1 | 10 | 85 | 92 | 56 | 91 | 28 | 11 | 75 | 81 | 27 | 1.495 | 49 |
| | 2 | 14 | 93 | 95 | 61 | 88 | 35 | 08 | 74 | 83 | 09 | 89 | 48 |
| | 3 | 17 | 0.601 | 99 | 66 | 86 | 41 | 05 | 73 | 86 | 1.792 | 83 | 46 |
| | 4 | 21 | 09 | 0.202 | 70 | 84 | 47 | 02 | 74 | 89 | 75 | 77 | 44 |
| | 5 | 24 | 0.617 | 06 | 75 | 82 | 54 | 2.99 | 74 | 91 | 58 | 71 | 43 |
| | 6 | 27 | 25 | 10 | 80 | 80 | 60 | 97 | 75 | 94 | 42 | 65 | 41 |
| | 7 | 30 | 34 | 14 | 85 | 77 | 66 | 94 | 76 | 97 | 26 | 59 | 40 |
| | 8 | 34 | 42 | 17 | 90 | 75 | 72 | 91 | 77 | 99 | 10 | 53 | 38 |
| | 9 | 37 | 50 | 21 | 95 | 73 | 77 | 88 | 79 | 0.502 | 1.675 | 48 | 37 |
| 14 | 0 | 0.940 | 0.658 | 0.225 | 1.300 | 3.71 | 0.3884 | 2.86 | 0.5881 | 05 | 80 | 42 | 1.36 |
| | 1 | 43 | 66 | 29 | 05 | 69 | 89 | 83 | 84 | 08 | 65 | 37 | 34 |
| | 2 | 46 | 74 | 33 | 10 | 67 | 94 | 80 | 87 | 11 | 51 | 31 | 33 |
| | 3 | 48 | 83 | 37 | 15 | 66 | 99 | 78 | 90 | 13 | 36 | 25 | 32 |
| | 4 | 51 | 91 | 41 | 21 | 64 | 0.3904 | 75 | 93 | 16 | 22 | 20 | 31 |
| | 5 | 54 | 99 | 45 | 27 | 62 | 09 | 73 | 98 | 19 | 09 | 15 | 29 |
| | 6 | 56 | 0.708 | 49 | 32 | 60 | 13 | 70 | 0.5902 | 21 | 1.595 | 09 | 28 |
| | 7 | 59 | 16 | 53 | 38 | 58 | 18 | 68 | 06 | 24 | 82 | 04 | 27 |
| | 8 | 61 | 24 | 57 | 43 | 57 | 22 | 65 | 10 | 27 | 69 | 1.399 | 26 |
| | 9 | 64 | 33 | 61 | 49 | 55 | 27 | 63 | 16 | 30 | 56 | 93 | 25 |
| 15 | 0 | 0.966 | 0.741 | 0.265 | 1.355 | 3.53 | 0.3931 | 2.61 | 0.5922 | 0.533 | 44 | 88 | 1.24 |
| | 1 | 68 | 50 | 69 | 61 | 52 | 35 | 58 | 28 | 35 | 31 | 83 | 23 |
| | 2 | 70 | 58 | 73 | 67 | 50 | 39 | 56 | 35 | 38 | 19 | 78 | 22 |
| | 3 | 72 | 67 | 77 | 73 | 48 | 43 | 54 | 41 | 41 | 08 | 73 | 20 |
| | 4 | 74 | 75 | 81 | 79 | 47 | 46 | 51 | 46 | 44 | 1.495 | 68 | 1.195 |
| | 5 | 76 | 84 | 85 | 85 | 45 | 49 | 49 | 55 | 47 | 84 | 63 | 85 |
| | 6 | 78 | 92 | 89 | 92 | 44 | 53 | 47 | 64 | 50 | 73 | 58 | 77 |
| | 7 | 80 | 0.801 | 94 | 98 | 42 | 56 | 45 | 73 | 53 | 62 | 54 | 66 |
| | 8 | 82 | 09 | 98 | 1.405 | 41 | 59 | 43 | 81 | 56 | 51 | 49 | 57 |
| | 9 | 83 | 18 | 0.302 | 11 | 39 | 61 | 40 | 90 | 59 | 40 | 44 | 48 |
| 16 | 0 | 0.985 | 0.826 | 06 | 18 | 3.38 | 0.3964 | 2.38 | 0.5999 | 0.562 | 30 | 40 | 1.140 |
| | 1 | 86 | 35 | 11 | 25 | 37 | 67 | 36 | 0.601 | 65 | 20 | 35 | 31 |
| | 2 | 88 | 44 | 15 | 31 | 35 | 69 | 34 | 02 | 68 | 09 | 30 | 23 |
| | 3 | 89 | 52 | 19 | 38 | 34 | 71 | 32 | 03 | 71 | 00 | 26 | 15 |
| | 4 | 90 | 61 | 23 | 45 | 33 | 73 | 30 | 04 | 74 | 1.390 | 21 | 07 |
| | 5 | 91 | 70 | 28 | 52 | 31 | 75 | 28 | 05 | 77 | 80 | 17 | 1.099 |
| | 6 | 93 | 78 | 32 | 59 | 30 | 77 | 26 | 06 | 80 | 71 | 12 | 91 |
| | 7 | 94 | 87 | 36 | 67 | 29 | 79 | 24 | 07 | 84 | 61 | 08 | 84 |
| | 8 | 0.9945 | 96 | 41 | 74 | 27 | 80 | 22 | 09 | 87 | 52 | 03 | 76 |
| | 9 | 54 | 0.904 | 45 | 82 | 26 | 82 | 20 | 10 | 90 | 43 | 1.299 | 69 |
| 17 | 0 | 0.9962 | 13 | 0.349 | 1.489 | 3.25 | 0.3983 | 2.18 | 0.611 | 0.593 | 34 | 95 | 1.062 |
| | 1 | 69 | 22 | 54 | 97 | 24 | 84 | 16 | 13 | 96 | 26 | 90 | 55 |
| | 2 | 76 | 30 | 58 | 1.505 | 23 | 85 | 14 | 14 | 0.600 | 17 | 86 | 49 |
| | 3 | 81 | 39 | 62 | 13 | 22 | 86 | 13 | 16 | 03 | 09 | 82 | 42 |
| | 4 | 86 | 48 | 67 | 21 | 20 | 87 | 11 | 17 | 06 | 00 | 78 | 36 |
| | 5 | 91 | 56 | 71 | 28 | 19 | 88 | 09 | 18 | 10 | 1.292 | 74 | 29 |
| | 6 | 94 | 65 | 75 | 37 | 18 | 89 | 07 | 20 | 13 | 84 | 70 | 23 |
| | 7 | 97 | 74 | 80 | 45 | 17 | 89 | 05 | 22 | 16 | 76 | 65 | 17 |
| | 8 | 99 | 83 | 84 | 53 | 16 | 89 | 04 | 23 | 20 | 68 | 61 | 11 |
| | 9 | 1 | 91 | 88 | 62 | 15 | 89 | 02 | 25 | 23 | 61 | 57 | 06 |
| 180 | | 1 | 1 | 0.393 | 1.571 | 3.14 | 89 | 2.00 | 0.627 | 0.627 | 1.253 | 1.253 | 1 |

# How to Specify Precision Knurls

GEORGE BARROW
Westinghouse Electric Corp.

**It takes more than a single adjective or short note on a drawing to specify a straight knurl on which tolerances are critical. An accurate method of designation such as the dimensioning system proposed by the author herein is required.**

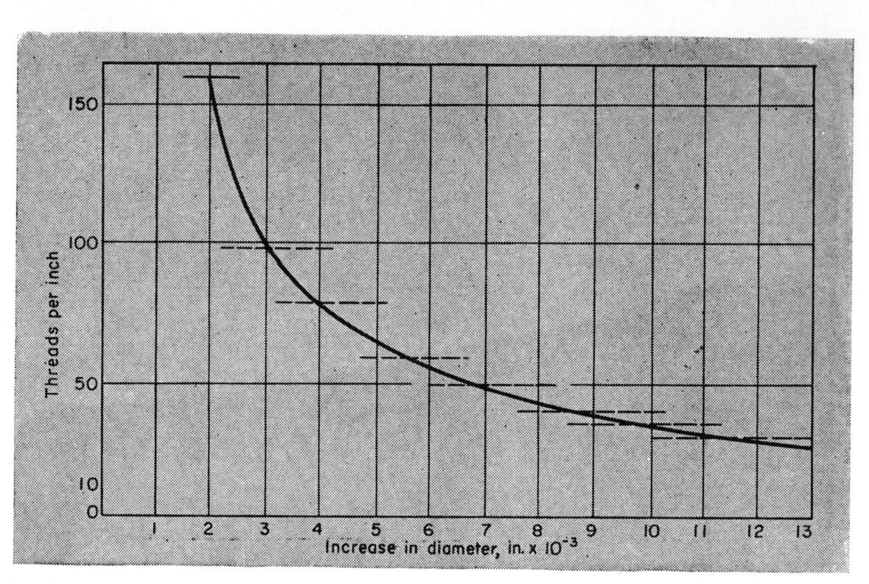


**Fig. 1—Data for steel, brass and aluminum for selecting a tool pitch to give the desired increase in outside diameter or buildup.**

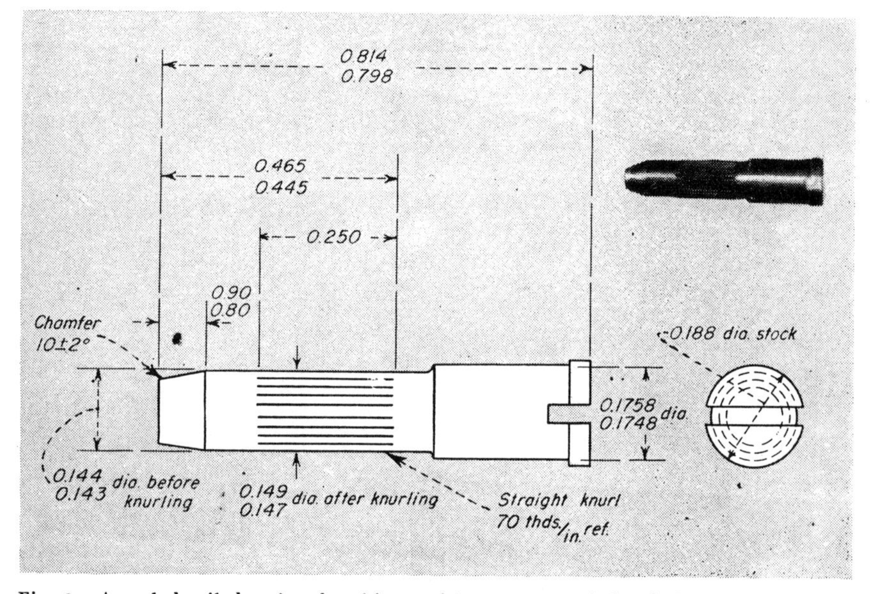


Fig. 3—Actual detail drawing for this precision parts used the designation "Medium Straight Knurl." Preferred system of specifying the information is shown.

FREQUENTLY, knurls are specified on drawings by the general classifications of fine, medium or coarse. Since knurled parts are seldom trade items and there is no standardization as such, this system of designation tends to put the responsibility for the end-product on the machine operator rather than on the designer. Furthermore, what is considered a medium knurl by one operator can be a fine, or even a coarse one, to an operator in a different shop or another machine operator in the same shop. This leads to non-uniformity of products, and does not provide control over critical dimensions.

Of course, precision is not always required. Some types of knurls, such as diagonal tooth or diamond shaped, are widely used to hold studs or inserts in die castings and molded plastic products. Close tolerances are unnecessary in these cases since the material will always mate intimately with the surface. The same condition exists if the serrations are used for decorative or grip purposes.

But some products, particularly small machine elements that employ straight knurling to obtain a press fit assembly, require a rigid system of designation. The degree of this minimal fit must be sufficient to meet performance requirements and forms the basis for the maximum press fit that occurs with the maximum diameter of the shaft and the minimum diameter of the hole in the hub or bushing. The following specification procedure was developed primarily for applications of this type.

## Specification Procedure

Establishing a designation system that is both accurate and practical from a production standpoint involves two factors. One is the relation between the diameter of stock and pitch of the teeth to prevent double knurling or splitting of serrations on repeated revolutions of the work against the tool. This requires an approximately even division of the circumference of the stock by the tooth pitch. Except in special cases where conditions warrant tool development it is unnecessary to compute the exact pitch size and supply a special tool for each diameter of stock encountered. A more practical solution is for the designer to concentrate on the necessary buildup, or increase in diameter, and allow the shop freedom in select-

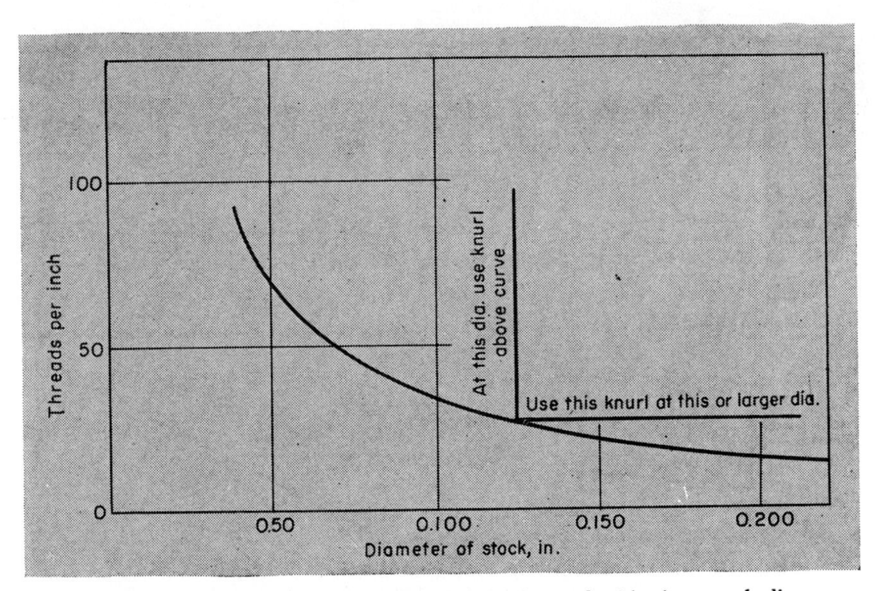


**Fig. 2—Approximate range of tool pitches that can be used with given stock diameters. Combinations that fall beneath the curve are to be avoided.**

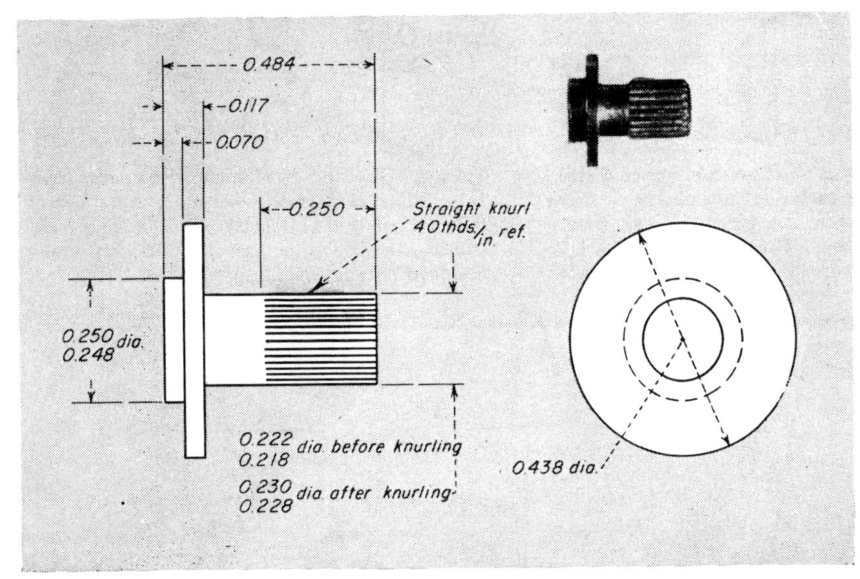


**Fig. 4—This knurl was originally specified as "Standard Fine." Following the authors suggestions, the part is detailed above to eliminate such arbitrary classifications.**

ing the final pitch. An approximate pitch can be suggested if modifications are allowed to prevent double-knurling. Thus, a change from 40 to 41 or from 47 to 50 threads per inch, as determined by the setup man, is usually better assurance of satisfactory results than the use of an untried calculated pitch for a given diameter.

The other factor, which was referred to previously, is the final diameter of the knurled area. For this, the classifications fine, medium and coarse, are entirely inadequate. From a practical basis, twelve is the minimum number of serrations that can be made on a given diameter of stock and yet retain good workmanship. This gives the maximum increase in diameter and can be designated the coarsest knurl. Reducing the pitch reduces the buildup and increases the number of serrations for a given diameter stock until at some point the result can be termed medium or fine. However, the buildup in diameter over the final surface is the important factor and the resulting number of serrations is secondary.

Data on the buildup of steel, brass and aluminum stock is given in Fig 1 to guide in the selection of a knurl tool. For a given pitch, the buildup increases slightly with increase in diameter of stock. The range is indicated by the short lines through the curve and parallel to the abscissa. The left end of the line represents the minimum of twelve serrations and indicates the buildup on a diameter of stock derived by dividing a constant 3.82 by the threads per inch. The other end of the line indicates the maximum buildup for the same knurl on a diameter of stock approximately 6.4 divided by threads per inch. Further increases in the diameter of the stock will not affect the amount of buildup.

Knowing the diameter of stock to be worked and the buildup required, in thousandths, an approximate tool pitch can be determined from the curve. The derivation of the constant 3.82 is as follows:

$$\frac{C}{N} = p$$

$$P = \frac{1}{p} = \frac{N}{C} = \frac{N}{\pi D}$$

When $N = 12$

$PD = 3.82$

$N$ = number of serrations
$D$ = diameter of stock before knurling
$C$ = circumference of stock before knurling
$P$ = $1/p$, threads per inch
$p$ = pitch

The use of constant 3.82 permits further enlargement of data as demonstrated by Fig. 2. This curve shows the approximate minimum limit for a given diameter of stock or a given pitch of tool to produce the twelve-serration type of area. For example, with a tool having 30 threads per inch the minimum diameter of stock on which it can be used is 0.127 in., but it can be used on any larger diameter stock, in which case it will make more than twelve serrations. Likewise it is seen that, at 0.127-in. diameter, 30 is the largest pitch that can be used, but tools with a smaller pitch can be used and if they are, more than twelve serrations will be made. Therefore, the area under the curve represents combinations of threads per inch and stock diameter that should be avoided.

A suggested designation for specifying this information on engineering drawings is shown in Figs. 3 and 4. The nominal increases in the diamaters of the knurled areas for these two precision parts are 0.0045 and 0.009 in., respectively. From Fig. 1, the suggested pitches are 70 and 40 threads per inch, respectively. These sizes are shown on the drawing for reference only. Alternatively, these values could have been specified by the designer with an appropriate tolerance. The diameter of the knurled area, which in the final analysis is the important detail, is specified definitely in both the before and after conditions. Sufficient information and latitude are given to permit manufacture and inspection of the critical areas without further questions or possibility of misinterpreting the designer's needs and specifications.

# how much HORSEPOWER to pipe liquids?

Douglas C. Greenwood

**Charts give answers for flow from 50 to 100,000 gpm for pipe-roughness ratios of 0.02 to smooth.**

The charts on these two pages will make it easier to design pumps and other equipment for handling liquid flowing in pipes.

First, it is often helpful to know the theoretical horsepower required to raise the liquid to various heads. This is obtained from the horsepower-gpm charts. They are plotted for a 10-ft head of water: For other heads, multiply hp by $H/10$ where $H$ is the revised head; for other liquids, multiply by the corresponding specific gravity.

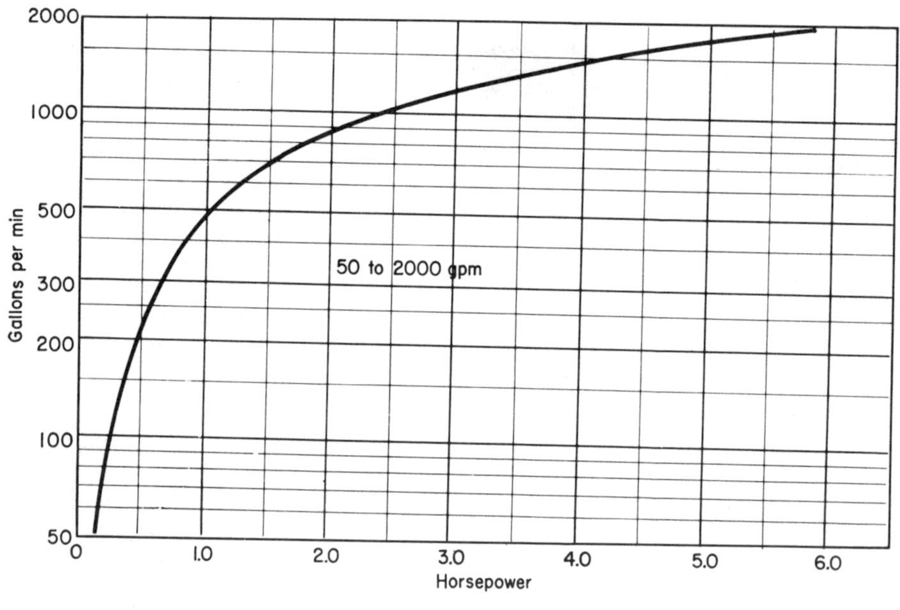


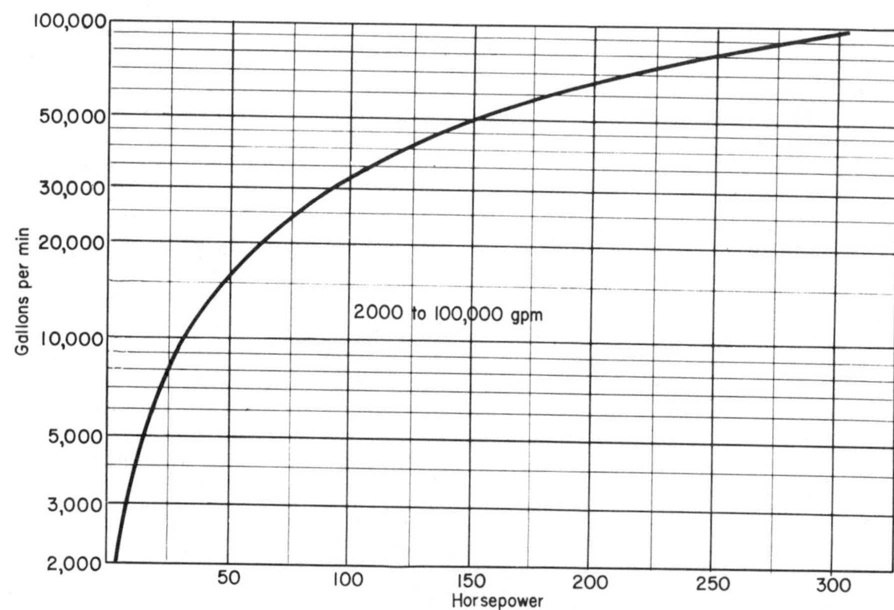


**Hp-gpm Chart . . .** shows how much hp is required to pump water against a 10-ft head. Full-pipe flow is assumed.

**SYMBOLS**

| | | |
|---|---|---|
| $G$ = flow rate, gpm | $R$ = Reynold's number $= 0.0833\ Vd/\nu$ | $w$ = fluid density, lb per cu ft |
| $H_1$ = head loss, ft | $V$ = velocity, fps | $r$ = relative roughness of pipe = $\epsilon/d$ |
| $L$ = length of pipe, ft | $d$ = pipe dia, in. | $\epsilon$ = effective height of roughness particle, in. |
| $P$ = pressure, psi | $f$ = friction factor | $\nu$ = kinematic viscosity, $ft^2/sec$ |

Next, for practical results friction losses must be accounted for. These vary and should be known for each individual case.

Much used in liquid-flow calculations is the Darcy formula

$$H_1 = f\,\frac{6LV^2}{d32.16}$$

which can be modified, if velocity is in gpm, to

$$H_1 = 0.0312f\,\frac{LG^2}{d^5}\,w$$

or for head loss in psi units

$$P = 0.000217f\,\frac{LG^2}{d^5}\,w$$

Practical values of $f$ vary from about 0.01 to 0.06 depending on pipe smoothness and dia. For laminar flow, $f = 64.4/R$. The flow chart gives $f$ for various values of $R$ and pipe roughness. Values $\epsilon$ for various pipes are: 0.00006 in. for smooth drawn tubing; 0.0018 in. for wrought iron; 0.01 in. for cast iron. Curves for relative roughness values of 0.0005 to 0.01 are plotted. Most of these lie in the transition zone between laminar flow and complete turbulence.

**Fluid-flow Chart . . .**

gives friction for various pipe conditions and values of Reynold's Number.

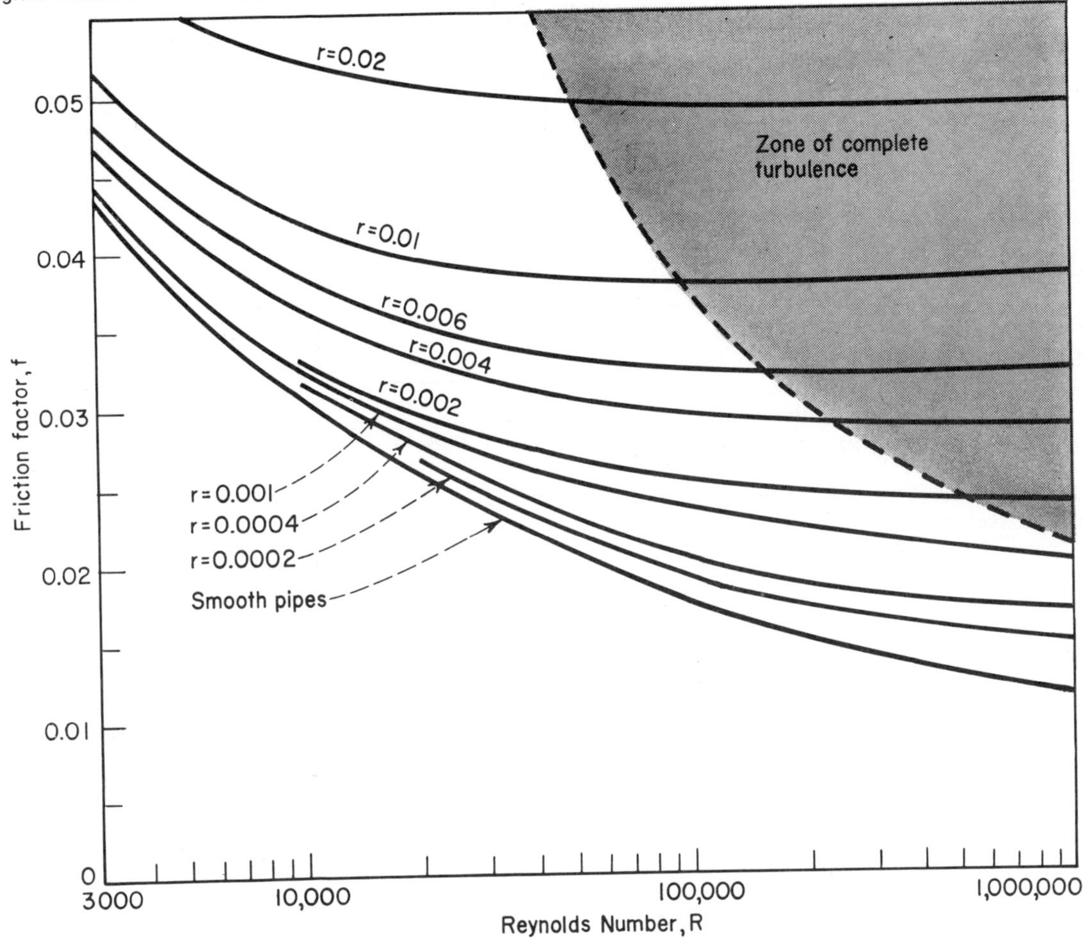

# Design Fundamentals for

• Principles that can be applied to reduce mold cost, avoid uneven shrinkage, and minimize internal stresses.

## MOLDING

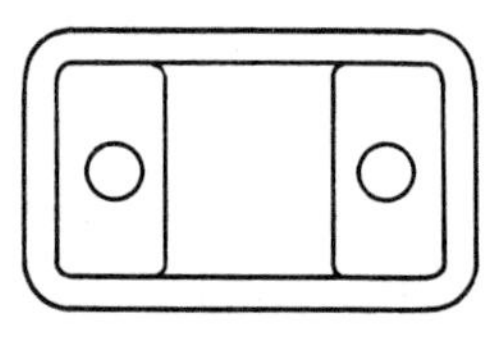
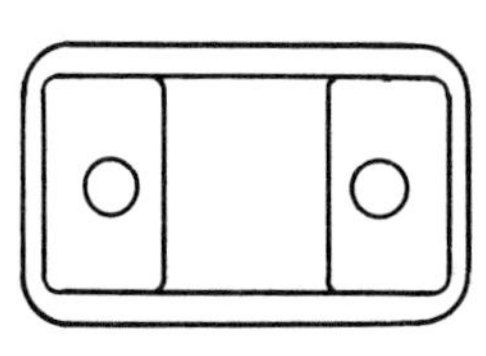
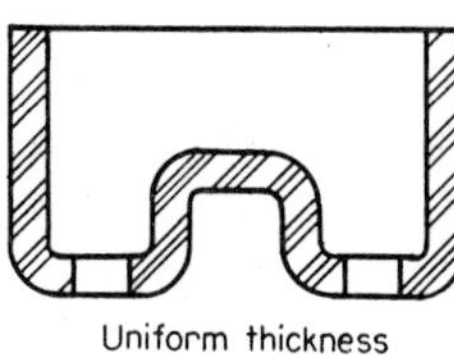
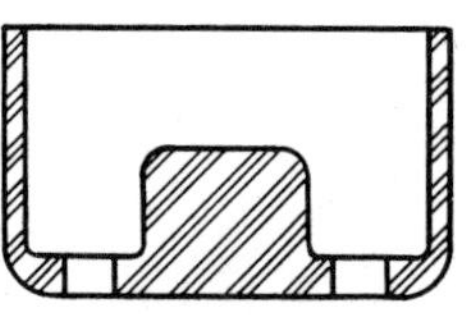

Uniform thickness preferred

Varying thickness not recommended

***Minimum wall thickness*** for a compression molded part is 1/16 in.; for injection molded part a minimum wall thickness of 0.050 in. should be maintained unless paths of plastic flow are short and high pressure is available. Walls of uniform thickness decrease the chance of uneven shrinkage and resultant internal stresses.

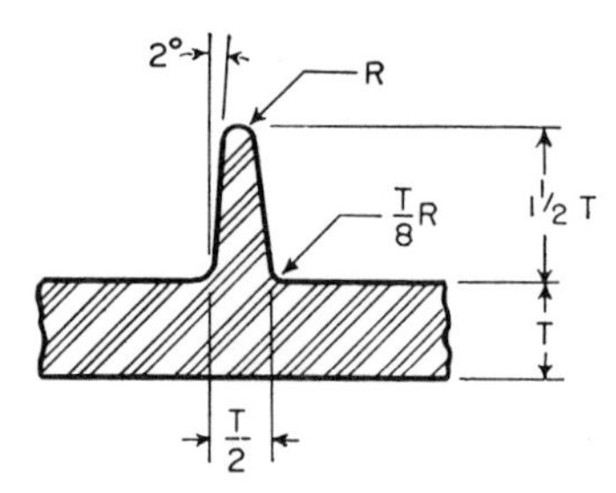

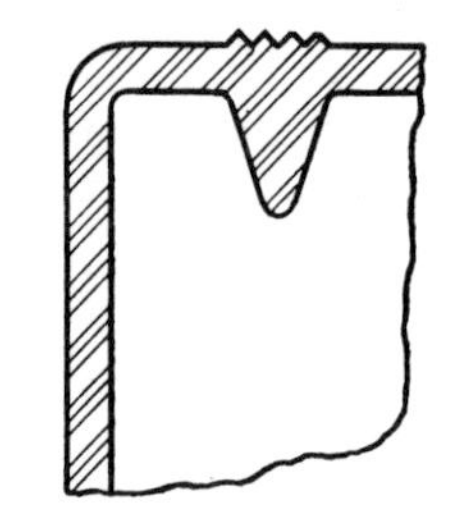

***Ribs*** should be designed with at least two to five deg taper per side and with radii at the top and base. The rib width at the base should be ½ the wall thickness; rib height should be 1½ the wall thickness. Internal ribs and bosses on thin wall sections that cause flow lines or sink marks on the outer surface can be concealed by decorative ribbing, as shown in right sketch, or by using two smaller ribs.

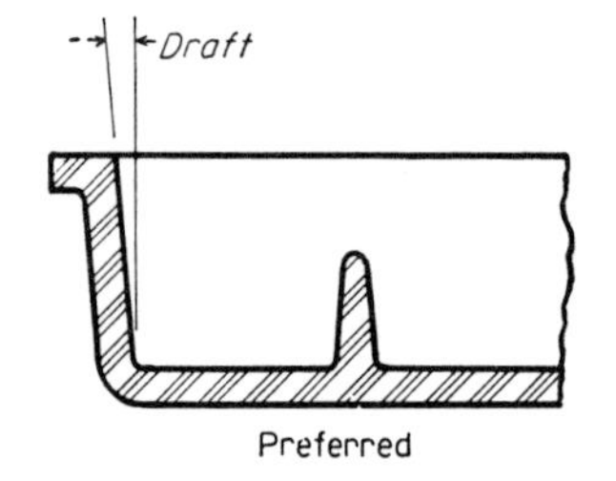

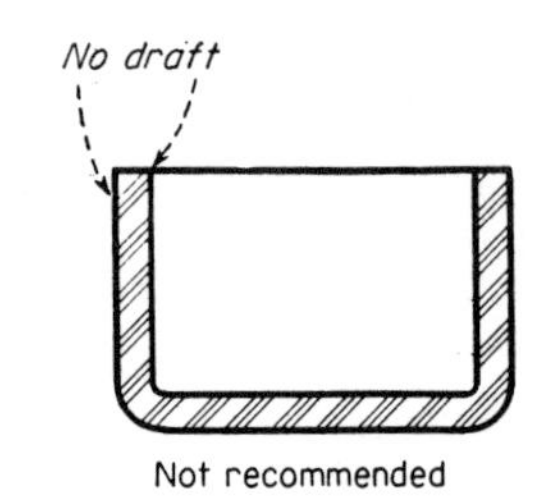


Preferred

Not recommended

***Draft*** may vary from ¼ to 4 deg per side depending upon the length of draw, surface area, type of material, and method of ejection. Sufficient draft must be allowed on all surfaces in the direction of the draw to ease ejection of the part from the mold and to protect the surface. In special instances where no draft is permitted, it should be specified by a note on the drawing to the affected surface.

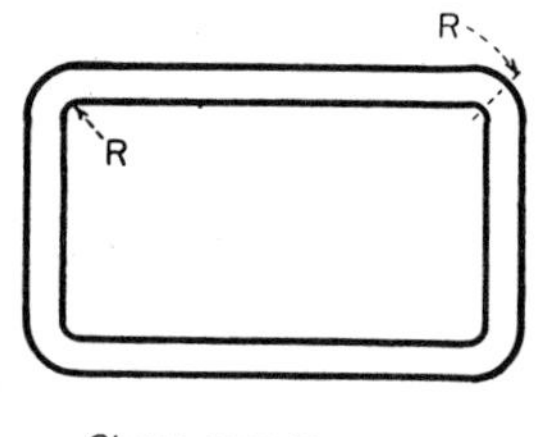

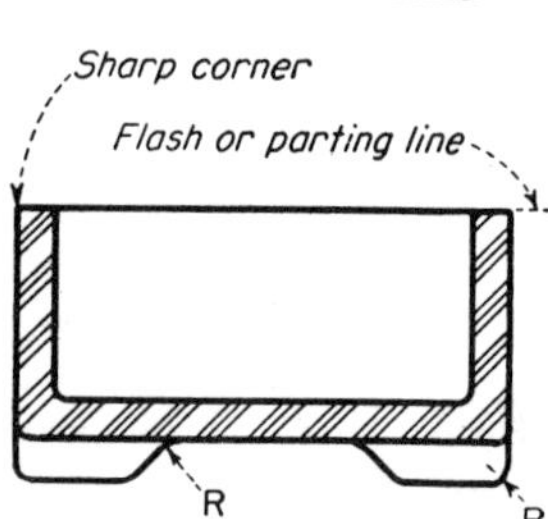


***Radii*** on all corners should be one-half section thickness but not less than 1/32 in. except for outside edges at the parting line, which are usually sharp. Fillets at the base of ribs and bosses and on inside corners facilitate the flow of plastic and reduce the concentration of stresses at these points. The use of radii and fillets in the design may reduce the mold cost, usually improve the appearance, and eliminate dust traps that make a part difficult to clean. Fillets at the base of small protruding pin type inserts reduce bending and breakage of pins. Fillets minimize the shadows and flow lines of inside ribs and bosses of parts molded of the translucent plastics.

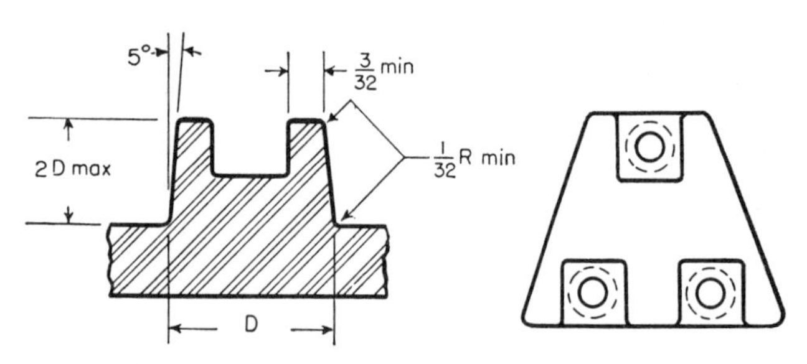


***Boss height*** should not be more than twice its diameter. Where high bosses are required, they should have at least 5 deg taper per side and radii at the top and base. High bosses formed in the upper part of a mold tend to trap gases that reduce density and strength of the part. To eliminate machining the faces of bosses to obtain a flat mounting surface, three bosses can be used instead of four.

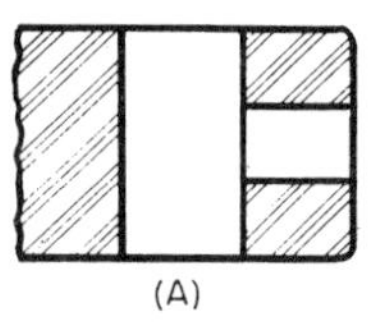

(A)

*Requires movable side core pin*

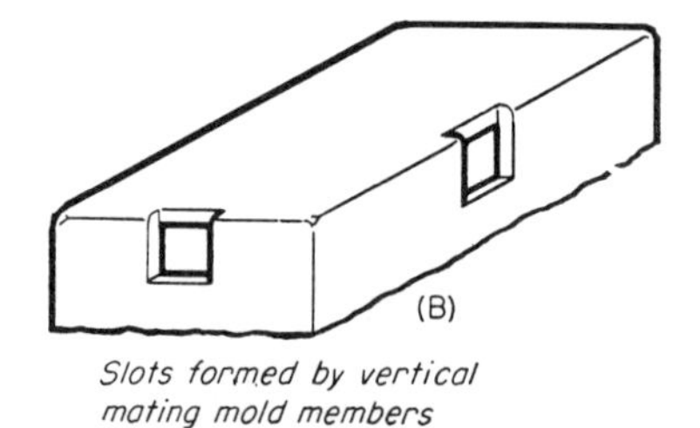

(B)

*Slots formed by vertical mating mold members*

***Holes*** should be located perpendicular to the parting line to permit easy removal from the mold. Where possible, holes that are to be molded at right or oblique angles to each other should be avoided because split molds or molds with removable side core pins are more costly. The drilling of side holes, after the molding of the part, may be more economical. Forming side openings or slots as shown at *B*, can be used in the molding of thin hollow sections.

# MOLDED PLASTIC PARTS

• Techniques that can be adopted to speed assembly time and make possible production of complex shapes.

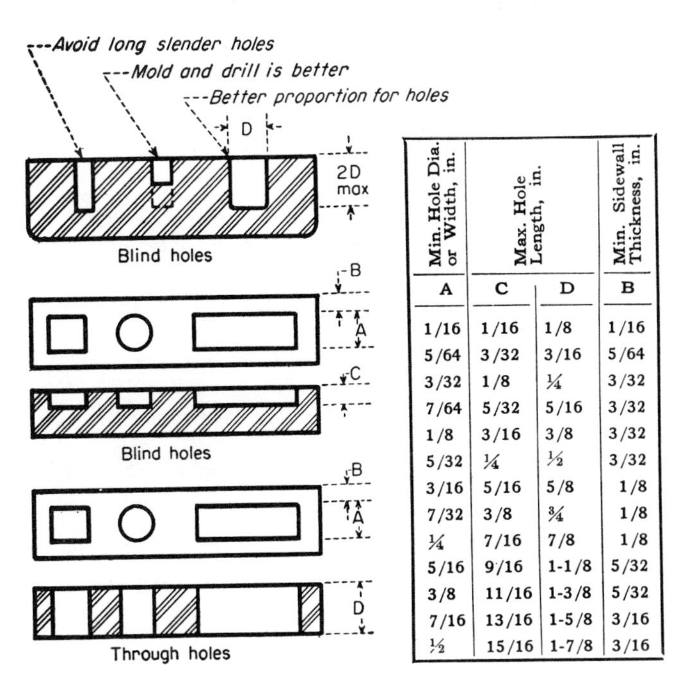


| Min. Hole Dia. or Width, in. | Max. Hole Length, in. | | Min. Sidewall Thickness, in. |
|---|---|---|---|
| A | C | D | B |
| 1/16 | 1/16 | 1/8 | 1/16 |
| 5/64 | 3/32 | 3/16 | 5/64 |
| 3/32 | 1/8 | ¼ | 3/32 |
| 7/64 | 5/32 | 5/16 | 3/32 |
| 1/8 | 3/16 | 3/8 | 3/32 |
| 5/32 | ¼ | ½ | 3/32 |
| 3/16 | 5/16 | 5/8 | 1/8 |
| 7/32 | 3/8 | ¾ | 1/8 |
| ¼ | 7/16 | 7/8 | 1/8 |
| 5/16 | 9/16 | 1-1/8 | 5/32 |
| 3/8 | 11/16 | 1-3/8 | 5/32 |
| 7/16 | 13/16 | 1-5/8 | 3/16 |
| ½ | 15/16 | 1-7/8 | 3/16 |

***Holes*** less than 1/16 in. dia are difficult to mold and are generally drilled. Long slender holes require thin mold pins. Long holes of small diameter may be molded for a short length and then drilled to the required depth.

Recommended proportions for through and blind hole designs are given in the accompanying table. Core pins for through holes can be supported at each end, therefore, through holes are easier to mold than blind holes. Depth of blind holes should be limited to twice their diameter or to once their diameter for holes less than 1/16 inch.

***Section thickness*** between any two holes should be as large as design permits but not less than 1/8 in. Between a hole and a molded insert, section thickness should not be less than 1/16 in. Thickness of section between the surface of a wall and a hole of 3/32 in. dia or less should be at least one hole diameter; larger holes may have a lower ratio.

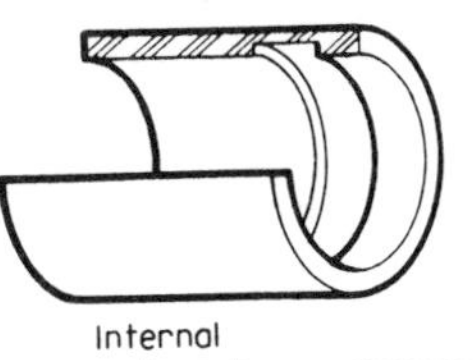


***Undercuts*** should be avoided in molded parts as they require split molds or removable core sections which are costly. Where undercuts are required, the designs should be reviewed to see if molding in two separate parts and assembling with fasteners is not more economical. If essential to mold undercuts, their design and location should be considered in conjunction with the ease of removal of the part from the mold.

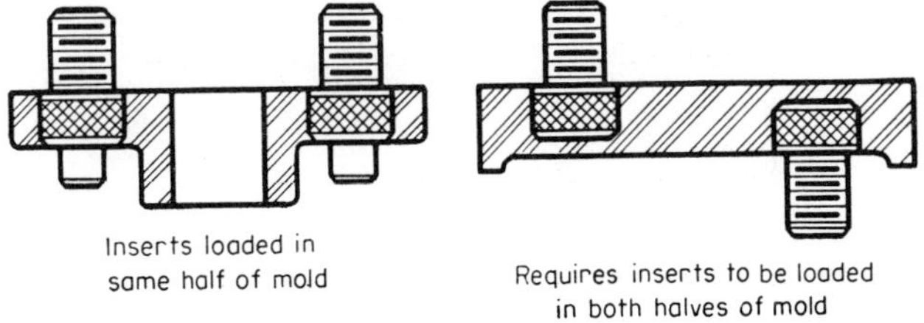


***Inserts*** should be of a design that permits both ends to be supported firmly in the mold. Inserts at right angles to the parting line simplify mold design. Where possible, all inserts should be placed on the same side of the parting line as it is more costly to load inserts in both halves of a mold. Long slender inserts tend to bend or break off during the closing of the mold and should be avoided, especially in compression molded parts. When adequately supported in the mold, however, delicate inserts may be used in transfer and injection molded parts.

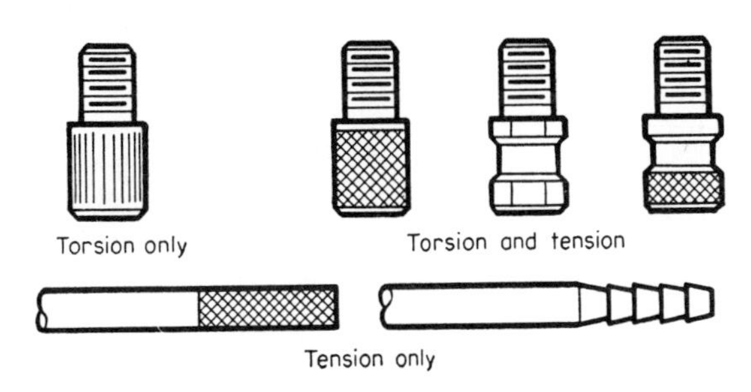


***Inserts*** should be firmly anchored in the molded part. A diamond knurled surface is good as it will withstand both tension and torsion loading. Grooved inserts should take tension only, while serrated units withstand only torsion. To ease the flow and to reduce stresses caused by shrinkage, the ends and grooves of inserts can be chamfered or rounded.

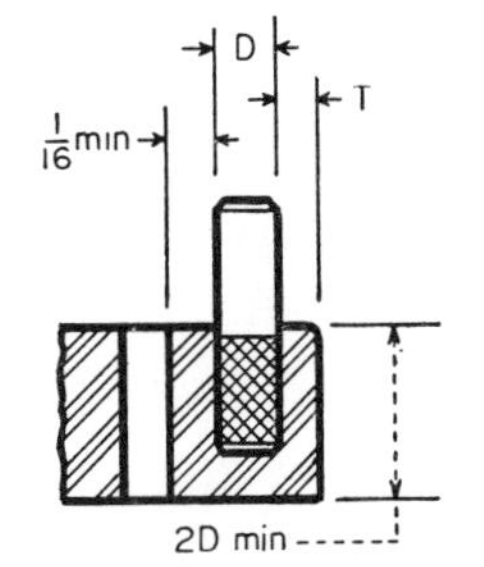


| D | T |
|---|---|
| $\frac{1}{4}$ and smaller | $\frac{3}{32}$ min |
| Over $\frac{1}{4}$ to $\frac{1}{2}$ | $\frac{1}{8}$ min |
| Over $\frac{1}{2}$ | $\frac{1}{4}$ min |

***Support*** of an insert by the plastic material requires that the amount of material around the insert be as large as the design permits and not less than that shown. Rod and wire types should be imbedded to a length equal to twice the diameter. The blind end should not be too close to a wall as a blister may form. Minimum section thickness between a hole and a molded insert should be 1/16 inch.

**MOLDING** —*continued*

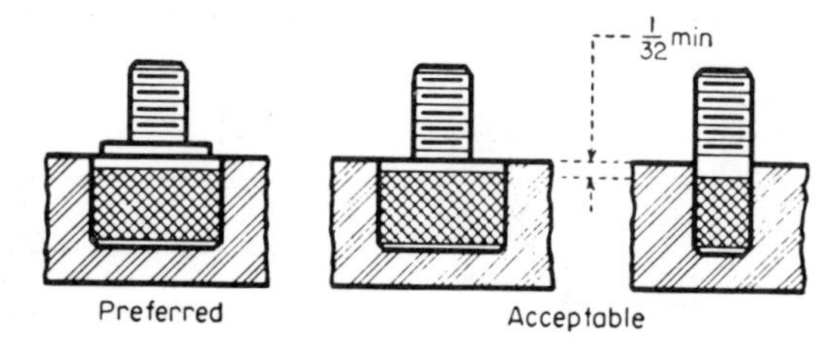


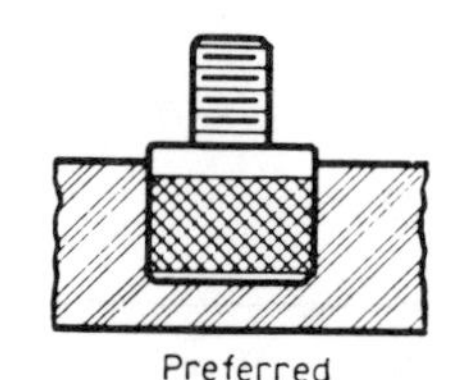


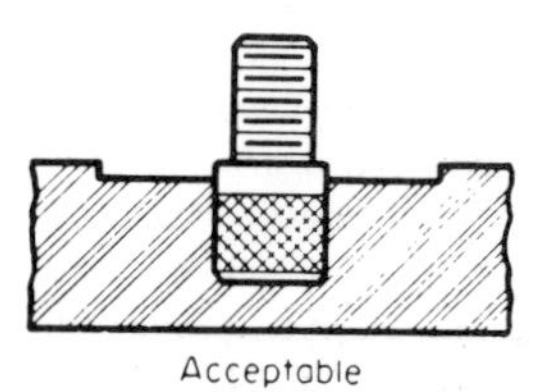


***Protruding portions*** of an insert should be round. This design is easier to clean of flash, and its corresponding hole in the mold is less expensive to form than square or hexagonal shapes. Plain shoulders of 1/32 in. between the edge of the knurl and the edge of the part will reduce or eliminate flash. The preferred designs shown give the best seal against the flow of plastic when molded.

## Parts Made of Laminated Plastic Stock

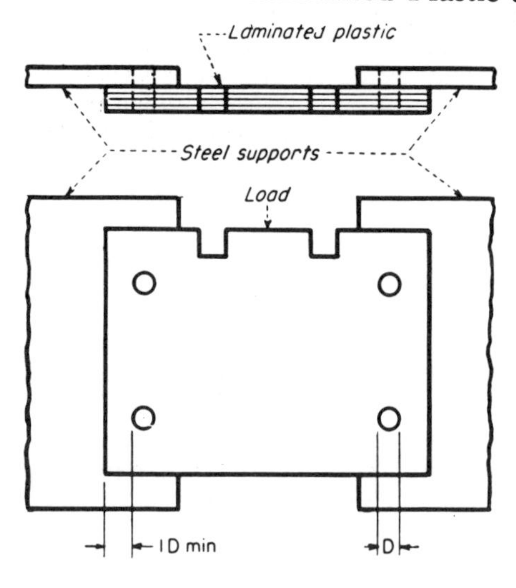


***In laminated parts*** (left), the load carrying member should be supported on each end. Holes should be placed as far as practicable from the edge of the laminated material to prevent breaking or fracturing. The one diameter spacing, as shown, is the minimum recommended for edge clearance.

***Where cantilever loading*** (right), cannot be avoided, the distance from the edge to holes should be as great as the design will permit to prevent the laminated material from tearing or breaking. A minimum of one inch is recommended for edge clearance.

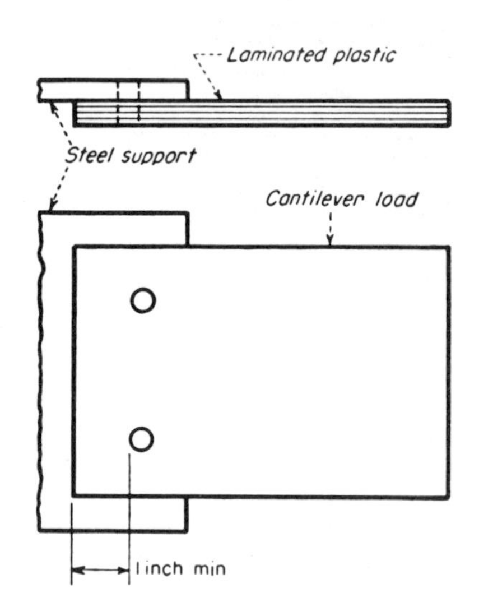


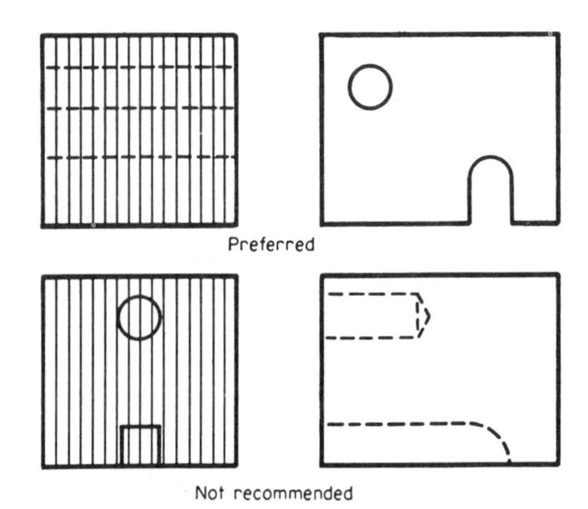


***Drilling and machining*** at right angles to the laminations avoids splitting the layers. Where these operations are unavoidable, and are parallel to the laminations, a cloth base laminate should be used.

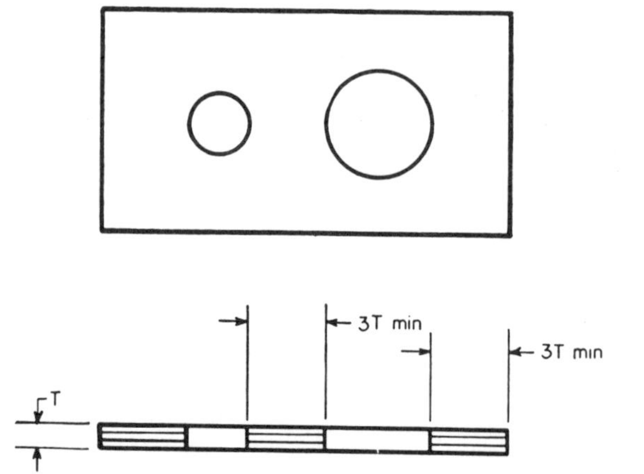


***Laminated material*** can be punched in thicknesses up to ⅛ in. Paper base laminates can be sheared in thickness up to 1/16 in. and most fabric base materials up to ⅛ in. Minimum hole clearance should be 3 times the stock thickness for all materials.

## ASSEMBLY

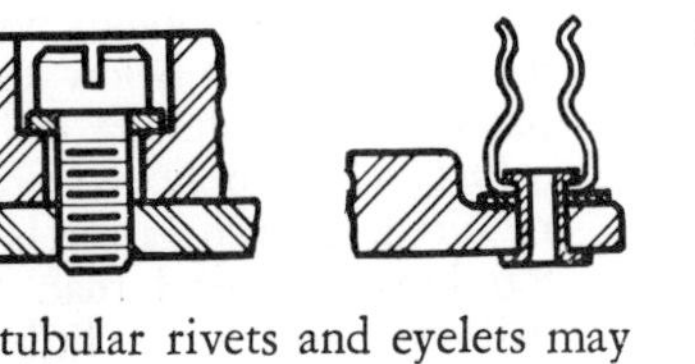

***Through holes*** in the molded part will allow screws, bolts, or studs to be inserted and secured in tapped holes of adjacent parts or with nuts or other fasteners. Solid and tubular rivets and eyelets may be used for permanent assemblies.

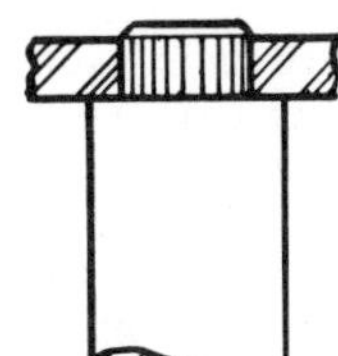

***Shrink fit*** of two plastic parts, one on the other, is another assembly method. Close control of the molding technique is required. Plastic parts may be shrunk on metal parts. The metal part is usually knurled where shrink fitted parts must not be free to turn relative to each other.

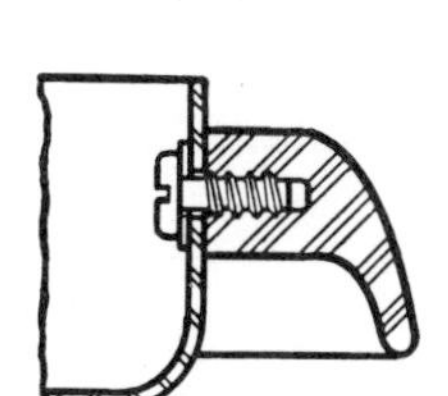
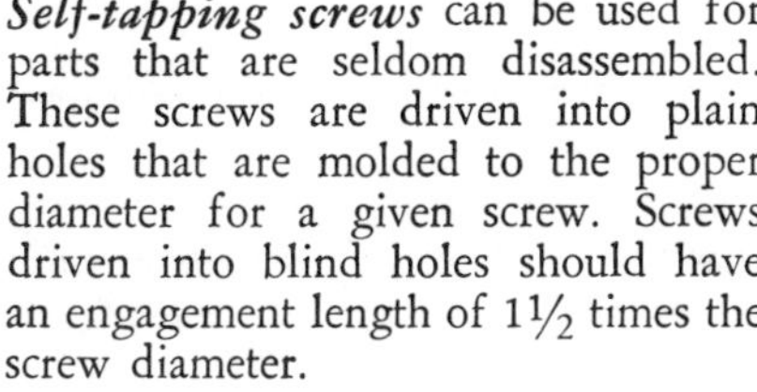

***Self-tapping screws*** can be used for parts that are seldom disassembled. These screws are driven into plain holes that are molded to the proper diameter for a given screw. Screws driven into blind holes should have an engagement length of $1\frac{1}{2}$ times the screw diameter.

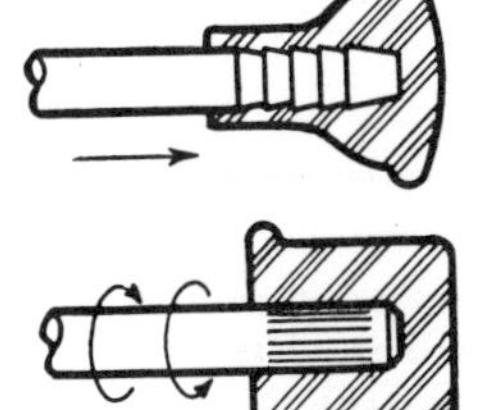

***Press-in inserts.*** These can be used to assemble parts that are molded of thermoplastic materials that are not too rigid. They should be considered only if other assembly methods and molded-in units are not practical or economical. For good anchorage, the insert should have a serrated or knurled surface.

***Drive screws*** may be used for permanent installations. These are usually driven into molded holes of suitable diameter for a given screw. The drive screws used in plastics vary in diameter from 1/16 in. minimum to 5/16 in. maximum.

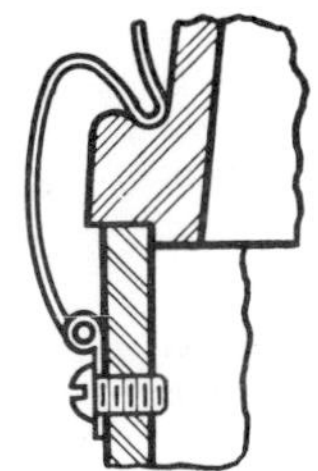
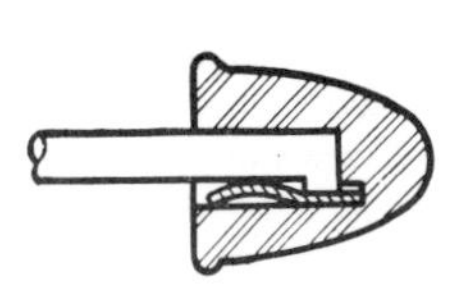

***Spring type*** metal hinges and clips are used to fasten two plastic parts or a molded part to one of metal. These items fit into recesses or over bosses of the molded part.

***Machine screws*** can be used for assemblies that are limited to light loads and infrequent disassembly. Threads must be tapped. Molded threads should be avoided for these assemblies.

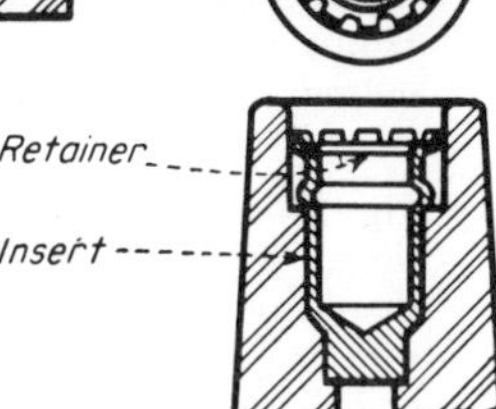


***Stud clips*** and retainers can also be used as fasteners. Pressing a spring clip over a boss will hold two parts together. A retainer may be pressed into a hole of a molded part to hold two parts together as shown.

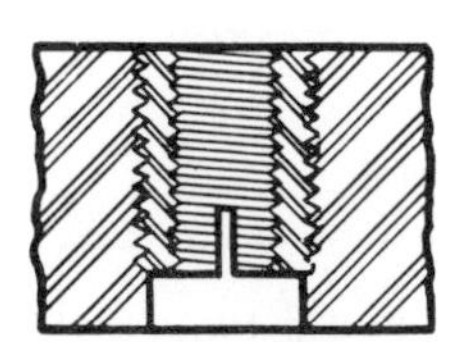
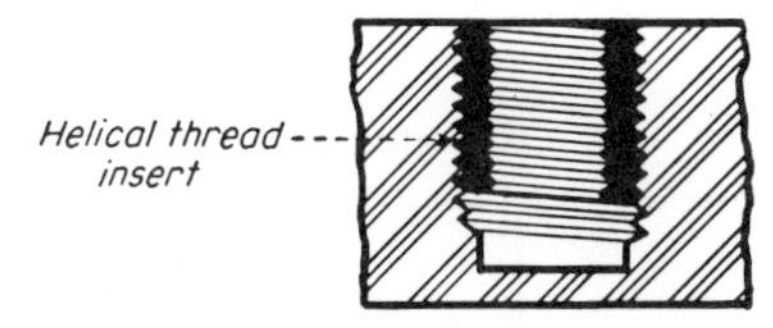


***Threaded inserts*** which lock in place instead of molded-in inserts, in certain applications, can be more economical. These inserts protect the plastic threads and permit disassembly. Types used are: A) self-tapping insert with a threaded hole, and B) special wire coiled to a given thread pitch before it is screwed into a hole tapped in the plastic.

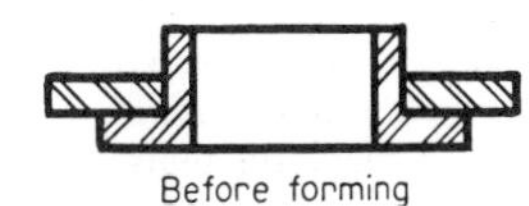

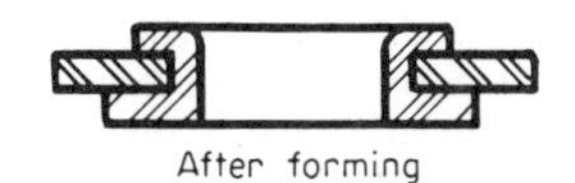


***Heat forming*** can be used to join certain thermoplastic molded parts to metal parts by reforming the plastic part under pressure with a hot punch at assembly.

***Either internal or external threads*** can be molded. These are generally more expensive to form than other threads because of the method of removing the part from the mold. Generally, only coarser pitches are molded.

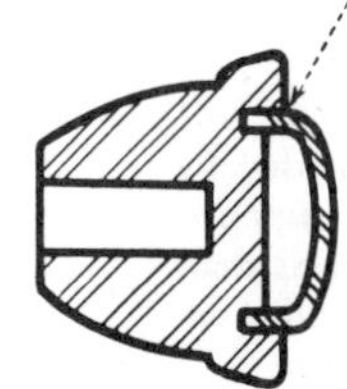


***Adhesives and cements*** can be used to fasten plastic parts where the loading is light. Care must be exercised in the selection of cement because certain solvents may cause cracking, crazing, or discoloration of the bonded plastic parts.

# Recommended Practices for Designing

EXPERIENCE IN THIS FIELD has proved that faulty design is responsible for more difficulties than faulty processing. Considerations that do not enter into ordinary sheet metal fabrication are vital to good porcelain enameling. For example, to be certain that a part can safely withstand the high firing temperatures, such factors as kind of base metals, gage of metal, size and shape of piece, method of forming, rigidity, holes for hanging, etc. must be considered. Selection of the proper gage metal cannot be over-emphasized. Although no hard and fast rules can be made, the table below can be used as a guide for parts of moderate size to be finished in white or light colors, where only moderate rigidity and flatness are required. In general, it is better to err on the side of too heavy a metal than to use one too light. The first case will cause little or no trouble during the enameling or mounting, the only loss being the extra amount of metal required. When the metal is too light, however, warpage, sagging, hairlining

**Fig. 1**—Avoid square corners in openings to prevent porcelain from hairlining in corners. For white or light colors, min. radius of flanges in (c) should be 3/16 in. and min. flange depth ¼ in. With dark colors, 9/64 is min. flange radius.

**Fig. 2**—Raw edges are undesirable on finished parts since porcelain will burn off the edges. Flanges should be turned down as shown in Fig. 1 (*c*) whenever possible.

**Fig. 3**—Use equal cross sections so that when parts are heated and cooled, expansion and contraction will be uniform, and warping and chipping of porcelain because of strains in the metal will be avoided.

**Fig. 4**—Wrong method of making a flange cut-out is shown in (*a*). Chipping and hairline will occur on the face. Method illustrated in (*b*) is correct. Corners of rectangular parts should have generous radius with sharp corners, annealing may be required to prevent recrystallization during firing.

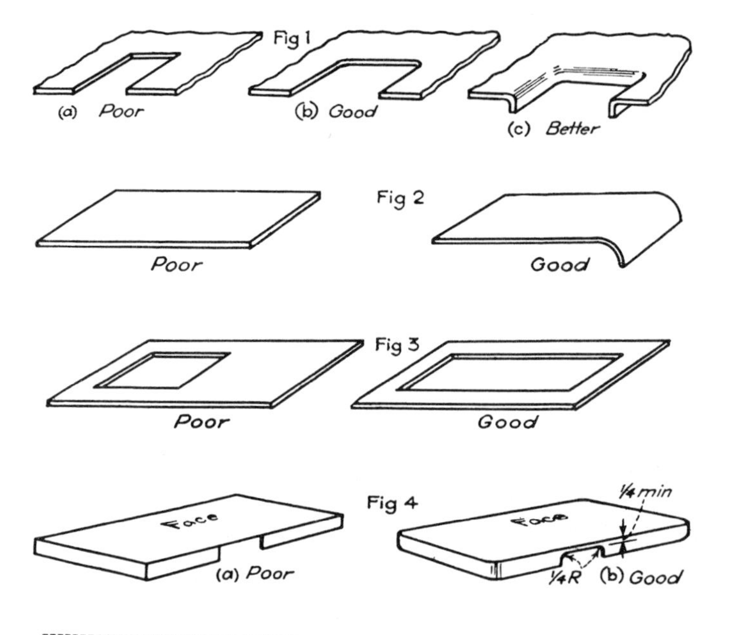


**Fig. 5**—Avoid welding on exposed surfaces if possible. When stiffening braces are required, they should be skeletonized to minimize weight and insure proper firing. Weld spots must be smoothed off or indentures will show.

**Fig. 6**—Do not use square notched corners in panels with flanges. Notching at 45 deg, (*b*), is preferred if parts are formed on press brake. When production volume warrants forming and blanking tools, the method in (*c*) can be used.

**Fig. 7**—Avoid turned-under flanges. If part is dipped, excess enamel cannot drain; if part is sprayed, under side of flange will be difficult to reach; and if the part is exposed to moisture, rusting might occur.

**Fig. 8**—In (*a*), the angle of the vertical flange will change during firing; in (*b*), support provided by end flange is inadequate. For proper support, *A-B = 1/3 F* as in (*c*) or if *A* and *B* are less than *1/3 F*, (*d*), additional support can be provided by the strap that is spot welded to the frame.

### Recommended Gages of Metal

| Gage No. | Approx. Thickness, in. | Width, in. | Total Area, $ft^2$ |
|---|---|---|---|
| 24 | 0.0239 | 6 | 0.5 |
| 24 | — | 12 | 3* |
| 24 | — | 18 | 5* |
| 22 | 0.0299 | 6 | 1 |
| 22 | — | 12 | 3.5 |
| 22 | — | 18 | 6* |
| 22 | — | 24 | 8* |
| 20 | 0.0359 | 6 | 1.5 |
| 20 | — | 12 | 5 |
| 20 | — | 18 | 8* |
| 20 | — | 24 | 10–15* |

* Should be embossed flanged or otherwise suitably reinforced. All parts larger than indicated should be made from 18 gage or heavier metal.

# Porcelain Enamel Products

PAUL F. METZ
*Crosley Div. Avco Manufacturing Corp.*

and/or chipping are frequently encountered. Using heavier gages will also help eliminate unequal stress distributions that often result from drawing operations.

The qualities that must be inherent in a good enameling stock are given as follows:

1. Very low carbon content.
2. Freedom from harmful solid or gaseous impurities.
3. Resistance to sagging and warping at enamel firing temperatures.
4. Good welding qualities.
5. Good drawing or working qualities.
6. Good surface texture.
7. Uniformity of composition.

The sketches shown below illustrate recommended industrial design practice and methods of fastening porcelain enamel plates to supporting structures; some of them were supplied by the Porcelain Enamel Institute, Inc. of Washington, D. C.

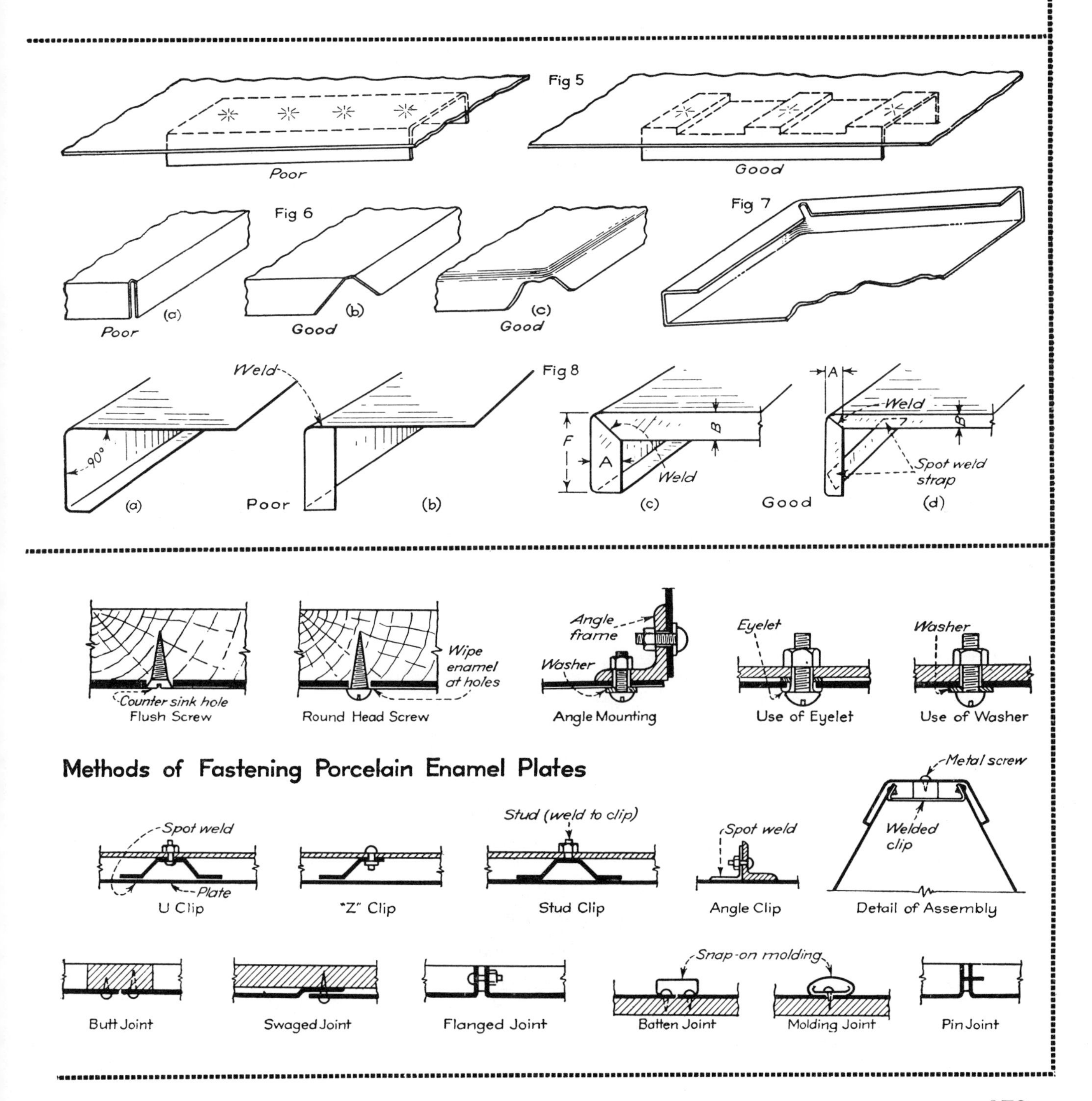

# how to calculate STRESSES in PRESS-FIT BUSHINGS

**Here is a family of formulas, each helpful for a typical application.**

E. H. SCHUETTE,

*Chief of Applications Research, Dow Chemical Co, Midland, Mich.*

**KEY TO FORMULAS**

$$A = \frac{b^2 + a^2}{b^2 - a^2}$$

$$B = \frac{b^2}{c^2 - b^2}$$

$$C = \frac{c^2}{c^2 - b^2}$$

$$D = \frac{d^2 + c^2}{d^2 - c^2}$$

$E$ = modulus of elasticity, psi
$S$ = stress, psi
$e$ = interference, in.
$\mu$ = Poisson's ratio

Subscripts:
$B$ = bushing
$M$ = matrix metal
$S$ = sleeve
1 = matrix-sleeve interface
2 = sleeve-bushing interface

With light metals, press-fit bushings greatly increase the strength of bearings and give only a small increase in weight. Many light-metal alloys, however, are sensitive to stress corrosion. Tensile stresses produced by an interference fit should, therefore, be kept to a safe level. Here, beginning on this page, are formulas that will help do this.

The formulas for cases I and II are extensions of the formulas for pressurized, thick-wall cylinders.

In cases III and IV the derived formulas were too long to be of practical value. To simplify them, therefore, 237 specific examples were calculated on an electronic computer. All probable designs were covered. Formulas III and IV, using arbitrary increase of dia c, were developed as a result of the electronic computation, and give results that are 99% accurate.

### Assumptions Behind Formulas

Assumption for cases I and II is that the outer bushing-face must deflect to a position coincident with the inner matrix-face. Since pressures on these two faces are equal (one reacting against the other) it is possible to say that matrix deflection minus bushing deflection equals the amount of interference. In cases III and IV, conditions governing analysis are:

1. At any position along the length, an unbalance may exist between internal matrix-pressure and the external bushing-pressure.
2. The total of unbalanced forces up to the given position, over a unit circumferential width, is balanced by total shear force on unit circumferential width of cross-section at that position.
3. Slope of the deflection curve at any position must be related, by rigidity modulus, to shear stress.

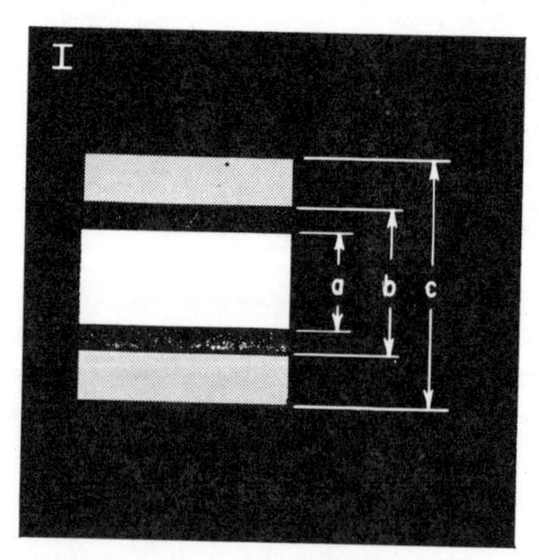


$$S_M = \frac{E_M \frac{e}{b}(B + C)}{(B + C + \mu_M) + \frac{E_M}{E_B}(A - \mu_B)}$$

**Matrix metal, bushing . . .**
have the form of a cylinder. This formula is an extension of those for thick-walled cylinders with internal pressure.

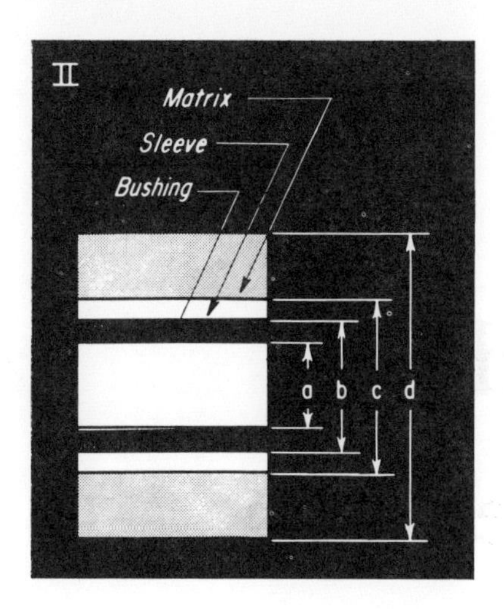


**Sleeve between matrix . . .**
and bushing complicates formulas. But though lengthy, the formulas involve only simple arithmetic calculations.

$$S_M = p_2 D \qquad S_S = p_1(B + C) - 2p_2 D$$

$$p_1 = E_M \frac{\frac{e_1}{b}\left[(D + \mu_M) + \frac{E_M}{E_S}(B + C - \mu_S)\right] + \frac{e_2}{b}\left[2\frac{E_M}{E_S}C\right]}{\left[\frac{E_M}{E_S}(B + C + \mu_S) + \frac{E_M}{E_B}(A - \mu_B)\right]\left[(D + \mu_M) + \frac{E_M}{E_S}(B + C - \mu_S)\right] - 4\left[\frac{E_M}{E_S}\right]^2 BC}$$

$$p_2 = E_M \frac{\frac{e_1}{b}\left[2\frac{E_M}{E_S}B\right] + \frac{e_2}{b}\left[\frac{E_M}{E_S}(B + C + \mu_S) + \frac{E_M}{E_B}(A - \mu_B)\right]}{\left[\frac{E_M}{E_S}(B + C + \mu_S) + \frac{E_M}{E_B}(A - \mu_B)\right]\left[(D + \mu_M) + \frac{E_M}{E_S}(B + C - \mu_S)\right] - 4\left[\frac{E_M}{E_S}\right]^2 BC}$$

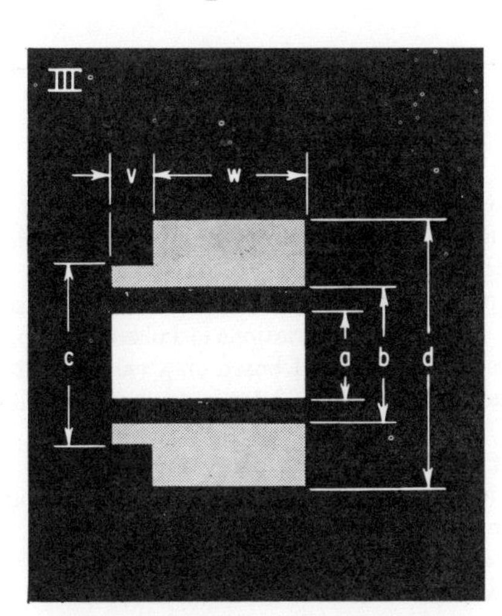


**Stepped matrix . . .**
causes stresses, pressures and deflections to vary over total length. Formula here derives value of $\Delta c$, which gives 99% accurate results when combined with formula for I.

Use formula for case I, substituting $(c + \Delta c)$ for c in $\left(C = \frac{c^2}{c^2 - b^2}\right)$

$$\frac{\Delta c}{b} = \left[\frac{1 + \frac{2a}{b}}{3}\right]\left[\frac{\frac{d-c}{b}}{\frac{d-c}{b} + 1}\right]\left[\frac{\frac{w}{b}}{\frac{w}{b} + \frac{v}{b}}\right]^{1 + \left(\frac{2w}{b}\right)}$$

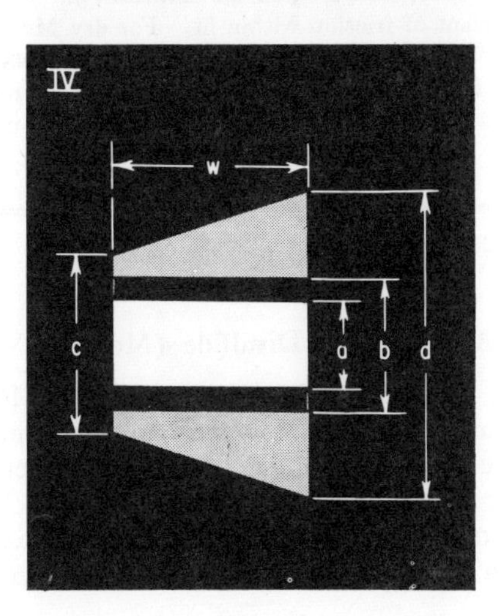


**Conical-matrix . . .**
formula gives $\Delta c$, which is then also incorporated into formula for I, giving same degree of accuracy as III.

Use formula for case I, substituting $(c + \Delta c)$ for c

$$\frac{\Delta c}{b} = \frac{1}{2}\left[\frac{d-c}{b}\right]\left[1 - \frac{(w/b)^2}{(w/b)^2 + 2}\right]$$

# Increasing the Holding Power of Press-Fits . . .

. . . can replace more expensive methods of fastening. A combination of charts and formulas simplifies the analysis of press-fit variables.

RICHARD HAZELETT

Research Engineer, The Alpha Corporation, Greenwich, Conn.

THE USE of $MoS_2$ lubricants in press-fitted assemblies not only reduces scoring, galling and distortion, but also eases the assembly and disassembly of parts. This permits larger interferences between mating parts and allows the use of press fits to replace more expensive methods of fastening such as shrink and taper fits and keyways. For example, by using the maximum permissible bores in standard unhardened gears and a minimum interference of 0.001 in. per in. of dia, nearly all standard gears up to 6 in. pitch dia will sustain their maximum running load before the fit will slip.

The holding power, or resistance to slippage under running load, the induced tensile strength, and the expansion or contraction of the mating parts can be calculated quickly by using the combination of graphs and formulas given below. They are not restricted to any one lubricant.

## SYMBOLS

$a, b, c$ = Diameters, in. (for a solid shaft, $a$ = zero)
$E$ = Modulus of elasticity, lb/sq in., of part $ab$ or $bc$
$L$ = Length of press fit, in.
$N_{ab}$, $N_{bc}$ = Dimensionless functions of $a/b$ or $b/c$ and Poisson's ratio. (Because of variations in Poisson's ratio, two curves appear in Fig. 1 based on a value of 0.28 for ferrous metals, and 0.33 for copper and aluminum alloys).
$P$ = Unit pressure within fit, psi. Does not apply near shouldered ends where local pressure will be higher.
$Q$ = Dimensionless function of $b/c$.
$T$ = Resistance torque available to prevent slippage.
$U_{ab}$, $U_{bc}$ = Dimensionless functions of $a/b$, and $b/c$.
$\Delta b$ = Interference, measured on the diameter, in.
$\mu$ = Coefficient of friction within fit. For dry $MoS_2$ on press-fitted, hardened and ground steel parts, $\mu$ varies from 0.06 to 0.08. In determining the minimum available torque, a safer value of 0.05 is used.

## EXAMPLE

Given: $b = 4.38$, $a/b = 9.14$, $b/c = 0.667$, $L = 3$, $\Delta_b = 0.002$ minimum, $\mu = 0.05$, parts ab and bc are steel.

Since both parts are steel, the graph lines marked "Fe" in Fig. 1 are used, and values of $N_{ab} = 11 \times 10^6$, $N_{bc} = 2.8 \times 10^6$, and $Q = 2.6 \times 10^6$ are obtained. Thus, from the formulas in Fig. 1

$$P = 10^6\left(\frac{0.002}{4.38}\right)\left[\frac{1}{\dfrac{11 \times 10^6}{29.5 \times 10^6} + \dfrac{2.8 \times 10^6}{29.5 \times 10^6}}\right] = 975 \text{ psi}$$

$$T = 1/2(975)(4.38)^2(3)(0.05)(\pi) = 4{,}400 \text{ in.-lb}$$

and

$$\text{Tensile stress} = (975)(2.6 \times 10^6)(10^{-6}) = 2{,}540 \text{ psi}$$

The graph in Fig. 2 is used to determine the elastic change in diameter.

### What is Molybdenum Disulfide ($MoS_2$)?

Experts believe that the lubricating properties of $MoS_2$, which is a solid lubricant similar in appearance and feel to graphite, is the result of its laminar type crystal structure where each layer of molybdenum atoms is sandwiched between two layers of sulfur atoms Because of the strong affinity that sulfur has for metal, the layer of sulfur atoms

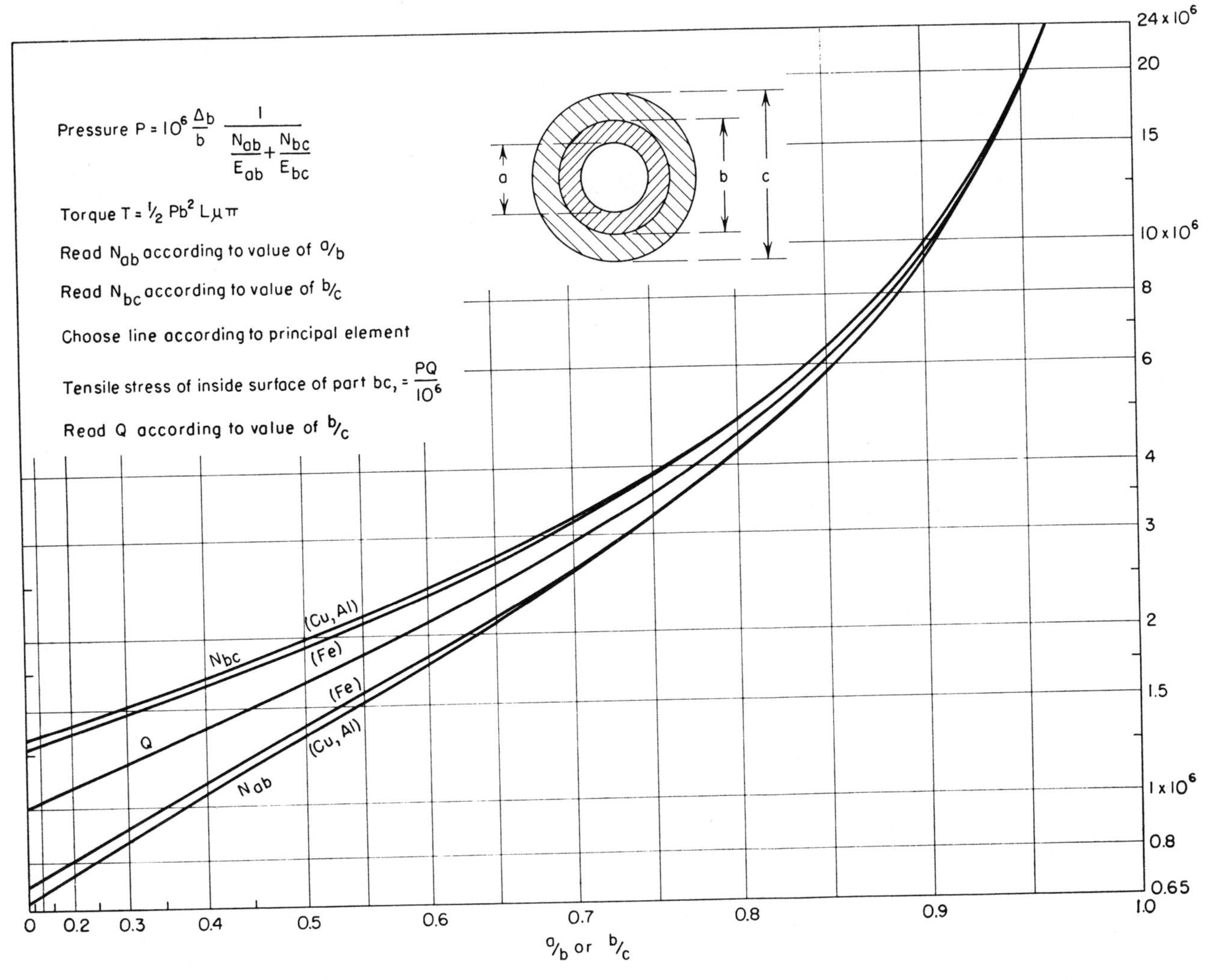


Fig. 1—Press and shrink fit variables.

adheres strongly to metal upon contact. On the other hand, the bond between two adjacent layers of sulfur atoms is weak, so they slip easily over one another. Since no free sulfur is present it does not attack metals. Unlike graphite, the lubricating action does not depend on the presence of a film of moisture and can be used in a vacuum, while its low oxidation tendency permits its use at temperatures from about −40 to 700 F. Since it is also a dielectric lubricant, it is a nonconductor and does not attract dust.

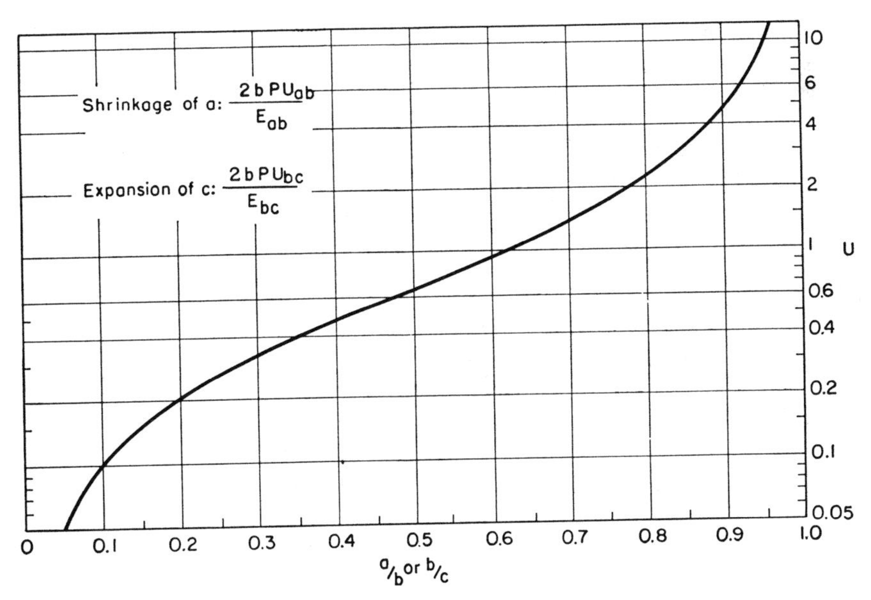


Fig. 2—Elastic change in diameter.

# RATCHET LAYOUT ANALYZED

**Here, in a brief but comprehensive rundown, are generally unavailable formulas and data for precise ratchet layout.**

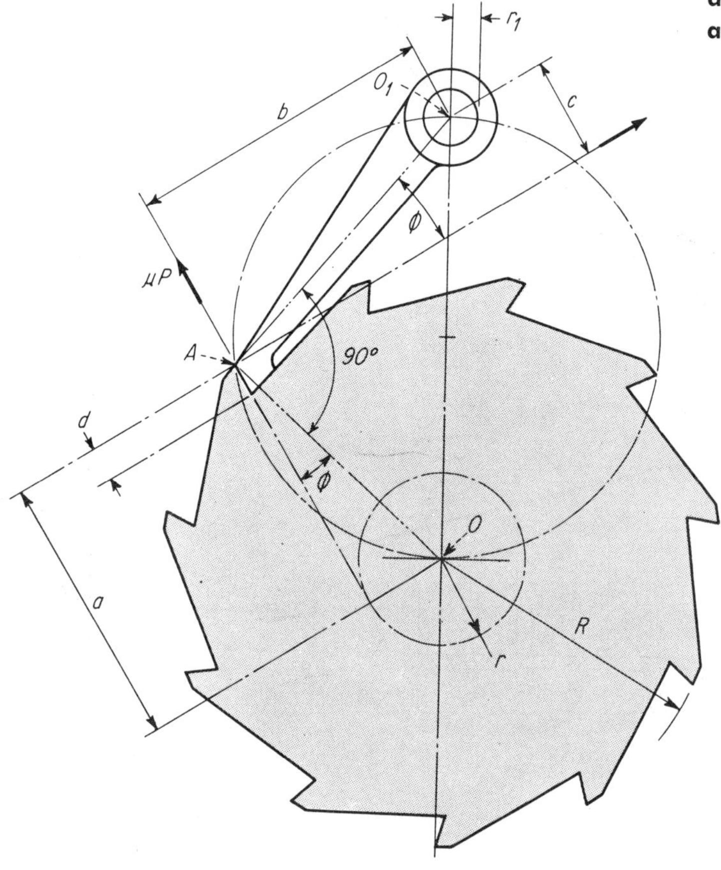


**Pawl in compression . . .**
has tooth pressure $P$ and weight of pawl producing a moment that tends to engage pawl. Friction-force $\mu P$ and pivot friction tend to oppose pawl engagement.

### Symbols

$a$ = moment arm of wheel torque
$M$ = moment about $O_1$ caused by weight of pawl
$O$, $O_1$ = ratchet and pawl pivot centers respectively
$P$ = tooth pressure = wheel torque/$a$
$P\sqrt{(1+\mu^2)}$ = load on pivot pin
$\mu$, $\mu_1$ = friction coefficients
Other symbols as defined in diagrams


**EMERY E. ROSSNER**
*New York, N. Y.*


The ratchet wheel is widely used in machinery, mainly to transmit intermittent motion or to allow shaft rotation in one direction only. Ratchet-wheel teeth can be either on the perimeter of a disc or on the inner edge of a ring.

The pawl, which engages the ratchet teeth, is a beam pivoted at one end; the other end is shaped to fit the ratchet-tooth flank. Usually a spring or counterweight maintains constant contact between wheel and pawl.

It is desirable in most designs to keep the spring force low. It should be just enough to overcome the separation forces—inertia, weight and pivot friction. Excess spring force should not be relied on to bring about and maintain pawl engagement against the load.

To insure that the pawl is automatically pulled in and kept in engagement independently of the spring, a properly layed out tooth flank is necessary.

The requirement for self-engagement is

$$Pc + M > \mu Pb + P\sqrt{(1+\mu^2)}\,\mu_1 r_1$$

Neglecting weight and pivot friction

$$Pc > \mu Pb$$

or

$$c/b > \mu$$

but $c/b = r/a = \tan\phi$, and since $\tan\phi$ is approximately equal to $\sin\phi$

$$c/b = r/R$$

Substituting in term (1)

$$rR > \mu$$

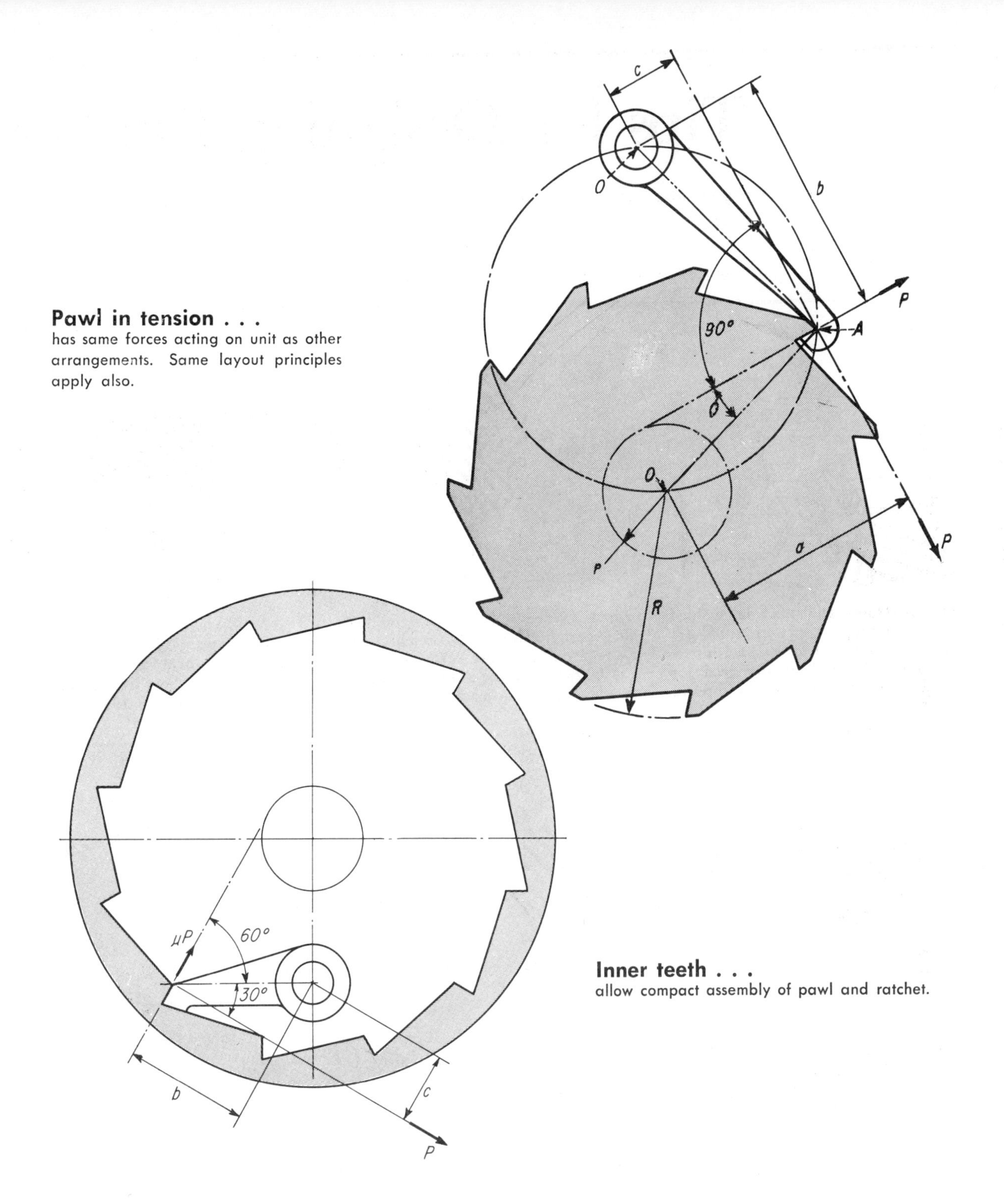


**Pawl in tension . . .**
has same forces acting on unit as other arrangements. Same layout principles apply also.

**Inner teeth . . .**
allow compact assembly of pawl and ratchet.

For steel on steel, dry, $\mu = 0.15$. Therefore, using

$$r/R = 0.20 \text{ to } 0.25$$

the margin of safety is large; the pawl will slide into engagement easily. For internal teeth with $\phi$ of 30°, $c/b$ is tan 30° or 0.577 which is larger than $\mu$, and the teeth are therefore self engaging.

When laying out the ratchet wheel and pawl, locate *points* O, A and $O_1$ on the same circle. AO and $AO_1$ will then be perpendicular to one another; this will insure that the smallest forces are acting on the system.

Ratchet and pawl dimensions are governed by design sizes and stress. If the tooth, and thus pitch, must be larger than required in order to be strong enough, a multiple pawl arrangement can be used. The pawls can be arranged so that one of them will engage the ratchet after a rotation of less than the pitch.

A fine feed can be obtained by placing a number of pawls side by side, with the corresponding ratchet wheels uniformly displaced and interconnected.

# Tips for Designing Rubber

For economical production, parts should be designed for die cutting in preference to molding wherever possible. When selecting the method of fastening rubber parts to other components, the following factors should be considered: hardness of rubber, operating loads, disassembly requirements, fabricating equipment availability, and production requirements.

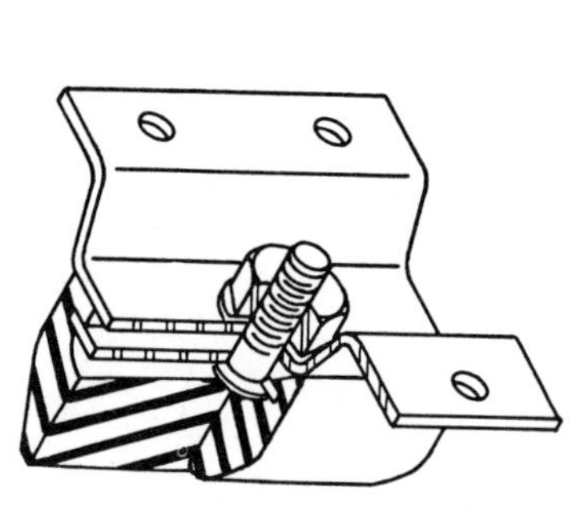

Fig. 1—Holes and slots in molded parts requiring screws or studs are generally preferred to threaded inserts because they eliminate the need for cleaning the rubber from the threads.

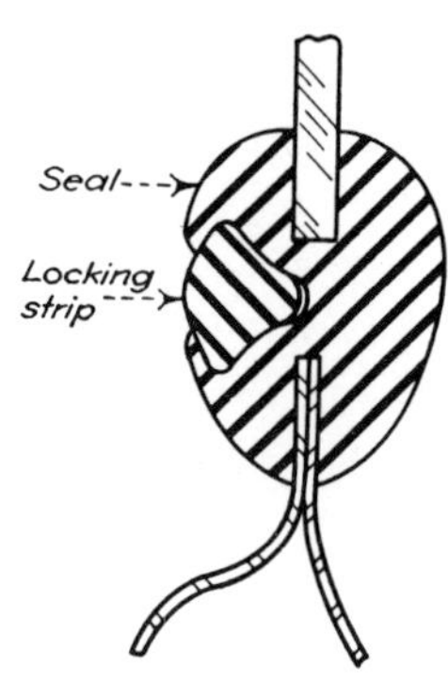


Fig. 2—When it is necessary to seal long peripheries and disassembly may be required, extruded soft rubber seals can be held in place effectively with a locking strip, also made of rubber.

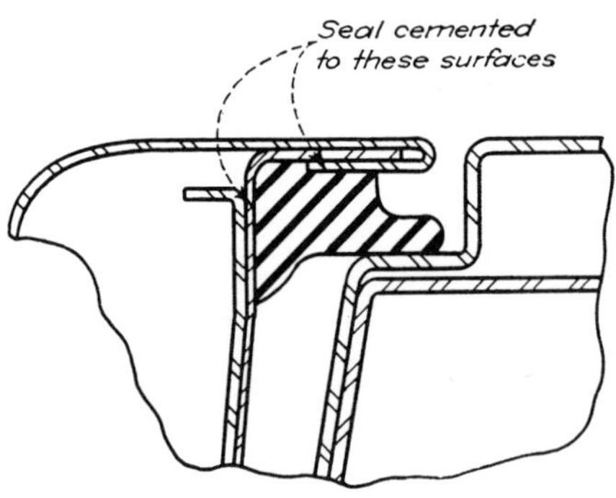


Fig. 3—When the member is subject only to compression loads, soft rubber items can be held to other surfaces with adhesives described in Product Engineering, February 1952.

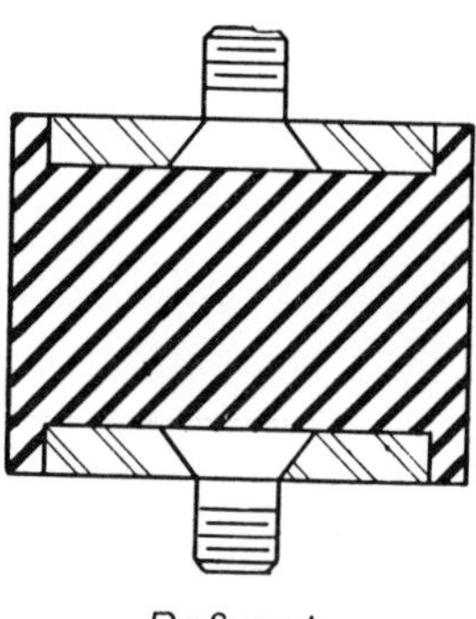

Fig. 6—Uniform sections will reduce adhesion failures where rubber is bonded to inserts and the assembly is subjected to both axial and shear loads. Uniform thicknesses also facilitate molding and eliminate uneven shrinkages.

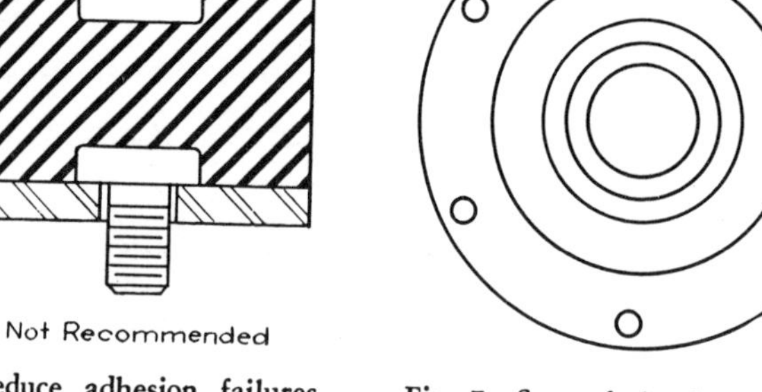

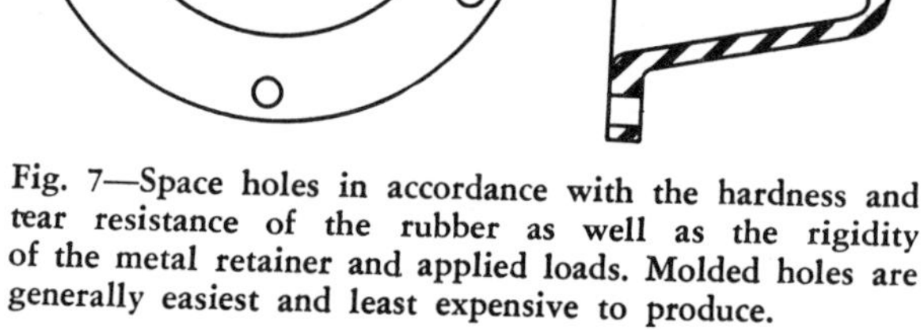

Fig. 7—Space holes in accordance with the hardness and tear resistance of the rubber as well as the rigidity of the metal retainer and applied loads. Molded holes are generally easiest and least expensive to produce.

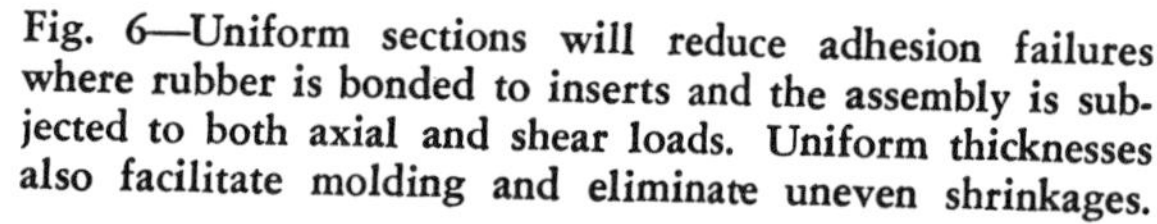

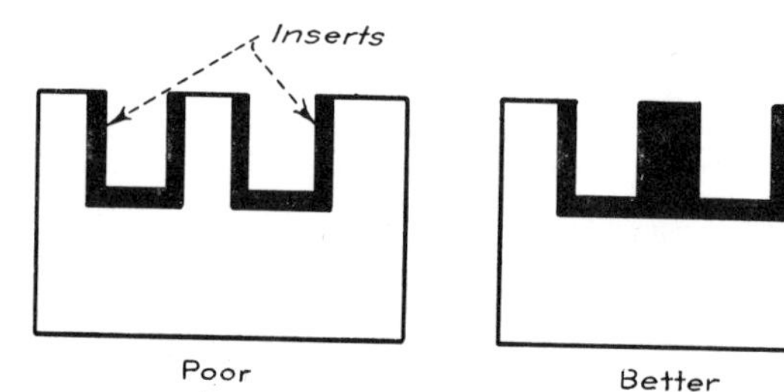


Fig. 11—Small and multiple inserts can increase costs as well as production problems. The single insert shown at the right is preferred to the two smaller inserts at left.

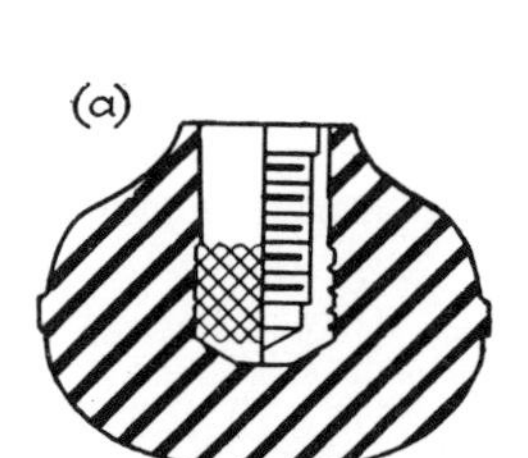


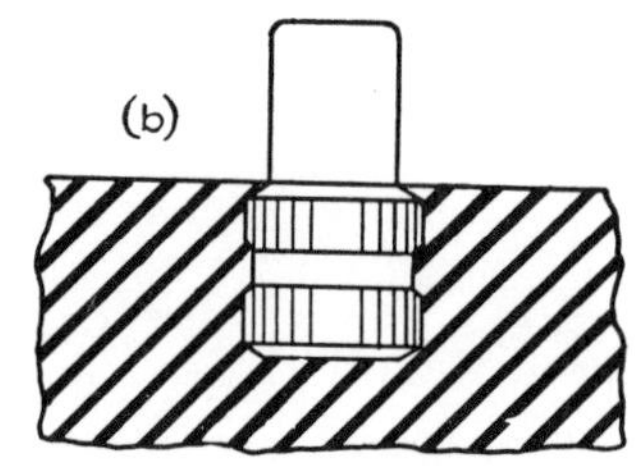


Fig. 12—Coarse diamond knurl (a) is good for most cylindrical inserts. Straight serrations resist twisting but require shoulder (b) to keep from pulling out.

# Parts and Assemblies

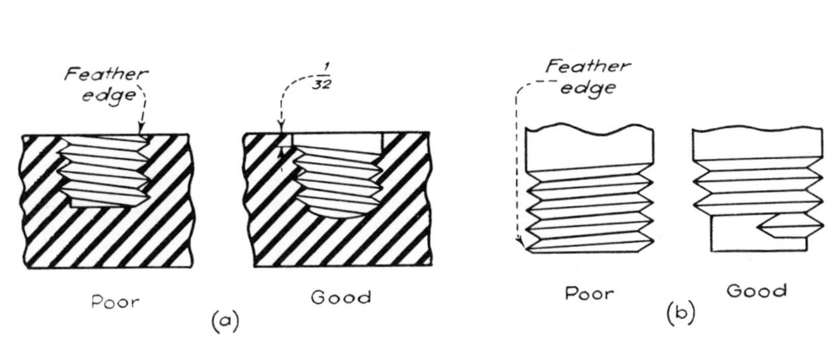


Fig. 4—Molded threads are usually limited to hard rubber items. Unless production requirements are high, they are expensive because of mold costs and the curing time required. On female threads, (a), a recess about 1/32 in. deep extending to the major thread dia will make assembly easier and prevent chipping. Male threads, (b), should start 1/32 in. from the end of the part.

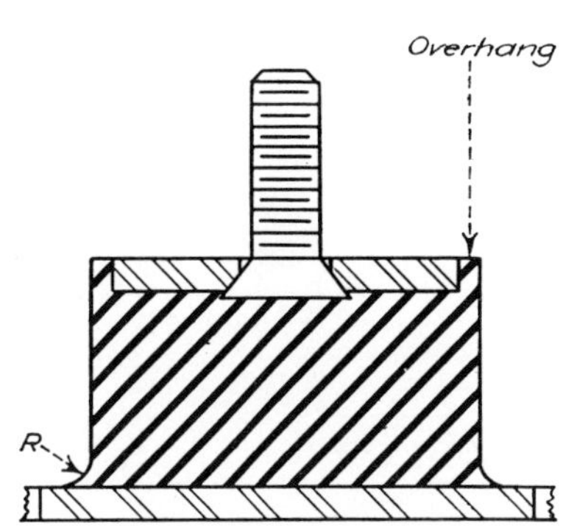


Fig. 5—To reduce adhesion failures in general, all fillet radii should be generous and the rubber should overhang the edges of the inserts whenever possible.

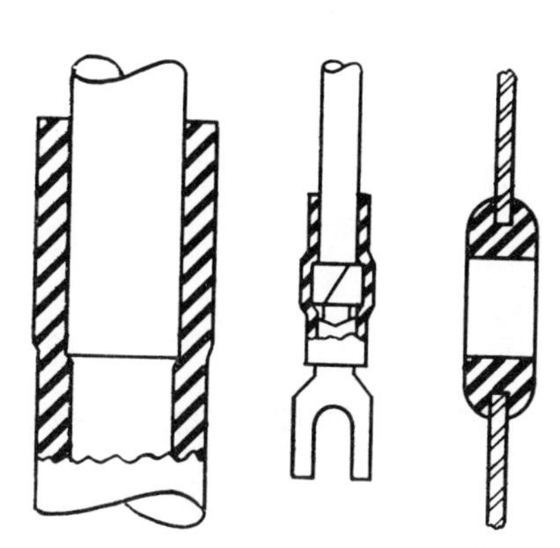

Fig. 8—Select rubbers to take advantage of elasticity to hold parts without fasteners or cement as well as provide leakproof seals in some cases.

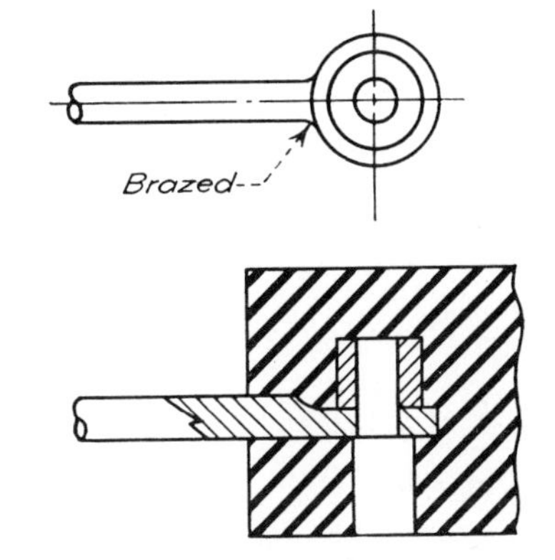


Fig. 9—Projecting ends of inserts should be standard shapes to facilitate pre-positioning in mold. Thin or unsupported inserts may shift in molding.

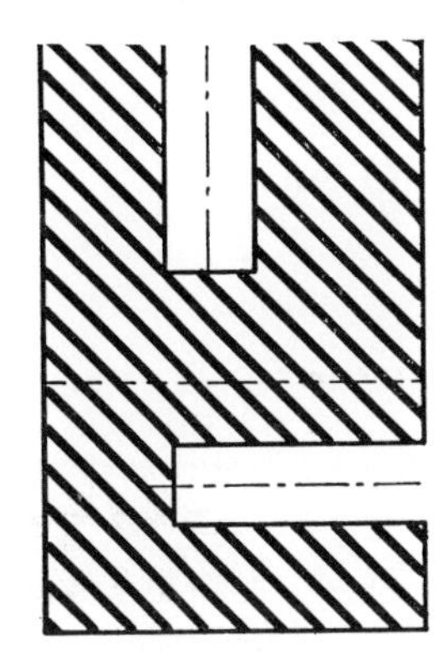

Fig. 10—Avoid holes or slots in two directions to cut production costs. It would be better to make this part in two sections as per the dotted line.

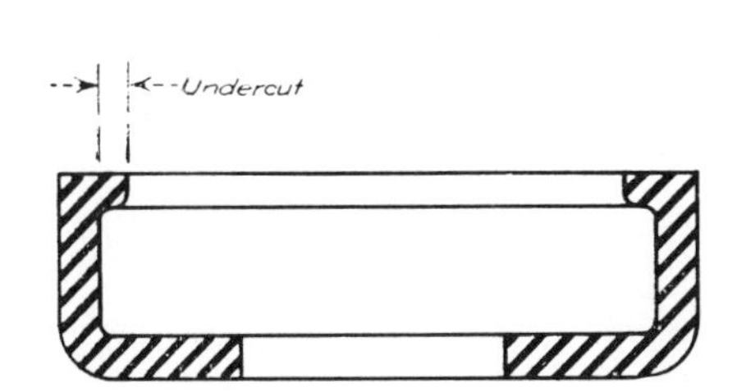


Fig. 13—Avoid undercuts that result in molding difficulties. Size of undercut that will strip easily depends on part size and rubber type.

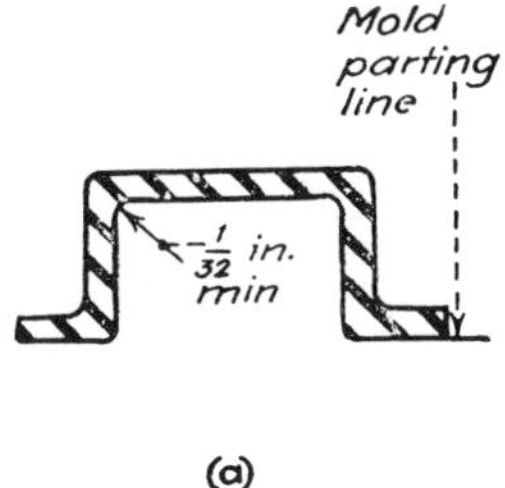


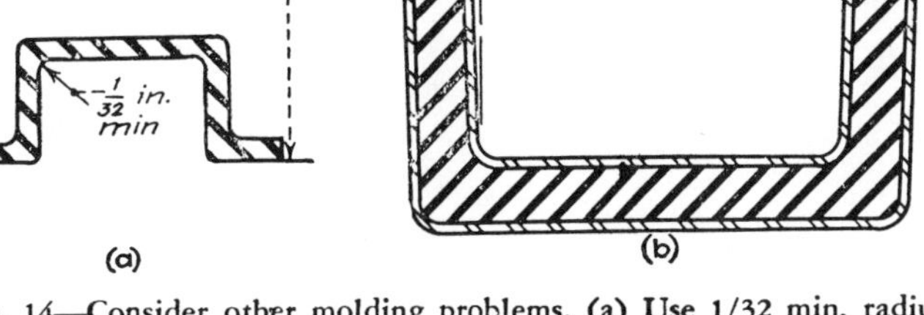


Fig. 14—Consider other molding problems. (a) Use 1/32 min. radius for edges or corners and avoid radii at the parting line. (b) Use draft of 1 degree minimum for surfaces perpendicular to parting line.

# SHRINK-FIT NOMOGRAPH

## *For steel and cast-iron rings of equal length.*

**SIGMUND RAPPAPORT,**
*Project Supervisor,*
*Ford Instrument Co.,*
*Long Island City, N. Y.*
*Adjunct Professor of Kinematics,*
*Polytechnic Institute of Brooklyn*

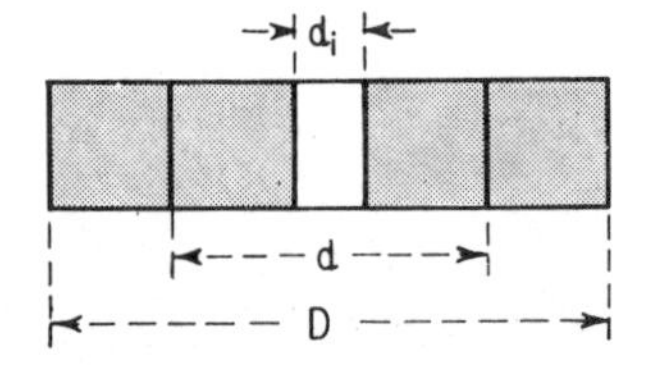


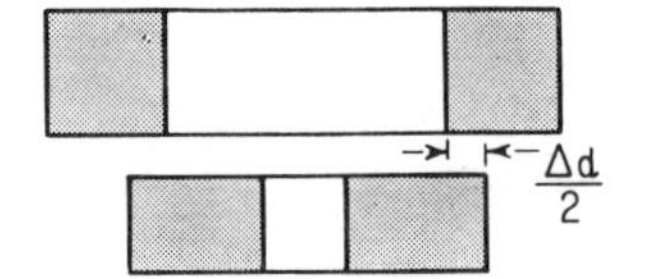


### SYMBOLS

$A$ = mating-surface area, in.$^2$
$E$ = 30,000,000 psi for steel
 = 15,000,000 psi for cast iron
$P$ = desired transmitted force, lb
$p$ = pressure of mating surfaces, psi
$S_f$ = safety factor
$\sigma$ = max. resultant stress, psi
$\epsilon$ = relative shrinkage, mils per in.
$\mu$ = friction coefficient

**STEP 1**

Find $\epsilon/p$ (1000) from nomograph A for whichever combination of steel and cast iron is being considered. In the construction shown this is 0.175, for case 1 (both rings made of steel).

**STEP 2**

It is then necessary to find $p$. This is found from

$$p = PS_f/A\mu$$

It's value must be high enough to allow the desired torque to be transmitted safely. This torque may be that required to drive a shaft; or it may be merely a test requirement for a

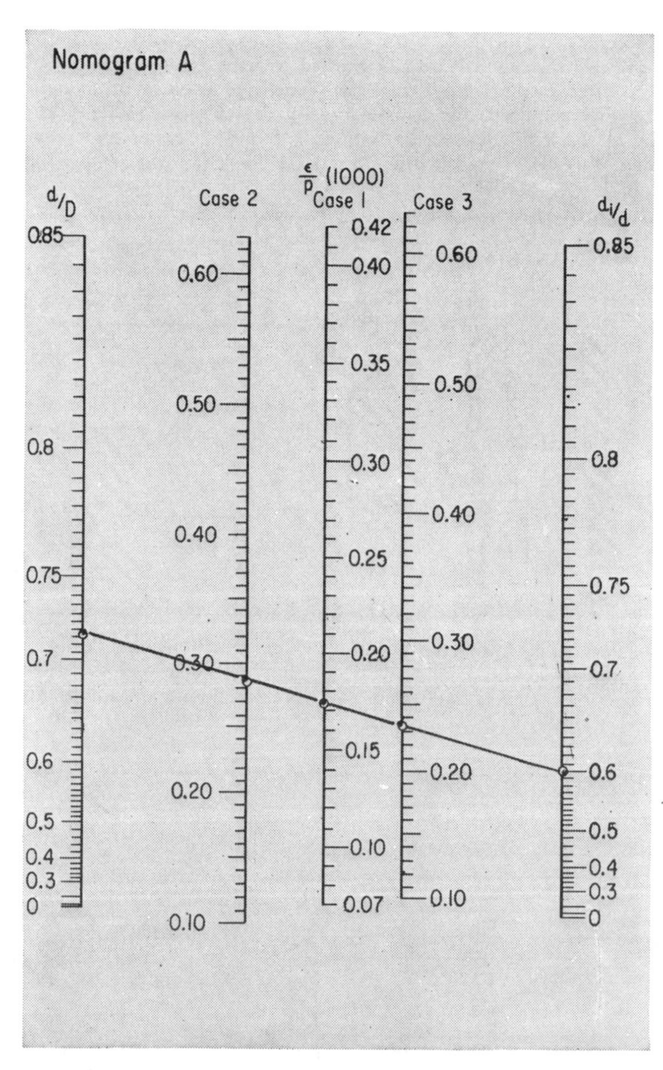


**Auxiliary nomograph . . .**
has scales for different combinations of cast-iron and steel rings.
Case 1: Both rings steel.
Case 2: Outer ring cast iron, inner ring steel.
Case 3: Outer ring steel, inner ring cast iron.

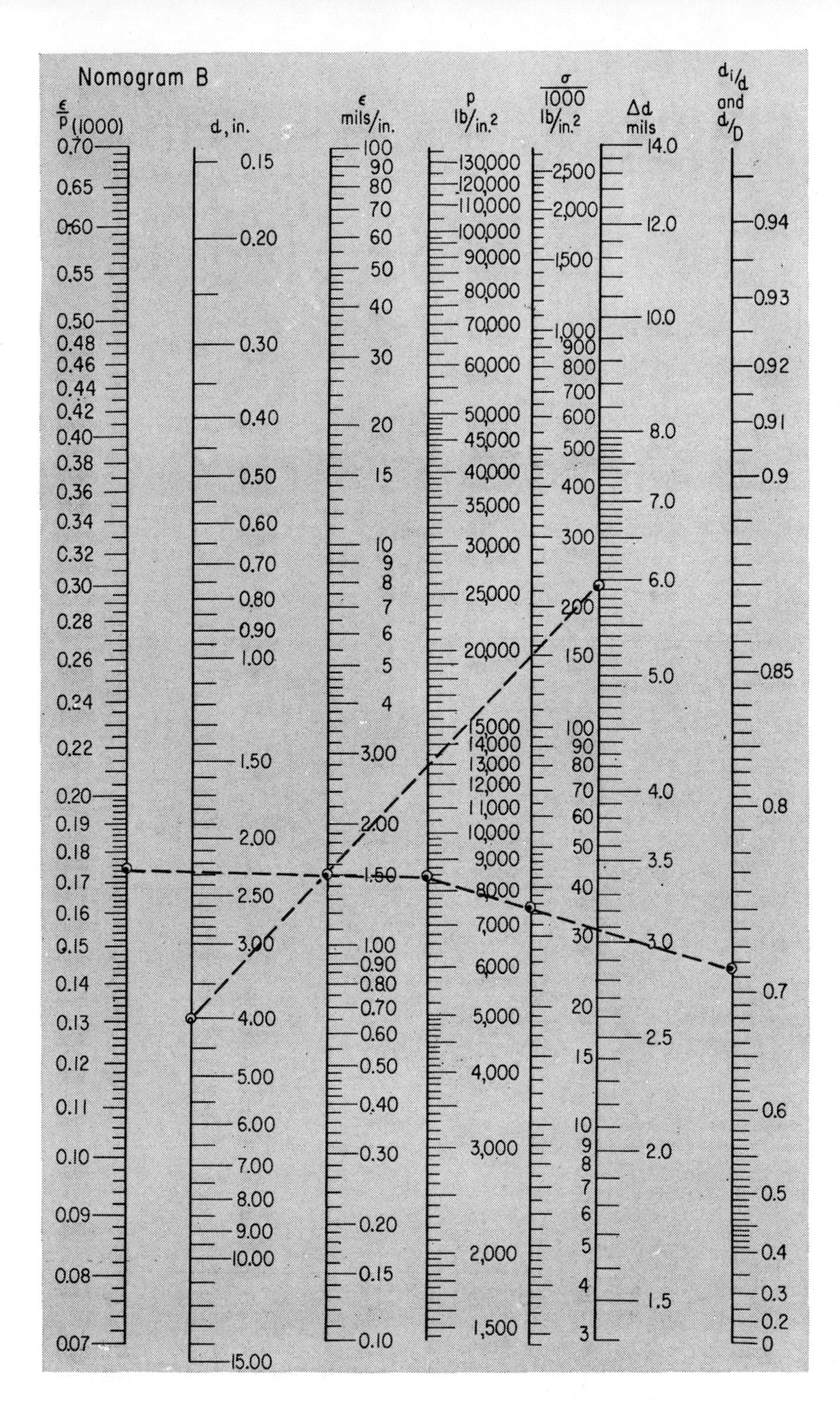


static structure such as a shrunk flange used as a support.

**STEP 3**

Enter $p$ in the $B$ nomograph at the value found from above formula. (8500 psi in the sample construction.) Connect this value with the previously found $\epsilon/p$ (1000) value, 0.175. At the point where this line cuts the $\epsilon$ scale draw a secant from the $d$ scale ($d = 4$ in. in sample.) This line cuts $\Delta d$ at 0.006 in. The rings would, therefore, be machined to have a 0.006 in. interference on the mating diameter.

**MAX STRESS . . .**

will occur at the bores for either ring. The ring with the larger diameter will have the larger max stress. It is better to choose diameter-ratios such that the strength of both rings is fully employed—even if the rings are made of different materials. Ninety percent of yield point is considered permissible for $\sigma$. In the sample nomograph, $\sigma$ has a value of 35,000 psi, obtained by drawing a secant through $p = 8500$ and $d_i/d = 0.715$. Since the yield point for the steel is about 40,000 to 50,000 psi, the value for $\sigma$ thus found would be satisfactory.

# the RIGHT FINISHES for STAINLESS STEEL

**These tables cover electropolishing and other finishing methods in terms of applications and the appearance you want.**

RICHARD E PARET, *Committee of Stainless Steel Producers, American Iron and Steel Institute*

## MILL FINISHES . . .

### present a wide choice to designers

| FINISH NO. | TYPE | SURFACE APPEARANCE | THESE ARE APPLICATIONS | |
|---|---|---|---|---|
| | | | Final Finish For | Preliminary Finish When |
| 1 | Hot-rolled, annealed & pickled | Frosty white | Hi-temp. conditions<br>Industrial needs<br>Other applications where appearance is not important | Surface will be damaged as in annealing |
| 2D | Dull cold-rolled & pickled | Similar to #1 but brighter | Curtain walls<br>Industrial applications | Severe drawing to be followed by polishing |
| 2B | Bright cold-rolled | Bright, dense | Curtain walls<br>Industrial, commercial & transportation equipment | Light forming to be followed by polishing |
| 3 | Intermediate polished | Bright, fairly good luster | A semifinished, polished surface is needed for subsequent finishing operations | |
| 4 | Polished | Bright, good luster; most commonly used polished finish | Architectural trim<br>Restaurant kitchen & sanitary equipment | Where forming operations will not mar surface or can be blended easily. As base for #6, #7, and 8 finishes. |
| 6 | Tampico brushed | Soft, satiny | Decorative without high reflectivity | Follows finish #4 |
| 7 | Polished | High luster | Decorative | |
| 8 | Polished | "Mirror-finish" | Decorative | |

## ELECTROPLATING BATHS . . .

### give high-luster finish not achieved by mechanical means (but will not remove scratches or similar blemishes)

| BATH | COMPOSITION | BATH TEMP., °F | CURRENT DENSITY AMP/SQ IN. | REMARKS |
|---|---|---|---|---|
| Phosphoric acid | 4 parts—85% orthophosphoric acid<br>1 part—water with wetting agent | 140–160 | 1–4<br>(2–12 v) | Removed metal remains in solution—ultimately requires replacement of bath |
| Glycolic acid—sulphuric acid | 55% glycolic acid; 30% sulphuric acid; 15% water (by volume) | 170–190 | 0.05–4.0<br>(4–9 v) | Removed metal settles out. Bath replenished with original solution plus conc. sulphuric acid |
| Citric acid—sulphuric acid | 55–60% citric acid; 15–20% sulphuric acid; rest, water (by wt.) | 180–200 | 0.05–4.0<br>(6–12 v) | Metal settles out. Replenish bath with equal portions of acids. |

## HOW TO GET A NO. 4 FINISH . . .

### no matter how rough the original surface

*(Abrasive: aluminum oxide or silicon carbide—must be iron-free.)*

| OPERATION | GRIT[b] | WHEEL | LUBRICANT | SURFACE FT. PER MIN. | REMARKS |
|---|---|---|---|---|---|
| 1 Grind[a] (rough) | 20–60 | Rubber or bakelite-bonded | Dry | 5–6000 | Starting operation for very rough weld beads |
| 2 Grind (finish) | 60–80 | Rubber or bakelite-bonded | Dry | 5–6000 | Starting operation for heavy sheet or plate; hot-rolled finishes; light welds |
| 3 Polish (rough) | 80–100 | Cloth | Polishing tallow or grease stick | 7500 | Starting operation for cold-rolled sheet or strip |
| 4 Polish (finish) | 120–150 | Cloth | Polishing tallow or grease stick | 7500 | Similar to No. 4 mill finish |

[a]Operation #1 is a preliminary required only for very rough work
[b]Number depends on quality of finish desired or grit used in prior operation—should be increased by 20 to 40 numbers for each operation

## FOR HIGHER FINISHES FROM STANDARD NO. 4 FINISH . . .

### These are the methods

| MILL FINISH NUMBER | OPERATION | GRIT | WHEEL | ABRASIVE | LUBRICANT | SURFACE FT. PER MIN. | REMARKS |
|---|---|---|---|---|---|---|---|
| 6 | Brushing | . . . . | Tampico or similar brush | Powdered pumice with oil applied as paste | | 1–5000 | Finish varies with abrasive and wheel speed |
| 7 | Polishing | 180 | Cloth | Aluminum oxide or silicon carbide* | Tallow or grease stick | 7500 | May be followed by second polishing with 220–240 grit |
| | Buffing (cutting) | . . . . | Bias type or sewed-piece buff | Commercial cutting compound (supplied in stick or cake with lubricant) | | 8–10,000 | Use only iron-free compounds; color buffing is for much shorter periods than cutting |
| | Buffing (coloring) | . . . . | Loose disk or bias type | Commercial coloring compound (supplied in stick or cake with lubricant) | | 8–10,000 for high finish | |
| 8 | Polishing | 180 | Cloth | Aluminum oxide or silicon carbide | Tallow or grease stick | 7500 | |
| | Polishing | 240–320 | Cloth | Aluminum oxide or silicon carbide | Tallow or grease stick | 7500 | |
| | Buffing (for both cutting and coloring) | Same as for #7 finish | | | | | |

*Manufactured-type abrasives—must be iron-free

# How to find
# Final Dimensions of Stretch-formed Parts

**BRUCE RUSSELL** *Chance Vought Aircraft, Dallas, Tex.*

**Serious errors can result when thickness, flange-depth, and web-width changes are ignored in the design of stretch-formed parts. Here is the way to calculate these dimensional changes.**

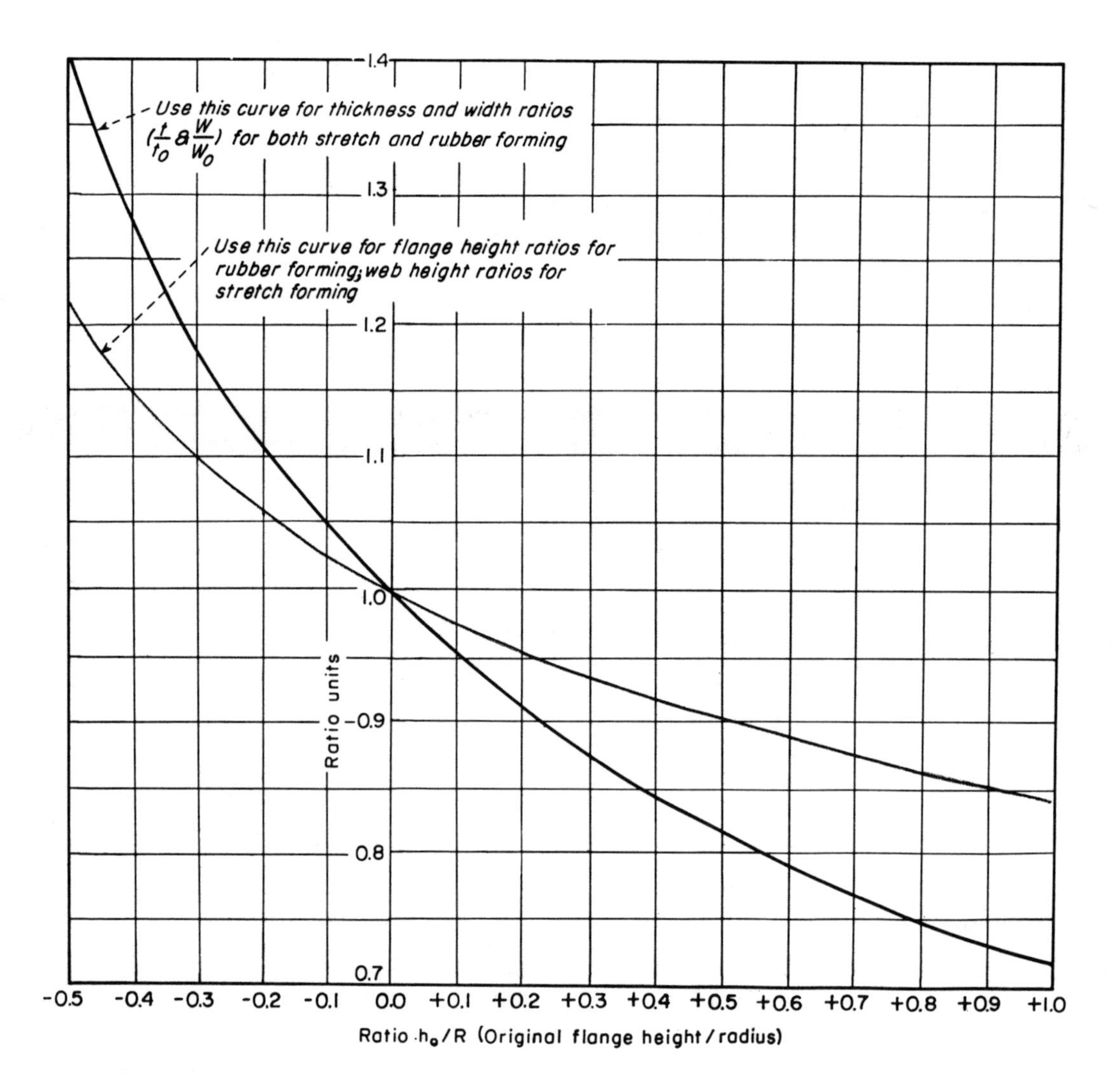


The final dimensions of stretch-formed parts can be found from the curves and formulas presented here. These aids get around the problem that contoured parts undergo dimensional changes in cross-section when formed on stretch press or rubber press. When stress calculations ignore such changes, unexpected failure of the part may occur. The problem is aggravated by new materials and production techniques that increase maximum elongation during forming; this magnifies error when strength calculations are based on the original dimension. For flat sheet or plate the thickness-elongation formula can also be employed to analyze failures in parts formed from sheet stock. All formulas give results accurate within 2%.

Negative values in the curves arise from the geometry used in deriving the equations; the numerical value of $R$ is negative when other terms are measured in a reverse direction. Note also that $h$, $h_o$ equal web height for stretch formed sections, and flange height for rubbed formed sections.

## STRETCH-PRESS

(entire sections stretched)

$$t = \frac{t_o}{\sqrt{1 + h_o/R}}$$

$$\frac{h}{h_o} = \frac{h_o}{4(R - h_o)}$$

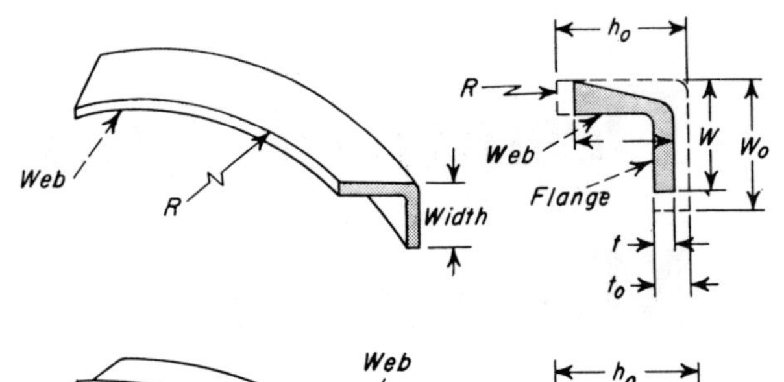


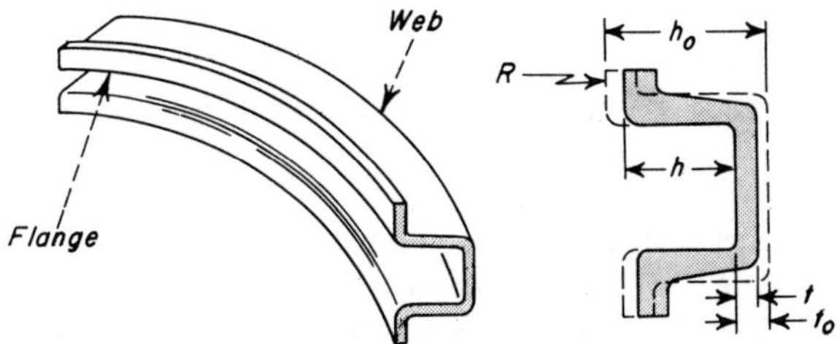


## RUBBER-PRESS

(flange stretched)

$$t = \frac{t_o}{\sqrt{1 + h_o/R}}$$

$$\frac{h}{h_o} = 1 - \frac{h_o}{4(R - h_o)}$$

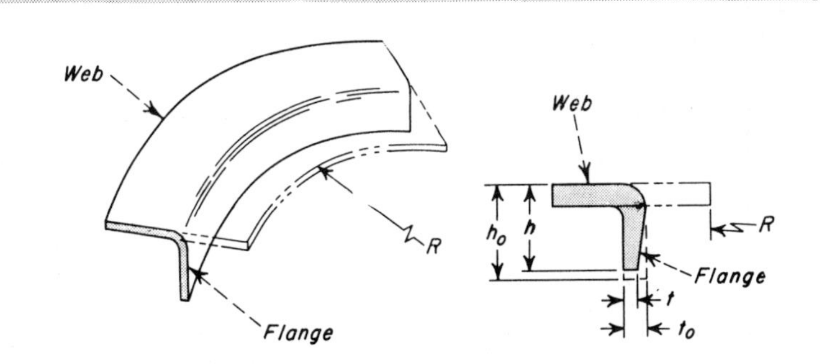


## RUBBER-PRESS

(flange shrunk)

$$t = \frac{t_o}{\sqrt{1 - h_o/R}}$$

$$\frac{h}{h_o} = 1 + \frac{h_o}{4(R - h_o)}$$

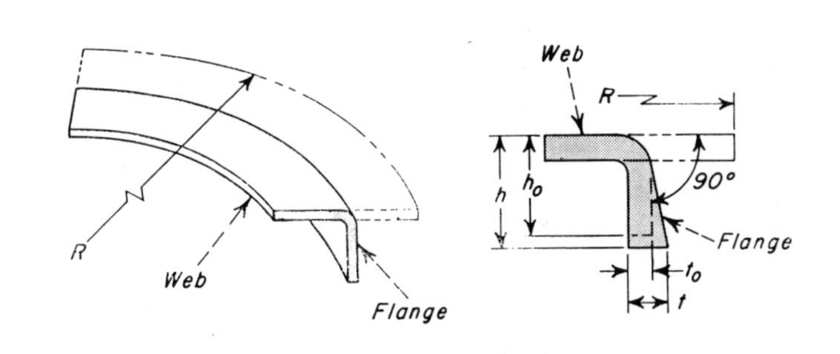


## Plate or Sheet Under Triaxial Strain

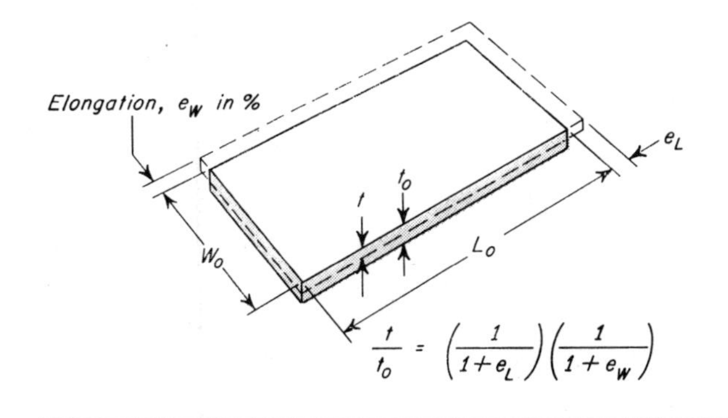


$$\frac{t}{t_o} = \left(\frac{1}{1+e_L}\right)\left(\frac{1}{1+e_W}\right)$$

## Unrestrained Along Width

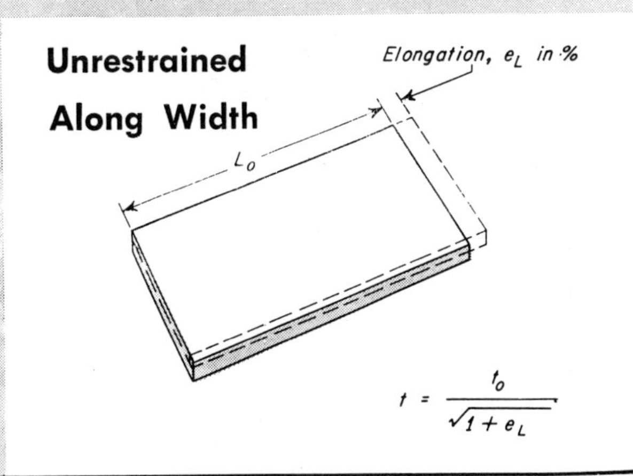


$$t = \frac{t_o}{\sqrt{1 + e_L}}$$

# Design of Toothed Mechanical Components

L. N. CANICK, SERVOMECHANISMS, INC.

TOOTHED COMPONENTS such as clutches and detents are designed to transmit a given torque. The basic relationship is $T = Rf$, where $f$ is the tangential force acting at an effective radius $R$ and $T$ is torque. In turn, force $f$ is produced indirectly as a result of an axial engaging force, $F$. This is illustrated in Fig. 1(A) for a toothed clutch and in (B) for a detent.

Equating the horizontal and vertical forces in Fig. 1(A) to zero

$$F + \mu N \cos\theta - N \sin\theta = 0 \qquad (1)$$

$$N \cos\theta + N \sin\theta - T/R = 0 \qquad (2)$$

where $\mu$ is the coefficient of friction

Eliminating $N$,

$$T = RF\,\frac{\cos\theta - \mu \sin\theta}{\sin\theta - \mu\cos\theta}$$

In the design of these elements the effective radius is

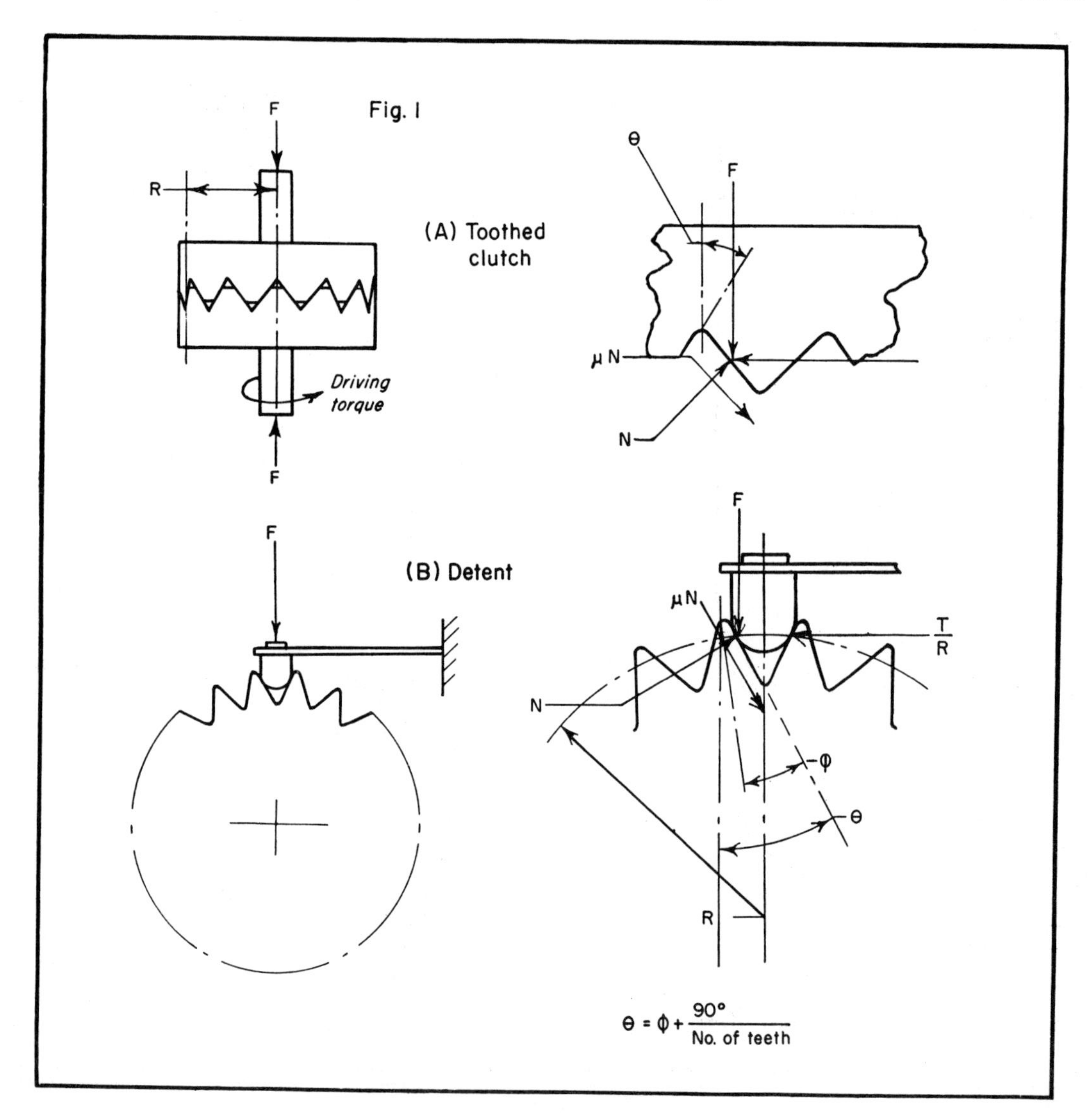

usually set by size considerations, and the required torque then will result from the values assigned to the engaging force, tooth angle, and coefficient of friction. The nomogram shown below solves the equation

$$K = \frac{\cos\theta - \mu\sin\theta}{\sin\theta - \mu\cos\theta}$$

Thus, it is a convenient means for considering the effect of variation in $\theta$ and $\mu$. For a given friction coefficient there is a tooth angle below which $K$ becomes infinite. This means that the clutch or detent is self-locking and will transmit torque limited only by its structural strength.

EXAMPLE—Find the tooth angle and best K for a self-locking clutch, taking the coefficient of friction as 0.4 minimum. Line I gives a tooth angle of slightly less than 22 deg for the self-locking condition under high friction coefficient conditions. Thus, the tooth angle must be larger than 22 deg if no disengaging force is to be used; 30 deg is chosen so that a standard cutter can be used. The minimum K to be expected, under low friction conditions, $\mu = 0.2$, is given by Line II as slightly less than 3. Line III shows the K for a flat-faced friction clutch. Thus, for equal effective radii and engaging forces, the toothed clutch will transmit 3/0.2 or 15 times the torque of the flat-faced clutch.

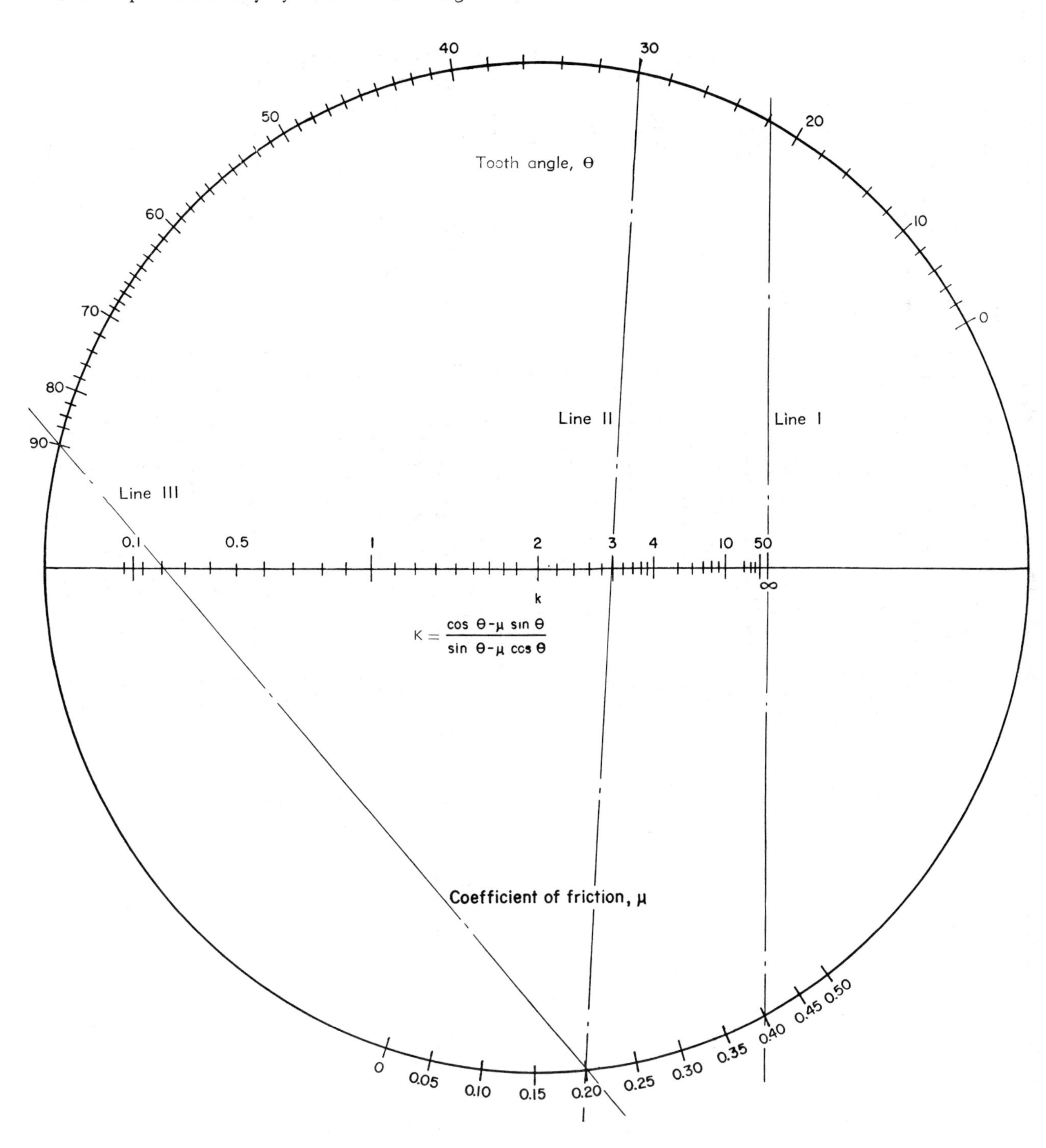

# Design Data for Metal

## TYPES OF STITCHES

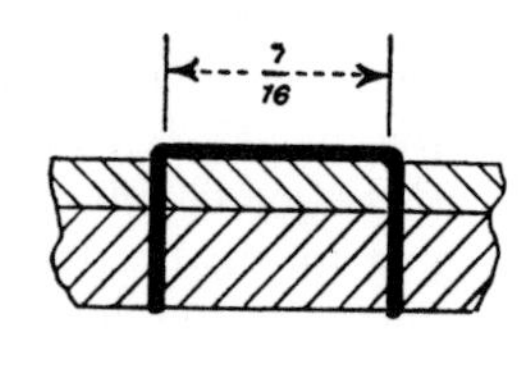


Unclinched

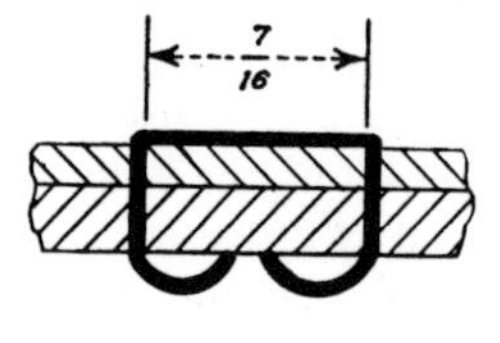


Standard loop

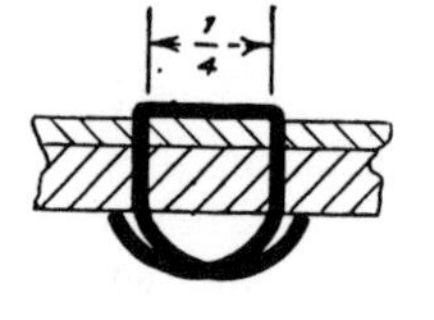


By-pass loop

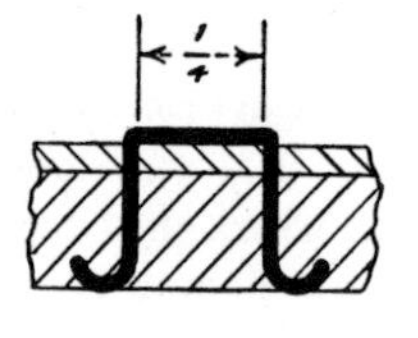


Outside loop

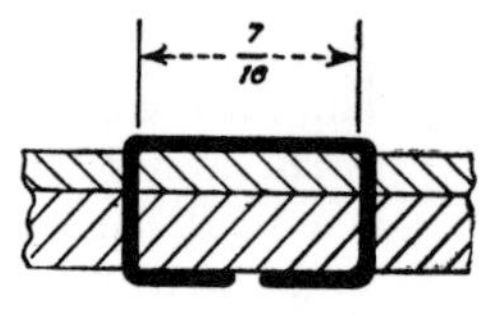


Flat clinch

## MAXIMUM STITCH PENETRATION

**Loop-clinch stitch**

| Metals | Metal to Metal thicknesses (in.) | | Metal to Non-metal thicknesses (in.) | | Non-metals |
|---|---|---|---|---|---|
| Aluminum (soft) 3SO, 52SO, 61SO, Alclad 24SO, R301-O, R301-W, 75SO, 75SW | 0.093 | 0.093 | 0.125 | 1/2 | Sheet cork |
| Aluminum (1/2 hard) 3S-H14, 52S-H34 | 0.064 | 0.064 | 0.080 | 3/8 | Leather |
| Aluminum (hard) 61ST6, Alclad 24ST3, Alclad 24ST36, R301-T, 75ST6 | 0.040 | 0.040 | 0.064 | 1/4 | Sheet asbestos |
| Aluminum extrusion | 0 062 | 0 062 | 0 093 | 1/2 | Fiberboard |
| 1010 Cold-rolled steel | 0.0475 | 0.0475[1] | 0.080 | 1/2 | Sponge rubber |
| Hot-rolled steel | 0 0475 | 0 0348[1] | 0 0625 | 1/4 | Solid rubber |
| Galvanized sheet | 0 0348 | 0.0348[1] | 0 0475 | 1/8 | Phenol[2] |
| Stainless, Type 302, full hard | 0.010 | 0.010 | 0.020 | 3/16 | Plastics[2] |
| Stainless, 1/2 hard | 0.012 | 0.012 | 0.025 | 3/8 | Standard Masonite |
| Stainless, 1/4 hard | 0.015 | 0.015 | 0.030 | 1/4 | Tempered Masonite |
| Stainless, annealed | 0.020 | 0 020 | 0.040 | 3/8[3] | Solid wood |
| Sheet brass, soft | 0.030 | 0.030 | 0.050 | 3/8[3] | Plywood |
| Sheet copper | 0.035 | 0.035 | 0.064 | | |

[1] 50RB or softer.
[2] Must be soft enough to allow penetration without cracking.
[3] Grain structure may cause leg wander in thicknesses over 3/8 in.
[4] Stitching full-hard or half-hard stainless to itself not recommended.

**Flat-clinch stitch**

| Metals | Metal to Metal thicknesses (in.) | | Metal to Non-metal thicknesses (in.) | | Non-metals |
|---|---|---|---|---|---|
| Aluminum (soft) 3SO, 52SO, 61SO, Alclad 24SO, R301-O, R301-W, 75SO, 75SW | 0.093 | 0.093 | 0.125 | 1/8 | Sheet cork |
| Aluminum (1/2 hard) 3S-H14, 52S-H34 | 0.064 | 0.064 | 0.080 | 3/16 | Leather |
| Aluminum (hard) 61ST6, Alclad 24ST3, Alclad 24ST36, R301-T, 75ST6 | 0.040 | 0.040 | 0.064 | 3/16 | Sheet asbestos |
| Aluminum extrusion | 0.062 | 0.062 | 0.093 | 1/4 | Fiberboard |
| 1010 Cold-rolled steel | 0.0348 | 0.0348 | 0.0348 | 3/8 | Sponge rubber |
| Hot-rolled steel | 0.0348 | 0.0348 | 0.0348 | 1/4 | Solid rubber |
| Galvanized sheet | 0.0312 | 0.0312 | 0.0312 | 1/8 | Phenolics[2] |
| Stainless, Type 302, full hard | 0.010 | [4] | 0.010 | 3/16 | Plastics[2] |
| Stainless, 1/2 hard | 0 012 | [4] | 0.012 | 3/16 | Standard Masonite |
| Stainless, 1/4 hard | 0.015 | 0.015 | 0.015 | 1/4 | Tempered Masonite |
| Stainless, annealed | 0.020 | 0 020 | 0.020 | 3/16 | Solid wood |
| Sheet brass, soft | 0.030 | 0.030 | 0.040 | 1/4 | Plywood |
| Sheet copper | 0.035 | 0.035 | 0.045 | | |

Example: A sheet of 0 093 in. 24S0 Alclad can be stitched to a 0.020 in. sheet of annealed steel with either type of stitch. This same Alclad sheet can be stitched to 1/4 in. thick tempered Masonite.

## RECOMMENDED MINIMUM SPACING (based on maximum holding capacity)

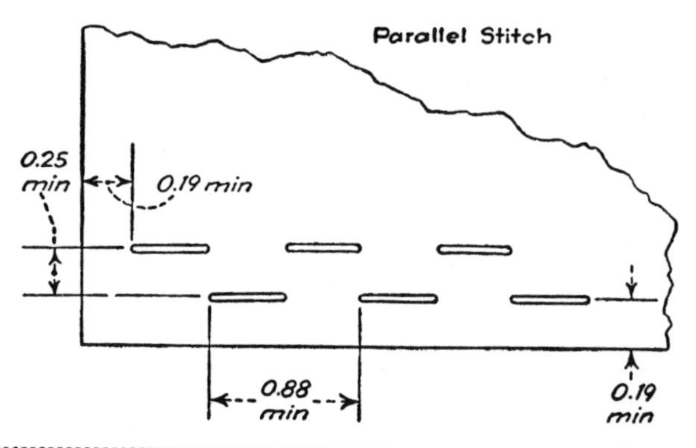


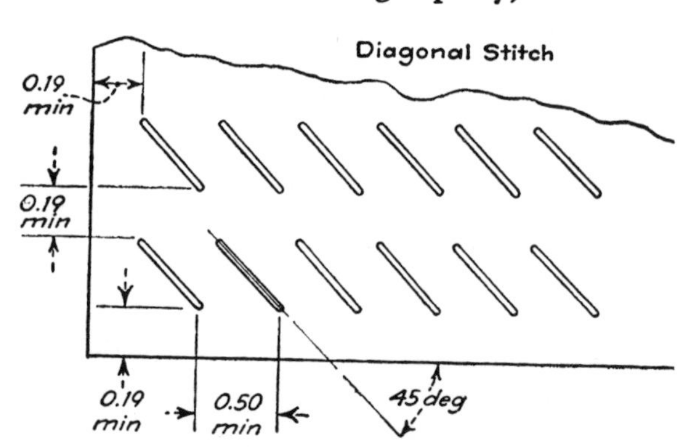

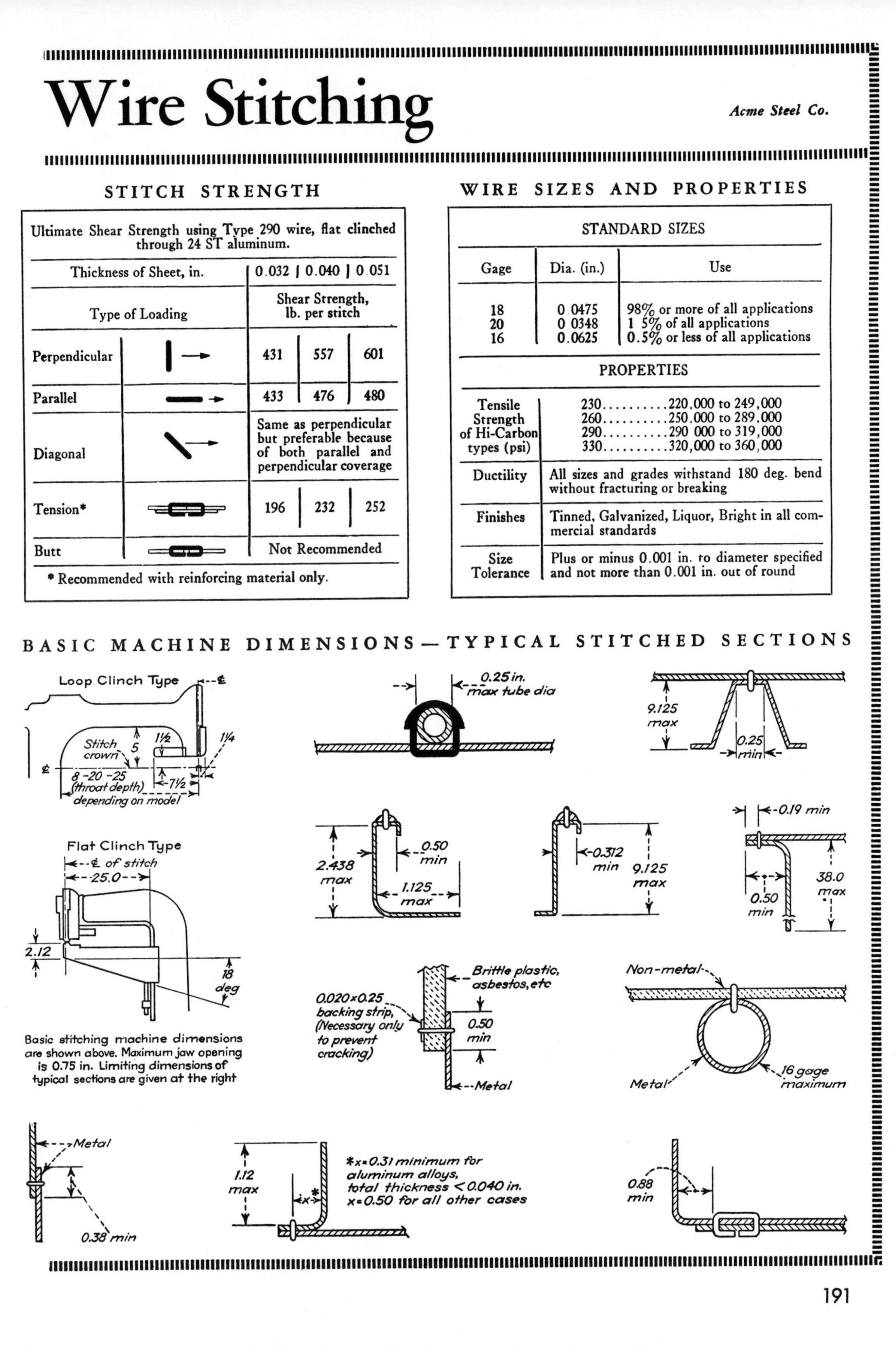

# Wire Stitching

*Acme Steel Co.*

## STITCH STRENGTH

Ultimate Shear Strength using Type 290 wire, flat clinched through 24 ST aluminum.

| Type of Loading | | Shear Strength, lb. per stitch — 0.032 | 0.040 | 0.051 |
|---|---|---|---|---|
| Thickness of Sheet, in. | | 0.032 | 0.040 | 0.051 |
| Perpendicular | | 431 | 557 | 601 |
| Parallel | | 433 | 476 | 480 |
| Diagonal | | Same as perpendicular but preferable because of both parallel and perpendicular coverage | | |
| Tension* | | 196 | 232 | 252 |
| Butt | | Not Recommended | | |

* Recommended with reinforcing material only.

## WIRE SIZES AND PROPERTIES

STANDARD SIZES

| Gage | Dia. (in.) | Use |
|---|---|---|
| 18 | 0 0475 | 98% or more of all applications |
| 20 | 0 0348 | 1 5% of all applications |
| 16 | 0.0625 | 0.5% or less of all applications |

PROPERTIES

| | |
|---|---|
| Tensile Strength of Hi-Carbon types (psi) | 230..........220,000 to 249,000<br>260..........250,000 to 289,000<br>290..........290 000 to 319,000<br>330..........320,000 to 360,000 |
| Ductility | All sizes and grades withstand 180 deg. bend without fracturing or breaking |
| Finishes | Tinned, Galvanized, Liquor, Bright in all commercial standards |
| Size Tolerance | Plus or minus 0.001 in. to diameter specified and not more than 0.001 in. out of round |

## BASIC MACHINE DIMENSIONS—TYPICAL STITCHED SECTIONS



Basic stitching machine dimensions are shown above. Maximum jaw opening is 0.75 in. Limiting dimensions of typical sections are given at the right

# 8

# DRIVES

| | |
|---|---|
| Ten Types of Belt Drives | 194 |
| Mechanisms for Adjusting Tension of Belt Drives | 196 |
| Methods of Reducing Pulsations in Chain Drives | 198 |
| Leather Belts – Hp Loss and Speeds | 200 |
| Power Capacity of Spur Gears | 201 |
| Friction Wheel Drives Designed for Maximum Torque | 204 |
| Accurate Solution for Disk-clutch Torque Capacity | 207 |
| Torque of Slip Couplings | 208 |
| Torsional Strength of Shafts | 210 |
| Basic Types of Variable Speed Friction Drives | 212 |

# Ten Types of Belt Drives

GEORGE R. LEDERER

Raybestos-Manhattan, Inc.

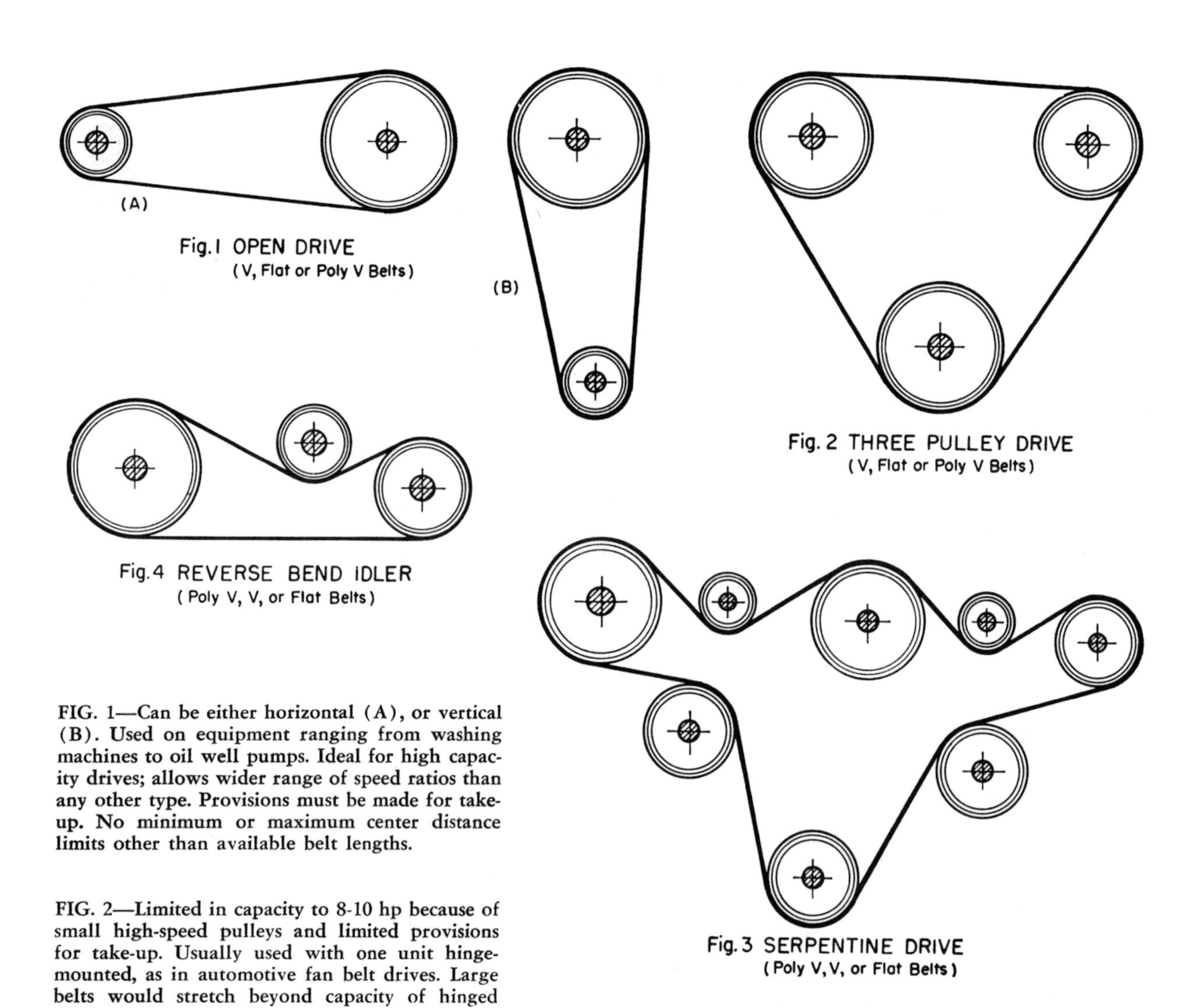


FIG. 1—Can be either horizontal (A), or vertical (B). Used on equipment ranging from washing machines to oil well pumps. Ideal for high capacity drives; allows wider range of speed ratios than any other type. Provisions must be made for take-up. No minimum or maximum center distance limits other than available belt lengths.

FIG. 2—Limited in capacity to 8-10 hp because of small high-speed pulleys and limited provisions for take-up. Usually used with one unit hinge-mounted, as in automotive fan belt drives. Large belts would stretch beyond capacity of hinged unit to take up slack.

FIG. 3—Useful where several units are driven from a central shaft. The Vv belt that resembles two v-belts joined back-to-back was developed especially for this drive. For Vv operation all pulleys must be grooved. For regular V-, Poly-V or flat-belt operation, only those driven by the belt face are grooved, others are flat and are driven by the back of the belt. Driving capacity range from 15-25 hp. Sheaves are small, speed is slow and belt flex is extremely high, which affects belt life.

FIG. 4—Used where driver and driven sheaves are fixed and there is no provision for take-up. Idler is placed on the slack side of the belt near the point where the belt leaves the driver sheave. Idler also gives increased wrap and increased arc of contact. Applications range from agricultural jackshaft drives to machine tools and large oil field drives. Idler can be spring loaded to keep belt tight if drive is subject to shock. For maximum belt life, the larger the idler, the better.

FIG. 5—Driver and driven sheaves are at right angles; belt must travel around horizontal sheave, turn, go over vertical sheave and return. Bend must be gradual to prevent belt from leaving sheave. Minimum center distance for V-belts is 5.5 in. × (pitch dia of largest sheave + width of sheave).

For Poly-V, minimum center distance = 13 × pitch dia of small sheave or 5.5 × (pitch dia + belt width). For flat belts it is 8 ×

Although countless types of belt drives are possible, these ten will solve most industrial applications. These pertain to power transmission only; the tooth type of timing belt is not included. For each drive are given: design pitfalls; speed and capacity ranges; and suggestions for application.

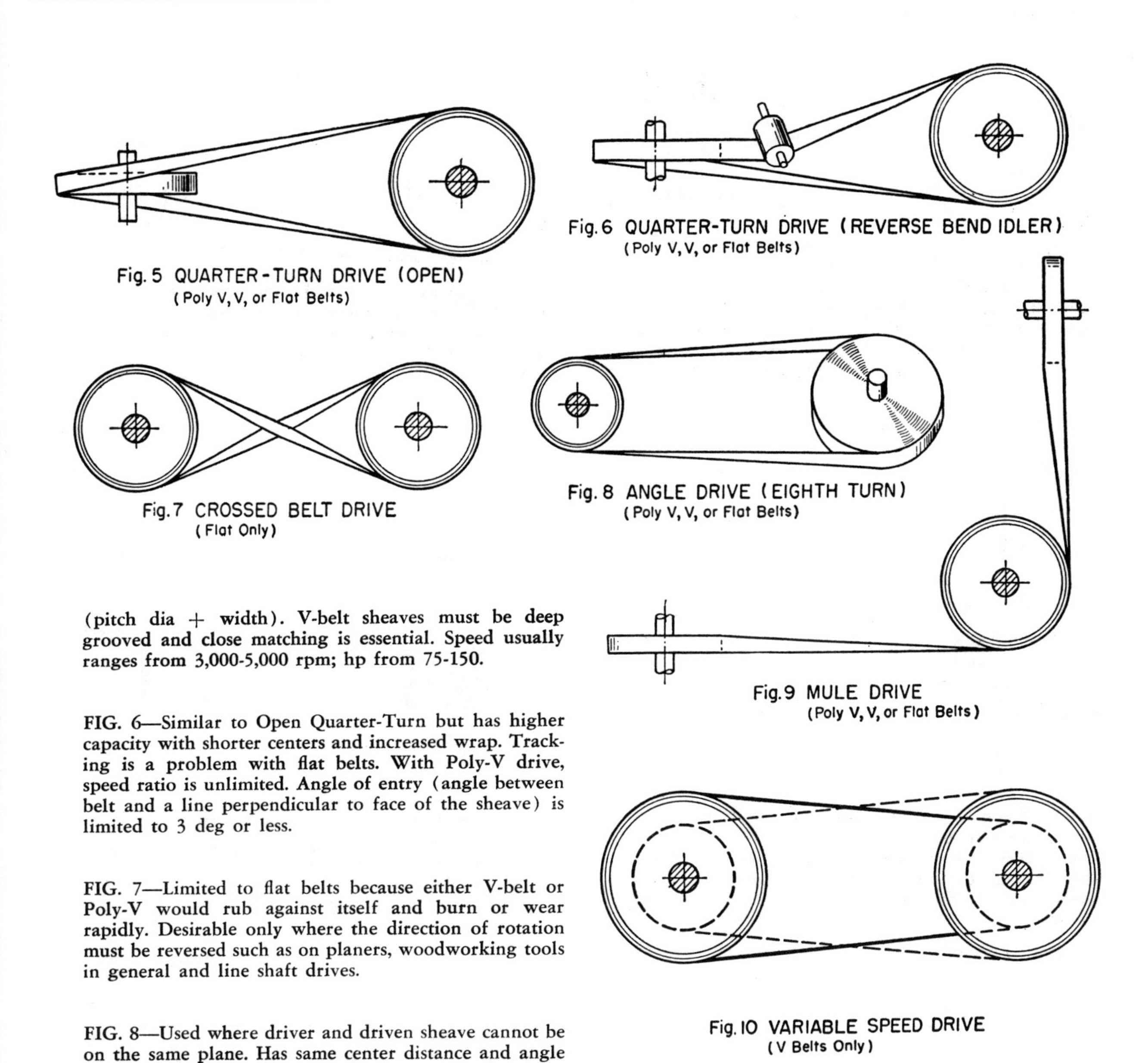

Fig. 5 QUARTER-TURN DRIVE (OPEN) (Poly V, V, or Flat Belts)

Fig. 6 QUARTER-TURN DRIVE (REVERSE BEND IDLER) (Poly V, V, or Flat Belts)

Fig. 7 CROSSED BELT DRIVE (Flat Only)

Fig. 8 ANGLE DRIVE (EIGHTH TURN) (Poly V, V, or Flat Belts)

Fig. 9 MULE DRIVE (Poly V, V, or Flat Belts)

Fig. 10 VARIABLE SPEED DRIVE (V Belts Only)

(pitch dia + width). V-belt sheaves must be deep grooved and close matching is essential. Speed usually ranges from 3,000-5,000 rpm; hp from 75-150.

FIG. 6—Similar to Open Quarter-Turn but has higher capacity with shorter centers and increased wrap. Tracking is a problem with flat belts. With Poly-V drive, speed ratio is unlimited. Angle of entry (angle between belt and a line perpendicular to face of the sheave) is limited to 3 deg or less.

FIG. 7—Limited to flat belts because either V-belt or Poly-V would rub against itself and burn or wear rapidly. Desirable only where the direction of rotation must be reversed such as on planers, woodworking tools in general and line shaft drives.

FIG. 8—Used where driver and driven sheave cannot be on the same plane. Has same center distance and angle of belt entry limitations as Quarter-Turn. Drive can be open if take-up can be accomplished at either end or it can be fitted with a reverse bend idler, but not an inside idler. Angle between shaft can be from zero to 90 deg.

FIG. 9—Especially developed for drill presses and special applications where driver and driven sheaves are at right angles to each other and yet on the same plane. Can operate around a corner or from one floor to another. The center sheave is 90 deg from the driver and driven sheave and acts as an idler. Twists affect belt life.

FIG. 10—Sheaves must be grooved to change the pitch diameter for variable or adjustable pitch operation. With two sheaves and one belt, it is possible to have a range of four different speeds. Widely used on propulsion drives and cylinder drives on agricultural combines and machine tools. Drive has same high capacity and advantage as standard open drive, with wider speed range. They are mostly single belt drives 1¼ to 2 in. wide. Small pulleys are not advisable. Most applications require special vari-speed cylinder or traction belt.

# Mechanisms for Adjusting

Sketches show devices for both manual and automatic take-up as required by wear or stretch. Some are for installations having fixed

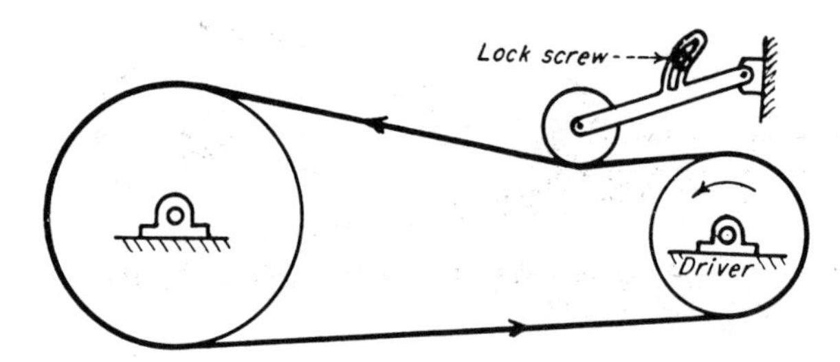


Fig. 1—Manually adjusted idler run on slack side of chain or flat belt. Useful where speed is constant, load is uniform and the tension adjustment is not critical. Can be adjusted while drive is running. Horsepower capacity depends upon belt tension.

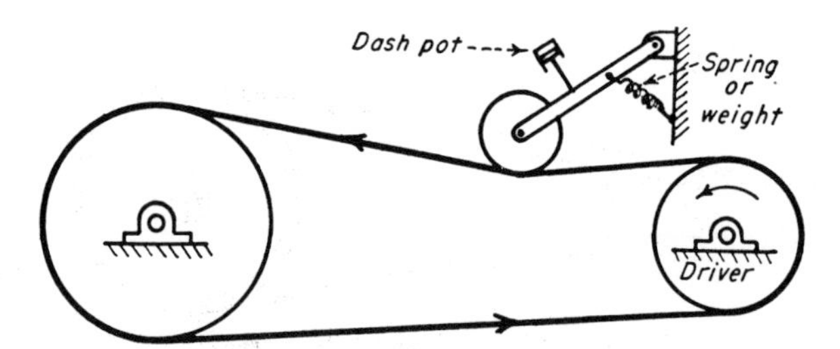


Fig. 2—Spring or weight loaded idler run on slack side of flat belt or chain provides automatic adjustment. For constant speed but either uniform or pulsating loads. Adjustments should be made while drive is running. Capacity limited by spring or weight value.

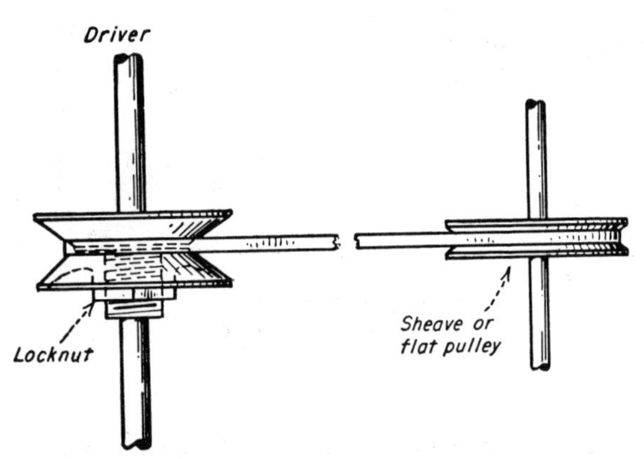


Fig. 5—Screw type split sheave for V-belts when tension adjustment is not critical. Best suited for installations with uniform loads. Running speed increases with take up. Drive must be stopped to make adjustments. Capacity depends directly upon value of belt tension.

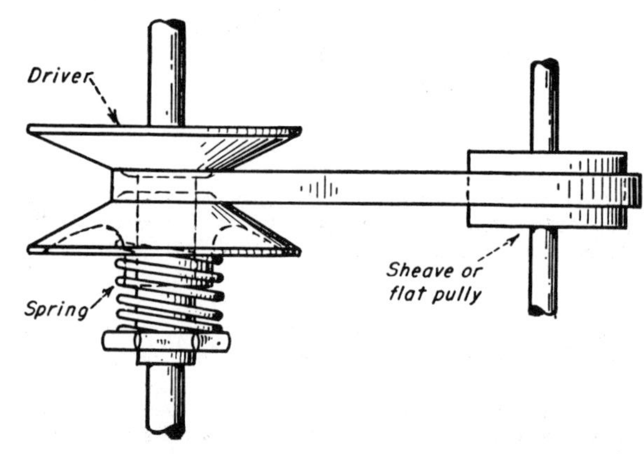


Fig. 6—Split sheave unit for automatic adjustment of V-belts. Tension on belt remains constant; speed increases with belt take up. Spring establishes maximum torque capacity of the drive. Hence, this can be used as a torque limiting or overload device.

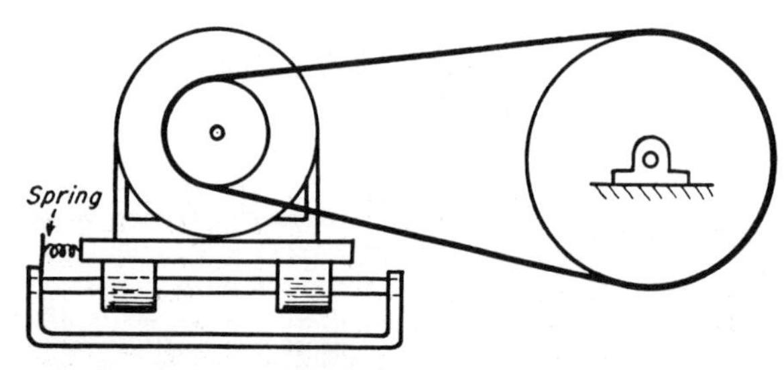


Fig. 9—Spring actuated base for automatic adjustment of uniformly loaded chain drive. With belts, it provides slipping for starting and suddenly applied torque. Can also be used to establish a safety limit for the horsepower capacity of belts.

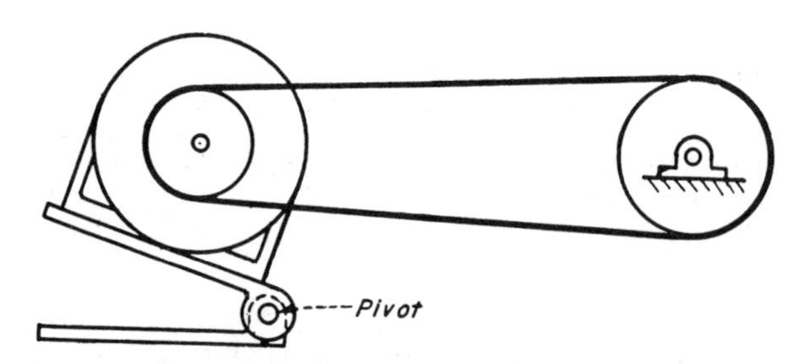


Fig. 10—Gravity actuated pivoting motor base for uniformly loaded belts or chains, only. Same safety and slipping characteristics as that of Fig. 9. Position of motor from pivot controls the proportion of motor weight effective in producing belt tension.

# Tension of Belt Drives

JOSEPH H. GEPFERT
Reeves Pulley Company

center distances; others are of the expanding center take-up types. Many units provide for adjustment of speed as well as tension.

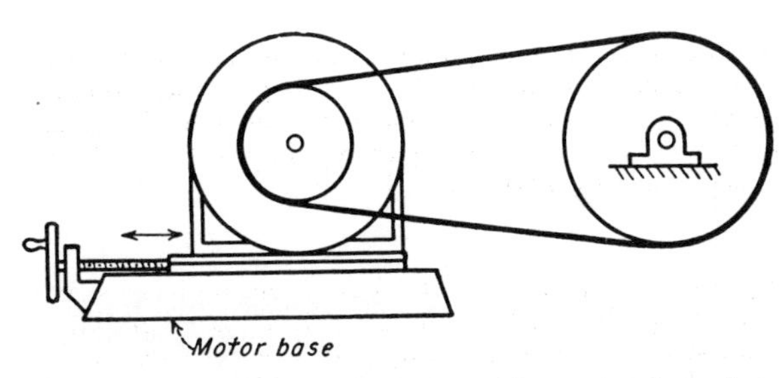


**Fig. 3—Screw-base type unit provides normal tension control of belt or chain drive for motors. Wide range of adjustments can be made either while unit is running or stopped. With split sheaves, this device can be used to control speed as well as tension.**

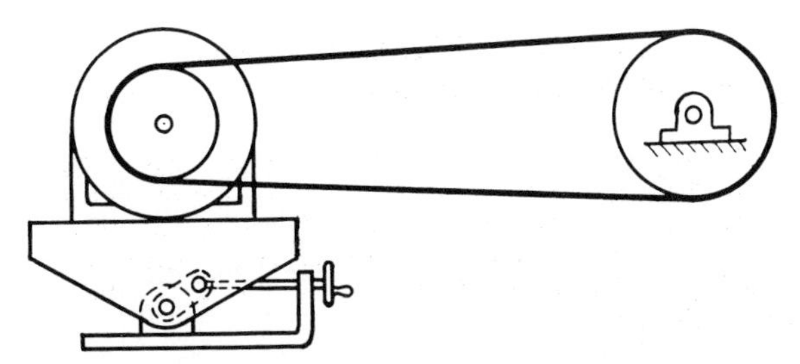

**Fig. 4—Pivoting screw base for normal adjustment of motor drive tension. Like that of Fig. 3, this design can be adjusted either while running or stopped and will provide speed adjustment when used with split sheaves. Easier to adjust than previous design.**

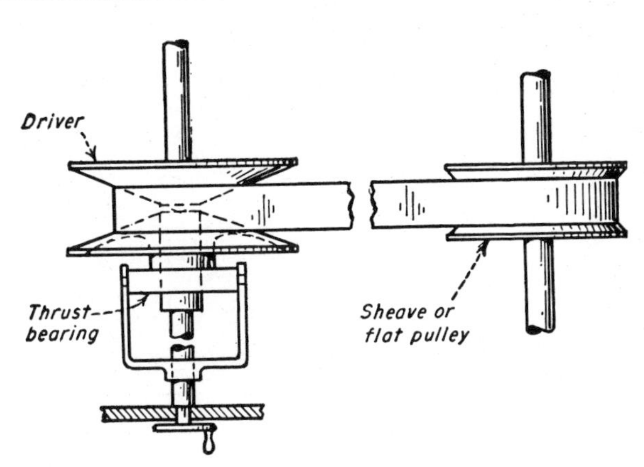


**Fig. 7—Another manually adjusted screw type split sheave for V-belts. However, this unit can be adjusted while the drive is running. Other characteristics similar to those of Fig. 6. Like Fig. 6, sheave spacing can be changed to maintain speed or to vary speed.**

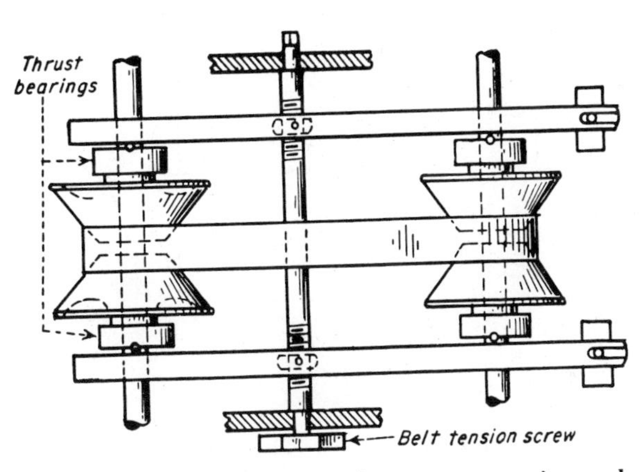


**Fig. 8—Special split sheaves for accurate tension and speed control of V-belts or chains. Applicable to parallel shafts on short center distances. Manually adjusted with belt tension screw. No change in speed with changes in tension.**

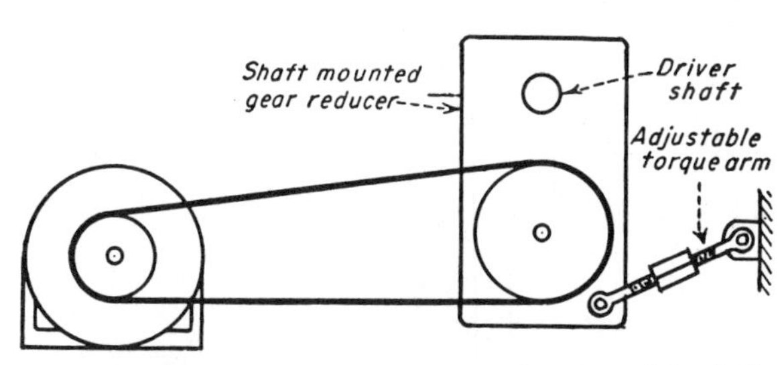


**Fig. 11—Torque arm adjustment for use with shaft mounted speed reducer. Can be used as belt or chain take up for normal wear and stretch within the swing radius of reducer; or for changing speed while running when spring type split sheave is used on motor.**

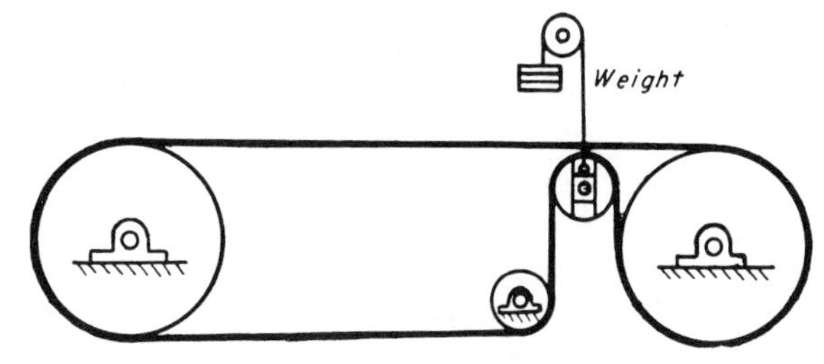


**Fig. 12—Wrapping type automatic take-up for flat ånd wire belts of any width. Used for maximum driving capacity. Size of weight determines tension put on belt. Maximum value should be established to protect the belt from being overloaded.**

# Methods for Reducing

Pulsations in chain motion created by the chordal action of chain and sprockets can be minimized or avoided by introducing a compensating cyclic motion in driving sprocket.

EUGENE I. RADZIMOVSKY
Ass't. Prof. of Mechanical Engineering, University of Illinois

Fig. 1—The large cast-tooth non-circular gear, mounted on the chain sprocket shaft, has wavy outline in which number of waves equals number of teeth on sprocket. Pinion has a corresponding noncircular shape. Although requiring special-shaped gears, drive completely equalizes chain pulsations.

Fig. 2—This drive has two eccentrically mounted spur pinions (1 and 2). Input power is through belt pulley keyed to same shaft as pinion 1. Pinion 3 (not shown), keyed to shaft of pinion 2, drives large gear and sprocket. However, mechanism does not completely equalize chain velocity unless the pitch lines of pinions 1 and 2 are noncircular instead of eccentric.

Fig. 3—Additional sprocket 2 drives noncircular sprocket 3 through fine-pitch chain 1. This imparts pulsating velocity to shaft 6 and to long-pitch conveyor sprocket 5 through pinion 7 and gear 4. Ratio of the gear pair is made same as number of teeth of spocket 5. Spring-

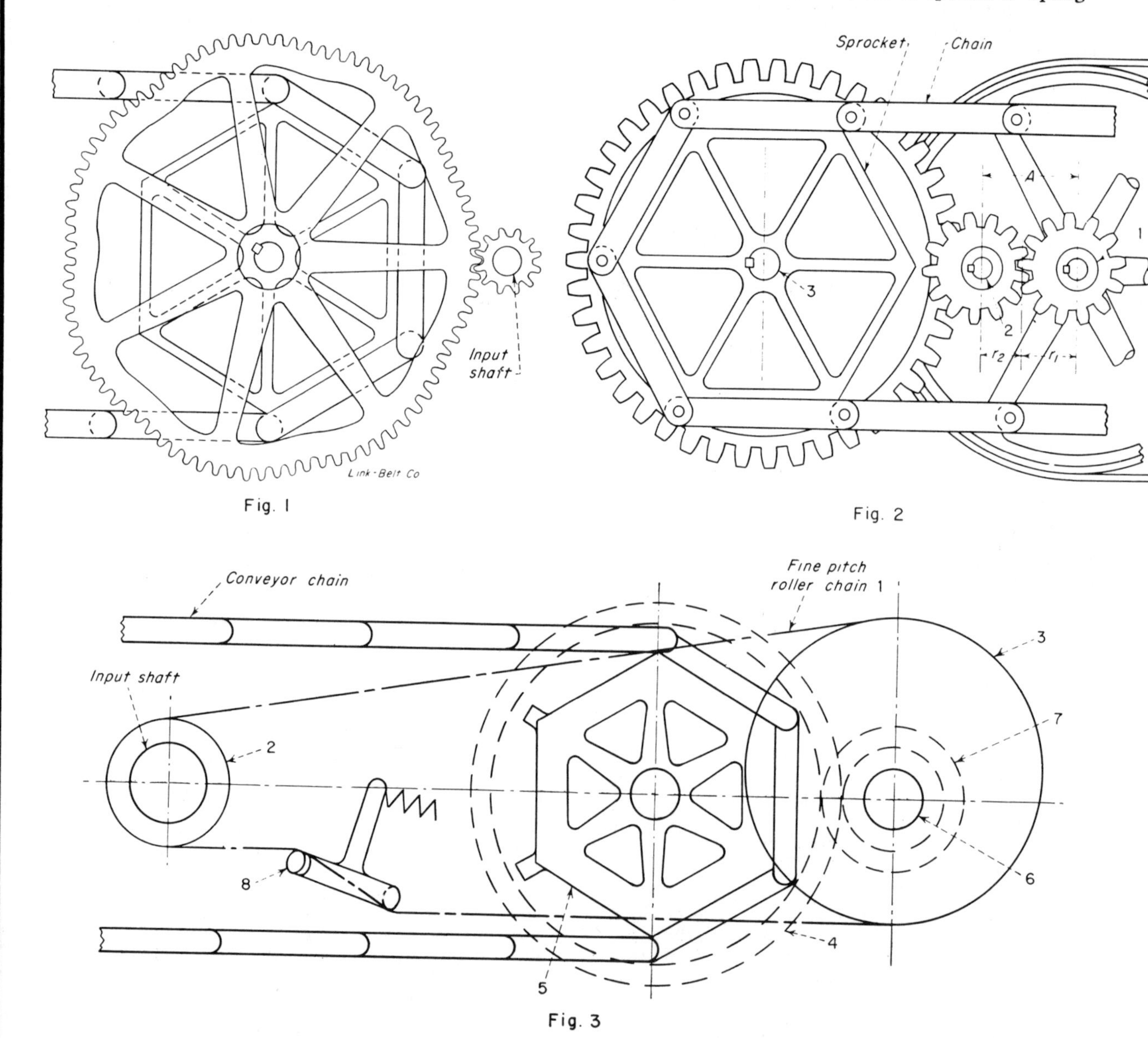


Fig. 1

Fig. 2

Fig. 3

# Pulsations in Chain Drives

Mechanisms for reducing fluctuating dynamic loads in chain drives and the pulsations resulting therefrom include non-circular gears, eccentric gears, and cam activated intermediate shafts.

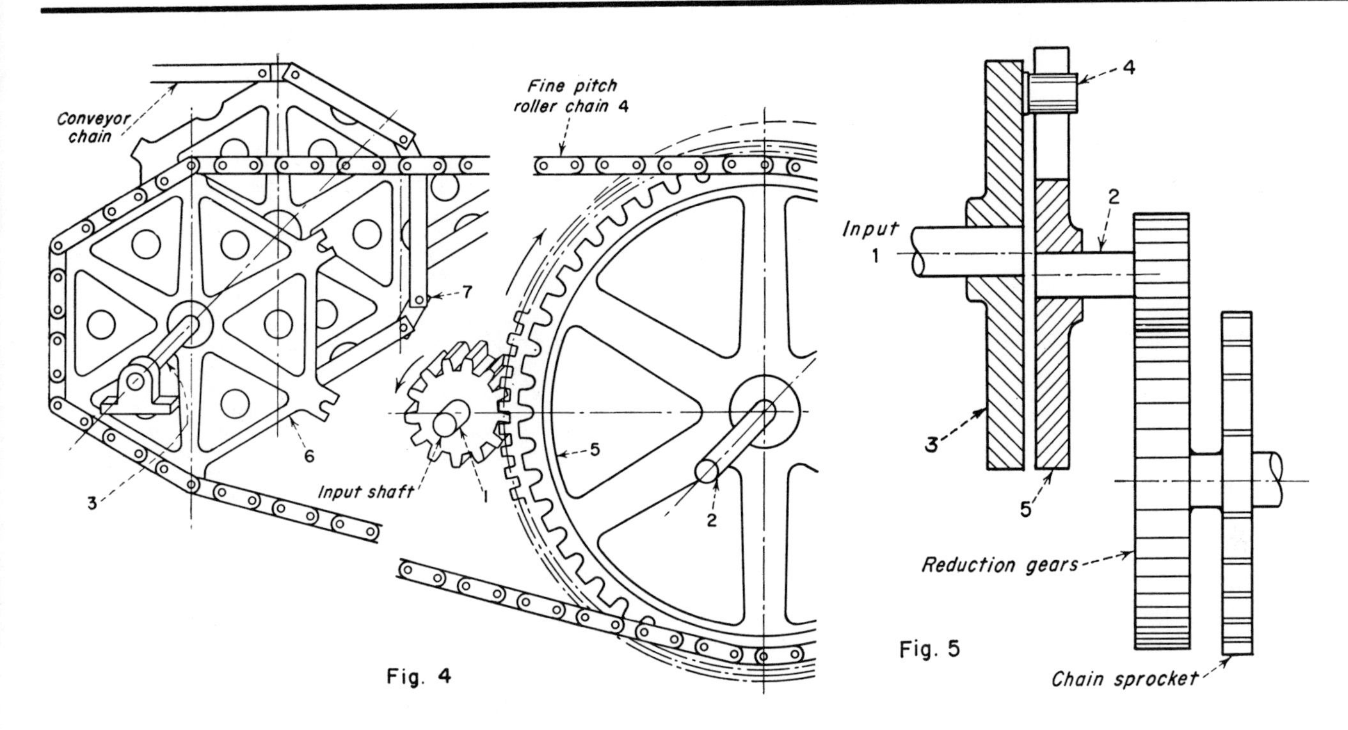


actuated lever and rollers 8 take up slack. Conveyor motion is equalized but mechanism has limited power capacity because pitch of chain 1 must be kept small. Capacity can be increased by using multiple strands of fine-pitch chain.

Fig. 4—Power is transmitted from shaft 2 to sprocket 6 through chain 4, thus imparting a variable velocity to shaft 3, and through it, to the conveyor sprocket 7. Since chain 4 has small pitch and sprocket 5 is relatively large, velocity of 4 is almost constant which induces an almost constant conveyor velocity. Mechanism requires rollers to tighten slack side of chain and has limited power capacity.

Fig. 5—Variable motion to sprocket is produced by disk 3 which supports pin and roller 4, and disk 5 which has a radial slot and is eccentrically mounted on shaft 2. Ratio of rpm of shaft 2 to sprocket equals number of teeth in sprocket. Chain velocity is not completely equalized.

Fig. 6—Integrated "planetary gear" system (gears 4, 5, 6 and 7) is activated by cam 10 and transmits through shaft 2 a variable velocity to sprocket synchronized with chain pulsations thus completely equalizing chain velocity. The cam 10 rides on a circular idler roller 11; because of the equilibrium of the forces the cam maintains positive contact with the roller. Unit uses standard gears, acts simultaneously as a speed reducer, and can transmit high horsepower.

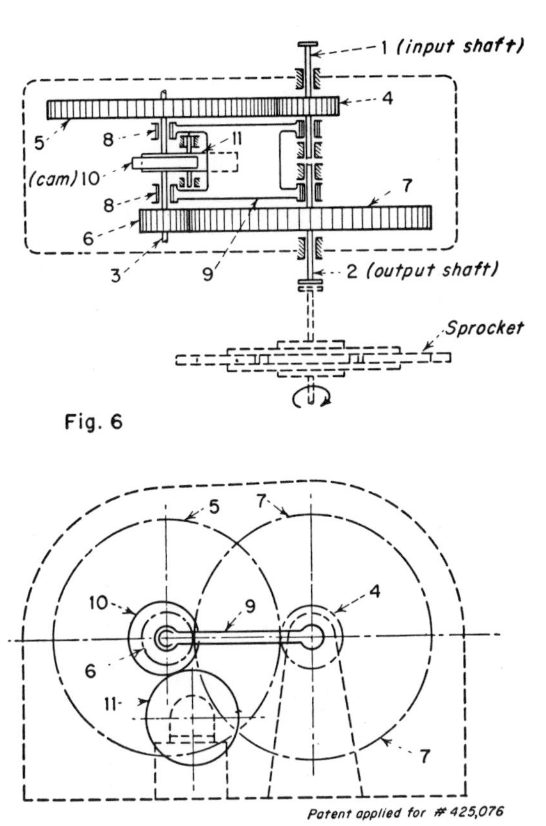

# Leather Belts—Hp Loss and Speeds...

Douglas C. Greenwood

**from 0-10,000 ft/min and 435-3450 rpm, for pulley diameters up to 30 in.**

Horsepower ratings and correction factors for various leather belt sizes, tensions, and operating conditions are given by most engineering handbooks or manufacturers' catalogs. Such data, however, are usually not corrected for centrifugal force. This chart may be entered at any axis or pulley-speed curve. As shown, secants parallel to the axes connect any four values in correct relationship. In the sample construction, a 12 in. dia pulley at 1150 rpm gives a belt velocity of about 3620 fps at which speed there is a 12% hp reduction. Consult belt manufacturer regarding suitability, efficiency and other factors in high-speed applications.

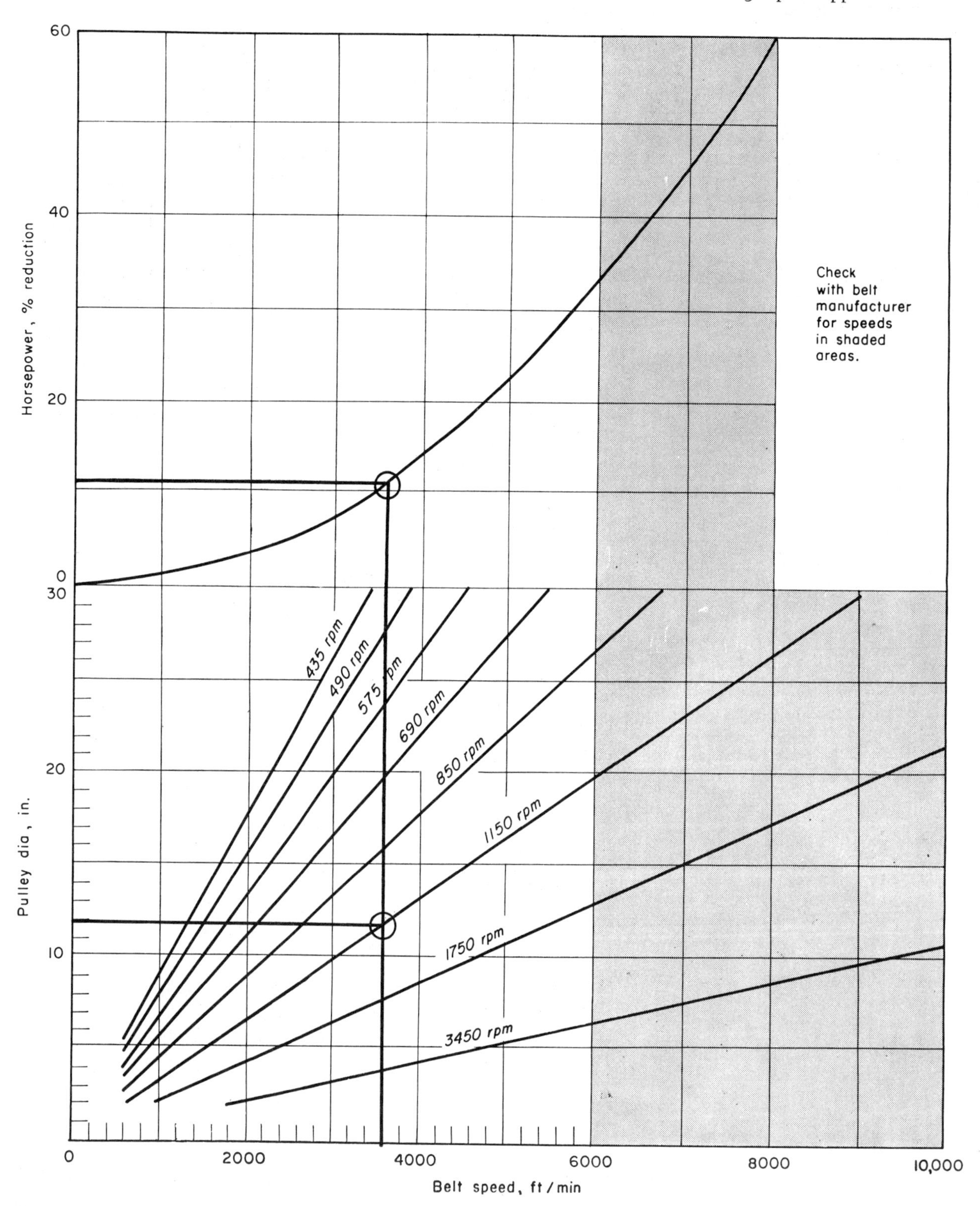

# Power Capacity of Spur Gears

CHARLES TIPLITZ
*Chief Engineer, J. M. Lehmann Company, Inc.*

MAXIMUM RATED HORSEPOWER that can safely be transmitted by a gear depends upon whether it runs for short periods or continuously. Capacity may be based on tooth strength if the gear is run only periodically; durability or wear governs rated horsepower for continuous running.

Checking strength and surface durability of gears can be a lengthy procedure. The following charts simplify the work and give values accurate to 5 to 10%. They are based on AGMA standards for strength and durability of spur gears.

**Strength Nomograph** is used first. Apart from the

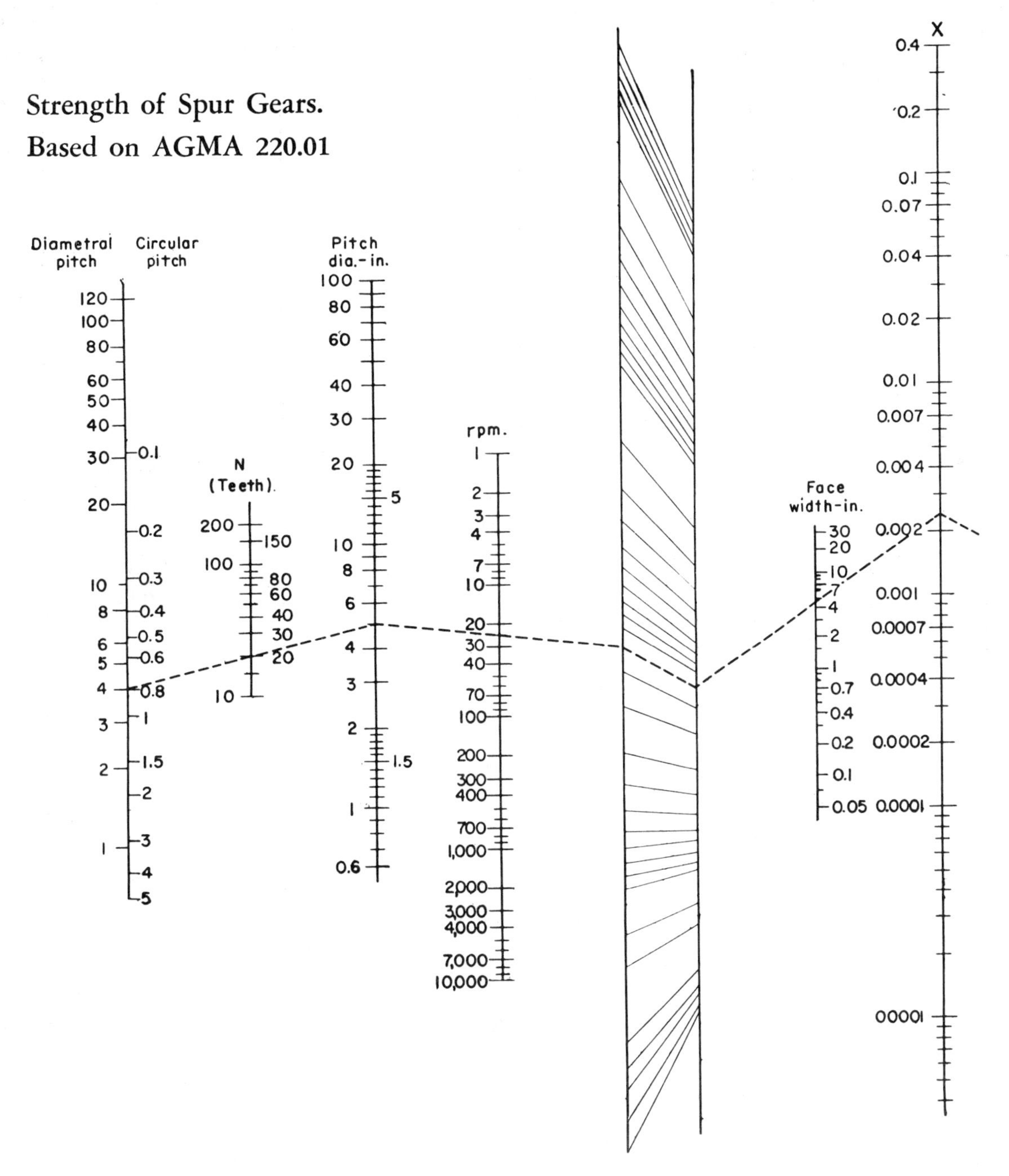

# Power Capacity of Spur Gears

usual design constants only two of the following three need be known: pitch, number of teeth and pitch diameter. To use the charts connect the two known factors by a straight line, cutting the third scale. From this point on the scale continue drawing straight lines through known factors, cutting the pivot scales. Between the double pivot scales the line should be drawn parallel to the adjacent lines.

**Durability nomograph** must be entered on scale X at the same value that was cut on the X scale on the strength chart. Both pinion and gear should be checked if made of different materials and the smaller of the values obtained should be used.

## Strength of Spur Gears (cont.)

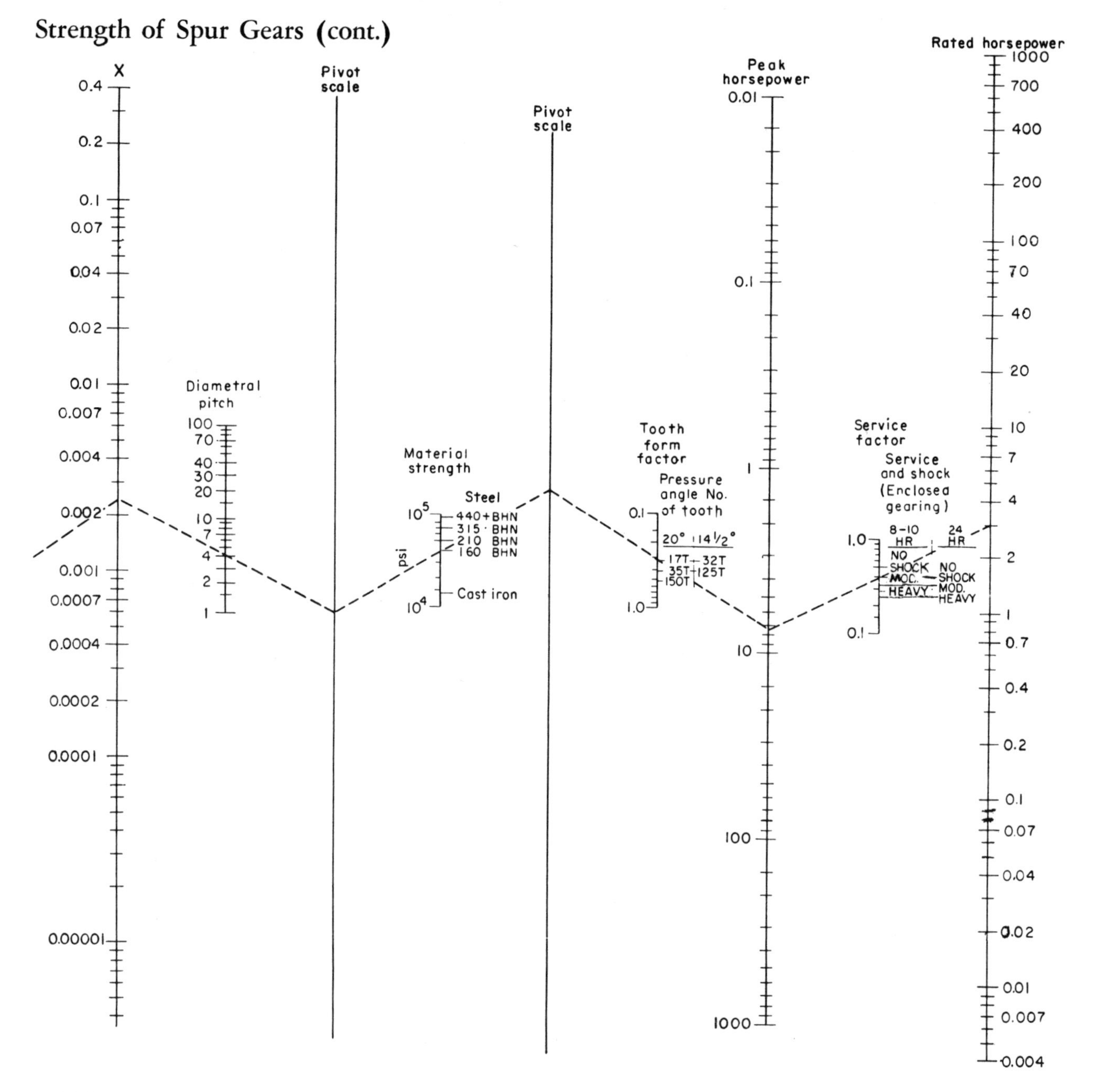

# Power Capacity of Spur Gears

## Surface Durability of Spur Gears. Based on AGMA 210.01

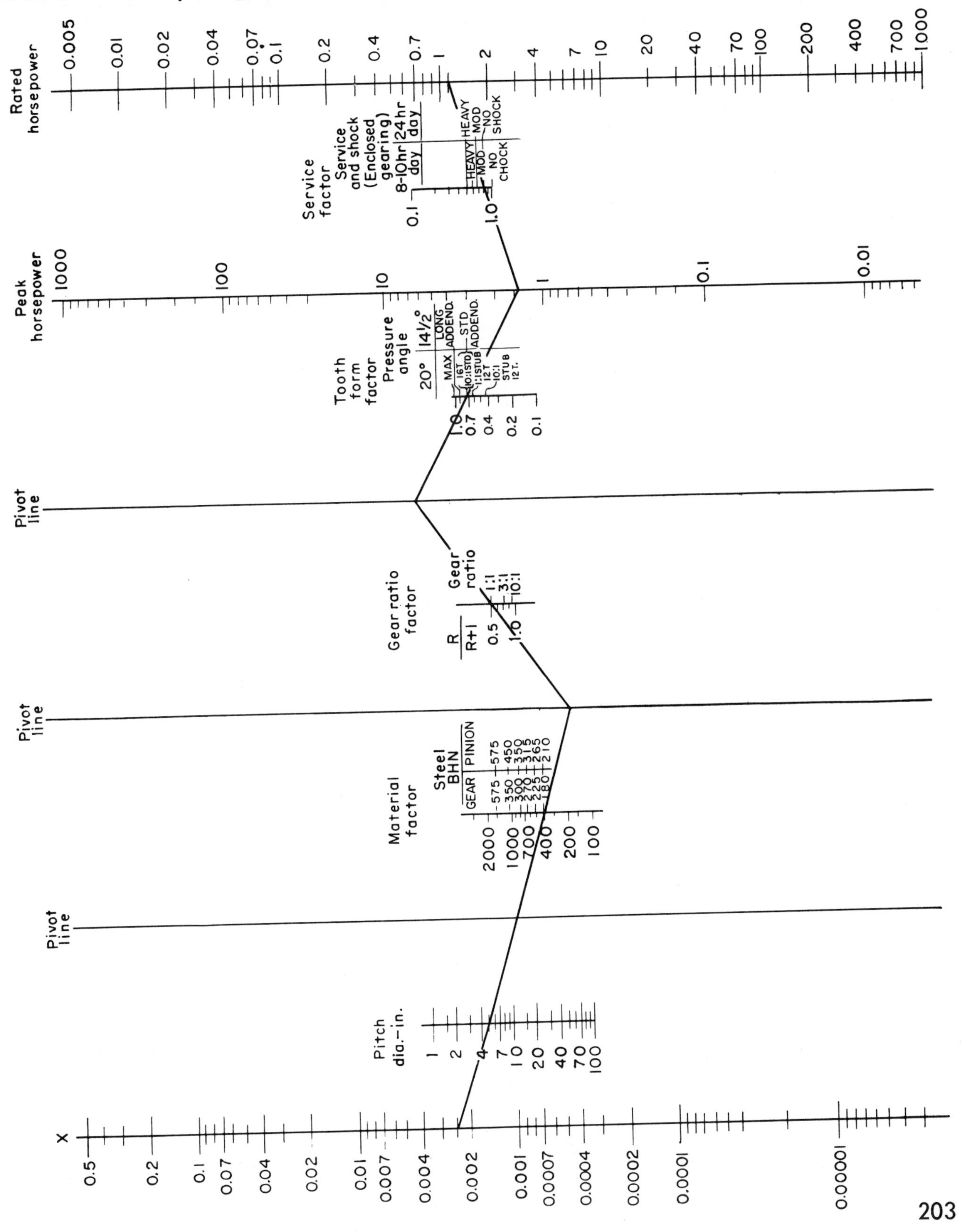

# Friction Wheel Drives Designed for Maximum Torque

Analysis of forces present in a friction wheel drive. Development of a design for drives that can be utilized to protect machines from excessive load torques.

RUDOLF KROENER

DURING THE POST WAR YEARS, as a consequence of the general scarcity of leather, rubber, and metals, in Germany efforts were made to replace belt drives and gear drives with friction wheel drives. The satisfactory results obtained are ascribed to:

1. Ability to manufacture a facing material having a high coefficient of friction. Cellulose type of materials have been developed possessing coefficients of friction ranging up to 0.5. When compared with materials having friction coefficients ranging from 0.15 to 0.2, the new materials offer an opportunity to reduce bearing pressures 60 to 70 percent.

2. Development of designs in which the contact or normal force between the friction wheels is varied automatically with changes in load torque of the driven machine. These designs make it possible to apply friction wheel drives to serve as disconnect clutches for limiting the transmission of torque before it becomes excessive.

When the faces of two friction wheels are pressed together, and where

$\mu$ = coefficient of friction of the materials in contact
$P$ = radial force pressing the wheels together, lb
$T$ = force transmitted tangentially to the wheels at their point of contact without slip, lb
$r$ = radius of driving wheel, in.
$n$ = speed of driving wheel, rpm
$H$ = horsepower transmitted by the friction wheel drive

$$T = \mu P \tag{1}$$

$$H = \frac{r\,n\,T}{63{,}025} \tag{2}$$

Since the radial force $P$ is equal and opposite to the force exerted by the wheels on their supporting bearings, it is evident that for constant values of $T$ the bearing pressures increase as the coefficient of friction decreases. Low coefficients of friction, therefore, are conducive to resultant power loss and bearing wear.

In a friction wheel drive where the wheel centers are adjusted and fixed to obtain a radial force sufficient to transmit a desired torque, the bearing pressure remains at a constant value regardless of variations in the transmitted torque. When the load torque varies to a large extent, such an arrangement compares unfavorably with gear drives and belt drives, since in such drives the bearing pressures vary with the torque transmitted.

This disadvantage is overcome in types of friction drives in which the driving wheel is mounted on a swinging center.

The swing drive shown in Fig. 1 is designed to change the radial pressure $P$ simultaneously and automatically with variations in load torque. In this arrangement the motor is fastened to a sub-base. The sub-base is free to swing on an axle at the right side. The opposite side of the base is supported by a spring. The driving friction wheel is pushed upward by the spring to maintain contact with the driven wheel. The driven wheel is mounted on a non-adjustable center.

When the torque load on the driven

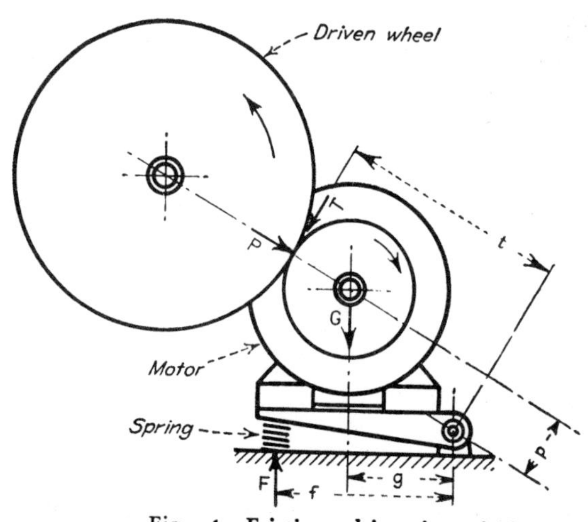


Fig. 1—Friction drive in which swing base is supported by a spring and axle.

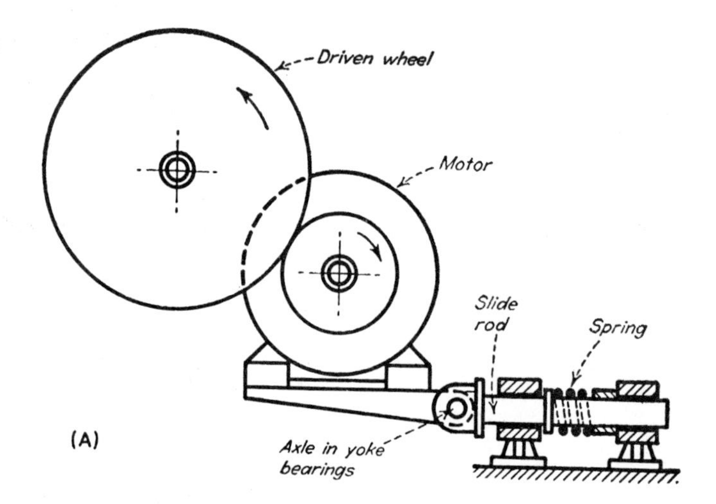


Fig. 2—(A) General arrangement of a maximum torque friction wheel drive with horizon-

wheel changes for any reason, the state of equilibrium is disturbed. The effect of an increase in torque load, until slipping occurs, is to cause the driving wheel to roll back or down on the driven wheel thus further compressing the spring. The spring force is thus increased, which results in an increased radial force $P$ and an increase in the transmitted torque. Where

$F$ = spring force, lb
$f$ = horizontal distance from center of axle to line of spring force, in.
$p$ = perpendicular distance from line of wheel centers to center of axle, in.
$G$ = resultant of motor weight, driving wheel weight, and subbase weight, referred to axis of the motor, lb
$g$ = horizontal distance from center of axle to vertical line passing through axis of motor, in.
$t$ = perpendicular distance from tangent through point of contact of wheel faces to center of axle, in.

then to satisfy conditions of equilibrium

$$Gg - Ff - Pp + Tt = 0 \tag{3}$$

and the spring force $F$ is found by substituting in Eq (3) the value of $P$ as given by Eq (1), or

$$F = \frac{Gg + T[t - (p/\mu)]}{f} \tag{4}$$

In the design shown in Fig. 1, the extent to which the radial pressure $P$ may build up, until slipping occurs, in response to increasing load torque is not limited. Excessive load torques may damage the friction facings, the driven machine, or the motor.

Any of many safety devices such as slip clutches, shear pins or keys, and breaking bolts, of course, can be used to protect the driven machine from excessive overloads. Fuses, overload relays, and thermal cut out devices can also be installed to protect the motor. Such protective devices are not necessary, however, when the friction wheel drive is designed to perform as a maximum torque clutch in which contact at the wheel faces ceases when a predetermined value of load torque is exceeded.

In the friction wheel drive shown in Fig. 2 (A), the drive motor $M$ is fastened to a swing plate, one side of which is supported on an axle. This axle is free to turn in yoke bearings on the ends of rods that are free to slide in fixed bearings. The spring $F$ is compressed between a shoulder and a spacer on each slide rod.

In this arrangement, an increase in load torque on the driven wheel causes the tangential force $T$ to increase, which in turn causes the driving wheel to ride at a lower position on the face of the driven wheel.

As the driving wheel drops to a lower position, the cosine of the angle included between the line of centers of the axle and motor and the horizontal centerline of the slide rods increases, thus compressing the spring $F$ and increasing the contact force $P$. With an increasing load torque, the driving wheel will finally fall away from the driven wheel.

At the instant of last contact of the two wheels, the spring has its maximum compression. The maximum torque that the arrangement shown in Fig. 2 (A) can transmit, therefore, depends upon the spring rate of the spring.

The geometrical relations present in the drive shown in Fig. 2 (A) when operating under a normal load and under maximum load are shown in Figs. 2 (B) and (C), respectively. For normal load conditions, the notations for dimensions and angles carry the subscript 1; for maximum load conditions they carry the subscript 2. The geometrical relations existing are.

AT A NORMAL LOAD,

$$a_1 = (R + r) \cos \beta_1$$
$$h_1 = g - (R + r) \sin \beta_1$$
$$b_1 = \sqrt{s^2 - h_1^2}$$
$$p_1 = s \sin (\alpha_1 - \beta_1)$$
$$t_1 = r + s \cos (\alpha_1 - \beta_1)$$
$$c_1 = a_1 + b_1$$

$$\sin \alpha_1 = \frac{h_1}{s} = \frac{g - (R + r) \sin \beta_1}{s}$$

AT MAXIMUM LOAD,

$$\alpha_2 = \beta_2$$
$$a_2 = (R + r) \cos \beta_2$$
$$h_2 = g - (R + r) \sin \beta_2$$
$$b_2 = \sqrt{s^2 - h_2^2} = h_2/\tan \beta_2$$
$$p_2 = s \sin (\alpha_2 - \beta_2) = 0$$
$$t_2 = r + s \cos (\alpha_2 - \beta_2) = r + s$$
$$c_2 = a_2 + b_2$$
$$\sin \alpha_2 = \sin \beta_2 = g/(R + r + s)$$

Where

$F$ = spring force or horizontal component of the reaction load exerted by the axle, lb
$N$ = vertical component of the reaction load exerted by the axle, lb

the relations that satisfy conditions of equilibrium are

$$Gb + Tt - Pp = 0 \tag{5}$$
$$P \cos \beta - T \sin \beta = F \tag{6}$$
$$P \sin \beta + G + T \cos \beta = N \tag{7}$$
$$\sqrt{N^2 + F^2} = S \tag{8}$$
$$T = \mu P \tag{9}$$

Substituting Eq (9) in Eq (6)

$$T_2\left(\frac{\cos \beta_2}{\mu} - \sin \beta_2\right) = F_2 \tag{10}$$

The spring force $F_2$ required to maintain sufficient radial pressure $P_2$ to transmit a maximum horsepower $H_2$ from Eqs (2) and (10) is then

$$F_2 = \frac{63,025\ H_2}{rn}\left(\frac{\cos \beta_2}{\mu} - \sin \beta_2\right) \tag{11}$$

by similar analysis

$$F_1 = \frac{63,025\ H_1}{rn}\left(\frac{\cos \beta_1}{\mu} - \sin \beta_1\right) \tag{11A}$$

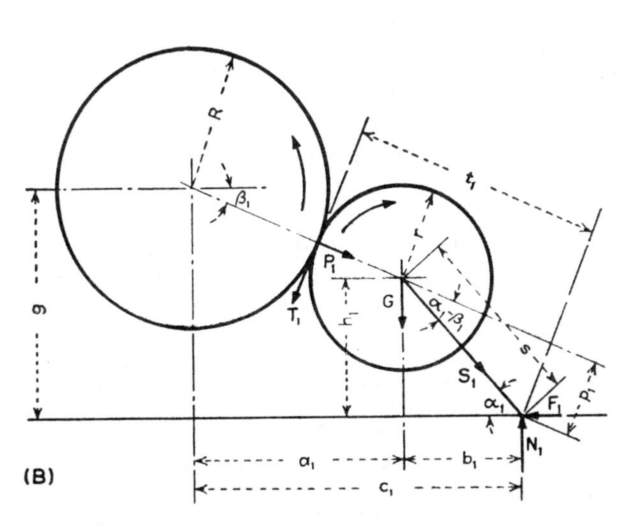


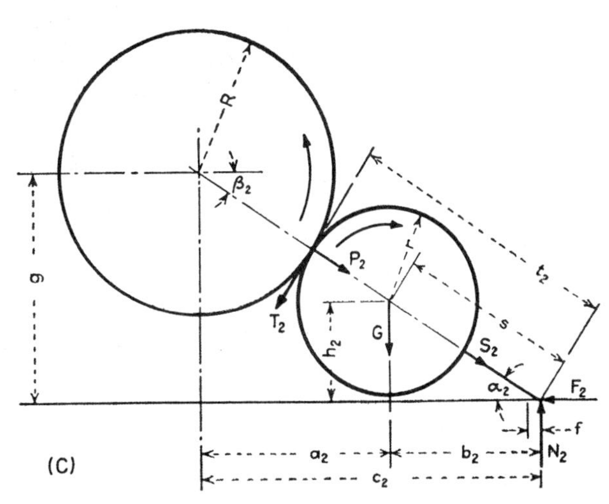


tal slide rods and compression springs. (B) Geometrical relation of parts under normal driving conditions. (C) Geometrical relation of parts under maximum torque driving conditions.

The angle $\beta_2$ that gives the minimum value of $F_2$ for a given driving wheel diameter and speed at a given horsepower is found by differentiating Eq (11), setting the derivative equal to zero and solving for $\beta_2$, which results in the relation

$$\beta_2 = \sin^{-1} (\mu / \sqrt{1 + \mu^2}) \qquad (12)$$

Substituting values of $\mu$ of in Eq (12)

For $\mu = 0.3$, $\beta_2 = 16$ deg 42 min
For $\mu = 0.4$, $\beta_2 = 21$ deg 48 min
For $\mu = 0.45$, $\beta_2 = 24$ deg 14 min

Employing an angle $\beta_2$ as determined by Eq (12) will assure full utilization of the coefficient of friction of the material up to the limit where sliding may occur, while simultaneously maintaining minimum radial pressure and bearing loads.

Where

$f$ = compression of the spring from spring load $F_1$ to spring load $F_2$, in.

$f = c_1 - c_2 = a_1 + b_1 - a_2 - b_2$

the spring rate $k$ is

$$k = (F_2 - F_1)/f \qquad (13)$$

The foregoing analysis of the drive shown in Fig. 2(A) can be criticized on the basis that the design is expensive and that the vertical component $N$ of the axle bearing reaction may introduce a frictional force that cannot be neglected. Furthermore, the analysis assumes that all the forces considered lie in the same vertical plane. Strictly speaking, the weights of motor, driving wheel, and swing base do not lie in the vertical planes that include radial pressure, tangential force, and spring force.

When the bearing rods are supported horizontally, however, these errors in assumption do not affect the spring calculations to a large extent when the friction wheel lies in a vertical plane that is central between the two bearing rods. These errors do, however, affect the magnitude of the vertical component $N$, which acts on the two bearing rods.

The friction wheel drive shown in Fig. 3 is supported by two bearing rods separated by the distance d. The axial thrusts on these rods are counter-balanced by the reaction of compression springs. Each spring, its adjusting nut, and its rod thrust plate are contained within a sleeve. When the motor and swing plate have fallen through, the sleeve can be removed by lifting it from its enclosure. By lifting the motor, the sleeve can be replaced in its enclosure without changing the spring setting.

By resolving all the forces present into their components acting parallel to the plane passing through the centerlines of the bearing rods and writing equations of equilibrium, the following relations are obtained:

$$G \sin \alpha_2 + T_1 \sin (\alpha_2 - \beta_1) + P_1 \cos (\alpha_2 - \beta_1) - F_{A1} - F_{B1} = 0 \qquad (14)$$

taking moments about $B$

$$(d/2)\, G \sin \alpha_2 - d\, F_{A1} - e\, [T_1 \sin (\alpha_2 - \beta_1) + P_1 \cos (\alpha_2 - \beta_1)] = 0 \qquad (15)$$

taking moments about $A$

$$(d/2)\, G \sin \alpha_2 + (d + e)\, [T_1 \sin (\alpha_2 - \beta_1) + P_1 \cos (\alpha_2 - \beta_1)] - d\, F_{B1} = 0 \qquad (16)$$

dividing Eq (15) through by $d$ and solving for $F_{A1}$

$$F_{A1} = \frac{G \sin \alpha_2}{2} - \frac{e}{d} \times [T_1 \sin (\alpha_2 - \beta_1) + P_1 \cos (\alpha_2 - \beta_1)] \qquad (15A)$$

dividing Eq (16) through by $d$ and solving for $F_{B1}$

$$F_{B1} = \frac{G \sin \alpha_2}{2} + \left(\frac{d + e}{d}\right) \times [T_1 \sin (\alpha_2 - \beta_1) + P_1 \cos (\alpha_2 - \beta_1)] \qquad (16A)$$

For the condition at the instant of maximum torque when the swing falls through

$$\alpha_2 = \beta_1 = \beta_2$$

therefore Eq (15A) becomes

$$F_{A2} = \frac{G \sin \alpha_2}{2} - \frac{e\, P_2}{d} \qquad (15B)$$

and Eq (16A) becomes

$$F_{B2} = \frac{G \sin \alpha_2}{2} + \frac{(d + e)\, P_2}{d} \qquad (16B)$$

The stroke or deflection $\Delta l$ of the spring between the initial load and the maximum load is

$$\Delta l = R + r + s - [(R + r) \times \cos (\beta_2 - \beta_1) + s \cos (\alpha_1 - \alpha_2)] \qquad (17)$$

The spring constant for spring $A$ is

$$k_A = (F_{A2} - F_{A1})/\Delta l \qquad (18)$$

The spring constant for spring $B$ is

$$k_B = (F_{B2} - F_{B1})/\Delta l \qquad (19)$$

In similar manner, by resolving the forces into their components acting normal to the plane passing through the centerlines of the bearing rods and writing equations of equilibrium, the bending and twisting loads acting on the eyes of the bearing rods can be determined.

The maximum torque friction wheel drive is suitable and advantageous in applications where it is desirable to limit the magnitude of torque that can be delivered to a machine. This drive protects the machine from excessive steady loads and shock loads.

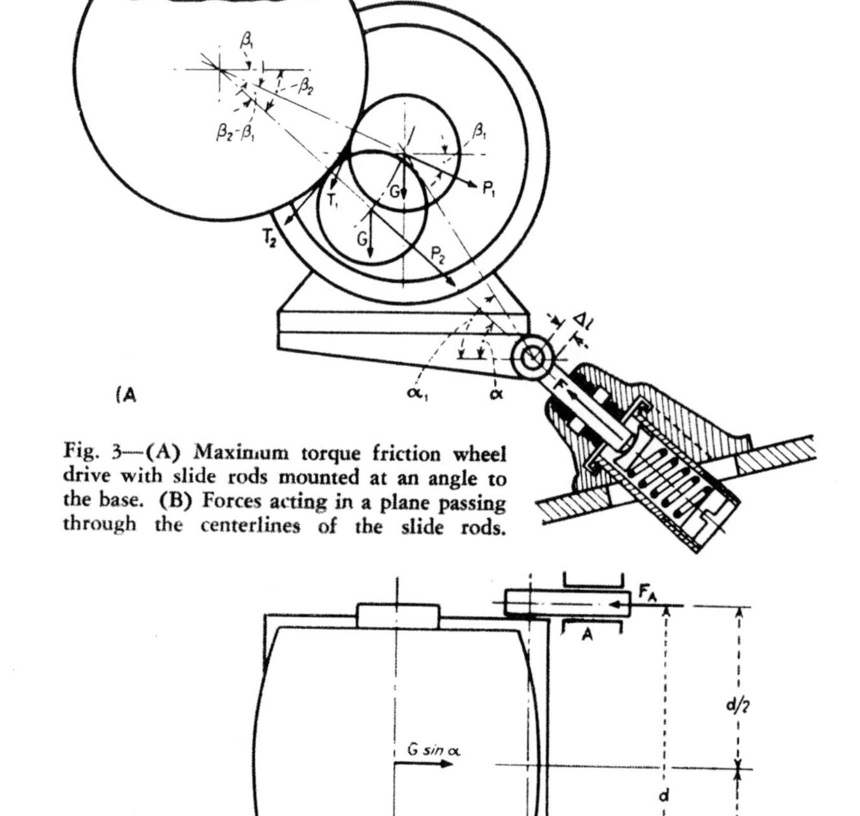


Fig. 3—(A) Maximum torque friction wheel drive with slide rods mounted at an angle to the base. (B) Forces acting in a plane passing through the centerlines of the slide rods.

# Accurate Solution for Disk-Clutch Torque Capacity

NILS M. SVERDRUP
Development Engineer,
AiResearch Manufacturing Company

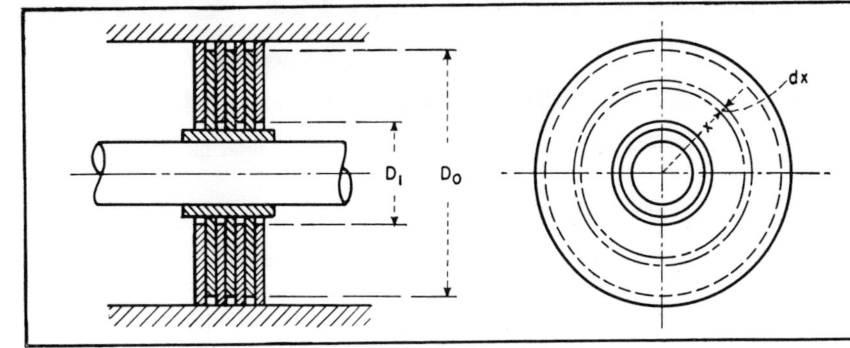


IN COMPUTING TORQUE CAPACITY, the mean radius $R$ of the clutch disks is often used. The torque equation then assumes the following form:

$$T = P\mu R n \qquad (1)$$

Where

$T$ = torque, in.-lb
$P$ = pressure, lb.
$\mu$ = coefficient of friction
$R$ = mean radius of disks, in.
$n$ = no. of friction surfaces

This formula, however, is not mathematically correct and should be used cautiously. The formula's accuracy varies with the ratio $D_1/D_0$. When $D_1/D_0$ approaches unity, the error is negligible; but as the value of this ratio decreases, the induced error will increase to a maximum of 33 percent.

By introducing a correction factor, $\phi$, Eq (1) can be written

$$T = P\mu R n\phi \qquad (2)$$

The value of the correction factor can be derived by the calculus derivation of Eq (2).

Sketch above represents a disk clutch with $n$ friction surfaces, pressure between plates being $p$ psi. Inside and outside diameters of effective friction areas are $D_1$ and $D_0$ in., respectively. Since the magnitude of pressure on an element of area, $dA$, at distance $x$ from center is $pdA$, the friction force is $pdA\mu$ and the moment of this force around the center is $pdA\mu x$.

Integrating within limits $D_1/2$ and $D_0/2$ and multiplying by $n$ friction surfaces, the expression for total torque in in.-lb is obtained.

Hence

$$T = \int_{D_1/2}^{D_0/2} pdA\mu xn \qquad (3)$$

but

$$dA = 2\pi x dx \qquad (4)$$

Substituting in Eq (3)

$$T = \int_{D_1/2}^{D_0/2} p\,(2\pi x dx)\,\mu x n$$

$$= (2/3)\pi p\mu n \left[\left(\frac{D_0}{2}\right)^3 - \left(\frac{D_1}{2}\right)^3\right]$$

$$\text{or} \quad T = 0.262\, p\mu n\,(D_0^3 - D_1^3) \qquad (5)$$

If the total pressure acting on clutch disks be $P$ lb, the expression for pressure per unit area is

$$p = \frac{P}{(\pi/4)\,(D_0^2 - D_1^2)}$$

Substituting this value for $p$ in Eq (5)

$$T = 0.333\, P\mu n \frac{D_0^2 + D_0 D_1 + D_1^2}{D_0 + D_1} \qquad (6)$$

Now let

$$\frac{D_1}{D_0} = m \text{ so that, } D_1 = m D_0 \qquad (7)$$

Substituting in Eq (6)

$$T = 0.333\, P\mu n D_0 \frac{1 + m^2 + m}{1 + m} \qquad (8)$$

Similarly, by substituting value of $D_1$ from Eq (7) in Eq (2), and having

$$R = \frac{D_0 + D_1}{4},$$

$$T = P\mu \frac{D_0 + m D_0}{4} n\phi$$

or

$$T = 0.25\, P\mu n\phi\, D_0 (1 + m) \qquad (9)$$

Equating expressions (8) and (9)

$$0.25\, P\mu n\phi\, D_0 (1 + m) =$$

$$0.333\, P\mu n\, D_0 \frac{1 + m^2 + m}{1 + m}$$

and solving for $\phi$, the result is

$$\phi = 1.333 \times \frac{1 + m^2 + m}{(1 + m)^2} \qquad (10)$$

With various diameter ratios, the values for $\phi$ were computed and represented in graph herewith. By using this graph and Eq (2), accurate values of torque can be easily determined.

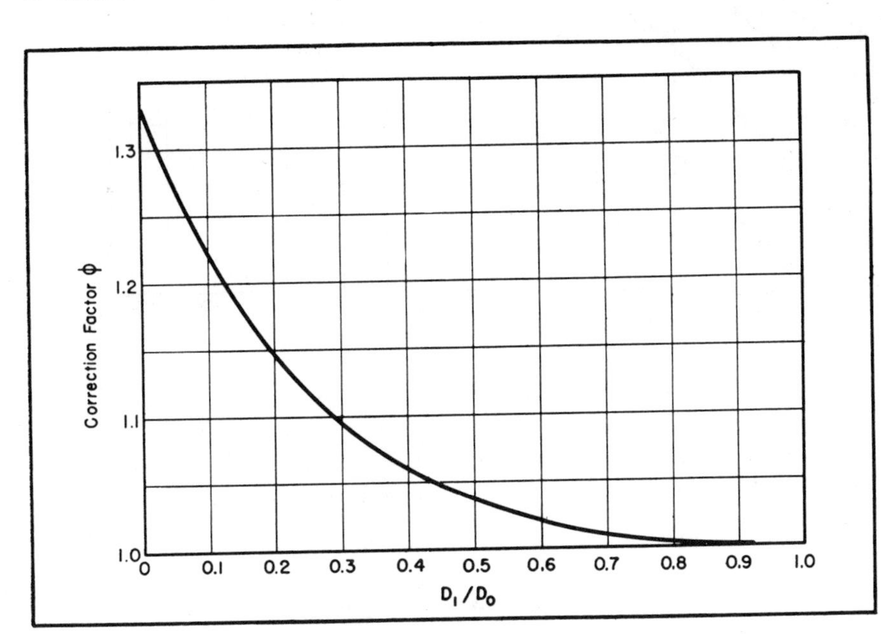

# Torque of Slip Couplings

Douglas C. Greenwood

**Here are formulas and charts of force, torque and friction conversion-factors for slip couplings from 0.50- to 5.0-in. mean radius.**

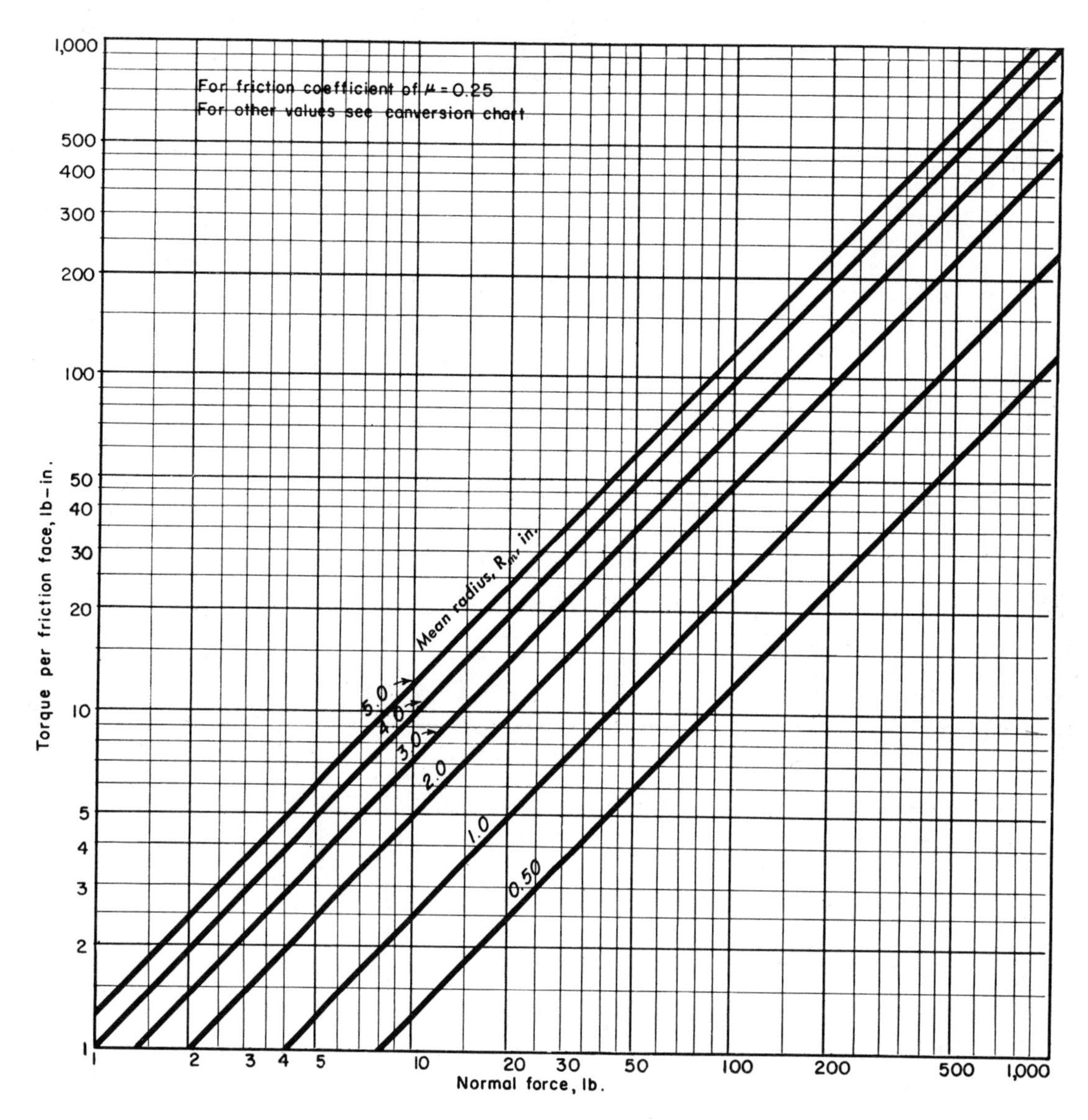


All slip-coupling designs consist, basically, of one member pressing against another. Friction face can be metal or some other material. When torque force exceeds the friction force between pressed-together faces, they slip relative to one another. Usually, pressure is adjustable so driving torque can be varied. Some form of flexible coupling should be mounted on one of the shafts close to the slip coupling.

For each friction face of a flat-disk coupling, torque, force and friction are related in

$$T = \mu F R_m$$

This formula is also true for cone couplings. If pressure is uniform throughout the friction face area the effective mean radius is

$$R_m = 2/3 \left( \frac{R_o^3 - R_i^3}{R_o^2 - R_i^2} \right)$$

When the friction face is between mating cones

$$F = \frac{F_a}{\text{Sin } \alpha}$$

If the cones are slipping, this becomes

$$F = \frac{F_a}{\text{Sin } \alpha + \mu \cos \alpha}$$

The face angle for cone couplings can range from about 7 to 30°. Usually it is about $12\frac{1}{2}$°.

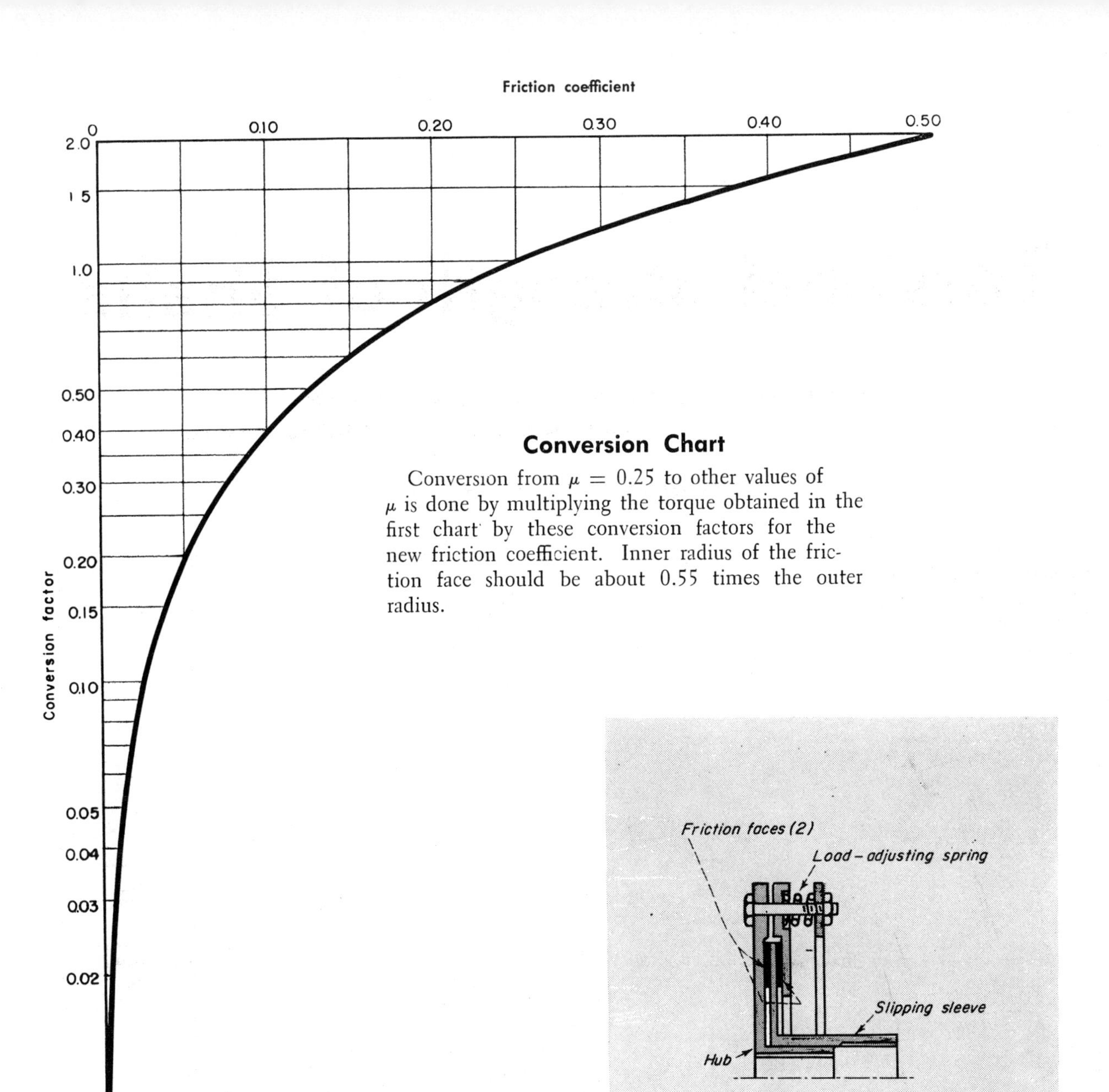


## Conversion Chart

Conversion from $\mu = 0.25$ to other values of $\mu$ is done by multiplying the torque obtained in the first chart by these conversion factors for the new friction coefficient. Inner radius of the friction face should be about 0.55 times the outer radius.

Maximum unit pressure for various friction materials should not be exceeded. While this is not so likely to occur in slip couplings as in clutches, it should nevertheless be checked. Maximum pressures for friction faces are available in most engineering handbooks.

If desirable to stop the drive when slipping occurs, a warning device such as a clacker can be mounted between the two members of the drive—movement of one relative to the other will be audible. An electrical contact similarly mounted can be connected through a slip ring to stop the drive automatically.

### Symbols

$F$ = force normal to friction face, lb
$F_a$ = axial force, lb
$R_m$ = mean radius of friction faces, in.
$R_o$ = outer radius of friction faces, in.
$R_i$ = inner radius of friction faces, in.
$T$ = torque, lb-in.
$n$ = number of friction faces
$a$ = cone angle relative to shaft axis
$\mu$ = friction coefficient

# Torsional Strength of Shafts

**Formulas and charts for horsepower capacity of shafts from ½ to 2½ in. dia. and 100 to 1000 rpm.**

Douglas C. Greenwood

For a maximum torsional deflection of 0.08° per foot, shaft length, diameter and horsepower capacity are related in

$$d = 4.6\sqrt[4]{\frac{hp}{R}}$$

where $d$ = shaft diameter, in.; hp = horsepower; $R$ = shaft speed, rpm. This deflection is recommended by many authorities as being a safe general maximum. The two charts are plotted from this formula, providing a rapid means of checking transmission-shaft strength for usual industrial speeds up to 20 hp. Although shafts under 1-in. dia are not transmission shafts, strictly speaking, lower sizes have been included.

When shaft design is based on strength alone, the diameter can be smaller than values plotted here. In such cases use the formula

$$d = \sqrt[3]{\frac{k\,hp}{R}}$$

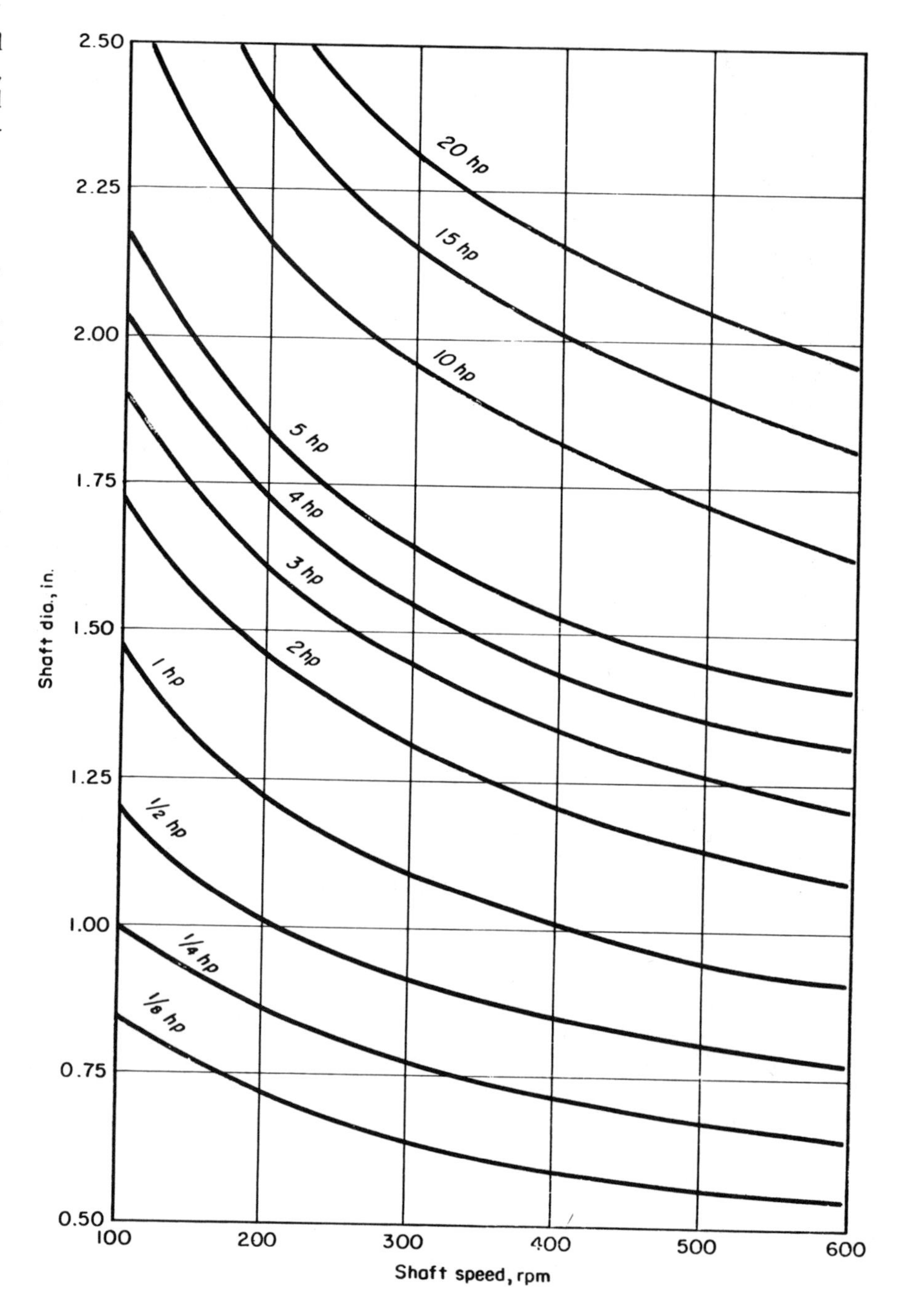

| k | LOADING CONDITION |
|---|---|
| 125 (2600 psi) | Head shafts subject to heavy strains. (Intermittent loads and slow speeds, clutches or gearing carried) |
| 100 (3200 psi) | Lineshafts 75–100 ft long, heavily loaded. Bearings 8 ft apart. |
| 90 (3550 psi) | Lineshafts 50–75 ft long, medium load, bearings 8 ft apart. |
| 75 (4300 psi) | Lineshafts 20–50 ft long, lightly loaded, bearings 6 ft apart. |

The value of $k$ varies from 125 to 38 according to allowable stress used. The figure accounts for members that introduce bending loads, such as gears, clutches and pulleys. But bending loads are not as readily determined as torsional stress. Therefore, to allow for combined bending and torsional stresses, it is usual to assume simple torsion and use a lower design stress for the shaft depending upon how it is loaded. For example, 125 represents a stress of approximately 2600 psi, which is very low and should thus insure a strong-enough shaft. Other values of $k$ for different loading conditions are shown in the table.

When bending stress is not considered, lower $k$ values can be used, but a value of 38 should be regarded as the minimum.

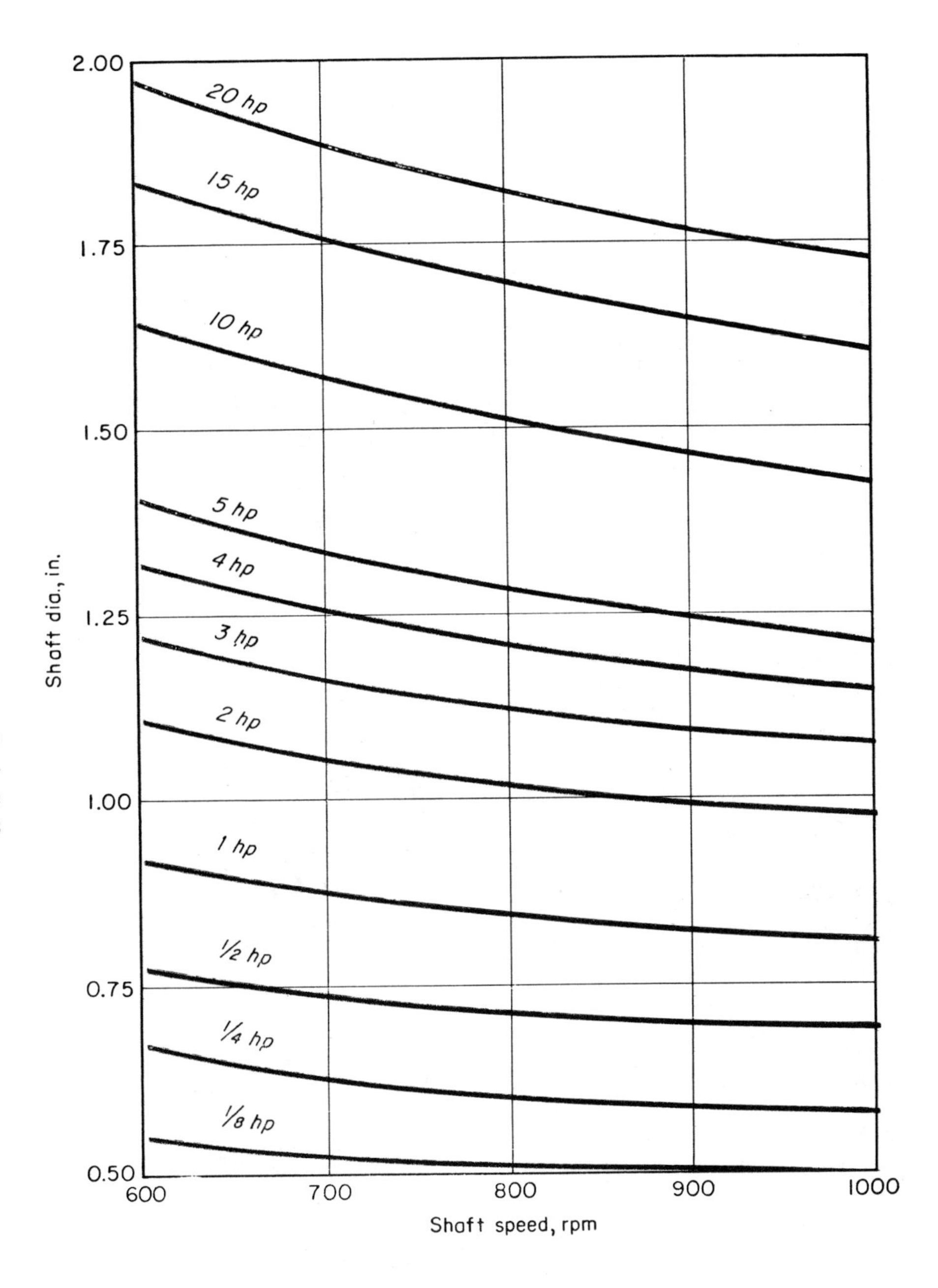

# Basic Types of Variable Speed

THESE DRIVES are used to transmit both high torque, as on power spindle presses, and low torque, as in laboratory instruments. All perform best if used to reduce and not to increase speed. All friction drives have a certain degree of slip due to imperfect rolling of the friction members, but by correct design this slip can be held constant, resulting in constant speed of the driven member. Variations in load should be compensated by inertia masses on the driven end. Springs or similar elastic members can be used to keep the friction parts in constant contact and exert the force necessary to create the friction. In some cases, gravity will take place of such members. Specially made friction materials are generally recommended, but leather or rubber are often satisfactory. Normally only one of the friction members is made or lined with this material, while the other is metal.

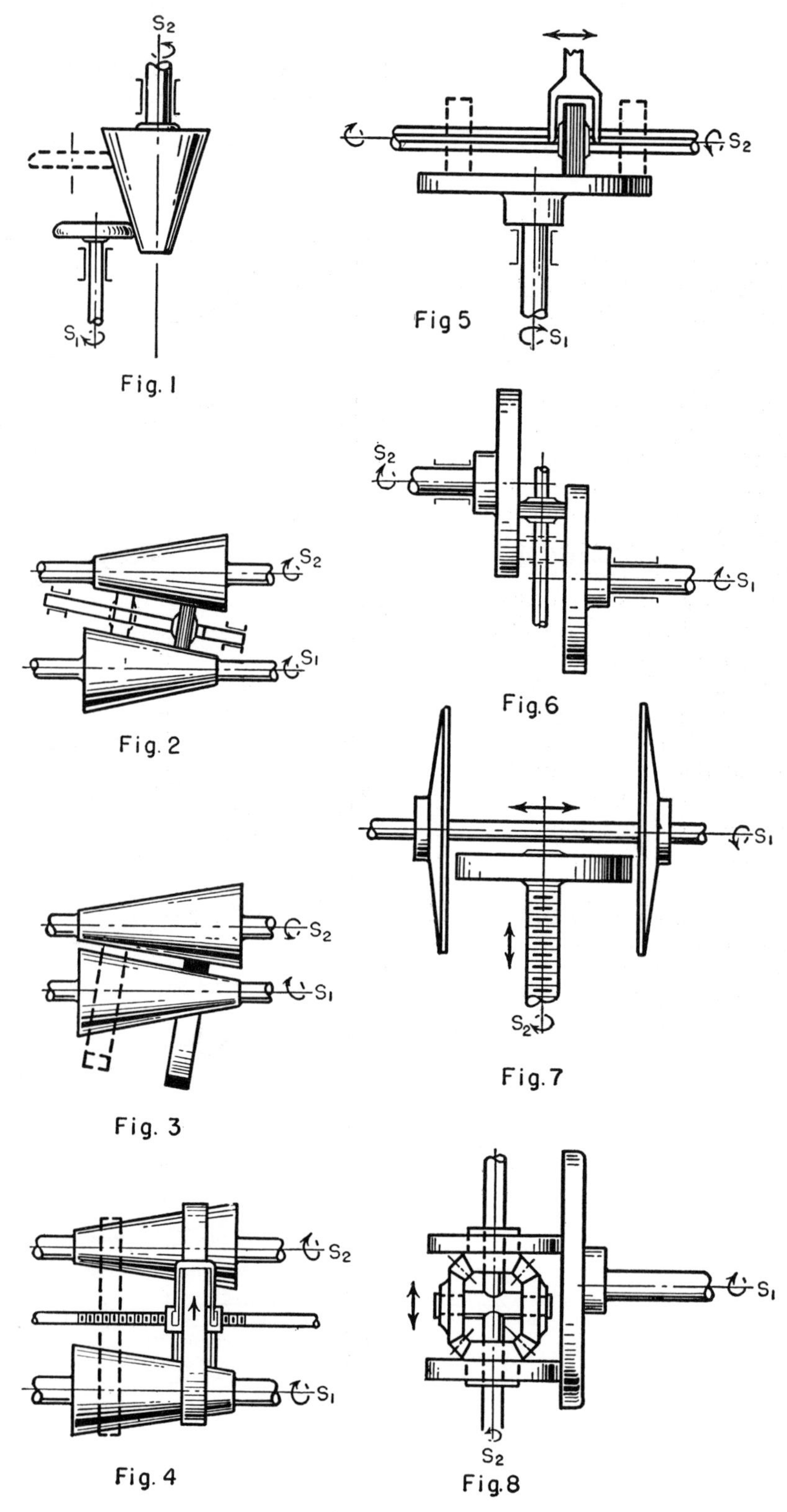


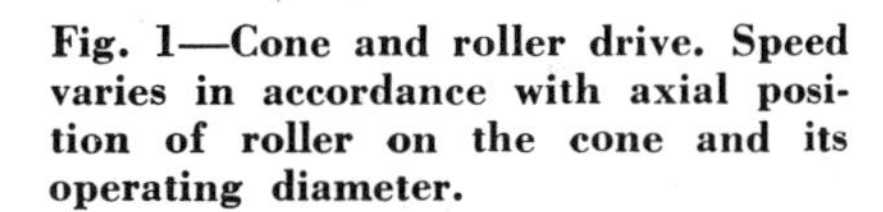

**Fig. 1—Cone and roller drive. Speed varies in accordance with axial position of roller on the cone and its operating diameter.**

**Fig. 2—Two cones and a free spinning roller that can be moved along the surfaces of the cones.**

**Fig. 3—Two cones and a free endless leather belt between them. The belt is shifted between cone faces to obtain the desired speed ratio.**

**Fig. 4—Two cones belted by a flat movable belt. Two metal fingers position belt along faces of the cones.**

**Fig. 5—A disk and roller drive. Roller is moved radially on the disk. Speed ratio depends upon the operating diameter of disk. Direction of relative rotation of shafts is reversed when roller is moved past the center of the disk as indicated by dotted lines.**

# Friction Drives

HAIM MURRO

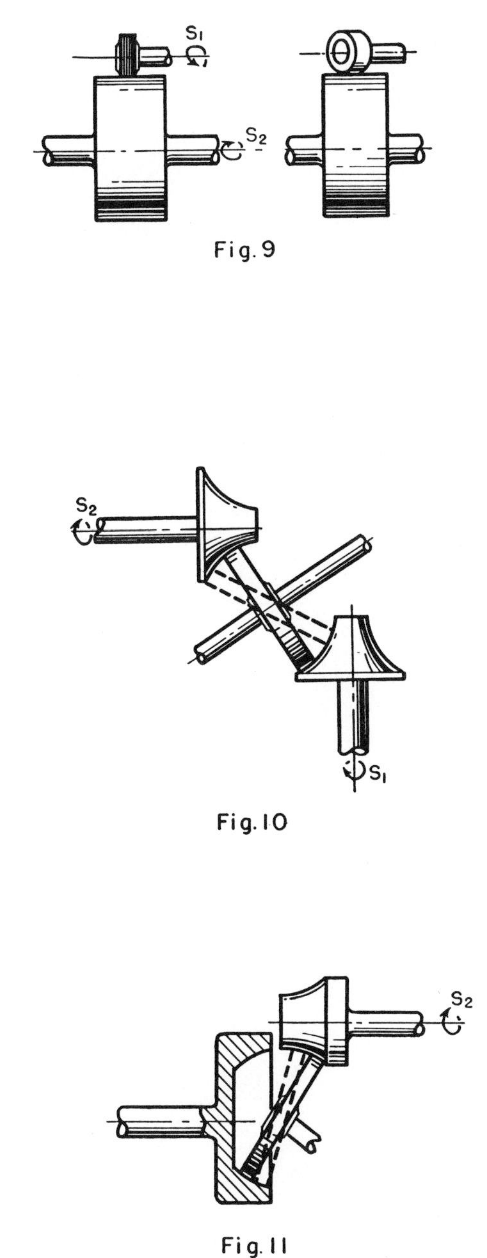


Fig. 9

Fig. 10

Fig. 11

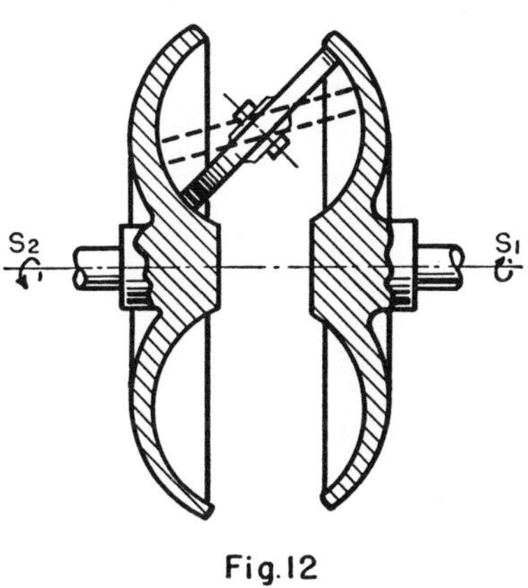


Fig. 12

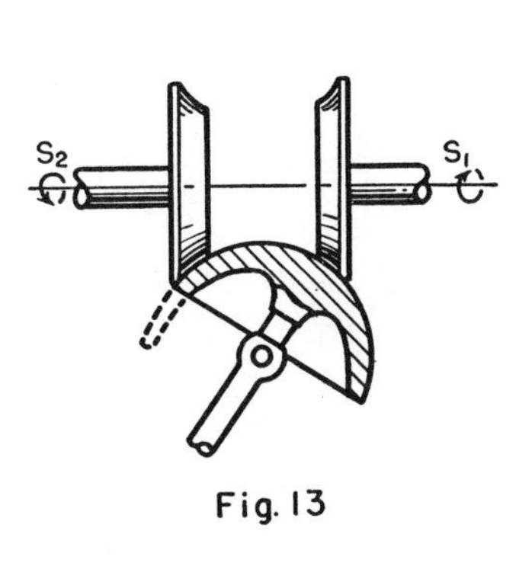


Fig. 13

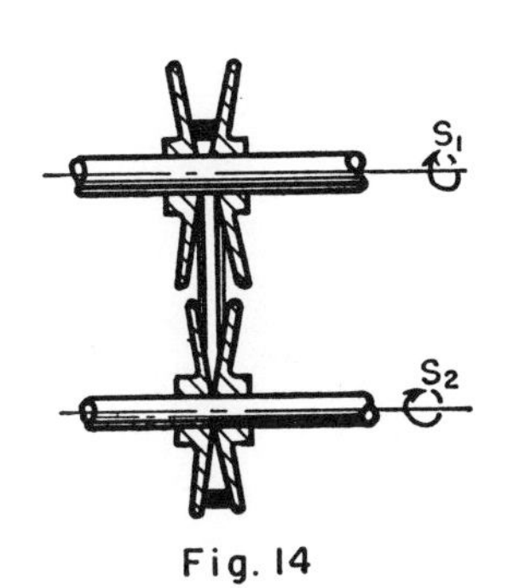


Fig. 14

**Fig. 6—Two disks and a free spinning roller, movable between them. Change of speed will be fast, because the operating diameters of the disks change in an inverse ratio.**

**Fig. 7—Two disks mounted on same shaft and a roller mounted on a threaded spindle. Contact of roller can be changed from one disk to the other to change direction of rotation. Rotation will be accelerated or decelerated with movement of the screw.**

**Fig. 8—Disk and two differential rollers. Rollers and their bevel gears are free to rotate on shaft, $S_2$. Other two bevel gears are free to rotate on pins connected by $S_2$. Good for high reduction and accurate adjustment of speed. $S_2$ will have the differential speed of the two rollers. Differential assembly is movable across face of disk.**

**Fig. 9—Drum and roller. Change of speed is effected by skewing the roller relative to the drum.**

**Fig. 10—Two spherical cones on intersecting shafts and a free roller.**

**Fig. 11—Spherical cone and groove with a roller. Suitable when small adjustments in speed are desired.**

**Fig. 12—Two disks with torus contours and a free rotating roller.**

**Fig. 13—Two disks with a spherical free rotating roller.**

**Fig. 14—Split pulleys for V belts. Effective diameter of belt grip can be adjusted by controlling distance between two parts of pulley.**

# 9

# ELECTRICAL, ELECTRONIC, AND MAGNETIC COMPONENTS


| | |
|---|---|
| Electrical Symbols and Standards | 216 |
| Fundamental Electronic Circuits: Power and Voltage Amplifiers | 218 |
| Tubes for Industrial Inspection and Control | 220 |
| Designing Electric Heating Elements | 222 |
| Typical Industrial Uses of Electric Heating Elements | 224 |
| Magnet Coil Design | 226 |
| Fundamental Types of Permanent Magnets | 230 |
| Applications for Permanent Magnets | 232 |
| Permanent Magnet Mechanisms, Their Design and Uses | 234 |
| Fractional Horsepower Motors | 236 |
| Torque Requirements of Various Motor Driven Loads | 238 |
| Thirty-seven Ideas for the Application of Precision Snap-acting Switches | 241 |
| Electrical Terminal Connections | 242 |
| 12 Ways to Retain Electron Tubes | 244 |

# ELECTRICAL SYMBOLS AND STANDARDS

There is no general industry agreement on graphical symbols for electrical components. Different groups advance different proposals. And the fact that the symbol must be simple is sometimes forgotten. Perhaps the following material will provide a start for that end. It is based in the main on the Engineering Drafting Standards of the Ford Motor Company, but has been supplemented by data from the American Standards Association and from the National Machine Tool Builders Association.

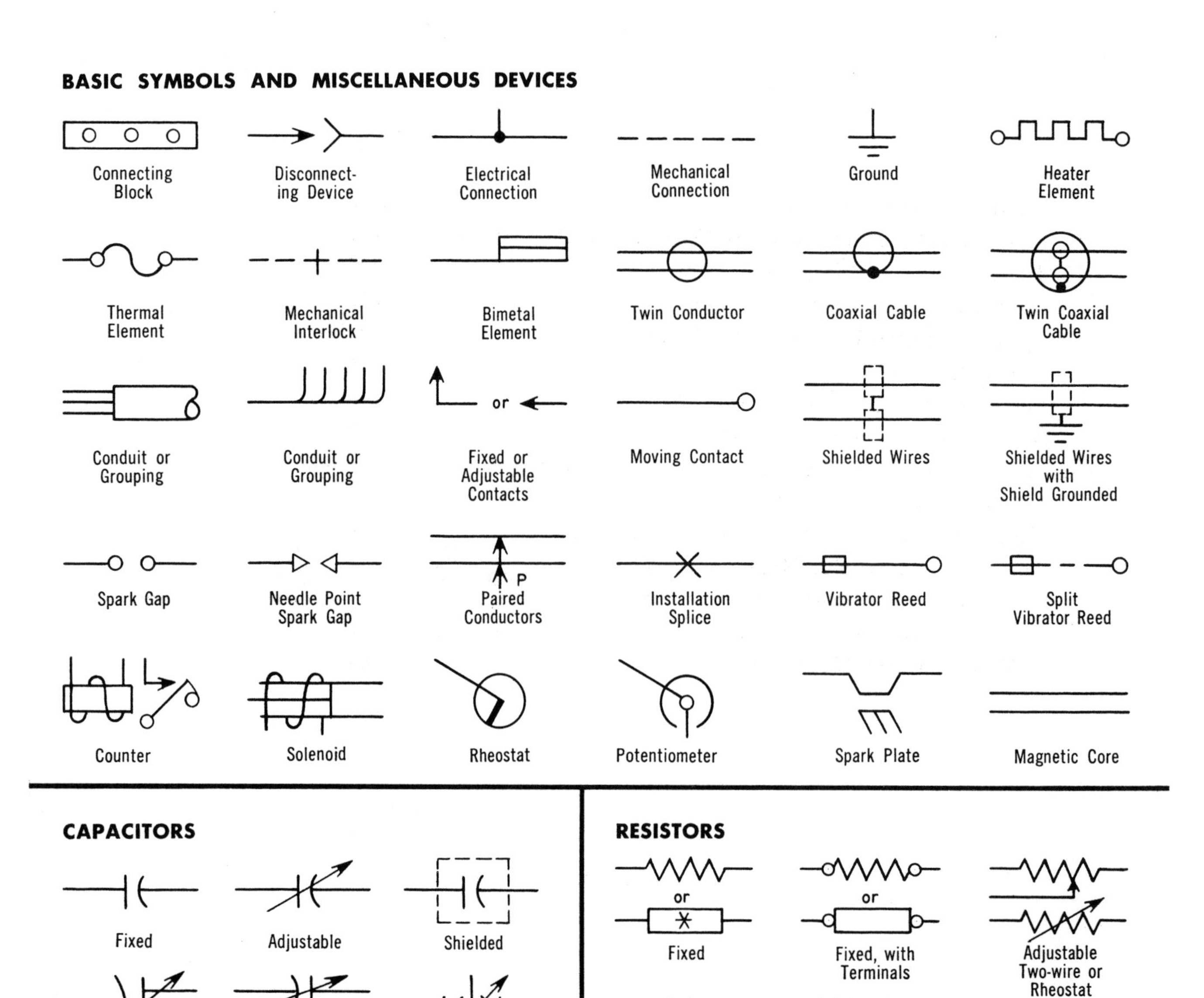

## BATTERIES

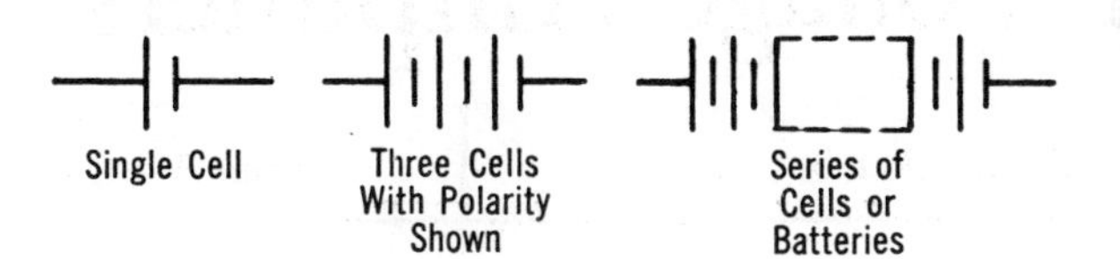

Single Cell — Three Cells With Polarity Shown — Series of Cells or Batteries

## COILS

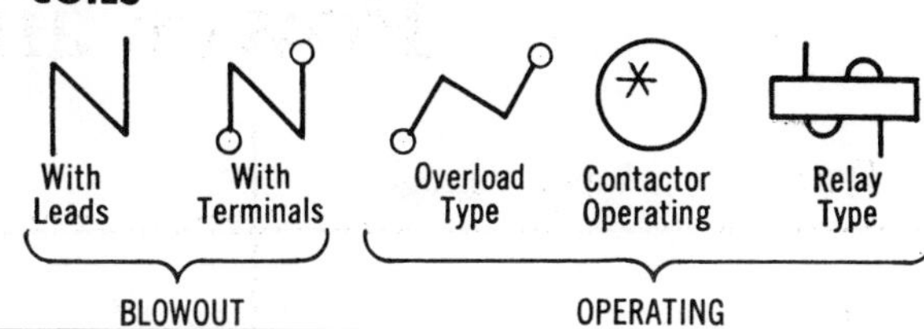

With Leads — With Terminals (BLOWOUT)

Overload Type — Contactor Operating — Relay Type (OPERATING)

## FUSES AND CIRCUIT BREAKERS

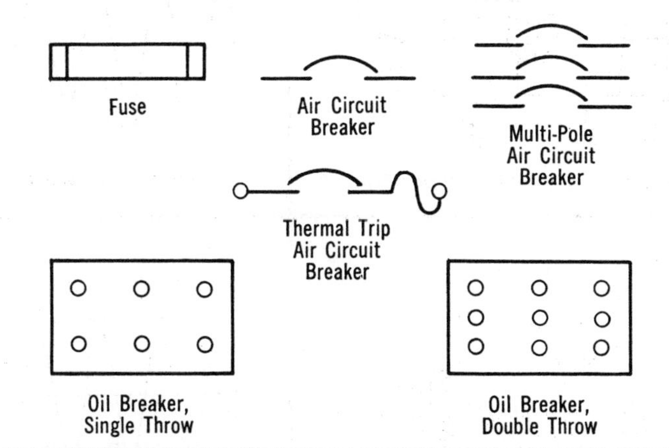

Fuse — Air Circuit Breaker — Multi-Pole Air Circuit Breaker — Thermal Trip Air Circuit Breaker — Oil Breaker, Single Throw — Oil Breaker, Double Throw

## CONTACTORS

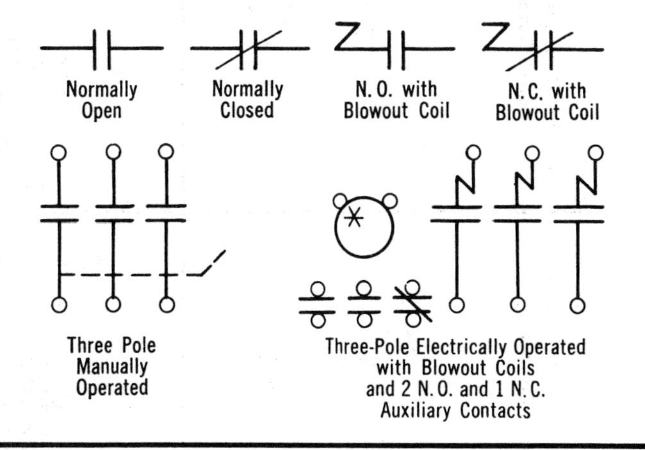

Normally Open — Normally Closed — N.O. with Blowout Coil — N.C. with Blowout Coil

Three Pole Manually Operated — Three-Pole Electrically Operated with Blowout Coils and 2 N.O. and 1 N.C. Auxiliary Contacts

## RELAYS

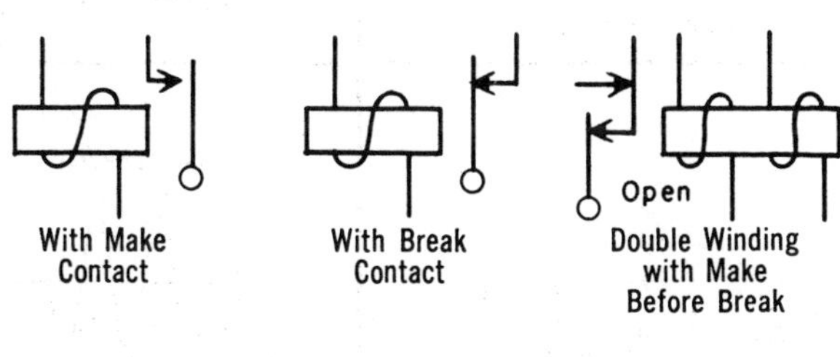



With Make Contact — With Break Contact — Double Winding with Make Before Break

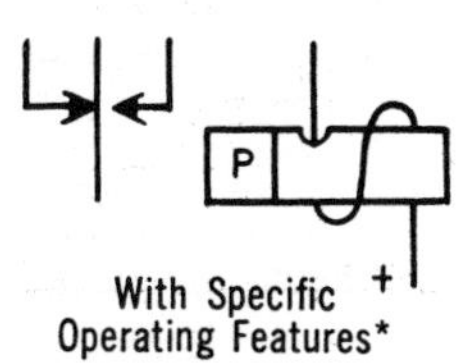



With Specific Operating Features*

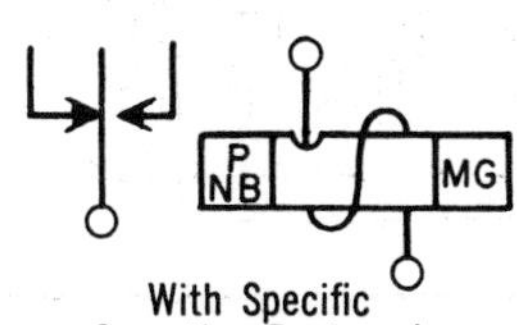



With Specific Operating Features*

*Signs designate proper poling. Current in the direction indicated will move or tend to move the armature toward the contact shown nearest the core. If the relay is equipped with numbered terminals, the proper terminal numbers should be shown.

AC — Alternating current or ringing
D — Differential
DB — Dashpot
EP — Electrically polarized
*FO — Fast operating
*FR — Fast release
MG — Marginal
NB — No bias
P — Magnetically polarized using biasing spring, or having magnet bias
SO — Slow operating
SR — Slow release
SW — Sandwich wound to improve balance to longitudinal currents

## JACKS

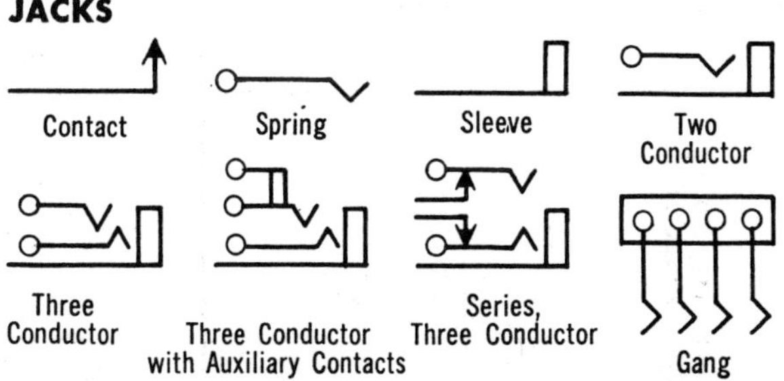

Contact — Spring — Sleeve — Two Conductor

Three Conductor — Three Conductor with Auxiliary Contacts — Series, Three Conductor — Gang

## SIGNALS

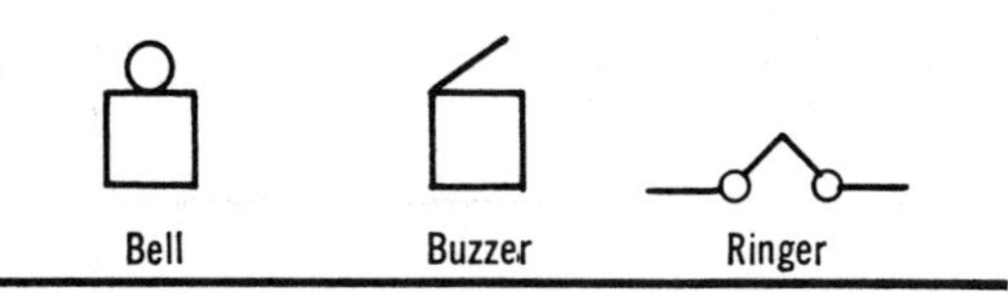

Bell — Buzzer — Ringer

## LAMPS

General — Indicating — Illuminating

Ballast, Resistance, or Heating

*Use letter or letters to indicate color.

A — Amber
B — Blue
C — Clear
G — Green
O — Orange
OP — Opalescent
P — Purple
R — Red
W — White
Y — Yellow
FL — Fluorescent

## THERMOSTATS

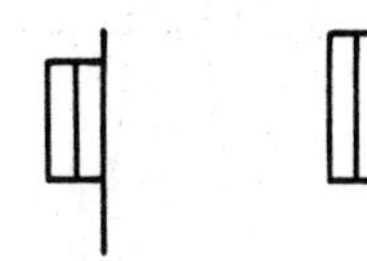

Self-Heated

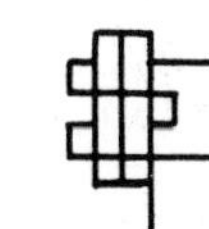

Externally Heated

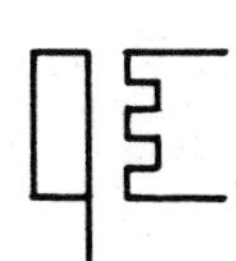

Integral Heater

External Heater

## RECTIFIERS

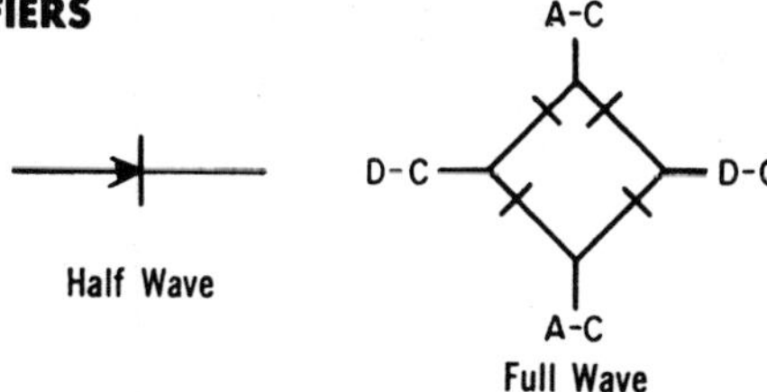



Half Wave — Full Wave

# Power and Voltage Amplifiers

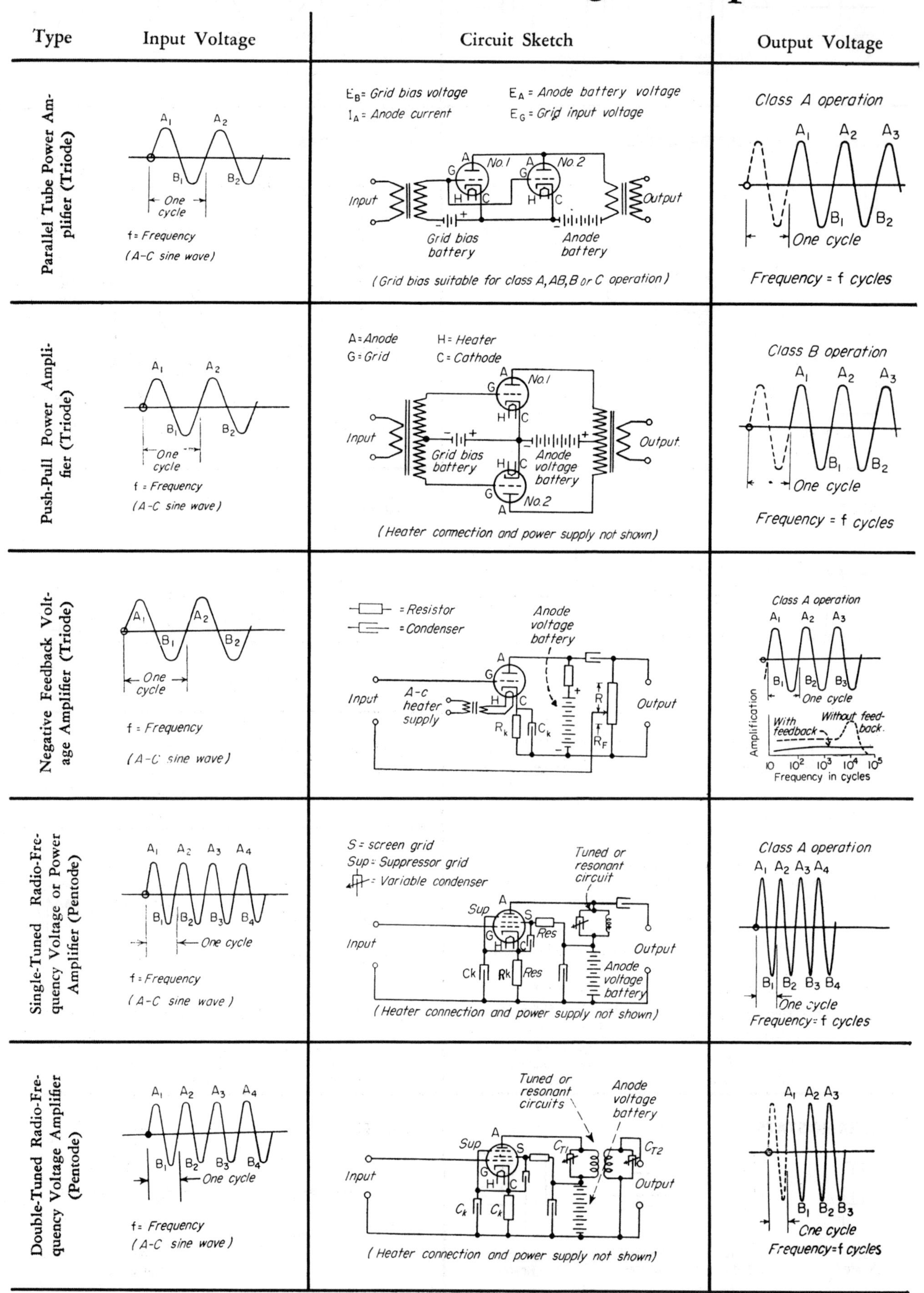

| Operation | Remarks |
|---|---|
| 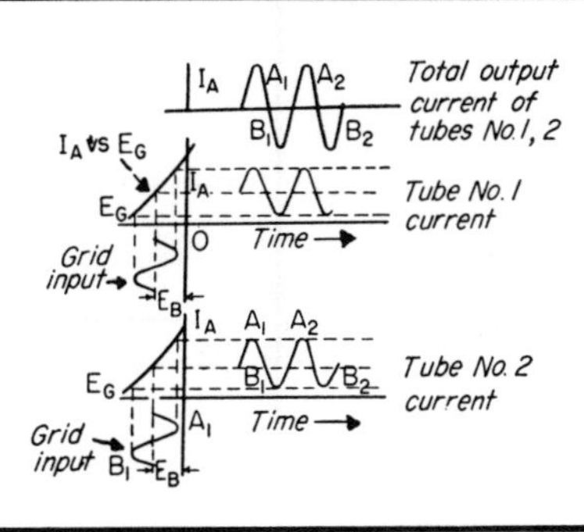  | Two or more tubes may be connected in parallel to provide a higher power output than is possible with one tube. The power output is proportional to the number of tubes employed in the circuit. The grid control potential required is the same as for one tube. If the tubes are operated for Class *C* amplification, or if power is consumed in the grid circuit, the grid power required is proportional to the number of tubes in the circuit. In this particular circuit, the d-c anode current saturates the transformer. The voltage phase shift can be reduced by reversing the polarity of the transformer output winding. |
| 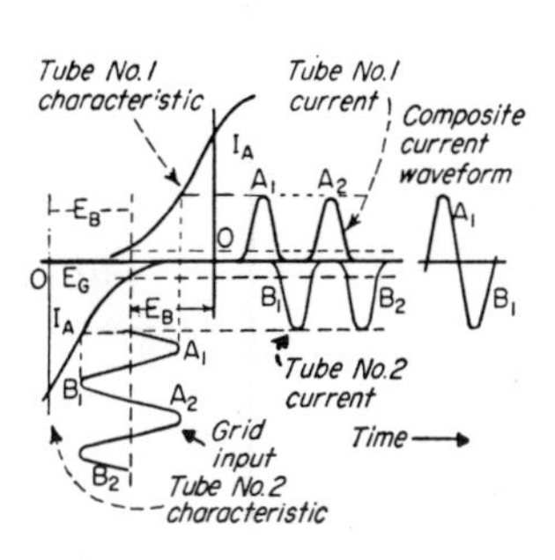  | The push-pull circuit for a power amplifier is superior to the parallel circuit shown above, because the wave form distortion is much less for the same power output. The term "push-pull" is used because the voltages and currents of the two tubes are out of phase. The current in one tube is increasing while the current in the other tube is decreasing. This characteristic can be illustrated very clearly in a graphical analysis for Class *AB* amplifier. This circuit is widely used because distortion is less (even harmonics cancel each other); there is no d-c transformer iron saturation (d-c currents cancel each other); and little or no filtering of the output is required. |
| 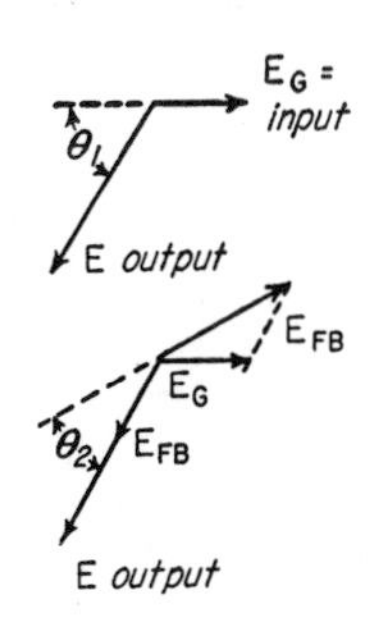  Victor diagram of amplifier with inductive load and without feedback. The angle $\theta$ is the phase distortion. With a feedback voltage $E_{fb}$ added to the grid voltage $E_g$, the phase distortion $\theta_2$ is reduced. The feedback factor is: $B = \frac{R_F}{R_F + R}$ | A negative feedback amplifier is one in which some of the amplified output energy is introduced into the input grid circuit in such a manner that the net grid input voltage is reduced. Although negative feedback reduces the gain or amplification, it is advantageous because it reduces the frequency, harmonic and phase distortion. It also provides greater stability than an amplifier without feedback. The performance characteristics are not as sensitive to changes in individual tubes or in the applied voltage. The amplification vs frequency characteristic is almost constant for a wide frequency range. Positive feedback is unstable and may cause the amplifier to oscillate. |
| 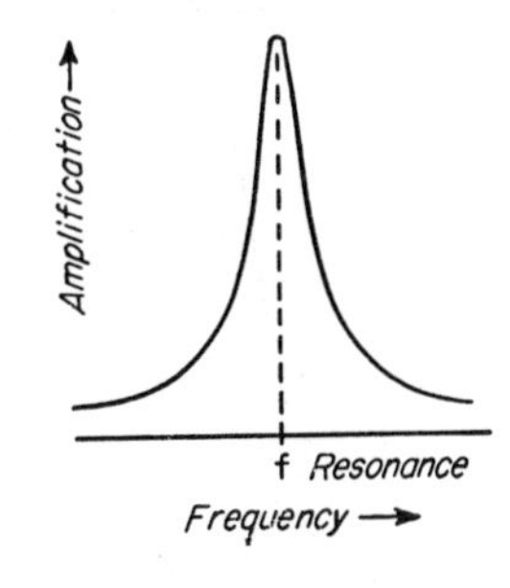  This type of amplifier operates similarly to those previously described except that it can be tuned to be more responsive to one particular frequency or frequency range. The value of inductance and capacitance determine the resonant frequency. | Five-element tubes are usually used in high frequency or radio-frequency amplifiers because the inter-electrode capacitances are considerably lower than those of a three-element tube. This particular type of amplifier circuit is seldom used as a voltage amplifier, since its characteristics are inferior to those of the double-tuned or transformer coupled types. Because of the inductance coil limitations, it is seldom possible to obtain more than 10 percent of the tube amplification factor. This circuit is frequently used for power amplification. |
| 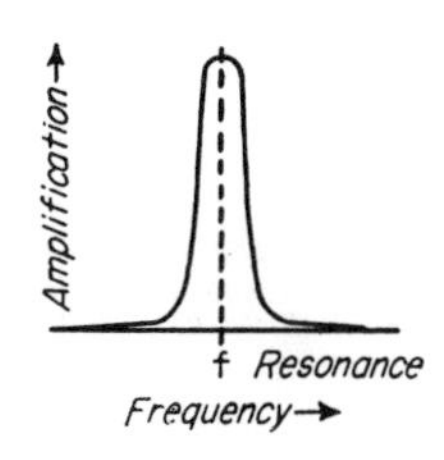  Both the primary and secondary of the transformer of this circuit are tuned. The frequency of resonance or maximum amplification can be varied by changing the capacitance value of condensers $C_{t1}$ and $C_{t2}$. | This type of circuit is widely used in broadcast receivers since it is responsive only to a certain frequency range. Also, as the cutoff characteristic is sharp and the maximum amplification range is broader and almost constant near the resonant frequency, it permits easy tuning or selection of the frequency to be amplified. A common application of this circuit is for intermediate frequency amplification in a superheterodyne radio receiver. |

| Type | Input Voltage | Circuit Sketch | Output Voltage |
|---|---|---|---|
| D-C Voltage Regulator | +, $E_{IN}$, 0 (Variable d-c) | A = *Anode* G = *Grid* C = *Cathode* P = *Potentiometer* H = *Heater*; + *Input (unregulated)* −; P; Grid bias battery $E_B$; A, G, H, C; Feedback resistor; + *Output (regulated)* −; *(Heater connection and power supply not shown)* | +, $E_{out}$, 0 *(Approximately steady d-c)* |
| Electronic Time Delay | +, $E_{IN}$, 0 (Steady d-c) | R = *Resistor* $C_G$ = *Grid condenser*; Control switch; − *Input* +; $R_1$, $R_2$, $R_3$; $C_G$; A, G, H, C; Relay coil; Output; Relay contacts | TIME DELAY PERIOD *(Period from instant that control switch is opened until relay operates)* |
| Generation of Rectangular Pulse | $A_1$, $A_2$, $B_1$, $B_2$; One cycle; f = frequency (a-c sine wave) | 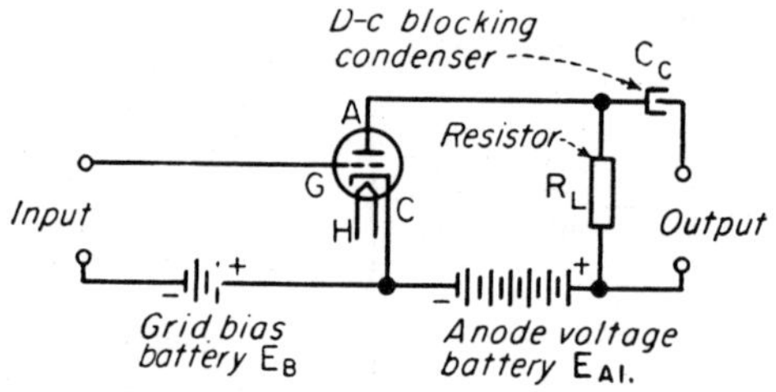 D-c blocking condenser $C_C$; Resistor $R_L$; A, G, H, C; Input; Output; Grid bias battery $E_B$; Anode voltage battery $E_{A1}$. (Grid bias for Class C operation) *(Heater connection and power supply not shown)* | $A_1$, $A_2$, $A_3$; One cycle *(Class C)*; f = frequency *(a-c periodic pulse)* |
| Frequency Multiplier | 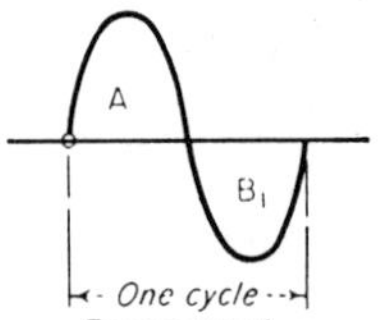 A, $B_1$; One cycle; Frequency = f (a-c sine wave) | 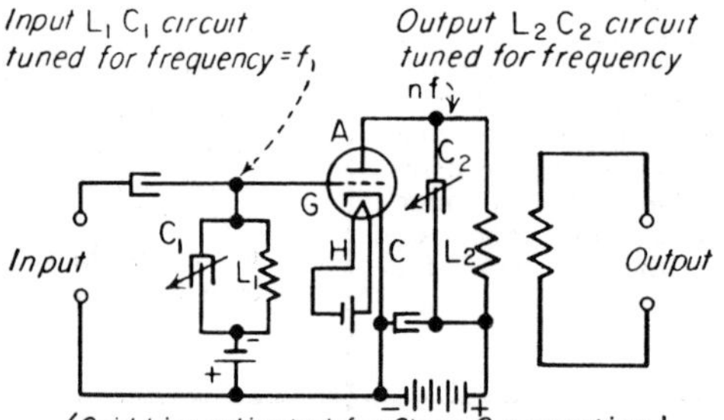 *Input $L_1$ $C_1$ circuit tuned for frequency = $f_1$*; *Output $L_2$ $C_2$ circuit tuned for frequency nf*; A, G, H, C; $C_1$, $L_1$, $C_2$, $L_2$; Input; Output; *(Grid bias adjusted for Class C operation)* | $A_1$, $B_1$; One cycle; nf = 2f = Frequency (a-c wave) |
| Production Of X-rays | +, $E_{IN}$, 0 (Steady d-c) or $A_1$, $A_2$, $B_1$, $B_2$ (a-c sine wave) | 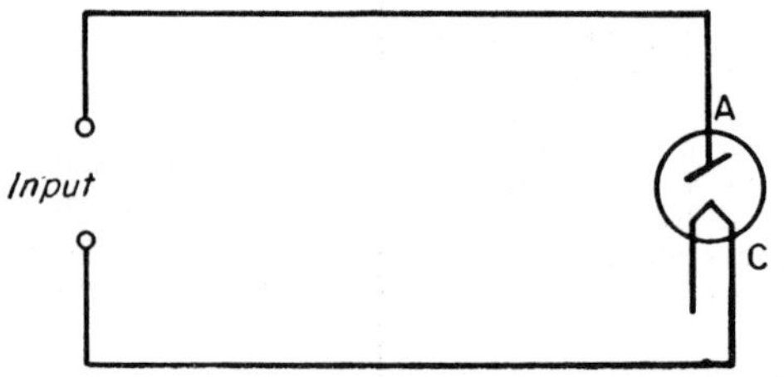 *Input*; A; C; *(Cathode connection and power supply not shown)* | X-RAYS *(Electromagnetic waves in a frequency range of approximately $10^{16}$ to $10^{20}$ cycles per second* |

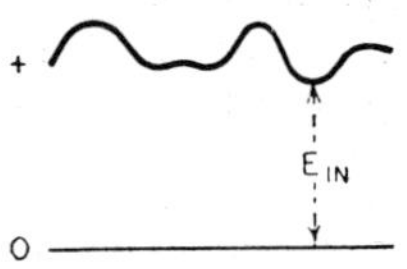

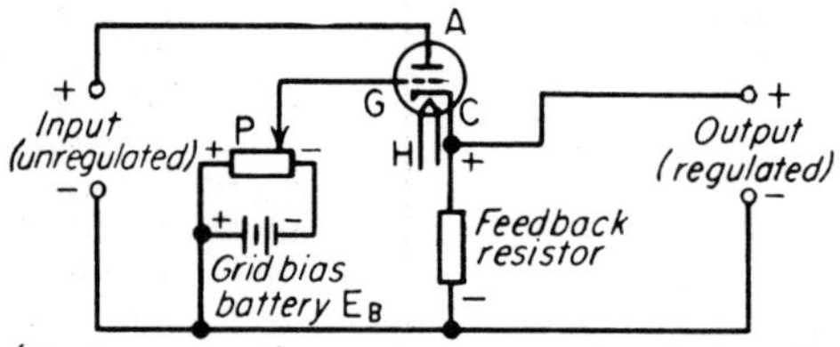

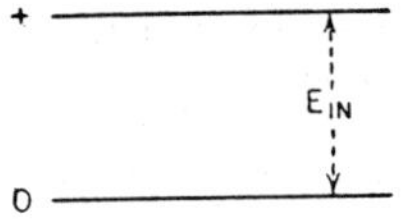

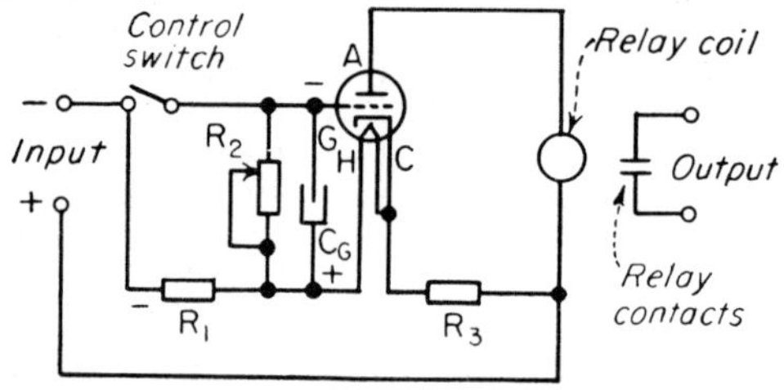

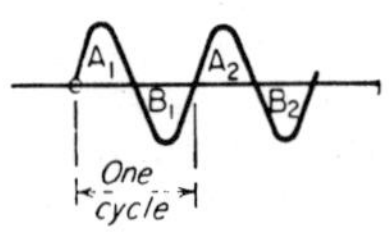

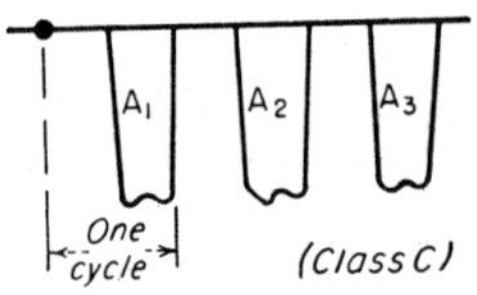

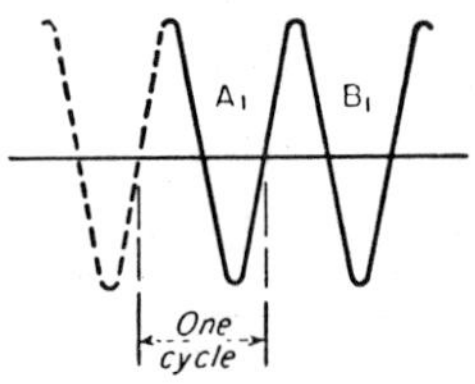

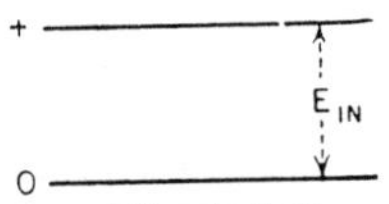

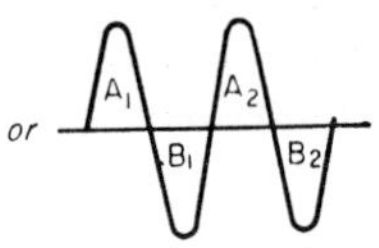

# Inspection and Control

RALPH B. IMMEL
Westinghouse Electric Corporation

| Operation | Remarks |
|---|---|

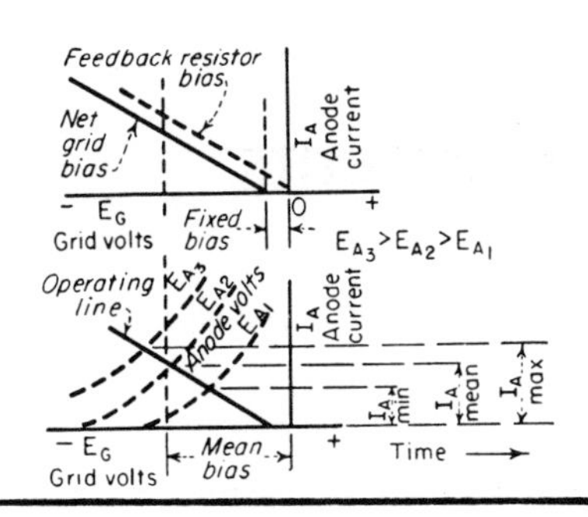


The battery and potentiometer circuit supplies a fixed value of bias to the grid and determines the mean value of the regulated anode current. If the unregulated voltage input exceeds the regulated output voltage value, the anode current increases. This increases the negative bias and holds the current constant. Conversely, a decrease in input voltage will decrease the bias so that the current will increase.

For a given input voltage variation the anode current $I_A$ and the voltage drop across the feedback resistor will be approximately constant. The regulation is determined by the resistor value and the amplification factor of the tube. The output voltage of this type may not vary more than one volt for a 20 v change in input voltage.

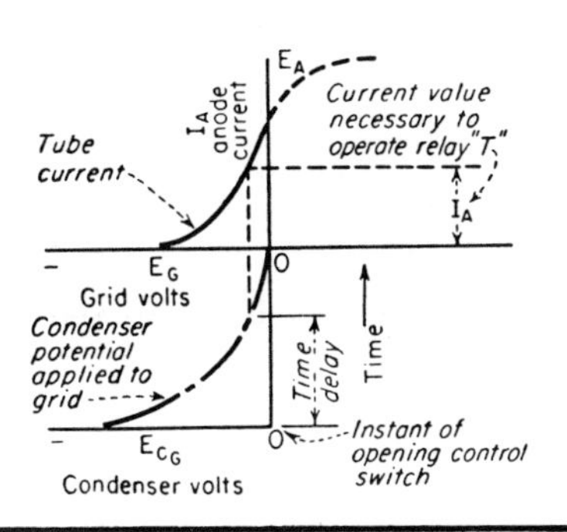


When the control switch is closed, the condenser $C_g$ charges to a potential that is the same as the voltage drop across resistor $R_1$. The time delay period is initiated by opening the control switch which permits the condenser to discharge through the potentiometer $R_2$. The negative potential on the condenser and grid decays until the anode current increases to a value sufficient to operate the relay.

The time period is the duration of time that elapses between the opening of the control switch and operation of relay. The time can be adjusted by varying $R_2$. The relay contacts can be either normally open or closed. It is seldom used for time periods exceeding one minute because of inaccuracies caused by condenser leakage. The reset time is almost instantaneous.

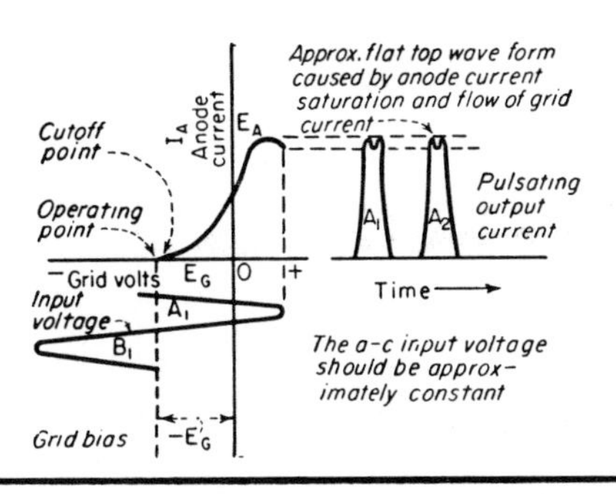


The voltage pulse may be used for triggering or initiating operation in control circuits of apparatus.

The circuit is similar to the pulse forming diode circuit except that the top of the wave form is flatter because of the flow of grid current.

Peak-clipping or the generation of a rectangular alternating current voltage wave form can be obtained by using two tubes. The square wave form may then be used to check the performance of an amplifier or transformer.

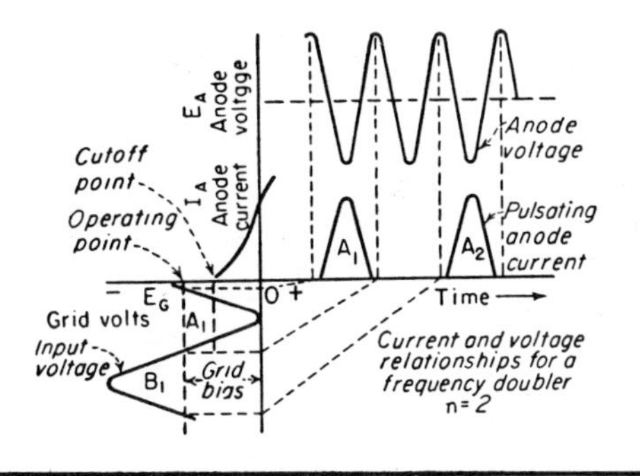


To obtain stable frequencies in excess of 10 megacycles, which is the practical limit for a crystal oscillator, a frequency-multiplier can be used. The input resonant circuit should be tuned for the input signal frequency and the anode resonant circuit should be tuned for an integral multiple of the input frequency. The anode circuit then offers a high impedance to the desired harmonic and a very low impedance to the fundamental and other harmonic frequencies. However, in the series resonant circuit the desired harmonic produces a voltage output that is predominant over that of any of the other frequencies. Frequencies of $n$ times the fundamental or input frequency can be obtained with approximately $1/n$ times the power output of a normally operated Class $C$ amplifier.

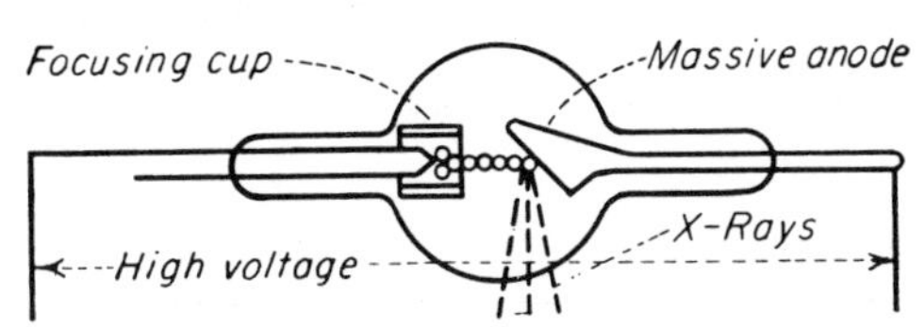


When electrons which are moving at a high velocity strike a metallic target (anode), the impact will result in a radiation of short electromagnetic waves which are known as x-rays.

$X$-ray wave lengths are so short that they will pass between the lattice-like structure of elements and compounds. However, when they strike an atom they are deflected. Denser substances cast shadows on photographic film or on a fluorescent screen.

$X$-rays provide a nondestructive means for medical examinations, inspection of industrial products, etc. The accelerating potential between the cathode and the anode is 300,000 v or more.

As the greater portion of the energy of the electron stream is converted into heat when the electrons impinge upon the anode, the anode must be massive and suitable for conducting away the heat.

# Designing Electric Heating Elements

JOSEPH GIALANELLA
Engineering Research Lab., Amersil Co., Inc.

When designing heaters for continuous operation, the curves shown in Figs. 1 and 2 can be used to determine the size and length of Nichrome V wire required for any given heat load.

In Fig. 1, the recommended B & S gage wire is given with a tolerance of $\pm 2$ sizes for 110 and 220 volt operation. Since the resistance of any conductor varies with temperature, cold or room temperature values have been used as the coordinate. This cold resistance is obtained by multiplying the resistance required at the operating temperature by a conversion factor from Table I. For materials other than Nichrome V, similar factors are available from the manufacturers.

When the heating element is formed into a helical coil, the curves in Fig. 2 can be used to find the required closed coil length. These curves are based on the following equation:

$$r = \frac{\pi D R'}{12d}$$

where

$r$ = ohms per in. of closed coil
$D$ = mean diameter of coil, in.
$R'$ = resistance per foot, ohms/ft.
$d$ = wire diameter, in.

As used above, a closed coil refers to a heating element that is compressed solid. In practice, a coil is stretched from $1\frac{1}{2}$ to 4 times its closed length to prevent shorting between adjacent turns and to supply sufficient space between turns for removing the heat.

EXAMPLE

A 2,500 watt immersion heater of Nichrome V wire is required for 110 volt operation. The largest outside coil diameter that will fit in the heater is $\frac{3}{4}$ inch.

From Fig. 1, the recommended B & S size is $11 \pm 2$. Using 13 B & S gage wire, the dashed line on Fig. 2 indicates that 16.0 in. of closed coil is required.

The curves in Fig. 2 can be used for materials other than Nichrome V, but the resistance or length of closed coil must be multiplied by the ratio of cold resistives. The resistivity of Nichrome V is 650 ohms per cir. mil-foot.

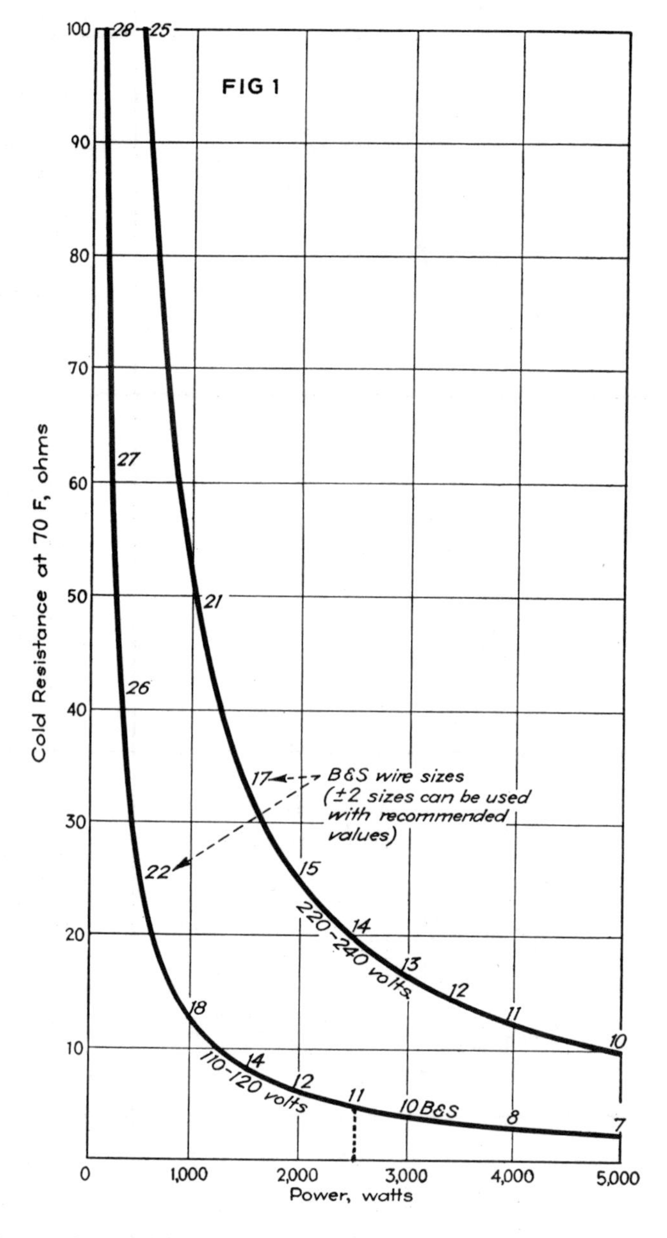


Table I—Conversion Factors For Nichrome V

| Operating Temp., F | 68 | 200 | 400 | 600 | 800 | 1000 | 1200 | 1400 | 1600 | 1800 | 2000 |
|---|---|---|---|---|---|---|---|---|---|---|---|
| Conversion Factors | 1.000 | 0.984 | 0.964 | 0.948 | 0.938 | 0.935 | 0.939 | 0.942 | 0.938 | 0.934 | 0.928 |

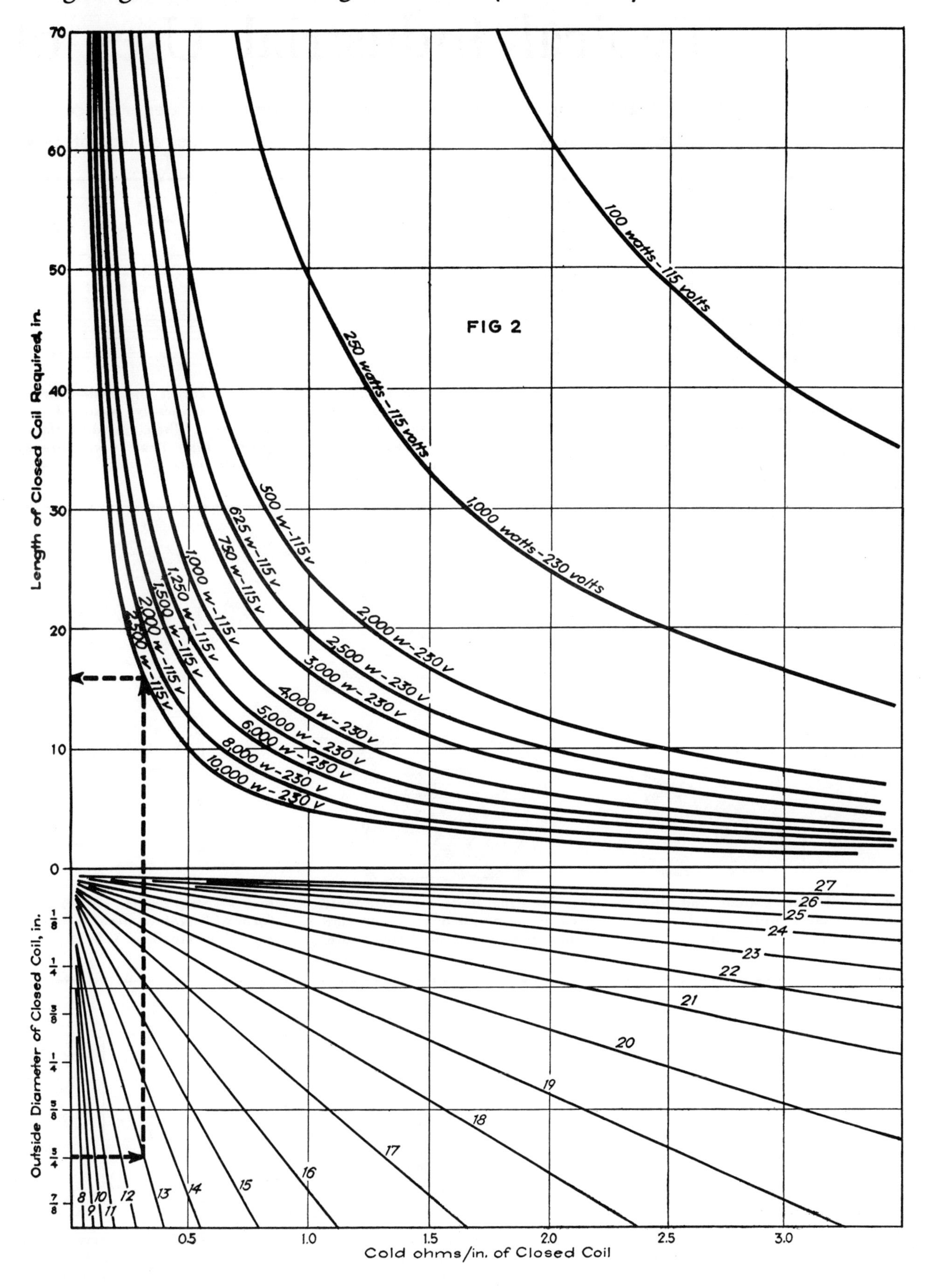
FIG 2
Length of Closed Coil Required, in.
Outside Diameter of Closed Coil, in.
Cold ohms/in. of Closed Coil
100 watts-115 volts
250 watts-115 volts
1,000 watts-230 volts
500 w-115 v
625 w-115 v
750 w-115 v
1,000 w-115 v
1,250 w-115 v
1,500 w-115 v
2,000 w-115 v
2,500 w-115 v
2,000 w-230 v
2,500 w-230 v
3,000 w-230 v
4,000 w-230 v
5,000 w-230 v
6,000 w-230 v
8,000 w-230 v
10,000 w-230 v
70
60
50
40
30
20
10
0
1/8
1/4
3/8
1/2
5/8
3/4
7/8
0.5
1.0
1.5
2.0
2.5
3.0
8
9
10
11
12
13
14
15
16
17
18
19
20
21
22
23
24
25
26
27

# Typical Industrial Uses of

Industrial heating applications can be divided into four categories: contact; immersion; radiant; and air. For these purposes a wide variety of types

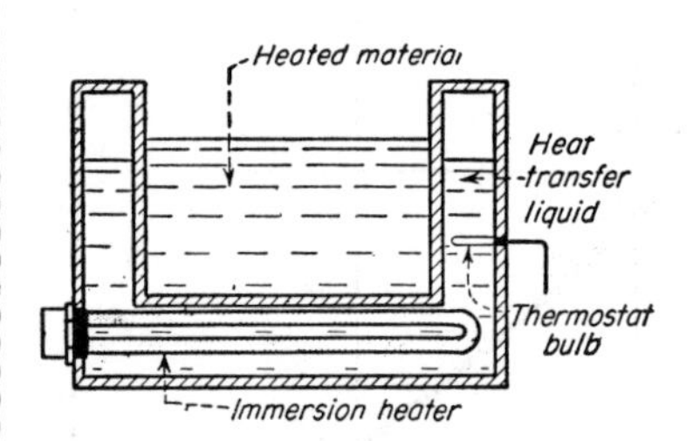


**Fig. 1—Indirect immersion heating with tubular type unit located in a transfer medium. Useful for adhesives or other materials that are easily damaged by overheating.**

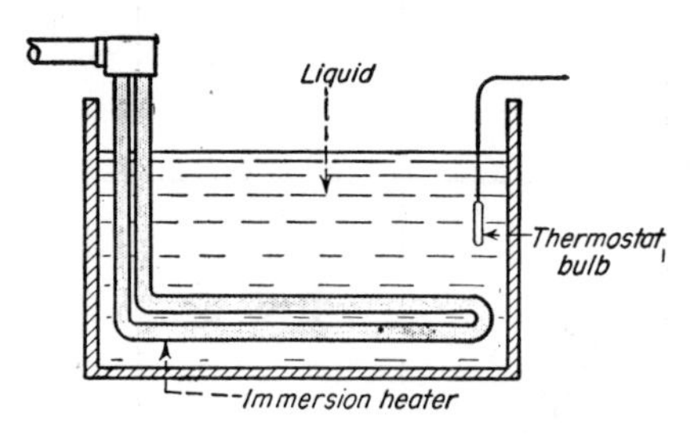


**Fig. 2—Direct immersion heating with portable tubular heaters. Can be used for molten salt, oil tempering baths, melting lead, solder, and stereo-type metal (not zinc).**

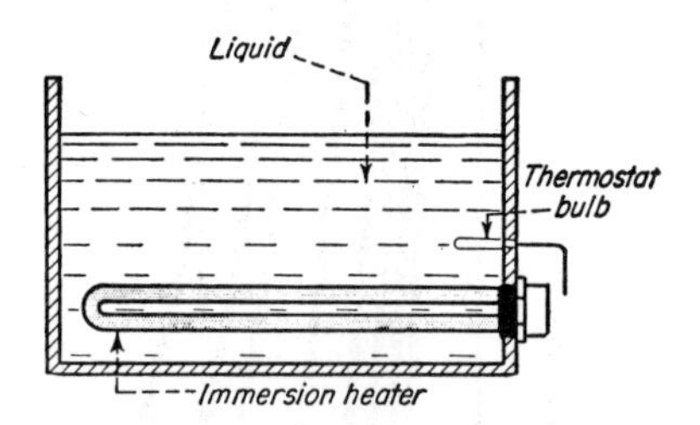


**Fig. 3—Direct immersion heating with permanently mounted tubular heater. When operating temperature is high, heaters with low watt densities must be used.**

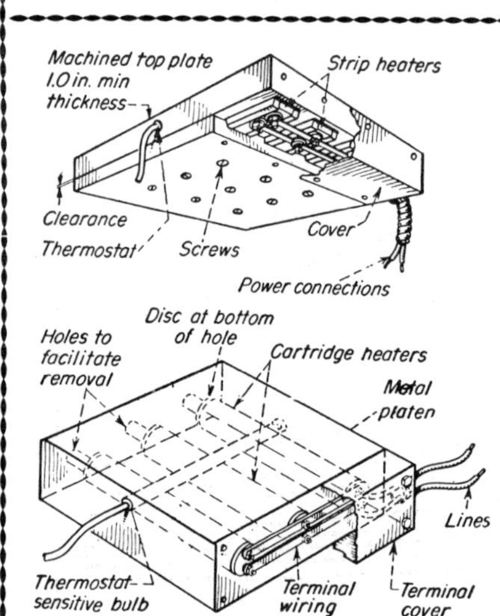


**Fig. 6—Moving platens and dies can be heated with strip heaters or with cartridge type units.**

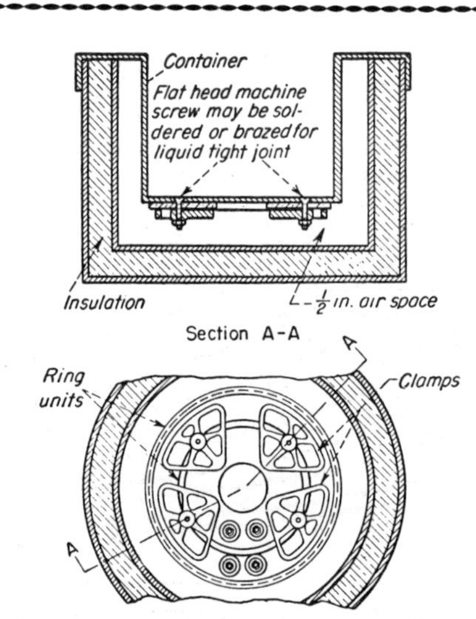


**Fig. 7—Ring units are commonly used in thin-walled tanks or containers as shown in the sketch above.**

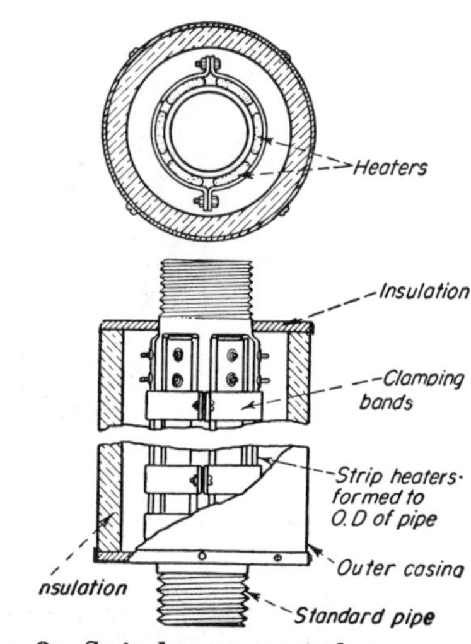


**Fig. 8—Strip heaters curved to conform with tank or pipe. Inside radii can be as small as 1 3/16 inch.**

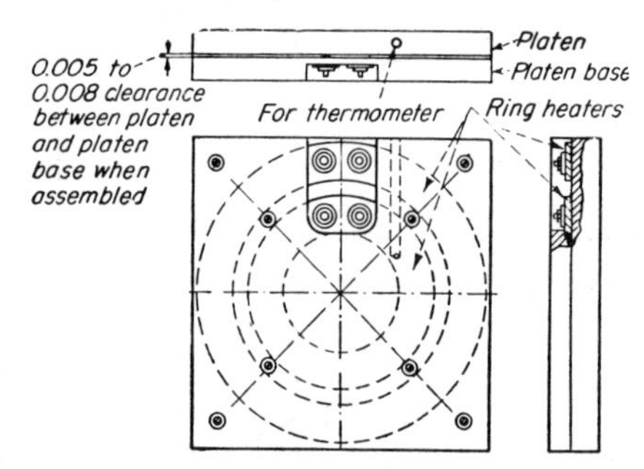


**Fig. 11—Ring type heaters can also be used as shown above in parts such as platens, dies and molds.**

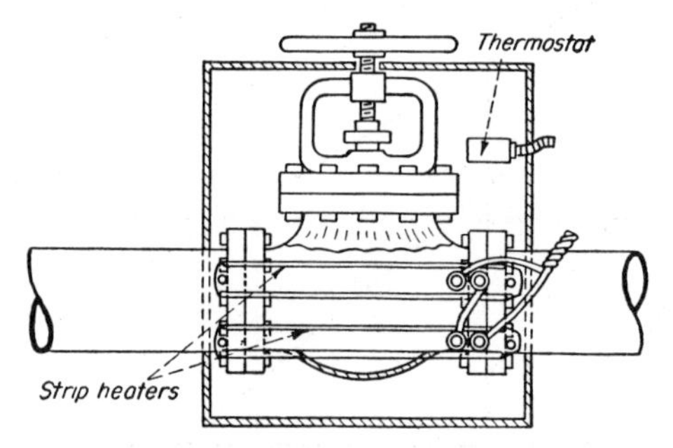


**Fig. 12—Strip heaters and thermostat are mounted in sheet metal casing to prevent valves freezing.**

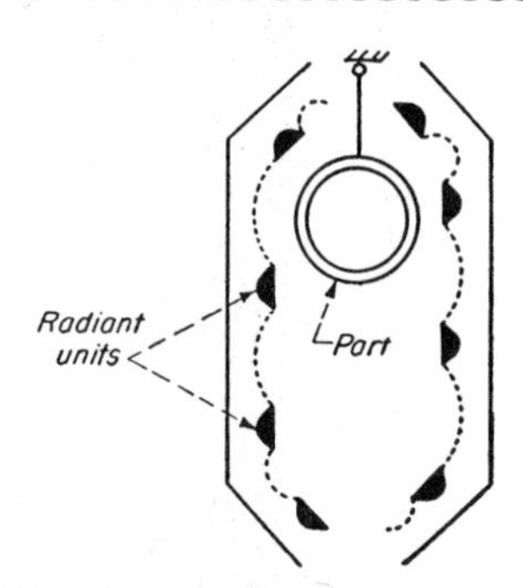


**Fig. 13—Radiant heaters for degreasing or for paint baking oven. Grease is removed by vaporizing.**

# Electric Heating Elements

and sizes of units have been developed. These sketches, which illustrate typical applications of each, were supplied by the Edwin L. Wiegand Company.

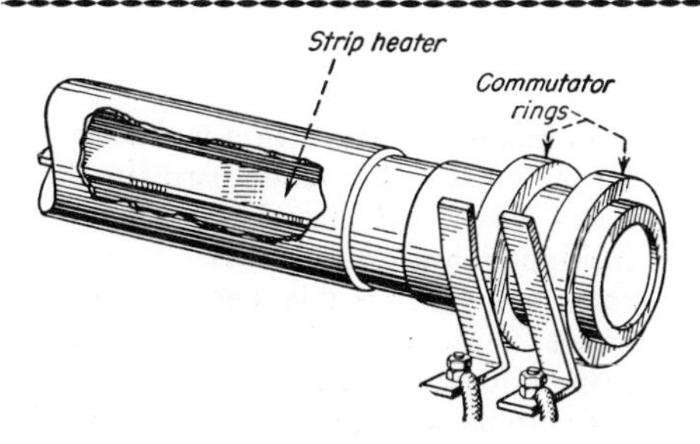


**Fig. 4—Strip heater used on machine parts in motion such as revolving rolls. One or more units can be placed within the roll, connected to commutator rings. Brushes wired to power supply contact the rings.**

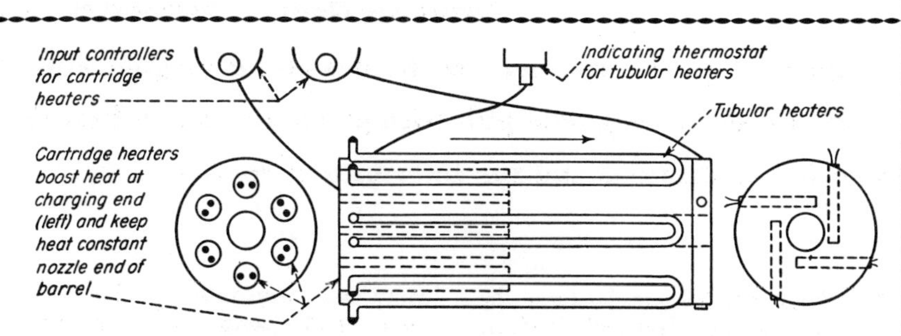


**Fig. 5—Six thermostatically controlled tubular heaters are used in conjunction with the cartridge units as shown to provide flexibly controlled zones of heat. Each set of cartridge units is separately controlled. Thus, heat and temperature can be regulated closely.**

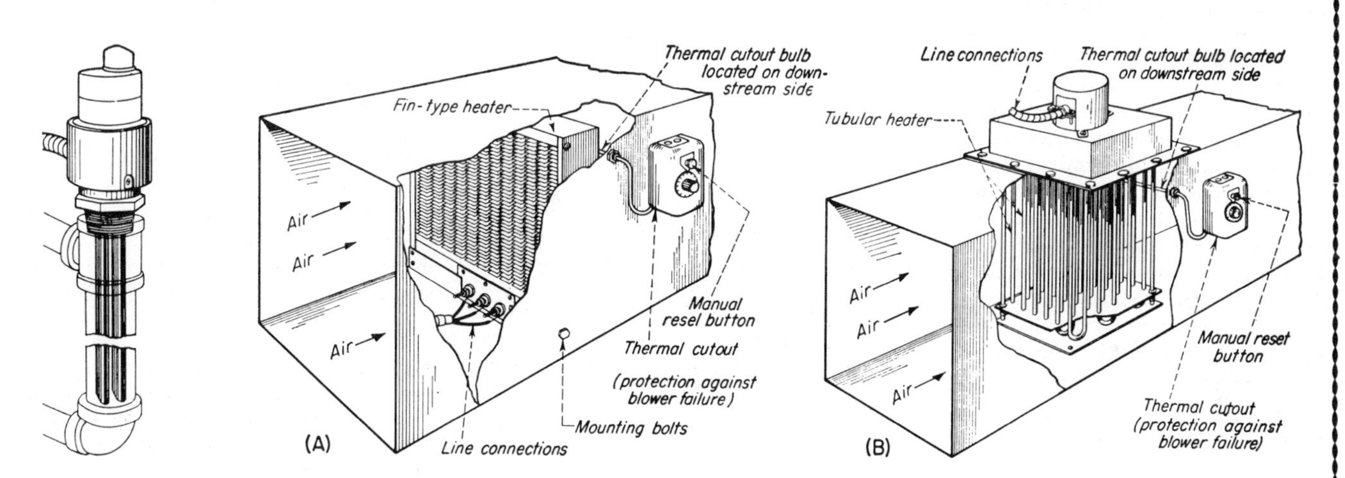


**Fig. 9 — Immersion heater screwed into standard tee fitting for heating liquids.**

**Fig. 10—Two types of heaters for forced air duct installations. Tubular type (*B*) is recommended for higher temperature uses than finned-units in (*A*). Operating temperatures can be over 1,000 F at air velocities of 6 ft per second.**

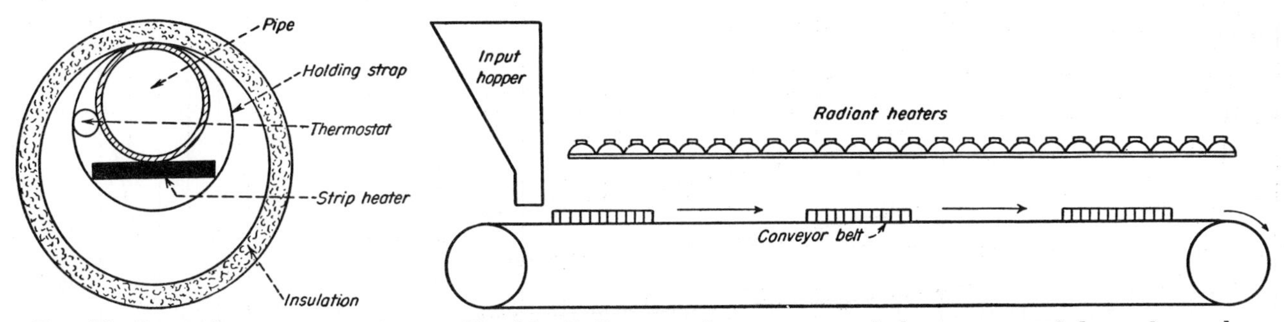


**Fig. 14—Long, low wattage strip heaters are strapped to pipe carrying viscous materials like tar.**

**Fig. 16—Radiant type heaters mounted above conveyor belt can be used for drying materials or parts. Wherever possible, radiant units should be staggered and reflectors placed opposite heaters to distribute heat evenly.**

# MAGNET COIL DESIGN

## *With Data for Estimating Essential Factors*

GRAHAM LEE MOSES
*Westinghouse Electric & Manufacturing Company*

**Discussion of problems involved in designing magnet coils to produce a desired magnetic pull with given air gap and imposed voltage for continuous and for intermittent service**

MAGNET COILS are in form the simplest of electrical devices. The design, manufacture and application of magnet coils, however, present complicated engineering and shop problems. From an engineering point of view, a magnet coil must produce sufficient magnetic pull to operate a particular device; dissipate the heat generated within itself without exceeding permissible temperature limits; and be insulated from the metallic parts of the apparatus. In addition, the winding must be insulated between turns and layers from terminal to terminal against both normal and maximum voltage surge peaks. Since each of these requirements presents a separate group of problems, each one will be discussed individually.

Magnet design problems usually involve the development of a coil to meet specific requirements of magnetic pull evaluated in terms of ampere-turns. The physical limits of the coil are generally established previously. The design of every coil starts with a preliminary estimate of the ampere-turns required to produce the desired force. It is preferable to obtain these data by testing the actual device and in any event this should be done before the final coil design is established as the factors involved are complicated. The accurate calculation of the required ampere-turns is difficult and is beyond the scope of this article.

### *Air Gap and Other Factors*

Since the air gap is ordinarily the determining factor, a set of curves is shown in Fig. 1 to help approximate the ampere-turns required for various pulls at different air gaps. The ampere-turns will, however, vary considerably with the shape and size of the magnetic circuit, and also with the shape of pole faces and type of iron used. The curves in Fig. 1 are intended to be used only for preliminary estimates of the ampere-turns.

Permissible temperature is determined by the class of insulation used. In Table I are listed and classified the kinds of insulation available for use on magnet coils with permissible temperatue rises based on 40 deg. C. as the maximum ambient temperature.

Continuous ratings are based on the maximum permissible temperature rises shown in Table I. Since most magnet coils, however, are not energized continuously some consideration must be given to the duty cycle. The simplest example is a coil which is energized only a part of the time on a constant voltage circuit where the equivalent voltage (square root of mean square), in so far as heating of the coil is concerned, is less than the intermittently applied voltage.

The curve in Fig. 2 shows the permissible increase in applied voltage beyond rated continuous voltage of the coil as determined by the percentage of the time energized. To explain further, if a coil is to be used on 250 volts and it is to be energized 60 per cent or less of the time,

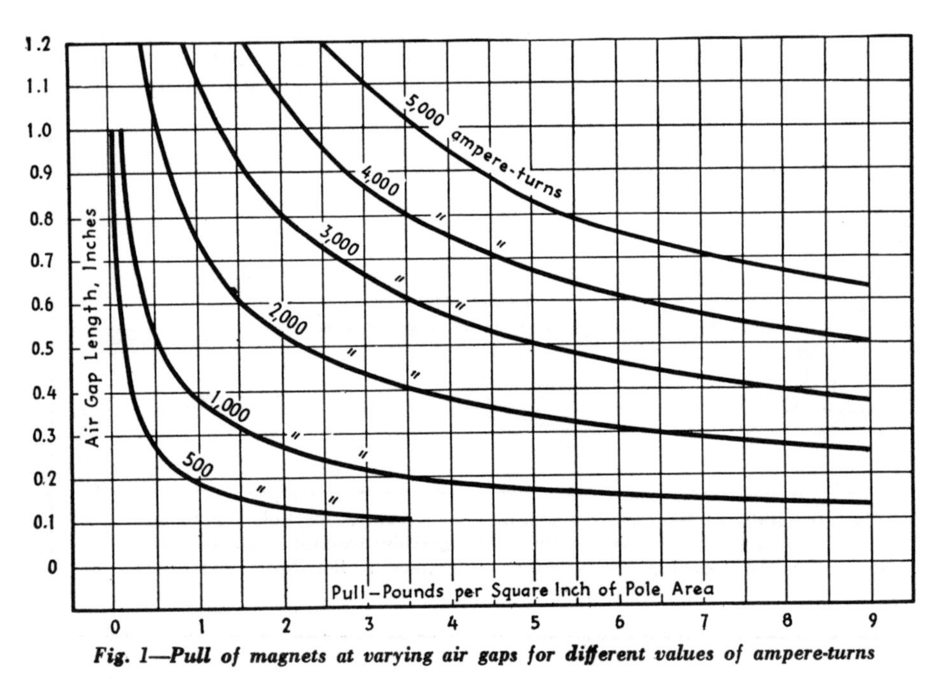


*Fig. 1—Pull of magnets at varying air gaps for different values of ampere-turns*

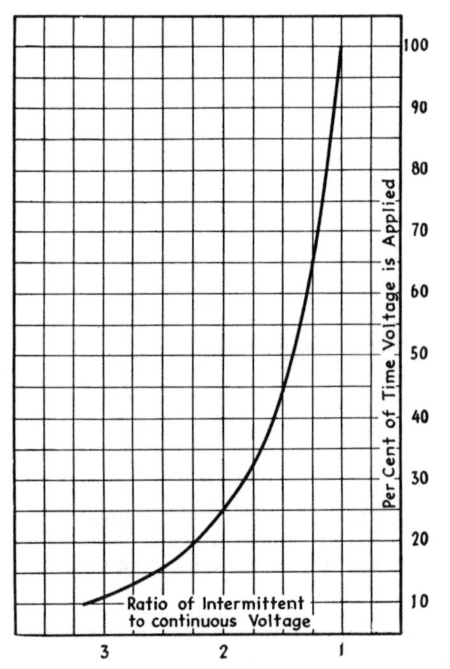


*Fig. 2—Permissible increase in applied voltage in terms of the continuous voltage rating of coil as determined by the percentage of the time energized on intermittent service*

**TABLE I—Limiting Temperature Rise in Deg. C. for Various Insulations**

| | *Class A Insulation* Organic, Impregnated (cotton, silk, paper, fullerboard, enamel) | | *Class B Insulation* Inorganic, Impregnated (mica, asbestos, glass textiles) | |
|---|---|---|---|---|
| | *Measured by Thermometer* | *Measured by Rise in Resistance* | *Measured by Thermometer* | *Measured by Rise in Resistance* |
| Wire wound coils.......... | 65 | 85 | 85 | 105 |
| Single layer series coils with exposed surfaces uninsulated or enamelled .... | 90 | — | 105 | — |

and the time that power is on does not exceed ½ hour periods, Fig. 2 shows that the coil can be operated at 1.3 times its normal voltage. For 250 volt service, therefore, a coil having a continuous rating of 192.5 volts will not overheat if energized 60 per cent of the time.

The continuous rating $E$ of a coil for any temperature rise can be obtained as follows:

$E$ = Continuous (square root of mean square) voltage

$R_c$ = Cold resistance at 25 deg.C.

$(t_2 - 25)$ = Temperature rise

$K$ = Constant for particular coil in deg. per watt.

This is dependent on coil surface, wall thickness and construction; and varies with size and shape. It must be determined for a particular coil size by test. It may be estimated from the expression

$K = 1/(AC)$

Where $A$ = Barrel surface of coil in sq.in.,

$C$ = Watts radiated per sq.in. of barrel surface per deg. C. rise as measured by the rise in resistance. Heat radiated varies from 0.01 for thick wall coils to 0.015 watts per sq.in. per deg. C. for thin wall coils.

$$E = \sqrt{\left[\frac{R_c\,(t_2 - 25)}{K}\right]\left[\frac{(t_2 - 25)}{259.5} + 1\right]}$$

Note: For a coil having the following approximate dimensions:

Length = 3¼ in., *O.D.* = 2¾ in., *I.D.* = 1⅛ in., Barrel Surface = 25.9 sq.in.

$K$ = 3 deg. per watt

$C$ = 0.01285 watts per sq.in. of barrel surface per deg. C.

There are in general two types of magnet coil construction: namely, coils which are wound in spools and the "self-sustained" coils which are wound on "universal" machines. Spool-wound coils may be wound with wire using any type of covering depending upon service conditions and the expense justified. This is the simplest type of coil to wind and the easiest to calculate. Provision must be made in the spool construction for a slot in one end-washer or a separate split washer to bring out the starting lead from the bottom layer to the suface.

"Self-sustained" or "universal" wound coils are usually wound with enamel-covered wire on a single tube in a close spiral so that the consecutive turns lie close together. At the same time, the machine winds one or more threads of cotton in an open spiral, so that the cotton lies diagonally along the barrel surface and overlaps the wire travel at each end of the coil. This builds up cotton walls at each end of the coil, laced together by diagonal threads. The cotton also provides channels through the body of the winding so that it can be thoroughly impregnated. On this type of coil the starting lead is usually insulated with varnished cloth and embedded in the cotton end wall so that split starting washers are not required.

Completely insulated coils should not exceed the diameter of the washers, generally they should be ¼ in. less in diameter than the end-washer to leave space for outer insulation. The determination of the various factors, with all dimensions expressed in inches should proceed as follows:

Uninsulated Coil O.D. = Washer O.D. − 1/4 in.

$$\text{Wall Thickness} = \frac{\text{Uninsulated Coil O.D.} - \text{Tube O.D.}}{2}$$

Mean Diameter = Wall Thickness + Tube O.D.

Mean Turn ($M$) = Mean Diameter × $\pi$

$$\text{Turns per layer } (T) = \frac{\text{Winding Space} \times \text{Factor}}{\text{Insulated Wire O.D.}}$$

**TABLE II—Data for Universal Wound Coils Using Enameled Wire and Cotton Yarn Layer Insulation**

| *AWG Size* | *Diam. in In.* | | *Per 1,000 Ft.* | | *Winding Factor* | *Conductors Per* | | *Per Cu. In.* | |
|---|---|---|---|---|---|---|---|---|---|
| | *Bare* | *Ins.* | *Res. Ohms* | *Wt. Lbs.* | | *Lin. In.* | *Sq. In.* | *Length Ft.* | *Resist.* |
| 20 | 0.032 | 0.0341 | 10.35 | 3.092 | 0.92 | 27 | 729 | 60.7 | 0.629 |
| 21 | 0.0285 | 0.0306 | 13.05 | 2.452 | 0.92 | 30 | 900 | 75.0 | 0.98 |
| 22 | 0.0254 | 0.0275 | 16.46 | 1.945 | 0.92 | 33 | 1,090 | 90.6 | 1.49 |
| 23 | 0.0226 | 0.0244 | 20.76 | 1.542 | 0.92 | 37 | 1,370 | 114.0 | 2.37 |
| 24 | 0.0201 | 0.0219 | 26.17 | 1.223 | 0.92 | 42 | 1,765 | 147.0 | 3.84 |
| 25 | 0.0179 | 0.0197 | 33.00 | 0.9699 | 0.91 | 46 | 2,120 | 177.0 | 5.84 |
| 26 | 0.0159 | 0.0175 | 41.62 | 0.7692 | 0.91 | 52 | 2,710 | 225.0 | 9.35 |
| 27 | 0.0142 | 0.0157 | 52.48 | 0.6100 | 0.91 | 58 | 3,370 | 281.0 | 14.73 |
| 28 | 0.0126 | 0.0141 | 66.17 | 0.4837 | 0.91 | 64 | 4,100 | 342.0 | 22.65 |
| 29 | 0.0113 | 0.0126 | 83.44 | 0.3836 | 0.91 | 72 | 5,190 | 432.0 | 36.25 |
| 30 | 0.0100 | 0.0113 | 105.2 | 0.3042 | 0.90 | 79 | 6,300 | 525.0 | 55.2 |
| 31 | 0.0089 | 0.0100 | 132.7 | 0.2413 | 0.89 | 89 | 7,930 | 660.0 | 87.5 |
| 32 | 0.008 | 0.0091 | 167.0 | 0.1913 | 0.88 | 96 | 9,210 | 768.0 | 128.0 |
| 33 | 0.0071 | 0.008 | 211.0 | 0.1517 | 0.88 | 110 | 12,100 | 1,008.0 | 213.0 |
| 34 | 0.0063 | 0.0072 | 266.0 | 0.1203 | 0.87 | 121 | 14,600 | 1,217.0 | 324.0 |
| 35 | 0.0056 | 0.0064 | 335.0 | 0.0954 | 0.86 | 134 | 17,950 | 1,495.0 | 501.0 |
| 36 | 0.0050 | 0.0058 | 423.0 | 0.0757 | 0.85 | 146 | 21,300 | 1,775.0 | 750.0 |
| 37 | 0.0045 | 0.0054 | 533.0 | 0.0600 | 0.85 | 163 | 26,550 | 2,210.0 | 1,177.0 |
| 38 | 0.0040 | 0.0047 | 672.0 | 0.0476 | 0.85 | 181 | 32,700 | 2,730.0 | 1,833.0 |
| 39 | 0.0035 | 0.0041 | 848.0 | 0.0377 | 0.85 | 207 | 42,800 | 3,570.0 | 3,010.0 |
| 40 | 0.0031 | 0.0037 | 1,069.0 | 0.0299 | 0.85 | 230 | 52,900 | 4,410.0 | 4,700.0 |

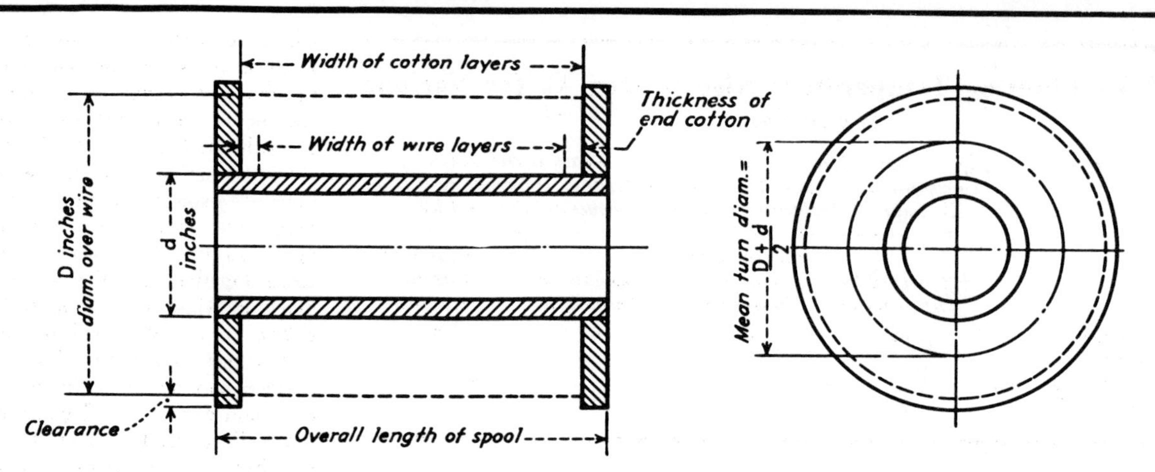


Size of wire number ________ Wire diam. inches ________

Weight of wire lb. per M. ft. ($w$) = ______ Winding factor ____

Turns per layer ($T$) $= \dfrac{\text{Winding space} \times \text{factor}}{\text{Insulated Wire O.D.}} =$

Number of layers ($L$) $= \dfrac{(D - d) \times \text{winding factor}}{2 \times \text{Insulated wire O.D.}} =$

Mean turn length in inches ($M$) $= \dfrac{(D - d) \times \pi}{2} =$

Total turns ($N$) $= T \times L =$

Length of wire ($B$) in M. ft. $= \dfrac{M \times N}{12 \times 1{,}000} =$

Total weight of copper in lb. $= w \times B =$

Resistance in ohms of wire per M. ft. at 25 deg. C. ($R_1$) =

Resistance in ohms of wire per M. ft. at $t_2$ deg. C. ($R_2$)

$$= R_1 \left[ 1 + 0.00386\,(t_2 - 25) \right] =$$

Resistance in ohms of mean turn length ($R_m$) $= \dfrac{M}{12} \times \dfrac{R_2}{1{,}000} =$

Resistance of coil at $t_2$ deg. C. ($R$) $= R_2 \times B =$

Watts radiated at $E$ volts at $t_2$ deg. C. $= \left( \dfrac{E^2}{R} \right) =$

Watts radiated per sq. in. of barrel surface at $E$ volts and $t_2$ deg. C.

$$= \left[ \left( \frac{E^2}{R} \right) \Big/ (\pi D \times \text{width of winding space}) \right] =$$

Ampere-turns $IN = \dfrac{E}{R_m} =$

*Fig. 3—Typical form of calculation sheet that can be used for estimating factors involved in magnet coil design*

$$\text{Layers } (L) = \frac{\text{Wall Thickness} \times \text{Factor}}{\text{Insulated Wire O.D.}}$$

Total Turns ($N$) = Turns per layer × Layers

$$\text{Coil Resistance} = N \times \frac{M}{12} \times \frac{R_2}{1{,}000}$$

Where $R_2$ = Resistance of wire in ohms per 1,000 ft. at temperature $t_2$ deg. C.

On spool-wound coils, where the wire is carefully wound in layers, a winding factor of 95 per cent should be used in determining the turns per layer, and 100 per cent in determining the number of layers. If the coils are wound promiscuously without regard to even layers, a winding factor of 95 per cent should be used for both turns and layers.

For self-sustained coils wound in universal machines, the calculation of the winding space and use of the winding factor is more complicated. In Fig. 3 is shown an excellent form of calculation sheet for estimating the various factors involved. This form may also be used for simpler types of coils by omitting some of the factors. The winding factor varies for different wire sizes, because of the cotton yarn occupying a differing percentage of the winding space. A list of winding factors of all sizes of wire from No. 20 to No. 40 AWG is given in Table II along with other useful data. These data are shown graphically in Fig. 4. From the conductors per sq.in. the number of turns for any given wall area can be quickly calculated. Also the resistance for any size of spool can be estimated from the active cubic inches (wall area multiplied by mean turn). The product of active cubic inches of winding and ohms per cubic inch for a given wire size is a close approximation of the resistance.

The ground insulation is usually supplied by the spool or insulating tube on which the coil is wound. For operating voltages below 500 it can usually be assumed that an insulating spool which is mechanically satisfactory will have suitable electrical characteristics. At 500 volts and above additional precautions must be taken to secure additional resistance to creepage by sealing any cracks between end-washers and tube. Spools made from fiber, micarta and various forms of molded plastic phenolics are generally found satisfactory. In the or-

## TABLE III—Analysis of Magnet Coil Design Data

*Specifications of Selected Coil*

| | | |
|---|---|---|
| Barrel Surface | = | 25.9 sq. in. |
| Active Winding | = | 12.75 cu. in. |
| Temperature Constant ($K$) | = | 3 deg. C. per watt |
| Total Turns | = | 30,500 |
| Resistance | = | 4,280 ohms at 25 deg. C. |

| *Watts* Total | *Watts* Per Cu. In. | Temp. Rise Deg. C | Hot Res. Ohms | Volts | Amp. | Amp. Turns | Amp. Turns Per Watt | Amp. Turns Per Cu. In. |
|---|---|---|---|---|---|---|---|---|
| 5 | 0.39 | 15 | 4,550 | 151 | 0.033 | 1,010 | 225 | 79 |
| 8 | 0.627 | 24 | 4,670 | 194 | 0.0414 | 1,260 | 157 | 99 |
| 10 | 0.78 | 30 | 4,770 | 218 | 0.046 | 1,400 | 140 | 110 |
| 15 | 1.18 | 45 | 5,010 | 275 | 0.0547 | 1,670 | 111 | 130 |
| 20 | 1.57 | 60 | 5,250 | 325 | 0.062 | 1,890 | 95 | 148 |
| 30 | 2.35 | 89 | 5,740 | 415 | 0.0725 | 2,210 | 74 | 174 |
| 40 | 3.14 | 121 | 6,260 | 500 | 0.08 | 2,440 | 61 | 192 |

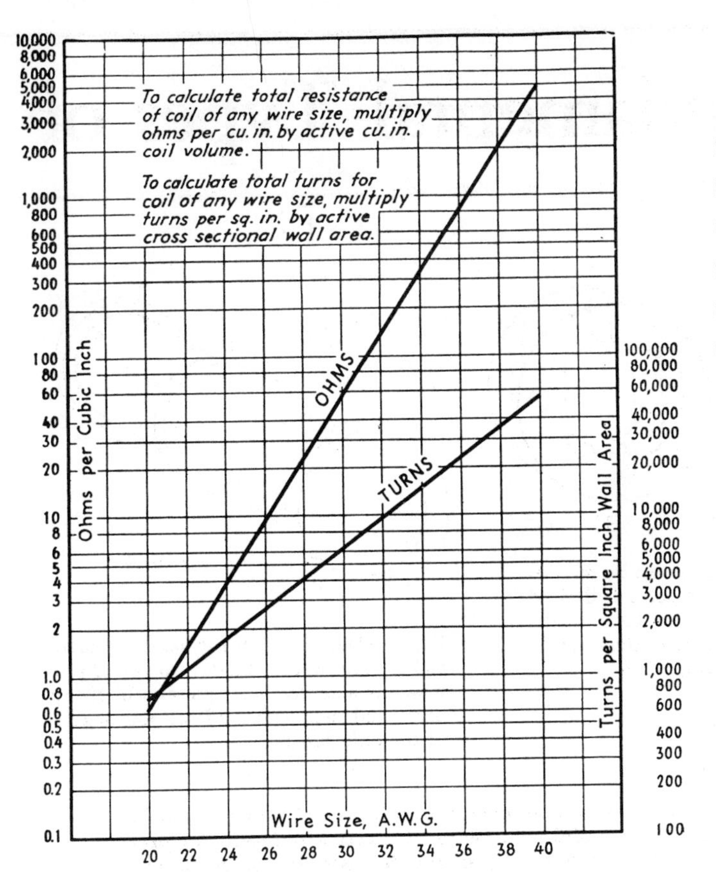


*Fig. 4—Resistance and turns constants for wire sizes from number 20 to number 40 A.W.G. on universal wound coils*

*Fig. 5—Magnet coil data presented graphically that shows the effect of coil volts on the temperature of surface and the internal temperature rise*

dinary sizes of magnet coil, tubes with 1/16 in. thick walls and 1/8 in. thick end washers are conventional.

Layer insulation as well as turn insulation is usually furnished by wire covering and, therefore, the kind of wire covering should be selected with care. For "universal wound" coils having interlaced cotton yarn, enameled wire is generally satisfactory. Spool wound coils may use enameled wire if the wall thickness is not so great as to build up enough pressure to damage the enamel, in which event paper and enamel covered wire is satisfactory. Where the operating temperature is in excess of that permitted for Class A materials, Class B insulated wire must be used. Asbestos covered wire has long been used for such applications and lately "Glasweve" insulated wire is coming into prominence. The latter has a better space factor than asbestos covered wire and has proven highly satisfactory.

An outer wrapper should be applied to most magnet coils to insulate the start and finish leads from each other and the coil surface. For coils rated up to 200 volts a single layer of 0.010 in. treated cloth is generally used. Above 200 volts 0.010 in. fishpaper and mica is preferable. For some coils, two or more layers of wrapper are required where line voltage surges may be 1,500 to 2,500 volts. A half lapped layer of cotton finishing tape should be used on Class A magnet coils. On Class B coils asbestos or "Glasweve" tape should be used.

Magnet coils for a.c. present problems that are in addition to those involved in d.c. coils. On a.c. magnets the current is not limited by resistance but by impedance which is a combination of resistance and inductive reactance. This latter factor changes as the magnet armature moves so that the current taken at a given voltage is much less with the air gap closed than the air gap open. The designer is primarily interested in the open gap current as it affects pick-up ampere-turns and in the closed gap as it affects heating.

### *Determining Impedance*

On most a.c. magnets the reactance is so much greater than the resistance that the resistance can be neglected as a factor of impedance. To determine the impedance $Z$ an existing coil should be tested on the magnetic circuit at open and closed gap. From this data the impedance can be calculated thus:

$Z$ = Impedance in ohms
$E$ = Volts applied to coil
$A$ = Amperes through coil
then $Z = E/A$

From this the magnetic constant $Q$ should be calculated for both conditions of gap as follows:

$Q$ = Magnetic constant of the device
$Z$ = Ohms impedance
$N$ = Turns
$F$ = Frequency in cycles per second
then $Q = Z/(N^2F)$

Once these magnetic constants have been established the impedance of a coil of any number of turns can be calculated from the same formula arranged in this form:

$$Z = QN^2F$$

Once a design has been worked out it will be interesting and instructive to analyze the component factors so as to obtain a clearer picture of the effect of the important factors; to illustrate this a coil has been selected having the characteristics given in Table III. The detail analysis of this coil is also shown in Table III. The same data are shown graphically in Fig. 5 which illustrate the effect of coil volts on surface and internal temperature rise, as well as total ampere-turns and ampere-turns per watt.

An important procedure in designing a magnet coil is to analyze its characteristics by drawing up its basic curves. Study of the characteristics of the finished design will disclose new angles which will help in producing better subsequent designs. Also, if data derived by studies of this nature are filed time can be saved later when calculating similar factors.

# Fundamental Types of

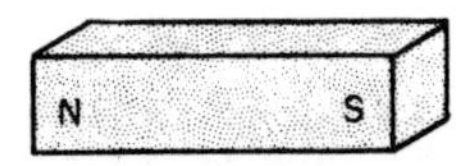


Fig. 1—Simplest form of a permanent magnet is a bar exhibiting two poles which may be of any cross section.

Fig. 2

Fig. 2—U or C-shaped permanent magnets consist of a bar "bent" to bring both of the pole faces into the same plane.

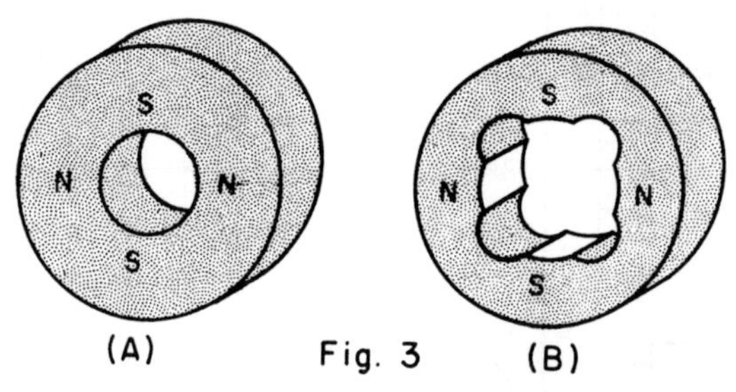


(A) Fig. 3 (B)

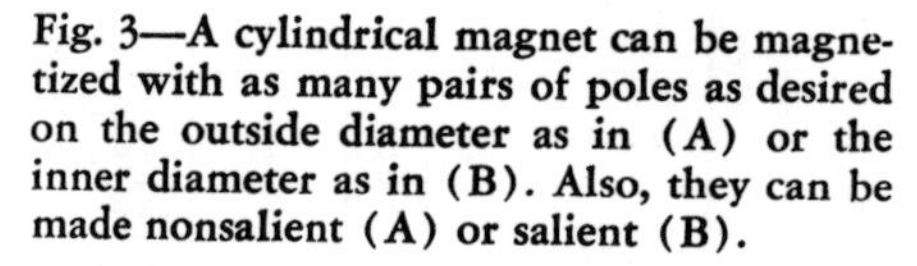

Fig. 3—A cylindrical magnet can be magnetized with as many pairs of poles as desired on the outside diameter as in (A) or the inner diameter as in (B). Also, they can be made nonsalient (A) or salient (B).

Fig. 4

Fig. 4—Magnetic roll for handling material (separators) made up of 2-pole magnets and steel pole pieces which supply equal magnetic field on 360 degrees of the pole surface.

Fig. 5—Magnets for generators and other devices may be assembled from multiple magnets with laminated (A) or solid pole pieces (E), cast with inserts for pressing in shafts, nonsalient (B) or salient (C) or cast for assembly on shafts (D).

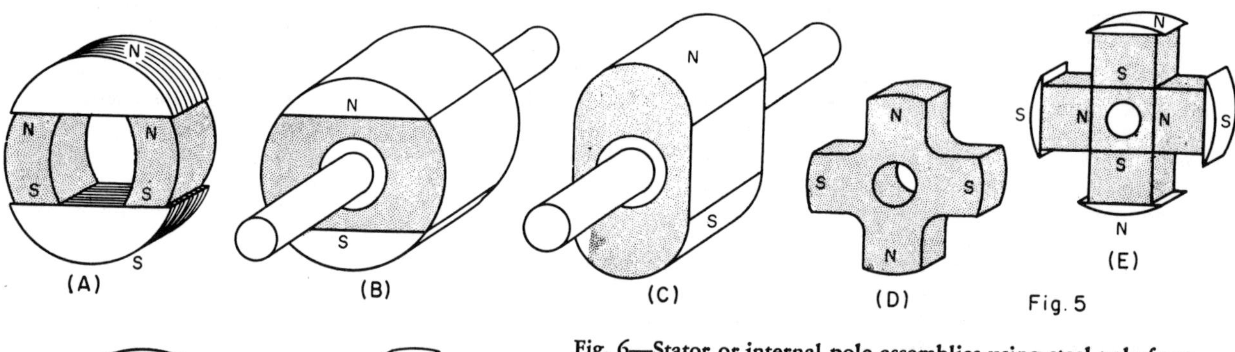


Fig. 5

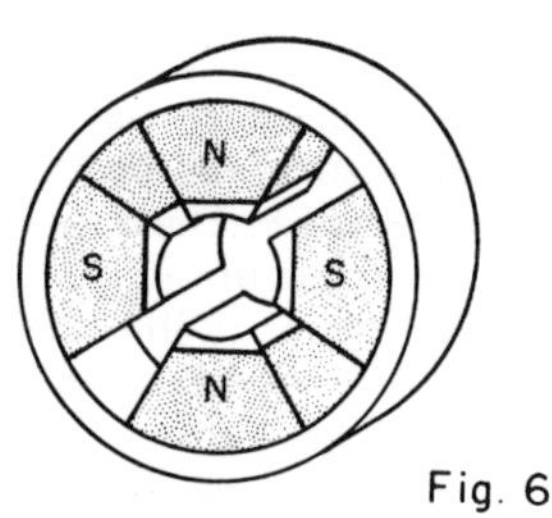


Fig. 6

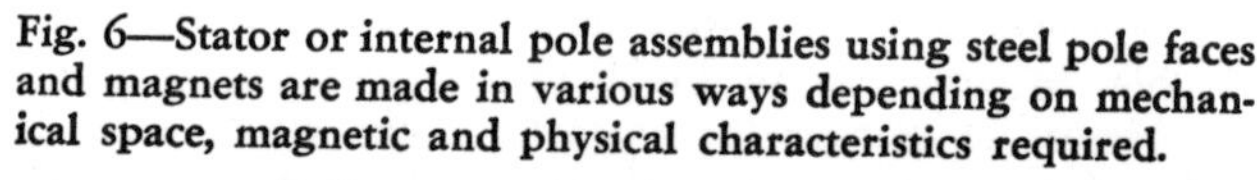

Fig. 6—Stator or internal pole assemblies using steel pole faces and magnets are made in various ways depending on mechanical space, magnetic and physical characteristics required.

Fig. 7—Another 4-pole magnet using soft steel. Although only a 4-pole unit is shown, it is possible to incorporate as many poles as required (A) using bar magnets. In style (B) it is possible to obtain several poles by using one 2-pole magnet.

Fig. 8—Double air-gap permanent magnet assemblies; one without steel poles (A), and one with steel poles (B).

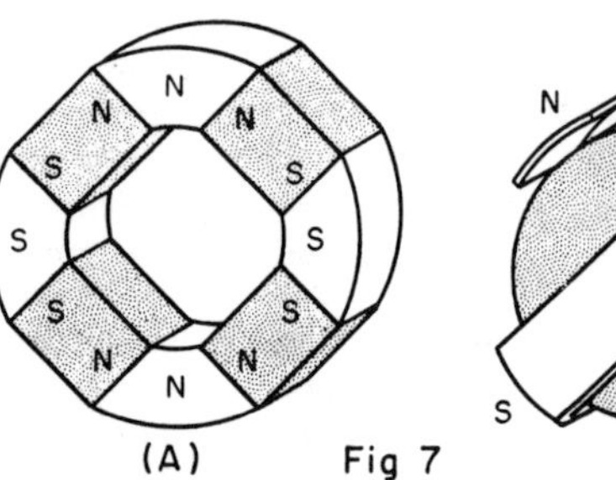


(A) Fig 7

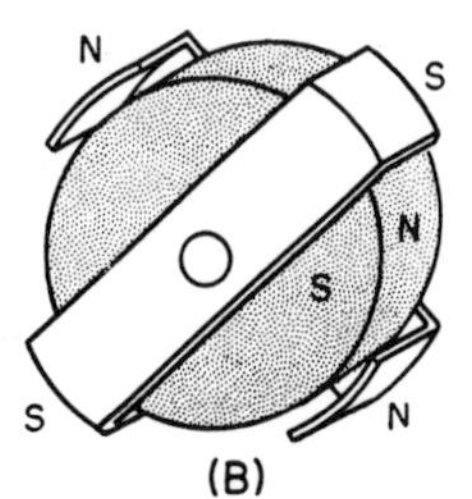


(B)

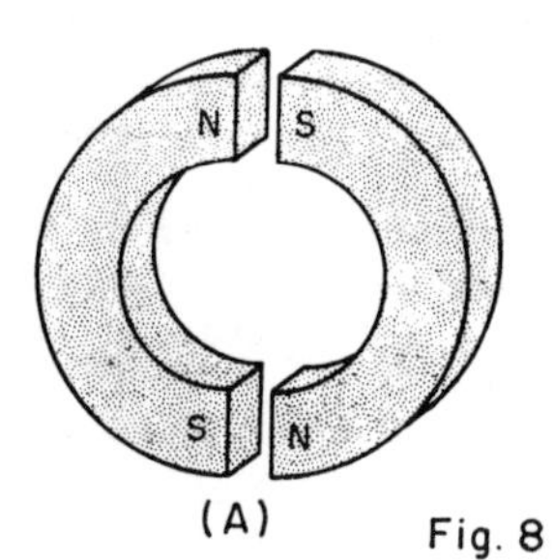


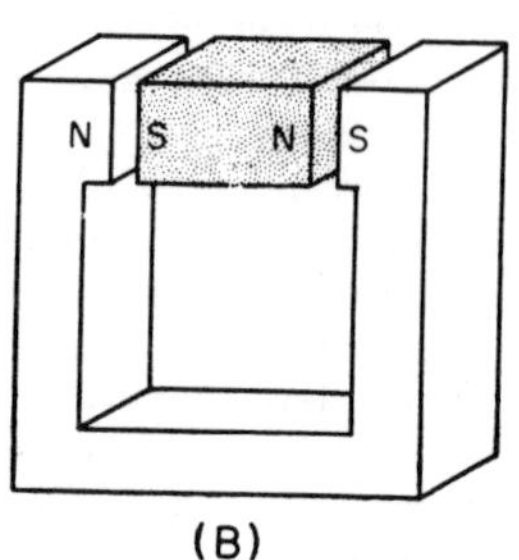


(A) Fig. 8 (B)

# Permanent Magnets

Carboloy Department
General Electric Company

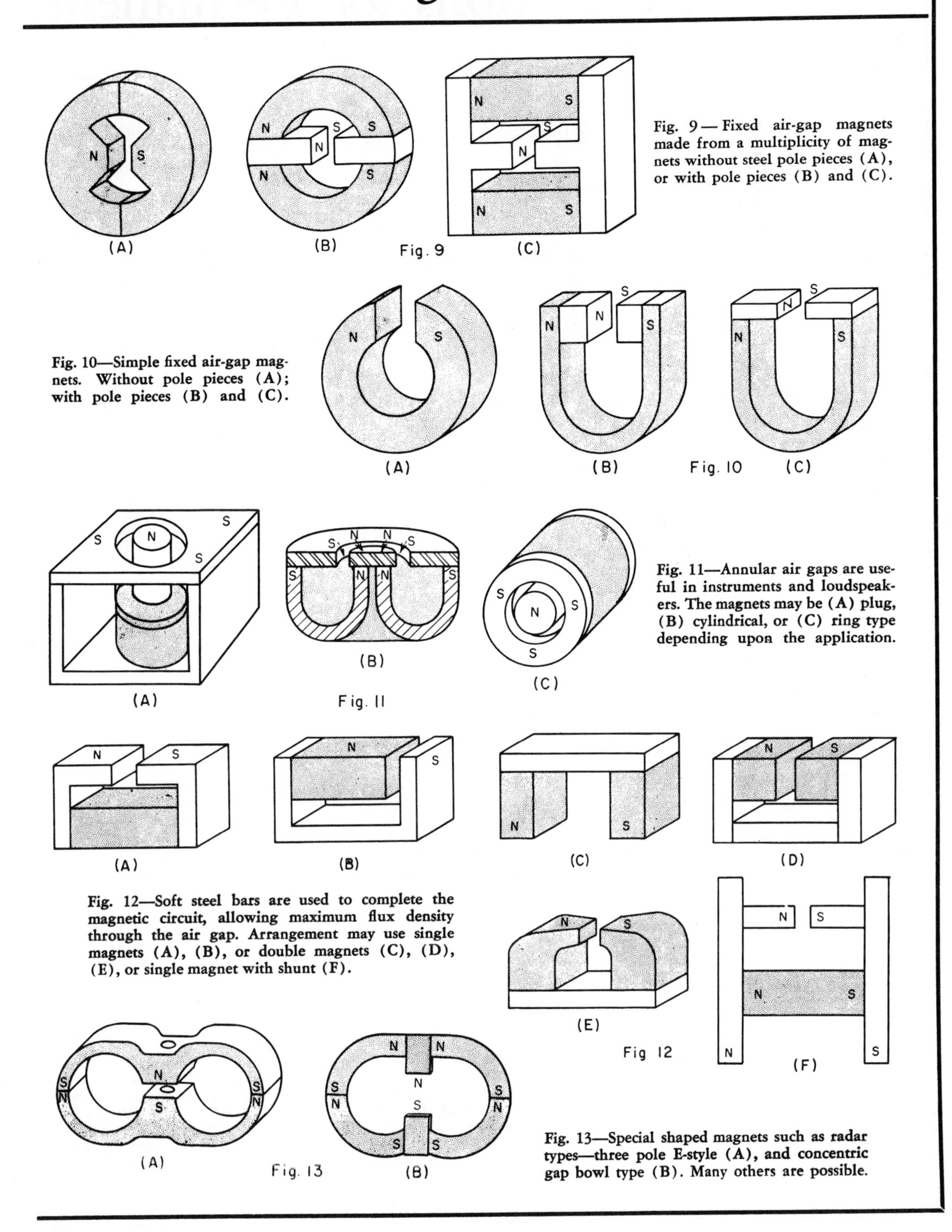


**Fig. 9—Fixed air-gap magnets made from a multiplicity of magnets without steel pole pieces (A), or with pole pieces (B) and (C).**

**Fig. 10—Simple fixed air-gap magnets. Without pole pieces (A); with pole pieces (B) and (C).**

**Fig. 11—Annular air gaps are useful in instruments and loudspeakers. The magnets may be (A) plug, (B) cylindrical, or (C) ring type depending upon the application.**

**Fig. 12—Soft steel bars are used to complete the magnetic circuit, allowing maximum flux density through the air gap. Arrangement may use single magnets (A), (B), or double magnets (C), (D), (E), or single magnet with shunt (F).**

**Fig. 13—Special shaped magnets such as radar types—three pole E-style (A), and concentric gap bowl type (B). Many others are possible.**

# Applications for Permanent

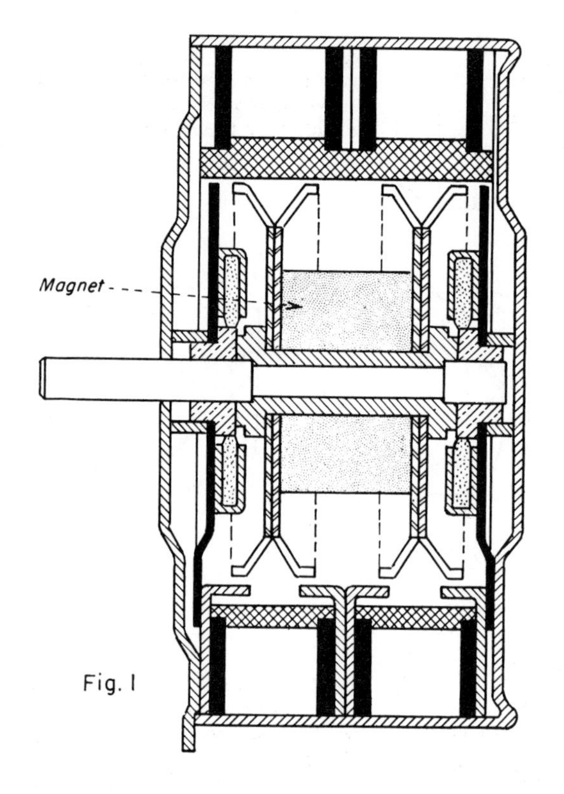


Fig. 1—Synchronous motor derives its d-c excitation from a permanent magnet mounted within the rotor structure. Outstanding features are self-start, low-speed and high torque.

Fig. 2—Eddy current set up by the rotating magnets is the only connection between speedometer cable and the dashboard pointer. Hairspring on cup spindle stabilizes needle.

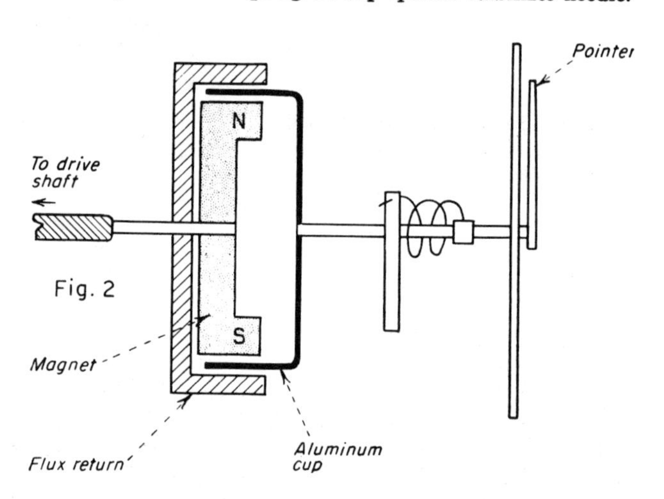


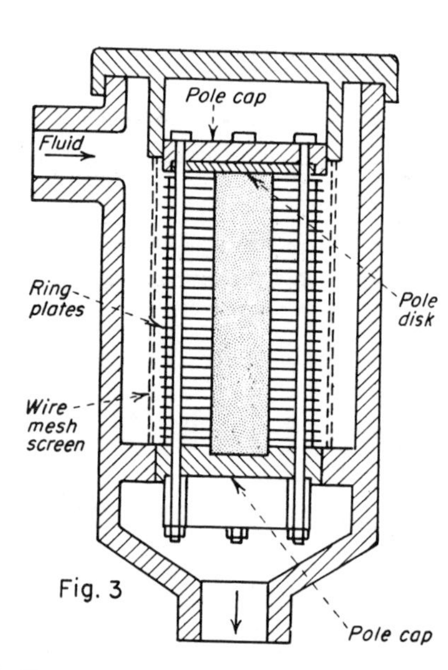


Fig. 3—Magnet filter, designed to remove minute ferrous particles from a liquid or gaseous medium by having a magnetic field of force across a multiplicity of air gaps.

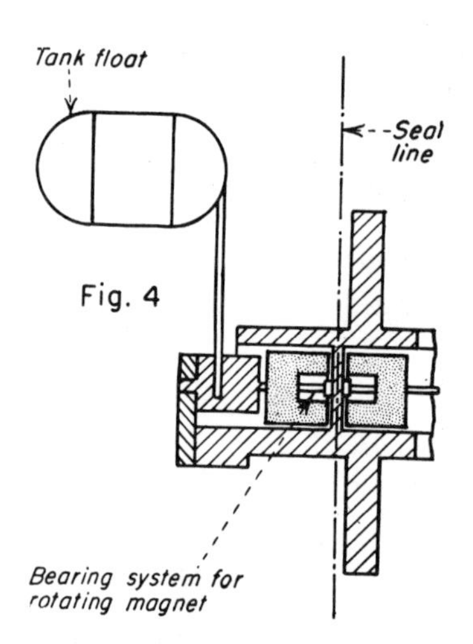


Fig. 4—Two magnets, separated by aluminum diaphragm, function as a leak-proof magnetic coupling to indicate accurately the level of the insulating liquid in transformer tanks. Portion of assembly at left of seal line can be located in a tank—magnet on outside then follows motion of magnet on inside.

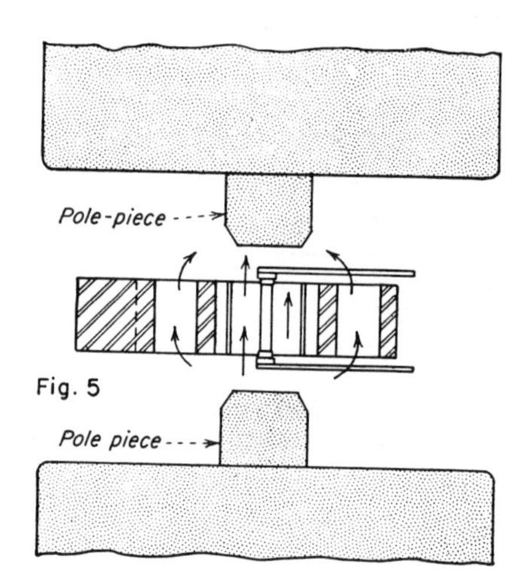


Fig. 5—In a Magnetron, a strongly concentrated field produces forces on the moving electrons between the two pole pieces in such a manner to cause an oscillating condition, creating electro-magnetic field energy. These pole pieces collect flux from permanent magnets several times their size, and concentrate it between their faces.

# Magnets

Carboloy Department
General Electric Company

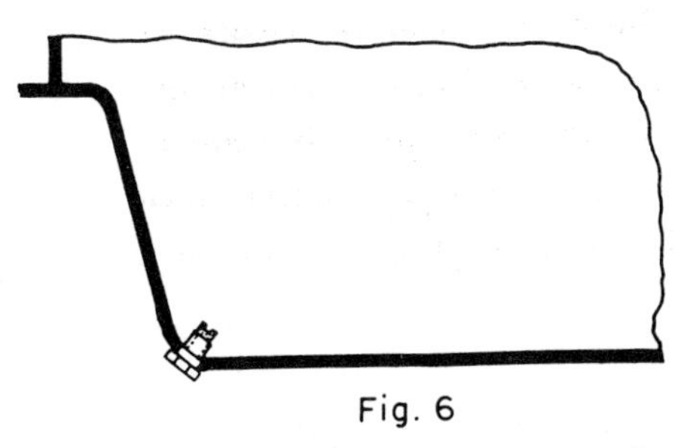


Fig. 6—Magnetic drain plug attracts and removes ferrous particles from a circulating oil or lubricant stream. The plug also helps to remove the particles that normally settle out.

Fig. 7—Permanent magnet switching device in a telephone. When handset is removed from instrument base, spring clips return to normal position, bringing contacts together and closing the circuit.

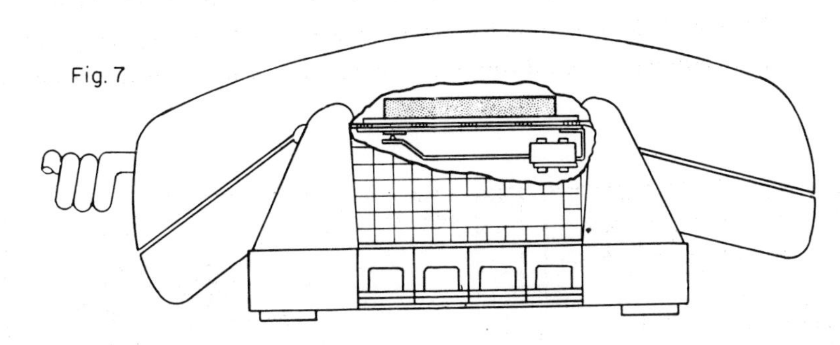


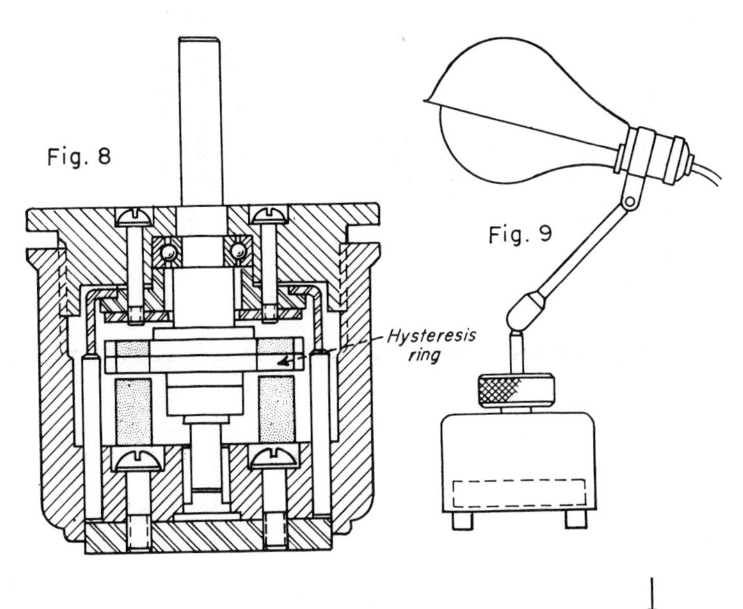


Fig. 8—Device uses magnets to provide smooth frictionless torque that is constant with speed. Torque is controlled by increasing or decreasing the distance between two magnets and a hysteresis ring also of permanent magnet material.

Fig. 9—Lamp socket and reflector is mounted on adjustable rod attached to magnetic base. Unit can be securely attached to a variety of metallic machine surfaces.

Fig. 10—Ring-section magnets mounted inside a conveyor drum. Ferrous contaminants are held in place until they pass barrier and drop down a disposal chute.

Fig. 11—Magnetic assembly acts as valve and holding agent. Working opposite to spring force, assembly moves to the right. When near the steel ring, it snaps against it and remains until mechanical component of the valve forces it away.

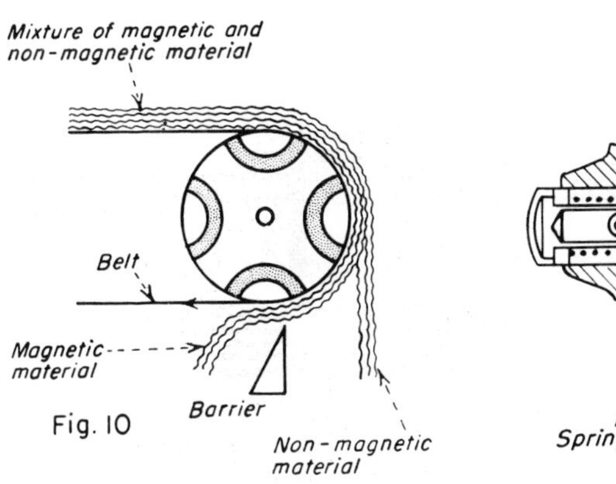


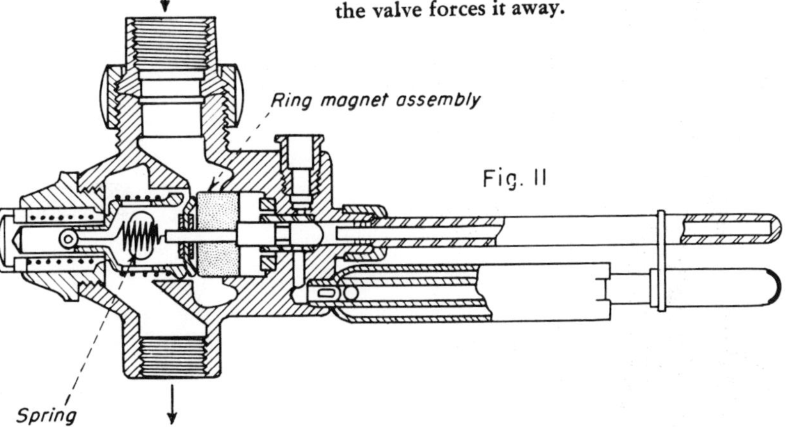

# Permanent Magnet Mechanisms,

The potential utility of permanent magnets in various mechanisms has been greatly increased in recent years by the development of higher and higher strength magnetic materials. These materials are, of course, best known to the electrical industry in which they were developed and to electrical engineers who have made most use of them. Many uses can be made of strong magnets in purely mechanical devices where electro-magnets are impractical or undesirable. Many such uses have already been made, as shown by the accompanying mechanisms and devices, most of which were developed by engineers in the employ of well-known manufacturers. Undoubtedly there are many other ways of using permanent magnets.

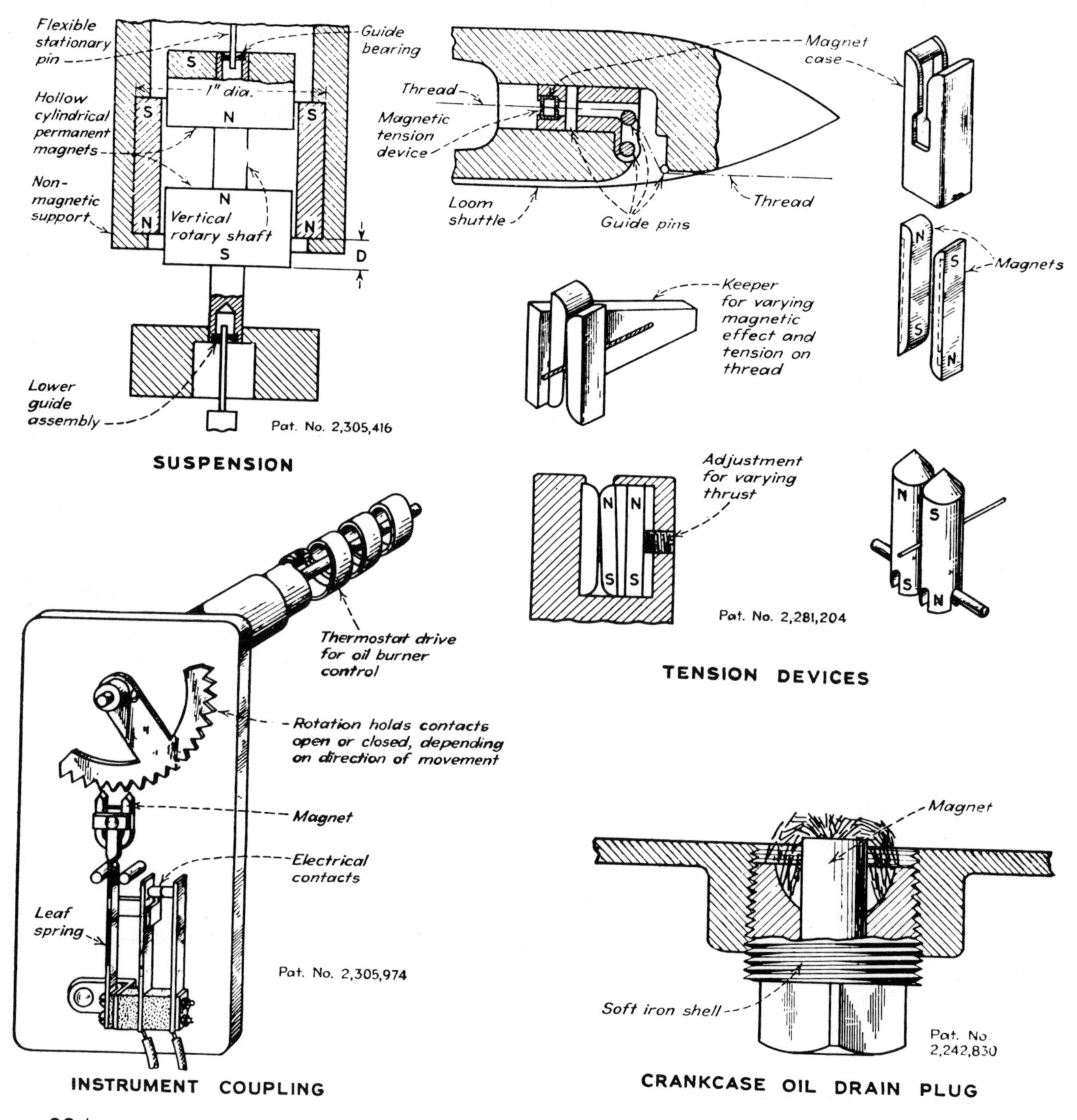

# Their Design and Uses

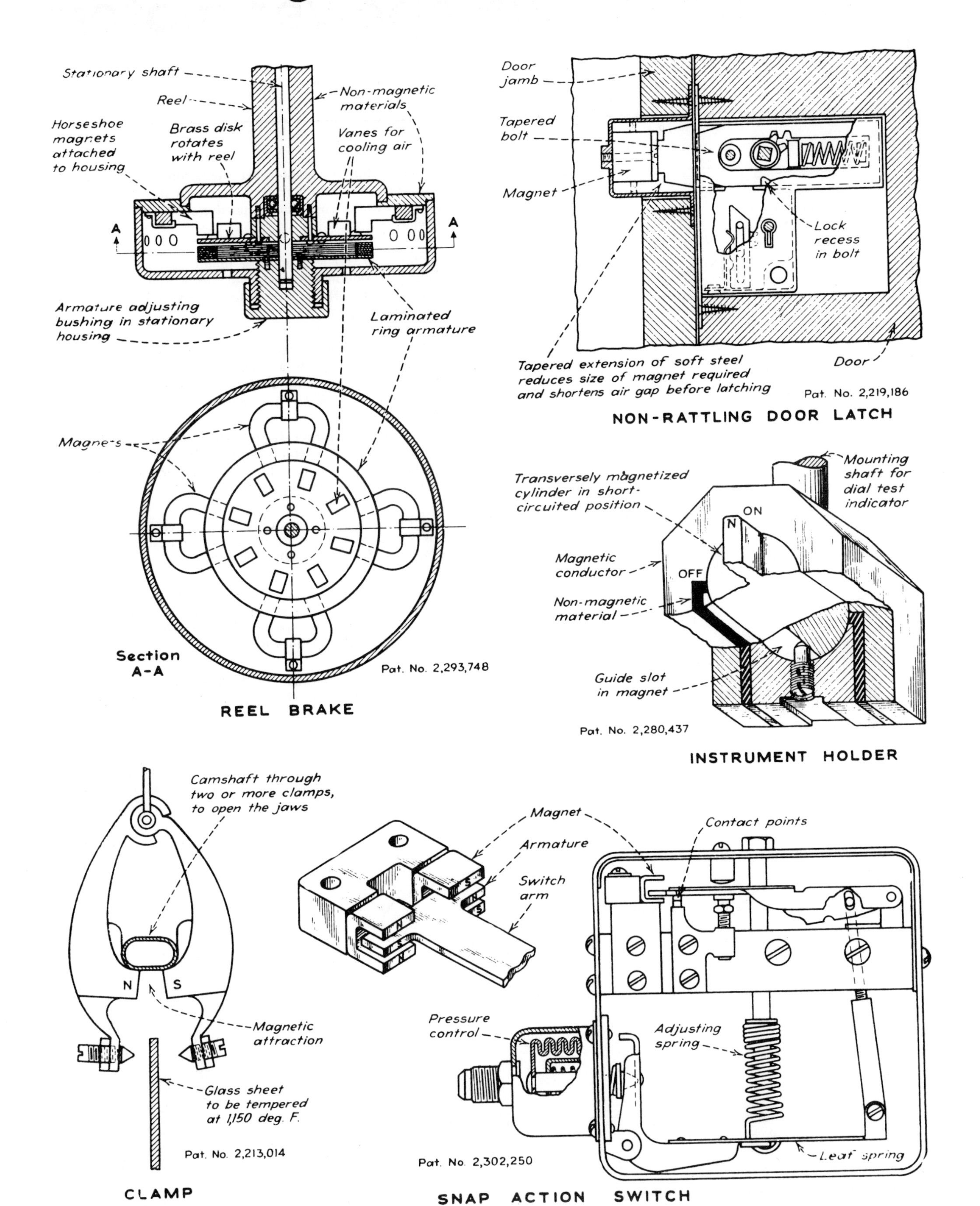


REEL BRAKE

NON-RATTLING DOOR LATCH

INSTRUMENT HOLDER

CLAMP

SNAP ACTION SWITCH

# Fractional horsepower motors

SIDNEY DAVIS, Servomechanisms, Inc.

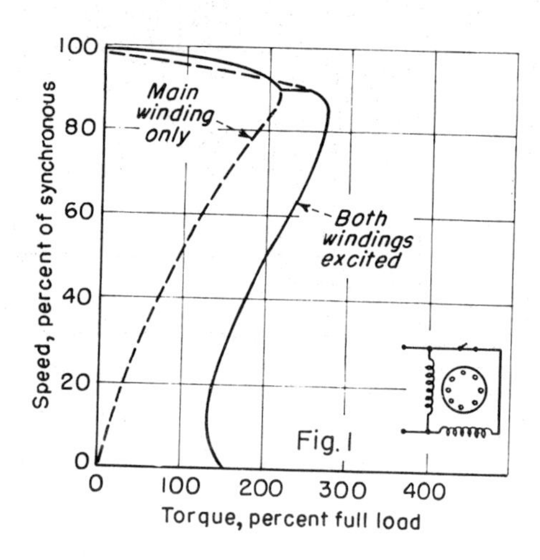


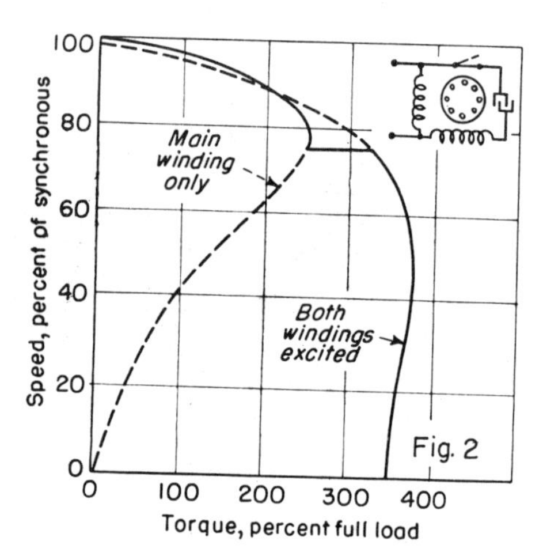


**Fig. 1—Split-Phase.** For constant-speed operations with moderate torque requirements: washing machines, oil burners, blowers, light machine tools. Lower in cost than capacitor-start motors. Range: 1/20-1/3 horsepower.

**Fig. 2—Capacitor-Start.** General-purpose motor suitable for heavy-duty applications requiring high locked-rotor torque and good running efficiency. Hoist service, fans, refrigerators, compressors, stokers. Range: 1/8-3/4 horsepower.

**Fig. 3—(A) Permanent-Split Capacitor.** Constant-speed motor for special-purpose applications such as: heaters, fans, blowers. Other speeds can be obtained by switching arrangement (B), or by using autotransformers. Range: 1/20-3/4 horsepower.

**Fig. 4—Repulsion-Start Induction.** High starting torque suitable for belt and chain drives in pumps, compressors, refrigerators. Becoming replaced by capacitor-start motor. Induction motor characteristics under load. Range: 1/8-3/4 horsepower.

**Fig. 5—Shaded Pole.** Generally under 1/20 hp. Least expensive of a-c motors but has low starting torque and efficiency. Fans, small blowers, unit heaters. Range: 1/2000-1/8 horsepower.

**Fig. 6—Synchronous Reluctance.** Constant speed regardless of load and voltage fluctuations. Teleprinters, sound recording, clocks. Synchronous hysteresis motors can synchronize high inertias such as gyro drives. Range: 1/3000-1/3 horsepower.

**Fig. 7—Universal.** Popular because of high starting torque, high speed, small size and weight for given output. Speed varies sharply with load. Portable tools, vacuum cleaners, mixers, sewing machines. Range: entire fractional horsepower.

**Fig. 8—Universal Governor-Controlled.** Speed is independent of voltage. Governor can be single speed or adjusted even while running. Business machines, typewriters, food mixers, movie projectors. Range: 1/50-3/4 horsepower.

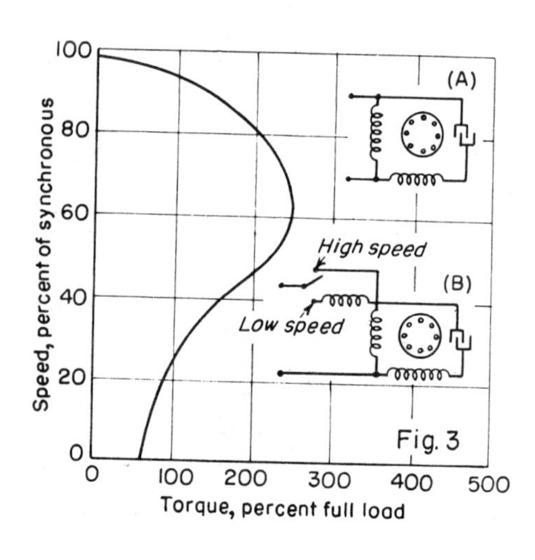


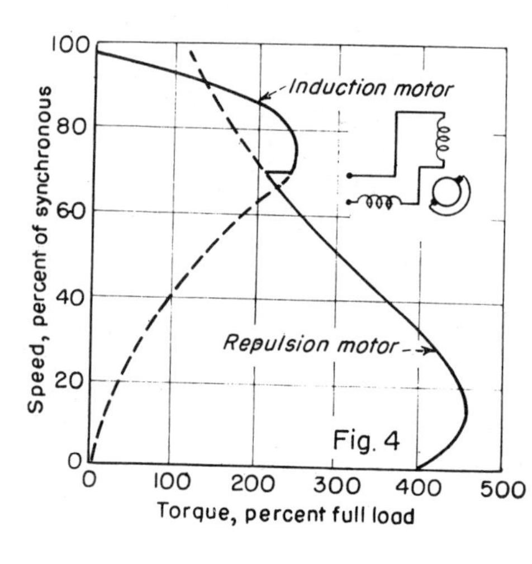


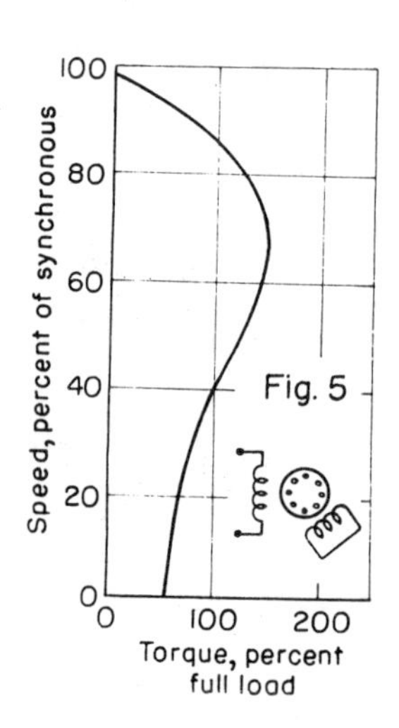

**Selection charts, schematic diagrams, speed-torque curves and typical applications for 12 types of fractional and subfractional motors.**

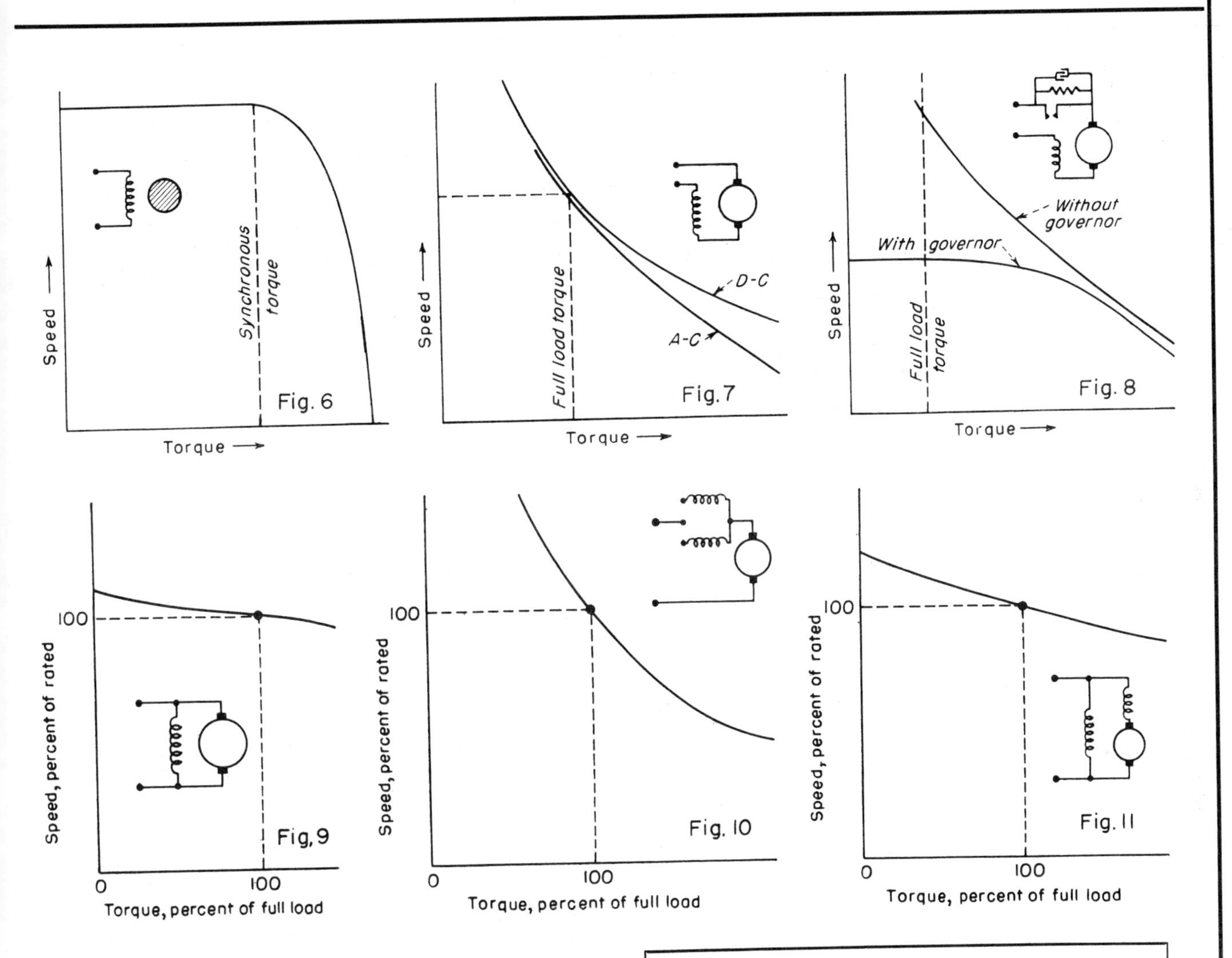


**Fig. 9—Shunt—General-purpose motor used where a-c is not available or where variable speed is desirable. Operates efficiently at many speeds by adjusting field rheostat. Applications similar to capacitor-start motor. Range: 1/20-3/4 horsepower.**

**Fig. 10—Split-Series Field. Used for ease of reversibility; requires only a single-pole double-throw switch for reversing. Races with light load and full voltage. Applications similar to universal motor. Range: 1/25-1/30 horsepower.**

**Fig. 11—Compound. Speed and torque characteristics between shunt and series motor depending upon degree of compounding. Applications similar to capacitor-start motor. Range: 1/100-3/4 horsepower.**

**SELECTION CHART**

**High Operating efficiency—Figs. 3, 4, 9, 10, 11**
**Constant speed—Figs. 1, 2, 3, 5, 6, 8, 9, 11**
**Synchronous—Fig. 6**
**Varying speed—Figs. 7, 10**
**Adjustable speed—Fig. 9**
**Good starting torque—Figs. 2, 4, 7, 8, 10**
**A-c only—Figs. 1, 2, 3, 4, 5, 6**
**D-c only—Figs. 9, 10, 11**
**Both a-c, d-c—Figs. 7, 8**
**Reversible—Figs. 3, 6, 9, 10, 11**
**Reversible from rest—Figs. 1, 2, 4**
**Minimum cost—Figs. 1, 5, 6**
**Medium cost—Figs. 2, 3, 7, 9, 10**

# Torque Requirements of Various Motor Driven Loads

G. BYBERG and E. H. FREDRICK
Allis-Chalmers Manufacturing Company

| Class of Load | Characteristics | Application | Torque Requirements, percent of Full Load Torque: Locked-Rotor (or Starting) Un-loaded | Locked-Rotor (or Starting) Loaded | Pull-in Un-loaded | Pull-in Loaded | Normal Pull-out or Break-down | Max. $Wk^2$ Ratio (Approx.) | Suitable or Typical Motors |
|---|---|---|---|---|---|---|---|---|---|
| 1(a) | Start—partially unloaded<br>Breakaway torque—medium<br>Pull-up torque—constant<br>$Wk^2$—relatively low | Blowers, Positive Displacement | 40 | — | 30 | — | 140 | 8 | Squirrel Cage Induction or Synchronous |
| | | Edgers | 40 | — | 30 | — | 250 | 5 | Squirrel Cage Induction or Synchronous |
| | | D-C Generators, Electroplating | 40 | — | 15 | — | 150 | 1 | Synchronous |
| | | , Cont. Rated | 12 | — | 8 | — | 150 | 1 | Synchronous or Squirrel Cage Induction |
| | | , 25% OL 2 hours | 15 | — | 10 | — | 200 | 1 | Synchronous or Squirrel Cage Induction |
| | | Jordans, Paper mill, Plug out | 50 | — | 50 | — | 150 | 1 | Synchronous or Sq. Cg. |
| | | Pulp Grinders—Magazines | 50 | — | 50 | — | 140 | 5 | Synchronous |
| | | —3 or 4 pocket | 40 | — | 30 | — | 140 | 4 | Synchronous |
| | | Metal-Rolling Mills | | | | | | | |
| | | Blooming & Slabbing, 3-high Structural Rail Roughing & Finishing; Plate Mills | 35 | — | 25 | — | 300 | 1 | Direct Current |
| | | Merchant Mill trains; Billet, Skelp and Sheet Bar mills continuous with lay-shaft; Tube reeling mills | 60 | — | 40 | — | 250 | 1 | Direct Current, Wound Rotor or Synchronous |
| | | Hot Strip mill Roughing Stand | 50 | — | 40 | — | 300 | 1 | Synchronous or W. R. |
| | | Tube Piercing; Expanding and Rolling mills | 60 | — | 40 | — | 350 | 1 | Synchronous |
| | | Saws—Gang | 60 | — | 30 | — | 200 | | Squirrel Cage or Synch. |
| | | —Trimmer | 40 | — | 30 | — | 200 | 1 | Squirrel Cage or Synch. |
| | | Vacuum Pumps (Hytor)......... | 40 | — | 60 | — | 150 | 4 | Squirrel Cage or Synch. |
| 1(b) | Start—partially unloaded<br>Breakaway torque—low<br>Pull-up torque—function of speed<br>$Wk^2$—relatively low | Centrifugal Type Machines | | | | | | | |
| | | Blowers & Compressors, High-Speed | 30 | — | 60 | — | 150 | 15 | Squirrel Cage Induction or Synchronous |
| | | Pumps, starting with discharge valve closed—High-speed | 40 | — | 50–60 | — | 150 | 1 | Squirrel Cage Induction or Synchronous |
| | | —Low-speed | 40 | — | 70–100 | — | 150 | 1 | Squirrel Cage Induction or Synchronous |
| | | Starting with discharge valve open— | — | 40 | — | 100 | 150 | 1 | Squirrel Cage or Synchronous |
| | | Pumps—Screw type—started dry | 40 | — | 30 | — | 150 | 1 | Squirrel Cage or Synch. |
| | | Started primed, disch. open | — | 40 | — | 100 | 150 | 1 | Squirrel Cage, Synchronous or W. R. |
| | | Pumps—Adjustable blade | 50 | — | 40–75 | — | 150 | | Squirrel Cage or Synch. |
| | | Shaker screens | 50 | — | — | 100 | 150 | 1 | Squirrel Cage or Synch. |
| 1(c) | Start—partially unloaded<br>Breakaway torque—high<br>Pull-up torque—constant average<br>$Wk^2$—relatively low | Reciprocating Type Machines | | | | | | | All 1(c) types:<br>High Speed; Squirrel Cg. or Synch.<br>Low Speed; Synchronous |
| | | Compressors | | | | | | | |
| | | Air and gas | 40 | — | 20–30 | — | 150 | 10 | |
| | | Ammonia, discharge pressure 25 lb per sq in. | 40 | — | 20–30 | 150 | 150 | 7 | |
| | | Freon | 40 | — | 40–60 | — | 150 | 4 | |
| | | Carbon Dioxide, Single-cyl., double acting | 40–120 | — | 30–40 | — | 150 | 5–10 | |
| | | Carbon Dioxide, 2-cylinder, double acting | 40–90 | — | 30–40 | — | 150 | 4–7 | |
| | | Pumps—started dry | 40 | — | 30 | — | 150 | 1 | |
| | | —by-passed | 40 | — | 40 | — | 150 | 1 | |
| | | Vacuum Pumps | 40 | — | 60 | — | 150 | 10 | |

## Torque Requirements of Various Motor Driven Loads

| Class of Load | Characteristics | Application | Torque Requirements, percent of Full Load Torque: Locked-Rotor (or Starting) Un-loaded | Locked-Rotor (or Starting) Loaded | Pull-in Un-loaded | Pull-in Loaded | Normal Pull-out or Break-down | Max. $Wk^2$ Ratio (Approx.) | Suitable or Typical Motors |
|---|---|---|---|---|---|---|---|---|---|
| 2 | Start—must be unloaded; Breakaway torque—high; Pull-up torque—constant average; $Wk^2$—varies with machine | Attrition Mills | 100 | — | 60 | — | 175 | 12 | Squirrel Cage, Wound Rotor or Synchronous |
| | | Crushers—B & W | 200 | — | 100 | — | 250 | 3 | Wound Rotor or High Torque Squirrel Cage |
| | | —Cone Type | 100 | — | 70 | — | 250 | 6 | Wound Rotor or High Torque Squirrel Cage |
| | | —Gyratory | 100 | — | 70 | — | 250 | 4 | Wound Rotor Induction |
| | | —Jaw | 100 | — | 70 | — | 250 | 2 | Wound Rotor Induction |
| | | —Roll | 100 | — | 70 | — | 250 | 3 | Wound Rotor Induction |
| | | Hammer Mills | 120 | — | 80 | — | 250 | 15-40 | Wound Rotor Induction |
| 3(a) | Start—partially unloaded; Breakaway torque—variable; Pull-up torque—variable; $Wk^2$—very high | Attrition Mills | 100 | — | 60 | — | 175 | 12 | Squirrel Cage, Wound Rotor or Synchronous |
| | | Bandsaw Mills | 125 | — | 30 | — | 250 | 100 | Synchronous Direct Connected Wound Rotor Belted |
| | | Blowers—Centrifugal | 40 | — | 60 | — | 150 | 10-15 | Squirrel Cage Induction or Synchronous |
| | | Compressors—Centrifugal | 40 | — | 60 | — | 150 | 15-25 | Sq. Cg., W. R., Synch. |
| | | —Reciprocating | 60-120 | — | 30 | — | 150 | 5-10 | Synchronous |
| | | Fans—Centrifugal | 40 | — | 60 | — | 150 | 20-50 | Squirrel Cage or Synch. |
| | | —Centrifugal (Sintering—with inlet gates closed) | 60 | — | 100 | — | 150 | 40-60 | Squirrel Cage, Synchronous or W. R. |
| | | Fans—Propeller type, with discharge open | 40 | — | 100 | — | 150 | 25 | |
| | | Hammer Mills | 120 | — | 80 | — | 250 | 15-40 | Squirrel Cage, Wound Rotor, Synchronous |
| | | Shredders | 50 | — | 50 | — | 250 | 30 | Squirrel Cage, Wound Rotor, Synchronous |
| | | Vacuum Pumps (Reciprocating, including sliding vane type) | 40 | — | 60 | — | 150 | 10 | Squirrel Cage or Synchronous |
| | | Wood Hogs | 60 | — | 30 | — | 225 | 30 | Sq. Cg., W. R., Synch |
| 4(a) | Start—fully loaded; Breakaway torque—high; Pull-up torque—high; $Wk^2$—generally low | Ball Mills—Rock | — | 150 | — | 110 | 150 | 2 | Synchronous |
| | | —Ore | — | 160 | — | 110 | 150 | 2 | Usually synchronous. Also Wound Rotor |
| | | Bowl Mills—(Coal Pulverizer) | — | 150 | — | 125 | 150 | | Squirrel Cage, Wound Rotor, Synchronous |
| | | Beaters-Pulp-Standard | — | 125 | — | 100 | 150 | 5 | Wound Rotor or Sq. Cg. |
| | | -Breaker | — | 125 | — | 100 | 200 | 5 | Wound Rotor or Sq. Cg. |
| | | Hydrapulpers | — | 125 | — | 125 | 150 | | Synchronous |
| | | Rod Mills-Ore grinding | — | 160 | — | 110 | 150 | 1 | Wound Rotor or Synch. |
| | | Line Shafts—Flour mills | — | 175 | — | 110 | 150 | 15 | Synchronous or W. R. |
| | | Reciprocating Pumps—Not by-passed (3-cylinder) | — | 150 | — | 110 | 150 | 1 | Squirrel Cage or Synchronous |
| 4(b) | Start—unload except in emergency; Breakaway torque—high; Pull-up torque—high; $Wk^2$—generally low | Rubber Mills—Banbury Mixers | — | 125 | — | 100 | 250 | 1 | Synchronous or W. R. |
| | | —Plasticators | — | 125 | — | 100 | 250 | 1 | Synchronous or W. R |
| | | Rolling Mills—Cold-rolling | — | 200 | — | 150 | 250 | 1 | Direct Current |
| | | Brass and Copper finishing mills | — | 150 | — | 125 | 250 | 1 | Direct Current |
| 4(c) | Start—cannot be unloaded; Breakaway—low; Pull-up torque—varies with speed; $Wk^2$—variable | Pumps-Axial-flow with discharge open— | | 40 | — | 100 | 150 | 1 | Squirrel Cage, Wound Rotor, Synchronous |
| | | discharge closed— | — | 40 | — | 200-300 | 150 | | Not practicable |

## PROCESS CONTROL

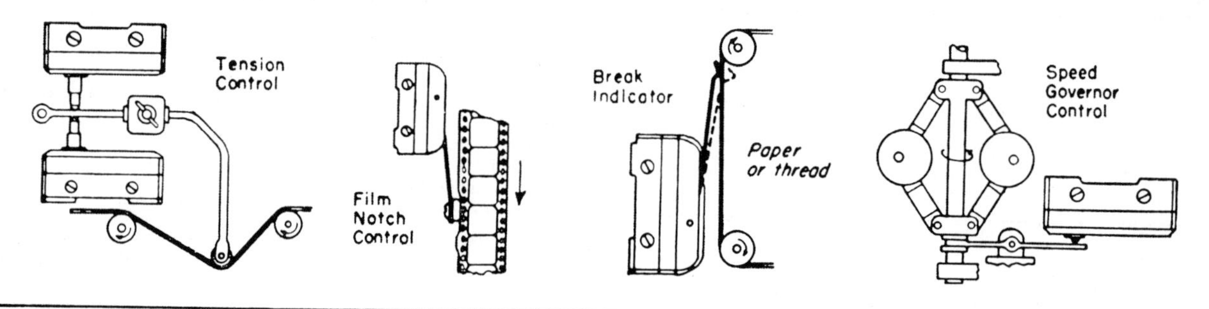


## MOTION CONTROL

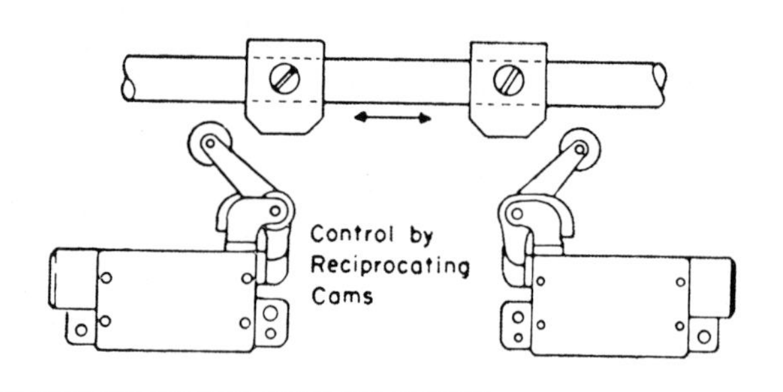


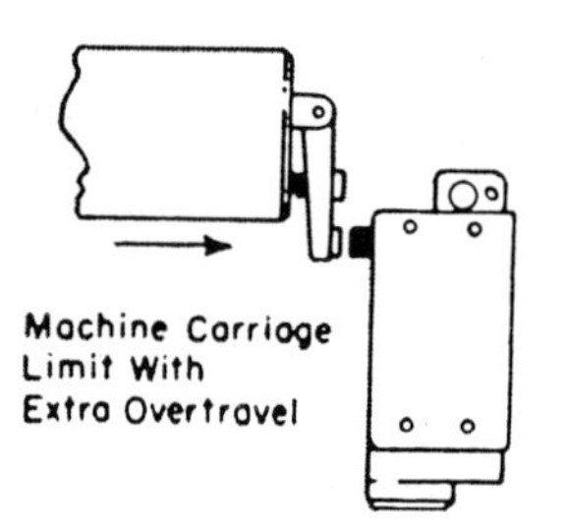


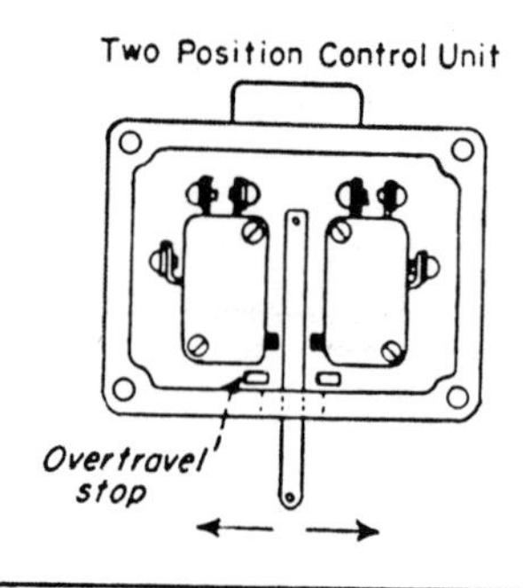


## ENCLOSURE PROTECTION

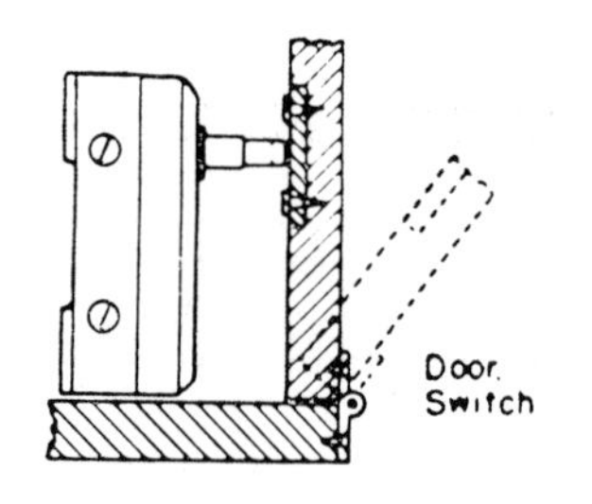

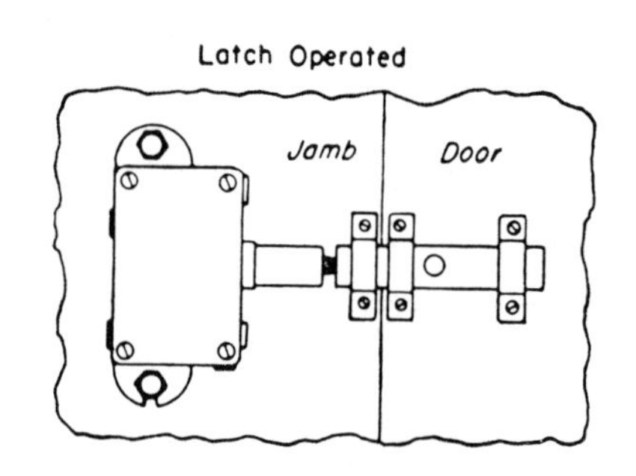


## THERMOSTATIC AND PRESSURE CONTROL

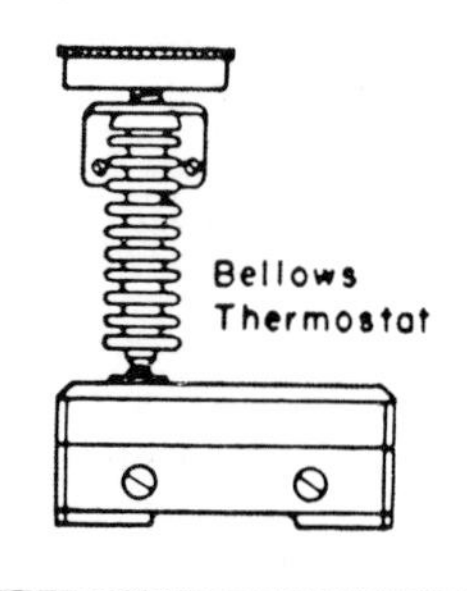


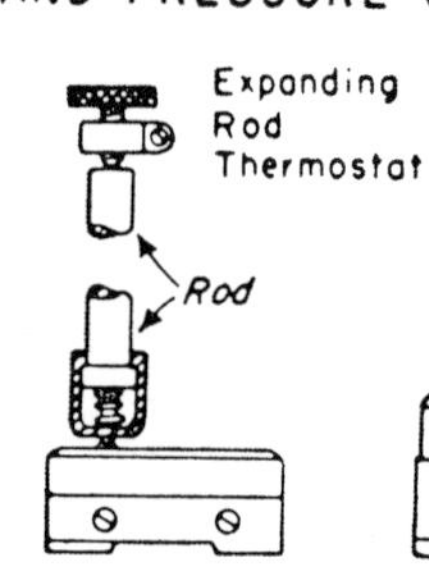


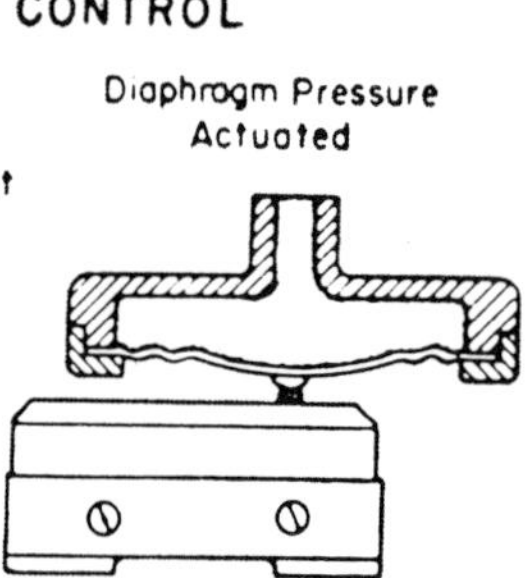


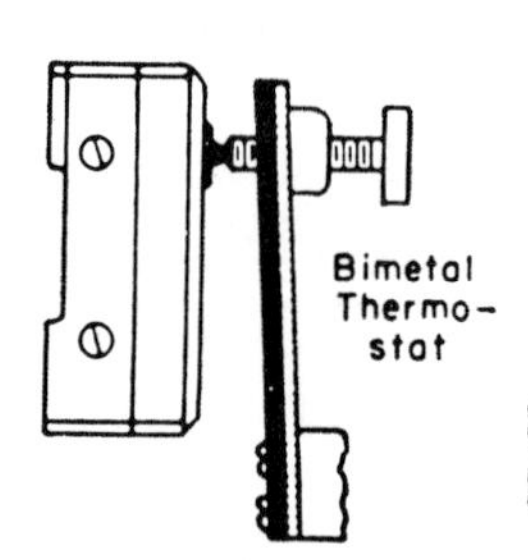

## COUNTING, SORTING AND FEEDING DEVICES

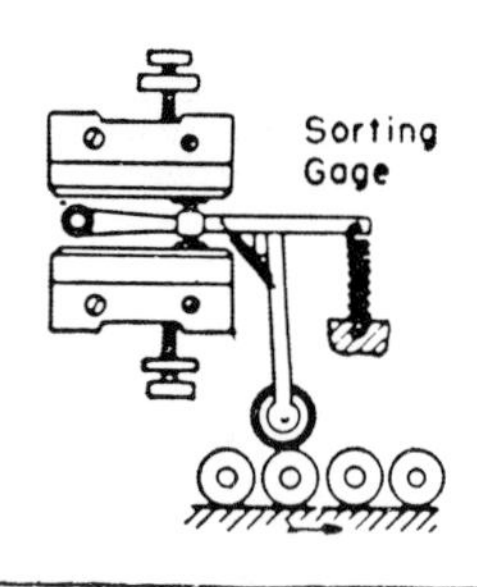

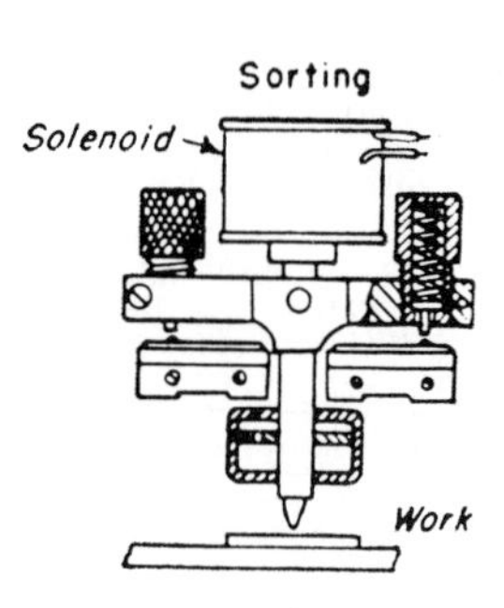

# Application of Precision Snap-Acting Switches

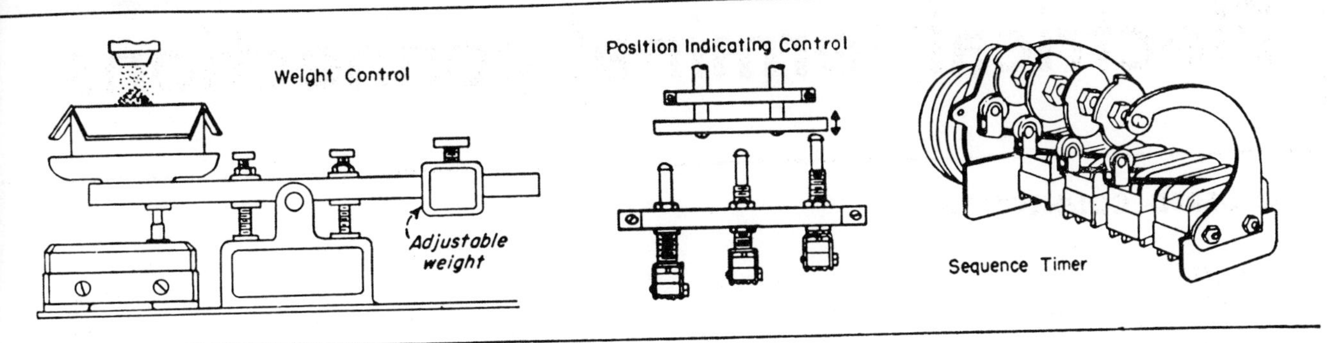


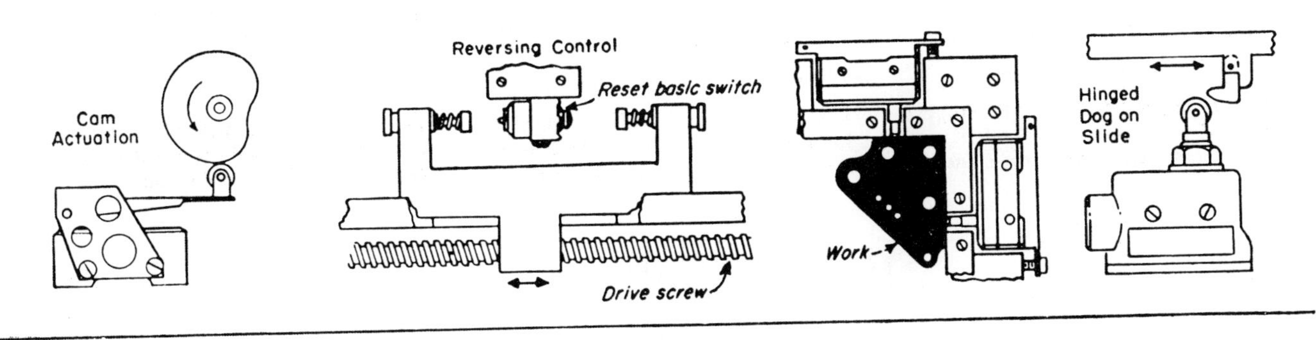


GAGING AND THICKNESS CONTROL

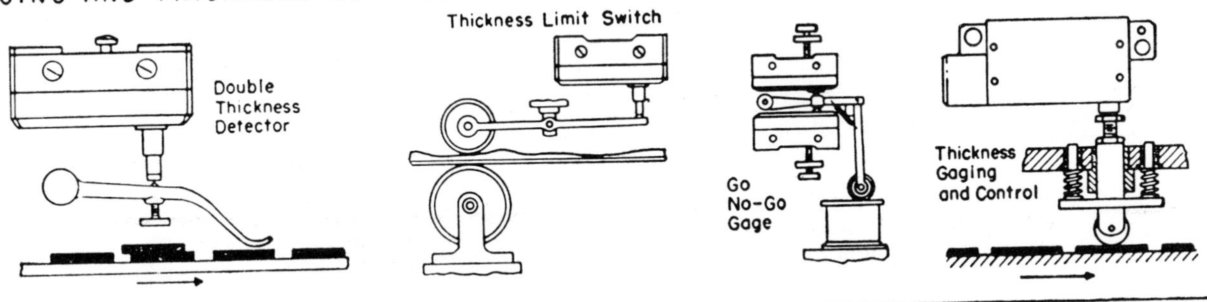


LEVEL CONTROL

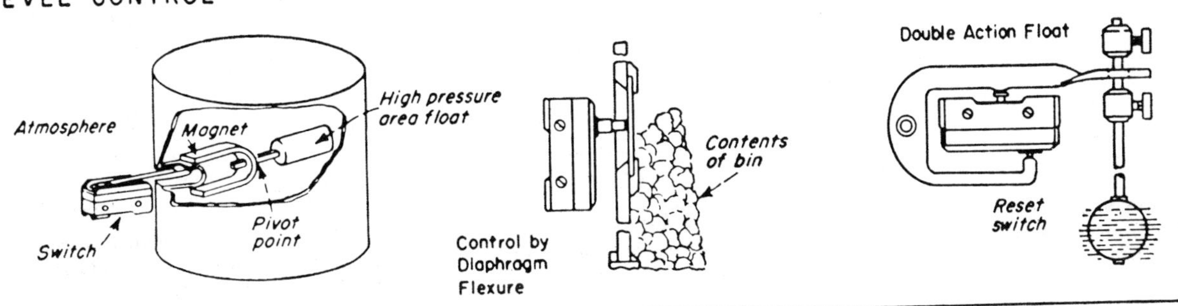


MISCELLANEOUS

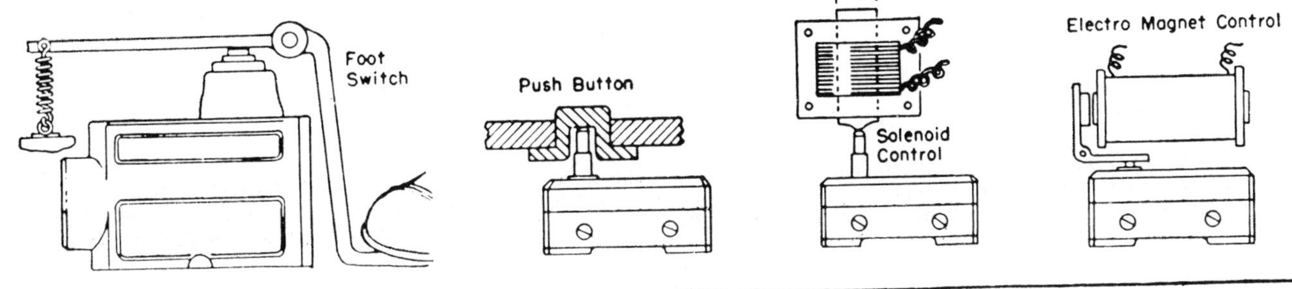

# Electrical terminal connections

## SOLDERED CONNECTIONS

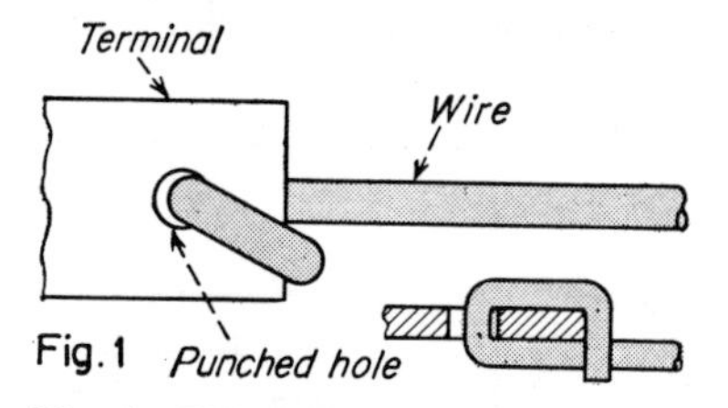


**Fig. 1—PUNCHED HOLE.** Stripped end of wire inserted into hole and bent around terminal before soldering.

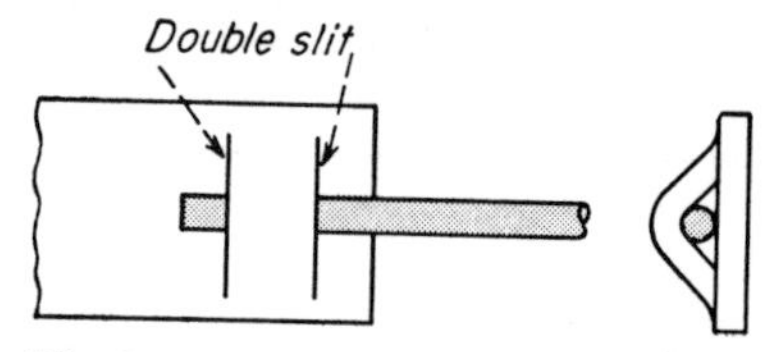


**Fig. 2—DOUBLE SLIT** stamped to form channel for wire. Wire is inserted; channel is pressed closed.

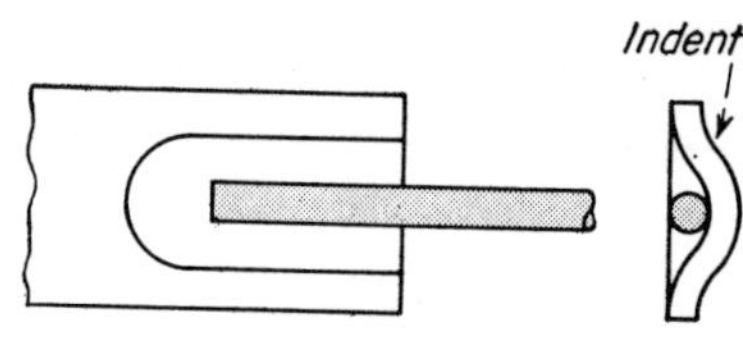


**Fig. 3—INDENT FORMED** in the terminal gives good electrical contact area and facilitates soldering.

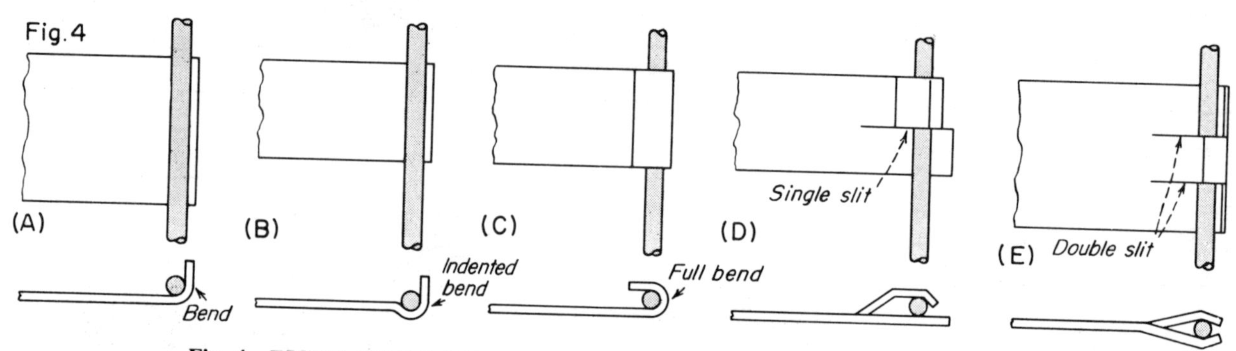


**Fig. 4—RIGHT-ANGLE CONNECTION** between wire and terminal. (A) Simple bend, (B) indented bend for greater contact area, (C) full bend for maximum contact area, (D) single slit for firm grip before soldering, (E) double slit for balanced grip.

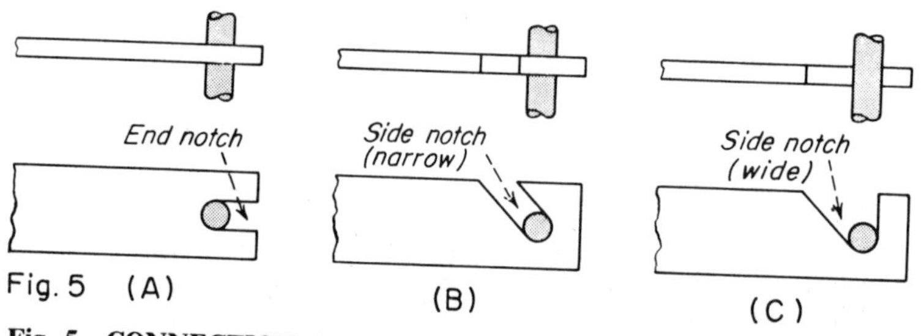


**Fig. 5—CONNECTION AT RIGHT-ANGLE** to terminal surface. (A) End notched, (B) side notched to resist end thrust, (C) modified side notch to avoid forming a sharp edge when stamping the notch in terminal.

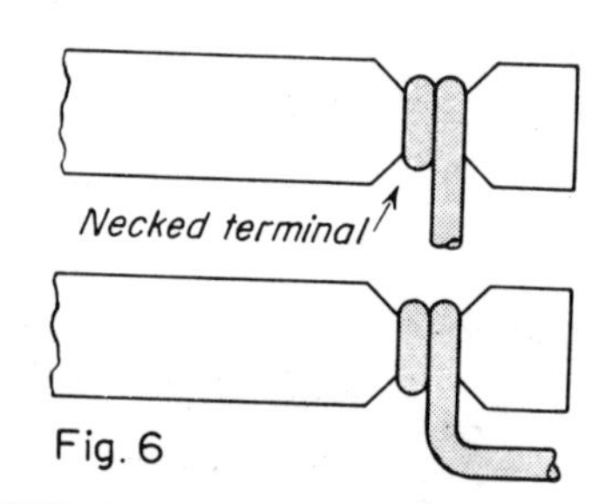


**Fig. 6—NECKS** cut in terminal permit wrapping the wire tightly before soldering to obtain a strong joint connection.

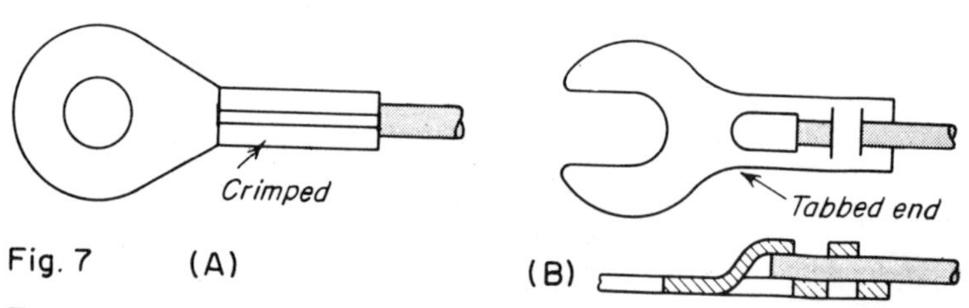


**Fig. 7—SHAPED TERMINALS.** (A) crimped around wire, (B) tabbed and then pressed after wire is inserted. Solder not needed in (A) but should be applied in (B) to obtain the maximum of electrical conductivity.

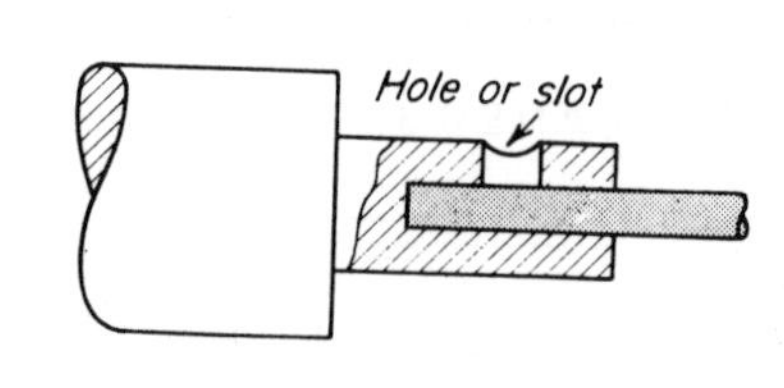


**Fig. 8—AXIAL CONNECTION** to a rod or shaft made by drilling two holes; one for wire, second to permit a good flow of solder.

FEDERICO STRASSER, Santiago de Chile

29 methods of fastening wire to terminals. Soldered connections are usually considered permanent. Screwed connections can be assembled and disassembled without special tools.

## SCREWED CONNECTIONS

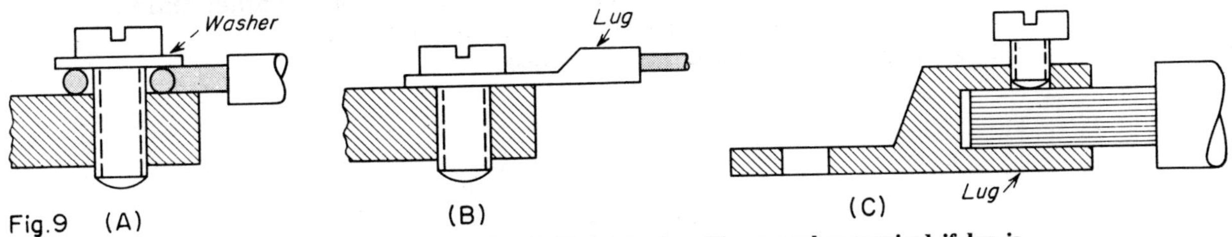


Fig. 9—(A) Screw and washer holds bent wire. (B) no washer required if lug is used. (C) Lug should be screwed if connection is too large for soldering.

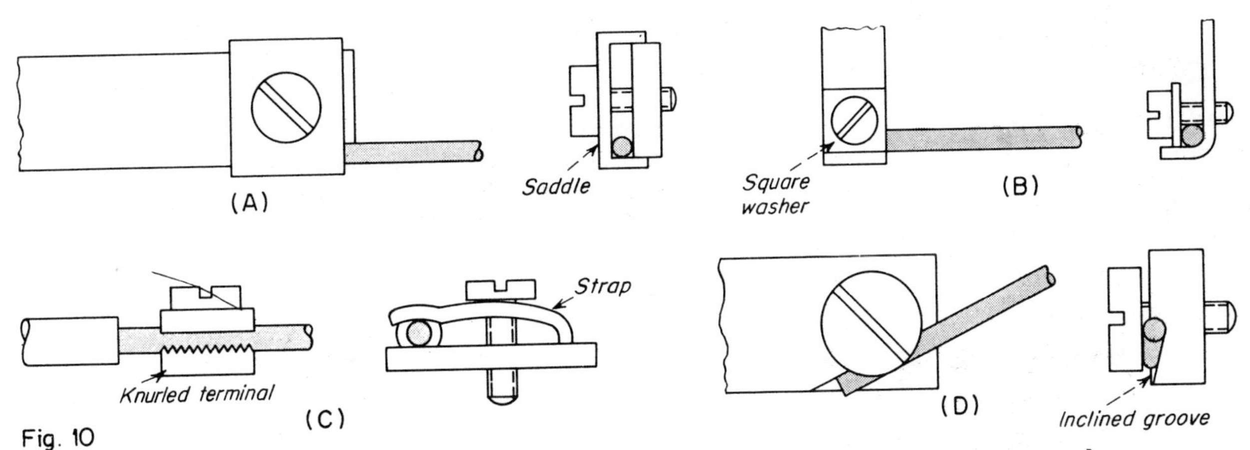


Fig. 10—CLAMPED WIRE held between (A) terminal and saddle, (B) terminal and square washer, (C) strap and knurled terminal for better grip, (D) nut and inclined groove cut in terminal.

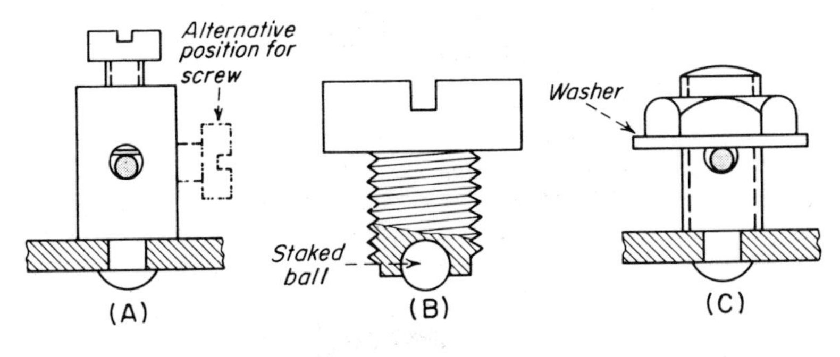


Fig. 11—TERMINAL POSTS allow quick and easy connections to be made. Screw has alternate position (A). Special screw (B) has ball staked in end to reduce turning force on wire. Pressure of nuts acts through washer (C). Side slot (D) allows for conditions where axial movement of wire is not possible. Pressure acts through round or square bar (E). Buttress threads (F) prevent terminal opening up and wire becoming loose.

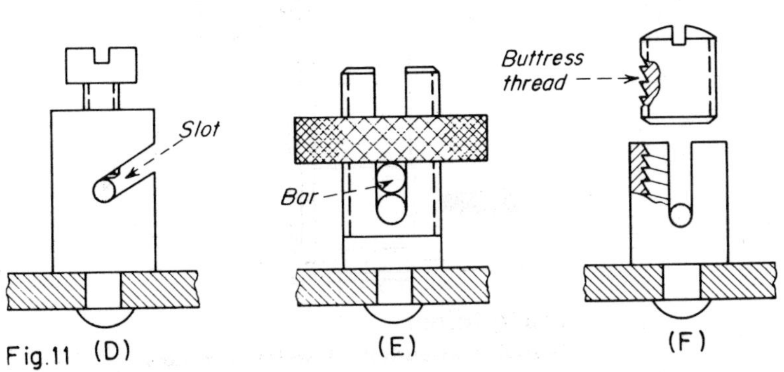


Fig. 12—WIRE CLAMPED between two nuts is usual method of connecting leads to male receptacle terminals.

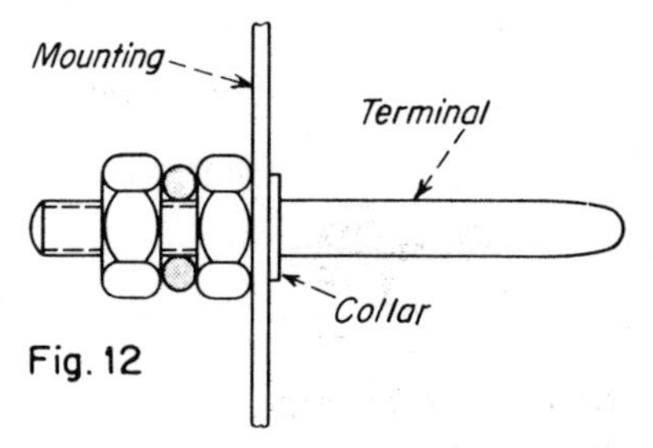

# 12 WAYS to RETAIN ELECTRON

## with their advantages and disadvantages

**Advantages are in boldface—**
*Disadvantages are in italics*

**Finding ways to hold electron tubes in their sockets is a problem that is often left to the mechanical designer rather than the electronic-circuit designer. Most holding devices in this survey are available for purchase.**

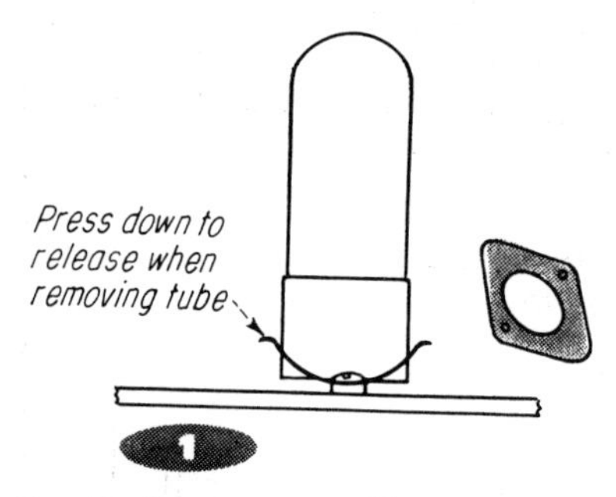


### Bowed spring collar

**Inexpensive**
**Simple to install**
*Not applicable to military equipment*
*Requires hand room to release clamp-action*

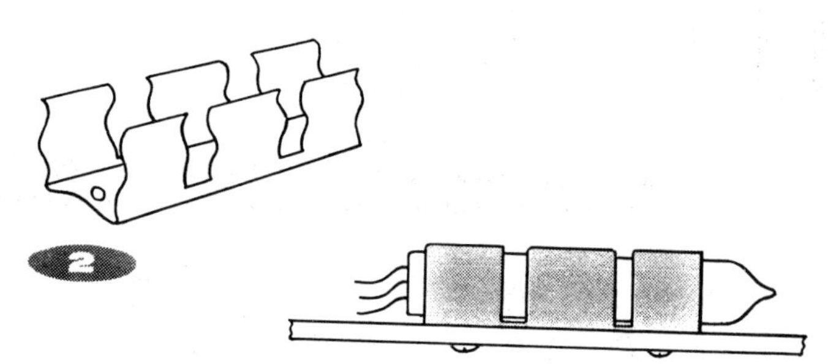


### Spring clamp

**Available in variety of styles and sizes**
**Dissipates heat from tube to chassis**
**Meets military specifications**
*Takes up much horizontal space*

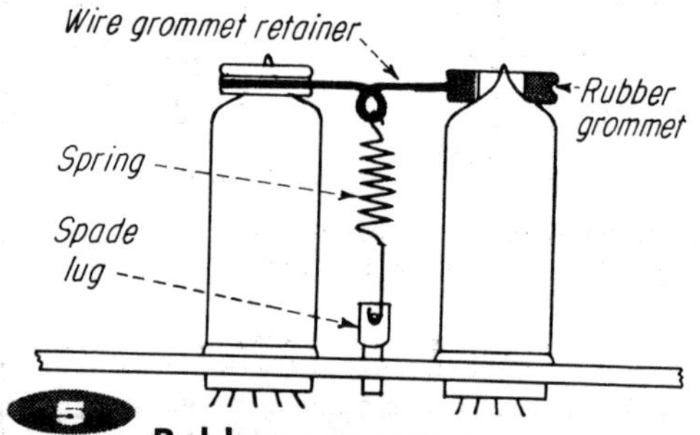


### Rubber grommet

**Protects tube from shock and vibration**
**Inexpensive**
**No loose parts to handle**
*Must be used with two equal-height tubes*

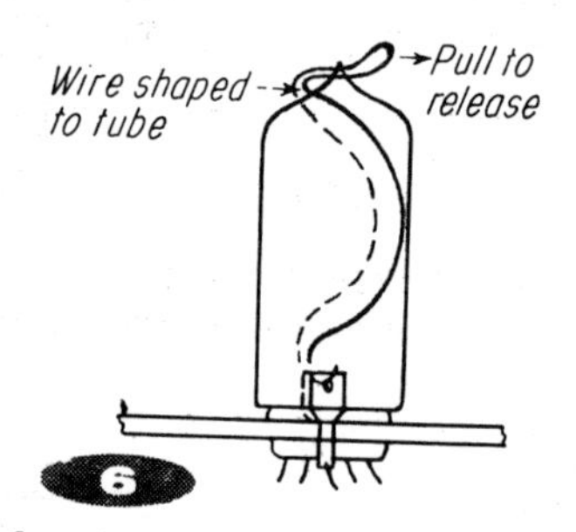


### Formed wire

**Inexpensive**
**Releases from top**
*Cannot be used on military equipment*

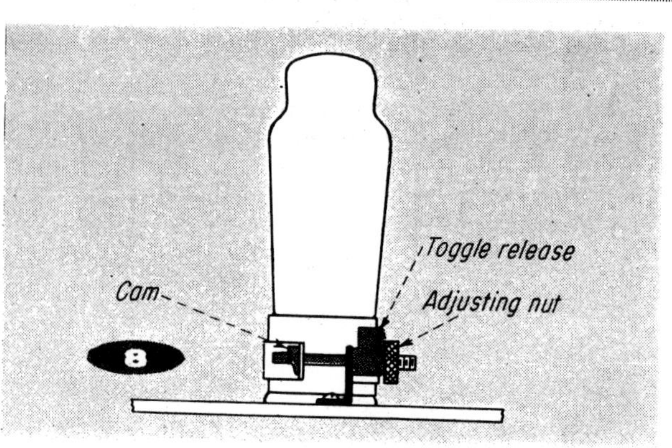


### Base cam-clamp

**Meets military specs**
**Adjustable to different base diameters**
*Must be located for toggle accessibility*

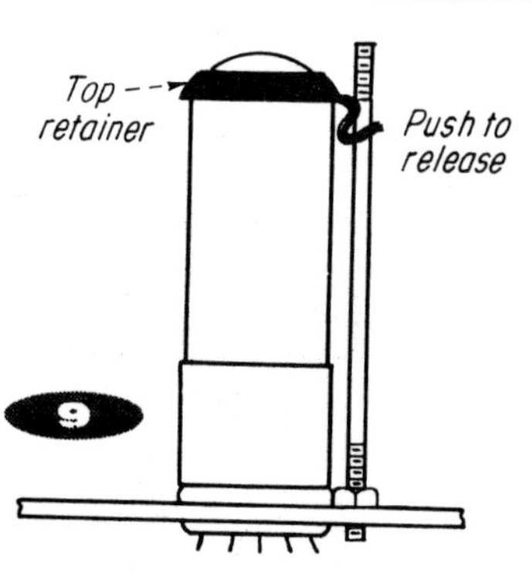


### Top retainer

**Available in variety of styles and sizes**
**Inexpensive**
**Meets military specs**
*Retainer is loose part when tube is removed*

# TUBES

**FRANK WILLIAM WOOD Jr.,**
*Design Engineer*
*Servonics Inc.,*
*Alexandria, Va.*

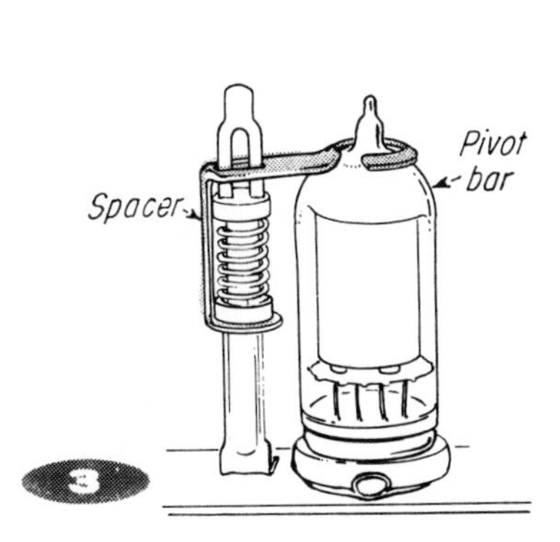


## Pivot bar

**Releases from top**
**Can easily accommodate different tube heights**
*Occupies valuable chassis area*

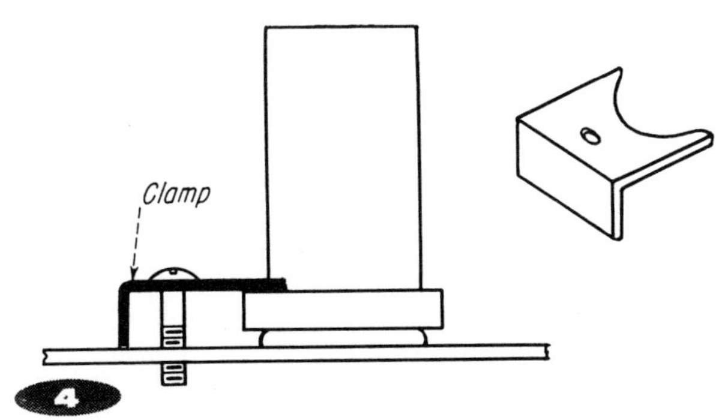


## Angle clamp

**Inexpensive**
**Suitable for military equipment**
*Screw must be completely removed before removing tube*
*Occupies much chassis area*
*Limited to small variety of tubes*

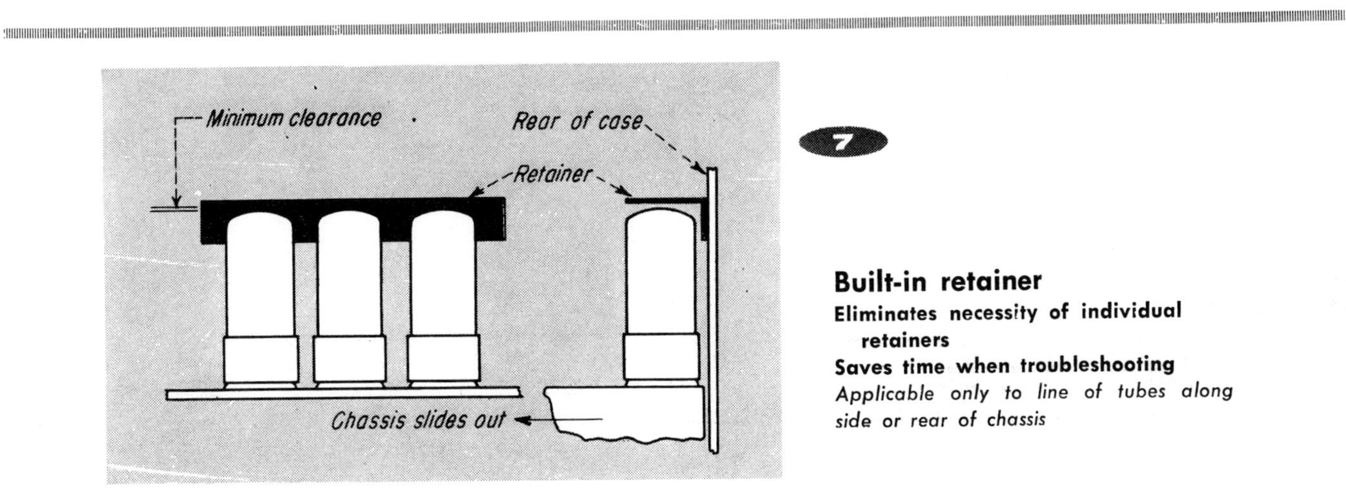

## Built-in retainer

**Eliminates necessity of individual retainers**
**Saves time when troubleshooting**
*Applicable only to line of tubes along side or rear of chassis*

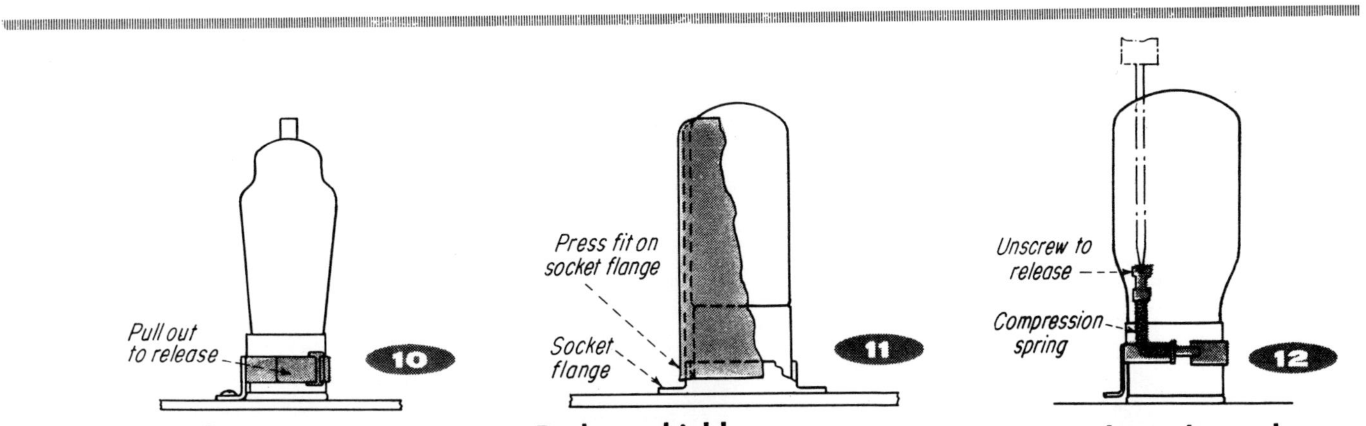


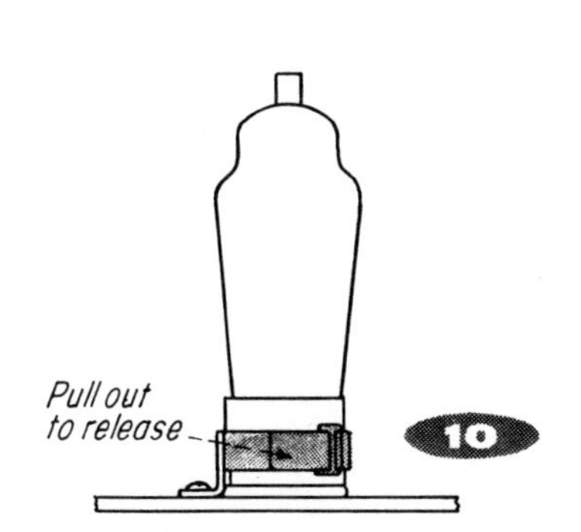


## Base clamp

**Meets military specs**
**Available in variety of sizes**
*Takes up much chassis area*
*Release space required around tube*

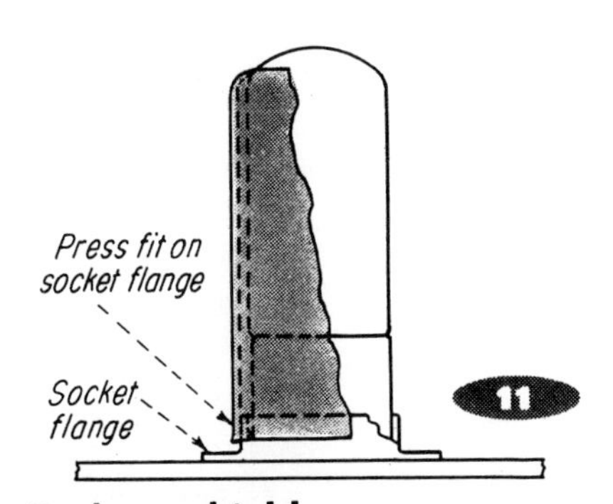


## Push-on shield

**Combination retainer and shield**
**Occupies no extra chassis area**
*No positive retention*
*Not applicable to military equipment*

## Top-release base clamp

**Available for purchase**
**Operates from top**
*Screwdriver required*

# 10

# MECHANICAL MOVEMENTS AND LINKAGES


| | |
|---|---|
| Mechanisms Actuated by Air or Hydraulic Cylinders | 248 |
| Computing Mechanisms I & II | 250 |
| Modified Geneva Drives and Special Mechanisms | 254 |
| Intermittent Movements and Mechanisms | 256 |
| Mechanisms for Providing Intermittent Rotary Motion | 258 |
| Friction Devices for Intermittent Rotary Motion | 260 |
| Kinematics of Intermittent Mechanisms | |
| I The External Geneva Wheel | 262 |
| II The Internal Geneva Wheel | 265 |
| III The Spherical Geneva Wheel | 269 |
| Kinematics of the Crank and Slot Drive | 272 |
| Linear to Angular Conversion of Gear-tooth Index Error | 275 |
| Accelerated and Decelerated Linear Motion Elements | 276 |
| Power Thrust Linkages and Their Applications | 278 |
| Transmission Linkages for Multiplying Short Motions | 280 |
| Special Purpose Mechanisms | 282 |
| Mechanisms for Producing Specific Types of Motions | 284 |
| Toggle Linkage Applications in Different Mechanisms | 286 |
| Traversing Mechanisms Used on Winding Machines | 288 |

# Mechanisms Actuated by Air or

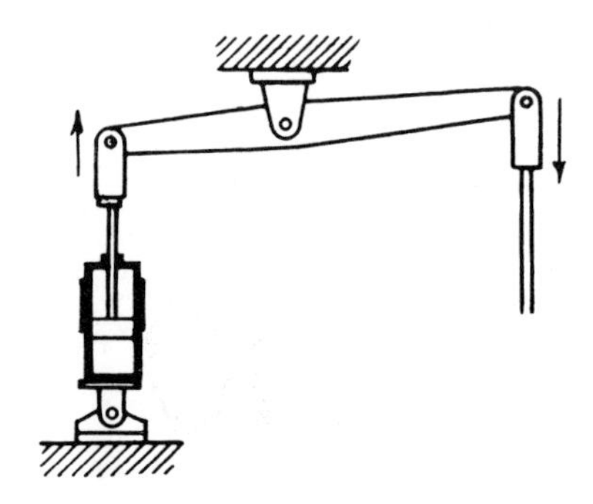

Fig. 1—Cylinder can be used with a first class lever.

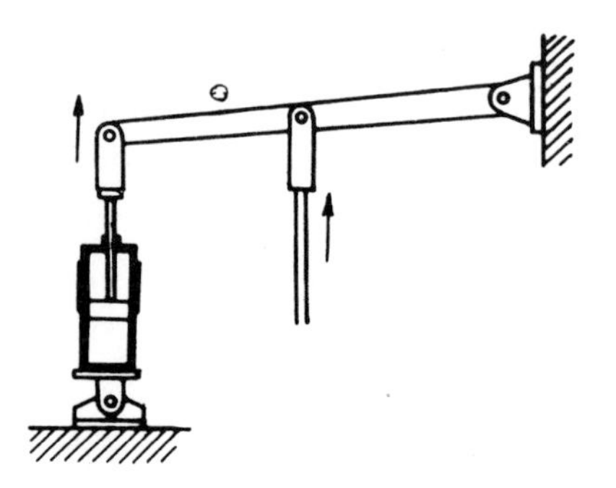

Fig. 2—Cylinder can be used with a second class lever.

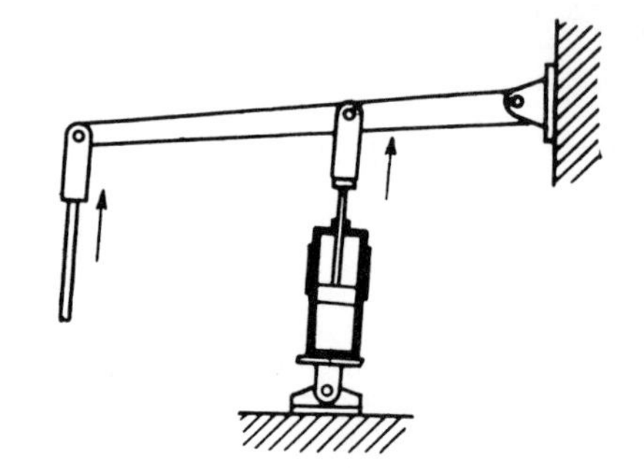

Fig. 3—Cylinder can be used with a third class lever.

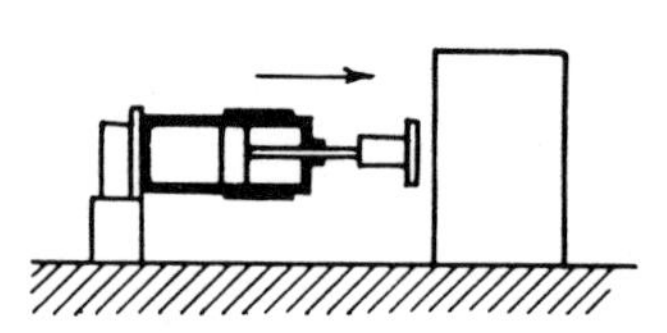

Fig. 4—Cylinder can be linked up directly to the load.

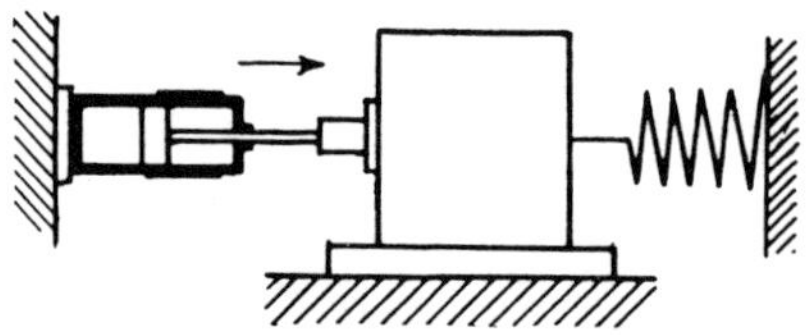

Fig. 5—Spring reduces the thrust at the end of the stroke.

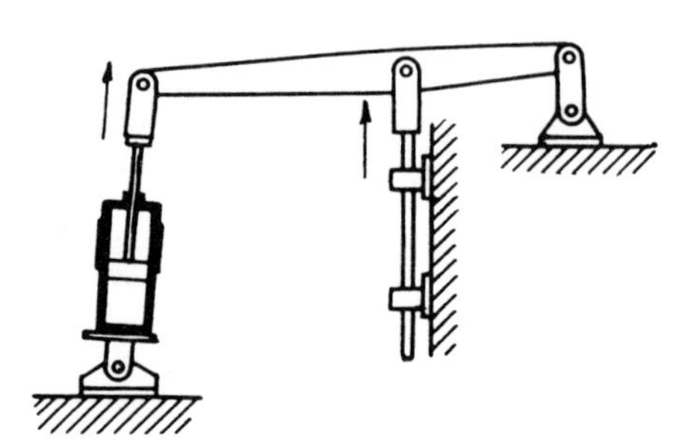

Fig. 6—Point of application of force follows the direction of thrust.

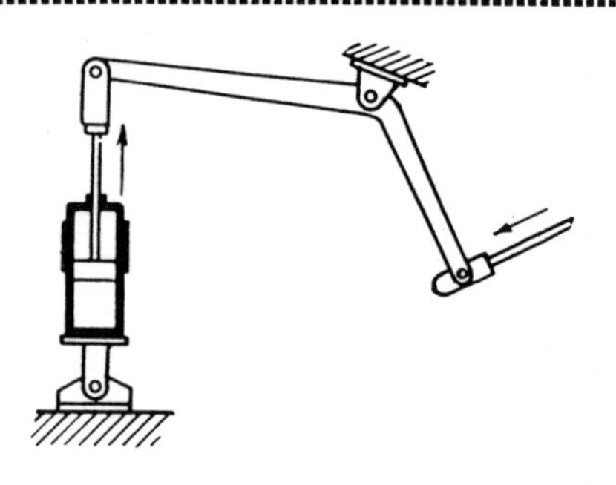

Fig. 7—Cylinder can be used with a bent lever.

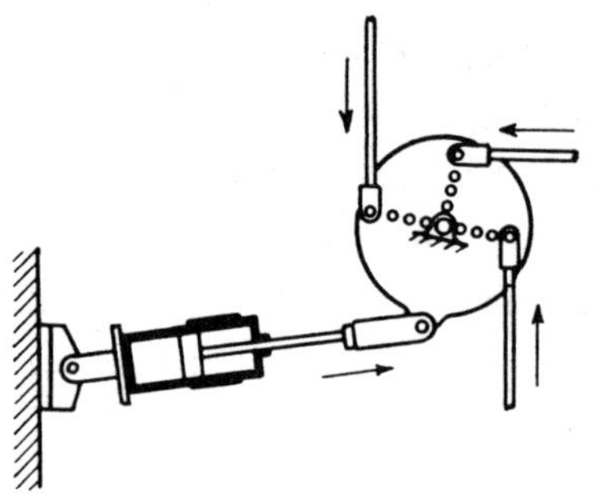

Fig. 8—Cylinder can be used with a trammel plate.

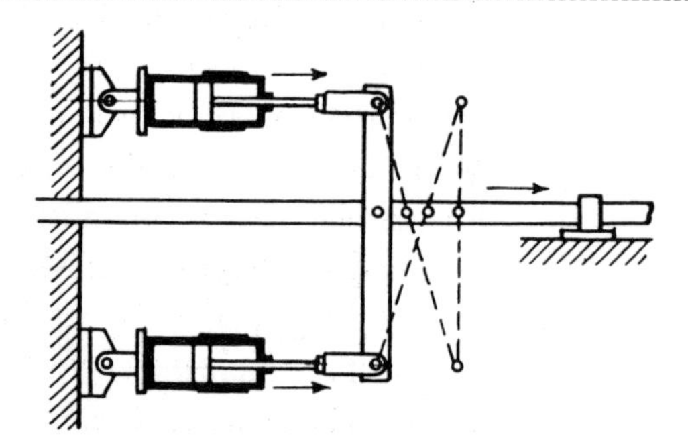

Fig. 9—Two pistons with fixed strokes position load in any of four stations.

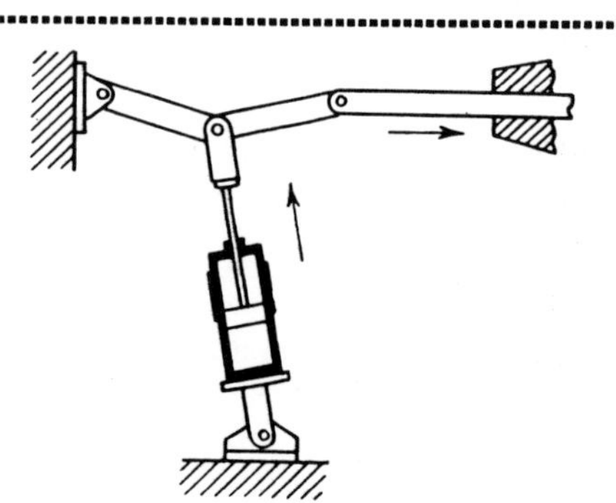

Fig. 10—A toggle can be actuated by the cylinder.

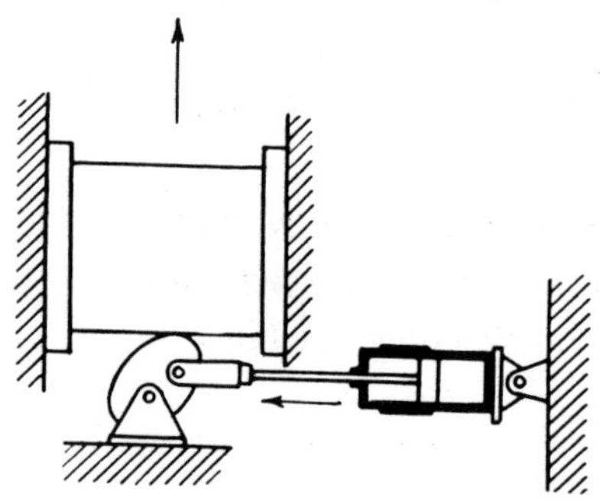

Fig. 11—The cam supports the load after completion of the stroke.

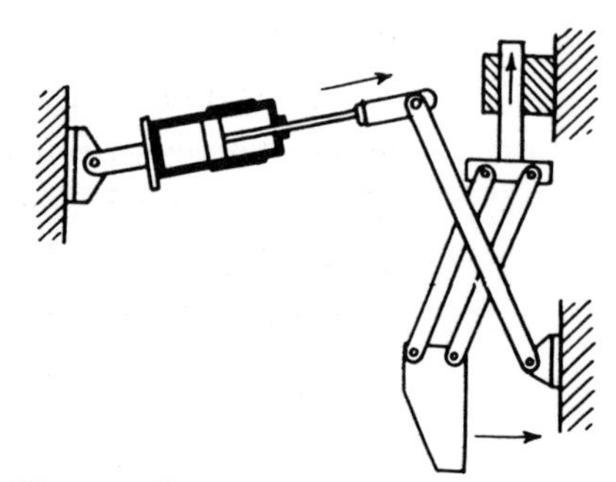

Fig. 12—Simultaneous thrusts in two different directions are obtained.

# Hydraulic Cylinders

Acknowledgment is made to Adel Precision Product Corporation, Blackhawk Manufacturing Company, Hydraulic Equipment Company, Mead Specialties Company, Westinghouse Air Brake Co., and especially to Hanna Engineering Works.

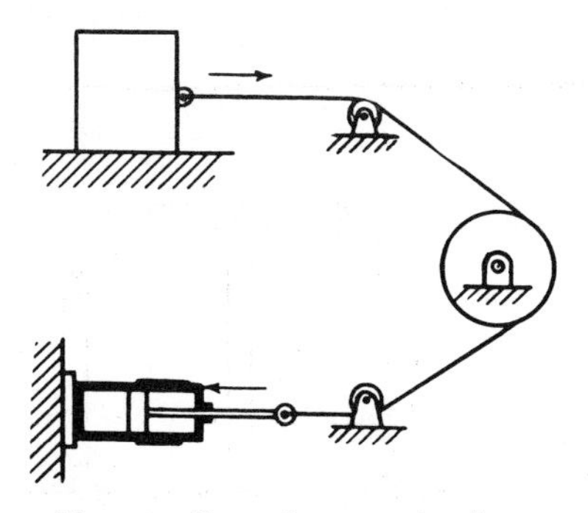

Fig. 13—Force is transmitted by a cable.

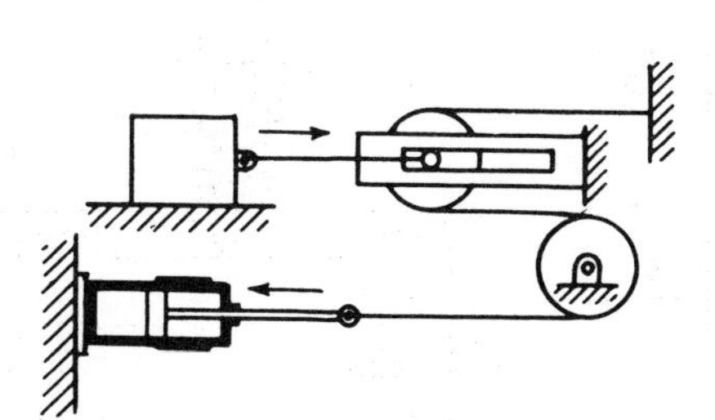

Fig. 14—Force can be modified by a system of pulleys.

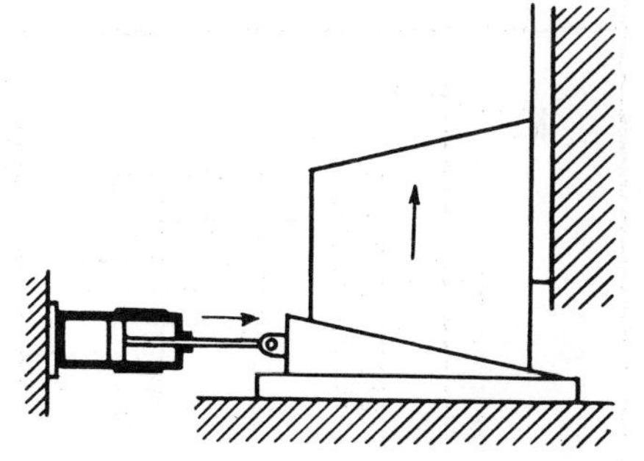

Fig. 15—Force can be modified by wedges.

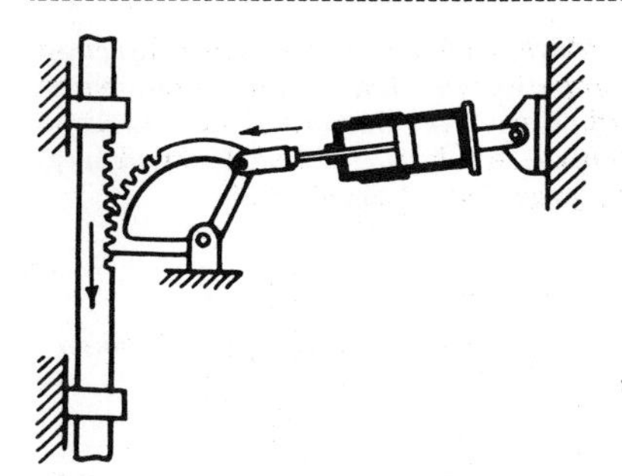

Fig. 16—Gear sector moves rack perpendicular to stroke of piston.

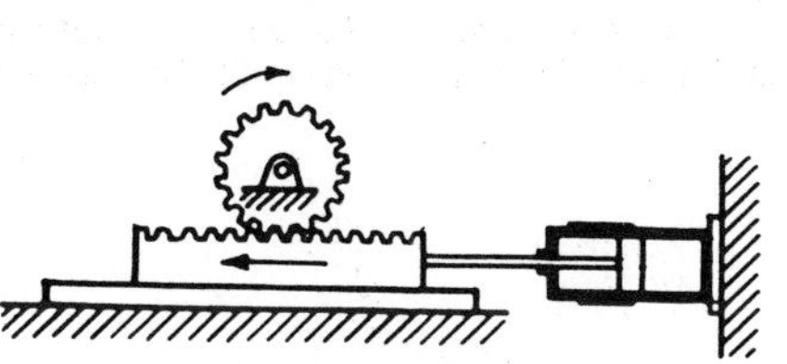

Fig. 17—Rack turns gear sector.

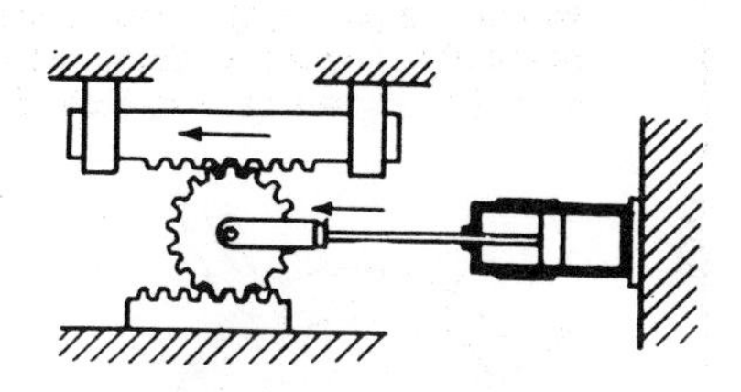

Fig. 18—Motion of movable rack is twice that of piston.

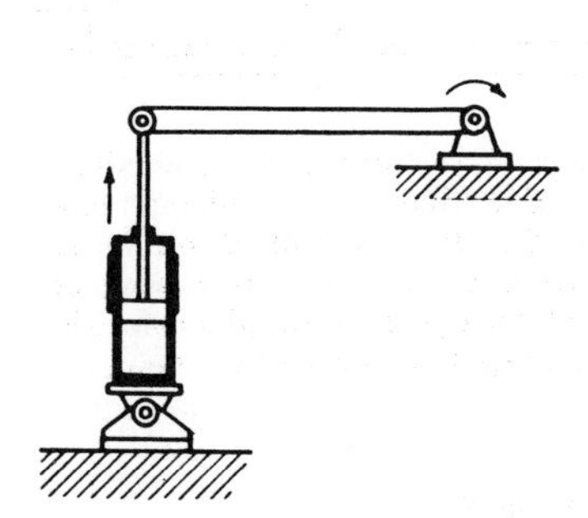

Fig. 19—Torque applied to the shaft' can be transmitted to a distant point.

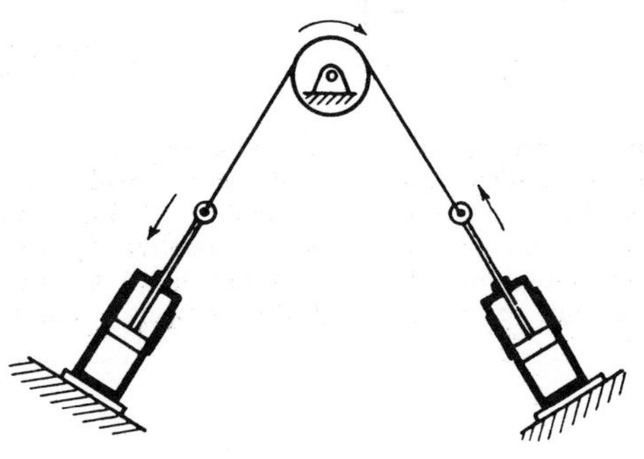

Fig. 20—Torque can also be applied to a shaft by a belt and pulley.

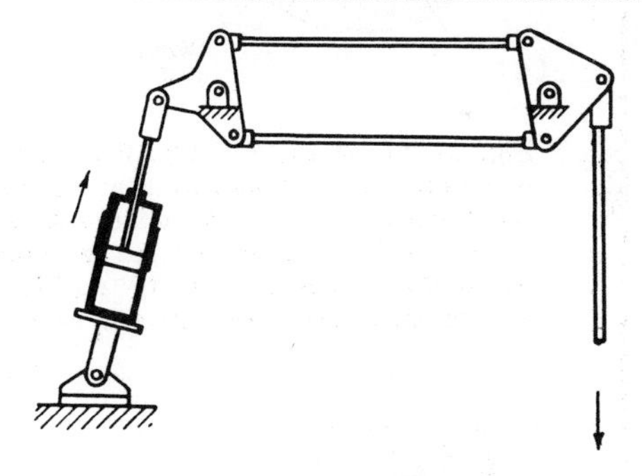

Fig. 21—Motion is transmitted to a distant point in the plane of motion.

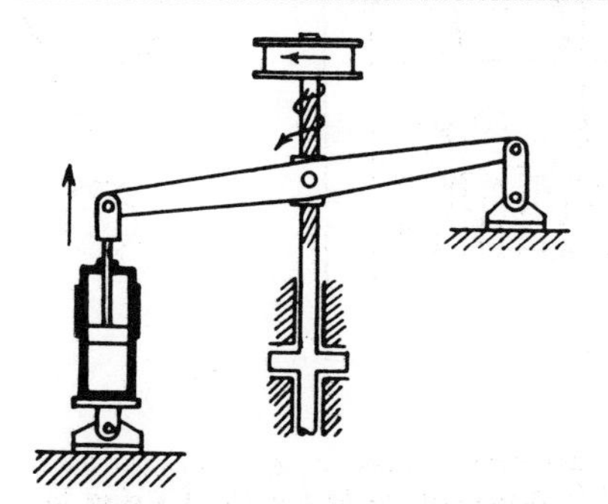

Fig. 22—A steep screw nut produces a rotation of shaft.

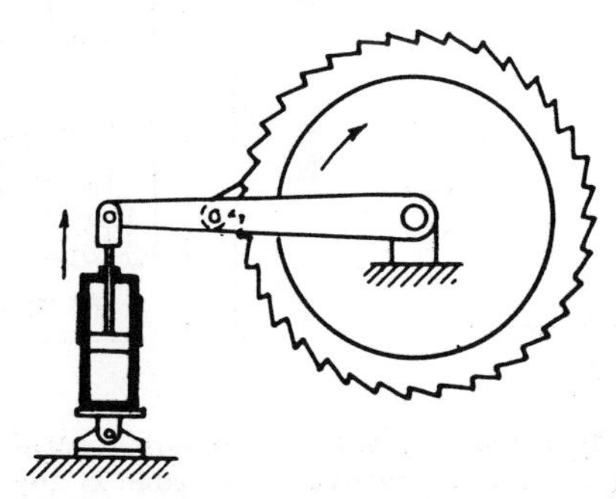

Fig. 23—Single sprocket wheel produces rotation in the plane of motion.

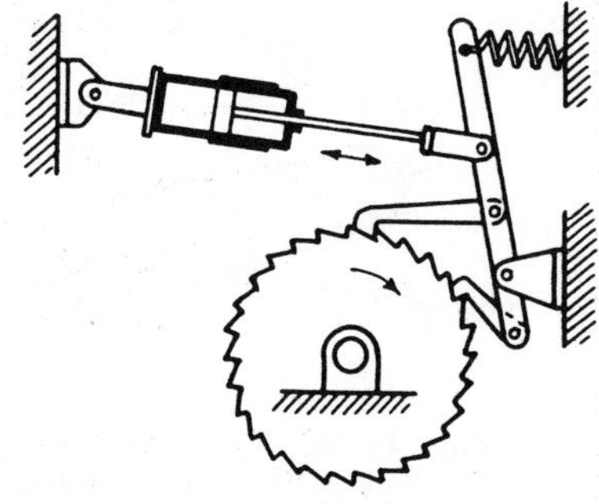

Fig. 24—Double sprocket wheel makes the rotation more nearly continuous.

# Computing mechanisms—I

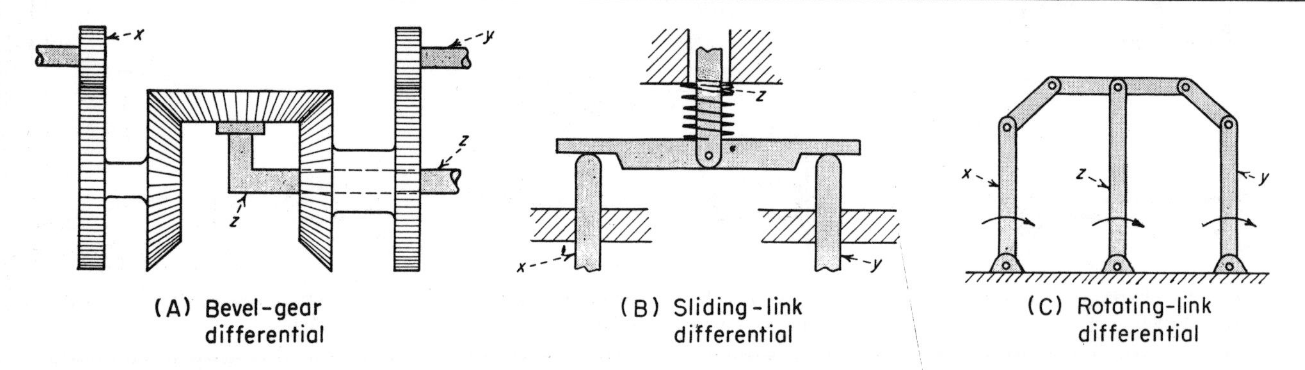


**Fig. 1—ADDITION AND SUBTRACTION.** Usually based on the differential principle; variations depend on whether inputs: (A) rotate shafts, (B) translate links, (C) angularly displaced links. Mechanisms solve equation: $z=c\ (x \pm y)$, where $c$ is scale factor, $x$ and $y$ are inputs, and $z$ is the output. Motion of $x$ and $y$ in same direction results in addition; opposite direction—subtraction. Nineteen additional variations are illustrated in "Linkage Layouts by Mathematical Analysis," Alfred Kuhlenkamp, Product Engineering, page 165, August 1955.

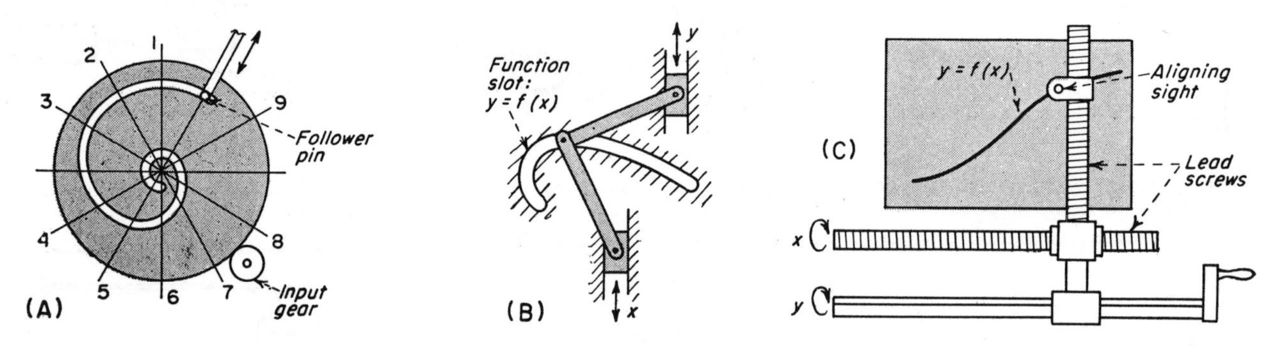


**Fig. 2—FUNCTION GENERATORS** mechanize specific equations. (A) Reciprocal cam converts a number into its reciprocal. This simplifies division by permitting simple multiplication between a numerator and its denominator. Cam rotated to position corresponding to denominator. Distance between center of cam to center of follower pin corresponds to reciprocal. (B) Function-slot cam. Ideal for complex functions involving one variable. (C) Input table. Function is plotted on large sheet attached to table. Lead screw for $x$ is turned at constant speed by an analyzer. Operator or photoelectric follower turns $y$ output to keep sight on curve.

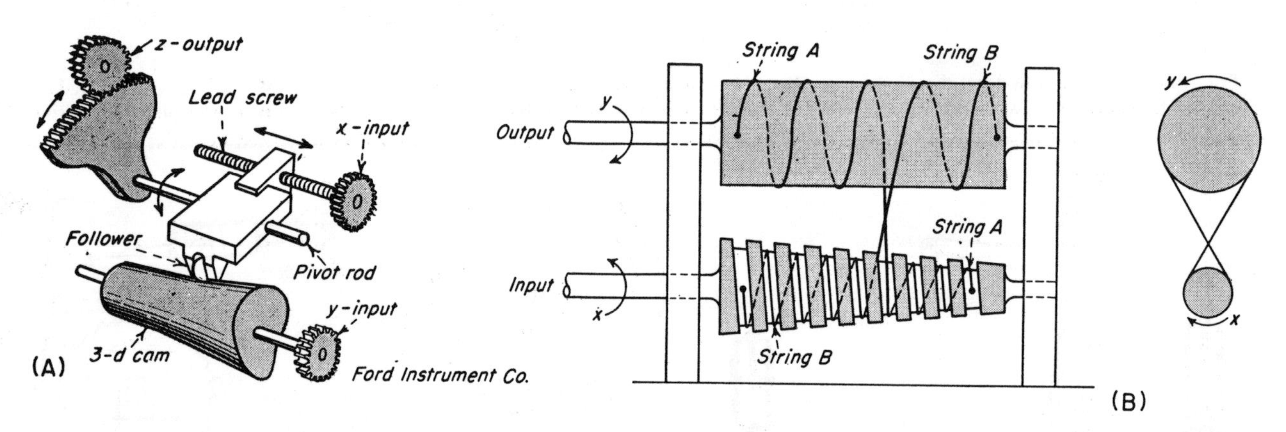


**Fig. 3—(A) THREE-DIMENSIONAL CAM** generates functions with two variables: $z=f\ (x,\ y)$. Cam rotated by $y$-input; $x$-inputs shifts follower along pivot rod. Contour of cam causes follower to rotate giving angular displacement to $z$-output gear. (B) Conical cam for squaring positive or negative inputs: $y=c\ (\pm x)^2$. Radius of cone at any point is proportional to length of string to right of point; therefore, cylinder rotation is proportional to square of cone rotation. Output is fed through a gear differential to convert to positive number.

Analog computing mechanisms are capable of virtually instantaneous response to minute variations in input. Basic units, similar to the types shown, are combined to form the final computer. These mechanisms add, subtract, resolve vectors, or solve special or trigonometric functions. Other computing mechanisms for multiplying, dividing, differentiating or integrating are presented on the next page.

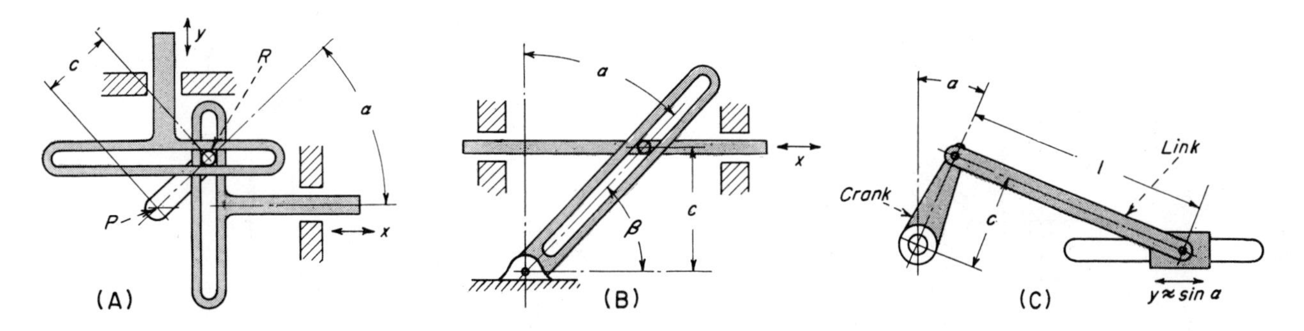


**Fig. 4—TRIGONOMETRIC FUNCTIONS.** (A) Scotch-yoke mechanism for sine and cosine functions. Crank rotates about fixed point *P* generating angle $\alpha$ and giving motion to arms: $y = c \sin \alpha$; $x = c \cos \alpha$. (B) Tangent-cotangent mechanism generates: $x = c \tan \alpha$ or $x = c \cot \beta$. (C) Eccentric and follower is easily manufactured but sine and cosine functions are approximate. Maximum error is: $e \max = l - \sqrt{l^2 - c^2}$; error is zero at 90 and 270 deg. $l$ is the length of the link and $c$ is the length of the crank.

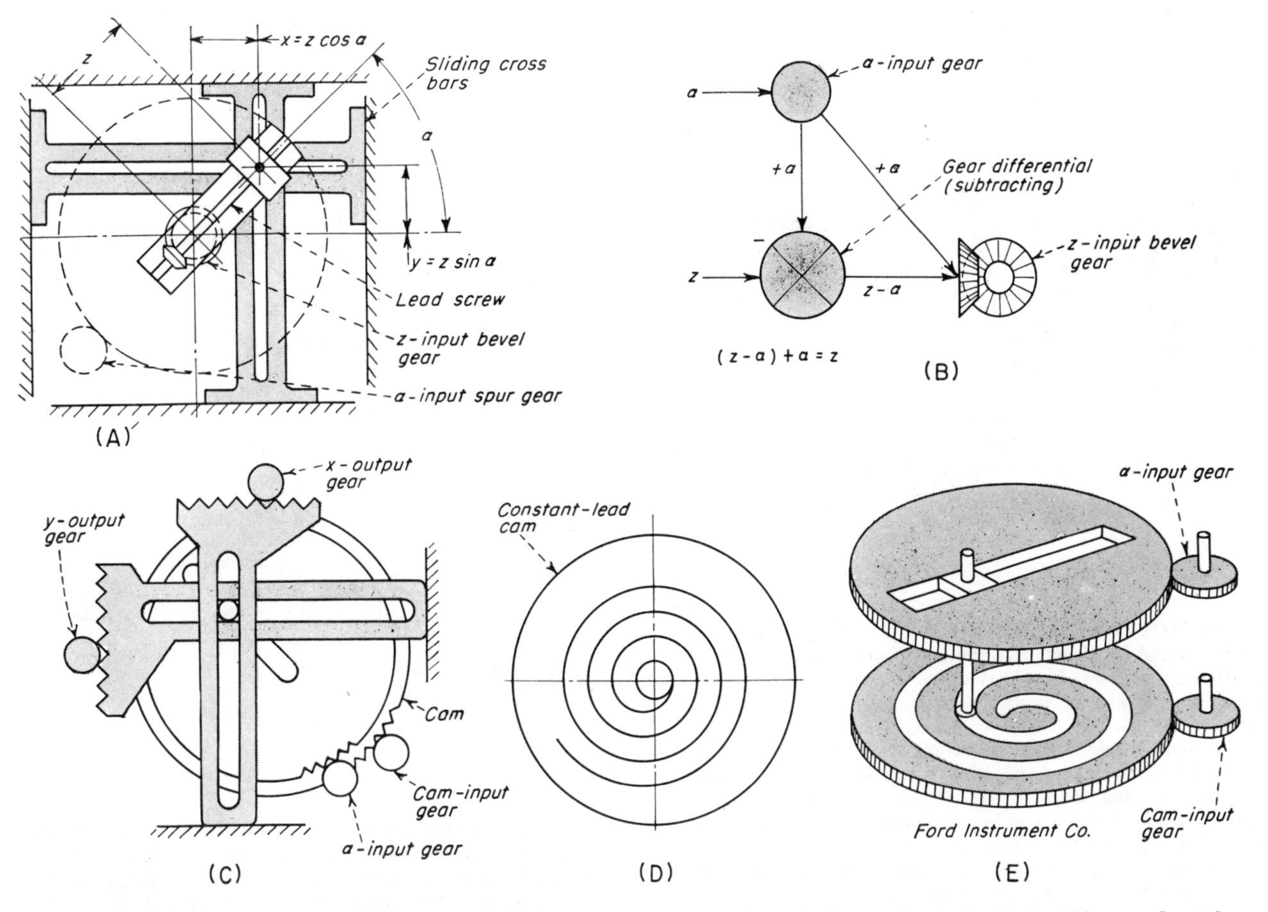


**Fig. 5—COMPONENT RESOLVERS** for obtaining $x$ and $y$ components of vectors that are continuously changing in both angle and magnitude. Equations are: $x = z \cos \alpha$, $y = z \sin \alpha$ where $z$ is magnitude of vector, and $\alpha$ is vector angle. Mechanisms can also combine components to obtain resultant. Input in (A) are through bevel gears and lead screws for $z$ input, and through spur gears for $\alpha$-input. Compensating gear differential (B) prevents $\alpha$-input from affecting $z$-input. This problem solved in (C) by using constant-lead cam (D) and (E).

# Computing mechanisms—II

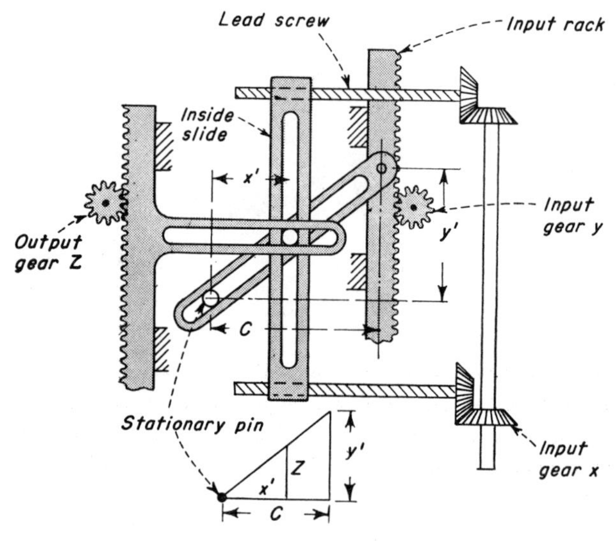


Fig. 1 (A)

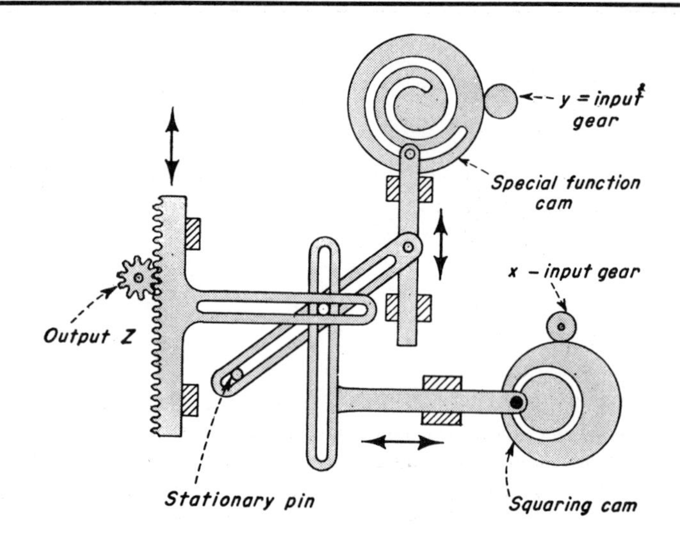


Fig. 2 (A)

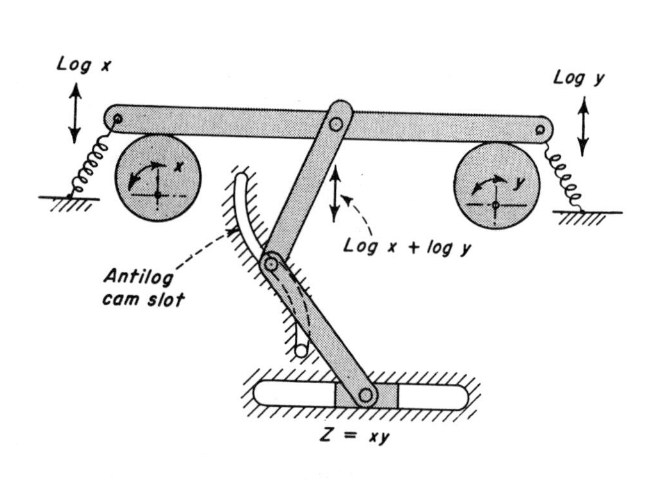


Fig. 1 (B)

**Fig. 1—MULTIPLICATION OF TWO TABLES $x$ and $y$ usually solved by either: (A) Similar triangle method, or (B) logarithmic method. In (A), lengths $x'$ and $y'$ are proportional to rotation of input gears $x$ and $y$. Distance $c$ is constant. By similar triangles: $z/x=y/c$ or $z=xy/c$, where $z$ is vertical displacement of output rack. Mechanism can be modified to accept negative variables. In (B), input variables are fed through logarithmic cams giving linear displacements of log $x$ and log $y$. Functions are then added by a differential link giving $z=\log x + \log y=\log xy$ (neglecting scale factors). Result is fed through antilog cam; motion of follower represents $z=xy$.**

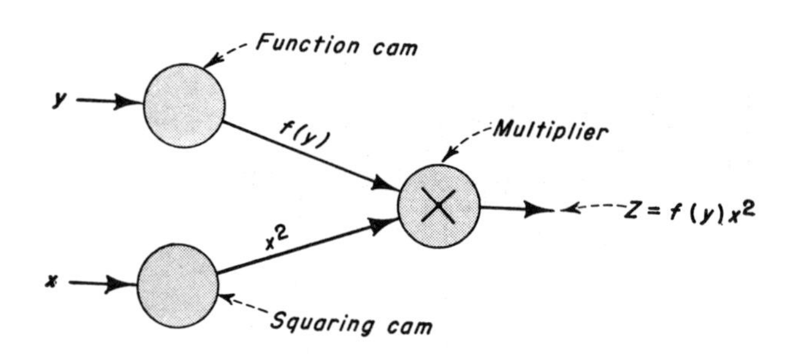


Fig. 2 (B)

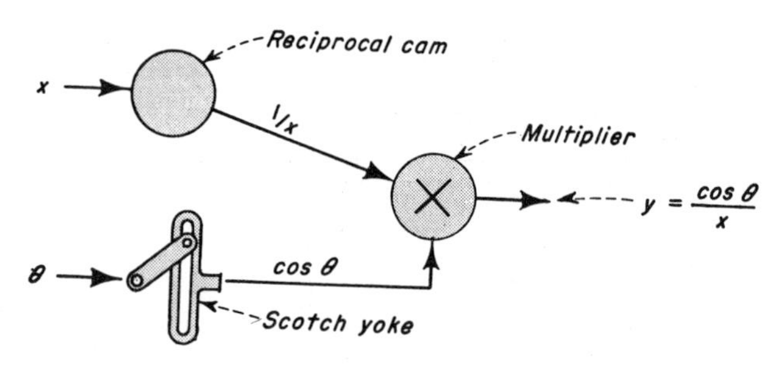


Fig. 2 (C)

**Fig. 2—MULTIPLICATION OF COMPLEX FUNCTIONS can be accomplished by substituting cams in place of input slides and racks of mechanism in Fig. 1. Principle of similar triangles still applies. Mechanism in (A) solves the equation: $z=f(y)\,x^2$. Schematic is shown in (B). Division of two variables can be done by feeding one of the variables through a reciprocal cam and then multiplying it by the other. Schematic in (C) shows solution of $y=\cos \phi/x$.**

Several typical computing mechanisms for performing the mathematical operations of multiplication, division, differentiation, and integration of variable functions. Analog computing mechanisms for adding and subtracting, for resolving vectors, and for trigonometric functions, are discussed in Part I.

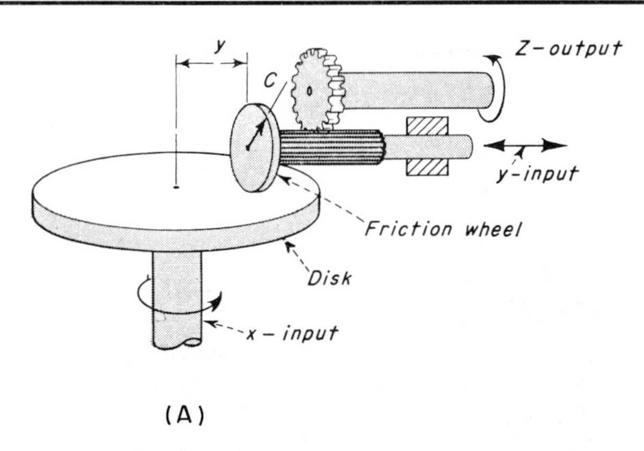


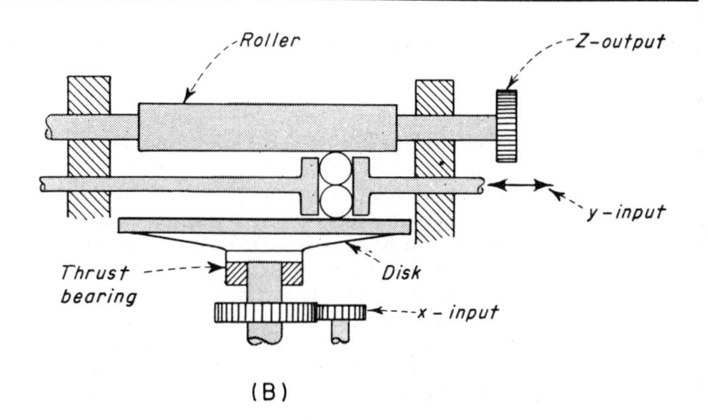


**Fig. 3—INTEGRATORS are basically variable speed drives. Disk in (A) is rotated by x-input which, in turn, rotates the friction wheel. Output is through gear driven by spline on shaft of friction wheel. Input y varies the distance of friction wheel from center of disk. For a wheel with radius c, rotation of disk through infinitesmal turn dx causes corresponding turn dz equal to: dz=(1/c) y dx. For definite x revolutions, total z revolutions will be equal to the integral of (1/c) y dx, where y varies as called for by the problem. Ball integrator in (B), gives pure rolling in all directions.**

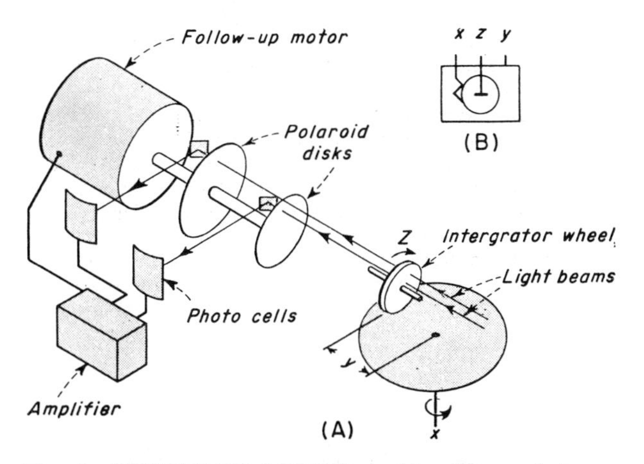


**Fig. 4—FOLLOW-UP MOTOR avoids slippage between wheel and disk of integrator in Fig. 3 (A). No torque is taken from wheel except to overcome friction in bearings. Web of integrator wheel is made of polaroid. Light beams generate current to amplifier which controls follow-up motor. Symbol at upper right corner is schematic representation of integrator. For more information see, "Mechanism," by Joseph S. Beggs, McGraw-Hill Book Co., N. Y., 1955.**

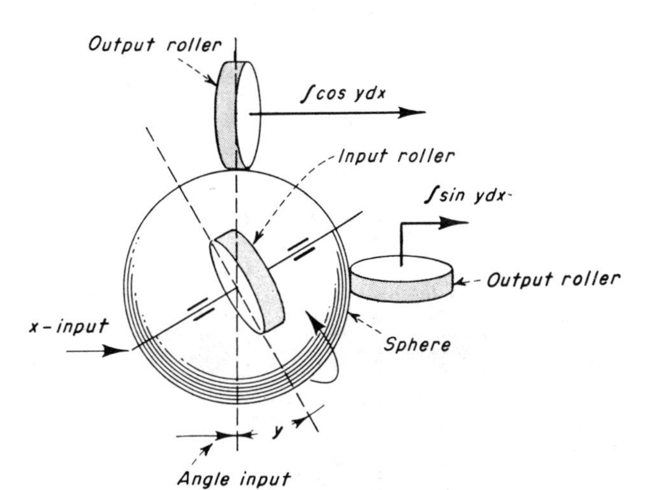


**Fig. 5—COMPONENT INTEGRATOR uses three disks to obtain $x$ and $y$ components of a differential equation. Input roller $x$ spins sphere; $y$ input changes angle of roller. Output rollers give integrals of components paralleling $x$ and $y$ axes. Ford Instrument Company.**

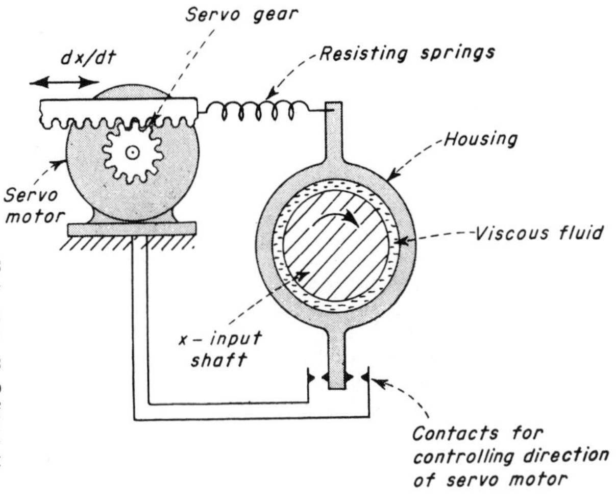


**Fig. 6—DIFFERENTIATOR uses principle that viscous drag force in thin layer of fluid is proportional to velocity of rotating x-input shaft. Drag force counteracted by spring; spring length regulated by servo motor controlled by contacts. Change in shaft velocity causes change in viscous torque. Shift in housing closes contacts causing motor to adjust spring length and balance system. Total rotation of servo gear is proportional to dx/dt. From "Mechanical Computing Mechanisms," Reid and Stromback, Product Engineering, October 1949 page 126.**

# Modified Geneva Drives and

These sketches were selected as practical examples of uncommon, but often useful mechanisms. Most of them serve to add a varying velocity component to the conventional Geneva motion. The data

Fig. 1—(Below) In the conventional external Geneva drive, a constant-velocity input produces an output consisting of a varying velocity period plus a dwell. In this modified Geneva, the motion period has a constant-velocity interval which can be varied within limits. When spring-loaded driving roller *a* enters the fixed cam *b*, the output-shaft velocity is zero. As the roller travels along the cam path, the output velocity rises to some constant value, which is less than the maximum output of an unmodified Geneva with the same number of slots; the duration of constant-velocity output is arbitrary within limits. When the roller leaves the cam, the output velocity is zero; then the output shaft dwells until the roller re-enters the cam. The spring produces a variable radial distance of the driving roller from the input shaft which accounts for the described motions. The locus of the roller's path during the constant-velocity output is based on the velocity-ratio desired.

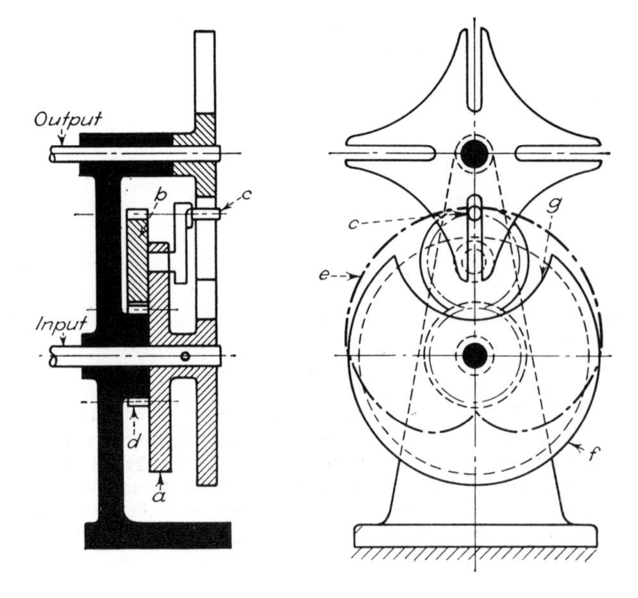


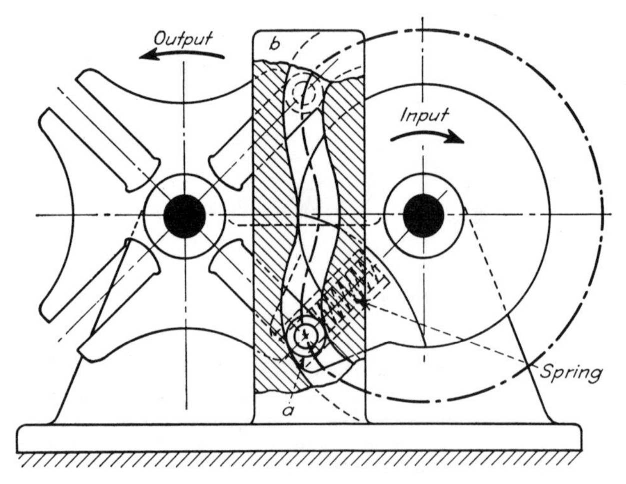


*Saxonian Carton Machine Co., Dresden, Germany*

Fig. 2—(Above) This design incorporates a planet gear in the drive mechanism. The motion period of the output shaft is decreased and the maximum angular velocity is increased over that of an unmodified Geneva with the same number of slots. Crank wheel *a* drives the unit composed of plant gear *b* and driving roller *c*. The axis of the driving roller coincides with a point on the pitch circle of the planet gear; since the planet gear rolls around the fixed sun gear *d*, the axis of roller *c* describes a cardioid *e*. To prevent the roller from interfering with the locking disk *f*, the clearance arc *g* must be larger than required for unmodified Genevas.

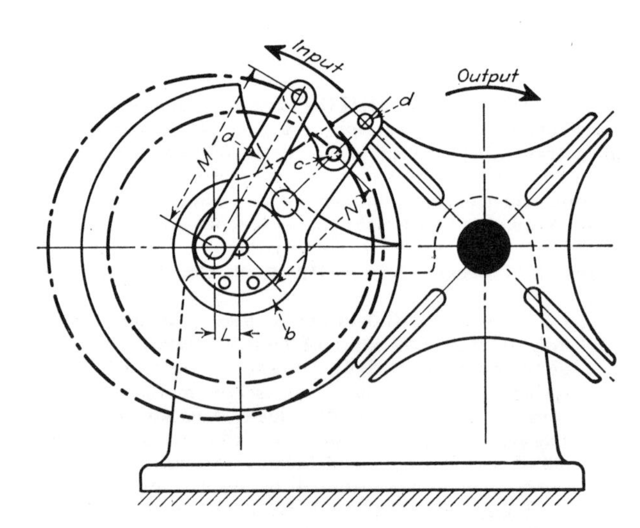


Fig. 3—A motion curve similar to that of Fig. 2 can be derived by driving a Geneva wheel by means of a two-crank linkage. Input crank *a* drives crank *b* through link *c*. The variable angular velocity of driving roller *d*, mounted on *b*, depends on the center distance *L*, and on the radii *M* and *N* of the crank arms. This velocity is about equivalent to what would be produced if the input shaft were driven by elliptical gears.

# Special Mechanisms

SIGMUND RAPPAPORT
Ford Instrument Company

**were based in part on material and figures in AWF und VDMA Getriebeblaetter, published by Ausschuss fuer Getriebe beim Ausschuss fuer wirtschaftiche Fertigung, Leipzig, Germany.**

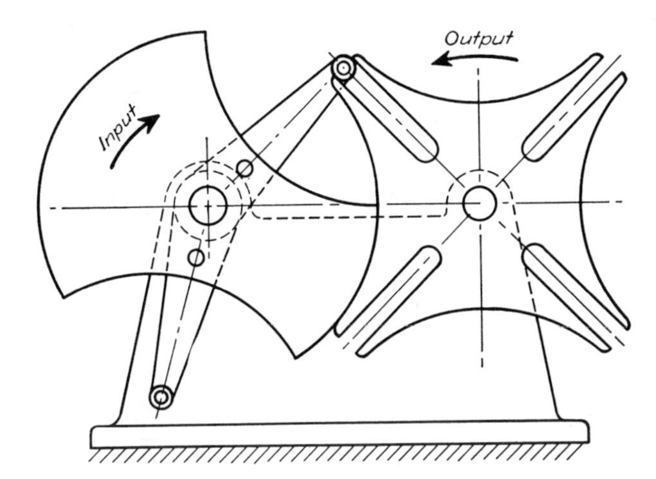


Fig. 4—(Left) The duration of the dwell periods is changed by arranging the driving rollers unsymmetrically around the input shaft. This does not affect the duration of the motion periods. If unequal motion periods are desired as well as unequal dwell periods, then the roller crank-arms must be unequal in length and the star must be suitably modified; such a mechanism is called an "irregular Geneva drive."

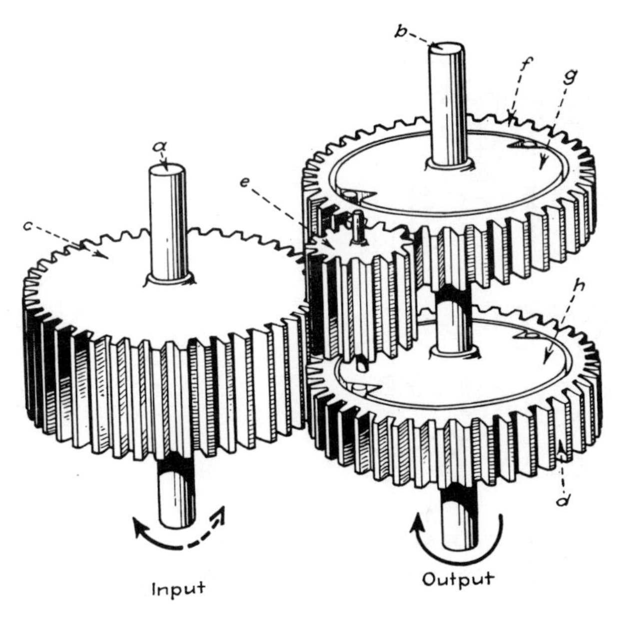


Fig. 5—(Below) In this intermittent drive, the two rollers drive the output shaft as well as lock it during dwell periods. For each revolution of the input shaft the output shaft has two motion periods. The output displacement $\phi$ is determined by the number of teeth; the driving angle, $\psi$, may be chosen within limits. Gear *a* is driven intermittently by two driving rollers mounted on input wheel *b*, which is bearing-mounted on frame *c*. During the dwell period the rollers circle around the top of a tooth. During the motion period, a roller's path *d* relative to the driven gear is a straight line inclined towards the output shaft. The tooth profile is a curve parallel to path *d*. The top land of a tooth becomes the arc of a circle of radius *R*, the arc approximating part of the path of a roller.

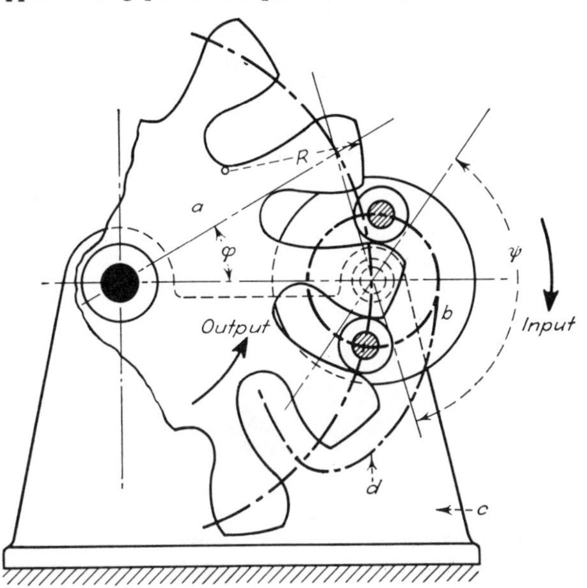


Fig. 6—This uni-directional drive was developed by the author and to his knowledge is novel. The output shaft rotates in the same direction at all times, without regard to the direction of the rotation of the input shaft; angular velocity of the output shaft is directly proportional to the angular velocity of the input shaft. Input shaft *a* carries spur gear *c*, which has approximately twice the face width of spur gears *f* and *d* mounted on output shaft *b*. Spur gear *c* meshes with idler *e* and with spur gear *d*. Idler *e* meshes with spur gears *c* and *f*. The output shaft *b* carries two free-wheel disks *g* and *h*, which are oriented uni-directionally.

When the input shaft rotates clockwise (bold arrow), spur gear *d* rotates counter-clockwise and idles around free-wheel disk *h*. At the same time idler *e*, which is also rotating counter-clockwise, causes spur gear *f* to turn clockwise and engage the rollers on free-wheel disk *g*; thus, shaft *b* is made to rotate clockwise. On the other hand, if the input shaft turns counter-clockwise (dotted arrow), then spur gear *f* will idle while spur gear *d* engages free-wheel disk *h*, again causing shaft *b* to rotate clockwise.

# INTERMITTENT MOVEMENTS

SEVERAL RECENTLY PATENTED mechanisms for producing intermittent motion are included in this group of devices. Many of the simpler and well-known mechanisms such as the common ratchet and pawl with its many modifications, escapements, numerous variations of the Geneva mechanisms and timed electrical contactors have been intentionally omitted in favor of the more ingenious combinations of mechanisms for accomplishing intermittent motion under unusual conditions.

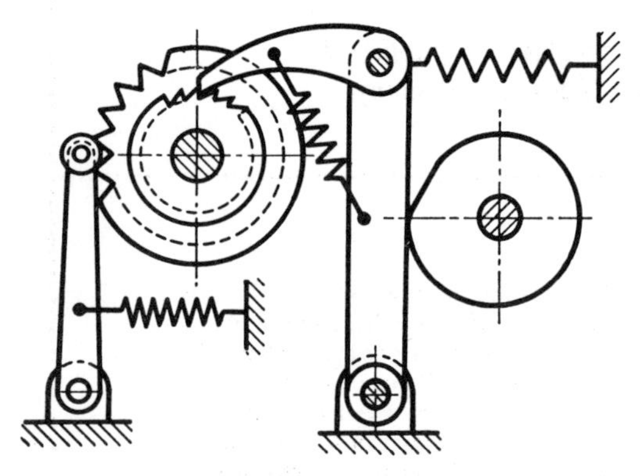

FIG. 1– CAM DRIVEN RATCHET

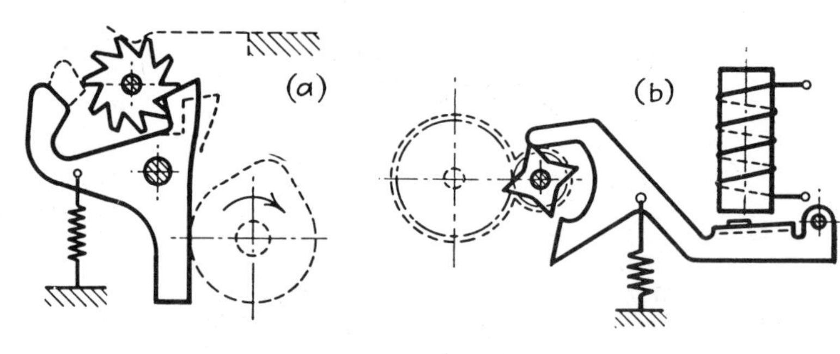


FIG. 3– (a) CAM OPERATED ESCAPEMENT ON A TAXIMETER
(b) SOLENOID OPERATED ESCAPEMENT

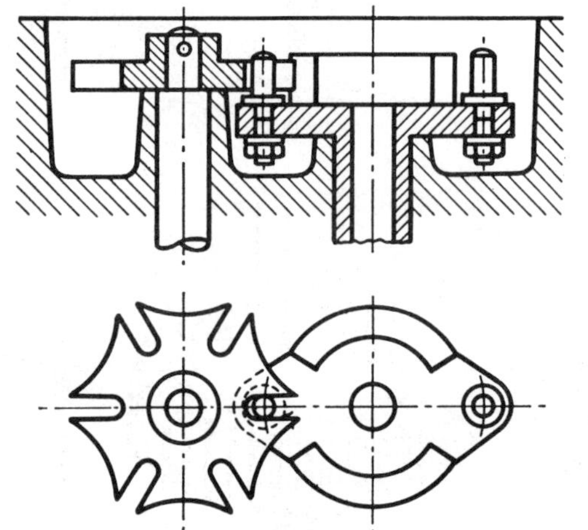

FIG. 2– SIX-SIDED MALTESE CROSS AND DOUBLE DRIVER GIVES 3:1 RATIO

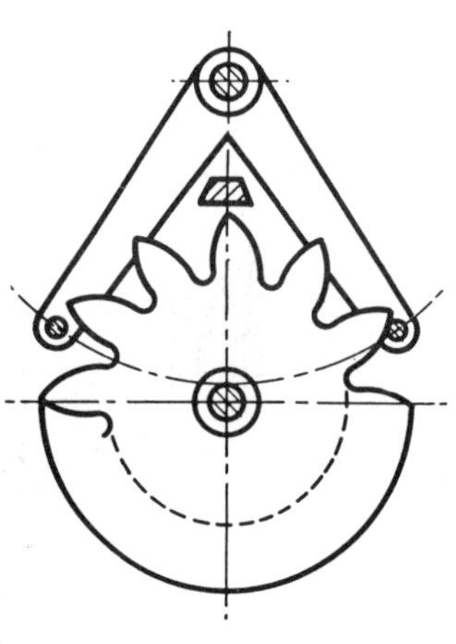

FIG. 4– ESCAPEMENT USED ON AN ELECTRIC METER

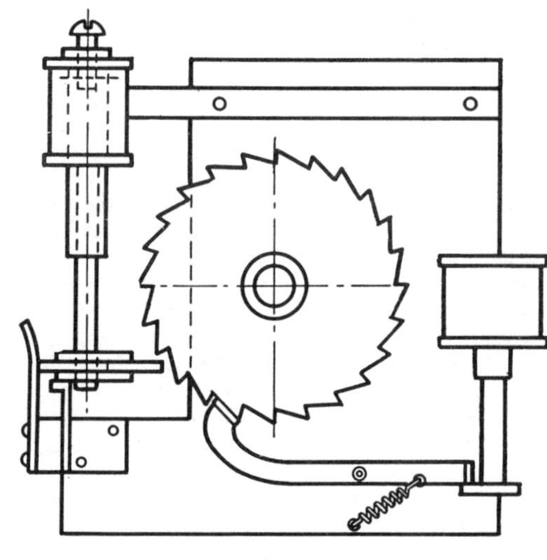

FIG. 5– SOLENOID-OPERATED RATCHET WITH SOLENOID RESETING MECHANISM. A SLIDING WASHER ENGAGES TEETH

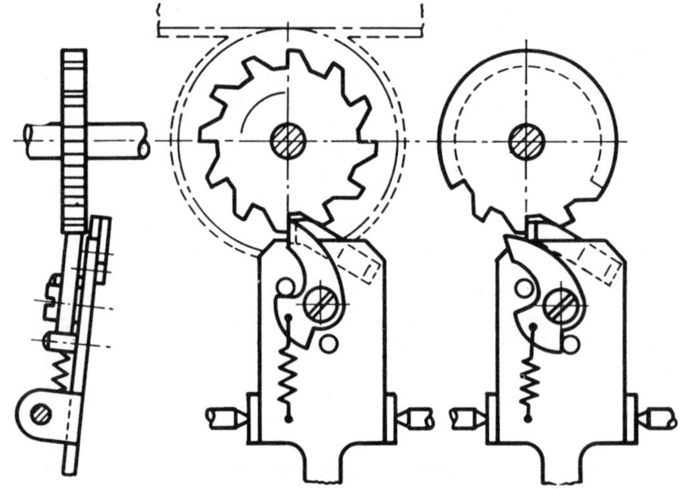

FIG. 6– PLATE OSCILLATING ACROSS PLANE OF RATCHET GEAR ESCAPEMENT CARRIES STATIONARY AND SPRING HELD PAWLS

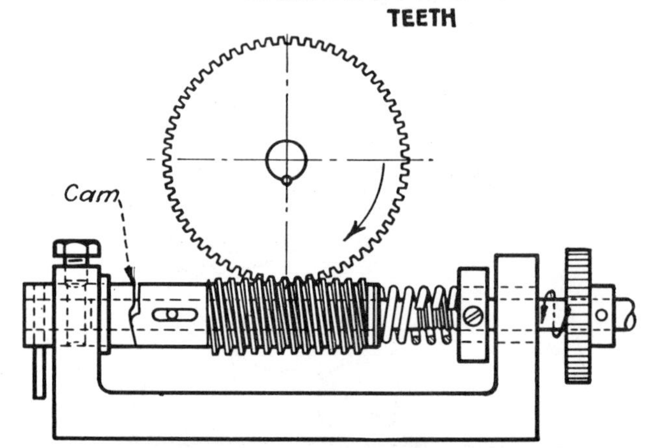


FIG. 7– WORM DRIVE COMPENSATED BY CAM ON WORK SHAFT, PRODUCES INTERMITTENT MOTION OF GEAR

# AND MECHANISMS

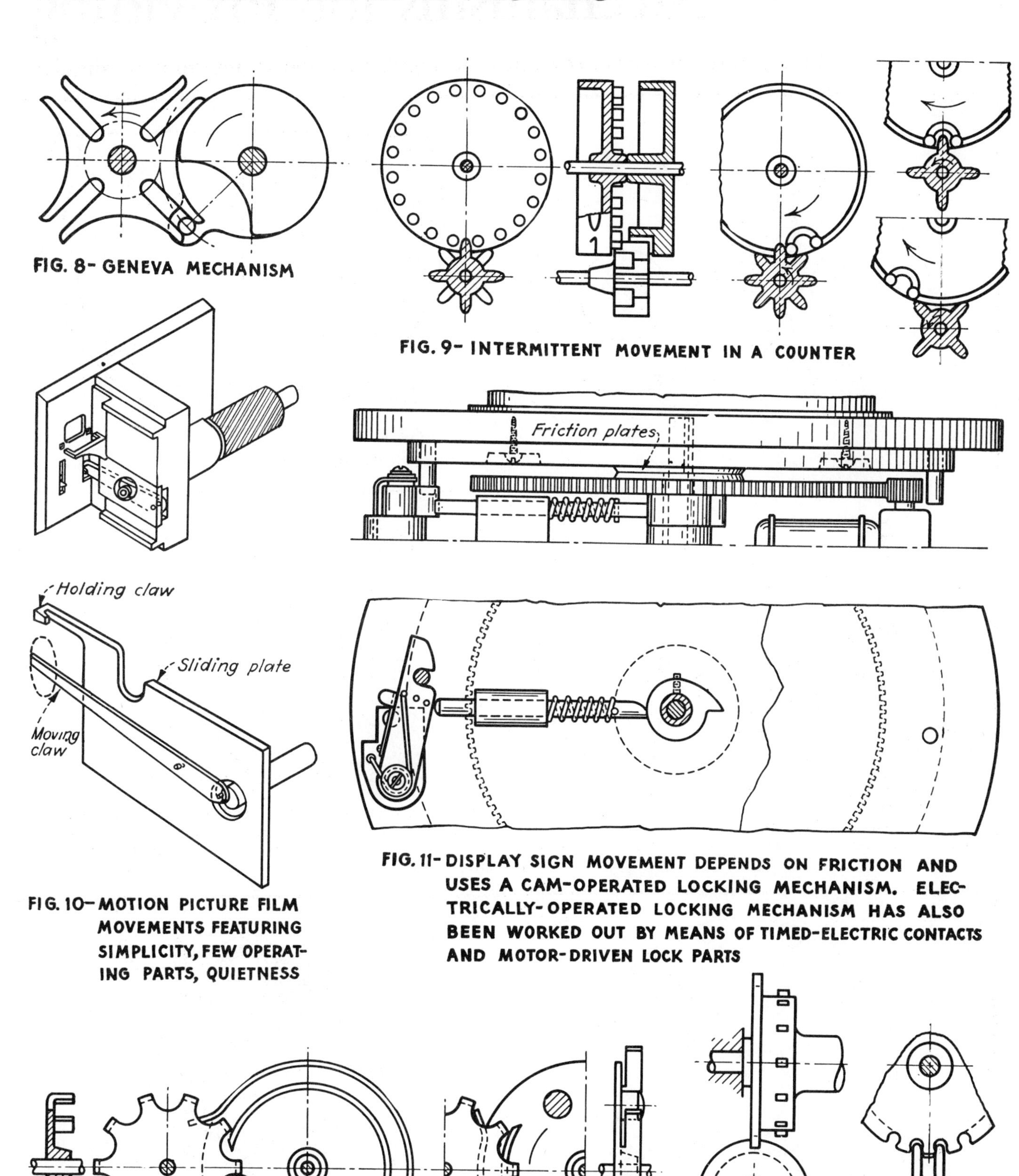


FIG. 8- GENEVA MECHANISM

FIG. 9- INTERMITTENT MOVEMENT IN A COUNTER

FIG. 10—MOTION PICTURE FILM MOVEMENTS FEATURING SIMPLICITY, FEW OPERATING PARTS, QUIETNESS

FIG. 11- DISPLAY SIGN MOVEMENT DEPENDS ON FRICTION AND USES A CAM-OPERATED LOCKING MECHANISM. ELECTRICALLY-OPERATED LOCKING MECHANISM HAS ALSO BEEN WORKED OUT BY MEANS OF TIMED-ELECTRIC CONTACTS AND MOTOR-DRIVEN LOCK PARTS

FIG. 12-MODIFICATION OF THE GENEVA MECHANISM

FIG. 13- SIMPLE WORM GEAR MECHANISM

# Mechanisms for Providing

**Eleven ways of converting uniform angular motion to intermittent angular**

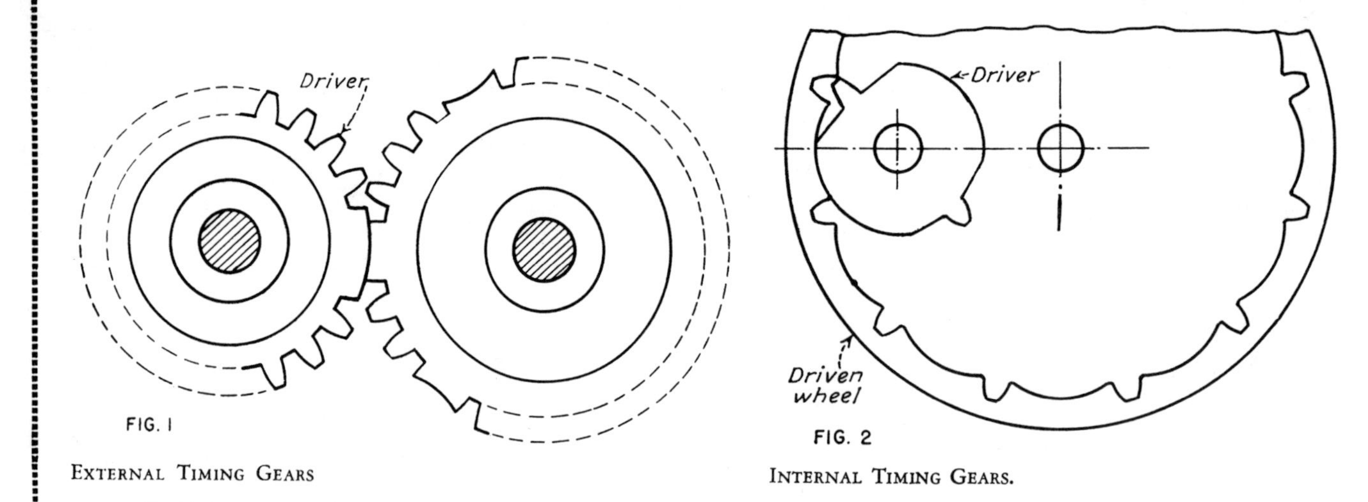


External Timing Gears

Internal Timing Gears.

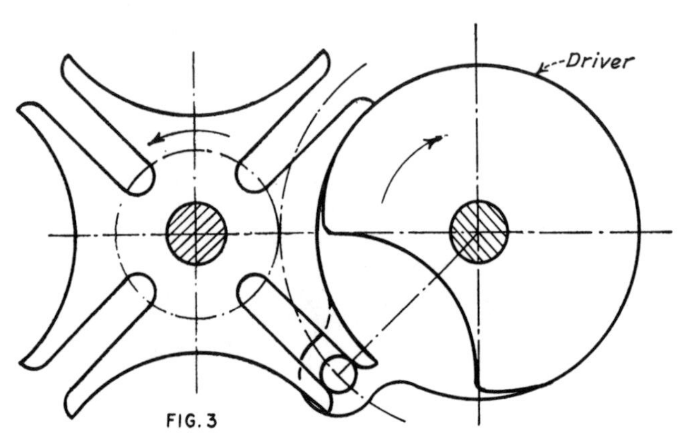


External Geneva Mechanism. Operation is smoother than timing gears. Practical number of slots is from 3 to 18. Duration of dwell is more than 180 deg of driver rotation.

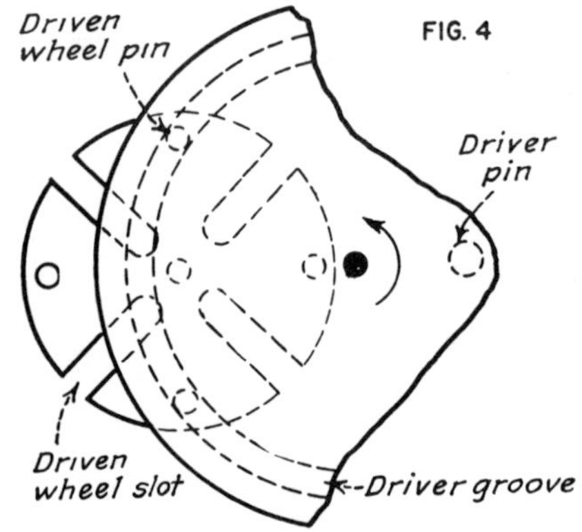


External Geneva Mechanism. Difference from mechanism of of Fig. 3 lies in method of locking. Driver grooves lock driven wheel pins during dwell. During movement, driver pin mates with driven wheel slot.

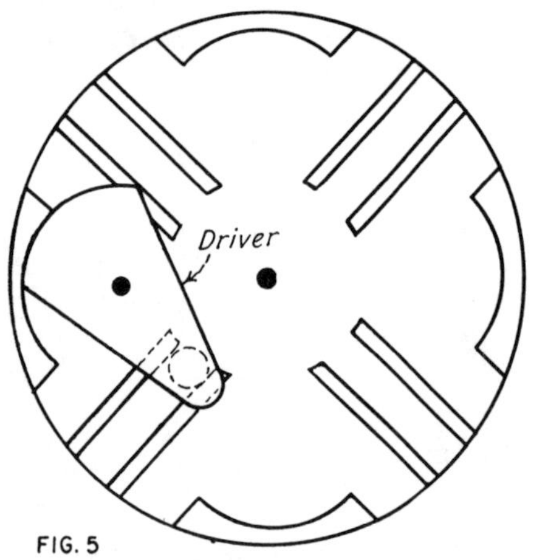


Internal Geneva Mechanism. Driver and driven wheel rotate in same direction. Duration of dwell is more than 180 deg of driver rotation.

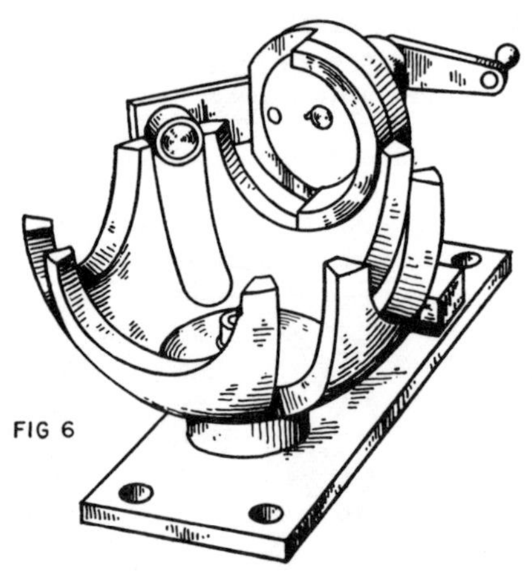


Spherical Geneva Mechanism. Driver and driven wheel are on perpendicular shafts. Duration of dwell is exactly 180 deg of driver rotation.

# Intermittent Rotary Motion

**motion and an explanation of two indirect ways to get this conversion.**

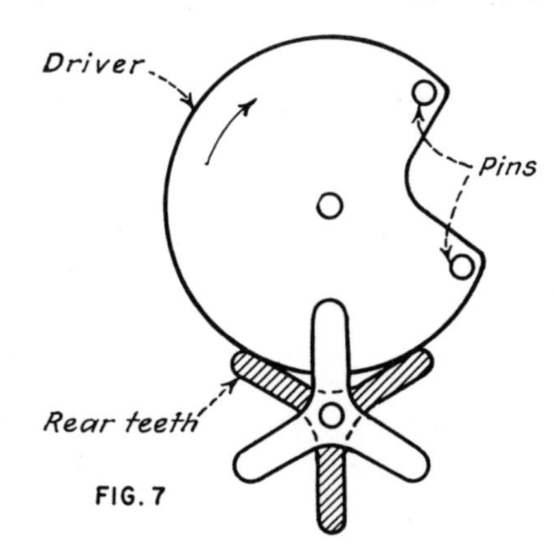


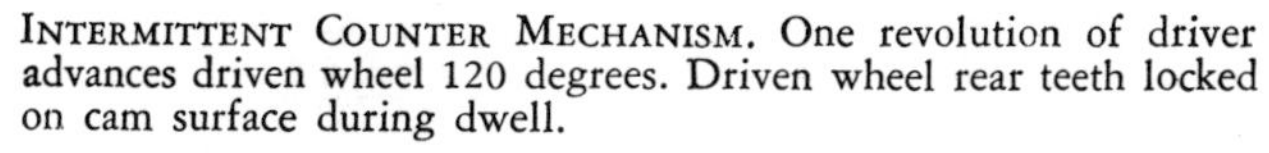

INTERMITTENT COUNTER MECHANISM. One revolution of driver advances driven wheel 120 degrees. Driven wheel rear teeth locked on cam surface during dwell.

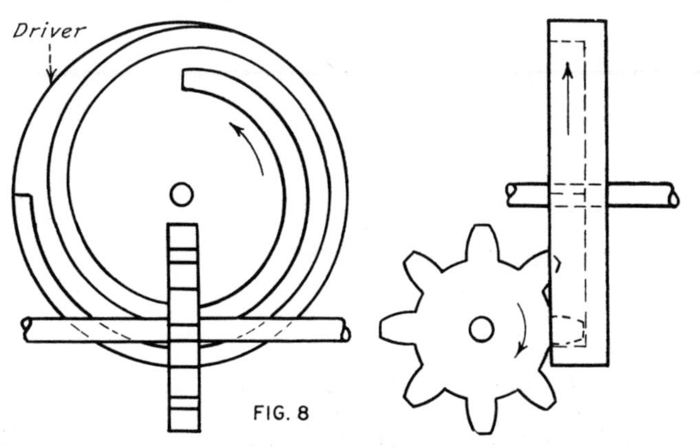


SPIRAL AND WHEEL. One revolution of spiral advances driven wheel 1 tooth. Driven wheel tooth locked in driver groove during dwell.

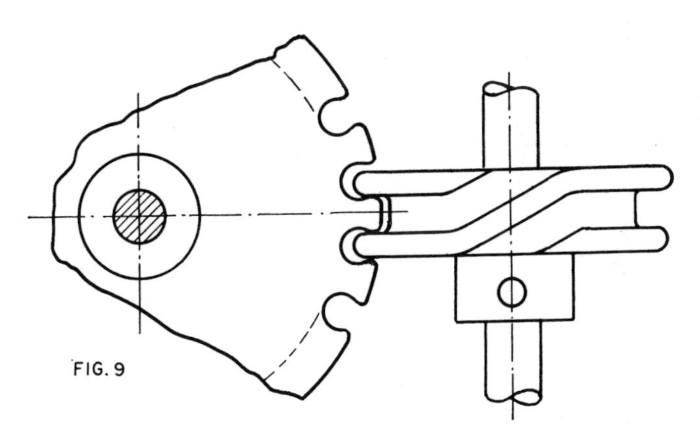


SPECIAL WORM AND WHEEL. Spiral of Fig. 8 is replaced by special worm.

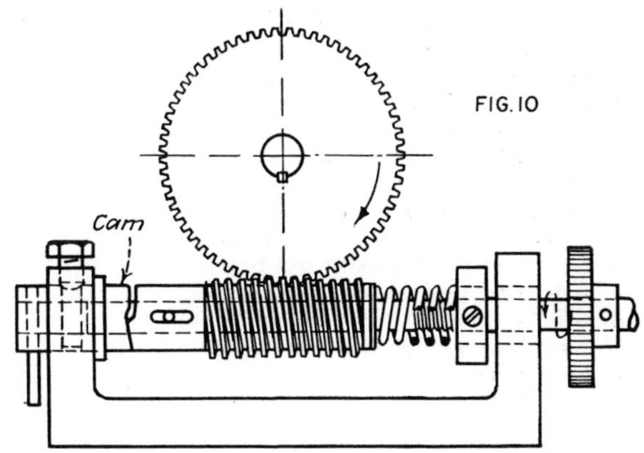


WORM, CAM AND WHEEL. Standard worm and cam replace special worm of Fig. 9.

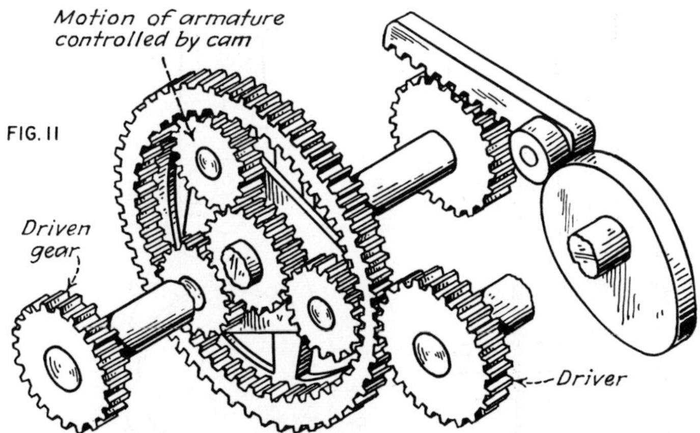


SPECIAL PLANETARY GEAR MECHANISM. Principle of relative motion of mating gears illustrated in Fig. 10 can be applied to spur gears in planetary system. Motion of normally fixed planet centers produces intermittent motion of sum gear.

THE CONVERSION of a uniform rotation to an intermittent rotation need not be performed in a single step. Two ways in which this can be carried out in two steps are:

1. Conversion of uniform rotation to reciprocating motion followed by conversion of reciprocating motion to intermittent motion.

2. Conversion of uniform rotation to angular oscillation followed by conversion of angular oscillation to intermittent rotation.

*Fig. 11—Courtesy G. J. Tarbourdet, ASME paper No. 48-SA-18.*

# FRICTION DEVICES
## For Intermittent Rotary Motion

W. M. HALLIDAY
Southport, England

FIG. 1—WEDGE AND DISK. Consists of shaft *A* supported in bearing block *J*; ring *C* keyed to *A* and containing an annular groove *G*; body *B* which can pivot around the shoulders of *C*; lever *D* which can pivot about *E*; and connecting rod *R* driven by an eccentric (not shown). Lever *D* is rotated counterclockwise about *E* by the connecting rod moving to the left until surface *F* wedges into groove *G*. Continued rotation of *D* causes *A*, *B* and *D* to rotate counterclockwise as a unit about *A*. Reversal of input motion instantly swivels *F* out of *G*, thus unlocking the shaft which remains stationary during return stroke because of friction induced by its load. As *D* continues to rotate clockwise about *E*, node *H*, which is hardened and polished to reduce friction, bears against bottom of *G* to restrain further swiveling. Lever *D* now rotates with *B* around *A* until end of stroke.

FIG. 2—PIN AND DISK. Lever *D*, which pivots around *E*, contains pin *F* in an elongated hole *K* which permits slight vertical movement of pin but prevents horizontal movement by means of set screw *J*. Body *B* can rotate freely about shaft *A* and has cut-outs *L* and *H* to allow clearances for pin *F* and lever *D*, respectively. Ring *C*, which is keyed to shaft *A*, has an annular groove *G* for clearance of the tip of lever *D*. Counterclockwise motion of lever *D* actuated by the connecting rod jams pin between *C* and the top of cut-out *L*. This occurs about seven degrees from the vertical axis. *A*, *B* and *D* are now locked together and rotate about *A*. Return stroke of *R* pivots *D* around *E* clockwise and unwedges pin until it strikes side of *L*. Continued motion of *R* to the right rotates *B* and *D* clockwise around *A* while the uncoupled shaft remains stationary because of its load.

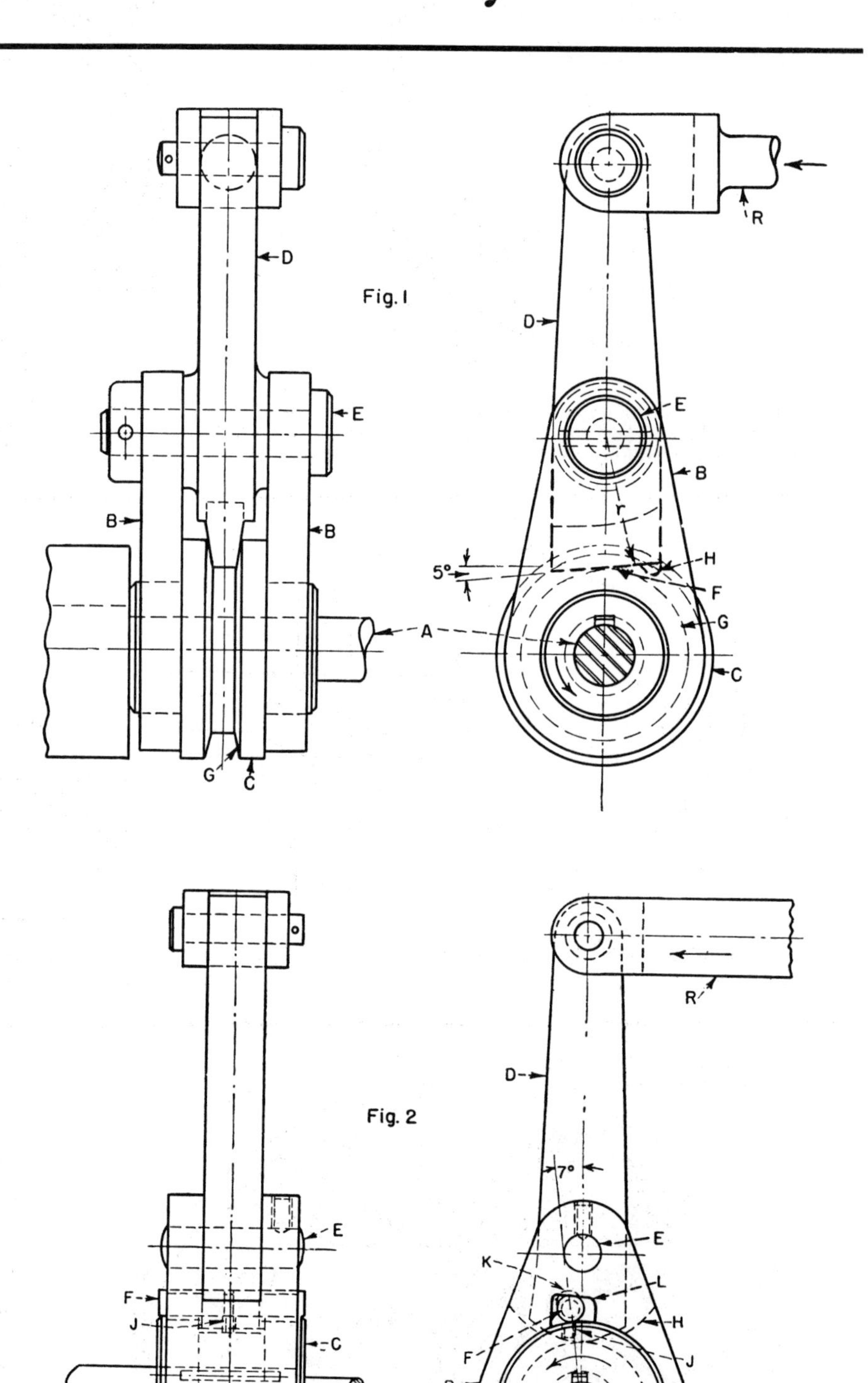

Friction devices are free from the common disadvantages inherent in conventional pawl and ratchet drives such as: (1) noisy operation; (2) backlash needed for engagement of pawl; (3) load concentrated on one tooth of the ratchet; and (4) pawl engagement dependent on an external spring.

The five mechanisms presented here convert the reciprocating motion of a connecting rod into intermittent rotary motion. The connecting rod stroke to the left drives a shaft counterclockwise; shaft is uncoupled and remains stationary during the return stroke of connecting rod to the right.

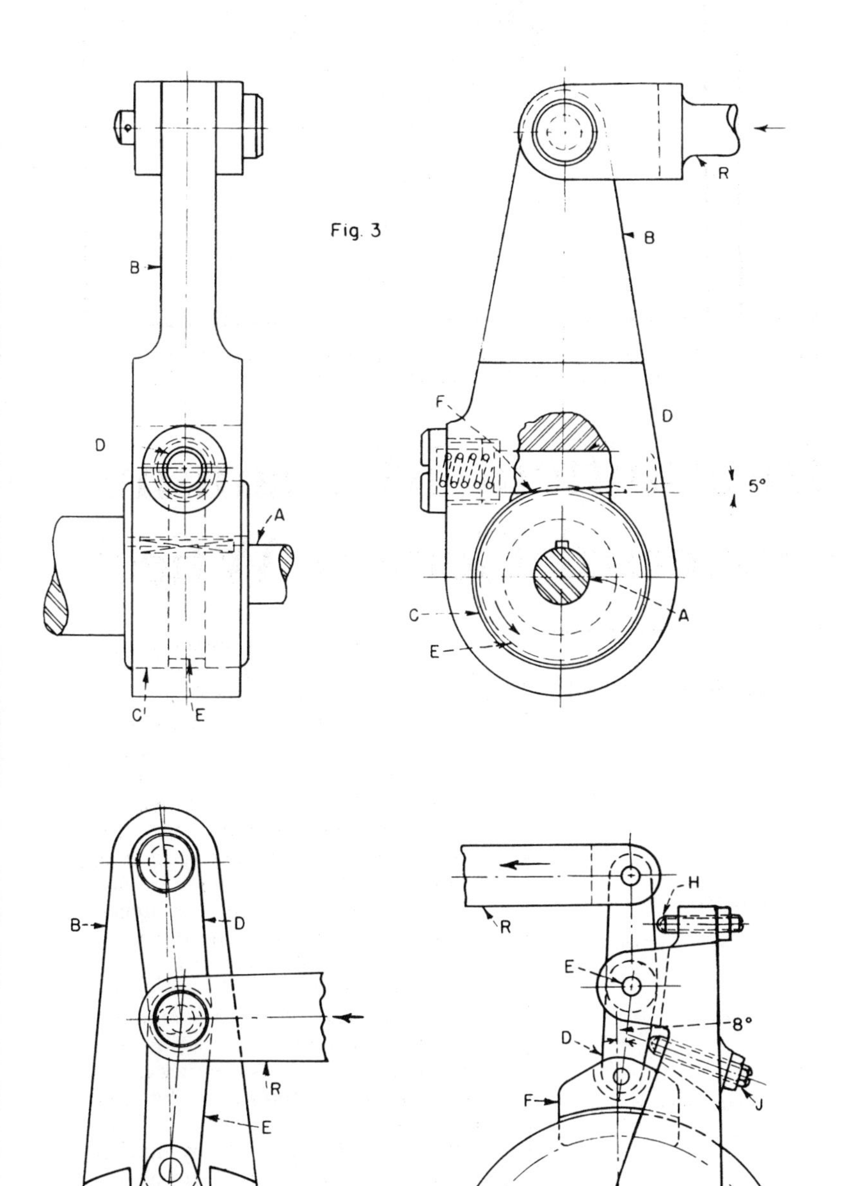


FIG. 3—SLIDING PIN AND DISK. Counterclockwise movement of body *B* about shaft *A* draws pin *D* to the right with respect to body *B*, aided by spring pressure, until the flat bottom *F* of pin is wedged against annular groove *E* of ring *C*. Bottom of pin is inclined about five degrees for optimum wedging action. Ring *C* is keyed to *A* and parts *A*, *C*, *D* and *B* now rotate counterclockwise as a unit until end of connecting rod's stroke. Reversal of *B* draws pin out of engagement so that *A* remains stationary while body completes its clockwise rotation.

FIG. 4—TOGGLE LINK AND DISK. Input stroke of connecting rod *R* to the left wedges block *F* in groove *G* by straightening toggle links *D* and *E*. Body *B*, toggle links and ring *C* which is keyed to shaft *A*, rotate counterclockwise together about *A* until end of stroke. Reversal of connecting rod motion lifts block, thus uncoupling shaft, while body *B* continues clockwise rotation until end of stroke.

FIG. 5—ROCKER ARM AND DISK. Lever *D*, activated by the reciprocating bar *R* moving to the left, rotates counterclockwise on pivot *E* thus wedging block *F* into groove *G* of disk *C*. Shaft *A* is keyed to *C* and rotates counterclockwise as a unit with body *B* and lever *D*. Return stroke of *R* to the right pivots *D* clockwise about *E* and withdraws block from groove so that shaft is uncoupled while *D*, striking adjusting screw *H*, travels with *B* about *A* until completion of stroke. Adjusting screw *J* prevents *F* from jamming in groove.

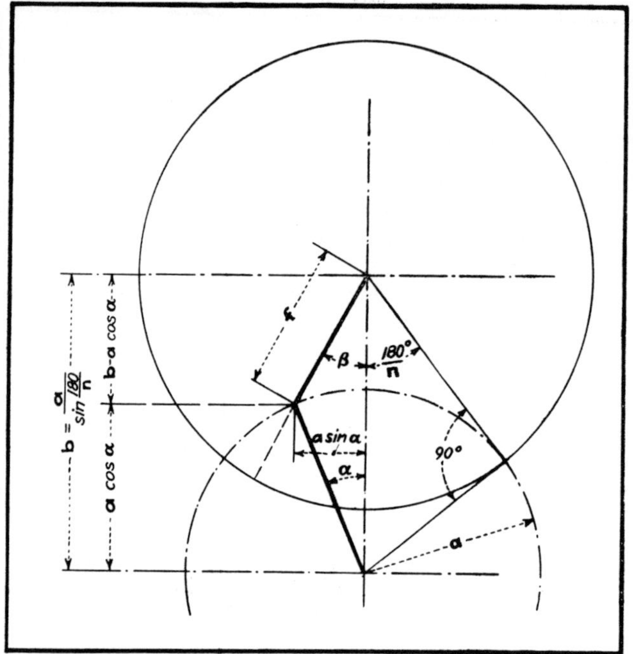


Fig. 1—Basic outline sketch for the external Geneva wheel. The symbols are identified for application in the basic equations.

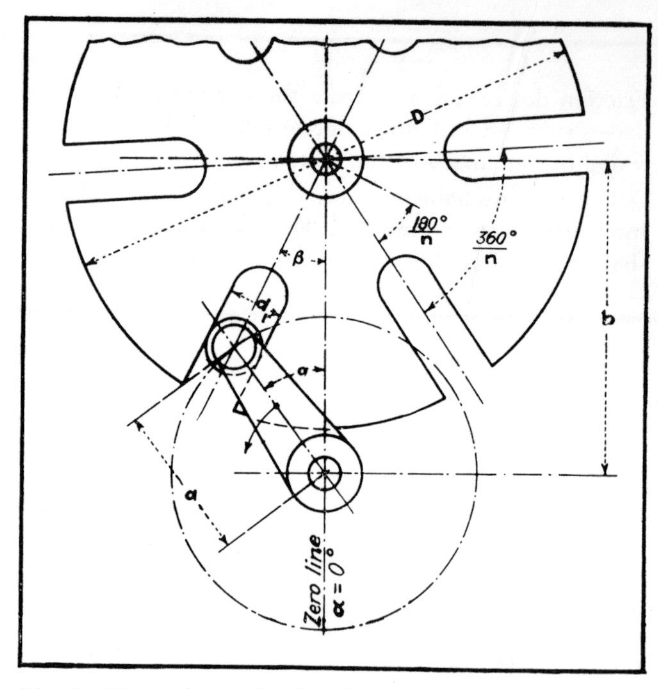


Fig. 2—Schematic sketch of a six slot Geneva wheel. Roller diameter, $d_r$, must be considered when determining $D$.

# Kinematics of Intermittent Mechanisms I—The External Geneva Wheel

**Table I—Notation and Formulas for the External Geneva Wheel**

Assumed or given: $a$, $n$, $d$ and $p$

$a$ = crank radius of driving member

$n$ = number of slots

$d_r$ = roller diameter

$p$ = constant velocity of driving crank in rpm

$$m = \frac{1}{\sin \frac{180}{n}}$$

$b$ = center distance = $am$

$$D = \text{diameter of driven member} = 2\sqrt{\frac{d_r^2}{4} + a^2 \cot^2 \frac{180}{n}}$$

$$\omega = \text{constant angular velocity of driving crank} = \frac{p\pi}{30} \text{ radians per sec}$$

$\alpha$ = angular position of driving crank at any time

$\beta$ = angular displacement of driven member corresponding to crank angle $\alpha$

$$\cos\beta = \frac{m - \cos\alpha}{\sqrt{1 + m^2 - 2m\cos\alpha}}$$

$$\text{Angular Velocity of driven member} = \frac{d\beta}{dt} = \omega\left(\frac{m\cos\alpha - 1}{1 + m^2 - 2m\cos\alpha}\right)$$

$$\text{Angular Acceleration of driven member} = \frac{d^2\beta}{dt^2} = \omega^2\left(\frac{m\sin\alpha\,(1 - m^2)}{(1 + m^2 - 2m\cos\alpha)^2}\right)$$

Maximum Angular Acceleration occurs when $\cos\alpha =$

$$\sqrt{\left(\frac{1 + m^2}{4m}\right)^2 + 2} - \left(\frac{1 + m^2}{4m}\right)$$

Maximum Angular Velocity occurs at $\alpha = 0$ deg, and equals

$$\frac{\omega}{m - 1} \text{ radians per sec}$$

ONE OF THE MOST commonly used mechanisms for producing intermittent rotary motion from a uniform input speed is the external Geneva wheel.

The driven member, or star wheel, contains a number of slots into which the roller of the driving crank fits. The number of slots determines the ratio between dwell and motion period of the driven shaft. Lowest possible number of slots is three, while the highest number is theoretically unlimited. In practice the 3 slot Geneva is seldom used because of the extremely high acceleration values encountered. Genevas with more than 18 slots also are infrequently used, since they necessitate wheels of comparatively large diameters.

In external Genevas of any number of slots, the dwell period always exceeds the motion period. The opposite is true of the internal Geneva, while for the spherical Geneva both dwell and motion periods are 180 degrees.

For proper operation of the external Geneva, the roller must enter the slot tangentially. In other words, the centerline of the slot and the line connecting roller center and crank rotation center must compose a right angle when the roller enters or leaves the slot.

Calculations that follow below are

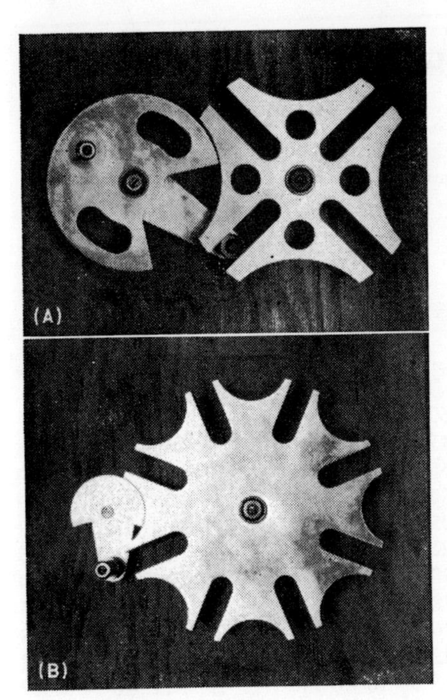


Fig. 3—Four slot Geneva (A) and eight slot (B). Both have locking devices.

S. RAPPAPORT
Consulting Engineer
Wright Machinery Company

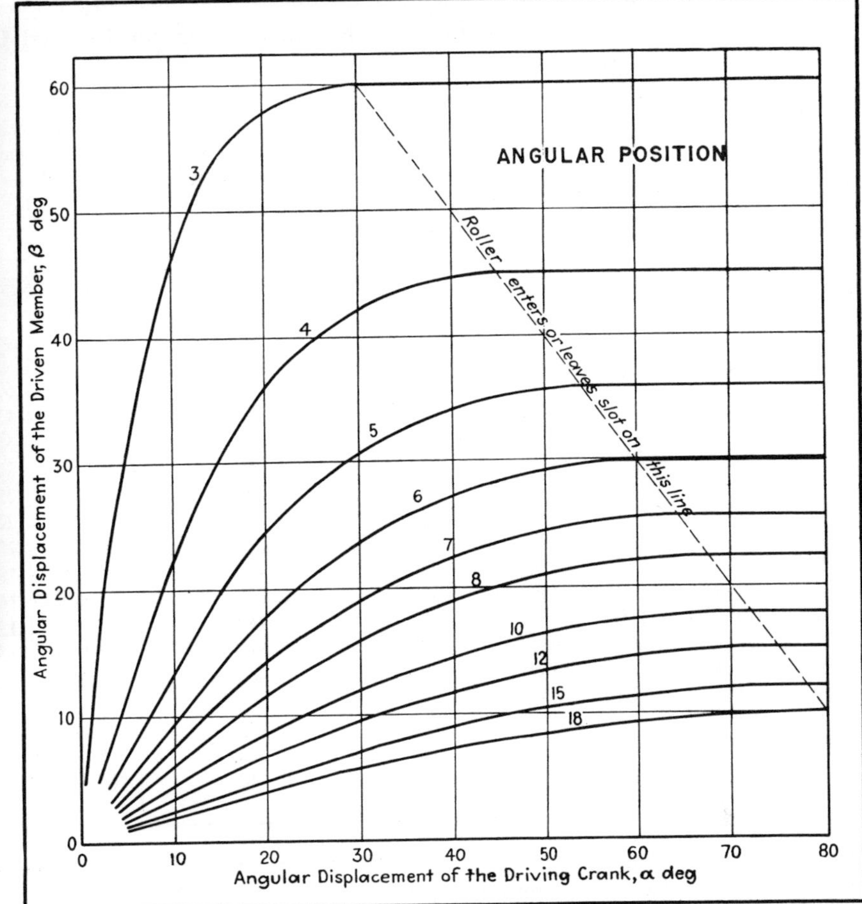


Fig. 4—Chart for determining the angular displacement of the driven member.

based upon these conditions stated.

Consider an external Geneva wheel, shown in Fig. 1, in which

$n$ = number of slots
$a$ = crank radius

From Fig 1, $b$ = center distance $= \dfrac{a}{\sin \dfrac{180}{n}}$

Let $\dfrac{1}{\sin \dfrac{180}{n}} = m$

then $b = a\,m$

It will simplify the development of the equations of motion to designate the connecting line of wheel and crank centers as the zero line. This is contrary to the practice of assigning the zero value of $\alpha$, representing the angular position of the driving crank, to that position of the crank where the roller enters the slot.

Thus, from Fig. 1, the driven crank radius $f$ at any angle is

$$f = \sqrt{(am - a\cos\alpha)^2 + a^2\sin^2\alpha} = a\sqrt{1 + m^2 - 2\,m\cos\alpha} \quad (1)$$

and the angular displacement $\beta$ can be found from

$$\cos\beta = \frac{m - \cos\alpha}{\sqrt{1 + m^2 - 2\,m\cos\alpha}} \quad (2)$$

A six slot Geneva is shown schematically in Fig. 2. The outside diameter

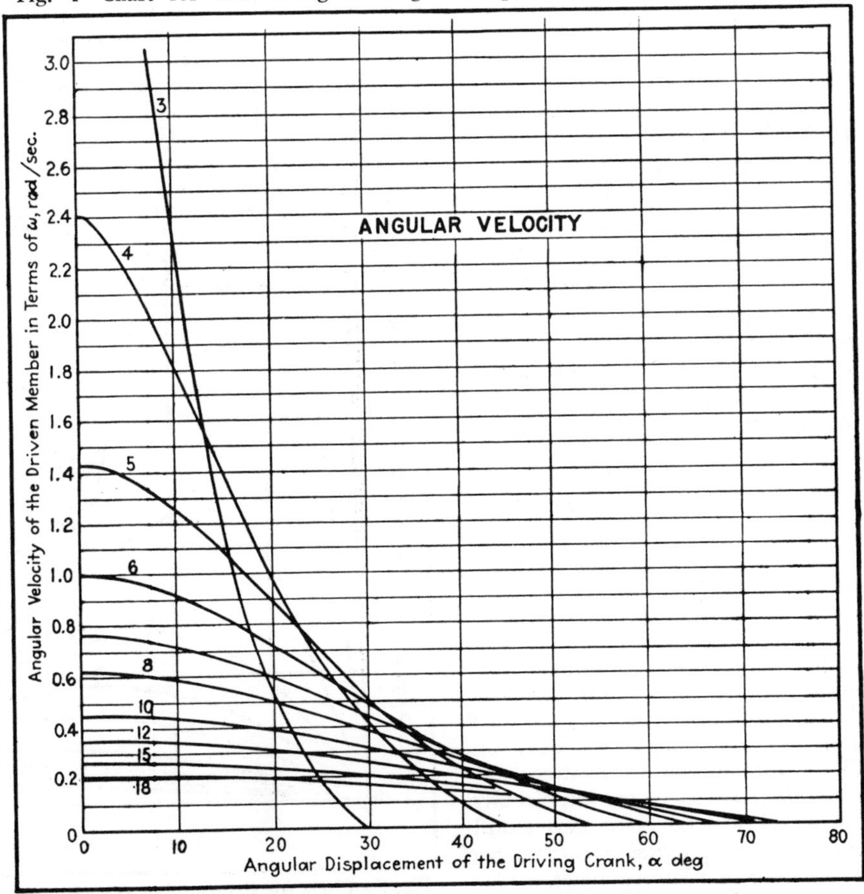


Fig. 5—Chart for determining the angular velocity of the driven member.

**Table II—Principal Kinematic Data for External Geneva Wheel**

| No. of Slots | $\frac{360°}{n}$ | Dwell period | Motion period | $m$ and center-distance for $a$ = 1 | Maximum angular velocity of driven member, radians per sec. equals $\omega$ multiplied by values tabulated. Crank at 0° position | Angular acceleration of driven member when roller enters slot, radians² per sec², equals $\omega^2$ multiplied by values tabulated. | | | Maximum angular Acceleration of driven member, radians² per sec², equals $\omega^2$ multiplied by values tabulated | | |
|---|---|---|---|---|---|---|---|---|---|---|---|
| | | | | | | $\alpha$ | $\beta$ | Multiplier | $\alpha$ | $\beta$ | Multiplier |
| 3 | 120° | 300° | 60° | 1.155 | 6.458 | 30° | 60° | 1.729 | 4° | 27° 58′ | 29.10 |
| 4 | 90° | 270° | 90° | 1.414 | 2.407 | 45° | 45° | 1.000 | 11° 28′ | 25° 11′ | 5.314 |
| 5 | 72° | 252° | 108° | 1.701 | 1.425 | 54° | 36° | 0.727 | 17° 31′ | 21° 53′ | 2.310 |
| 6 | 60° | 240° | 120° | 2.000 | 1.000 | 60° | 30° | 0.577 | 22° 55′ | 19° 51′ | 1.349 |
| 7 | 51° 25′ 43″ | 231° 30′ | 128° 30′ | 2.305 | 0.766 | 64° 17′ 8″ | 25° 42′ 52″ | 0.481 | 27° 41′ | 18° 11′ | 0.928 |
| 8 | 45° | 225° | 135° | 2.613 | 0.620 | 67° 30′ | 22° 30′ | 0.414 | 31° 38′ | 16° 32′ | 0.700 |
| 9 | 40° | 220° | 140° | 2.924 | 0.520 | 70° | 20° | 0.364 | 35° 16′ | 15° 15′ | 0.559 |
| 10 | 36° | 216° | 144° | 3.236 | 0.447 | 72° | 18° | 0.325 | 38° 30′ | 14° 16′ | 0.465 |
| 11 | 32° 43′ 38″ | 212° 45′ | 147° 15′ | 3.549 | 0.392 | 73° 38′ 11″ | 16° 21′ 49″ | 0.294 | 41° 22′ | 13° 16′ | 0.398 |
| 12 | 30° | 210° | 150° | 3.864 | 0.349 | 75° | 15° | 0.268 | 44° | 12° 26′ | 0.348 |
| 13 | 27° 41′ 32″ | 207° 45′ | 152° 15′ | 4.179 | 0.315 | 76° 9′ 14″ | 13° 50′ 46″ | 0.246 | 46° 23′ | 11° 44′ | 0.309 |
| 14 | 25° 42′ 52″ | 205° 45′ | 154° 15′ | 4.494 | 0.286 | 77° 8′ 34″ | 21° 51′ 26″ | 0.228 | 48° 32′ | 11° 3′ | 0.278 |
| 15 | 24° | 204° | 156° | 4.810 | 0.263 | 78° | 12° | 0.213 | 50° 30′ | 10° 27′ | 0.253 |
| 16 | 22° 30′ | 202° 30′ | 157° 30′ | 5.126 | 0.242 | 78° 45′ | 11° 15′ | 0.199 | 52° 24′ | 9° 57′ | 0.232 |
| 17 | 21° 10′ 35″ | 201° | 159° | 5.442 | 0.225 | 79° 24′ 43″ | 10° 35′ 17″ | 0.187 | 53° 58′ | 9° 26′ | 0.215 |
| 18 | 20° | 200° | 160° | 5.759 | 0.210 | 80° | 10° | 0.176 | 55° 30′ | 8° 59′ | 0.200 |

$D$ of the wheel (when taking the effect of the roller diameter $d_r$ into account) is found to be

$$D = 2\sqrt{\frac{d_r^2}{4} + a^2 \cot^2 \frac{180}{n}} \quad (3)$$

Differentiating Eq (2) and dividing by the differential of time, $dt$, the angular velocity of the driven member is

$$\frac{d\beta}{dt} = \omega\left(\frac{m \cos\alpha - 1}{1 + m^2 - 2m\cos\alpha}\right) \quad (4)$$

where $\omega$ represents the constant angular velocity of the crank.

By differentiation of Eq (4) the acceleration of the driven member is found to be

$$\frac{d^2\beta}{dt^2} = \omega^2\left(\frac{m \sin\alpha\,(1 - m^2)}{(1 + m^2 - 2m\cos\alpha)^2}\right) \quad (5)$$

All notations and principal formulas are given in Table I for easy reference. Table II contains all the data of principal interest for external Geneva wheels having from 3 to 18 slots. All other data can be read from the charts: Fig. 4 for angular position, Fig. 5 for angular velocity, and Fig. 6 for angular acceleration.

Similar analyses of the internal Geneva wheel will be given next month, and for the spherical Geneva wheel will appear in a later number.

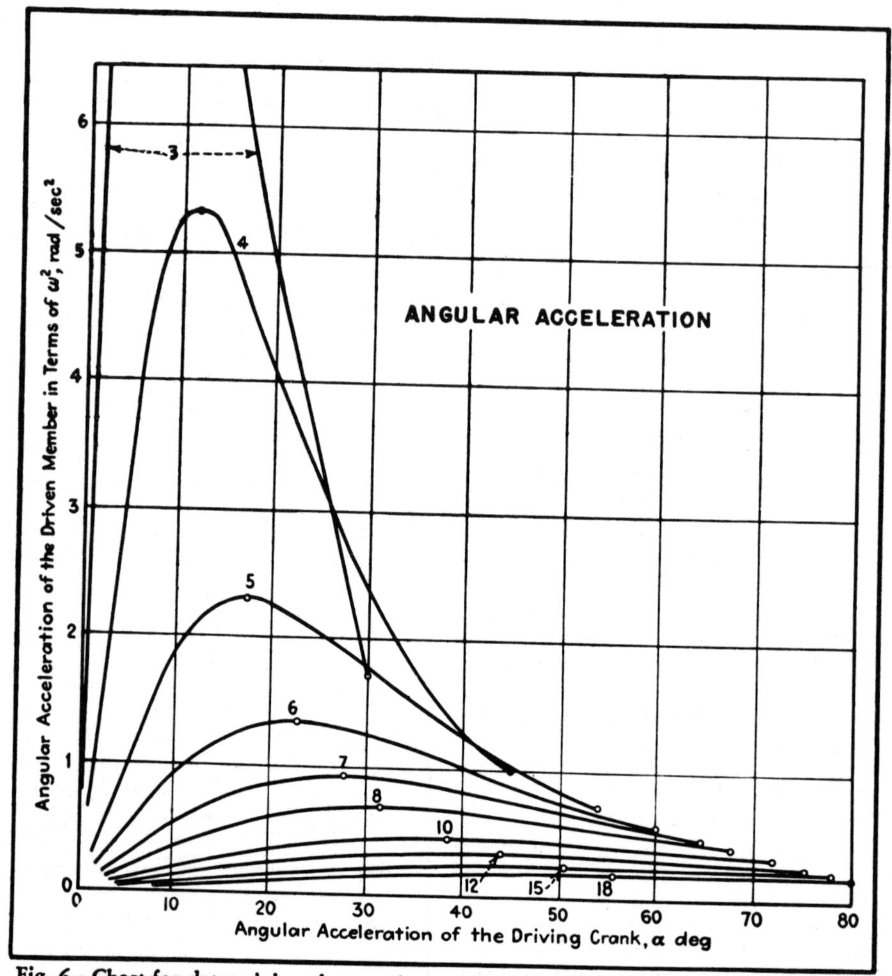


Fig. 6—Chart for determining the angular acceleration of the driven member.

Fig. 1—A four slot internal Geneva wheel incorporating a locking mechanism. The basic sketch is shown in Fig. 3.

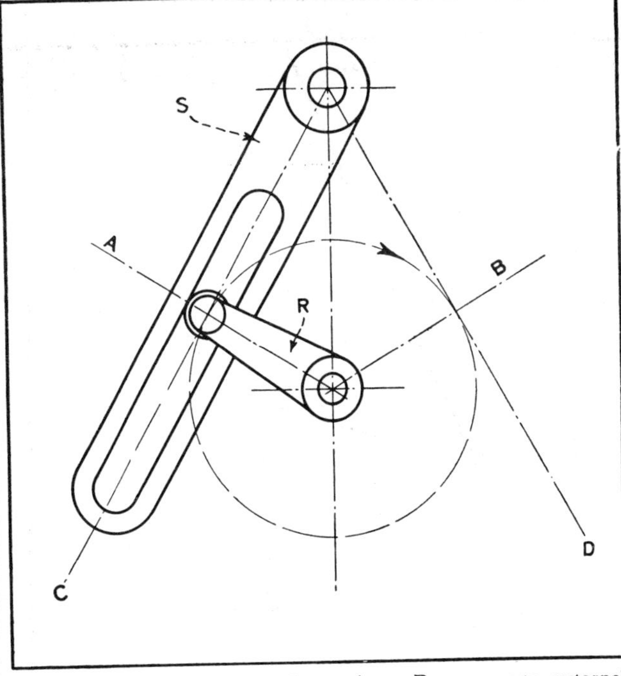


Fig. 2—Slot-crank motion from $A$ to $B$ represents external Geneva action; from $B$ to $A$ represents internal Geneva motion.

# Kinematics of Intermittent Mechanisms II—The Internal Geneva Wheel

S. RAPPAPORT
Consulting Engineer
Wright Machinery Company

WHERE INTERMITTENT DRIVES must provide dwell periods of more than 180 deg, the external Geneva wheel (PRODUCT ENGINEERING, July 1949, p 110) is quite satisfactory and is almost the standard device employed. But where the dwell period has to be less than 180 deg, other intermittent drive mechanisms must be used. The internal Geneva wheel is one way of obtaining this type of motion.

Dwell period of all internal Genevas is always smaller than 180 deg. Thus more time is left for the star to achieve maximum velocity, and acceleration is lower. The highest value of the angular acceleration occurs when the roller enters or leaves the slot. However, the acceleration curve does not reach a peak within the range of motion of the driven wheel. The geometrical maximum would occur in the continuation of the curve, but this continuation has no significance, since the driven member will have entered the dwell phase associated with the high angular displacement of the driving member.

The geometrical maximum lies in the continuation of the curve, falling into the region representing the mo-

**Table I—Notation and Formulas for the Internal Geneva Wheel**

Assumed or given: $a$, $n$, $d$ and $p$

$a$ = crank radius of driving member
$n$ = number of slots
$d$ = roller diameter
$p$ = constant velocity of driving crank in rpm

$$m = \frac{1}{\sin \frac{180°}{n}}$$

$b$ = center distance = $a\,m$

$$D = \text{inside diameter of driven member} = 2\sqrt{\frac{d^2}{4} + a^2 \cot^2 \frac{180°}{n}}$$

$\omega$ = constant angular velocity of driving crank in radians per sec = $\frac{p\pi}{30}$ radians per sec

$\alpha$ = angular position of driving crank at any time
$\beta$ = angular displacement of driven member corresponding to crank angle $\alpha$

$$\cos\beta = \frac{m + \cos\alpha}{\sqrt{1 + m^2 + 2m\cos\alpha}}$$

$$\text{Angular velocity of driven member} = \frac{d\beta}{dt} = \omega\left(\frac{1 + m\cos\alpha}{1 + m^2 + 2m\cos\alpha}\right)$$

$$\text{Angular acceleration of driven member} = \frac{d^2\beta}{dt^2} = \omega^2\left[\frac{m\sin\alpha\,(1 - m^2)}{(1 + m^2 + 2m\cos\alpha)^2}\right]$$

Maximum angular velocity occurs at $\alpha = 0°$ and equals $= \frac{\omega}{1 + m}$ radians per sec

Maximum angular acceleration occurs when roller enters slot and equals $= \frac{\omega^2}{\sqrt{m^2 - 1}}$ radians$^2$ per sec$^2$

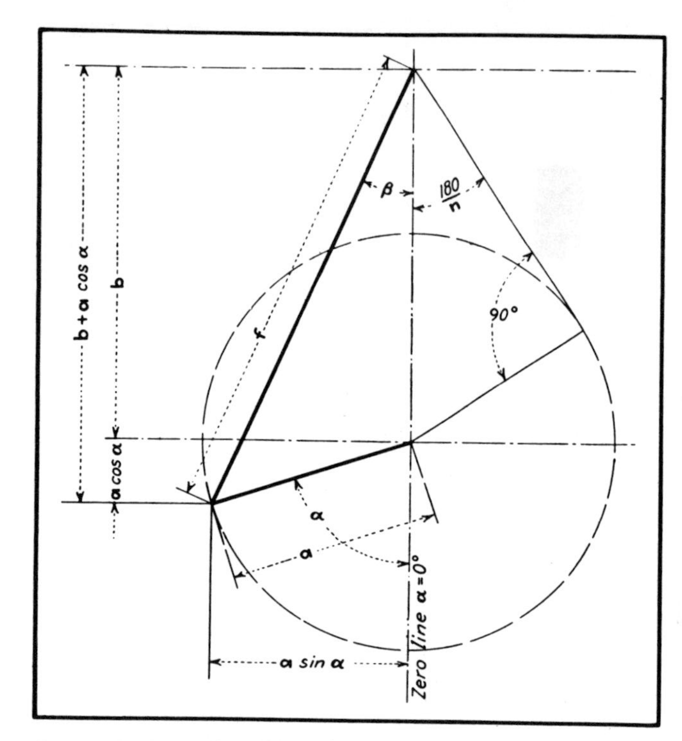


Fig. 3—Basic outline sketch for developing the equations of the internal Geneva wheel, using the notations as shown.

Fig. 4—Schematic sketch of a six slot internal Geneva wheel. Symbols are identified and motion equations given in Table I.

tion of the external Geneva wheel. This can be seen by the following considerations of a crank and slot drive, sketched in Fig. 2.

When the roller crank *R* rotates, slot link *S* will perform an oscillating movement, for which the displacement, angular velocity and acceleration can be given in continuous curves.

When the crank *R* rotates from *A* to *B*, then the slot link *S* will move from *C* to *D*, exactly reproducing all moving conditions of an external Geneva of equal slot angle. When crank *R* continues its movement from *B* back to *A*, then the slot link *S* will move from *D* back to *C*, this time reproducing exactly (though in a mirror picture with the direction of motion being reversed) the moving conditions of an internal Geneva.

Therefore, the characteristic curves of this motion contain both the external and internal Geneva wheel conditions; the region of the external Geneva lying between *A* and *B*, the region of the internal Geneva lying between *B* and *A*.

The geometrical maxima of the acceleration curves lie only in the region between *A* and *B*, representing that portion of the curves which belongs to the external Geneva.

Principal advantage of the internal Geneva, other than its smooth operation, is the sharply defined dwell period. A disadvantage is the relatively large size of the driven member, which increases the force resisting acceleration. Another feature, which is

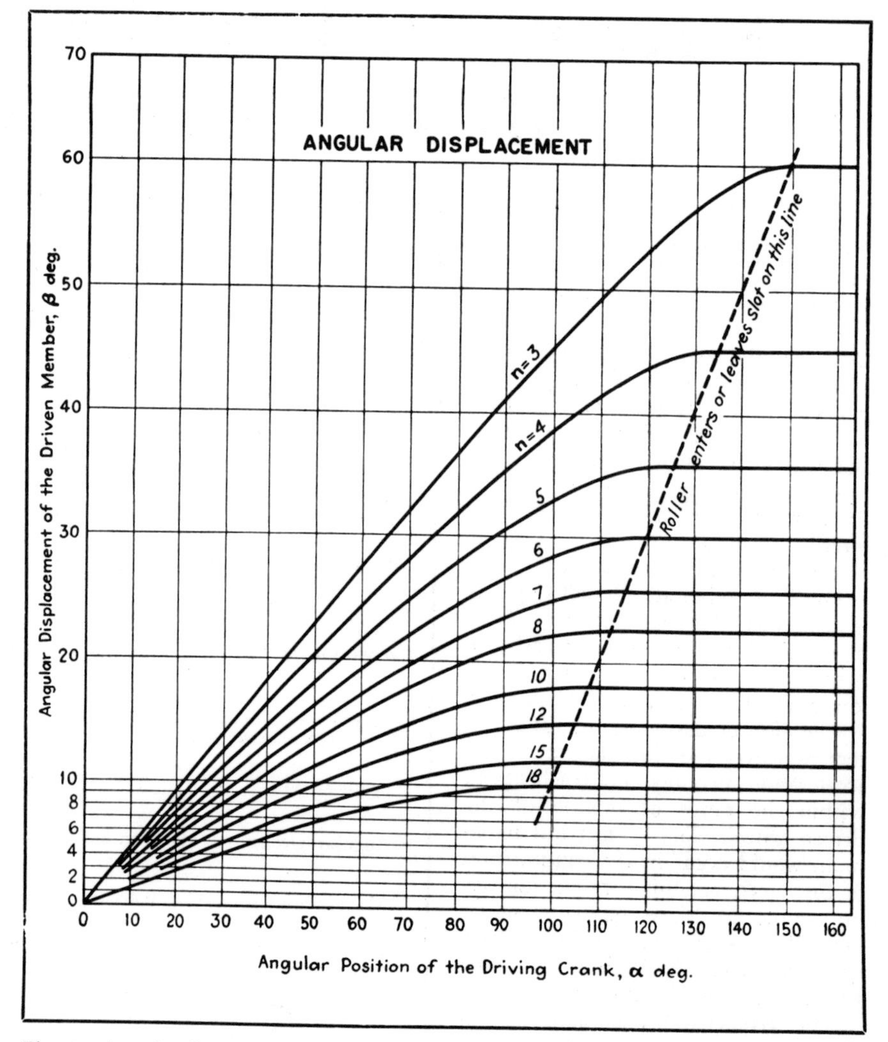


Fig. 5—Angular displacement of the driven member can be determined from this chart.

sometimes a disadvantage, is the cantilever arrangement of the roller crank shaft. This shaft cannot be a through shaft because the crank has to be fastened to the overhanging end of the input shaft.

To simplify the equations, the connecting line of wheel and crank centers is taken as the zero line. The angular position of the driving crank, $\alpha$, is zero when on this line. Then the following relations are developed, based on Fig. 3:

$n$ = number of slots
$a$ = crank radius

$$b = \text{center distance} = \frac{a}{\sin \frac{180°}{n}}$$

Let

$$\frac{1}{\sin \frac{180°}{n}} = m,$$

then

$$b = am$$

To find the angular displacement, $\beta$, of the driven member, the driven crank radius, $f$, is first calculated from

$$f = \sqrt{a^2 \sin^2 \alpha + (am + a \cos \alpha)^2} = a\sqrt{1 + m^2 + 2m \cos \alpha} \quad (1)$$

and since

$$\cos \beta = \frac{m + \cos \alpha}{f}$$

it follows:

$$\cos \beta = \frac{m + \cos \alpha}{\sqrt{1 + m^2 + 2m \cos \alpha}} \quad (2)$$

From this formula, $\beta$, the angular displacement, can be calculated for any angle $\alpha$, the angle of the driving member.

The first derivative of Eq (2) gives the angular velocity as

$$\frac{d\beta}{dt} = \omega \left( \frac{1 + m \cos \alpha}{1 + m^2 + 2m \cos \alpha} \right) \quad (3)$$

where $\omega$ designates the uniform speed of the driving crank shaft, namely

$$\omega = \frac{p\pi}{30}$$

if $p$ equals its number of revolutions per minute.

Differentiating Eq (3) once more develops the equation for the angular acceleration:

$$\frac{d^2\beta}{dt^2} = \omega^2 \left[ \frac{m \sin \alpha (1 - m^2)}{(1 + m^2 + 2m \cos \alpha)^2} \right] \quad (4)$$

The maximum angular velocity occurs, obviously, at $\alpha = 0$ deg. Its value is found by substituting 0 deg for $\alpha$ in Eq (3). It is

$$\frac{d\beta}{dt}_{\max} = \frac{\omega}{1 + m} \quad (5)$$

The highest value of the accelera-

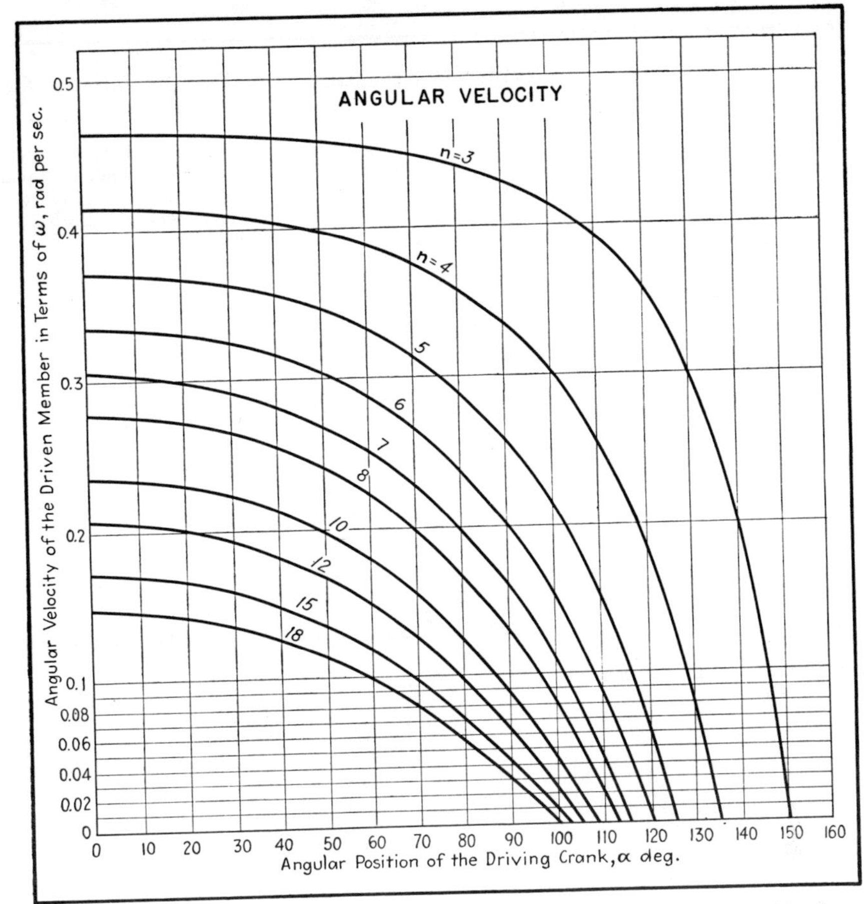


Fig. 6—Angular velocity of the driven member can be determined from this chart.

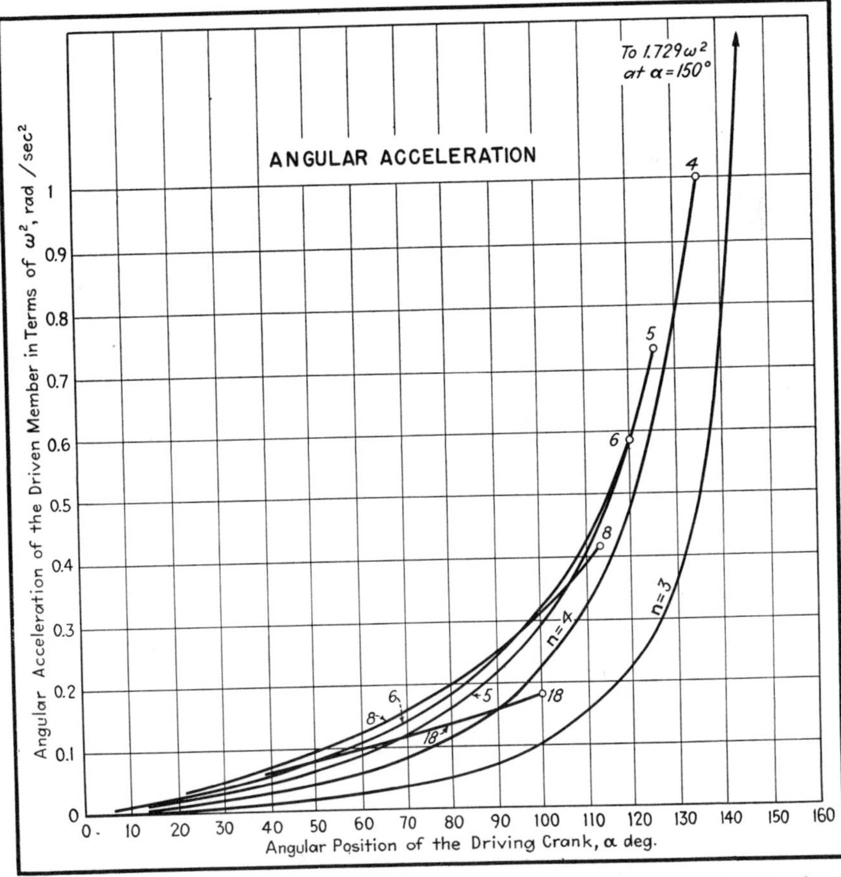


Fig. 7—Angular acceleration of the driven member can be determined from this chart.

## Table II—Kinematic Data For the Internal Geneva Wheel

| Number of slots, $n$ | $\frac{360°}{n}$ | Dwell period | Motion period | $m$ and center-distance for $a=1$ | Maximum angular velocity of driven member equals $\omega$ radians per sec, multiplied by values tabulated. Both $\alpha$ and $\beta$ in 0° position | Angular acceleration of driven member when roller enters slot equals $\omega^2$ radians² per sec² multiplied by values tabulated: $\alpha$ | $\beta$ | Multiplier |
|---|---|---|---|---|---|---|---|---|
| 3 | 120° | 60° | 300° | 1.155 | 0.464 | 150° | 60° | 1.729 |
| 4 | 90° | 90° | 270° | 1.414 | 0.414 | 135° | 45° | 1.000 |
| 5 | 72° | 108° | 252° | 1.701 | 0.370 | 126° | 36° | 0.727 |
| 6 | 60° | 120° | 240° | 2.000 | 0.333 | 120° | 30° | 0.577 |
| 7 | 51° 25′ 43″ | 128° 30′ | 231° 30′ | 2.305 | 0.303 | 115° 42′ 52″ | 25° 42′ 52″ | 0.481 |
| 8 | 45° | 135° | 225° | 2.613 | 0.277 | 112° 30′ | 22° 30′ | 0.414 |
| 9 | 40° | 140° | 220° | 2.924 | 0.255 | 110° | 20° | 0.364 |
| 10 | 36° | 144° | 216° | 3.236 | 0.236 | 108° | 18° | 0.325 |
| 11 | 32° 43′ 38″ | 147° 15′ | 212° 45′ | 3.549 | 0.220 | 106° 21′ 49″ | 16° 21′ 49″ | 0.294 |
| 12 | 30° | 150° | 210° | 3.864 | 0.206 | 105° | 15° | 0.268 |
| 13 | 27° 41′ 32″ | 152° 15′ | 207° 45′ | 4.179 | 0.193 | 103° 50′ 46″ | 13° 50′ 46″ | 0.246 |
| 14 | 25° 42′ 52″ | 154° 15′ | 205° 45′ | 4.494 | 0.182 | 102° 51′ 26″ | 12° 51′ 26″ | 0.228 |
| 15 | 24° | 156° | 204° | 4.810 | 0.172 | 102° | 12° | 0.213 |
| 16 | 22° 30′ | 157° 30′ | 202° 30′ | 5.126 | 0.163 | 101° 15′ | 11° 15′ | 0.199 |
| 17 | 21° 10′ 35″ | 159° | 201° | 5.442 | 0.155 | 100° 35′ 17″ | 10° 35′ 17″ | 0.187 |
| 18 | 20° | 160° | 200° | 5.759 | 0.148 | 100° | 10° | 0.176 |

tion is found by substituting $180/n + 90$ for $\alpha$ in Eq (4):

$$\frac{d^2\beta}{dt^2_{max}} = \frac{\omega^2}{\sqrt{m^2 - 1}} \tag{6}$$

A schematic sketch for a six slot internal Geneva wheel is shown in Fig. 4. All the symbols used in this sketch, and throughout the text, are compiled in Table I for easy reference.

Table II contains all the data of principal interest on the performance of internal Geneva wheels having from 3 to 18 slots. Other data can be read from the charts: Fig. 5 for angular position, Fig. 6 for angular velocity and Fig. 7 for angular acceleration.

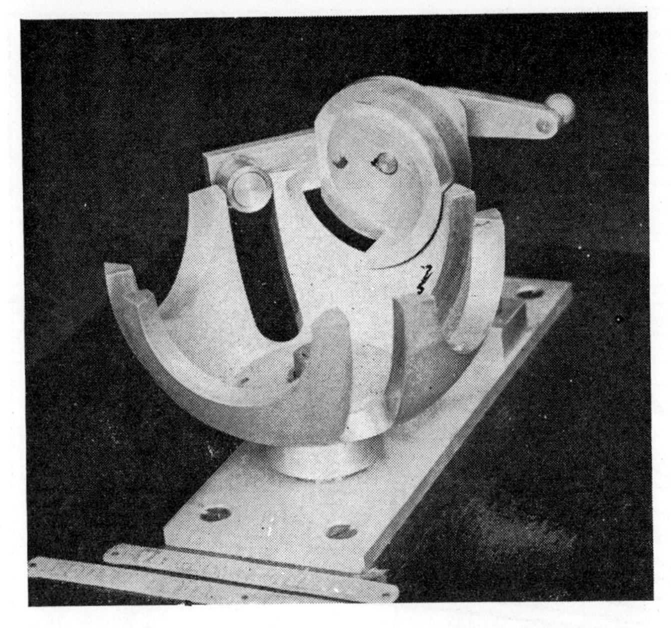

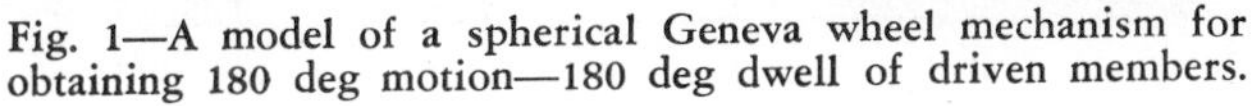

Fig. 1—A model of a spherical Geneva wheel mechanism for obtaining 180 deg motion—180 deg dwell of driven members.

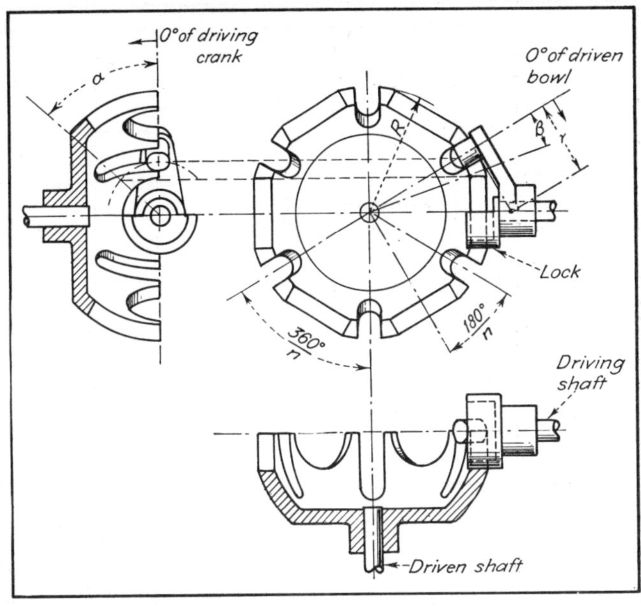


Fig. 2—Drawing of a six slot spherical Geneva wheel, indentifying the nomenclature used in the equations and the charts.

# Kinematics Of Intermittent Mechanisms III—The Spherical Geneva Wheel


S. RAPPAPORT
Consulting Engineer
Wright Machinery Company


A RARE, but useful member of the Geneva wheel group is the spherical Geneva. Where the external Geneva provides dwell periods greater than motion periods, and the internal Geneva provides the opposite, the spherical Geneva has a dwell period exactly equal to the motion period. Both motion and dwell periods occur over 180 deg of the driving shaft motion. In spite of its simplicity and of its smooth, shock free operation this device is hardly known to the mechanical engineer. The writer even applied for a patent on it, only to find that it was patented thirty years ago.

The spherical Geneva, Fig. 1, consists of a crank, attached to the driving shaft and carrying the driving roller, and the driven bowl, attached to the output shaft. Driving and driven shafts are not parallel as in the conventional Geneva drives, but at 90 deg to each other, with their axes intersecting. The bowl has the general shape of a hollow hemisphere, having slots along its meridian lines into which the driving roller fits.

The axis of the driving roller intersects the axis of the driving shaft at an angle determined by the number of slots on the bowl. While rotating, the drive roller axis describes the surface of a cone. The cone's apex lies in the intersection point of the axis of input shaft, output shaft and drive roller. This point is also the center of the hemispherical bowl.

Attached to the crank is the semicircular lock, fitting into semicircular cutouts on the bowl and working in much the same way as the conventional lock in the ordinary Genevas.

Motion period of this drive is always 180 deg of the cycle, independent of the number of slots, since the roller always engages and disengages on the equator of the bowl.

To determine the equations of motion the following notations (see Fig. 2) are used

$R$ = outside radius of bowl
$n$ = number of slots

With that position of the roller axis where the roller enters the slot established as the zero line, the crank radius is

$$r = R \tan \frac{180^\circ}{n}$$

if $r$ is measured tangentially to the bowl equator as Fig. 2 shows.

Calling

$$\tan \frac{180^\circ}{n} = m$$

then

$$r = Rm \tag{1}$$

It can easily be shown that

$$\frac{m - \tan \beta}{1 + m \tan \beta} = m \cos \alpha \tag{2}$$

where $\alpha$ designates the angular position of the driving crank at any time, and $\beta$ the corresponding angular displacement of the bowl. It follows from Eq (2) that

$$\tan \beta = m \left( \frac{1 - \cos \alpha}{1 + m^2 \cos \alpha} \right) \tag{3}$$

from which the angular displacement $\beta$ is found.

Differentiating Eq (3) the angular velocity of the bowl is

$$\frac{d\beta}{dt} = \omega \left( \frac{m \sin \alpha}{1 + m^2 \cos^2 \alpha} \right) \tag{4}$$

where the constant angular velocity of the driving shaft, rotating at $p$ revolutions per minute equals $\omega$ and

$$\omega = p\pi/30 \tag{5}$$

Differentiating Eq (4) the angular acceleration is found to be

$$\frac{d^2 \beta}{dt^2} = \omega^2 m \cos \alpha \left( \frac{1 + m^2 + m^2 \sin^2 \alpha}{(1 + m^2 \cos^2 \alpha)^2} \right) \tag{6}$$

Maximum velocity occurs at $\alpha$ equals 90 deg and must equal $\omega$ $m$.

To find the maximum acceleration, differentiate Eq (6). Making this third derivative zero, the maximum

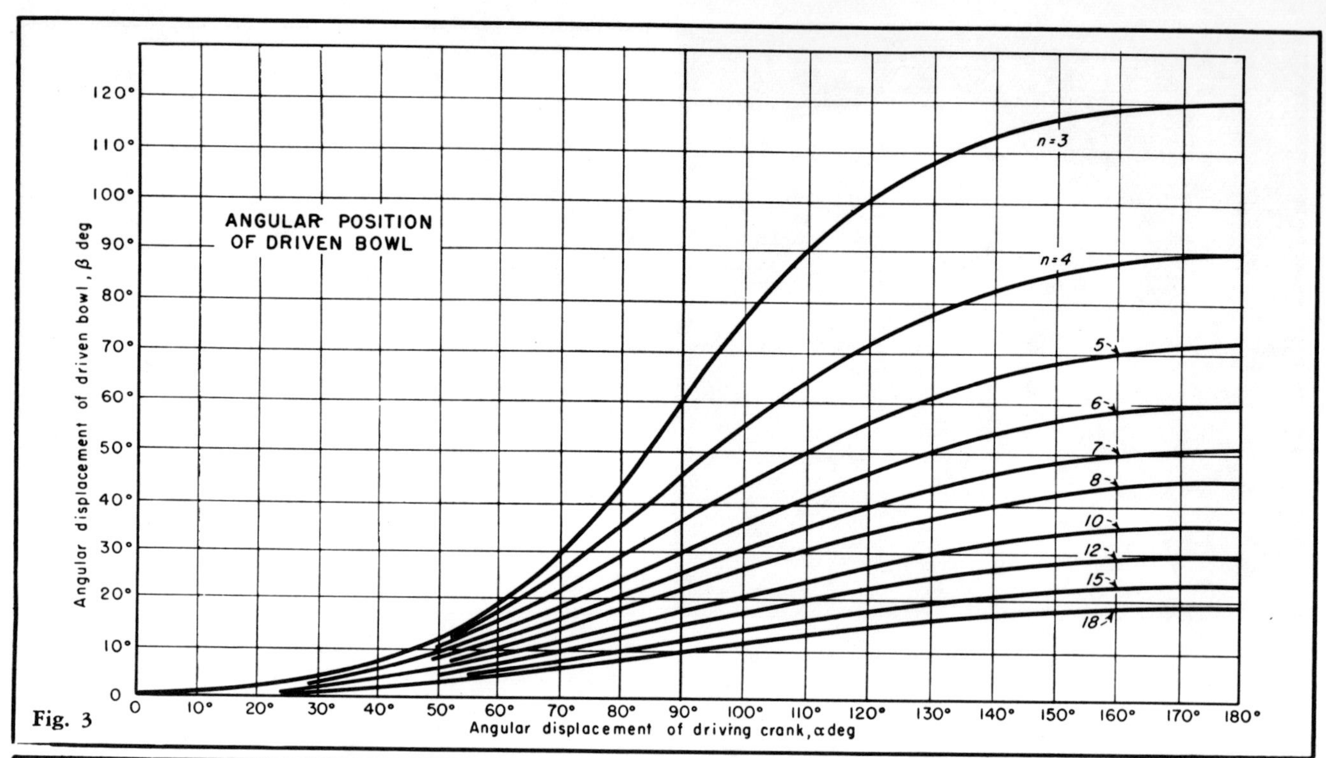


Fig. 3

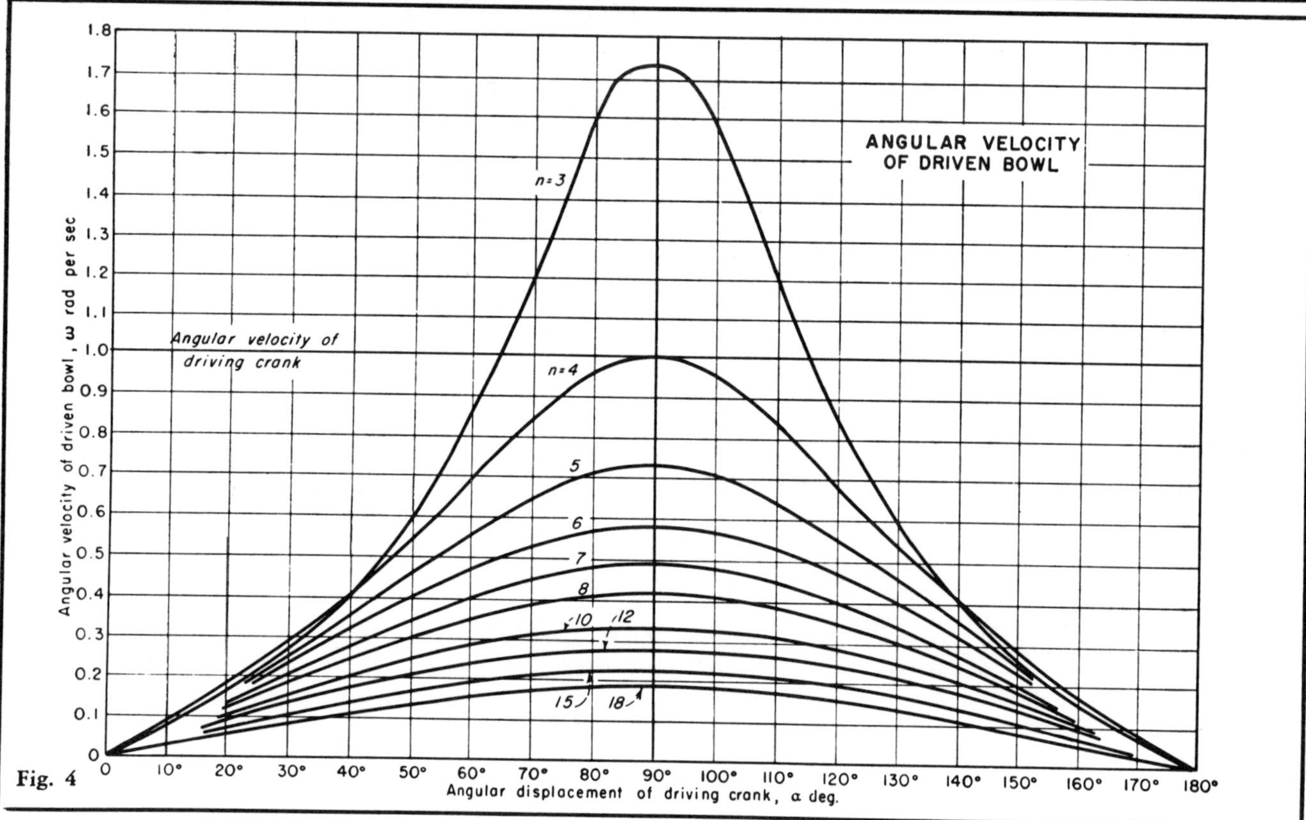


Fig. 4

acceleration is found to occur where

$$\cos \alpha = \frac{\sqrt{3 + 3m^2 - \sqrt{9m^4 + 16m^2 + 8}}}{m} \quad (7)$$

It is obvious that Eq (7) will produce irrational values of $\alpha$ for certain values of $m$. A closer investigation shows that a rational extreme outside of $\alpha$ equals 0 exists only for 3, 4, 5, 6 and 7 slots. If there are more than 7 slots in the bowl, then the highest acceleration occurs at $\alpha$ equals 0, as can be seen from Fig. 5.

To find the value of the highest acceleration for each given number of slots, substitute $\alpha$, found from Eq (7), into Eq (6), if the bowl contains less than 8 slots. If the bowl contains 8 or more slots, the maximum acceleration occurs at $\alpha$ equals 0 and this maximum acceleration will equal

$$\omega^2 \left( \frac{m}{1 + m^2} \right)$$

Table I shows the principal kinematic data for bowls with from 3 to 18 slots. Graphs in Figs. 3, 4 and 5 show respectively the angular displacement, velocity and acceleration.

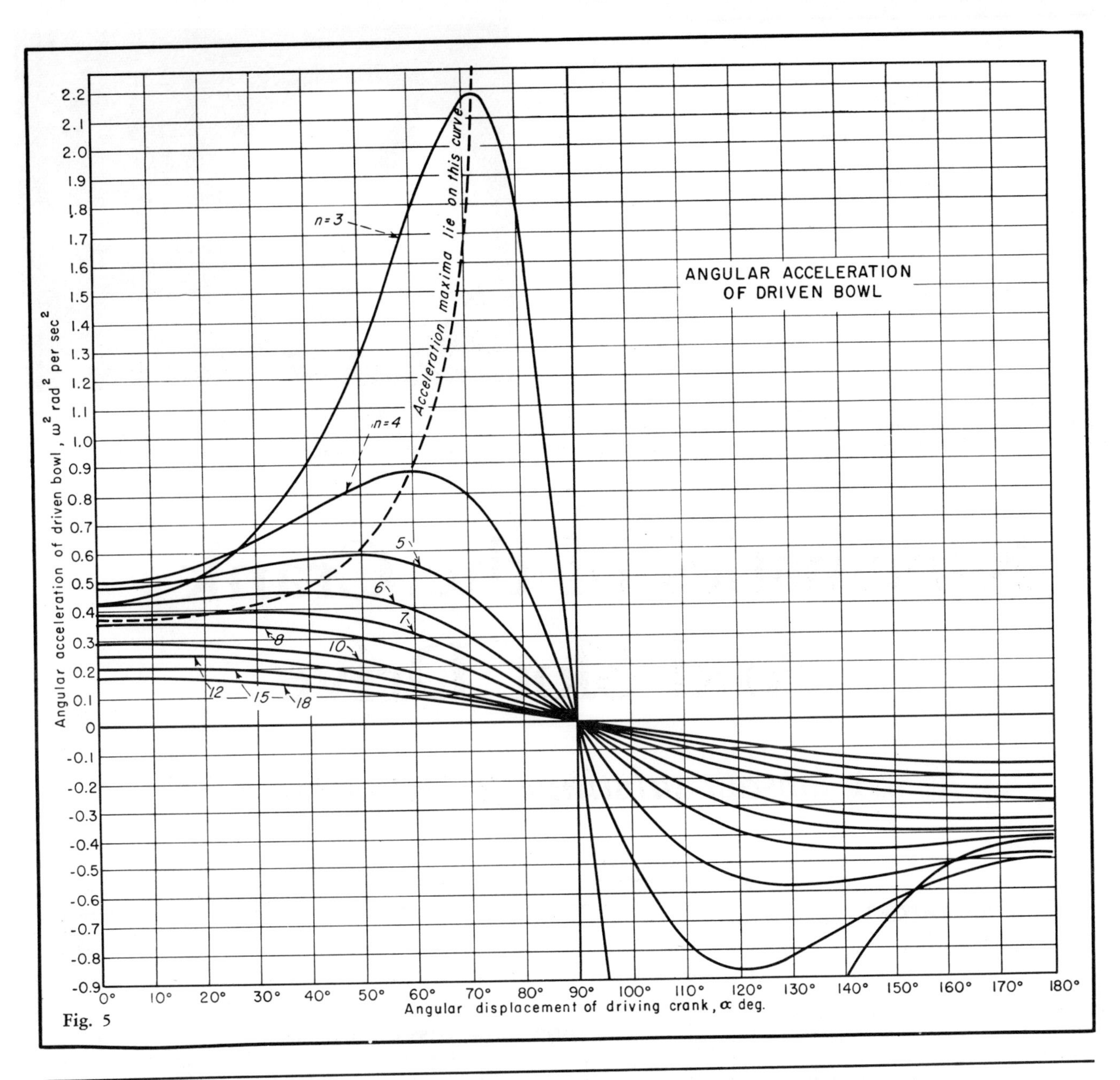

Fig. 5

## Table I—Principal Kinematic Data for Spherical Geneva Wheels

| Number of Slots, $n$ | Angle through which bowl rotates in one cycle | $\frac{180°}{n}$ | $\tan \frac{180}{n} = m$ and crank radius $r$ for $R = 1$ | Maximum angular velocity of driven bowl, rad per sec, equals $\omega$ multiplied by values tabulated | | | Acceleration of bowl when roller enters slot, rad² per sec² equals $\omega^2$ multiplied by values tabulated. $\alpha = 0$ | Maximum angular acceleration of driven bowl, rad² per sec², equals $\omega^2$ multiplied by values tabulated | | |
|---|---|---|---|---|---|---|---|---|---|---|
| | | | | $\alpha$ | $\beta$ | Multiplier | | $\alpha$ | $\beta$ | Multiplier |
| 3 | 120° | 60° | 1.73205 | 90° | 60° | 1.732 | 0.434 | 72° 13′ | 32° 8′ | 2.17 |
| 4 | 90° | 45° | 1.00000 | 90° | 45° | 1.000 | 0.500 | 59° 40′ | 18° 12′ | 0.873 |
| 5 | 72° | 36° | 0.72654 | 90° | 36° | 0.727 | 0.475 | 48° 35′ | 10° 16′ | 0.582 |
| 6 | 60° | 30° | 0.57735 | 90° | 30° | 0.577 | 0.432 | 37° 28′ | 5° 23′ | 0.455 |
| 7 | 51° 25′ 43″ | 25° 42′ 52″ | 0.48162 | 90° | 25° 43′ | 0.482 | 0.393 | 20° 42′ | 1° 28′ | 0.394 |
| 8 | 45° | 22° 30′ | 0.41421 | 90° | 22° 30′ | 0.414 | 0.354 | 0° | 0° | 0.354 |
| 9 | 40° | 20° | 0.36397 | 90° | 20° | 0.364 | 0.321 | 0° | 0° | 0.321 |
| 10 | 36° | 18° | 0.32492 | 90° | 18° | 0.325 | 0.294 | 0° | 0° | 0.294 |
| 11 | 32° 43′ 38″ | 16° 21′ 49″ | 0.29368 | 90° | 16° 22′ | 0.294 | 0.271 | 0° | 0° | 0.271 |
| 12 | 30° | 15° | 0.26795 | 90° | 15° | 0.268 | 0.251 | 0° | 0° | 0.251 |
| 13 | 27° 41′ 32″ | 13° 50′ 46″ | 0.24655 | 90° | 13° 51′ | 0.247 | 0.232 | 0° | 0° | 0.232 |
| 14 | 25° 42′ 52″ | 12° 51′ 26″ | 0.22842 | 90° | 12° 51′ | 0.228 | 0.216 | 0° | 0° | 0.216 |
| 15 | 24° | 12° | 0.21256 | 90° | 12° | 0.213 | 0.204 | 0° | 0° | 0.204 |
| 16 | 22° 30′ | 11° 15′ | 0.19891 | 90° | 11° 15′ | 0.199 | 0.192 | 0° | 0° | 0.192 |
| 17 | 21° 10′ 35″ | 10° 35′ 17″ | 0.18745 | 90° | 10° 35′ | 0.187 | 0.181 | 0° | 0° | 0.181 |
| 18 | 20° | 10° | 0.17633 | 90° | 10° | 0.176 | 0.171 | 0° | 0° | 0.171 |

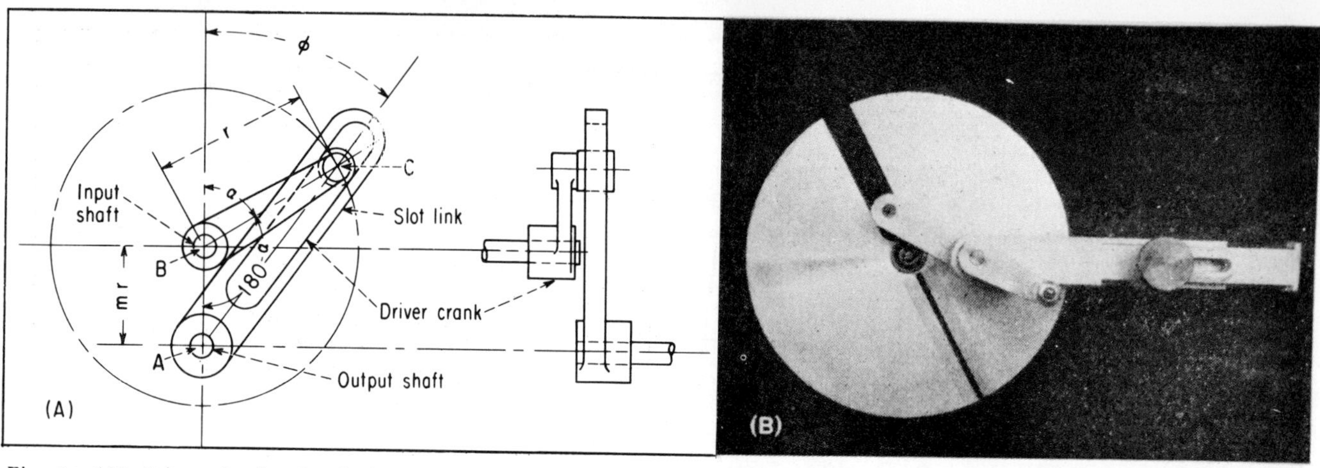


Fig. 1—(A) Schematic sketch of the crank and slot drive. By changing distance *mr*, the non-uniform rotary motion can be varied. The driven member will perform a complete revolution as long as *mr* is kept smaller than *r*. (B) Model of the crank and slot drive having an adjustable center distance. Slot link is replaced by a slot carrying disk for balancing purposes.

# Kinematics of the Crank And Slot Drive

## *A low-cost substitute for elliptical gearing*

**S. RAPPAPORT**
Consulting Engineer,
Wright Machinery Company

APPLICATION OF NON-UNIFORM ROTATION is sometimes unavoidable—and even to be desired. One frequent application is on rotary cross-cutting knives for continuously moving paper or foil webs.

Here, a fixed length of the web passes beneath the knife during one cycle of the machine, while the knife performs one rotation per cycle. The knife's peripheral speed, while cutting, should approximately equal the web speed, which in turn determines the cut length. Obviously, these conditions can be fulfilled for one particular cut length only if the knife diameter is kept constant. If the paper feed is set for a larger cut length, the web speed increases and the peripheral speed of the knife is relatively too slow.

To overcome this difficulty, elliptic gears are sometimes employed to drive the knife shaft. They cause the knife to accelerate in its cutting position to the approximate web speed and slow

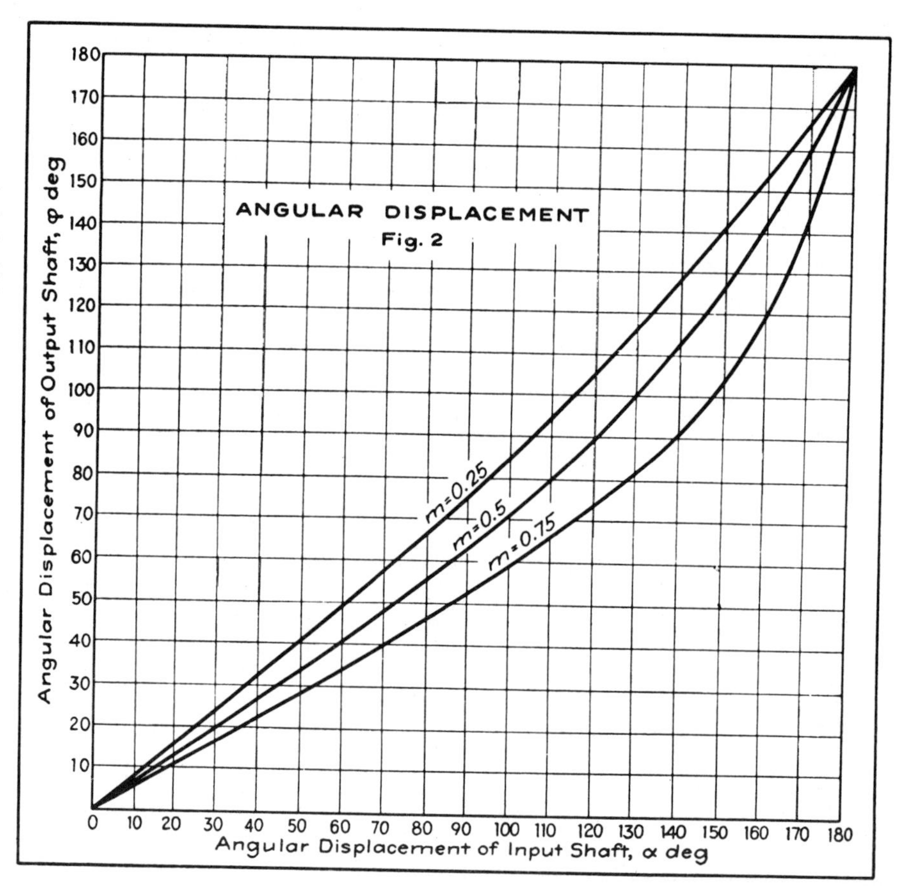

down in the remainder of its cycle.

There are other machines for which non-uniform output speed is desired, but elliptic gears are difficult to make and accordingly are expensive. There is, however, another solution for this kind of drive—the crank and slot drive.

It is a comparatively simple mechanism giving similar results. Fig 1(A) shows a schematic sketch.

It is obvious that the driven member, the slot link, will perform a complete revolution for each turn of the driving crank as long as their center distance is smaller than $r$, the crank length, or $m < 1$; (where $m$ equals a constant by which $r$ is multiplied to equal the distance between input and output shaft centers). For $m > 1$, the slot link will perform oscillations.

The equations of motion then will equal those of the Internal Geneva, PRODUCT ENGINEERING, Aug 1949, p 109, in that part of the oscillation where the direction of rotation is that of the driving crank. The equations will be those of the External Geneva, PRODUCT ENGINEERING, July 1949, p 110, when the slot link is on its backstroke, rotating in an opposite direction to the driving crank.

An advantage of this drive over the elliptic gear drive is the possibility of making the center distance adjustable. Thus it is possible to change the degree of circular irregularity to meet the particular requirement. Fig. 1(B) shows a photograph of a model with adjustable center distance.

For the development of the equations of motion refer to Fig. 1(A). The displacement of the output shaft, $\phi$, is found from triangle $ABC$, where $\alpha$ is the driving crank displacement:

$$\tan \phi = \frac{r \sin (180 - \alpha)}{mr - r \cos (180 - \alpha)} = \frac{\sin \alpha}{m + \cos \alpha} \quad (1)$$

Curves, Fig. 2, represent this equation for different values of $m$.

By differentiating Eq (1) the angular velocity of the driven shaft is found:

$$\frac{d\phi}{dt} = \omega \left( \frac{1 + m \cos \alpha}{1 + m^2 + 2m \cos \alpha} \right) \quad (2)$$

where $\omega$ is the uniform angular velocity of the driving crank in radians per second. This is plotted in Fig. 3. The minimum velocity occurs at $\alpha = 0$ deg and equals

$$\omega \left( \frac{1}{1 + m} \right).$$

The maximum velocity occurs at $\alpha =$ 180 deg and equals

$$\omega \left( \frac{1}{1 - m} \right).$$

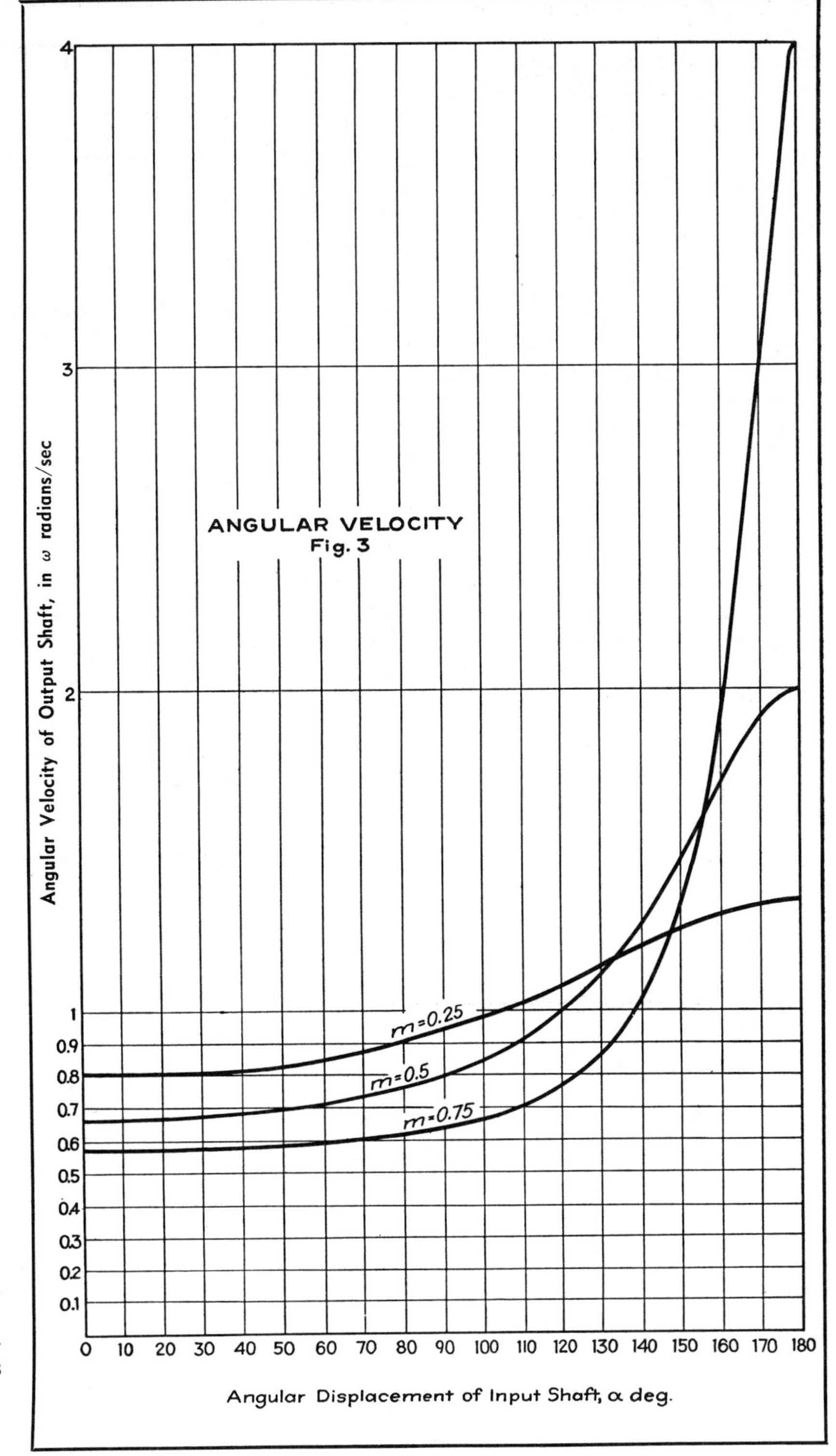


These relations can also be clearly seen from Fig. 1(A) taking the ratios between driving crank radius and the limits of the driven slot radius.

The degree of circular irregularity follows as

$$\frac{1 + m}{1 - m} \quad (3)$$

It is significant that Eq (2) closely resembles the velocity equation of elliptic gears which is

$$\frac{d\phi}{d_t} = \omega \left( \frac{1 - e^2}{1 + e^2 + 2e \cos \alpha} \right) \quad (4)$$

Where $e = \frac{1}{a} \sqrt{a^2 - b^2}$

Here $2a$ is the large axis, and $2b$ the small axis of the pitch ellipse.

The difference between the velocity outputs of elliptic gears and the crank and slot drive lies in the fact that the elliptic gear drive produces a lower velocity minimum than the crank and slot drive if the velocity maximum is kept alike in both drives. Accordingly, the degree of circular irregularity is larger for elliptic gears; namely:

$$\left(\frac{1+e}{1-e}\right)^2$$

as compared to Eq (3).

The acceleration curves, Fig. 4, are obtained by differentiation of Eq (2):

$$\frac{d^2\phi}{dt^2} = \omega^2\left[\frac{m(1-m^2)\sin\alpha}{(1+m^2+2m\cos\alpha)^2}\right] \quad (5)$$

The designer is interested mainly in the maximum acceleration, chiefly to calculate the highest force encountered.

To find at what position, $\alpha$, of the driver the maximum acceleration of the slot link occurs, let

$$\frac{d^3\phi}{dt^3} = 0$$

or:

$$(1+m^2+2m\cos\alpha)^2\, m(1-m^2)\cos\alpha + 4m^2(1-m^2)\sin^2\alpha\,(1+m^2+2m\cos\alpha) = 0 \quad (6)$$

from which follows:

$$\cos\alpha = \frac{1+m^2-\sqrt{(1+m^2)^2+32m^2}}{4m} \quad (7)$$

Substituting the value of $\alpha$ thus found into Eq (5) gives the maximum acceleration for the specified dimensions and input speed.

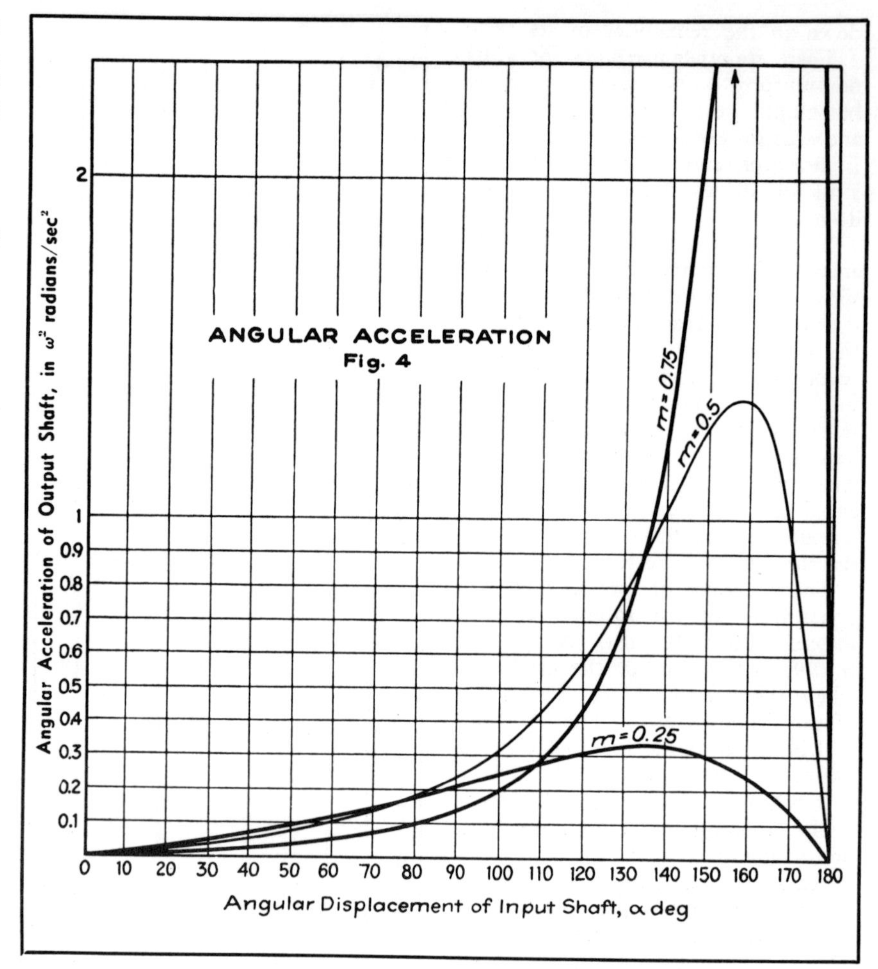

## linear to angular conversion of . . .

# GEAR-TOOTH INDEX ERROR

**For pitch diameters up to 200 in., chart quickly converts index error from ten-thousandths of an inch to seconds of arc.**

HAROLD R RONAN JR, *Research Engineer, Gould & Eberhardt Inc, Irvington, NJ*

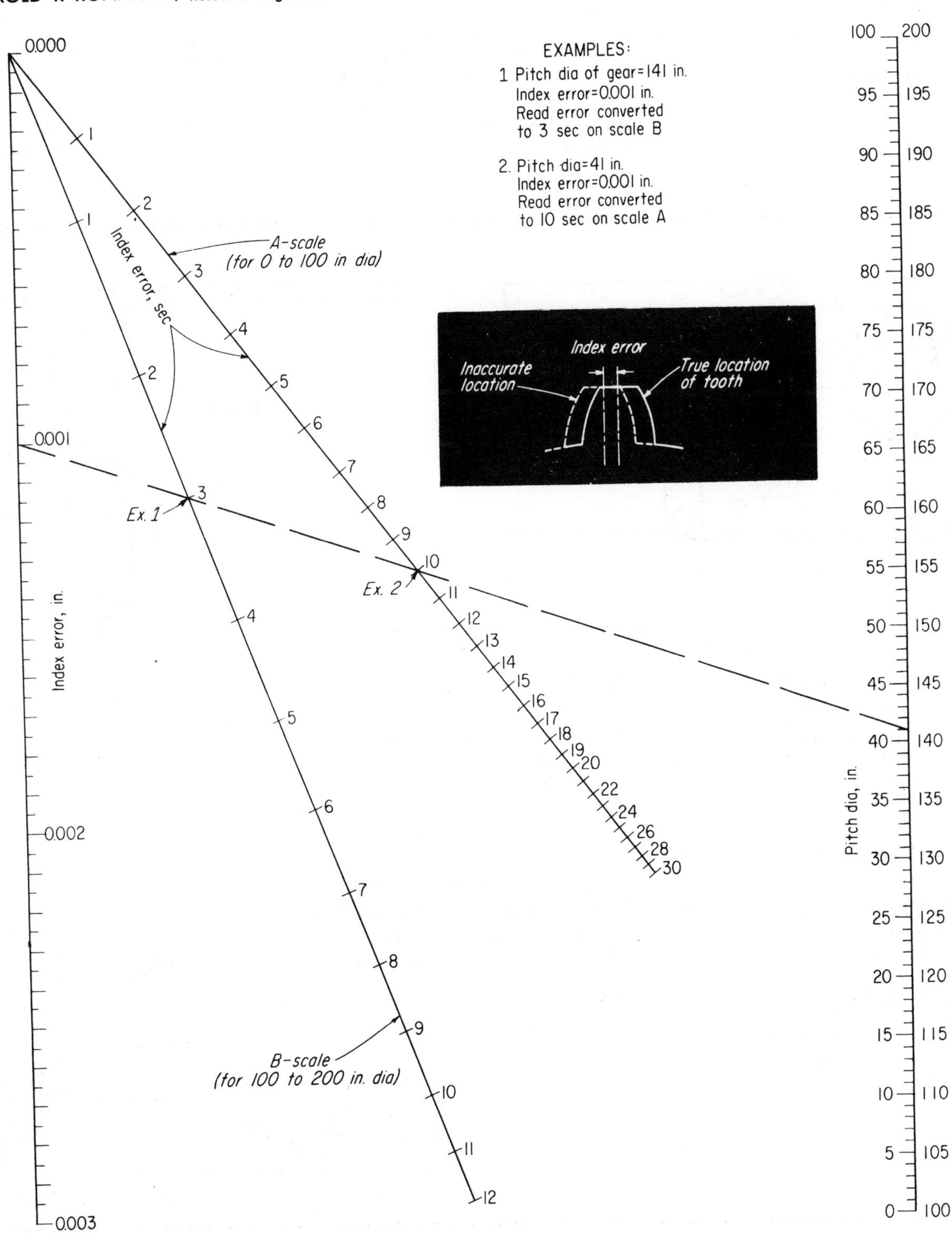

# ACCELERATED and DECELERATED

**When ordinary rotary cams cannot be conveniently applied, the mechanisms here presented, or adaptations of them, offer a variety of interesting possibilities for obtaining either acceleration or deceleration, or both**

JOHN E. HYLER
*Peoria, Ill.*

FIG. 1—Slide block moves at constant rate of reciprocating travel and carries both a pin for mounting link *B* and a stud shaft on which the pinion is freely mounted. Pinion carries a crankpin for mounting link *D* and engages stationary rack, the pinion may make one complete revolution at each forward stroke of slide block and another in opposite direction on the return—or any portion of one revolution in any specific instance. Many variations can be obtained by making the connection of link *F* adjustable lengthwise along link that operates it, by making crankpin radially adjustable, or by making both adjustable.

FIG. 2—Drive rod, reciprocating at constant rate, rocks link *BC* about pivot on stationary block and, through effect of toggle, causes decelerative motion of driven link. As drive rod advances toward right, toggle is actuated by encountering abutment and the slotted link *BC* slides on its pivot while turning. This lengthens arm *B* and shortens arm *C* of link *BC*, with decelerative effect on driven link. Toggle is spring-returned on the return stroke and effect on driven link is then accelerative.

FIG. 3—Same direction of travel for both the drive rod and the driven link

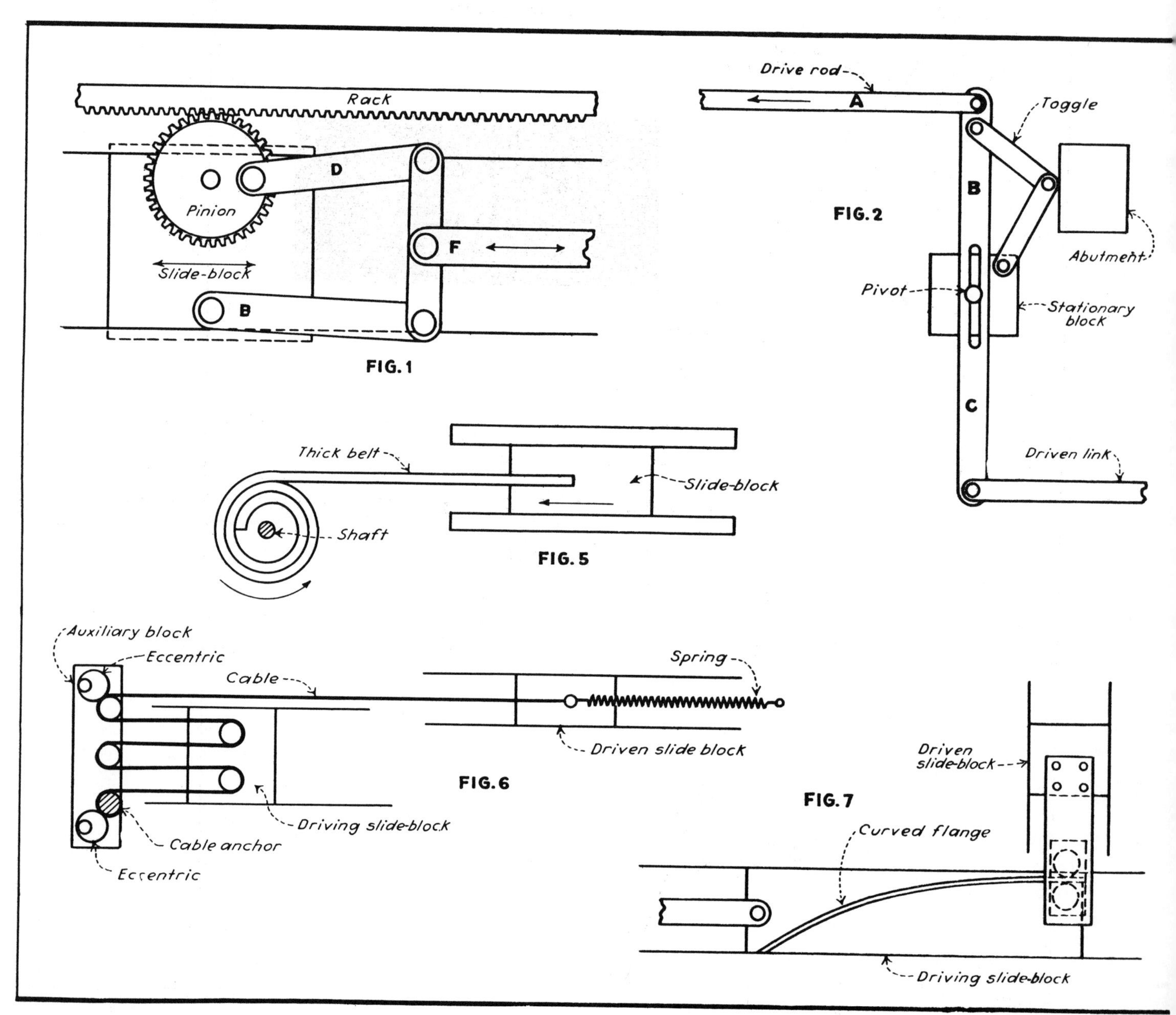

is provided by this variation of the preceding mechanism. Here, acceleration is in direction of arrows and deceleration occurs on return stroke. Accelerative effect becomes less as toggle flattens.

FIG. 4—Bellcrank motion is accelerated as rollers are spread apart by curved member on end of drive rod, thereby in turn accelerating motion of slide block. Driven elements must be spring-returned to close system.

FIG. 5—Constant-speed shaft winds up thick belt, or similar flexible member, and increase in effective radius causes accelerative motion of slide block. Must be spring- or weight-returned on reversal.

FIG. 6 — Auxiliary block, carrying sheaves for cable which runs between driving and driven slide blocks, is mounted on two synchronized eccentrics. Motion of driven block is equal to length of cable paid out over sheaves resulting from additive motions of the driving and the auxiliary blocks.

FIG. 7—Curved flange on driving slide block is straddled by rollers pivotally mounted in member connected to driven slide block. Flange can be curved to give desired acceleration or deceleration, and mechanism is self-returned.

FIG. 8—Stepped acceleration of the driven slide block is effected as each of the three reciprocating sheaves progressively engages the cable. When the third acceleration step is reached, the driven slide block moves six times faster than the drive rod.

FIG. 9—Form-turned nut, slotted to travel on rider, is propelled by reversing screw shaft, thus moving concave roller up and down to accelerate or decelerate slide block.

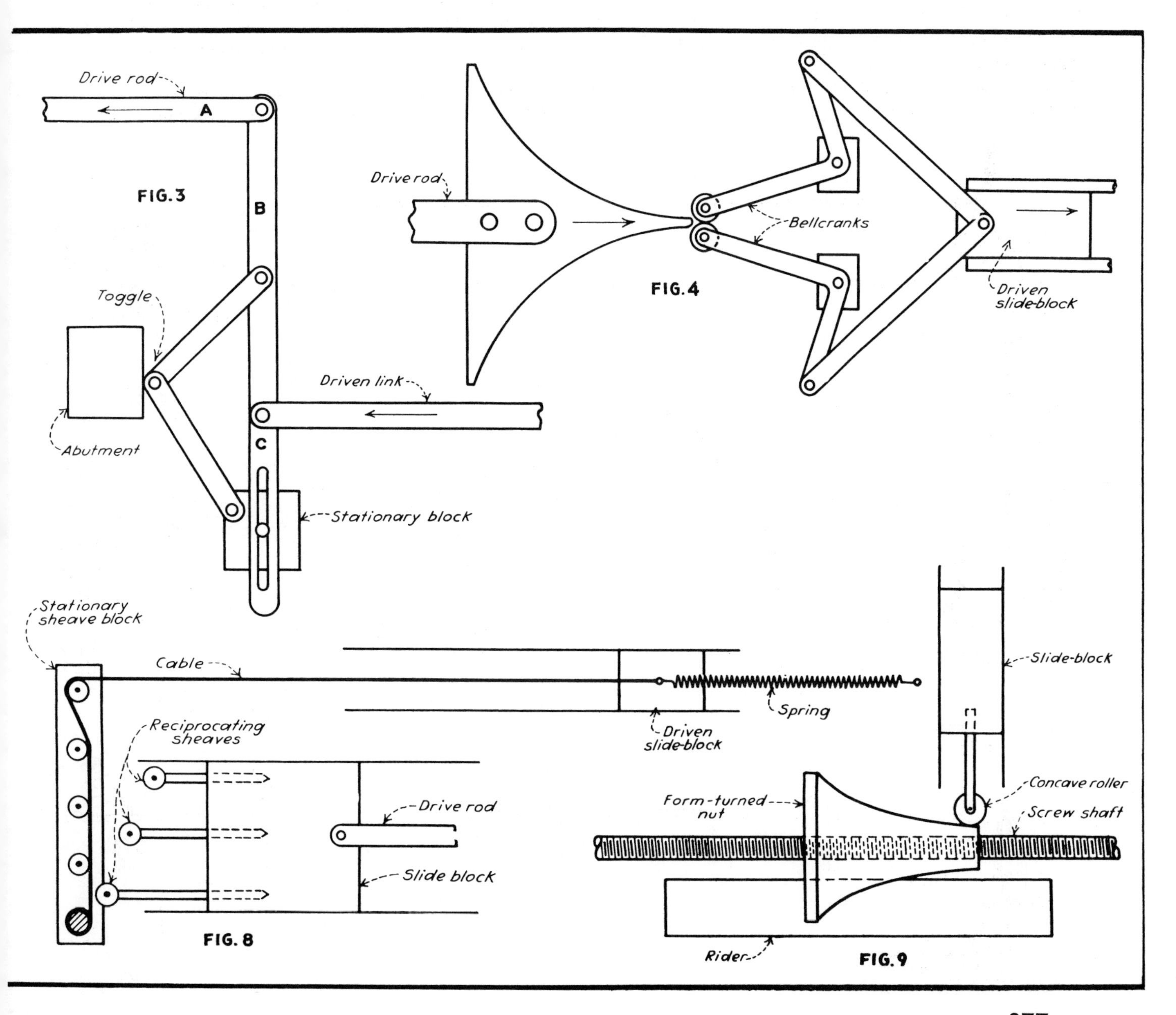

# POWER THRUST LINKAGES

POWERED STRAIGHT LINE MOTION over short distances is applicable to many types of machines or devices for performing specialized services. These motions can be produced by a steam, pneumatic or hydraulic cylinder, or by a self-contained electric powered unit such as the General Electric Thrustors shown herewith. These Thrustors may be actuated manually by pushbuttons, or automatically by mechanical devices or the photo-electric relay as with a door opener. These illustrations will suggest many other arrangements.

Fig. 1—Transfer motion to distant point.

Fig. 2—Double throw by momentary applications.

Fig. 3—Trammel plate divides effort and changes directions of motion.

Fig. 4—Constant thrust toggle. Pressure is same at all points of throw.

Fig. 5—Multiplying motion, 6 to 1, might be used for screen shift.

Fig. 6—Bell crank and toggle may be applied in embossing press, extruder or die-caster.

Fig. 7—Horizontal pull used for clay pigeon traps, hopper trips, and sliding elements with spring or counterweighted return.

Fig. 8—Shipper rod for multiple and distant operation as series of valves.

Fig. 9—Door opener. Upthrust of helical racks rotates gear and arm.

Fig. 10—Accelerated motion by shape of cam such as on a forging hammer.

Fig. 11—Intermittent lift as applied to lifting pipe from well.

Fig. 12—Straight-line motion multiplied by pinion and racks.

Fig. 13—Rotary motion with cylindrical cam. Operates gate on conveyor belt.

Fig. 14—Thrust motions and "dwells" regulated by cam.

Fig. 15—Four positive positions with two Thrustors.

Fig. 16—Toggle increasing thrust at right angle.

Fig. 17—Horizontal straight-line motion as applied to a door opener.

Fig. 18—Thrusts in three directions with two Thrustors.

Fig. 19—Fast rotary motion using step screw and nut.

Fig. 20—Intermittent rotary motion. Operated by successive pushing of operating button, either manually or automatically.

Fig. 21—Powerful rotary motion with worm driven by rack and pinion.

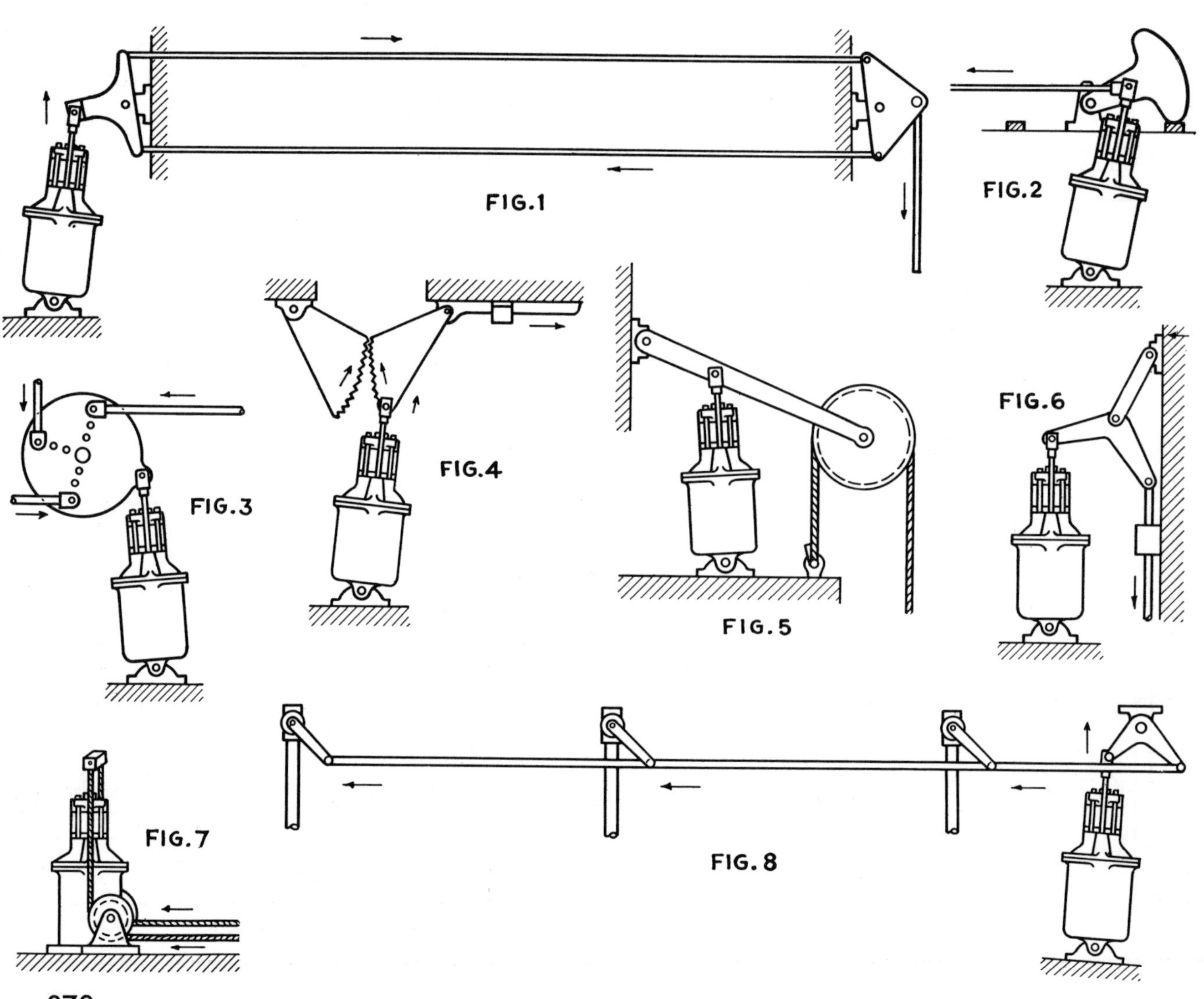

# AND THEIR APPLICATIONS

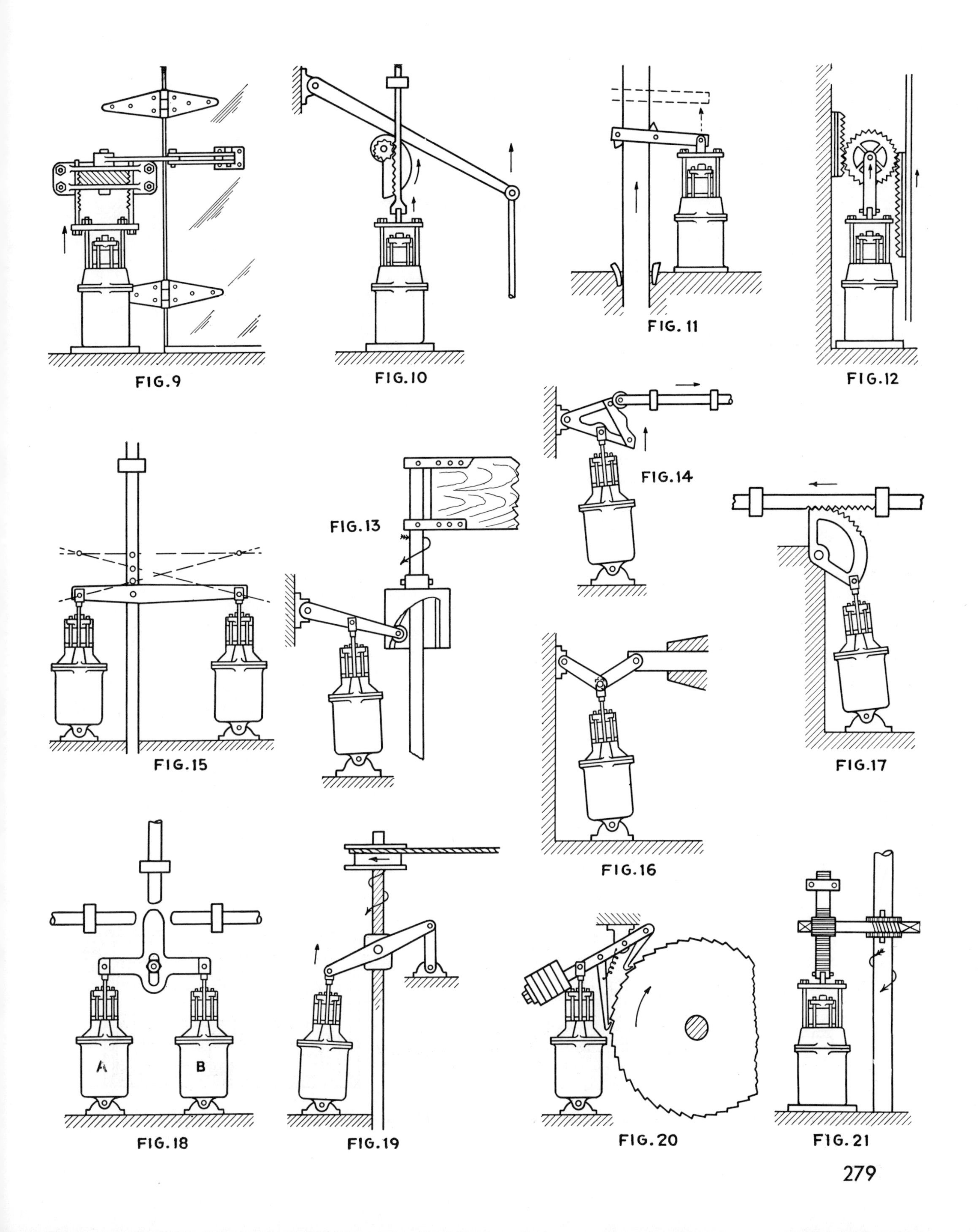

# TRANSMISSION LINKAGES FOR

THE ACCOMPANYING SKETCHES show typical mechanisms for multiplying short linear motions, usually converting the linear motion into rotation. Although the particular mechanisms shown are designed to multiply the movements of diaphragms or bellows, the same or similar constructions have possible applications wherever it is required to obtain greatly multiplied motions. These patented transmissions depend on cams, sector gears and pinions, levers and cranks, cord or chain, spiral or screw feed, magnetic attraction, or combinations of these devices.

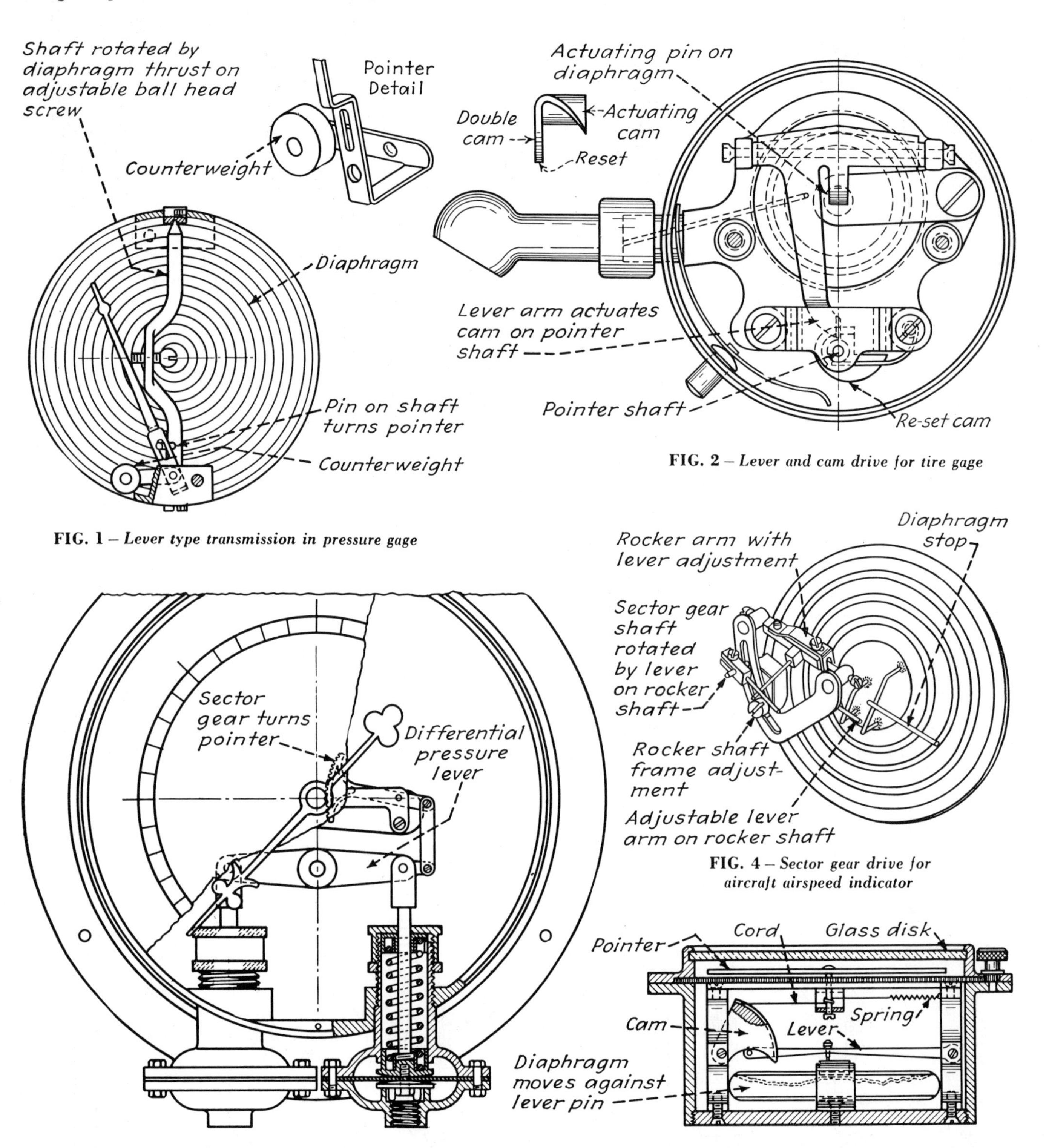


**FIG. 1** – *Lever type transmission in pressure gage*

**FIG. 2** – *Lever and cam drive for tire gage*

**FIG. 3** – *Lever and sector gear in differential pressure gage*

**FIG. 4** – *Sector gear drive for aircraft airspeed indicator*

**FIG. 5** – *Lever, cam and cord transmission in barometer*

# MULTIPLYING SHORT MOTIONS

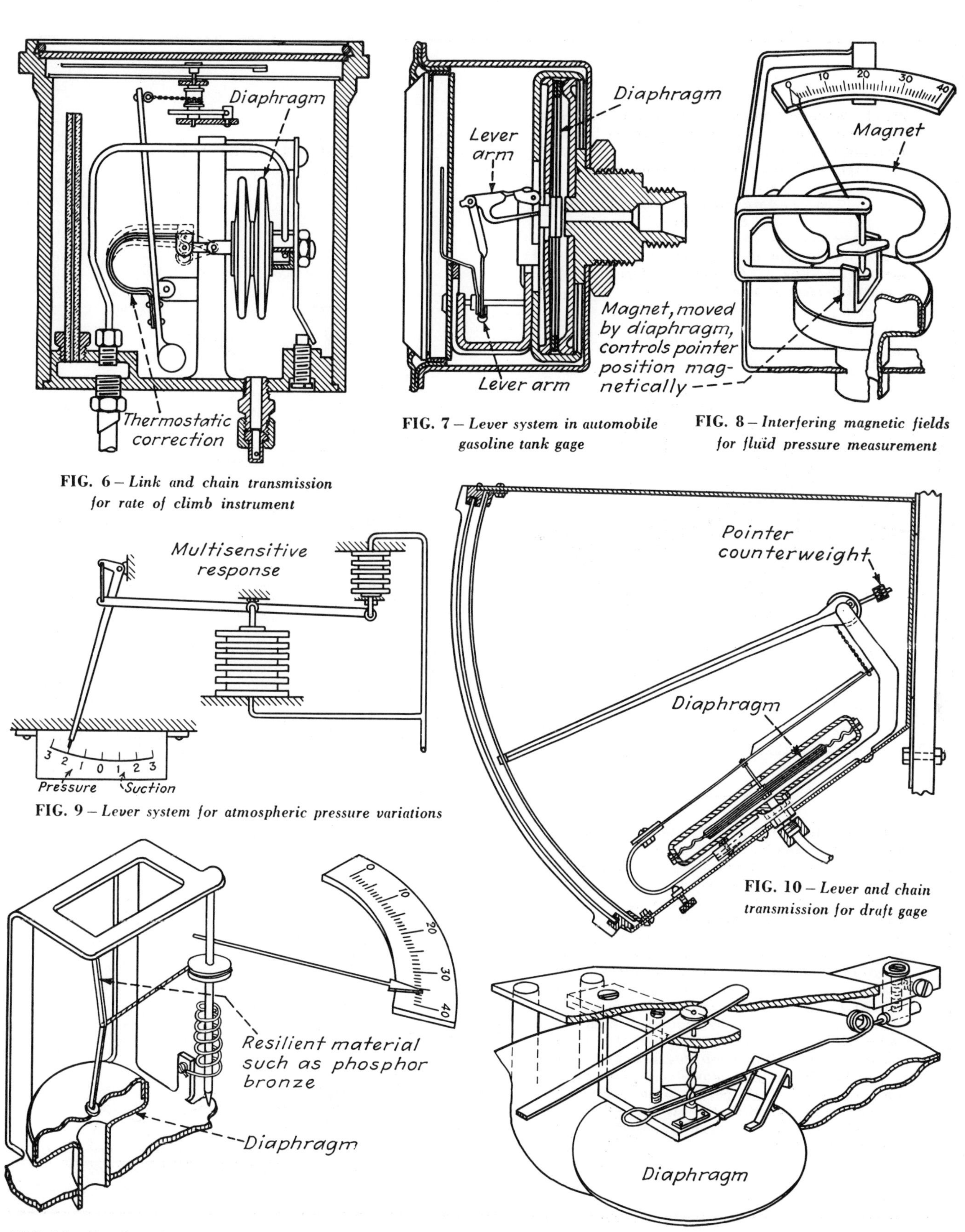


**FIG. 6** – *Link and chain transmission for rate of climb instrument*

**FIG. 7** – *Lever system in automobile gasoline tank gage*

**FIG. 8** – *Interfering magnetic fields for fluid pressure measurement*

**FIG. 9** – *Lever system for atmospheric pressure variations*

**FIG. 10** – *Lever and chain transmission for draft gage*

**FIG. 11** – *Toggle and cord drive for fluid pressure instrument*

**FIG. 12** – *Spiral feed transmission for general purpose instrument*

# Special Purpose Mechanisms

SIGMUND RAPPAPORT
Ford Instrument Company
(Div. of The Sperry Corporation)

**Fig. 1—Spherical linkage, produces an oscillating rotational output from a constant speed input for an input angular displacement.**
**where $\psi$=output angular displacement**
**$\rho$=tan $\alpha$**
**$\alpha$=constant angle between roller shaft *b* and input axis *c***
**$\tan \psi = \rho \sin \phi$**
**For mechanism to function properly, axes of semicircular link *d*, of roller *a*, and of input shaft *c* must intersect at point *M*.**

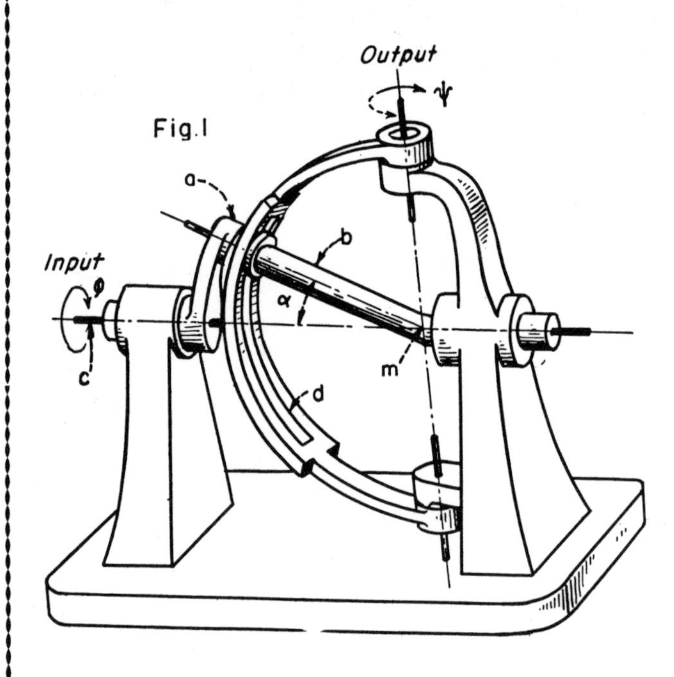


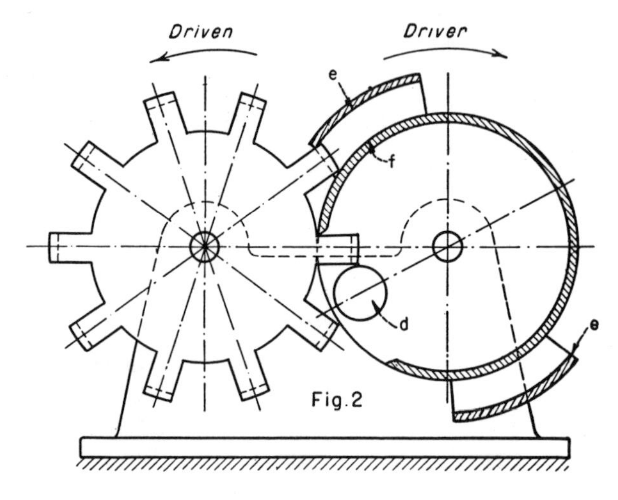


**Fig. 2—Intermittent drive with cylindrical lock. Shortly before and after engagement of two teeth with driving pin *d* at the end of the dwell period, the inner cylinder *f* is not adequate to effect positive locking of the driven gear. A concentric auxiliary cylinder *e* is therefore provided with which only two segments are necesary to get positive locking. Their length is determined by the circular pitch of the driven gear.**

**Fig. 3—"Multilated tooth" intermittent drive. Driver *b* is a circular disk of width *w* with a cut-out *d* on its circumference and carries a pin *c* close to the cutout. The driven gear, *a* of width *2w* has standard spur gear teeth, always an even number, which are alternately of full and of half width (mutilated). During the dwell period two full width teeth are in contact with the circumference of the driving disk, thus locking it; the multilated tooth between them is behind the driver. At the end of the dwell period pin *c* comes in contact with the mutilated tooth and turns the driven gear for one circular pitch. Then, the full width tooth engages the cutout *d* and the driven gear moves one more pitch, whereupon the dwell period starts again and the cycle is repeated. Used only for light loads primarily because of high accelerations encountered.**

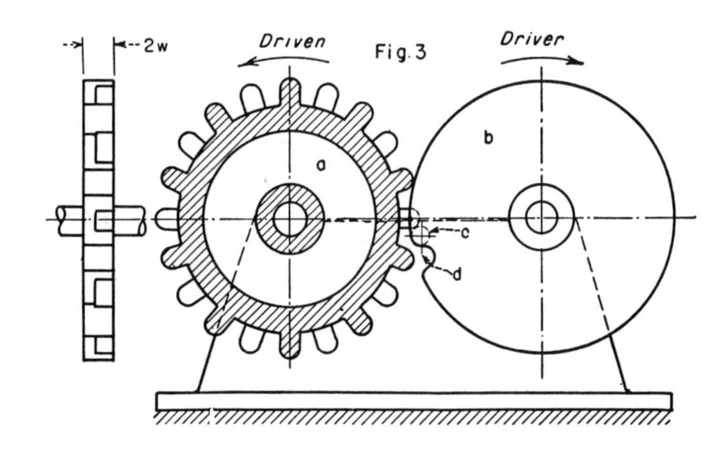

Data based on material and sketches in AWF and VDMA Getriebeblaetter, published by Ausschuss fuer Getriebe beim Ausschuss fuer wirtschaftiche Fertigung, Leipzig, Germany.

**Fig. 4—Mechanism for transmitting intermittent motion between two skewed shafts. The shafts need not be at right angles to one another. Angular displacement of the output shaft per revolution of input shaft equals the circular pitch of the output gear wheel divided by its pitch radius. The duration of the motion period depends on the length of the angular join *a* of the locking disks *b*. This drive was used extensively on motion picture projectors. Two other drives, used to move the film intermittently in projectors; are shown in Figs. 5 and 6.**

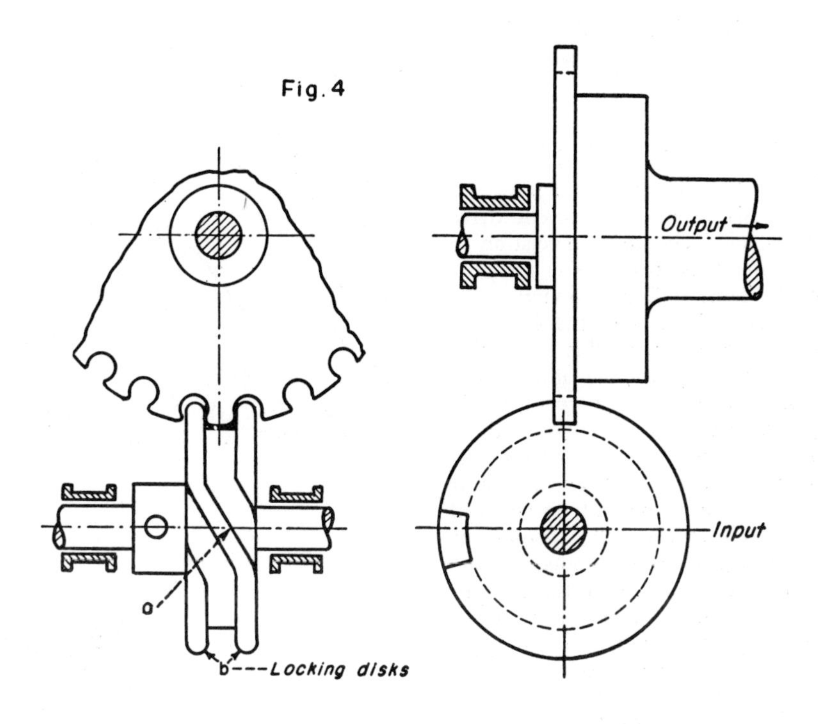


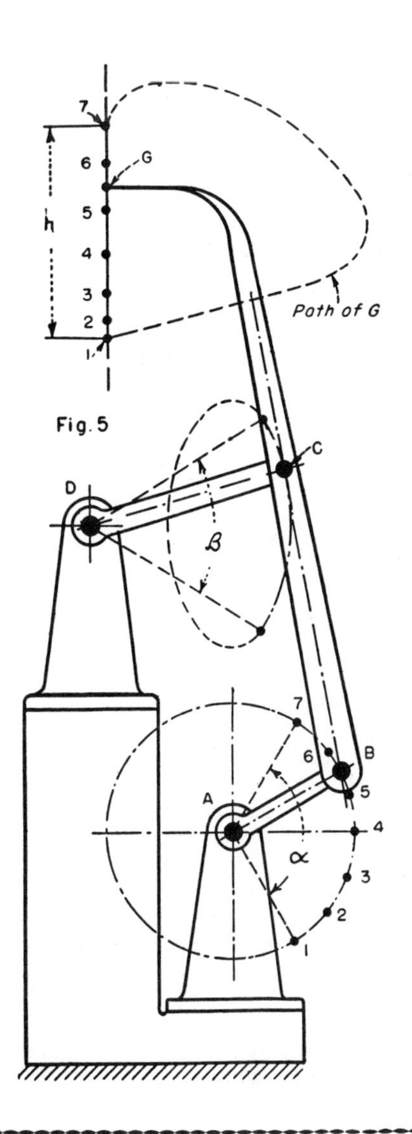


**Fig. 5—The task of describing a "D" curve, as accomplished by a mechanism of Fig. 6, can also be achieved by this four-bar linkage. Hint for designing: Start with a straight line portion of the path of *G*, replace the oval arc of *C* by an arc of the osculating circle, thus determining the length of link *DC*.**

**Fig. 6—This mechanism is intended to accomplish the following: (1) Film hook, while moving the film strip, must describe very nearly a straight line; (2) Engagement and disengagement of the hook with the perforation of the film must take place in a direction approximately normal to the film; (3) Engagement and disengagement should be shock free. Slight changes in the shape of the guiding slot *f* enable the designer to vary the shape of the output curve as well as the velocity diagram appreciably.**

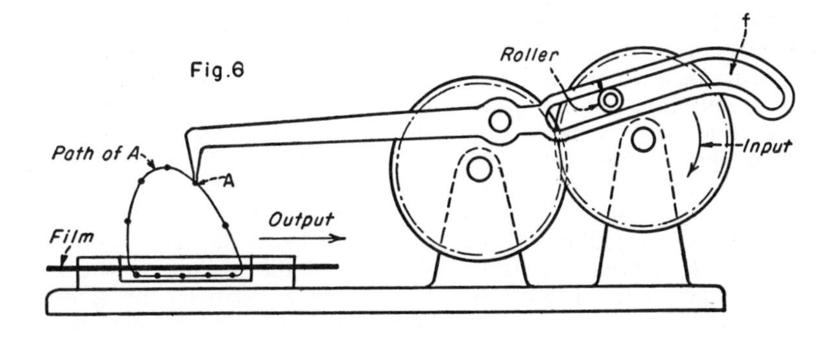

# Mechanisms for Producing

SIGMUND RAPPAPORT
Ford Instrument Company

## Straight Line Motion

FIG. 1—No linkages or guides are used in this modified hypocyclic drive which is relatively small in relation to the length of its stroke. The sun gear of pitch diameter $D$ is stationary. The drive shaft, which turns the T-shaped arm, is concentric with this gear. The idler and planet gears, the latter having a pitch diameter of $D/2$, rotate freely on pivots in the arm extensions. Pitch diameter of the idler is of no geometrical significance, although this gear does have an important mechanical function. It reverses the rotation of the planet gear, thus producing true hypocyclic motion with ordinary spur gears only. Such an arrangement occupies only about half as much space as does an equivalent mechanism containing an internal gear. Center distance $R$ is the sum of $D/2$, $D/4$ and an arbitrary distance $d$, determined by a particular application. Points $A$ and $B$ on the driven link, which is fixed to the planet, describe straight-line paths through a stroke of $4R$. All points between $A$ and $B$ trace ellipses, while the line $AB$ envelopes an astroid.

## Parallel Motion

FIG. 2—A slight modification of the mechanism in Fig. 1 will produce another type of useful motion. If the planet gear has the same diameter as that of the sum gear, the arm will remain parallel to itself throughout the complete cycle. All points on the arm will thereby describe circles of radius $R$. Here again, the position and diameter of the idler gear are of no geometrical importance. This mechanism can be used, for example, to cross-perforate a uniformly moving paper web. The value for $R$ is chosen such that $2\pi R$, or the circumference of the circle described by the needle carrier, equals the desired distance between successive lines of perforations. If the center distance $R$ is made adjustable, the spacing of perforated lines can be varied as desired.

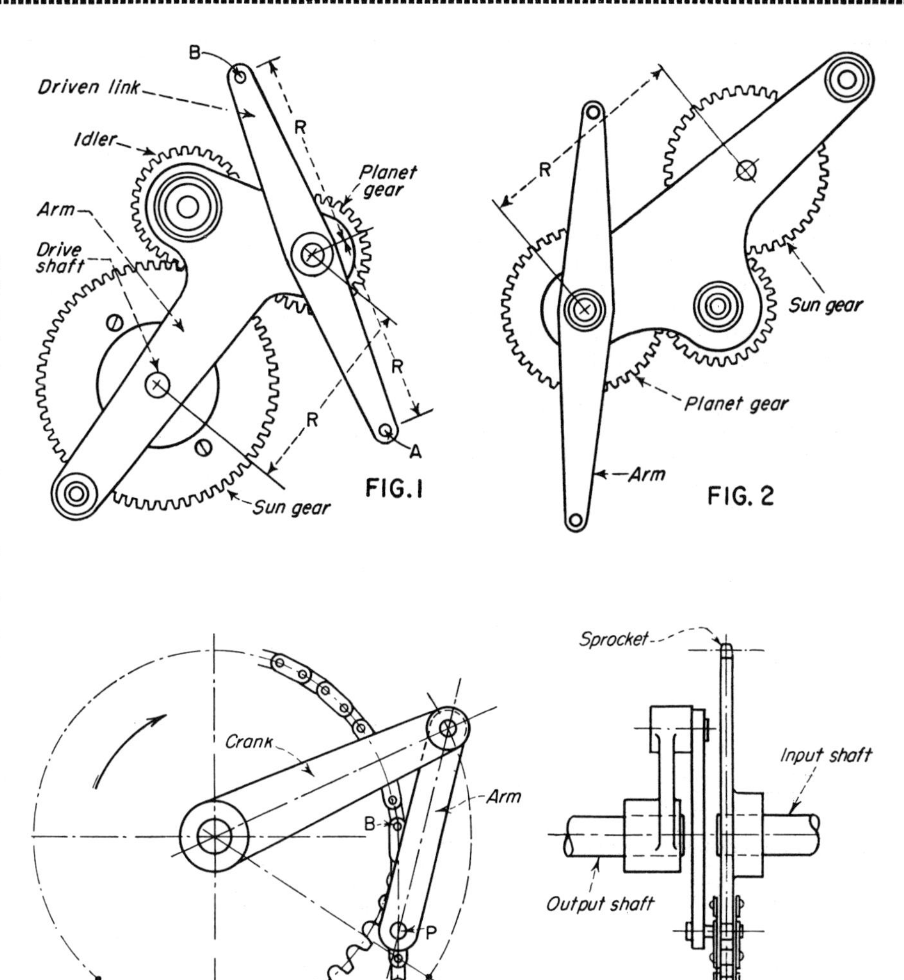


## Intermittent Motion

FIG. 3—This mechanism, developed by the author and to his knowledge novel, can be adapted to produce a stop, a variable speed without stop or a variable speed with momentary reverse motion. Uniformly rotating input shaft drives the chain around the sprocket and idler, the arm serving as a link between the chain and the end of the output shaft crank. The sprocket drive must be in the ratio $N/n$ with the cycle of the machine, where $n$ is the number of teeth on the sprocket and $N$ the number of links in the chain. When point $P$ travels around the sprocket from point $A$ to position $B$, the crank rotates uniformly. Between $B$ and $C$, $P$ decelerates; between $C$ and $A$ it accelerates;; and at $C$ there is a momentary dwell. By changing the size and position of the idler, or the lengths of the arm and crank, a variety of motions can be obtained. If in the sketch, the length of the crank is shortened, a brief reverse period will occur in the vicinity of $C$; if the crank is lengthened, the output velocity will vary between a maximum and minimum without reaching zero.

# Specific Types of Motions

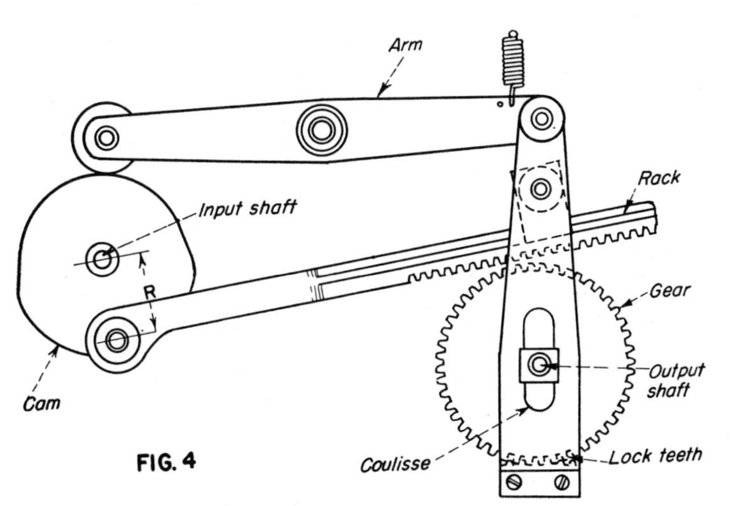


FIG. 4

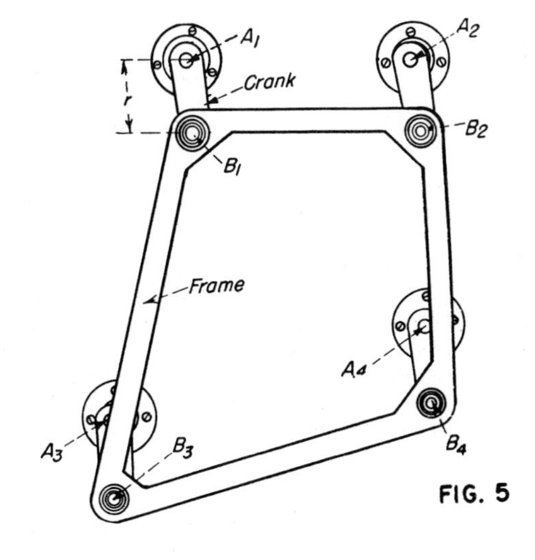


FIG. 5

## Intermittent Motion

FIG. 4—An operating cycle of 180 deg motion and 180 deg dwell is produced by this mechanism. The input shaft drives the rack which is engaged with the output shaft gear during half the cycle. When the rack engages, the lock teeth at the lower end of the coulisse are disengaged and, conversely, when the rack is disengaged, the coulisse teeth are engaged, thereby locking the output shaft positively. The change-over points occur at the dead-center positions so that the motion of the gear is continuously and positively governed. By varying *R* and the diameter of the gear, the number of revolutions made by the output shaft during the operating half of the cycle can be varied to suit requirements.

## Rotational Motion

FIG. 5—The absence of backlash makes this old but little used mechanism a precision, low-cost replacement for gear or chain drives otherwise used to rotate parallel shafts. Any number of shafts greater than two can be driven from any one of the shafts, provided two conditions are fulfilled: (1) All cranks must have the same length *r;* and (2) the two polygons formed by the shafts *A* and frame pivot centers *B* must be identical. The main disadvantage of this mechanism is its dynamic unbalance, which limits the speed of rotation. To lessen the effect of the vibrations produced, the frame should be made as light as is consistent with strength requirements.

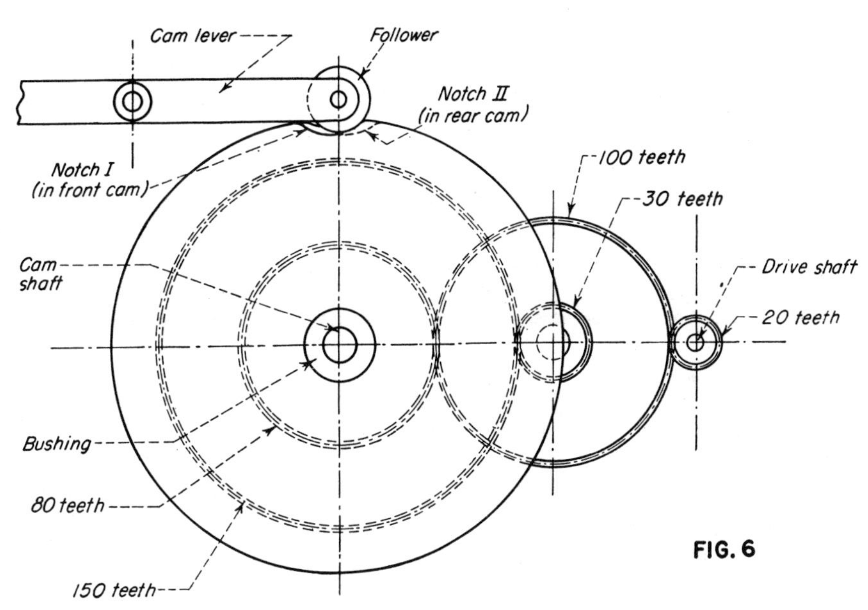


FIG. 6

## Fast Cam-Follower Motion

FIG. 6—Fast cam action every *n* cycles when *n* is a relatively large number, can be obtained with this manifold cam and gear mechanism. A single notched cam geared $1/n$ to a shaft turning once a cycle moves relatively slowly under the follower. The double notched-cam arrangement shown is designed to operate the lever once in 100 cycles, imparting to it a rapid movement. One of the two identical cams and the 150-tooth gear are keyed to the bushing which turns freely around the cam shaft. The latter carries the second cam and the 80-tooth gear. The 30- and 100-tooth gears are integral, while the 20-tooth gear is attached to the one-cycle drive shaft. One of the cams turns in the ratio of 20/80 or 1/4; the other in the ratio 20/100 times 30/150 or 1/25. The notches therefore coincide once every 100 cycles ($4 \times 25$). Lever movement is the equivalent of a cam turning in a ratio of 1/4 in relation to the drive shaft. To obtain fast cam action, *n* must be broken down into prime factors. For example, if 100 were factored into 5 and 20, the notches would coincide after every 20 cycles.

# Toggle Linkage Applications in

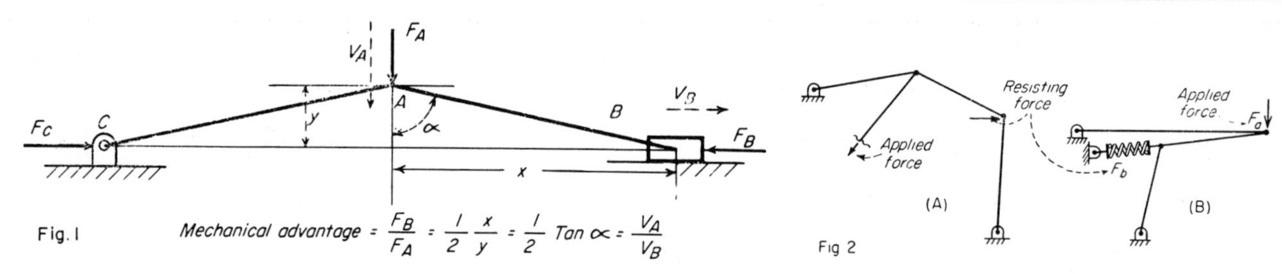


MANY MECHANICAL LINKAGES are based on the simple toggle which consists of two links that tend to line-up in a straight line at one point in their motion. The mechanical advantage is the velocity ratio of the input point $A$ to the output point $B$: or $V_A/V_B$. As the angle $\alpha$ approaches 90 deg, the links come into toggle and the mechanical advantage and velocity ratio both approach infinity. However, frictional effects reduce the forces to much less than infinity although still quite high.

FORCES CAN BE APPLIED through other links, and need not be perpendicular to each other. (A) One toggle link can be attached to another link rather than to a fixed point or slider. (B) Two toggle links can come into toggle by lining up on top of each other rather than as an extension of each other. Resisting force can be a spring force.

## HIGH MECHANICAL ADVANTAGE

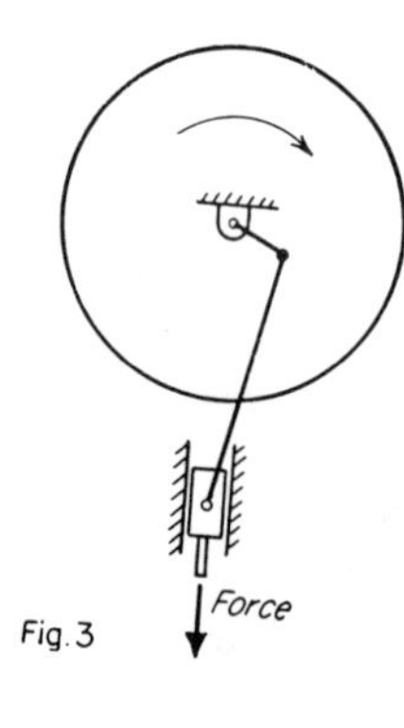


IN PUNCH PRESSES, large forces are needed at the lower end of the work-stroke, however little force is required during the remainder. Crank and connecting rod come into toggle at the lower end of the punch stroke, giving a high mechanical advantage at exactly the time it is most needed.

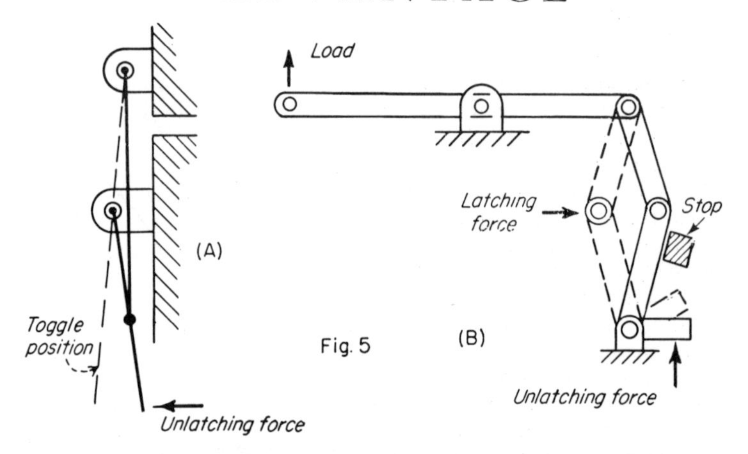


COLD-HEADING RIVET MACHINE is designed to give each rivet two successive blows. Following the first blow (point 2) the hammer moves upward a short distance (to point 3), to provide clearance for moving the workpiece. Following the second blow (at point 4), the hammer then moves upward a longer distance (to point 1). Both strokes are produced by one revolution of the crank and at the lowest point of each stroke (points 2 and 4) the links are in toggle.

LOCKING LATCHES produce a high mechanical advantage when in the toggle portion of the stroke. (A) Simple latch exerts a large force in the locked position. (B) For positive locking, closed position of latch is slightly beyond toggle position. Small unlatching force opens linkage.

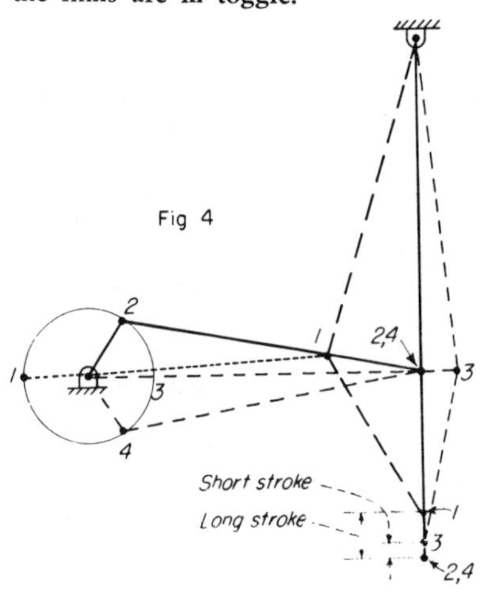


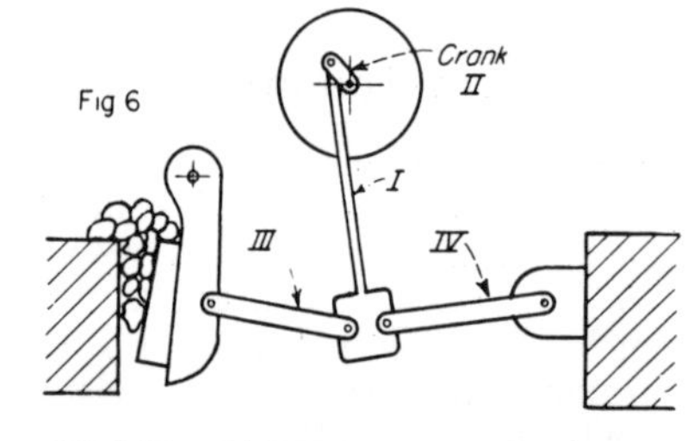


STONE CRUSHER uses two toggle linkages in series to obtain a high mechanical advantage. When the vertical link I reaches the top of its stroke, it comes into toggle with the driving crank II; at the same time, link III comes into toggle with link IV. This multiplication results in a very large crushing force.

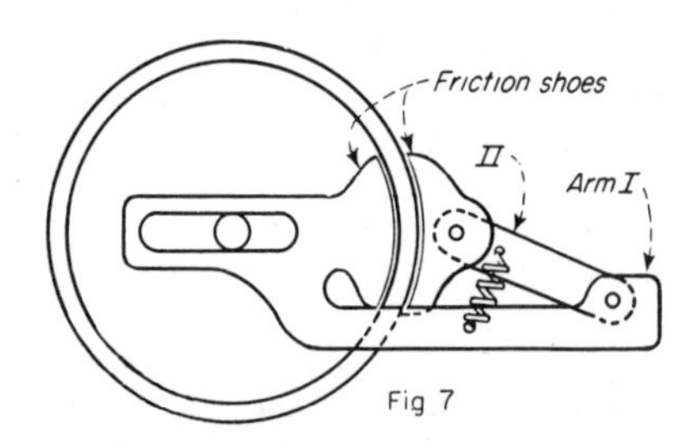


FRICTION RATCHET is mounted on a wheel; light spring keeps friction shoes in contact with the flange. This device permits clockwise motion of the arm I. However, reverse rotation causes friction to force link II into toggle with the shoes which greatly increases the locking pressure.

# Different Mechanisms

THOMAS P. GOODMAN
Westinghouse Electric Corporation

## HIGH VELOCITY RATIO

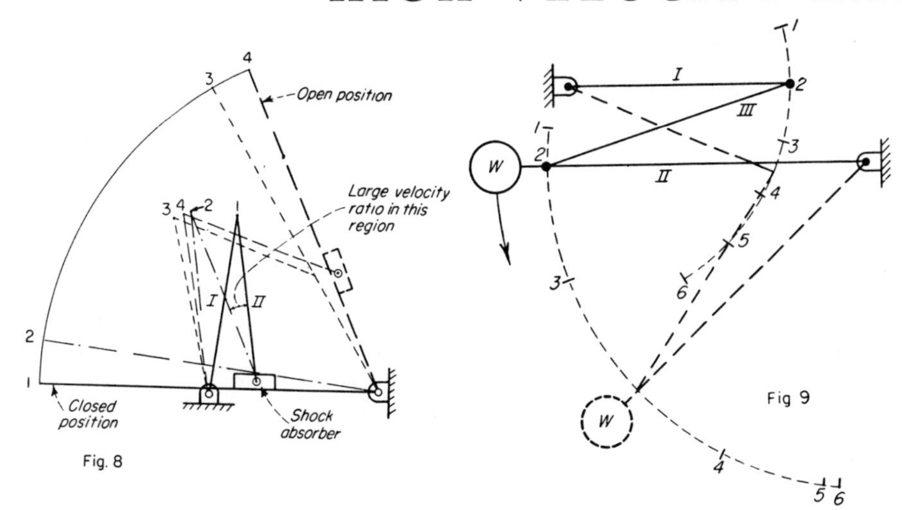


**DOOR CHECK LINKAGE** gives a high velocity ratio at one point in the stroke. As the door swings closed, connecting link I comes into toggle with the shock absorber arm II, giving it a large angular velocity. Thus, the shock absorber is more effective retarding motion near the closed position.

**IMPACT REDUCER** used on some large circuit breakers. Crank I rotates at constant velocity while lower crank moves slowly at the beginning and end of the stroke. It moves rapidly at the mid stroke when arm II and link III are in toggle. Falling weight absorbs energy and returns it to the system when it slows down.

## VARIABLE MECHANICAL ADVANTAGE

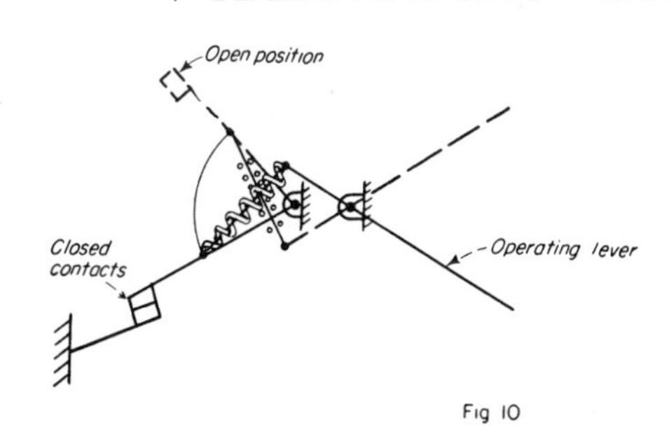


**TOASTER SWITCH** uses an increasing mechanical advantage to aid in compressing a spring. In the closed position, spring holds contacts closed and the operating lever in the down position. As the lever is moved upward, the spring is compressed and comes into toggle with both the contact arm and the lever. Little effort is required to move the links through the toggle position; beyond this point, the spring snaps the contacts closed.

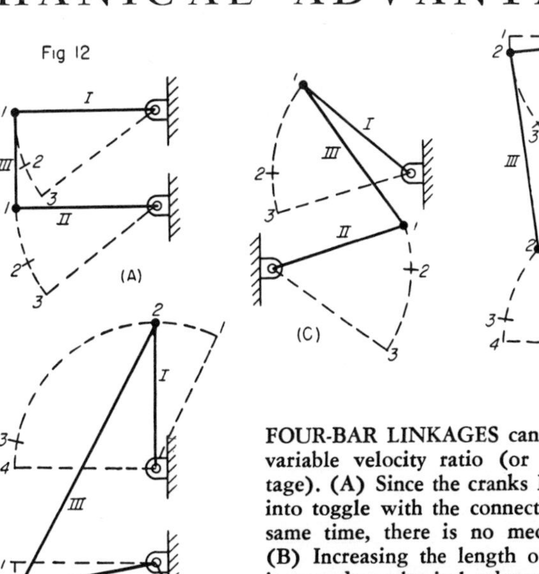


**FOUR-BAR LINKAGES** can be altered to give variable velocity ratio (or mechanical advantage). (A) Since the cranks I and II both come into toggle with the connecting link III at the same time, there is no mechanical advantage. (B) Increasing the length of link III gives an increased mechanical advantage between positions 1 and 2, since crank I and connecting link III are near toggle. (C) Placing one pivot at the left produces similar effects as in (B). (D) Increasing the center distance puts crank II and link III near toggle at position 1; crank I and link III approach toggle position at 4.

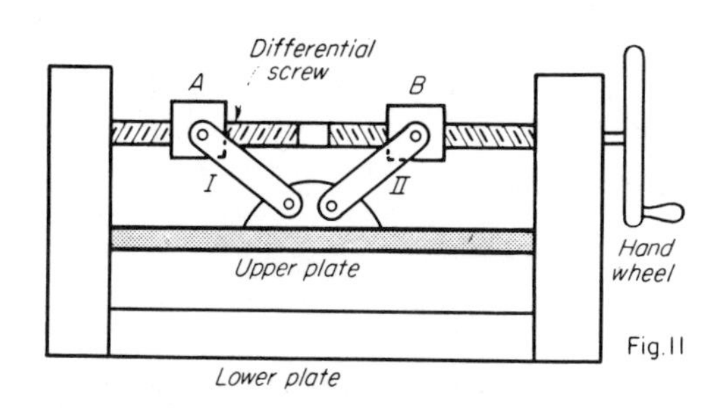


**TOGGLE PRESS** has an increasing mechanical advantage to counteract the resistance of the material being compressed. Rotating handwheel with differential screw moves nuts *A* and *B* together and links I and II are brought into toggle.

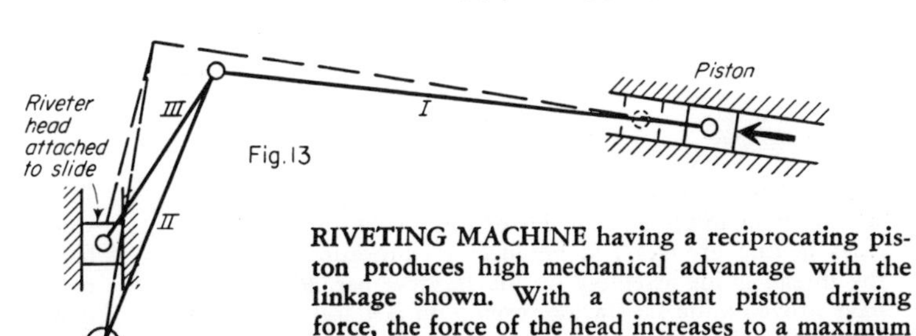


**RIVETING MACHINE** having a reciprocating piston produces high mechanical advantage with the linkage shown. With a constant piston driving force, the force of the head increases to a maximum value when links II and III come into toggle.

# Traversing Mechanisms Used

**The seven mechanisms shown below are used on different types of yarn and coil winding machines. Their fundamentals, however, may be applicable to other machines which require**

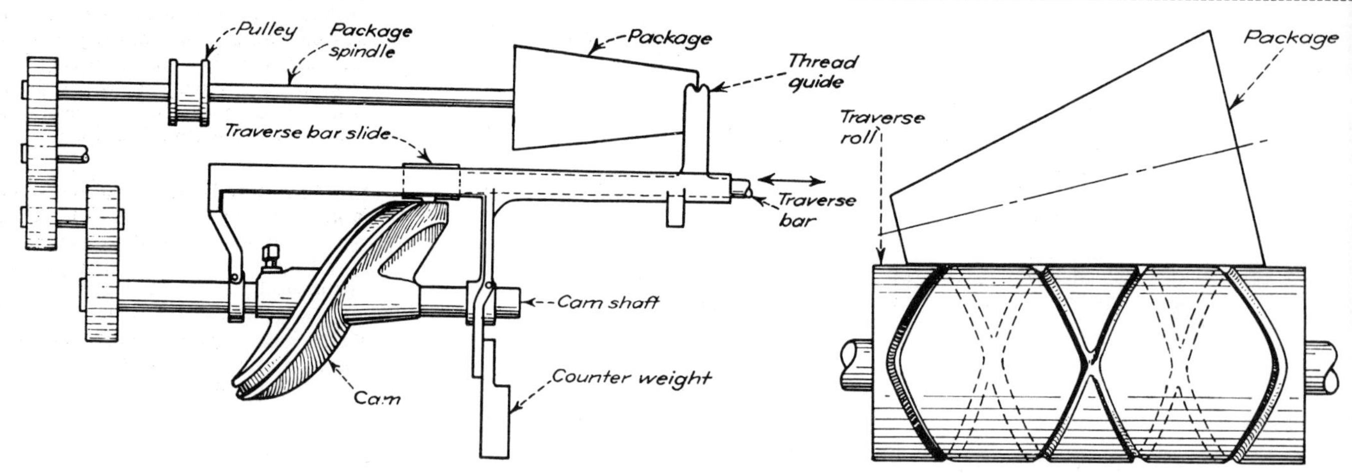


FIG. 1. Package is mounted on belt driven shaft on this precision type winding mechanism. Cam shaft imparts reciprocating motion to traverse bar by means of cam roll that runs in cam groove. Gears determine speed ratio between cam and package. Thread guide is attached to traverse bar. Counterweight keeps thread guide against package.

FIG. 2. Package is friction-driven from traverse roll. Yarn is drawn from the supply source by traverse roll and is transferred to package from the continuous groove in the roll. Different winds are obtained by varying the grooved path.

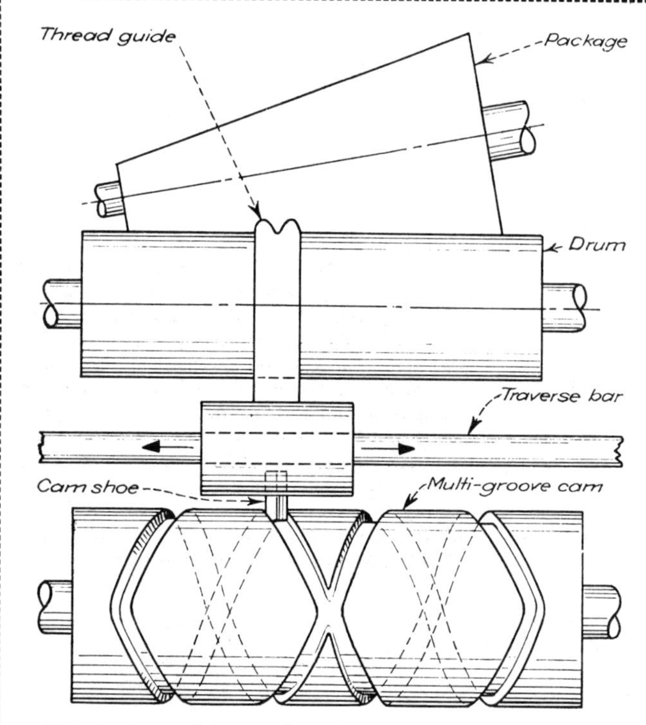


FIG. 4. Drum drives package by friction. Pointed cam shoe which pivots in the bottom side of the thread guide assembly rides in cam grooves and produces reciprocating motion of the thread guide assembly on the traverse bar. Plastic cams have proved quite satisfactory even with fast traverse speeds. Interchangeable cams permit a wide variety of winds.

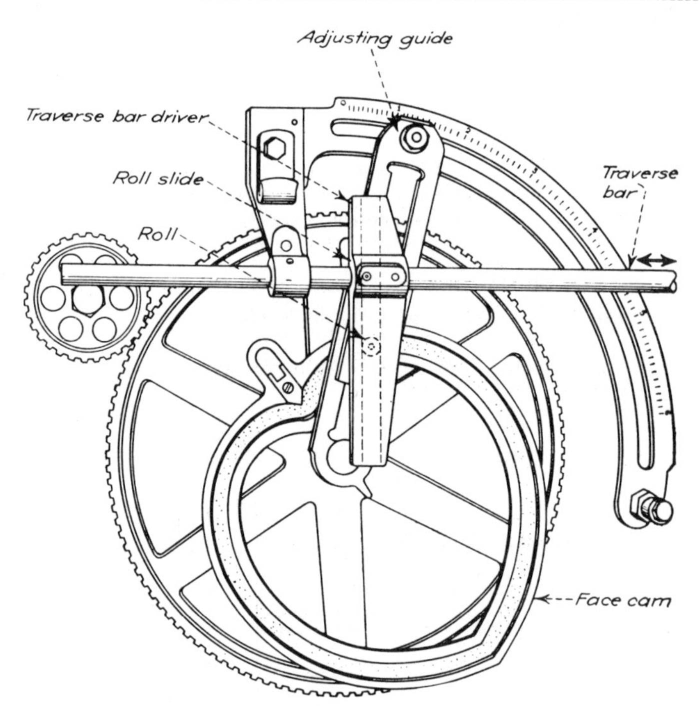


FIG. 5. Roll that rides in heart-shaped cam groove engages slot in traverse bar driver which is attached to the traverse bar. Maximum traverse is obtained when adjusting guide is perpendicular to the driver. As angle between guide and driver is decreased, traverse decreases proportionately. Inertia effects limit this type mechanism to slow speeds.

# on Winding Machines

E. R. SWANSON
Universal Winding Company

similar changes of motion. Except for the lead screw as used, for example on lathes, these seven represent the operating principles of all well-known, mechanical types of traversing devices.

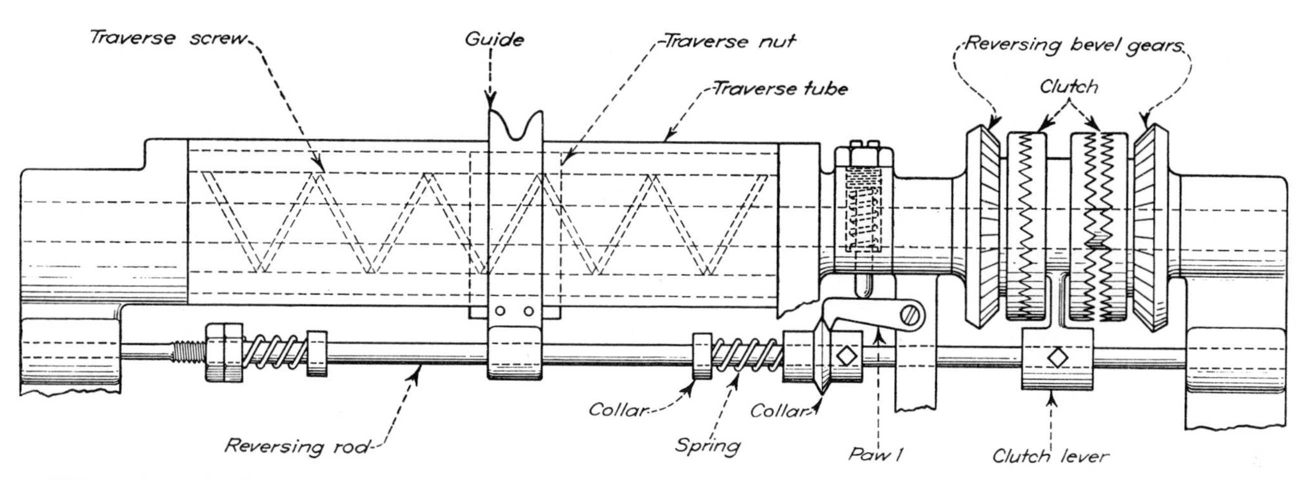


FIG. 3. Reversing bevel gears which are driven by a common bevel gear, drive the shaft carrying the traverse screw. Traverse nut mates with this screw and is connected to the yarn guide. Guide slides along the reversing rod. When nut reaches end of its travel, the thread guide compresses the spring that actuates the pawl and the reversing lever. This engages the clutch that rotates the traverse screw in the opposite direction. As indicated by the large pitch on the screw, this mechanism is limited to low speeds, but permits longer lengths of traverse than most of the others shown.

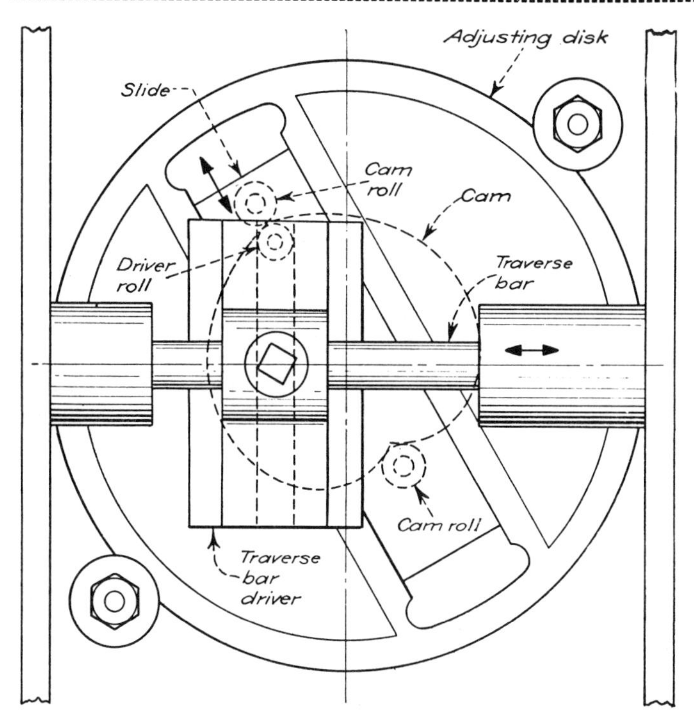


FIG. 6. Two cam rolls that engage heart shaped cam are attached to the slide. Slide has a driver roll that engages a slot in the traverse bar driver. Maximum traverse (to capacity of cam) occurs when adjusting disk is set so slide is parallel to traverse bar. As angle between traverse bar and slide increases, traverse decreases. At 90 deg traverse is zero.

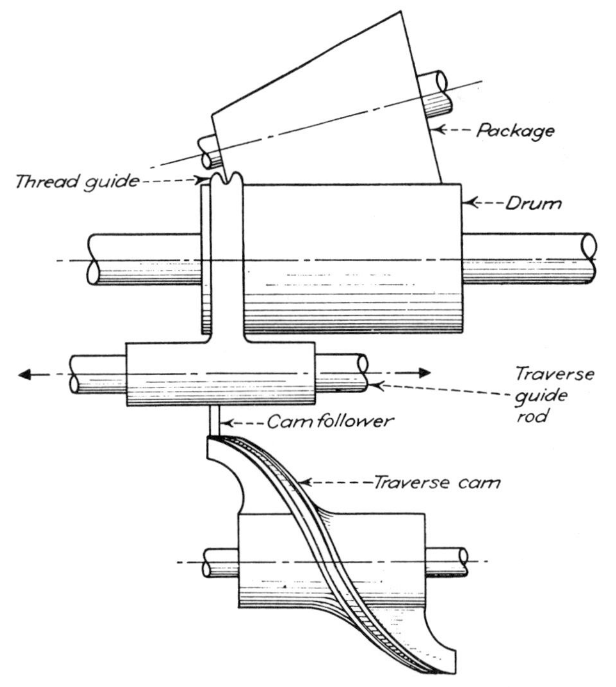


FIG. 7. Traverse cam imparts reciprocating motion to cam follower which drives thread guides on traverse guide rods. Package is friction driven from drum. Yarn is drawn from the supply source through thread guide and transferred to the drum-driven package. Speed of this type of mechanism is determined by the weight of the reciprocating parts.

# 11

# MISCELLANEOUS DESIGN AIDS


How to Provide for Backlash in Threaded Parts 292
Recommended Design Details to Reduce Corrosion 294
For Better Electroplating, These Answers to Trouble Spots 296
For Better Electroplating – Better Mounting, Drainage, Radii 298
Lubrication of Roller Chains 300
Design of Parts for Conditions of Variable Stress 302

# How to Provide for Backlash

These illustrations are based on two general methods of providing for lost motion or backlash. One allows for relative movement of the nut and screw in the plane

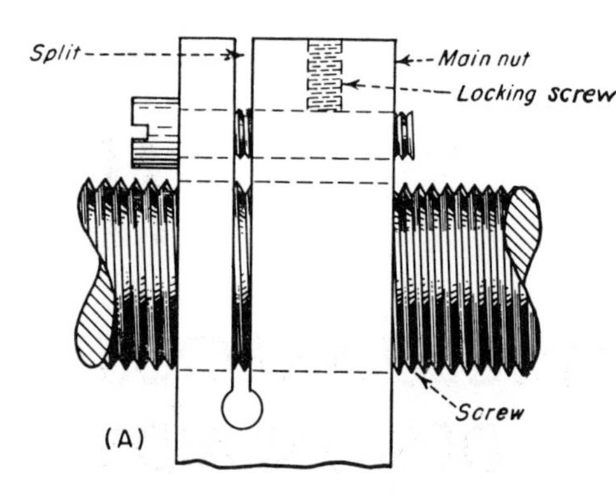


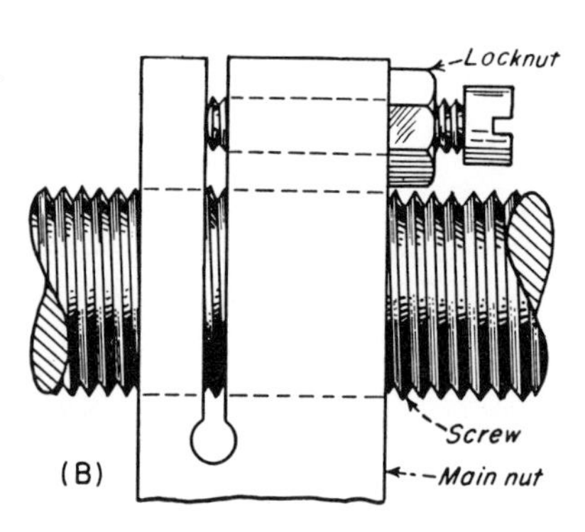


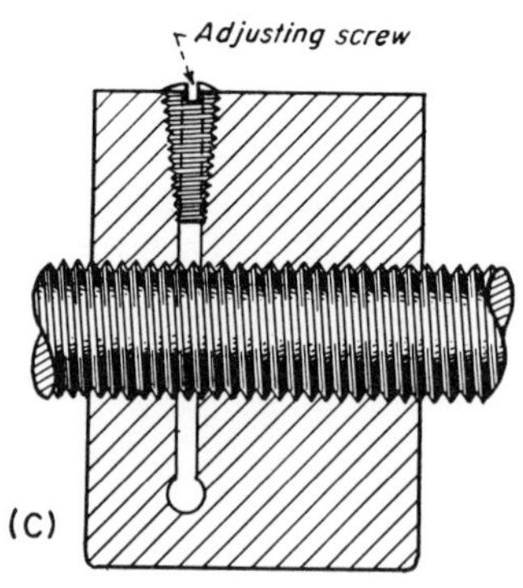


THREE METHODS of using slotted nuts. In *(A),* nut sections are brought closer together to force left-hand nut flanks to bear on right-hand flanks of screw thread and vice versa. In *(B),* and *(C)* nut sections are forced apart for same purpose.

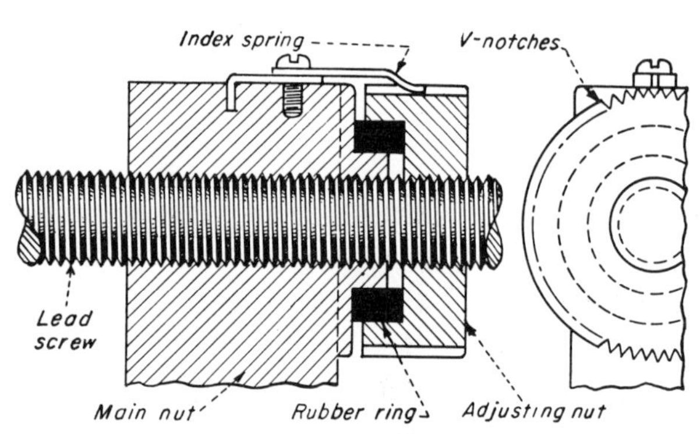


AROUND THE PERIPHERY of the backlash-adjusting nut are "v" notches of small pitch which engage the index spring. To eliminate play in the lead screw, adjusting nut is turned clockwise. Spring and adjusting nut can be calibrated for precise use.

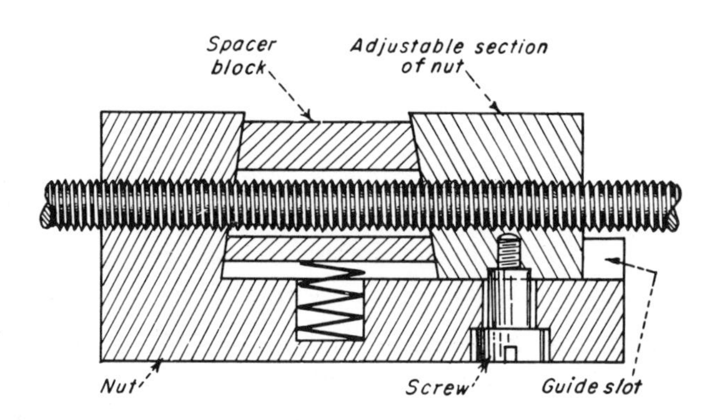


SELF-COMPENSATING MEANS of removing backlash. Slot is milled in nut for an adjustable section which is locked by a screw. Spring presses the tapered spacer block upwards, forcing the nut elements apart, thereby taking up backlash.

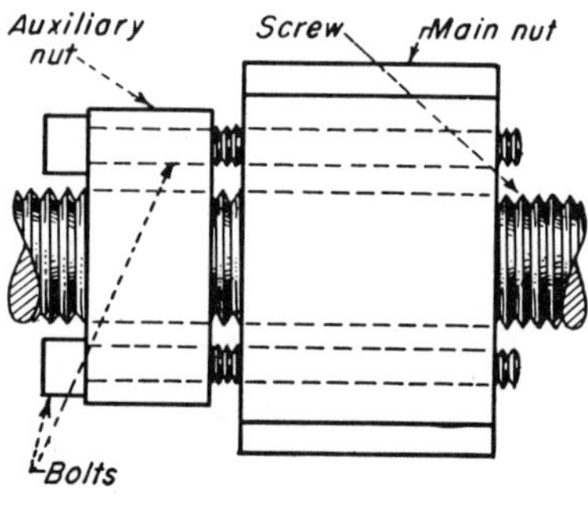


MAIN NUT is integral with base attached to part moved by screw. Auxiliary nut is positioned one or two pitches from main nut. The two are brought closer together by bolts which pass freely through the auxiliary nut.

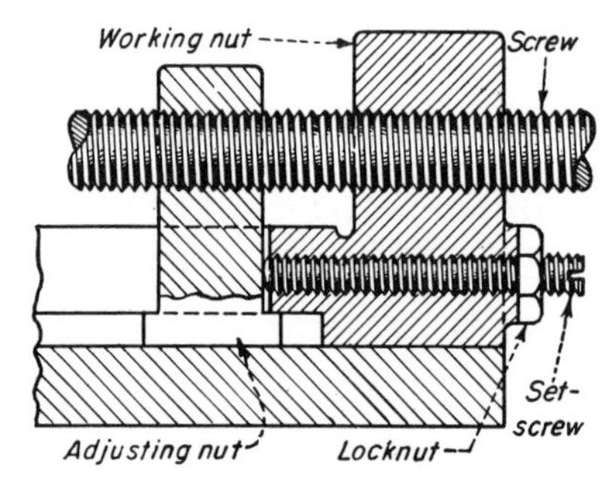


ANOTHER WAY to use an auxiliary or adjusting nut for axial adjustment of backlash. Relative movement between the working and adjusting nuts is obtained manually by the set screw which can be locked in place as shown.

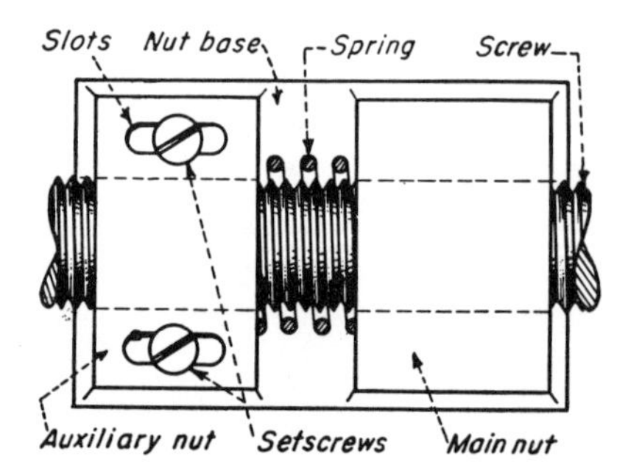


COMPRESSION SPRING placed between main and auxiliary nuts exerts force tending to separate them and thus take up slack. Set screws engage nut base and prevent rotation of auxiliary nut after adjustment is made.

# in Threaded Parts

CLIFFORD T. BOWER
London, England

parallel to the thread axis; the other method involves a radial adjustment to compensate for clearance between sloping faces of the threads on each element.

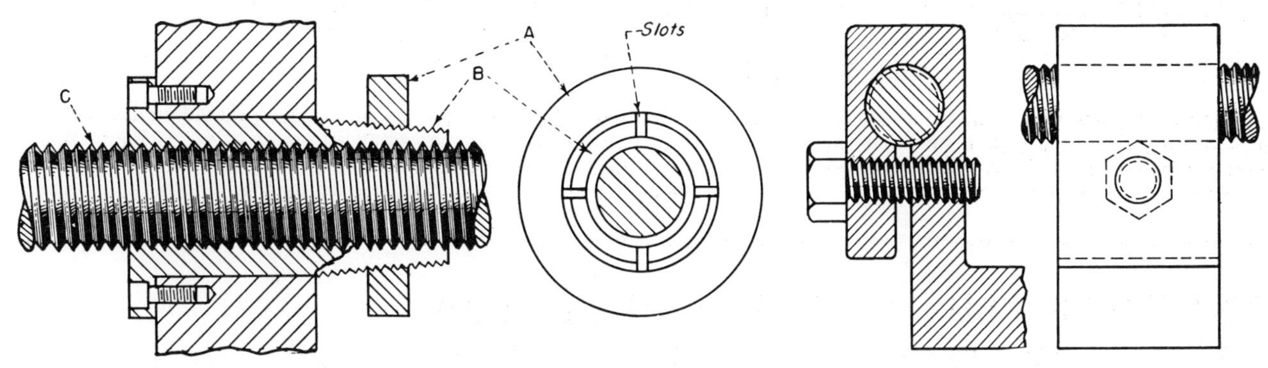


NUT *A* IS SCREWED along the tapered round nut, *B*, to eliminate backlash or wear between *B* and *C*, the main screw, by means of the four slots shown.

ANOTHER METHOD of clamping a nut around a screw to reduce radial clearance.

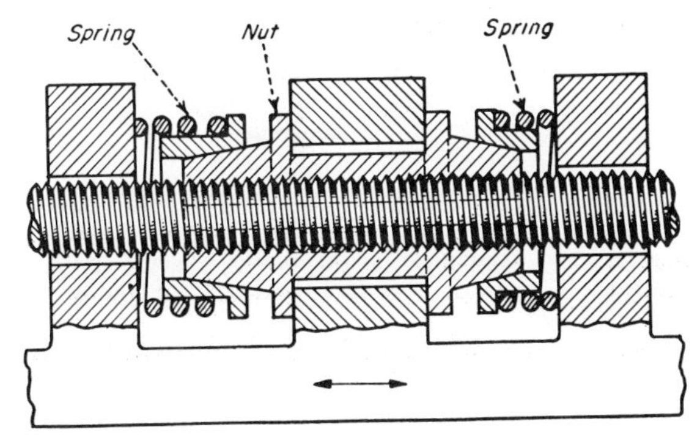


AUTOMATIC ADJUSTMENT for backlash. Nut is flanged on each end, has a square outer section between flanges and slots cut in the tapered sections. Spring forces have components which push slotted sections radially inward.

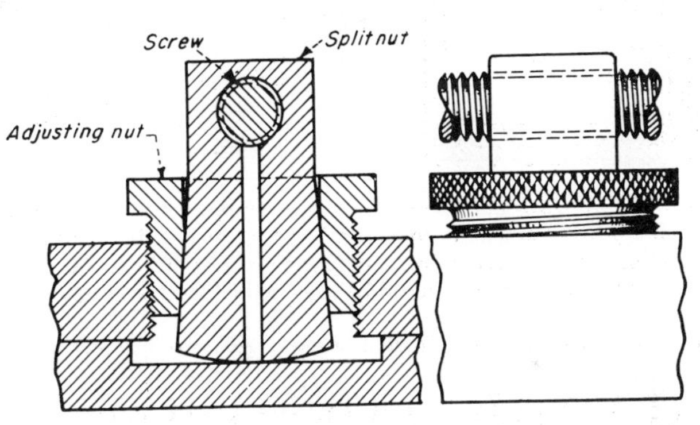


SPLIT NUT is tapered and has a rounded bottom to maintain as near as possible a fixed distance between its seat and the center line of the screw. When the adjusting nut is tightened, the split nut springs inward slightly.

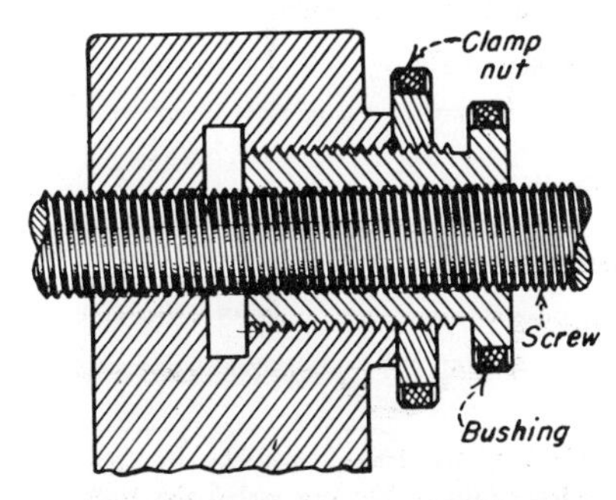


CLAMP NUT holds adjusting bushing rigidly. Bushing must have different pitch on outside thread than on inside thread. If outer thread is the coarser one, a relatively small amount of rotation will take up backlash.

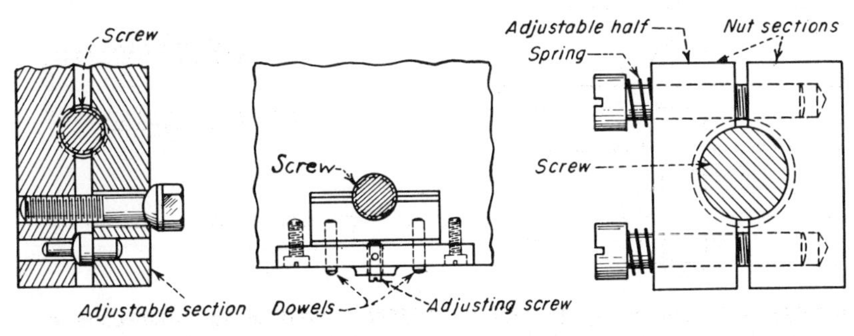


TYPICAL CONSTRUCTIONS based on the half nut principle. In each case, the nut bearing width is equal to the width of the adjustable or inserted slide piece. In the sketch at the extreme left, the cap screw with the spherical seat provides for adjustments. In the center sketch, the adjusting screw bears on the movable nut section. Two dowels insure proper alignment. The third illustration is similar to the first except that two adjusting screws are used instead of only one.

# Recommended Design Details

## ATMOSPHERIC CORROSION

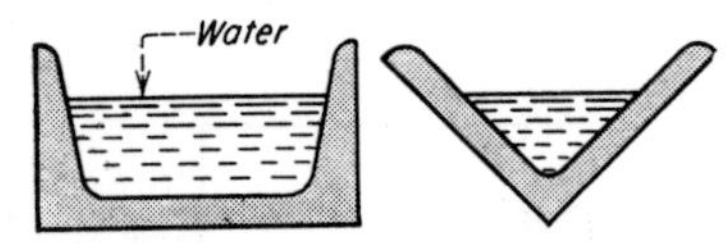

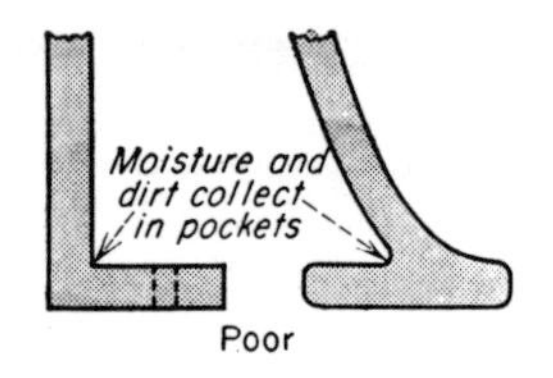

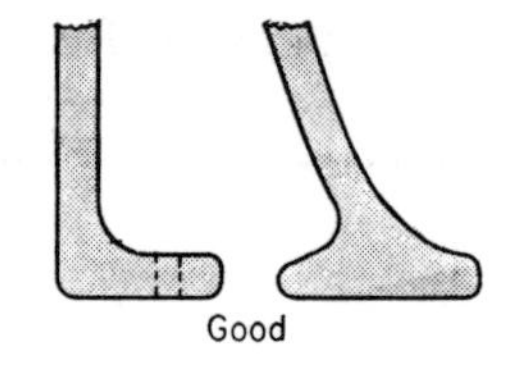


**STRUCTURAL MEMBERS** should be arranged so that moisture or liquids will not collect. Inverting angles and channels prevents this condition. Drilling holes is another method to insure proper drainage.

**ROUNDED CORNERS** and smooth contours should be used whenever possible to prevent the accumulation of moisture, liquids and solid matter. Using corrosion resistant materials is often found to be more economical due to greater service life.

## CONCENTRATION CELL CORROSION

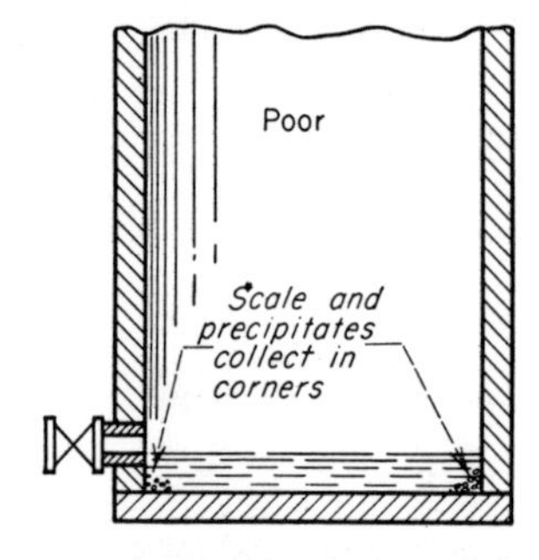

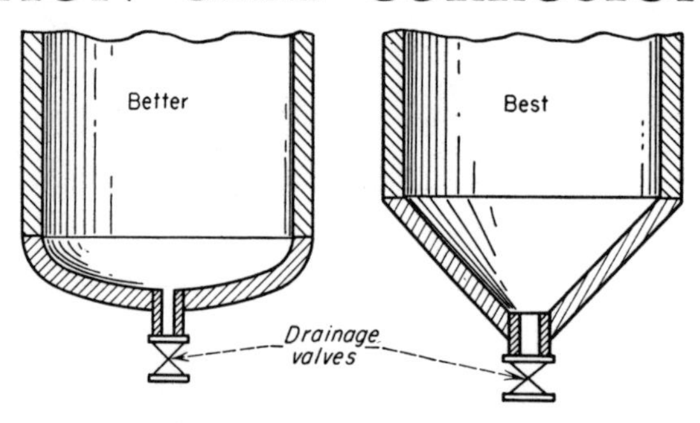

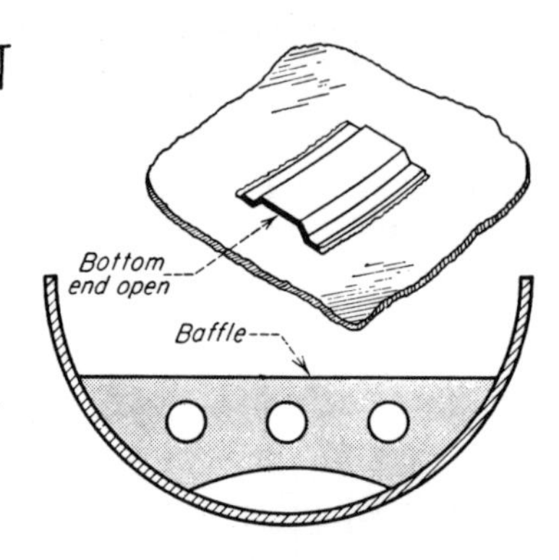


**ALL LIQUID CONTAINERS** should be designed with smooth and rounded corners. Sharp corners, stagnant areas and other such conditions are favorable to the accumulation of precipitates, solids and scale which promote concentration-cell attack. Sloping bottoms should be used with valves arranged for complete drainage.

**ALL BAFFLES** and internal stiffeners in tanks should have openings arranged to avoid liquid pockets and permit the free drainage of fluids.

## GALVANIC CORROSION

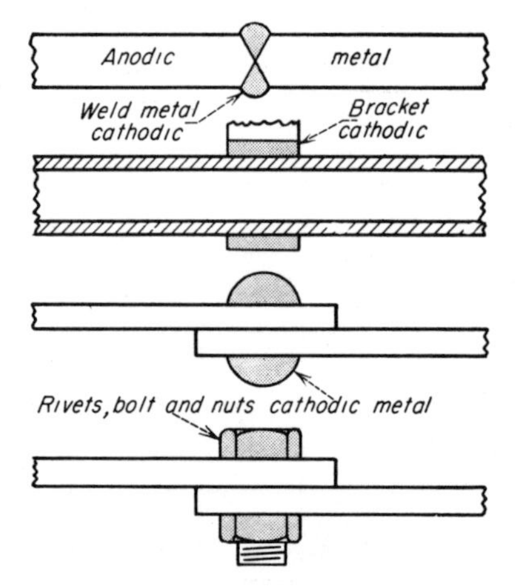

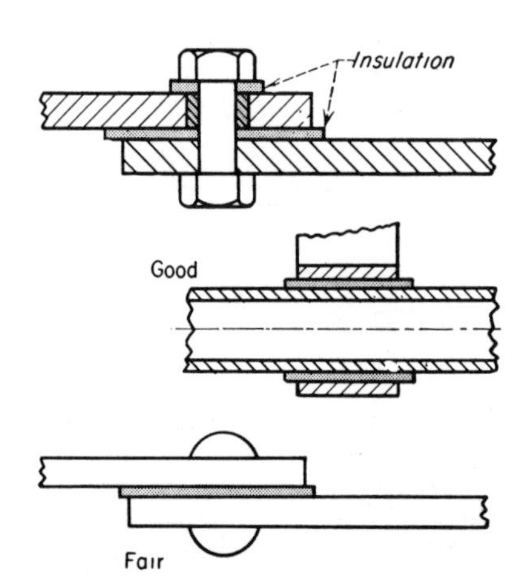

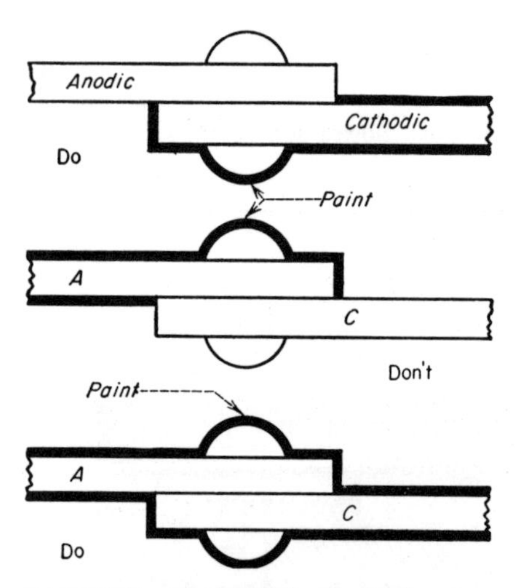


**IN JOINTS** and connections, the proportions of dissimilar metals should be chosen so the anodic or less noble metal has the greater exposed area. If fastenings such as bolts and rivets are required, they should be made of the more noble or cathodic type material.

**WHEN POSSIBLE** the connection of dissimilar metals should be separated by an insulating material to reduce or prevent the current flow in the galvanic circuit. Paint or plastic coatings serve to reduce the galvanic current by increasing the circuit resistance.

**PAINTED COATINGS** should be applied with caution. Do not paint the less noble material, otherwise greatly accelerated corrosion will occur at imperfections in the coating. If possible exposed surfaces should be painted. Commercial protectives are now available.

# to Reduce Corrosion

FRED M. REINHART
National Bureau of Standards
Washington, D. C.

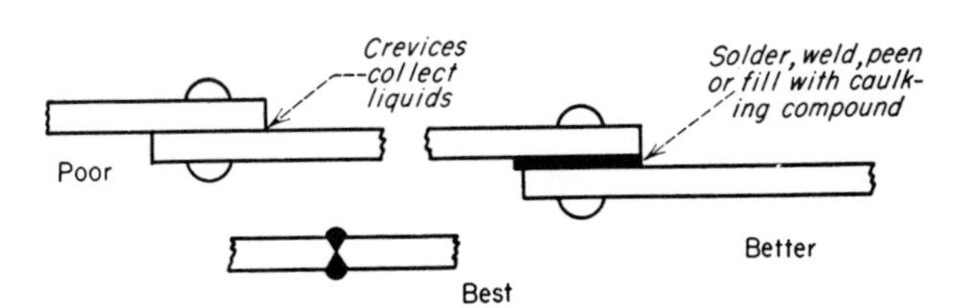


**BUTT WELDED JOINTS** are less likely to corrode and should be used. If lap joints are required, all crevices should be filled with a non-absorbing caulking compound or welded to prevent retention of liquids in crevices.

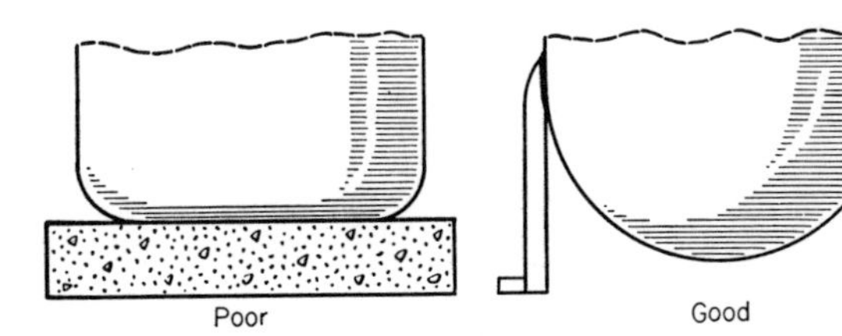


**STORAGE TANKS** and other containers should be supported on legs to allow a free circulation of air underneath. This prevents the possibility of any condensation and collection of moisture under the tank.

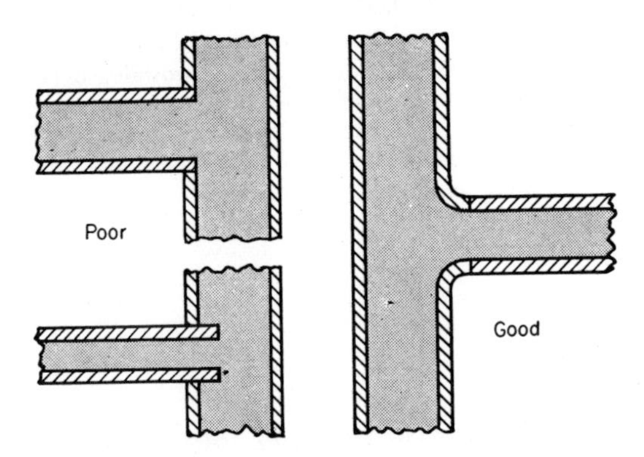


**IN THE DESIGN** of liquid passages, all pipes and connections should be constructed to insure uniform flow, with a minimum turbulence and air entrainment. This also reduces settling of solids.

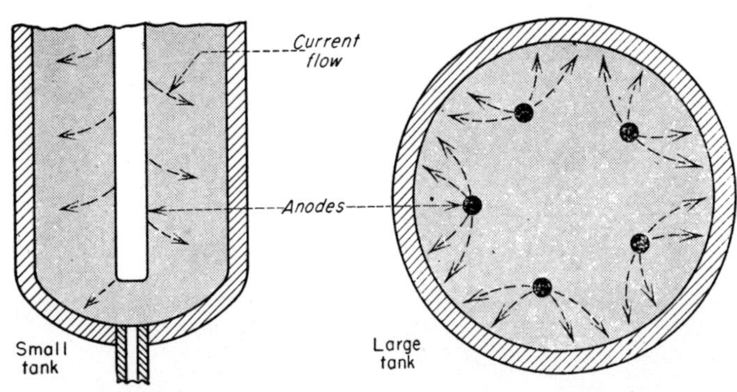


**CATHODIC PROTECTION** of containers with corrosive liquids can be done by immersing a rod of a more anodic material inside. This reverses the galvanic current and the container becomes cathodic and is less likely to corrode. Magnesium and zinc rods are often used.

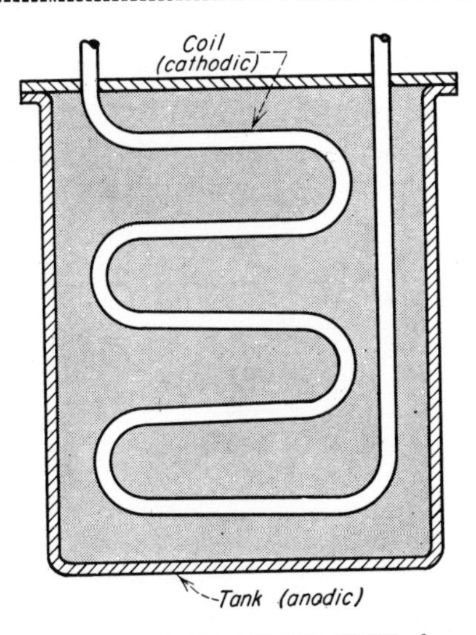


**IN DESIGNING EQUIPMENT,** keep different metals far apart in the solution; this increases the resistance of the electrolytic path. Chemical inhibitors are often added to corrosive solutions. Some bare zinc, magnesium or steel in the liquid will counteract corrosion.

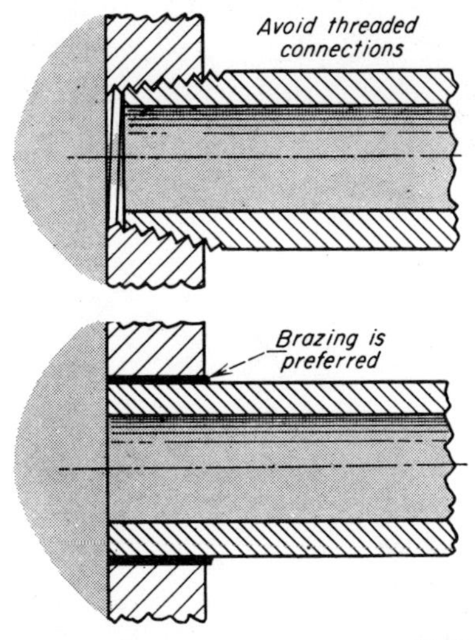


**IN JOINING DISSIMILAR MATERIALS,** well apart in the galvanic series, avoid threaded connections since the threads deteriorate rapidly. Brazed joints are preferred using a brazing alloy more noble than at least one of the metals that are being joined together.

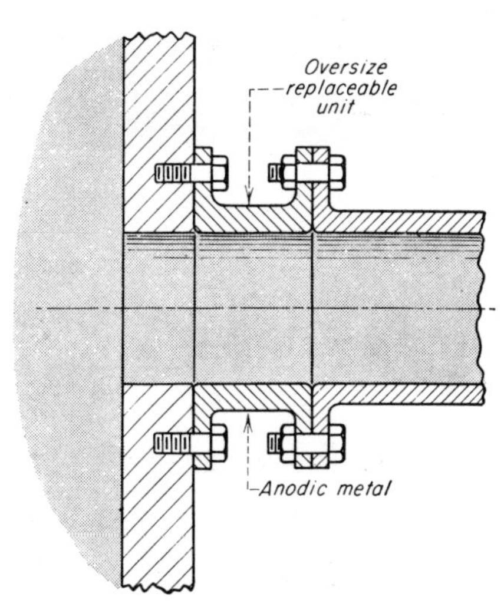


**AT CONNECTIONS** of dissimilar materials, consider using small replaceable sections made of the less noble metal. These expendable parts should be easy to replace and made oversize to increase their corrosive life. Nonmetallic gaskets increases the circuit resistance.

# For BETTER ELECTROPLATING . . .

**PETER C. NOY,** Manufacturing Engineer, Canadian General Electric Co., Ltd., Barrie, Ont.

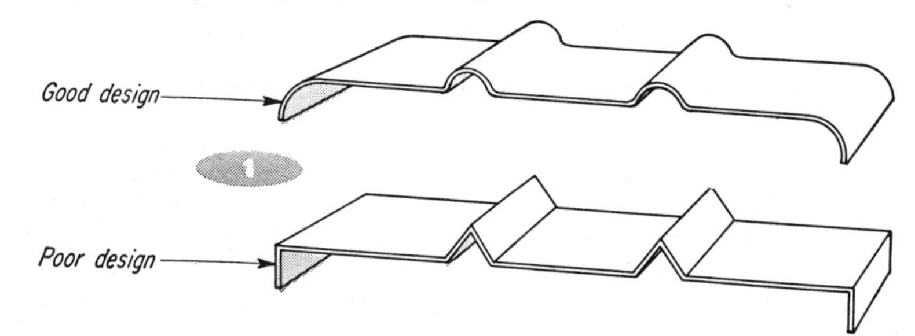


## Ridges and grooves . . .

on brightly finished presswork should be rounded as much as possible. The plating buildup that occurs on sharp ridges is not only wasteful, but—when abnormally thick—causes greater stress and often increases brittleness. Combined with buffing-wheel pressure in later operations, this condition can weaken the plate.

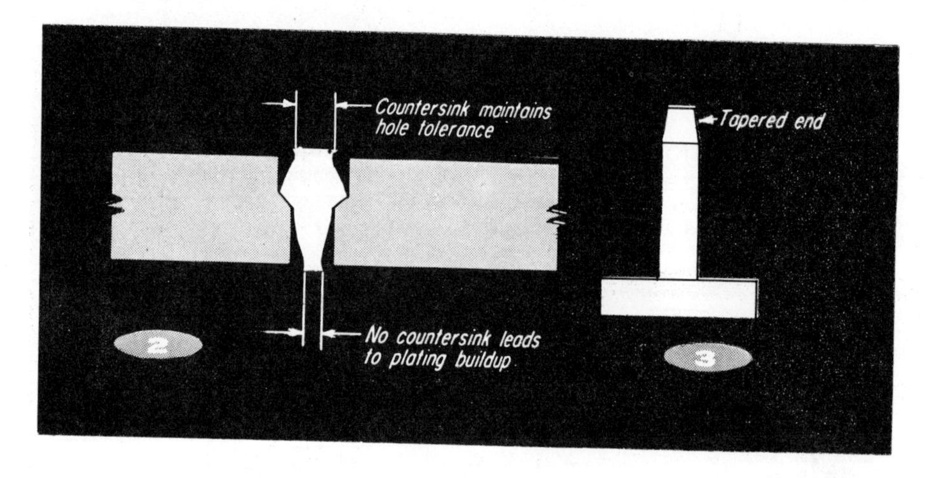


## Adequate hole tolerance . . .

is necessary to prevent the buildup on hole edges that reduces effective dia. Countersink (2) to keep hole fully open—deep in the hole, plating will be deposited either very thinly or not at all. Studs (3) should, for similar reasons, be provided with tapered entry; otherwise, excessive tolerances may be necessary.

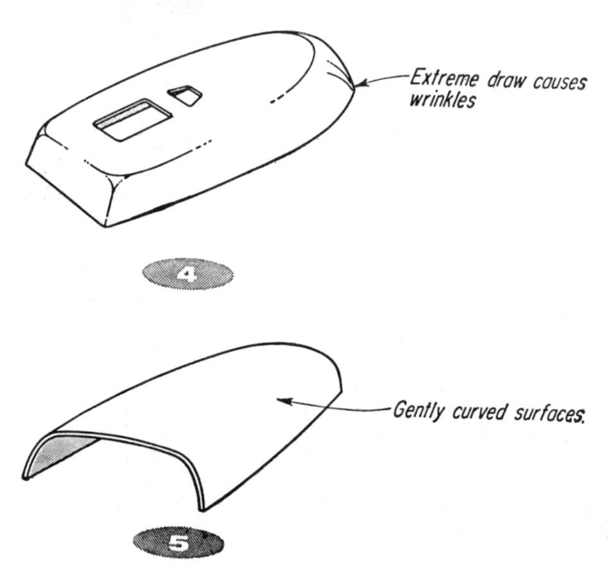


## Draws on presswork . . .

should be as gentle as possible. Extreme draws (4) often cause wrinkles. When these are polished out, the crystalline center of the metal, instead of the wrought surface, is bared to the plating solution. Gently curved surfaces on highly finished presswork (5) are easier to buff and polish than flat surfaces.

# . . . These answers to trouble spots

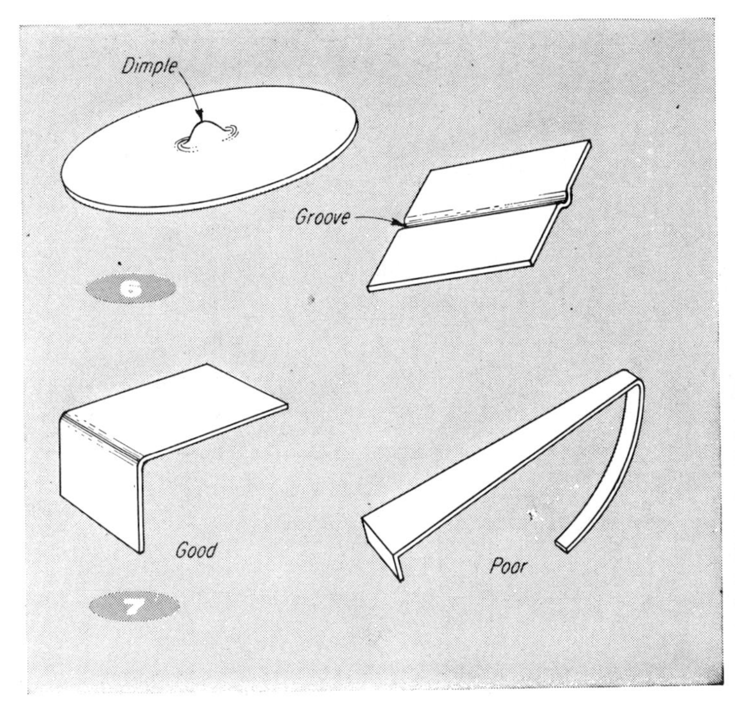


**Small flat parts . . .** that will be barrel plated should be grooved or dimpled (6). This prevents them from clinging together and blanking one another from the plating solution. Small parts (7) should also be stable enough to withstand the vigorous tumbling action of barrel plating without distorting permanently or becoming badly entangled.

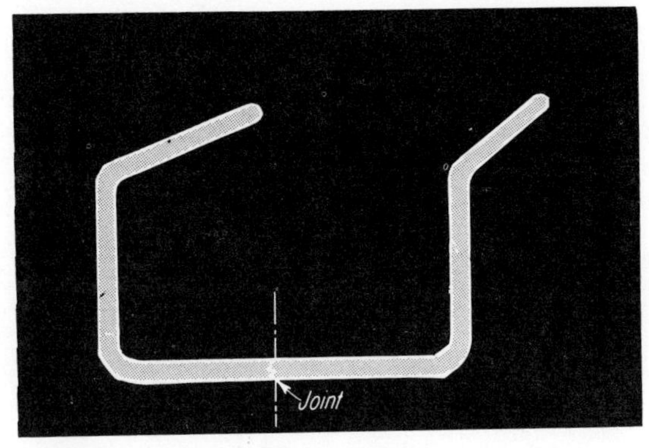


8 **Extra large parts . . .** such as metal furniture and other tubular units should be designed in sections convenient to plate. Few plating solutions are more than 3 ft, 6 in. deep.

## GENERAL HINTS

**Pure tin solder** (9) should be specified on all assemblies. Lead is hard to prepare for plating, contaminates adjacent metals and causes plate failure from lack of adhesion.

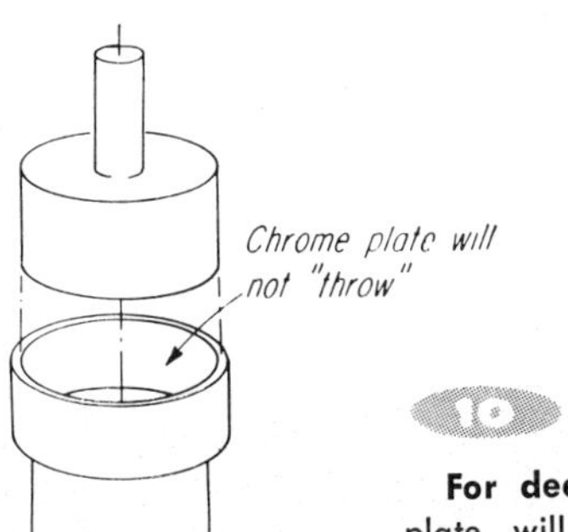


10

**For deeply recessed parts** (10) chrome plate will not "throw" in recesses unless auxiliary anodes are used. Nickel is better but real corrosion resistance may require special processes or designs.

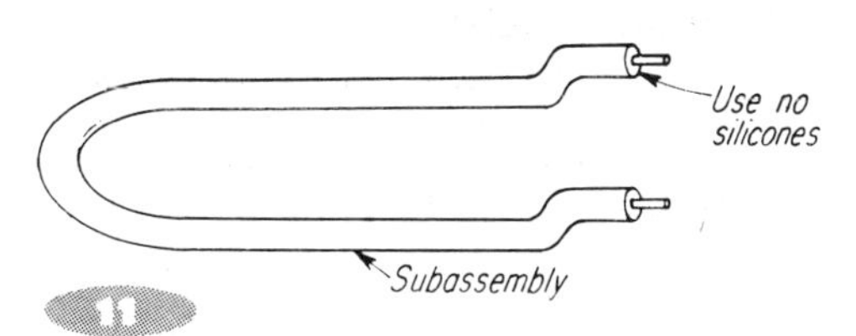


11

**Compatibility with plating solution** (11) is essential for materials used on plated subassemblies. Silicones, for example, will kill all foams purposely produced in cleaning and plating solutions; rubber gaskets will dissolve or thicken; aluminum dissolves.

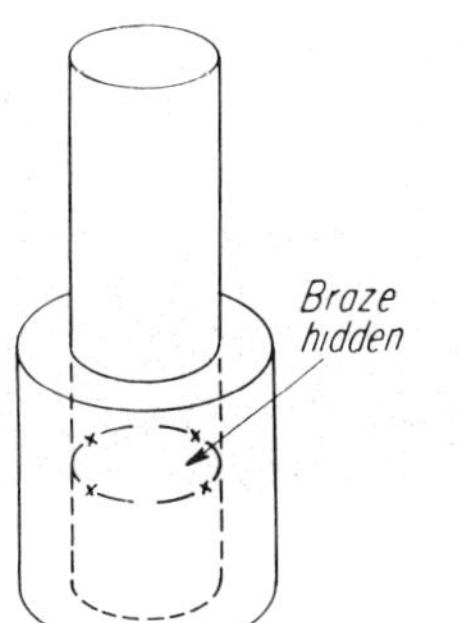


12

**Silver-brazed joints . . .** on decorative copper alloy work should be hidden from sight—the silver alloys will result in a matte finish.

# For BETTER

## . . . better mounting, drainage, rad

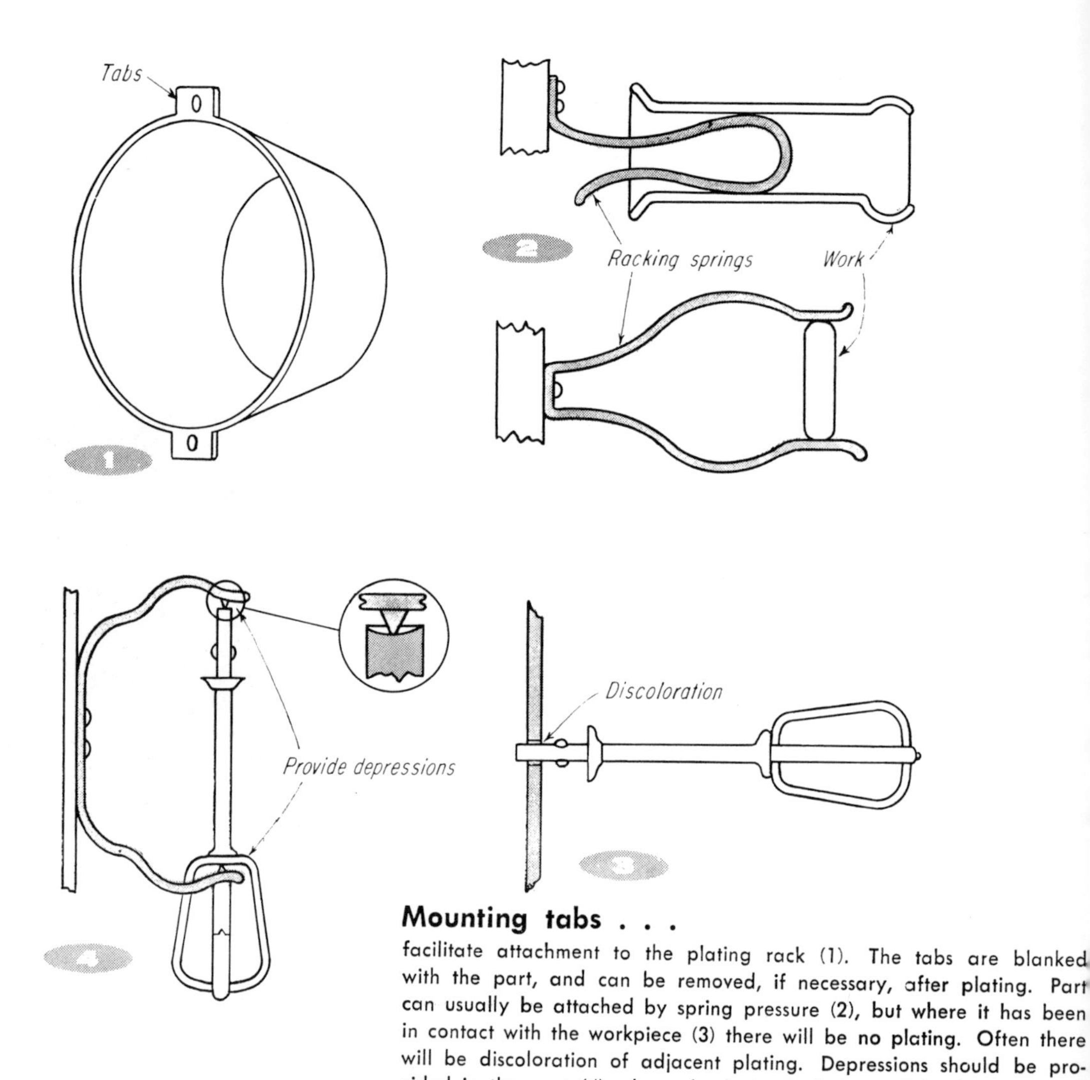


### Mounting tabs . . .

facilitate attachment to the plating rack (1). The tabs are blanked with the part, and can be removed, if necessary, after plating. Part can usually be attached by spring pressure (2), but where it has been in contact with the workpiece (3) there will be no plating. Often there will be discoloration of adjacent plating. Depressions should be provided in the part (4) where the lack of plating will be least noticed.

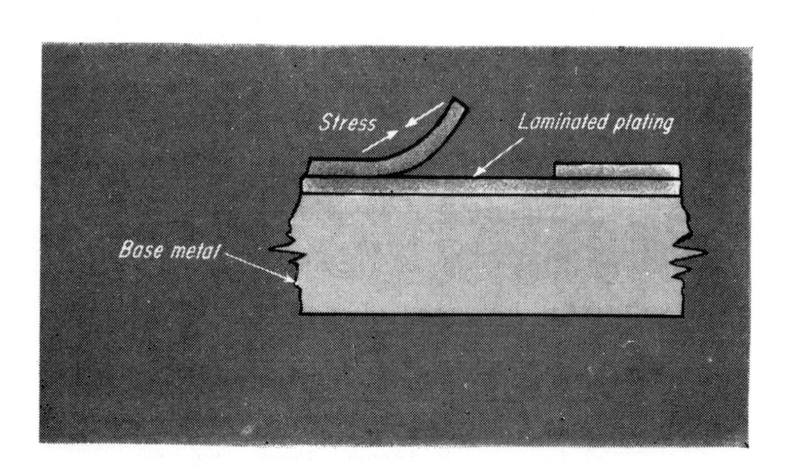

### Peeling results . . .

when point of attachment is not firm. Any wobble caused by solution agitation will result in laminated plating because of intermittent electrical contract. The laminations, under stress, will peel.

# ELECTROPLATING . . .

**PETER C. NOY,** *Manufacturing Engineer, Canadian General Electric Co., Ltd., Barrie, Ont.*

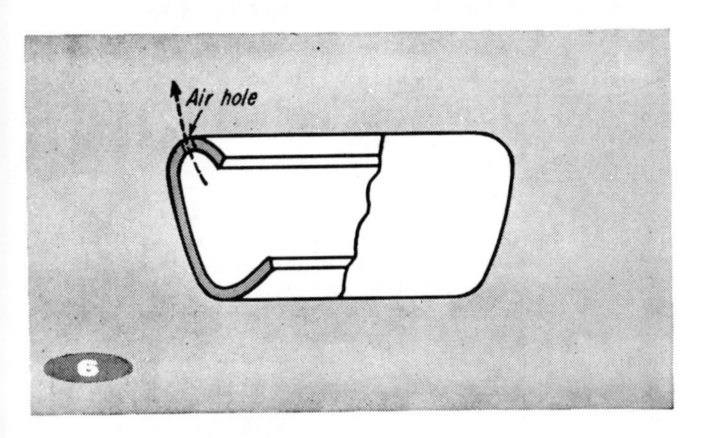


### Air-relief hole . . .

should be provided wherever an air trap is likely to occur. Plating would not be deposited inside parts similar to the one illustrated unless a hole allowed release of trapped air during submersion.

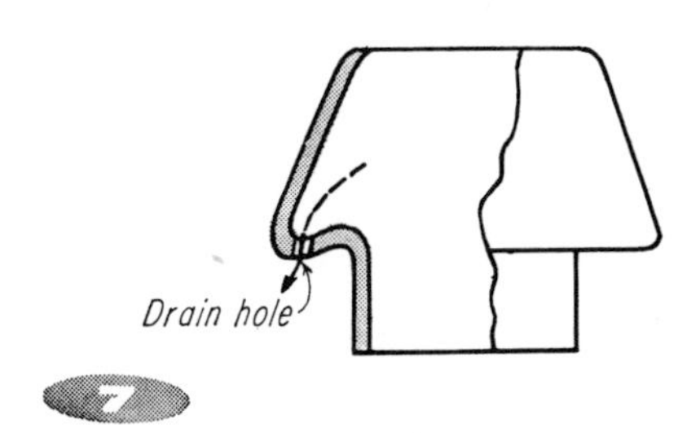


7

### Drain hole . . .

releases solution which would otherwise be trapped inside a part, causing solution "drag-out." Plating solutions are expensive and, in a well-regulated bath, drag-out losses are often the highest cost item. Undrained solution washed away by subsequent water rinses is literally money down the drain!

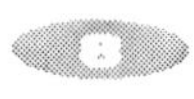

### Inside-corner . . .

radii should be large as possible. In theory, plate can not be deposited where an inside corner is less than 90°. Plating thickness at such corners is vastly improved by large radii. If radii are less than 1/4 in., auxiliary anodes may be necessary to achieve sufficient plate thickness.

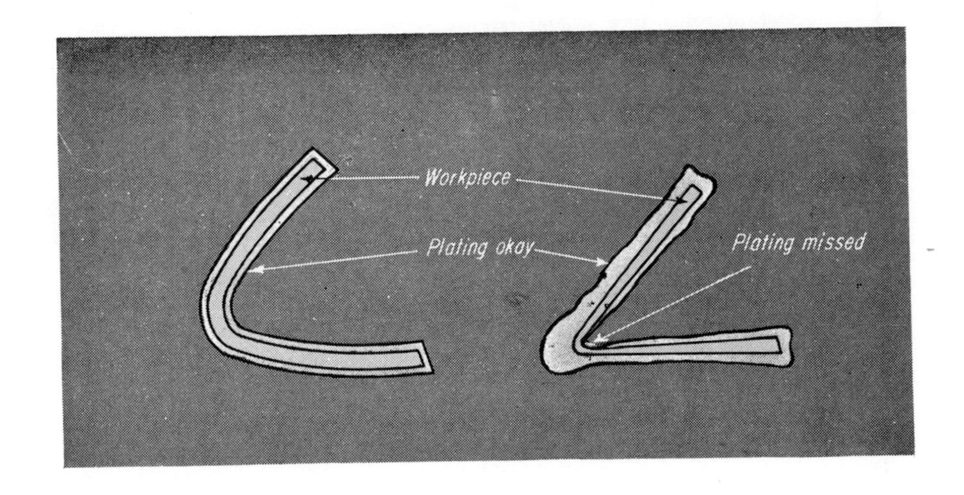

### Outside-corner radii . . .

should also be large. Nickel builds up on prominences because the current tends to be greater there. Often, the plater must provide metal fingers called "robbers" on his plating rack. Being cathodic, they help to distribute current density, for even plating.

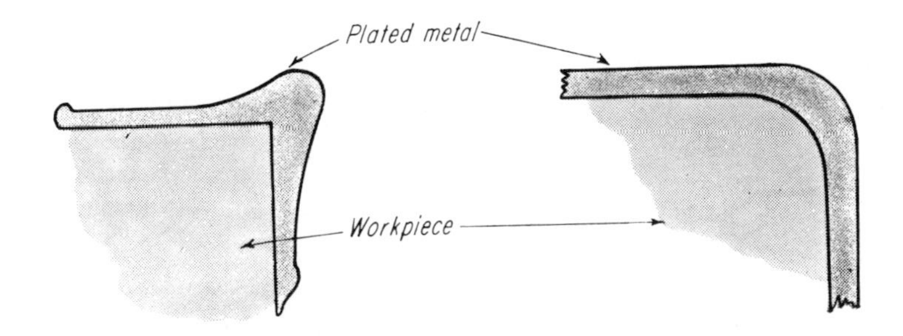

# Lubrication of roller chains

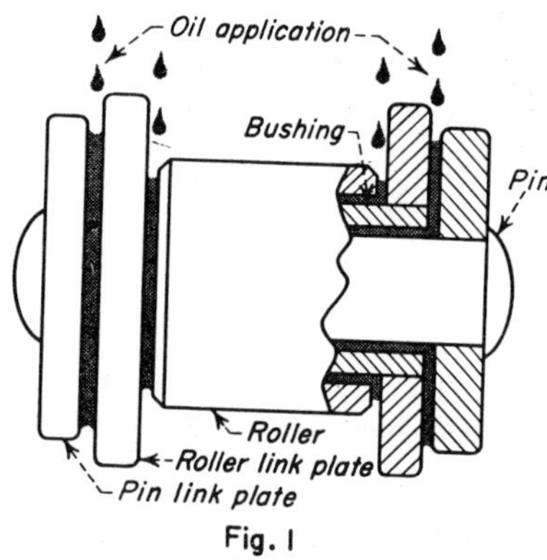


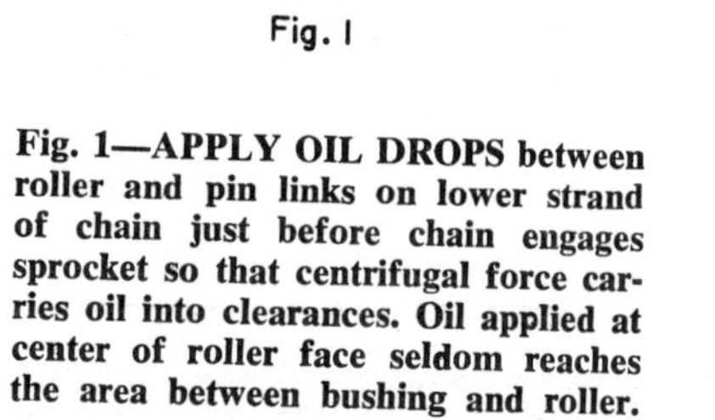

**Fig. 1—APPLY OIL DROPS** between roller and pin links on lower strand of chain just before chain engages sprocket so that centrifugal force carries oil into clearances. Oil applied at center of roller face seldom reaches the area between bushing and roller.

**Fig. 2—MANUAL APPLICATION OF LUBRICANT** by (A) flared-lip oil can, or (B) hand brush, is simplest method for low-speed applications not enclosed in casings. New chains should be lubricated daily until sufficiently "broken-in," after which weekly lubrication programs should suffice.

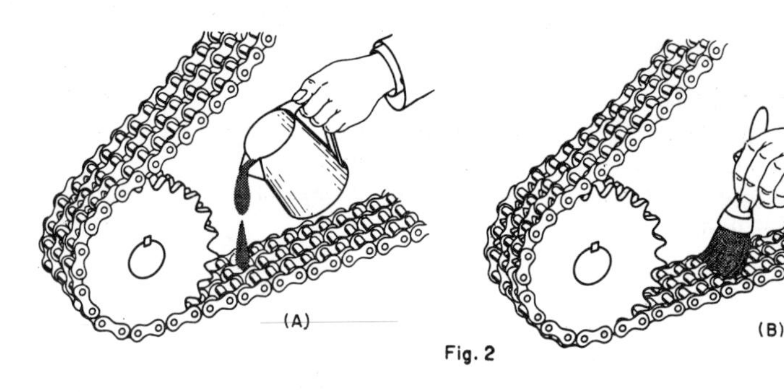


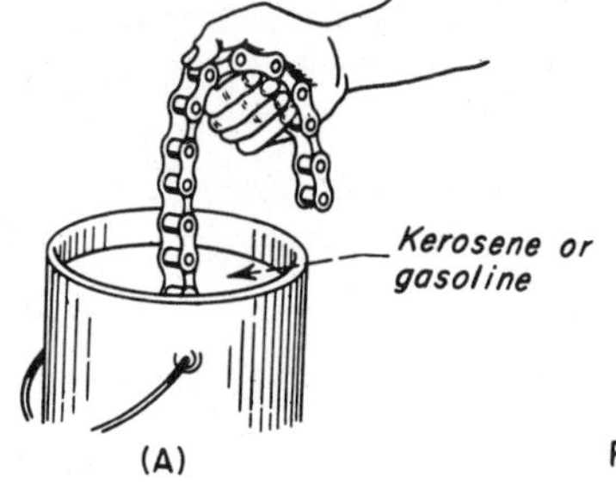


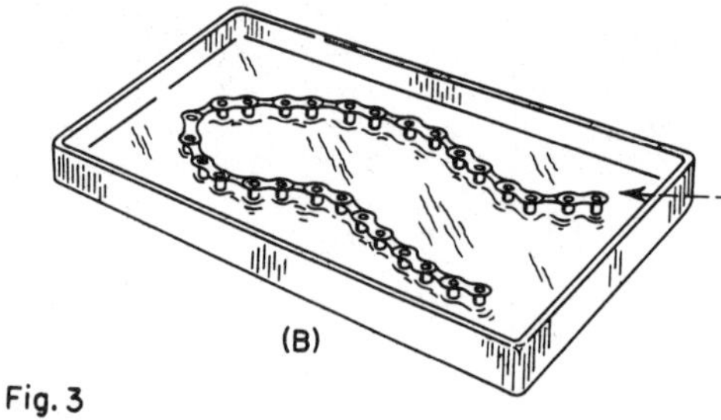


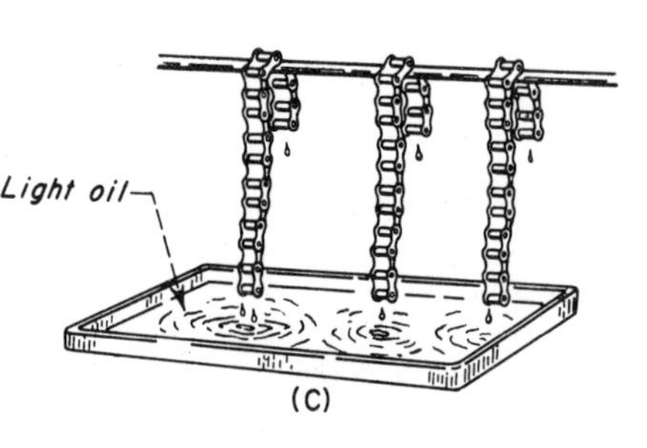


**Fig. 3—CHAINS WITHOUT CASING** should be: (A) removed periodically and washed in kerosene, (B) soaked in light oil after cleaning, and (C) draped to permit excess oil to drain.

**Fig. 4—DRIP LUBRICATION** can be adjusted to feed oil to edges of link plates at rate of 4 to 20 drops per minute depending on chain speed. Pipe contains oil-soaked wick to feed multiple-width chains.

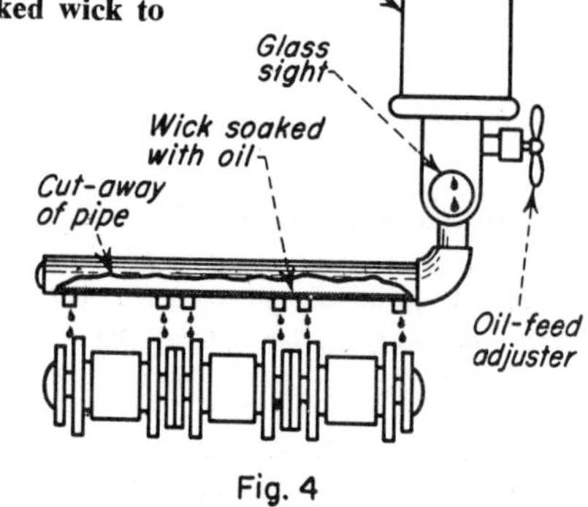


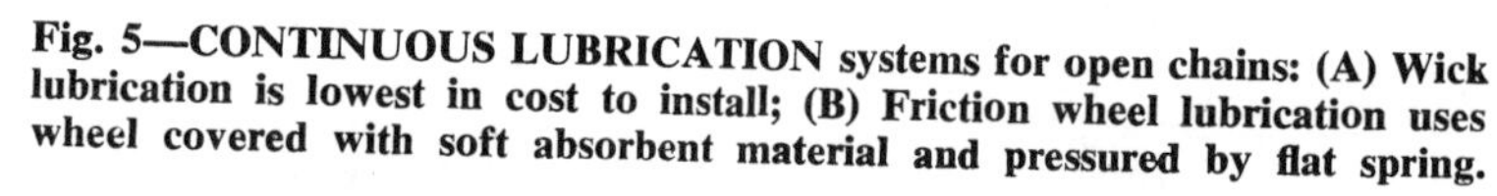

**Fig. 5—CONTINUOUS LUBRICATION** systems for open chains: (A) Wick lubrication is lowest in cost to install; (B) Friction wheel lubrication uses wheel covered with soft absorbent material and pressured by flat spring.

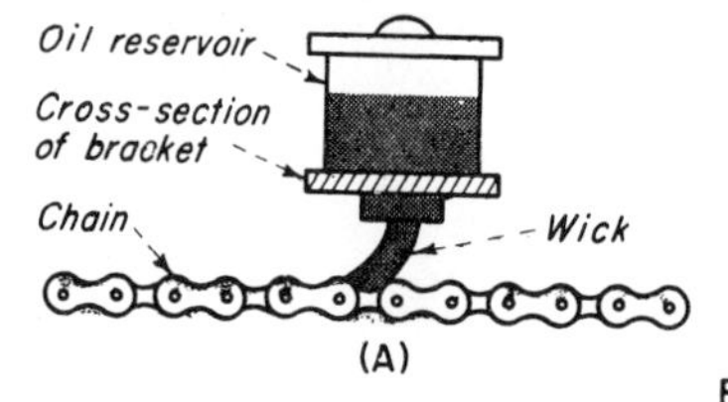


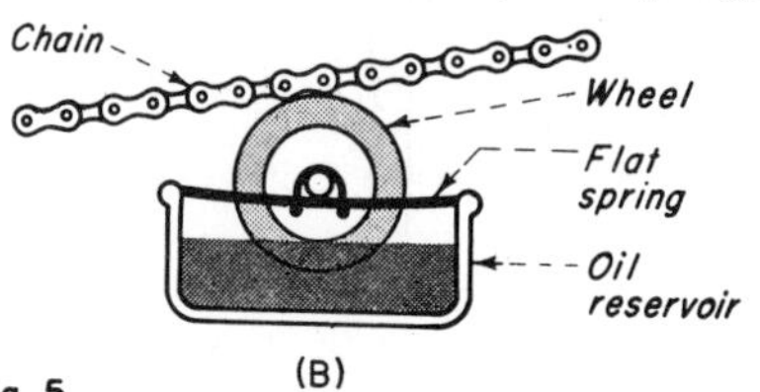


Fig. 5

Unsatisfactory chain life is usually the result of poor or ineffective lubrication. More damage is caused by faulty lubrication than by years of normal service. Illustrated below are 9 methods for lubricating roller chains. Selection should be made on basis of chain speed as shown in Table I. Recommended lubricants are listed in Table II.

**Table I—Recommended Methods**

| Chain Speed, ft/min | Method |
|---|---|
| 0–600 | Manual: brush, oil can<br>Slow Drip: 4–10 drops, min<br>Continuous: wick, wheel |
| 600–1500 | Rapid Drip—20 drops, min<br>Shallow Bath, Disk |
| over 1500 | Force Feed Systems |

**Table II—Recommended Lubricants**

| Pitch of chain, in. | Viscosity at 100 F, SUS | SAE No. |
|---|---|---|
| 1/4–5/8 | 240–420 | 20 |
| 3/4–1 1/4 | 420–620 | 30 |
| 1 1/2–up | 620–1300 | 40 |

Note: For ambient temperatures between 100 to 500 F use SAE 50.

**Fig. 6—SHALLOW BATH LUBRICATION** uses casing as reservoir for oil. Lower part of chain just skims through oil pool. Levels of oil must be kept tangent to chain sprocket to avoid excessive churning. Should not be used at high speeds because of tendency to generate excessive heat.

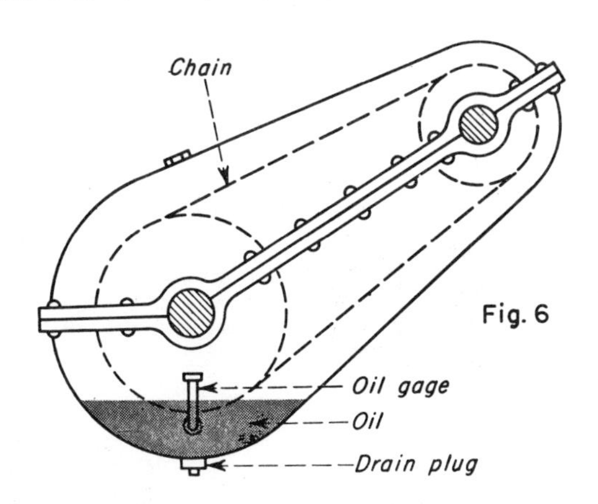


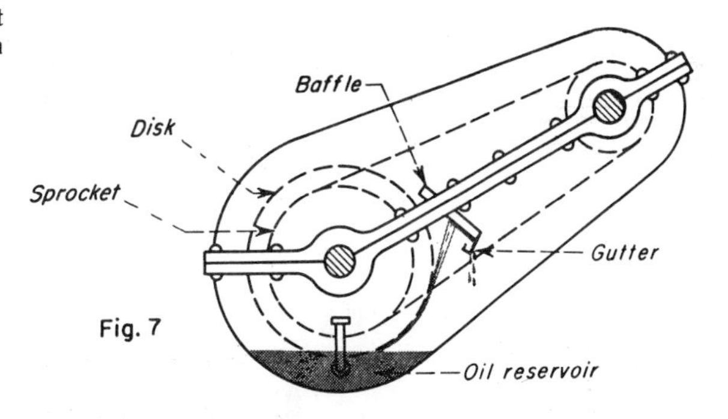


**Fig. 7—DISK OR SLINGER** can be attached to lower sprocket to give continuous supply of oil. Disk scoops up oil from reservoir and throws it against baffle. Gutter catches oil dripping down from baffle and directs it on to chain.

**Fig. 8—FORCE-FEED LUBRICATION** for chains running at extremely high speeds. Pump driven by motor delivers oil under pressure to nozzles that direct spray on to chain. Excess oil collects in reservoir which has wide area to cool oil.

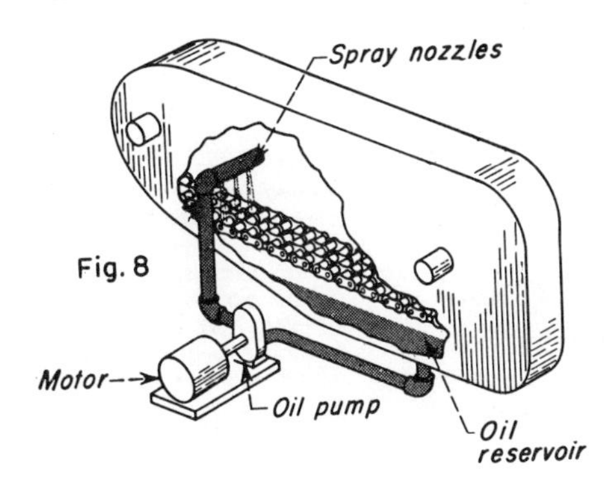


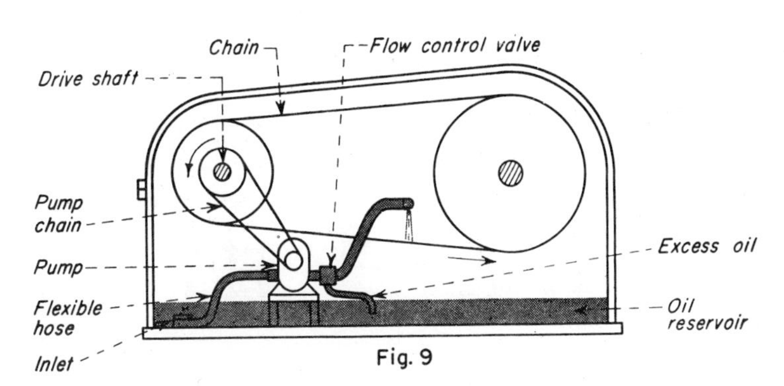


**Fig. 9—CHAIN-DRIVEN FORCE-FEED** system has pump driven by main drive shaft. Flow control valve, regulated from outside of casing, by-passes excess oil back to reservoir. Inlet hose contains filter. Oil should be changed periodically—especially when hue is brown instead of black.

# Design of Parts for Conditions

## BOLT HEADS

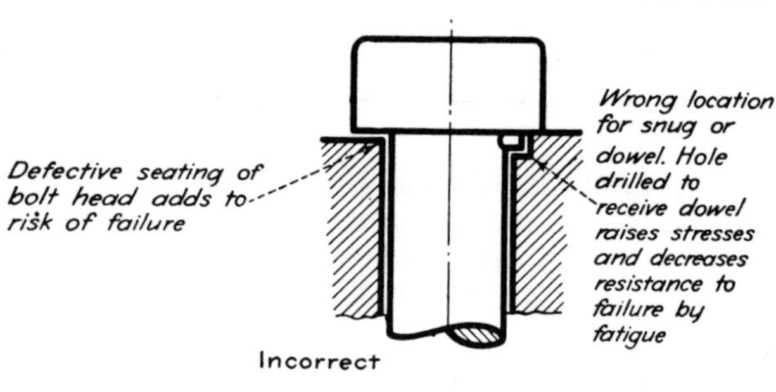

Defective seating of bolt head adds to risk of failure

Wrong location for snug or dowel. Hole drilled to receive dowel raises stresses and decreases resistance to failure by fatigue

Incorrect

Seating of bolt head uniform

Shank reduced to core diameter of thread permits less abrupt change of section from shank to head

Dowel located at point of low internal stress. No interruption of symmetrical flow of stress lines from shank into head

Correct

## NUTS

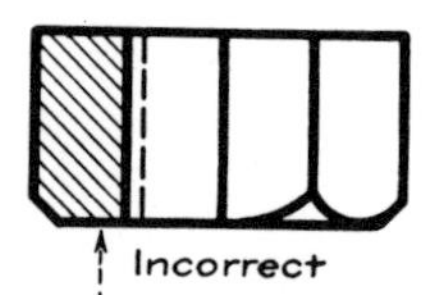

Incorrect

Nut too rigid. Strain incorrectly distributed and high stress localized

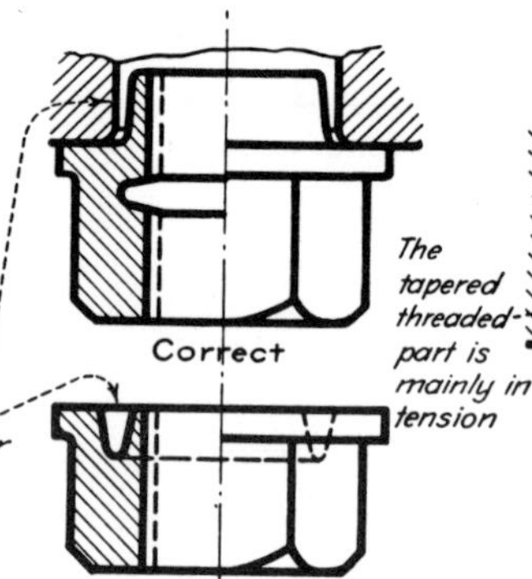

Correct

Internal recesses permit better load distribution over a number of threads. Designs relieve stress in thread portion of bolts

Correct

The tapered threaded part is mainly in tension

Correct

Reduced outside diameter allows better distribution of circumferential strain and better distribution of the load over all the threads

Groove of progressively decreasing depth improves load distribution over a number of threads. Groove makes thread more flexible under load. Permits better adjustment of load over greater thread length

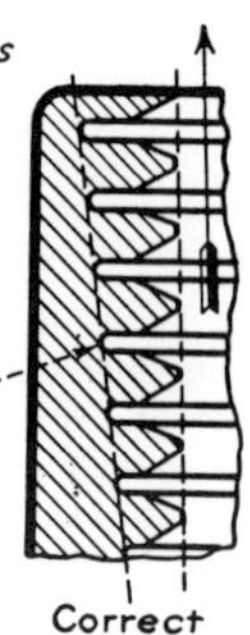

Correct

## STUDS

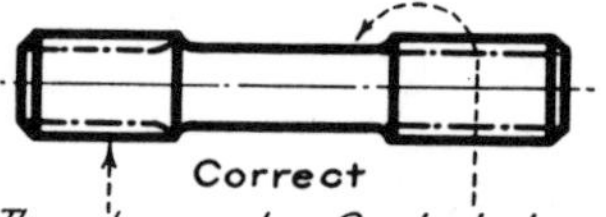

Correct

Thread on end which is fitted into tapped hole must not run through to shank, nor finish too far from the shank

Gradual change of section from core diameter to shank diameter reduces danger of fracture by producing gradual change of stress line direction

Incorrect

Abrupt change of section from the core diameter of the thread to the shank diameter produces abrupt change of stress line direction and increases danger fracture in vicinity of change of section

## SHOULDERS

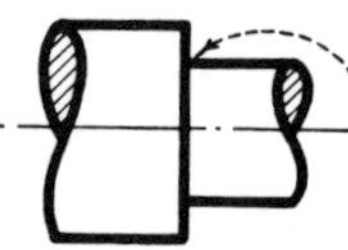

Abrupt change of section causes intense local stresses when shaft is loaded

Incorrect

Loose collar shaped to permit use of fillet radius in shaft and eliminate stress concentration. For use when sharp transverse change of section is required

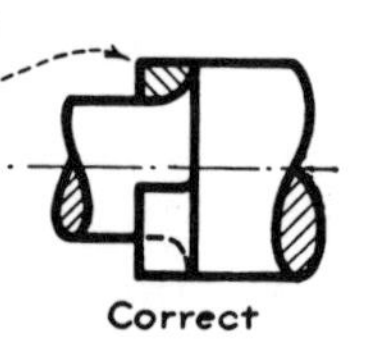

Correct

## SHOCK AND CREEP

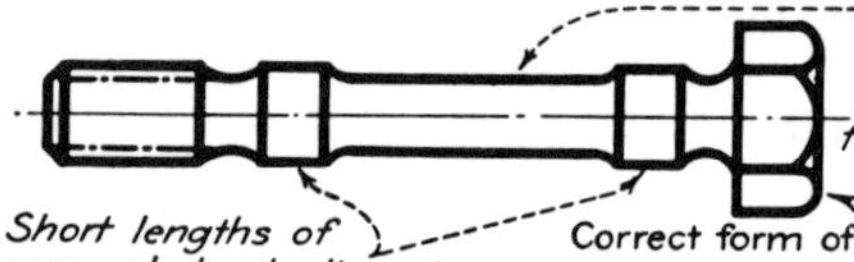

Shank of bolt reduced to core diameter of thread for greatest possible length to obtain high resilience

Short lengths of normal shank diameter provided for location in bolt holes

Correct form of bolt for shock resistance

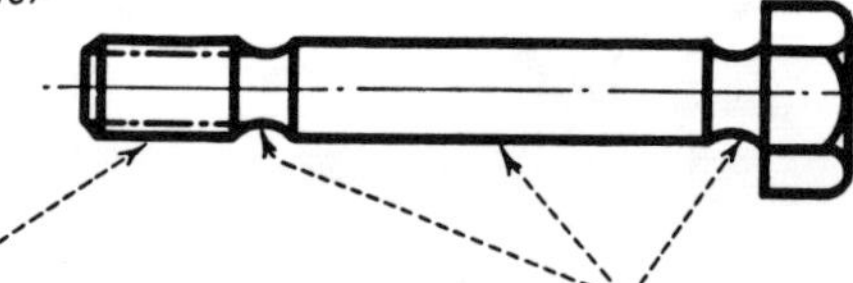

Correct form of bolt for resistance to creep at elevated temperatures

Smooth change of section and large shank diameter gives lower stresses and lower creep rate than a bolt with a reduced shank diameter

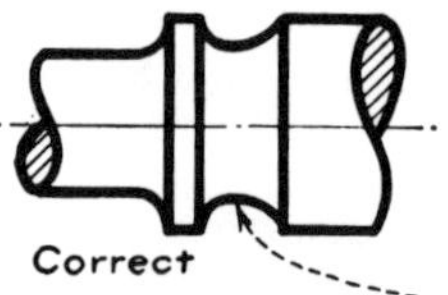

Correct

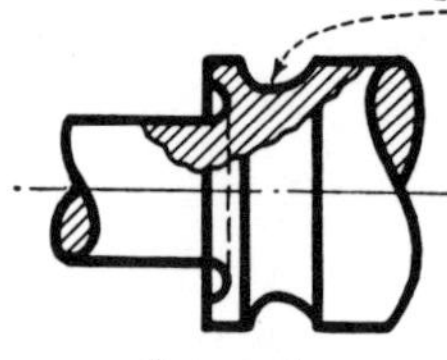

Correct

Stress relieving grooves turned in shaft behind shoulder. Grooves cause horizontal stress-lines in smaller shaft to follow a smooth change of direction upon entering larger part of shaft. Grooves are not semicircular in cross-section

# Of Variable Stress


J. SELWYN CASWELL

Engineering Department, University College of Swansea


**Load variations in machine parts are always accompanied by variations in internal stress, which may, under repeated loading, cause fatigue fracture of the part. Fractures in some machine parts can be avoided by these design modifications.**

## TAPERED ENDS AND COTTER HOLES

*Defective form of cotter hole. Square corners permit stress concentration and reduce strength across hole*

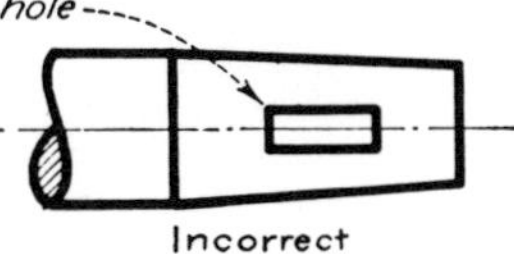

Incorrect

*Rounded ends of hole give more practical shape and greater strength through hole. Holes should be cut by end mill mortising, not by drilling and filing*

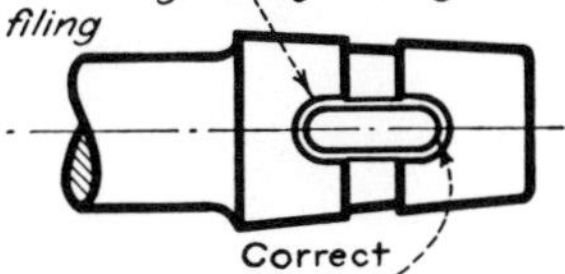

Correct

*Slight counterboring and recessing prevents interference by burrs*

*Effects of drilling, chipping, and filing. Sharp corners, produced by filing, induce high local stress and become starting points of fatigue fracture*

Incorrect

## HOLES IN SHAFTS

*Material around mouth of hole subjected to excessive stress. Resistance of material to fatigue failure is diminished*

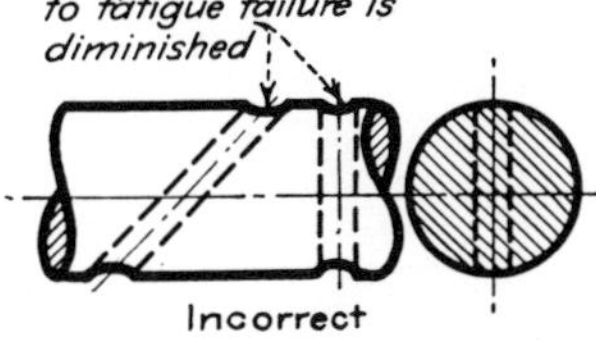

Incorrect

*Stress relieving grooves cut on each side of hole reduce local stresses and increase fatigue resistance of shaft*

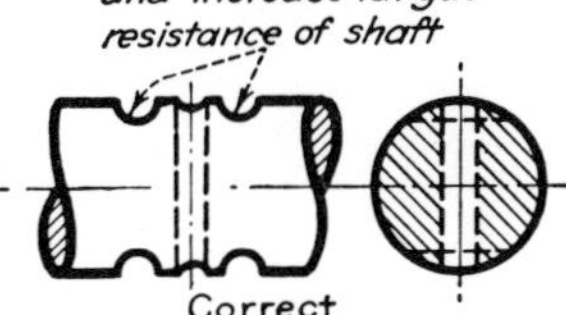

Correct

*Diameter of shaft enlarged around hole to reduce local stresses and increase fatigue resistance of shaft*

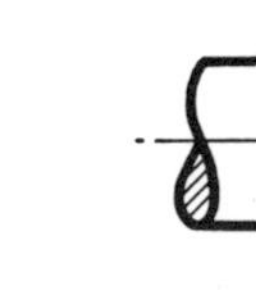

Correct

## KEYWAYS AND SPLINES

*Shaft weakened by keyway*

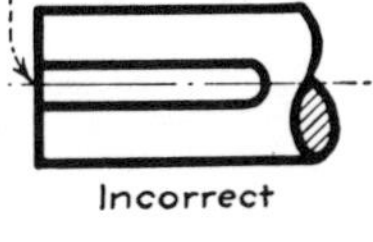

Incorrect

*Sharp corners at bottom of keyway weaken shaft*

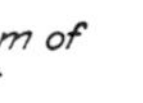

*Keyway seat on a diamter larger than shaft permits keyway to be cut without weakening shaft*

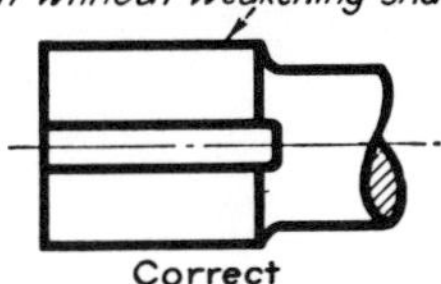

Correct

*Small fillet radius in corner at bottom of keyway reduces stress concentration*

*Shaft weakened by splines*

Incorrect

*Sharp corners at bottom of spline weaken shaft*

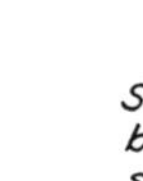

*Spline seat of larger diameter than shaft permits spline to be cut without weakening shaft*

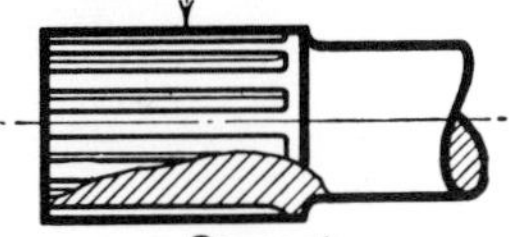

Correct

*Small fillet radius in corners at bottom of spline reduces stress concentration*

## SHRINK AND PRESS FITS

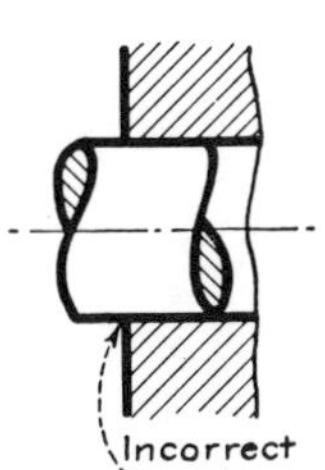

Incorrect

*Additional stresses set up in shaft by shrink or press fit may cause fatigue failure in shaft*

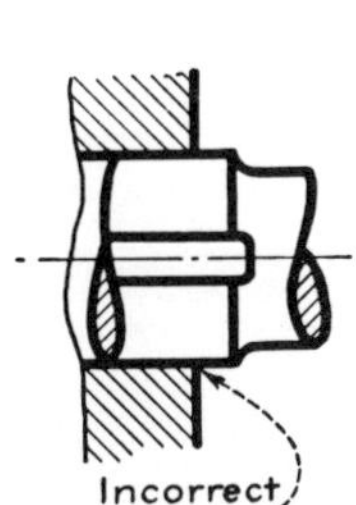

Incorrect

*Sharp change at edge of hub from high compressive stress to no compressive stress causes high local stress to start fatigue fracture*

*Seat and hub equal length*

*Stress relieved at ends of hubs. Shrinkage allowance diminished at these points*

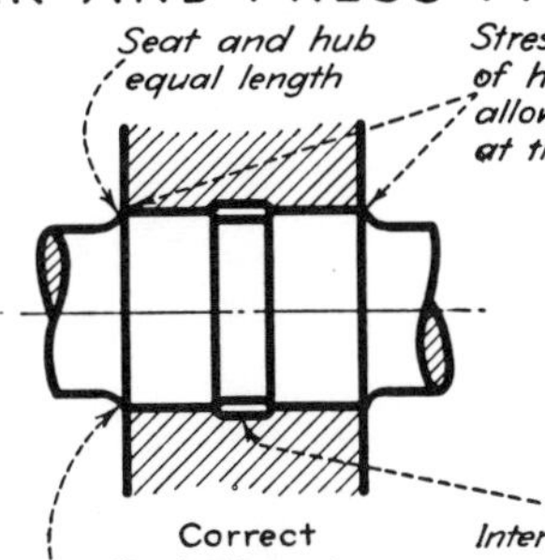

Correct

*Seat diameter increased to reduce pinch effects in adjacent parts of shaft*

*Internal recessing provides better symmetry of shrink fit stresses. Prevents a fulcrum or rocking point in imperfect fits*

*Seat and hub equal length*

Correct

*When keys are fitted, seat diameter is increased and gradual step-down to adjacent shaft diameter provided*

# 12

# SHAFT SEALS


Design Recommendations for O-ring Seals 306
Non-rubbing Seals for Oil Retention 308
Rubbing Seals for Oil Retention 310
Typical Methods of Sealing Rotating Shafts 312

# Design Recommendations

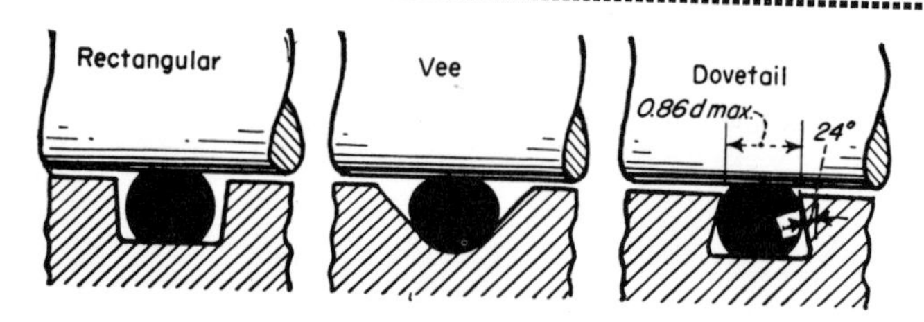


**1** Rectangular grooves are recommended for most applications, whether static or dynamic. Slightly sloping sides (up to 5 deg) facilitate machining with form tools. Where practical, all groove surfaces should have the same degree of finish as the rod or cylinder against which the O-ring operates. The Vee type groove is used for static seals and is especially effective against low pressures. The dovetail groove reduces operating friction and minimizes starting friction. The effectiveness of the seal with this groove is critical depending upon: pressure, ring squeeze and angle of undercut. In general, the groove volume should exceed the maximum ring volume by at least 15 percent.

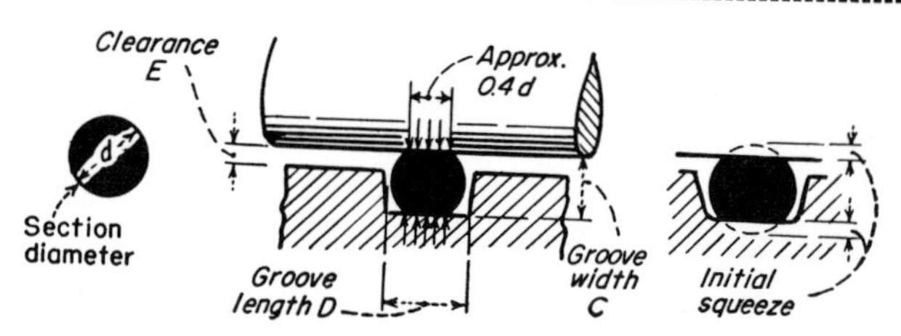


**2** To insure a positive seal, a definite initial squeeze or interference of the ring is required. As a rule, this squeeze is approximately 10 percent of the O-ring cross sectional diameter *d*. This results in a ring contact distance of approximately 40 percent under zero pressure and can increase as much as 80 percent of the cross section diameter depending on pressure and composition of the ring. Starting friction can be reduced somewhat by decreasing the amount of squeeze but such a seal would be only moderately effective at pressures above 500 psi. Table I lists the recommended dimensions and tolerances for O-ring grooves for both static and dynamic applications.

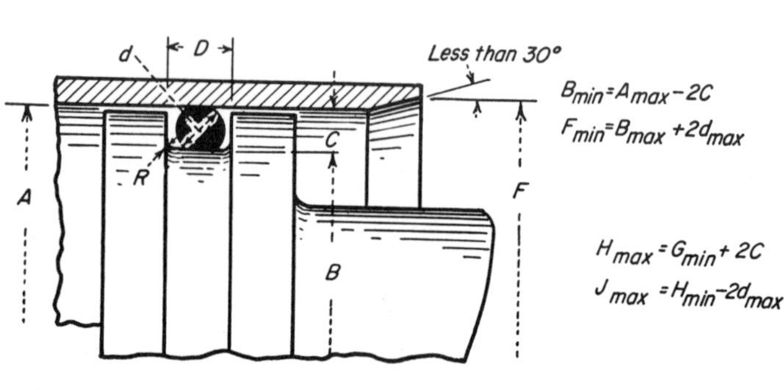


**3** On small diameters, to facilitate machining, O-ring grooves should be located on the ram or rod rather than on an inside surface. For larger diameters, grooves can be machined either way. One important factor is that the rubbing surfaces must be extremely smooth. The recommended dimensional data in Table I and listed under dynamic seals should be used for these applications. All cylinders and rods should have a gradual taper to prevent damage to the O-ring during assembly. Equations are listed for calculating limiting dimensions for both external and internal grooves.

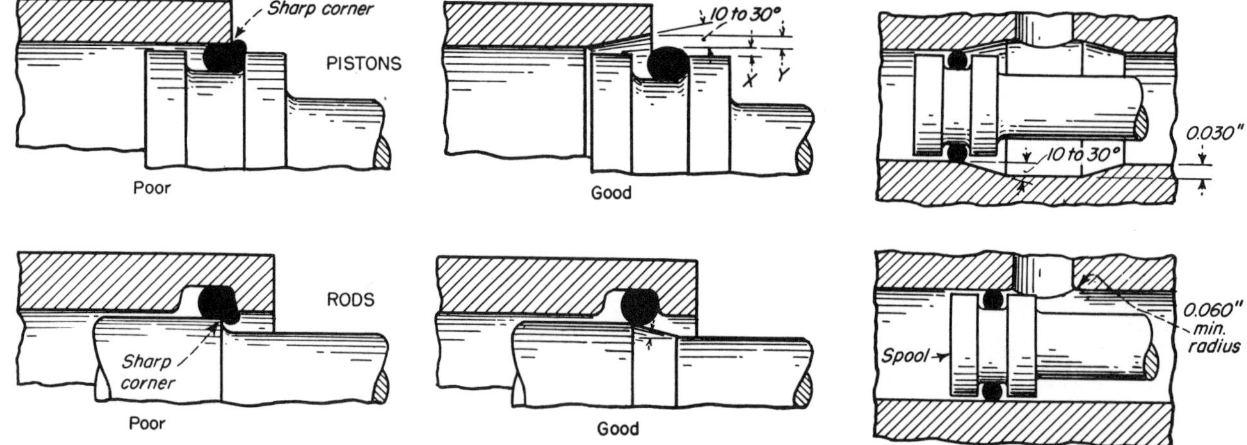


**6** To facilitate assembly, all members which slide over O-rings should be chamfered or tapered at an angle less than 30 degrees. An alternative method is to use a generous radius. Such details prevent any possibility of pinching or cutting the O-ring during assembly.

**7** Undercut all sharp edges, or cross-drilled ports over which O-rings must pass. While under pressure, rings should not pass over ports or grooves.

# for O-Ring Seals

J. H. SWARTZ
Linear, Inc., Philadelphia, Pa.

Table I—Dimensional Data for Standard AN or J.I.C. O-Rings and Gaskets

| Specification AN 6227 or J. I. C. O-Ring Dash Number | Nominal Ring Section Diameter | *d* Actual Section Diameter | For Static Seals | | For Dynamic Seals | | *D* Groove Length** | *R* Minimum Radius | 2*E* Diametral Clearance (maximum) | Eccentricity (maximum) |
|---|---|---|---|---|---|---|---|---|---|---|
| | | | Diametral Squeeze* (minimum) | *C* Groove Width +0.000 −0.005 | Diametral Squeeze* (minimum) | *C* Groove Width +0.000 −0.001 | | | | |
| 1 to 7 | 1/16 | 0.070±0.003 | 0.015 | 0.052 | 0.010 | 0.057 | 3/32 | 1/64 | 0.005 | 0.002 |
| 8 to 14 | 3/32 | 0.103±0.003 | 0.017 | 0.083 | 0.010 | 0.090 | 9/64 | 1/64 | 0.005 | 0.002 |
| 15 to 27 | 1/8 | 0.139±0.004 | 0.022 | 0.113 | 0.012 | 0.123 | 3/16 | 1/32 | 0.006 | 0.003 |
| 28 to 52 | 3/16 | 0.210±0.005 | 0.032 | 0.173 | 0.017 | 0.188 | 9/32 | 3/64 | 0.007 | 0.004 |
| 53 to 88 | 1/4 | 0.275±0.006 | 0.049 | 0.220 | 0.029 | 0.240 | 3/8 | 1/16 | 0.008 | 0.005 |
| AN 6230 or J. I. C. gaskets 1 to 52 | 1/8 | 0.139±0.004 | 0.022 | 0.113 | —— | —— | 3/16 | 1/32 | 0.006 | 0.003 |

Note: All dimensions are in inches.
* Diametral squeeze is the minimum interference between O-Ring cross section diameter *d* and gland width *C*.
** If space is limited, the groove length *D* can be reduced to a distance equal to the maximum O-Ring diameter *d* plus the static seal squeeze.

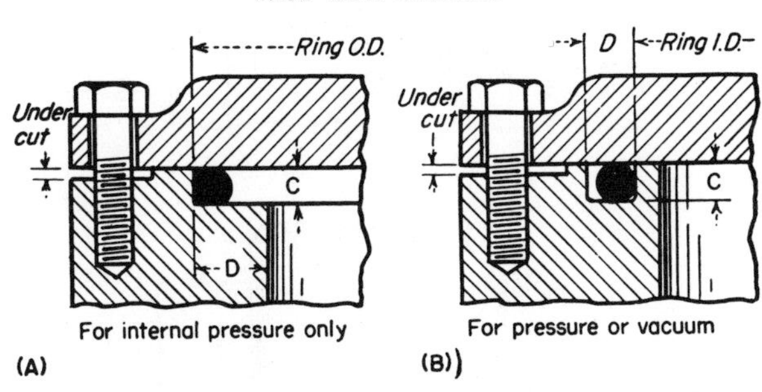


**4** For static face seals, two types of grooves are shown. Type (A) is more commonly used because of simpler machining. Groove depths listed in Table I under static seals apply to this application. In high pressure applications where steel flanges are used, slight undercutting of one face (not exceeding 0.010 in.) minimizes possible O-ring extrusion.

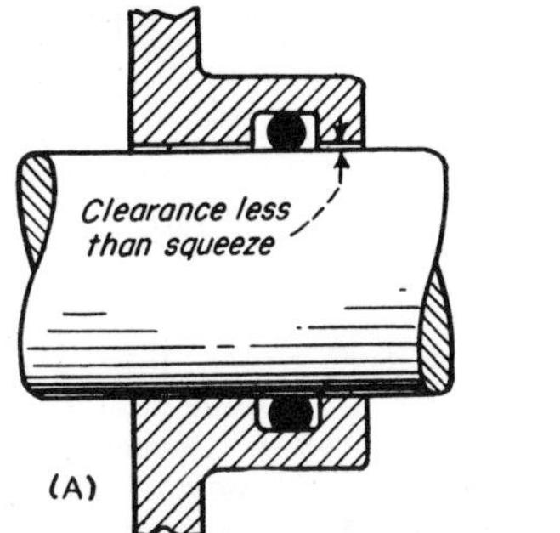


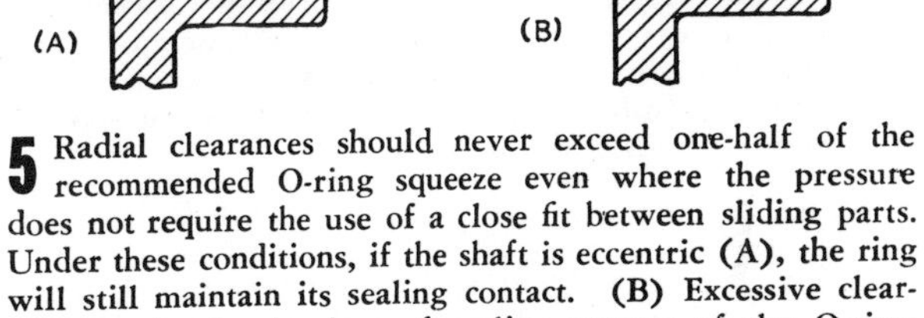


**5** Radial clearances should never exceed one-half of the recommended O-ring squeeze even where the pressure does not require the use of a close fit between sliding parts. Under these conditions, if the shaft is eccentric (A), the ring will still maintain its sealing contact. (B) Excessive clearance results in the loss of sealing contact of the O-ring.

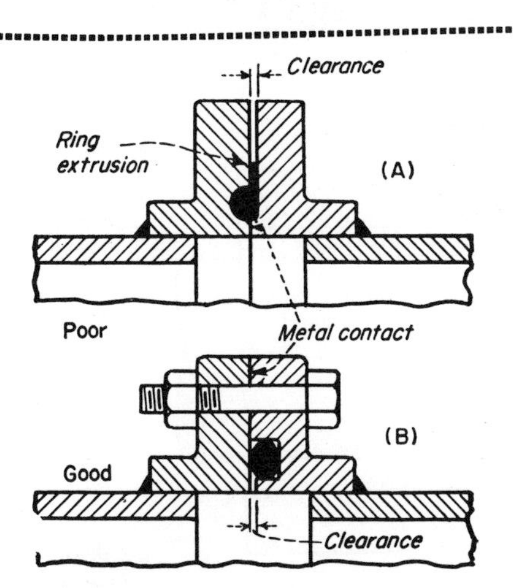


**8** Metal-to-metal contact of the inner mating surfaces (A) should be avoided. Clearances should be permitted only on inner surfaces (B).

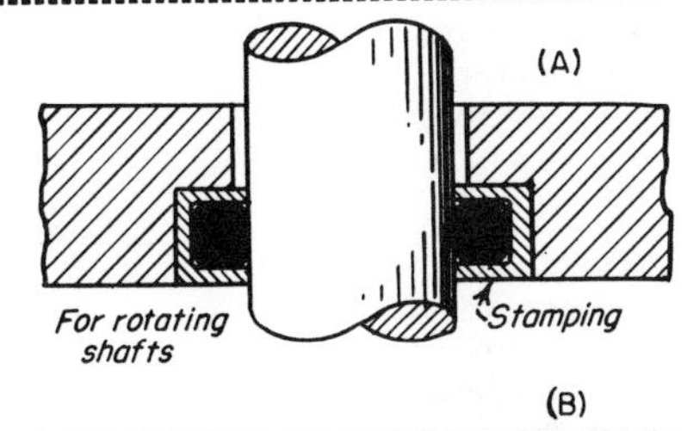


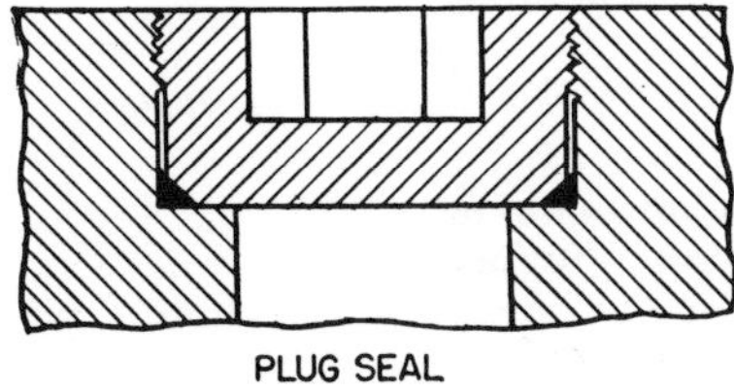


**9** Simple stamping (A) pressed in housing is for low speeds and pressures. (B) Chamfered corners of plug makes a recess for an O-ring.

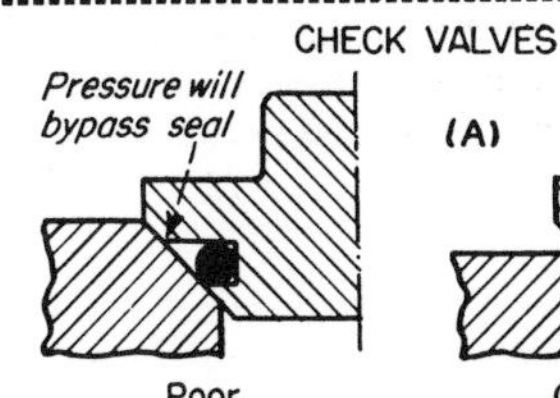


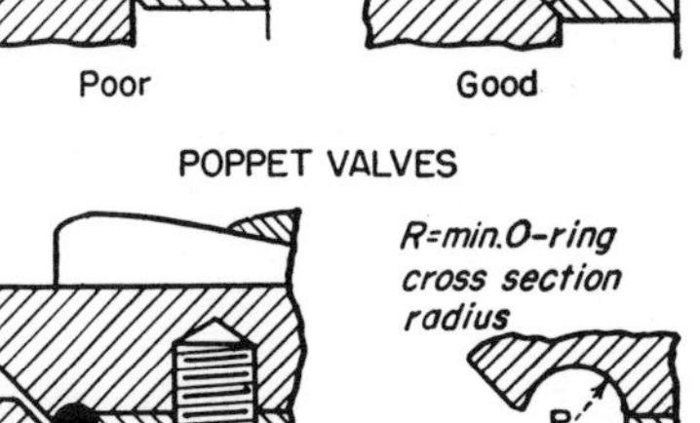


**10** Rectangular grooves (A) should be normal to the sealing surface. Special grooves (B) avoid the washout of O-rings during pressure surges.

# Non-Rubbing Seals for Oil

There are two general types of seals: rubbing and non-rubbing. Non-rubbing seals use oil or grease to lubricate mating surfaces and exclude foreign matter by forcing the lubricant out between the bearing surfaces. These seals are not limited by operation or speed since

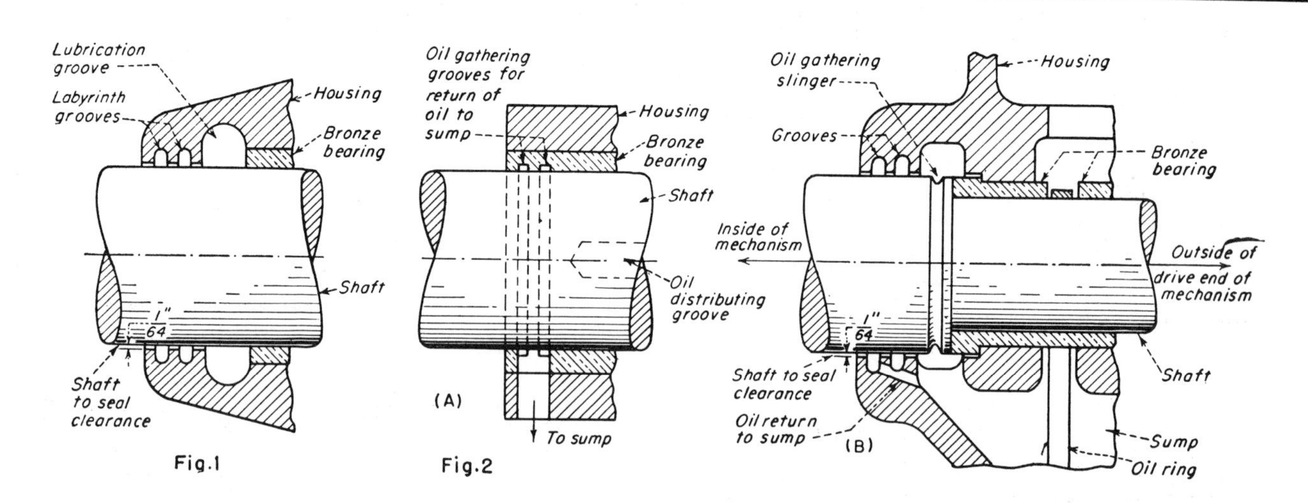


**Fig. 1—Non-rubbing seal for grease lubrication.** Grooves and housing are filled with grease at assembly or an automatic feed system can be incorporated. Seal offers protection against entrance of foreign matter because of the outward flow of grease through the labyrinth passages. Clearance between shaft and seal is about 1/64 inch.

**Fig. 2—Non-rubbing seal for oil lubrication.** Grooves can be located in bearing, (A), or in housing, (B). Grooves are connected to an oil return passage leading to a sump in housing. This keeps oil loss to a minimum and maintains a constant supply to the bearing. Design does not prevent entrance of foreign material. Radical clearance between shaft and seal is 1/64 inch.

**Fig. 3—Seal for vertical installation.** Circular groove picks up lubricant and feeds it to a spiral groove in bearing or shaft. Spiral feeds lubricant to top of bearing. Lubricant runs down between shaft and bearing. Design is effective when the shaft is rotating.

**Fig. 4—Similar to Fig. 3 but used for a horizontal installation to reduce leakage of lubricant.** Straight groove feeds oil to the bearing. Circular groove collects oil and the spiral groove, located in shaft or bearing, returns it to other end of bearing. Design prevents loss of oil by leakage at outside end.

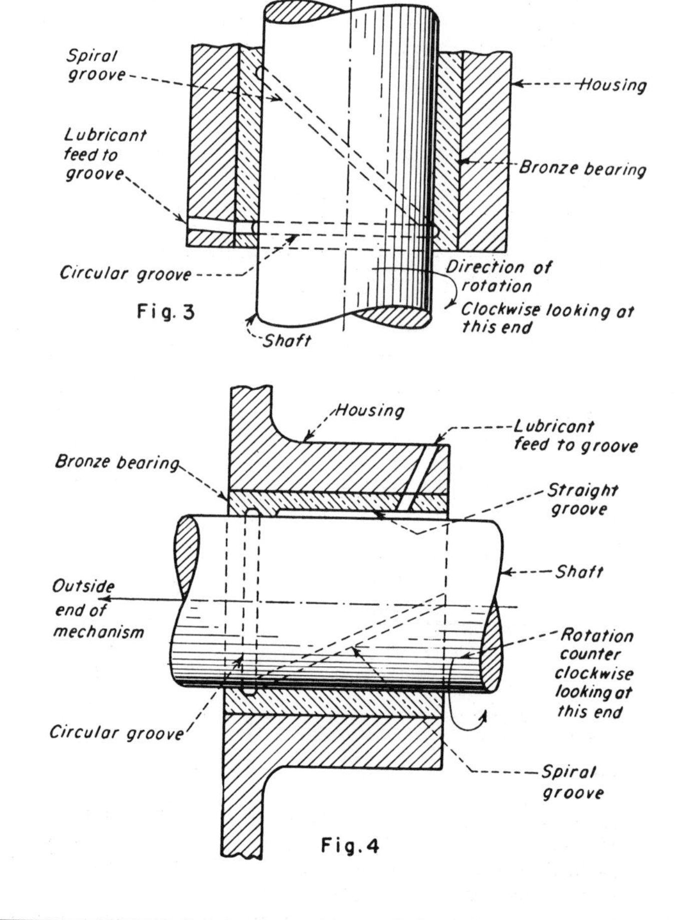

# Retention

DAVID C. SPAULDING, JR.
Product Engineer, The Bunting Brass and Bronze Co.
Toledo, Ohio

friction is negligible. They are, however, often more expensive than rubbing seals since the shaft, bearing, or housing must be grooved to distribute the lubricant. Distribution over the bearing area is, however, better. Lubricants may be forced-fed or gravity-fed.

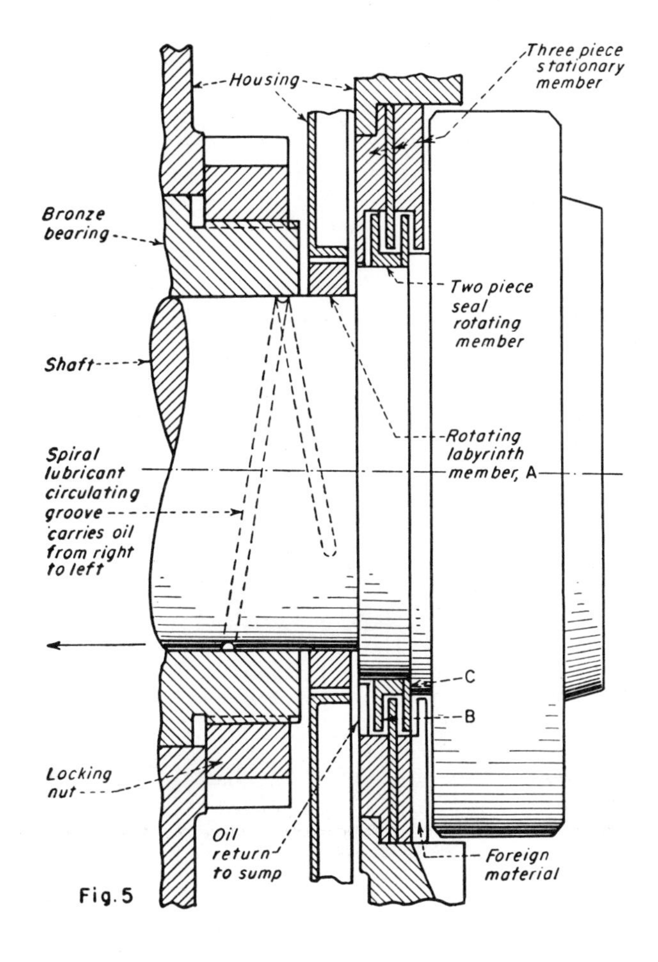


**Fig. 5—Labyrinth seals offer good protection especially at high speeds where the narrow zigzag passage is used in conjunction with centrifugal force. Oil and foreign matter are separated by slinger which limits oil flow past rotating member *(A)*. Inner member *(B)* throws oil back to sump. Member *(C)* throws out foreign matter.**

**Fig. 6—Non-rubbing seal for oil lubrication. Shaft rotation throws lubricant into the inboard groove *(B)* in the housing and is returned to the oil sump. First slinger *(D)* throws foreign material out of the assembly. Second slinger *(D)* feeds foreign material out through groove *(A)* and hole at *(C)*. Lubricant feeds between shaft and bearing to housing grooves.**

**Fig. 7—Reservoir type feed for grease lubrication. Grease is distributed in annular groove and feeds through holes in bearing to lubricate the shaft. Foreign matter can be excluded if clean grease is used. Grease will be lost through open bearing ends and assembly must be repacked periodically.**

**Fig. 8—Reservoir type feed for oil lubrication. Two bearings are used forming an oil reservoir between them. If porous wall type bearings are used oil will saturate and feed through bearings to the shaft. Outward flow of oil prevents entry of foreign material. Reservoir must be periodically refilled with new oil. Bearings are a press fit in the housing.**

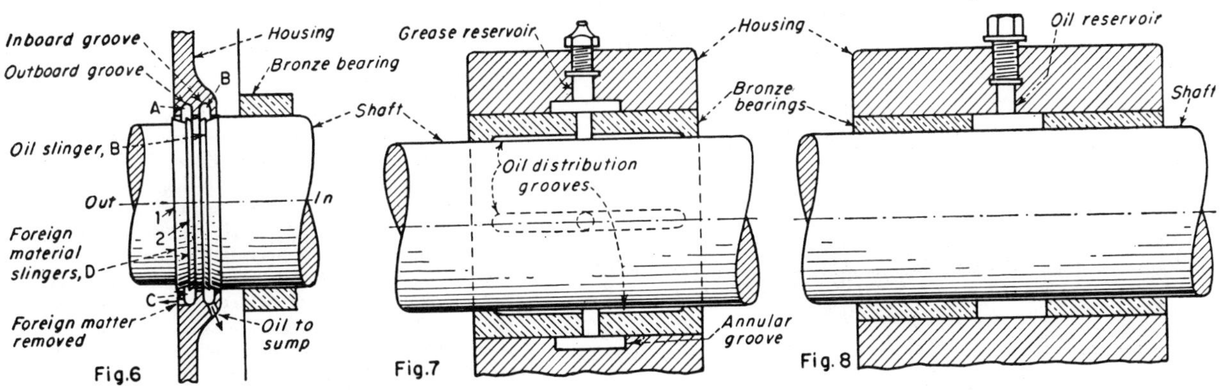

# Rubbing Seals for Oil

Rubbing seals cover all applications where a positive sliding contact exists between the seal and either the rotating or stationary member. They are limited as to type of operation and speed because of the friction between the contacting surfaces and they should not be used in

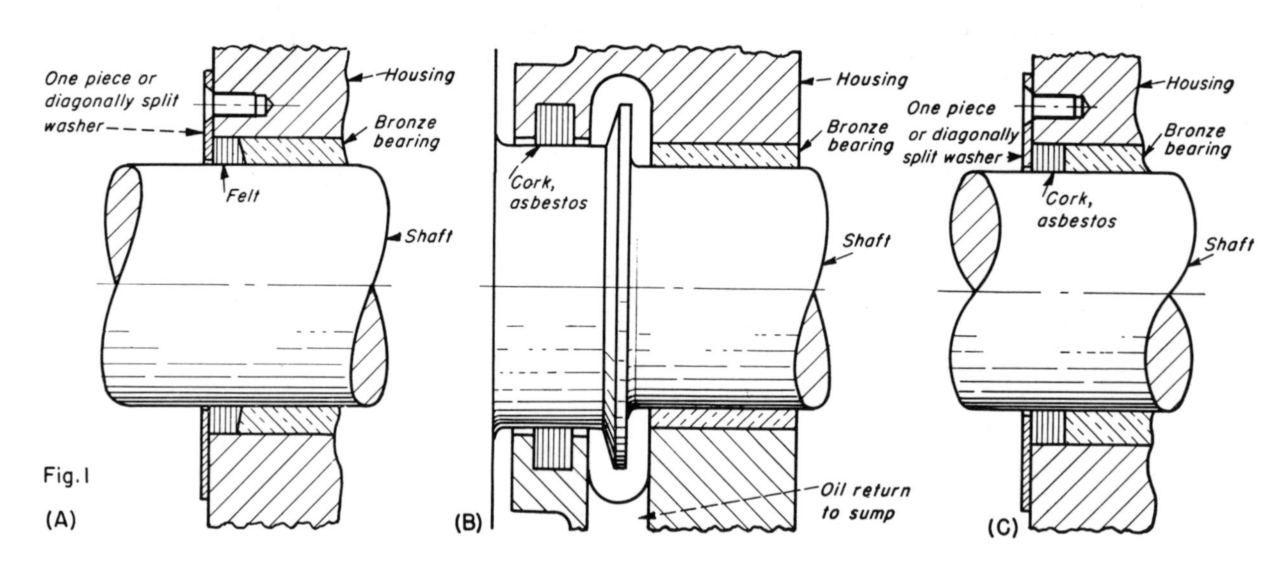


**FIG. 1—Rubbing seal for oil lubrication. Felt, cork, asbestos, natural or synthetic rubber or other materials can be used. The natural resiliency of felt provides a close contact between seal and shaft without the excessive pressures often encountered with other types. It also absorbs and retains oil providing for almost constant lubrication. For the retention of felt, design A is recommended, because the tapered sides insure close contact and the removable plate permits easy replacement. (B) and (C) may also be used. Cork and asbestos should be retained as shown in (B). Groove must be straight sided and narrow enough to compress the material slightly to prevent it from turning.**

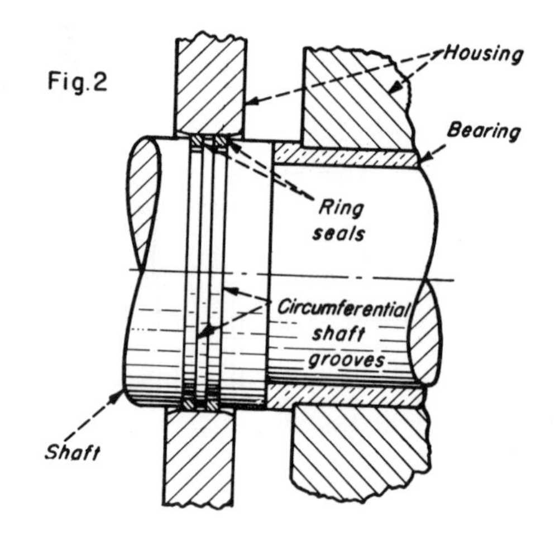


**FIG. 2—Bronze or cast iron rings are frequently used to seal bearings. This type of seal is equally effective for reciprocating and rotary motion. Circumferential grooves are cut in the shaft and the rings are compressed and inserted. They bear on the housing effectively sealing in the lubricant.**

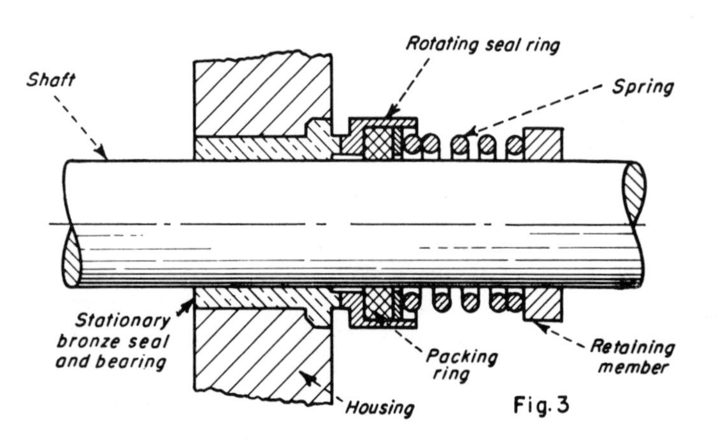


**FIG. 3—High pressure and high rotative speeds where leakage is critical use mechanical face type seals. The seal has low total friction, can withstand high misalignment and compensates for wear. Parts of the seal are: stationary seal ring, rotating seal ring, flexible type joint, (diaphragm, bellows, or packing ring) spring, and retaining members. The stationary member can be an integral part of the bearing when a cast bronze sleeve bearing is used. As illustrated, the rotating seal ring, packing spring and retaining member turn with the shaft and the spring keeps the seal ring in contact.**

# Retention

DAVID C. SPAULDING, JR.
Product Engineer, The Bunting Brass and Bronze Co.
Toledo, Ohio

abrasive surroundings. Types that are held against the rotating member by spring pressure can be used where there is a pressure head of fluids within the assembly or on the exterior. For high pressure stuffing box and O-ring type seals are used. O-rings are also used for zero leakage.

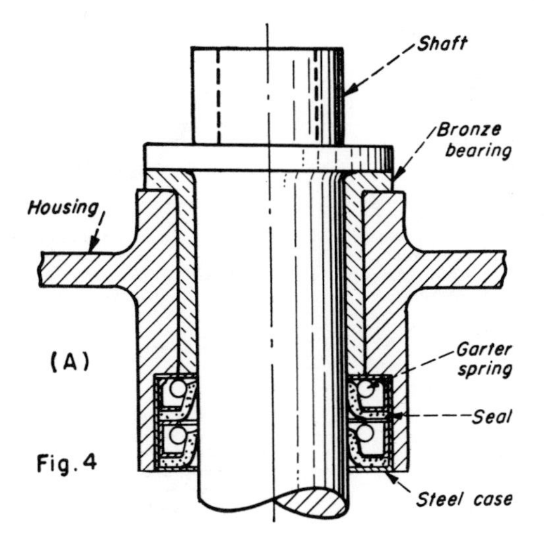


**FIG. 4—Rubbing seals, of the type shown, (A), have widespread use in all types of equipment. The spring tension and sealing ring material may be varied so that a variety of applications can be handled. Small units can be had where the O. D. is the same as the O. D. of the sleeve bearing, (B), thus eliminating the counterboring operation on the housing. The seal may be reversed and used to keep foreign matter out of the assembly. A drain hole may be provided to carry away surplus lubricant. Retention is by press fit on the outside diameter.**

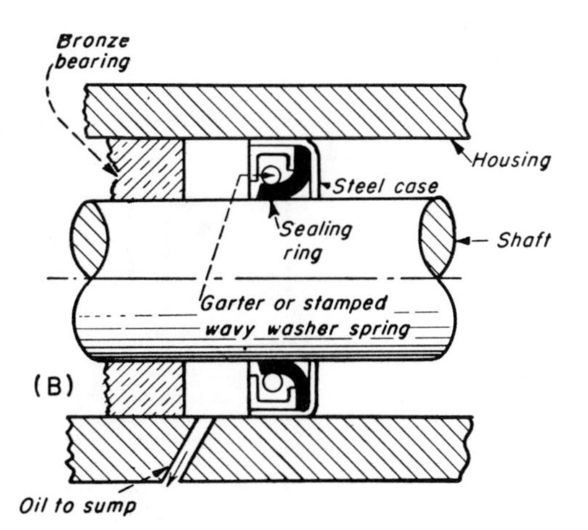


**FIG. 5—Rubbing seals of the stuffing box type (A), are used where high pressure are encountered. It can be used for all types of motion and the packing material can be varied depending upon the fluid to be sealed and the application. For rotating motion some leakage is necessary so it cannot be used when permissible leakage is zero. O-rings can also be used for rotary motion if the speed is slow. Special designs use O-ring seals (B), when zero leakage is demanded for either stationary or reciprocating motion.**

**This ring is made of natural rubber or synthetic rubber depending on the type of solution resistance required. Synthetic rubber, such as buna or neoprene, is resistant to aromatic hydrocarbons, while natural rubber resists the action of alcohol and glycerine.**

**O-rings can be located either in the shaft or in the housing and any movement or pressure forces the ring to one side, thereby forming a tight seal.**

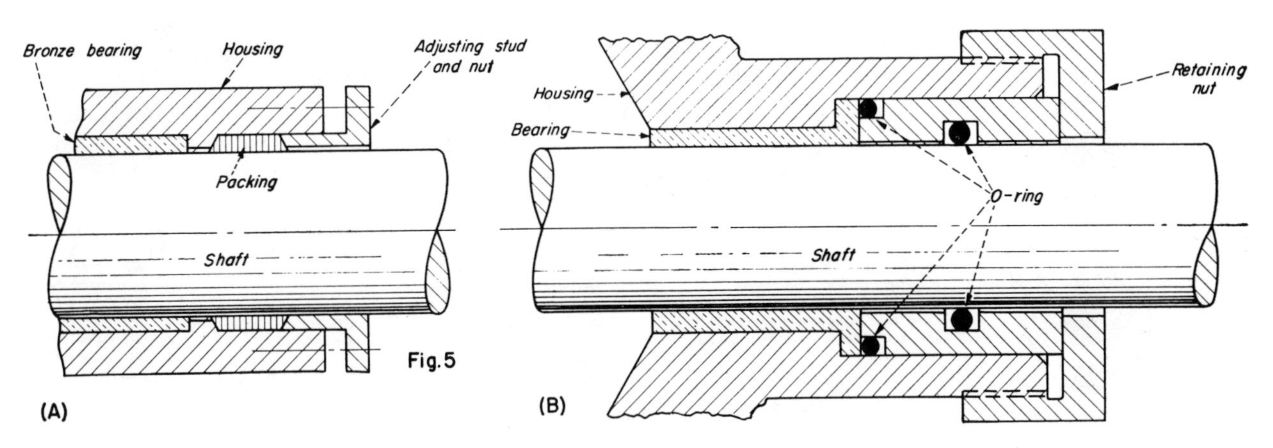

# TYPICAL METHODS OF SEALING

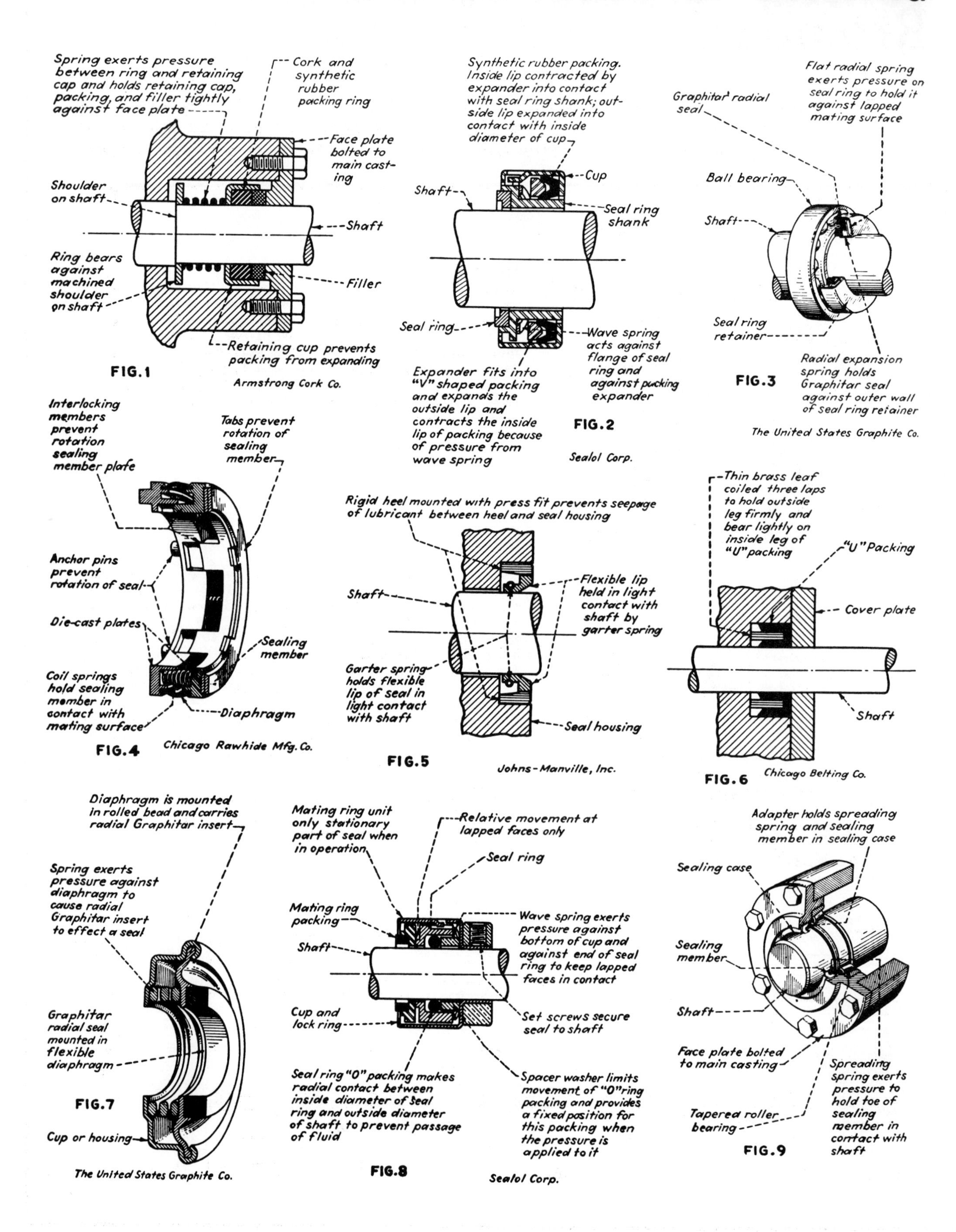

# ROTATING SHAFTS

Shaft seals, used to prevent seepage of lubricating oils or other liquids, range from simple flange packings to elaborate bellows and lapped-surface devices. Some of these, utilizing wave, coil, and garter springs, diaphragms, and locking rings, are described and illustrated.

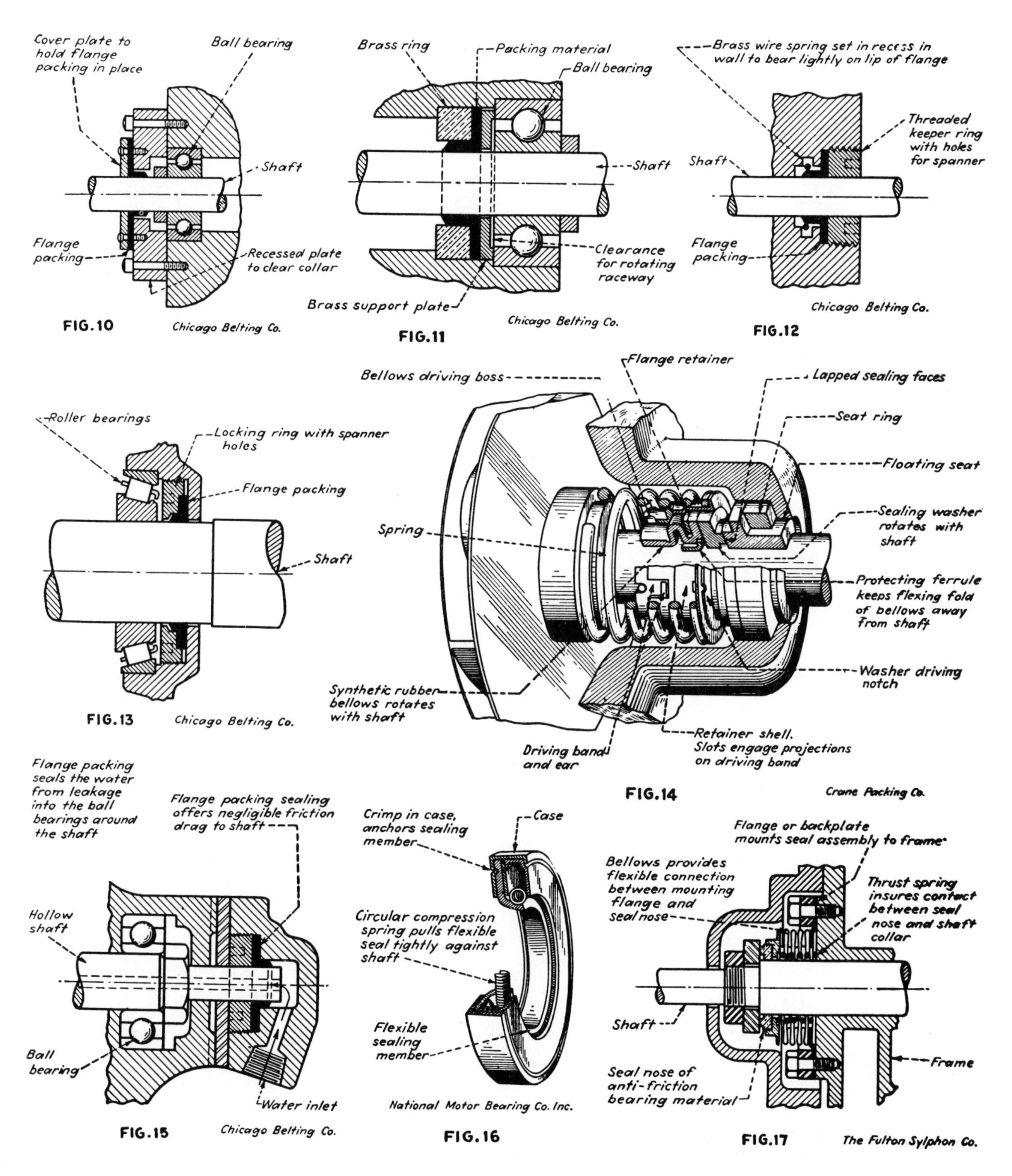

# 13

# SPRINGS


Compression Spring Adjusting Methods — I & II 316
Adjustable Extension Springs 320
How Much Force to Deflect a Spring Sideways? 322
Overriding Spring Mechanisms for Low-torque Drives 324
17 Ways of Testing Springs 326
Unusual Uses for Helical Wire Springs 328

# COMPRESSION SPRING

**In many installations where compression springs are used, adjustability of the spring tension is frequently required. The methods shown incorporate various designs of screw and nut adjustment with numerous types of spring-centering means to guard against buckling. Some designs incorporate frictional reducing members to facilitate adjustment especially for springs of large diameter and heavy wire**

HENRY MARTIN

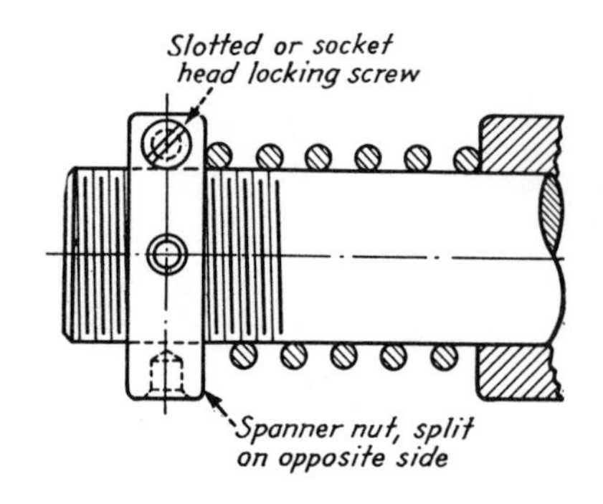


FIG.3

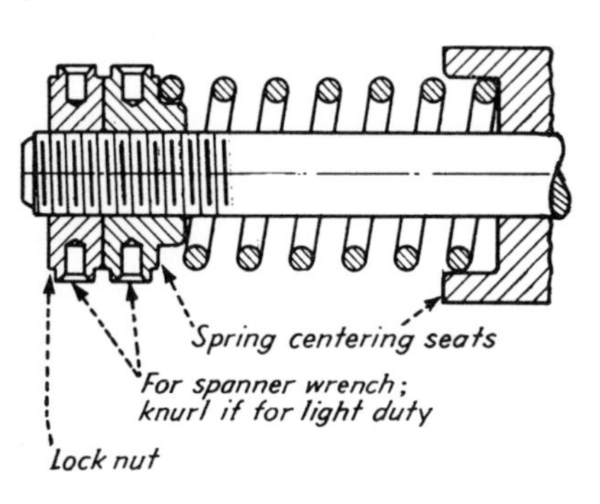


FIG.1

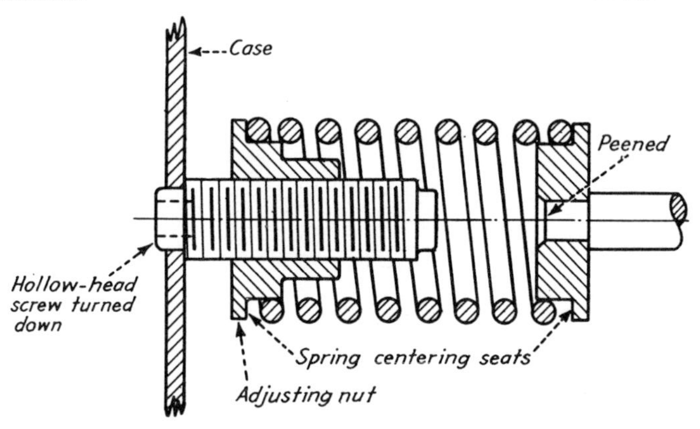


FIG.2

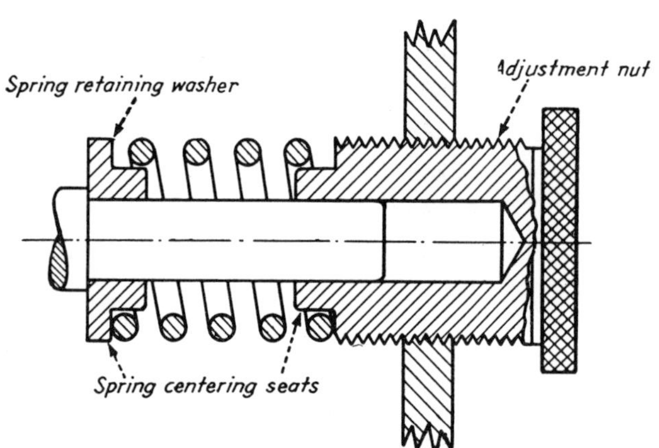


FIG.4

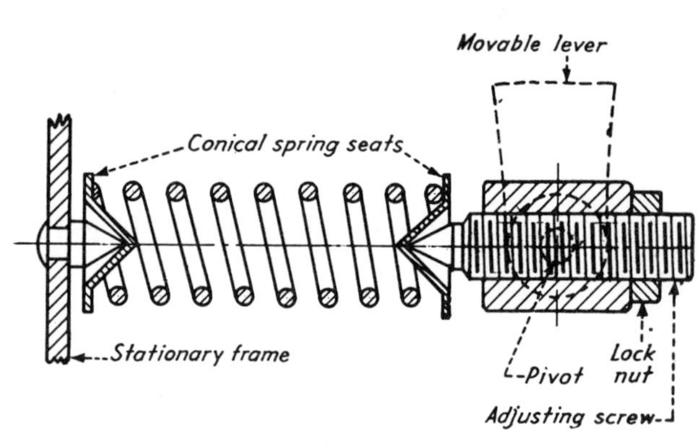


FIG.5

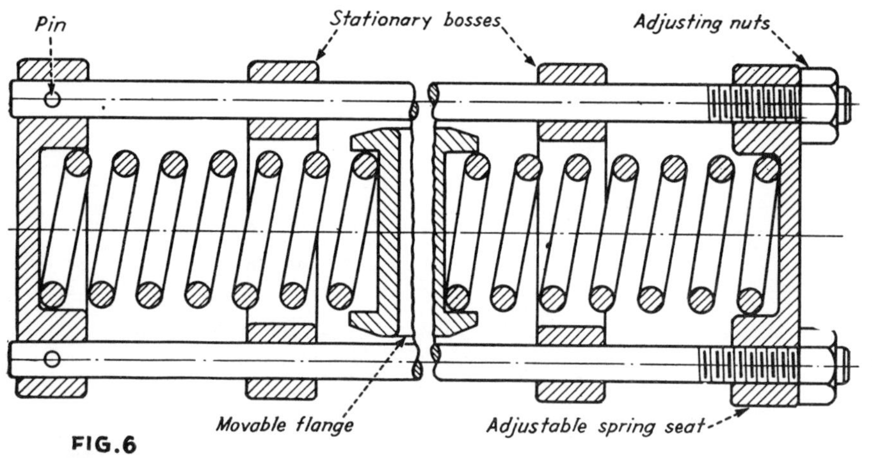

FIG.6

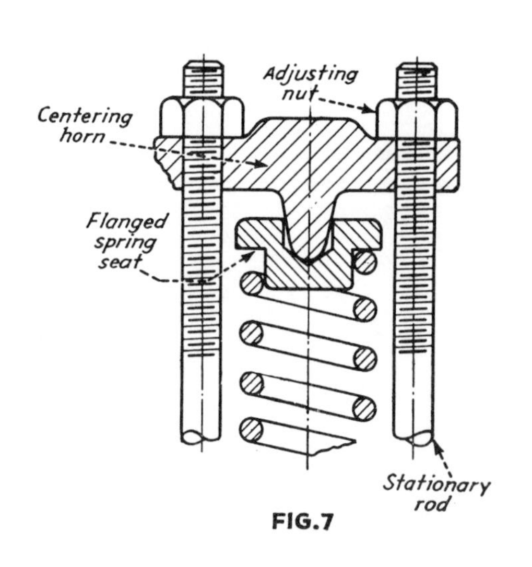


FIG.7

# ADJUSTING METHODS–I

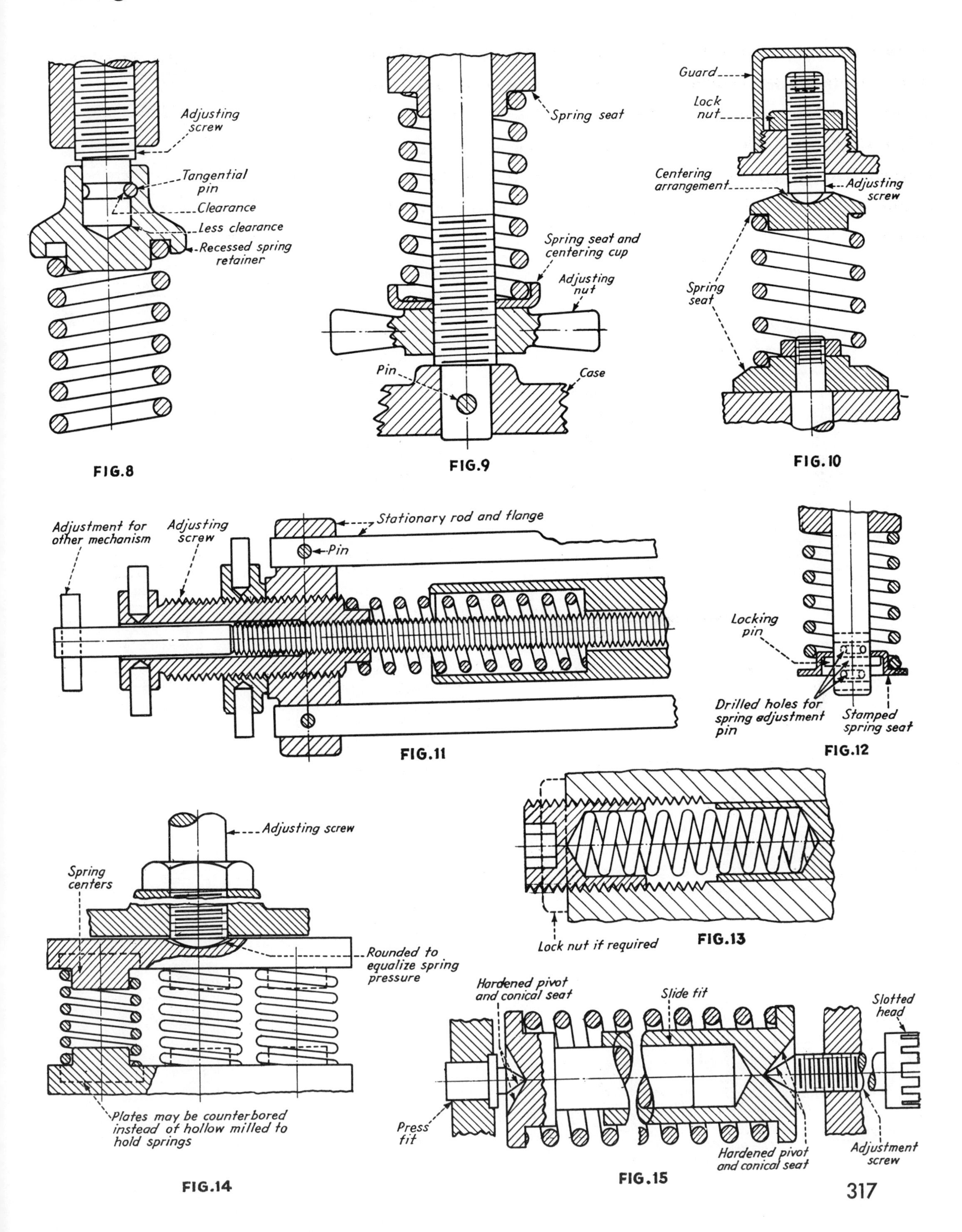

# COMPRESSION SPRING ADJUSTING

**In this concluding group of adjustable compression springs, several methods are shown in which some form of anti-friction device is used to make adjustment easier. Thrust is taken against either single or multiple steel balls, the latter including commercial ball thrust bearings. Adjustments of double spring arrangements and other unconventional methods are also illustrated**

HENRY MARTIN

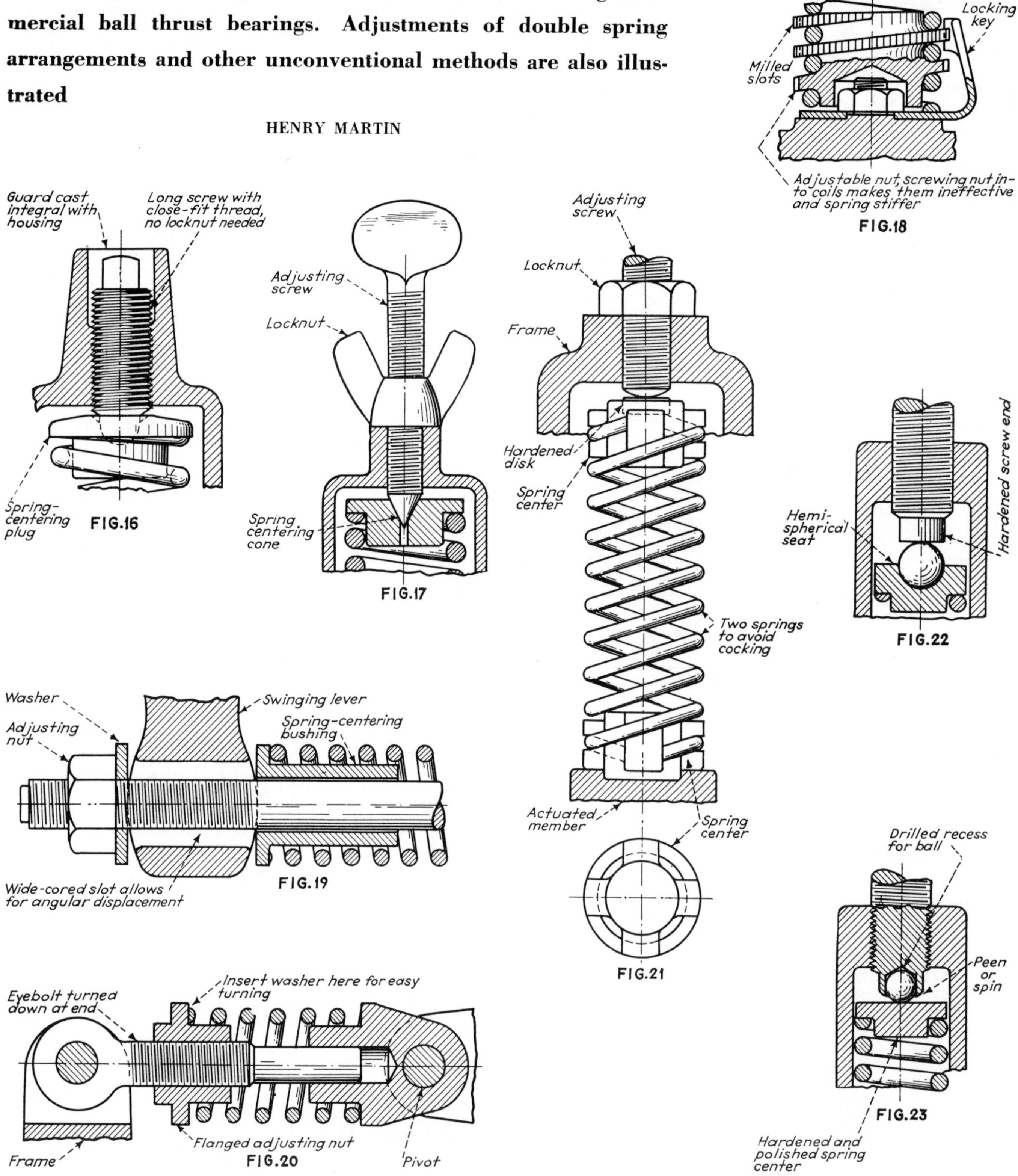

# METHODS—II

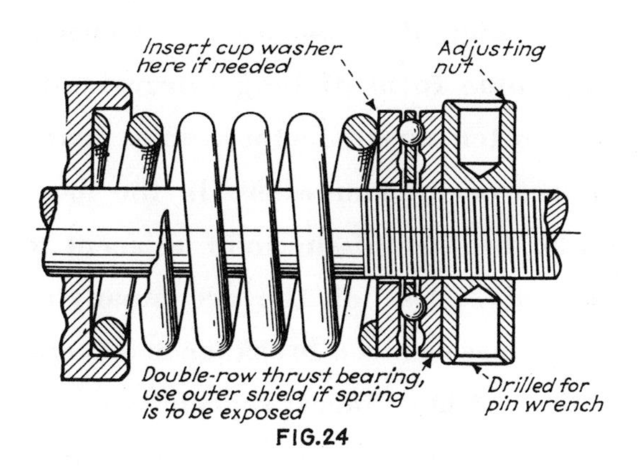

FIG.24

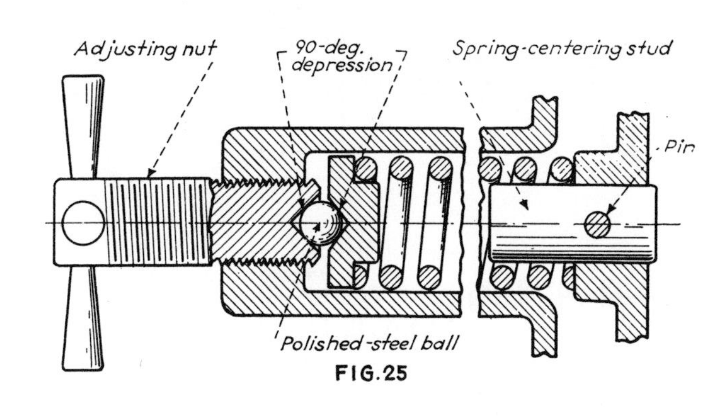

FIG.25

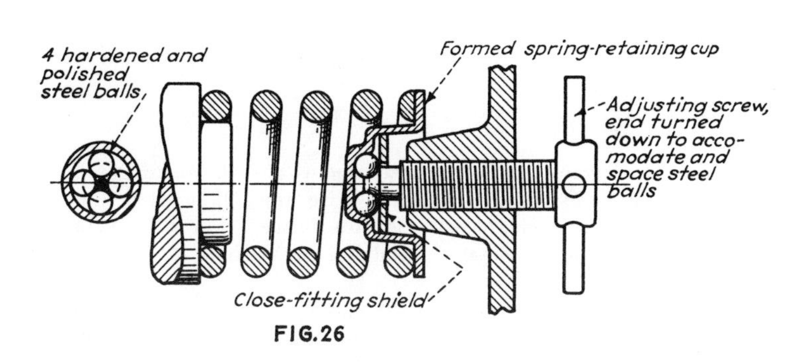

FIG.26

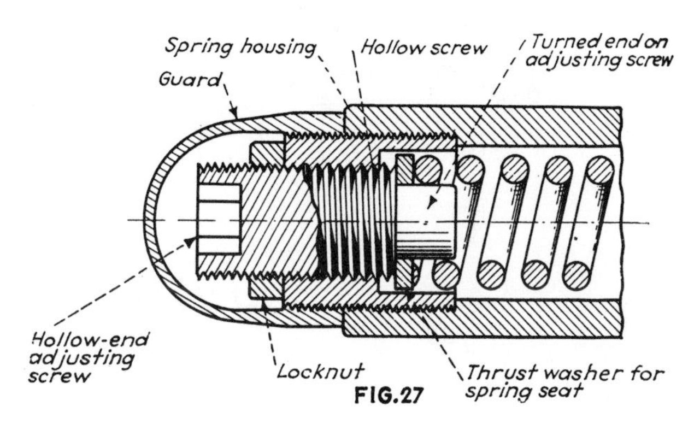

FIG.27

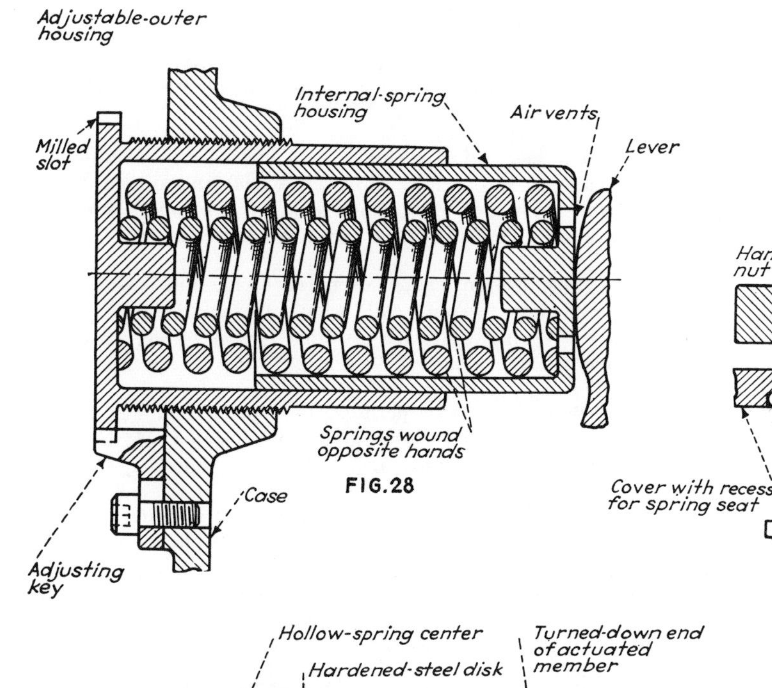

FIG.28

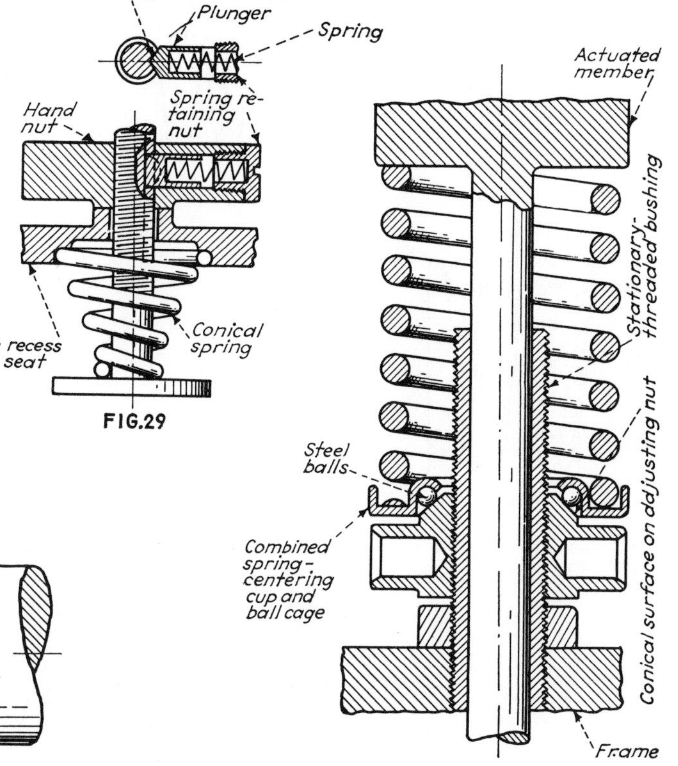

FIG.29

FIG.31


FIG.30

# ADJUSTABLE EXTENSION

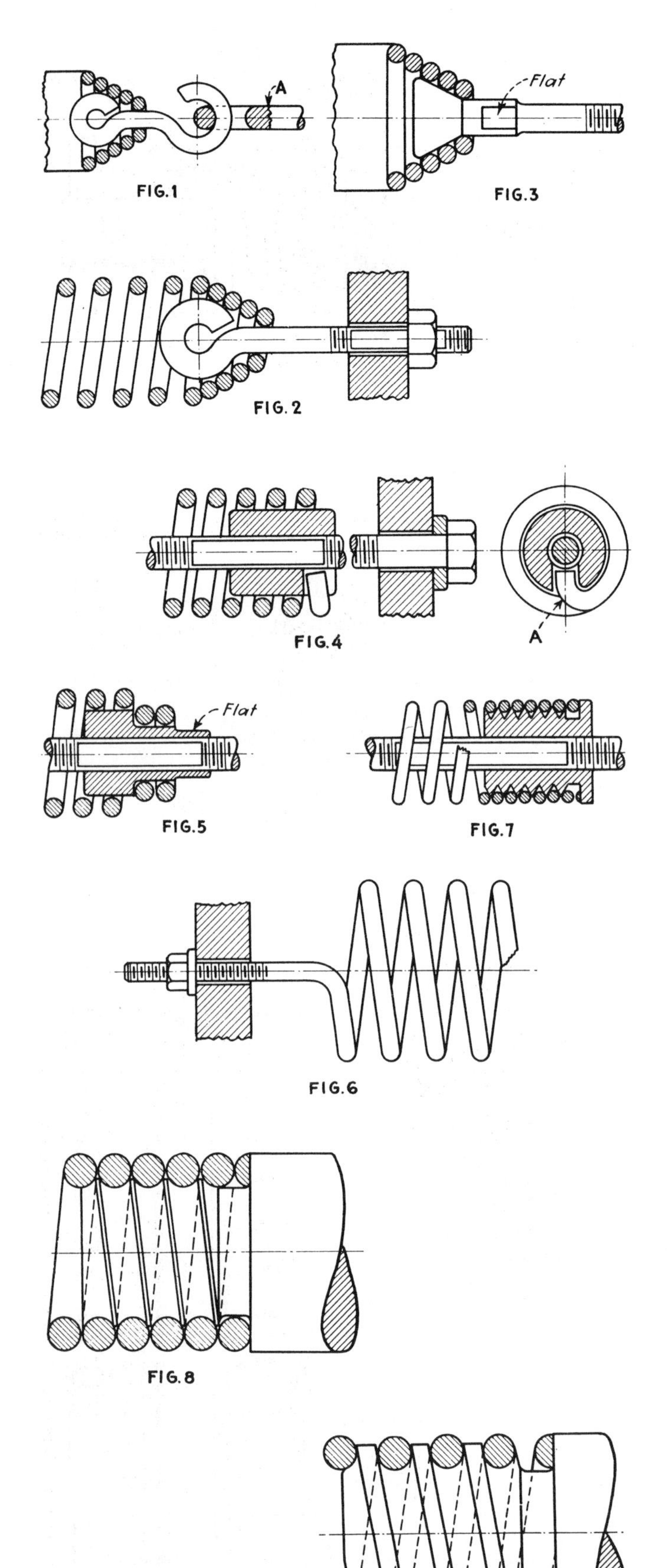


**Design of the end of a tension or extension spring using some form of loop integral with the spring is often unsatisfactory, since many spring failures occur somewhere in the loop, most often at the base of the loop adjacent to the spring body. Use of the accompanying tested methods has reduced breakage and therefore down-time of machinery, especially where adjustability of tension and length is required**

HENRY MARTIN

Fig. 1—Spring-end is tapered about a loop made of larger diameter and somewhat softer wire than that used for the spring. Upper end of wire is also formed into a loop, larger and left open to engage a rod-end or eye-bolt *A*.

Fig. 2—A loop is formed at the end of a soft steel rod threaded at the opposite end for a hex adjusting nut. Ordinary threaded rod-end may be substituted if desired.

Fig. 3—End of adjusting screw is upset in shape of a conical head to coincide with taper of spring-end. Unless initial tension of spring is sufficiently great a wrench flat on stem is provided to facilitate adjustment.

Fig. 4—The last coil of spring is bent inwardly to form hoop *A* which engages slot in nut. Although a neat and simple design, all spring tension is exerted on hook at one point, somewhat off-center of spring axis. Not recommended for heavy loads.

Fig. 5—An improved method over Fig. 4. The nut is shouldered to accommodate two end coils which are wound smaller than the body of spring. Flats are provided for use of wrench during adjustment.

Fig. 6—When wire size permits, the spring end can be left straight and threaded for adjustment. Because of the small size of nut a washer must also be used as shown.

Fig. 7—The shouldered nut is threaded with a coarse V-thread and is screwed into the end of the spring. The point of tangency between the 30-deg. side of thread and wire diameter should be such that the coils cannot pull off. The end of the spring is squared for sufficient friction so that nut need not be held when turning the adjusting screw.

Fig. 8—For close-wound extension springs, end of rod may be threaded with a shallow thread the root of which is the same curvature as that of the spring wire. This form of thread cut with the crests left sharp provides greater engagement contact.

Fig. 9—For more severe duty, the thread is cut deeper than that shown in Fig. 8. The whole spring is close-wound, but when screwed on adjusting rod, the coils are spread, thereby creating greater friction for better holding ability. Spring is screwed against the relieved shoulder of rod.

# SPRINGS

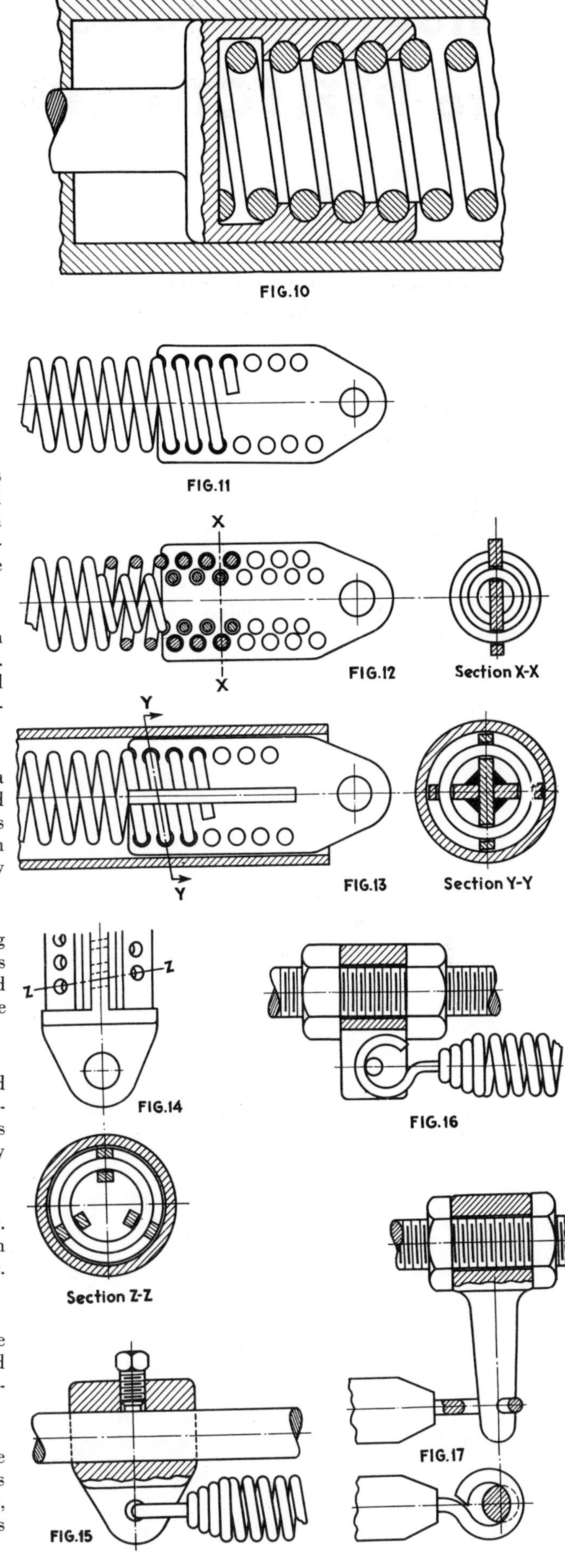


Fig. 10—When design requires housed spring, adjusting rod is threaded internally. Here also, the close-wound coils are spread when assembled. Unless housing bore is considerably larger than shouldered diameter of adjusting rod, or sufficient space is available for a covered spring, methods shown in Figs. 8 or 9 will be less expensive.

Fig. 11—A thin piece of cold-drawn steel is drilled to exact pitch of the coils with a series of holes slightly larger than spring wire. Three or four coils are screwed into the piece which has additional holes for further adjustment. It will be seen that all coils so engaged are inactive or dead coils.

Fig. 12—A similar design to that shown in Fig. 11, except that a smaller spring lies inside the larger one. Both springs are wound to the same pitch for ease of adjustment. By staggering the holes as shown, the outer diameter of the inner spring may approach closely that of the inner diameter of the outer spring, thereby leaving sufficient space for a third internal spring if necessary.

Fig. 13—When the spring is to be guarded, and to prevent binding of the spring attachment in the housing, the end is cross-shaped as shown in the section. The two extra vanes are welded to the solid vane. The location of the series of holes in each successive vane is such as to advance spring at one quarter the pitch.

Fig. 14—This spring end has three vanes and is turned, bored and milled from solid round stock where welding facilities are not convenient. In sufficient quantities, the use of a steel casting precludes machining bar stock. The end with the hole is milled approximately ¼ in. thick for the adjusting member.

Fig. 15—A simple means of adjusting tension and length of spring. The spring anchor slides on a plain round rod and is fastened in any position by a square head setscrew and brass clamping shoe. The eye in the end of the spring engages a hole in the anchor.

Fig. 16—A block of cold-drawn steel is slotted to accommodate the eye of the spring by means of a straight pin. The block is drilled slightly larger than the threaded rod and adjustment and positioning is by the two hex nuts.

Fig. 17—A similar arrangement to that shown in Fig. 16. The spring finger is notched at the outer end for the spring-eye as illustrated in the sectioned end view. In these last three methods, the adjustable member can be made to accommodate 2 or 3 springs if necessary.

# how much force to DEFLECT a SPRING SIDEWAYS?

**Formulas for force and stress when a side load deflects a vertically loaded spring.**

W H SPARING, *Mechanical Engineer*
*American Steel Foundries, Chicago*

There are many designs in which one end of a helical spring must be moved laterally relative to the other end. How much force will be required to do this? What deflection will the force cause? What stress will result from combined lateral and vertical loads? Here are formulas that find the answers.

It is assumed that the spring ends are held parallel by a vertical force $P$ (which does not appear in these formulas), and that the spring is long enough to allow overlooking the effect of closed end-turns.

**Lateral load** for a steel spring

$$Q = \frac{10^6 d^4 \delta_L}{AnD\,[0.204\,(L-d)^2 + 0.265\,D^2]}$$

where $n$ = number of turns = $(h/d) - 1.2$, $D$ = mean dia in., $A$ = correction factor.

**Lateral deflection**

$$\delta_L = \frac{AQnD\,[0.204\,(L-d)^2 + 0.265\,D^2]}{10^6 d^4}$$

The correction factor $A$ can never be unity (see chart on continuing page); also $P$ can never be zero. This is because there will always be some vertical deflection, and a side load will always cause a resultant vertical force if the ends are held parallel and at right angles to the original center line.

**Combined stress**

$$f_c = f\left\{1 + \frac{1}{D}\left[\delta_L + \frac{Q}{P}\,(L-d)\right]\right\}$$

where $f$ = vertical-load stress. Accurate within 10%, these formulas show that the nearer a spring approaches its solid position, the greater the discrepancy between calculated and actual load. This results from premature closing of the end-turns. It is best to provide stops to prevent the spring from being compressed solid. An example shows the combined stress at the stop position may even be higher than the solid stress caused by vertical load only.

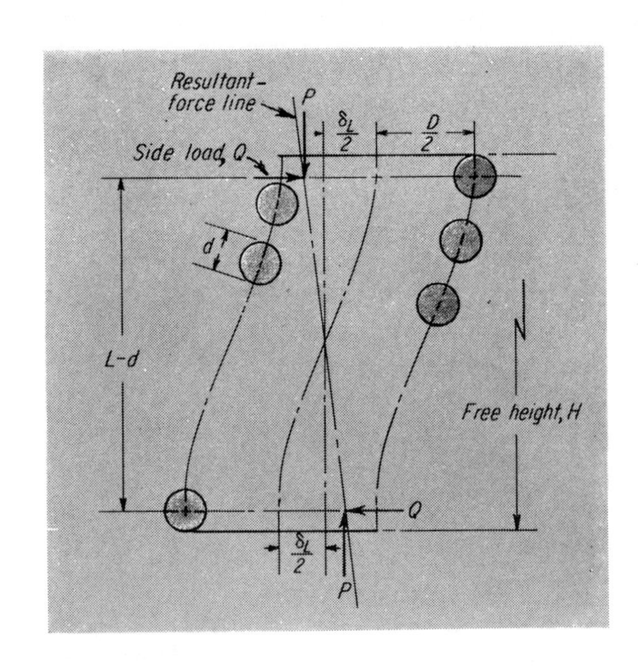


## A Working Example

A spring has the following dimensions, in inches:

| | |
|---|---|
| Outside dia | 9 |
| Bar dia ($d$) | 1 15/16 |
| Free height ($H$) | 16 1/2 |
| Solid height ($h$) | 13 |
| Loaded height ($L$) | 14 5/16 |
| Lateral deflection ($\delta_L$) | 1 1/2 |
| Stop height | 13 3/4 |

From these dimensions compute values at loaded height and stop height.

| | Loaded position | Stop position |
|---|---|---|
| $D$ | 7.0625 | 7.0625 |
| $n$ | 5.51 | 5.51 |
| $y$ (vertical deflection) | 2.1875 | 2.75 |
| $H - d$ | 14.563 | 14.563 |
| $L - d$ | 12.375 | 11.813 |
| $(H - d)/D$ | 2.06 | 2.06 |
| $y/(H - d)$ | 0.150 | 0.189 |
| $A$ | 1.30 | 1.40 |
| $Q$ | 9400 lb | 9310 lb |

From standard formulas for vertical loads only:

| | |
|---|---|
| Solid load | 36,300 lb |
| Load at stop | 28,500 lb |
| Stress $f$ when solid | 141,500 psi |
| Stress $f$ at stop | 111,200 psi |

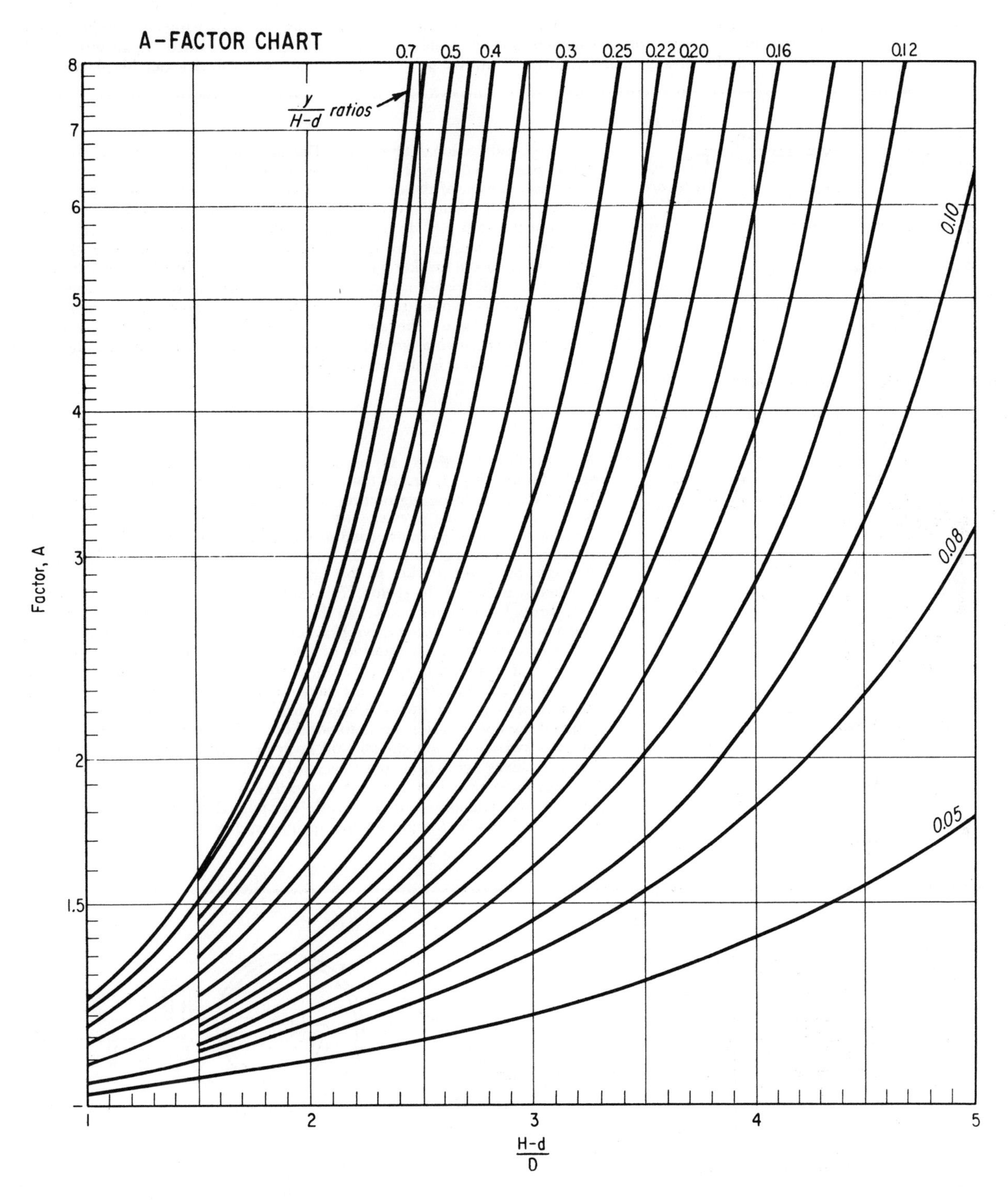


From the combined-stress formula

$$f_c = 111{,}200 \times 1.759$$
$$= 195{,}600 \text{ psi}$$

This stress is so high that settling in service would occur. This particular spring should be redesigned.

Generally, combined stress under the worst condition anticipated should not exceed the solid stress caused by vertical load only. A stop at a reasonable height above solid height is thus desirable—otherwise, spring may have to be modified.

# Overriding Spring Mechanisms

Extensive use is made of overriding spring mechanisms in the design of instruments and controls. Anyone of the arrangements illustrated allows an incoming motion to override the outgoing motion whose limit has been reached. In an instrument, for example, the spring device can be placed between

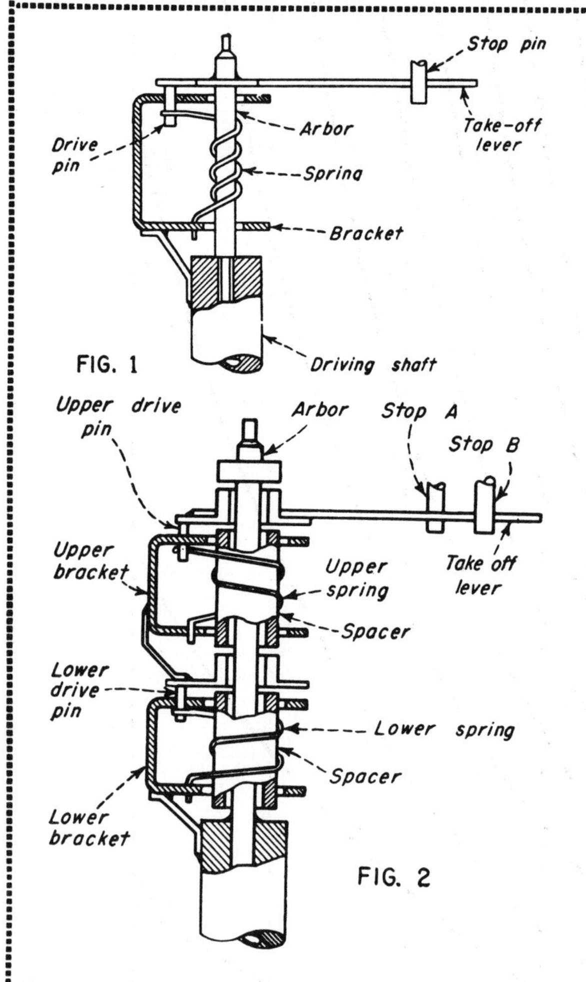


Fig. 1—Unidirectional Override. The take-off lever of this mechanism can rotate nearly 360 deg. It's movement is limited by only one stop pin. In one direction, motion of the driving shaft also is impeded by the stop pin. But in the reverse direction the driving shaft is capable of rotating approximately 270 deg past the stop pin. In operation, as the driving shaft is turned clockwise, motion is transmitted through the bracket to the take-off lever. The spring serves to hold the bracket against the drive pin. When the take-off lever has traveled the desired limit, it strikes the adjustable stop pin. However, the drive pin can continue its rotation by moving the bracket away from the drive pin and winding up the spring. An overriding mechanism is essential in instruments employing powerful driving elements, such as bimetallic elements, to prevent damage in the overrange regions.

Fig. 2—Two-directional Override. This mechanism is similar to that described under Fig. 1, except that two stop pins limit the travel of the take-off lever. Also, the incoming motion can override the outgoing motion in either direction. With this device, only a small part of the total rotation of the driving shaft need be transmitted to the take-off lever and this small part may be anywhere in the range. The motion of the driving shaft is transmitted through the lower bracket to the lower drive pin, which is held against the bracket by means of the spring. In turn, the lower drive pin transfers the motion through the upper bracket to the upper drive pin. A second spring holds this pin against the upper drive bracket. Since the upper drive pin is attached to the take-off lever, any rotation of the drive shaft is transmitted to the lever, provided it is not against either stop *A* or *B*. When the driving shaft turns in a counterclockwise direction, the take-off lever finally strikes against the adjustable stop *A*. The upper bracket then moves away from the upper drive pin and the upper spring starts to wind up. When the driving shaft is rotated in a clockwise direction, the take-off lever hits adjustable stop *B* and the lower bracket moves away from the lower drive pin, winding up the other spring. Although the principal uses for overriding spring arrangements are in the field of instrumentation, it is feasible to apply these devices in the drives of major machines by beefing up the springs and other members.

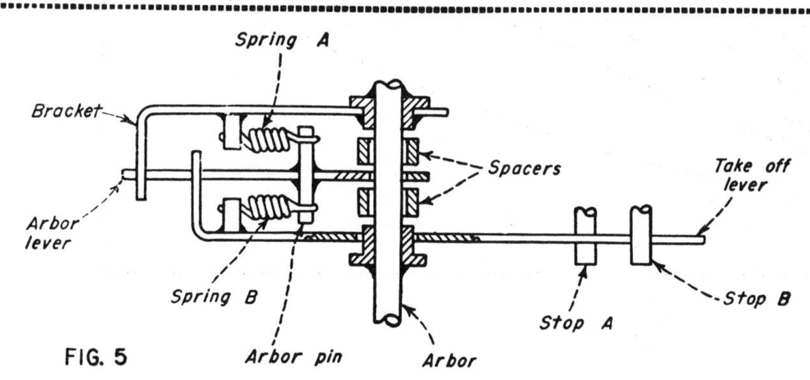


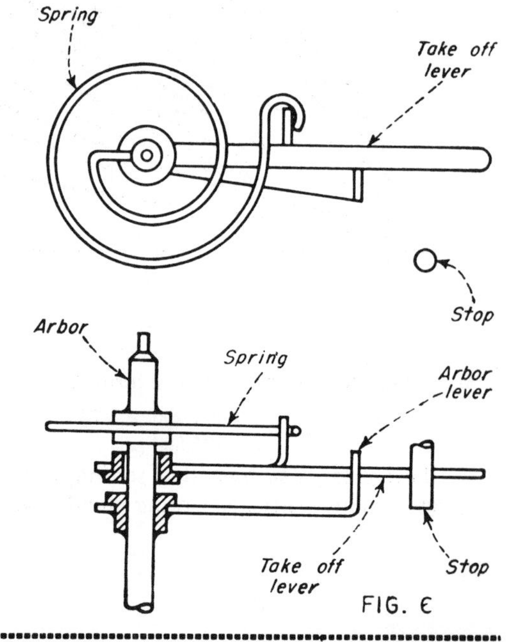


Fig. 5—Two-directional, 90 Degree Override. This double overriding mechanism allows a maximum overtravel of 90 deg in either direction. As the arbor turns, the motion is carried from the bracket to the arbor lever, then to the take-off lever. Both the bracket and the take-off lever are held against the arbor lever by means of springs *A* and *B*. When the arbor is rotated counterclockwise, the take-off lever hits stop *A*. The arbor lever is held stationary in contact with the take-off lever. The bracket, which is soldered to the arbor, rotates away from the arbor lever, putting spring *A* in tension. When the arbor is rotated in a clockwise direction, the take-off lever comes against stop *B* and the bracket picks up the arbor lever, putting spring *B* in tension.

# for Low-Torque Drives

HENRY L. MILO, JR.
Division Engineer,
The Foxboro Company

the sensing and indicating elements to provide over-range protection. The dial pointer is driven positively up to its limit, then stops; while the input shaft is free to continue its travel. Six of the mechanisms described here are for rotary motion of varying amounts. The last is for small linear movements.

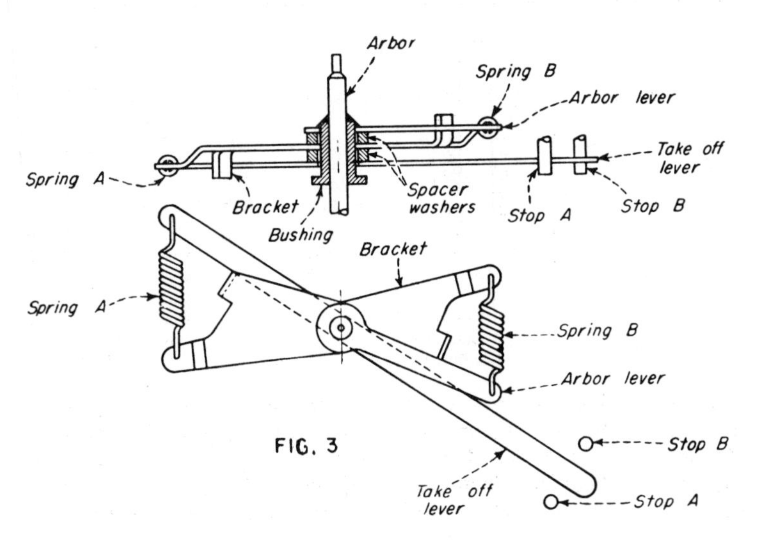


Fig. 3—Two-directional, Limited-Travel Override. This mechanism performs the same function as that shown in Fig. 2, except that the maximum override in either direction is limited to about 40 deg, whereas the unit shown in Fig. 2 is capable of 270 deg movement. This device is suited for uses where most of the incoming motion is to be utilized and only a small amount of travel past the stops in either direction is required. As the arbor is rotated, the motion is transmitted through the arbor lever to the bracket. The arbor lever and the bracket are held in contact by means of spring *B*. The motion of the bracket is then transmitted to the take-off lever in a similar manner, with spring *A* holding the take-off lever and the bracket together. Thus the rotation of the arbor is imparted to the take-off lever until the lever engages either stops *A* or *B*. When the arbor is rotated in a counterclockwise direction, the take-off lever eventually comes up against the stop *B*. If the arbor lever continues to drive the bracket, spring *A* will be put in tension.

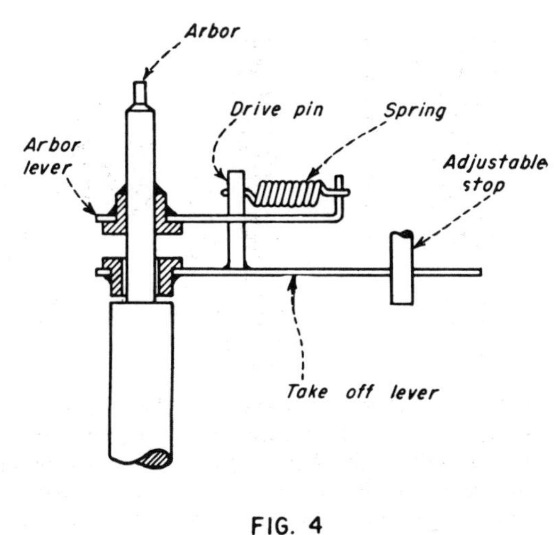


Fig. 4—Unidirectional, 90 Degree Override. This is a single overriding unit, that allows a maximum travel of 90 deg past its stop. The unit as shown is arranged for over-travel in a clockwise direction, but it can also be made for a counterclockwise override. The arbor lever, which is secured to the arbor, transmits the rotation of the arbor to the take-off lever. The spring holds the drive pin against the arbor lever until the take-off lever hits the adjustable stop. Then, if the arbor lever continues to rotate, the spring will be placed in tension. In the counterclockwise direction, the drive pin is in direct contact with the arbor lever so that no overriding is possible.

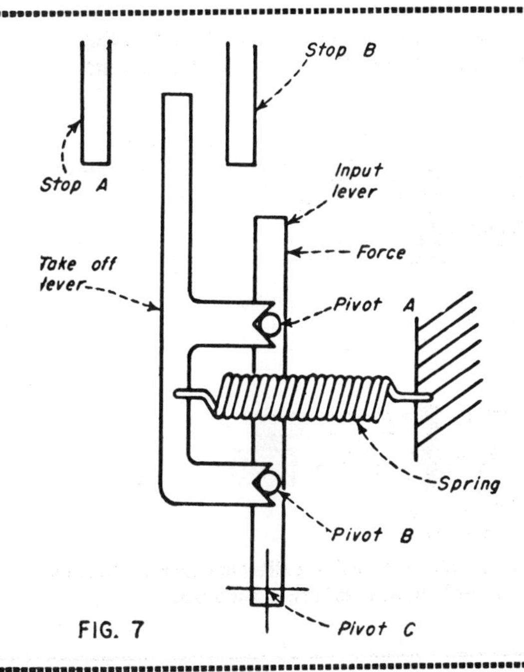


Fig. 6—Unidirectional, 90 Degree Override. This mechanism operates exactly the same as that shown in Fig. 4. However, it is equipped with a flat spiral spring in place of the helical coil spring used in the previous version. The advantage of the flat spiral spring is that it allows for a greater override and minimizes the space required. The spring holds the take-off lever in contact with the arbor lever. When the take-off lever comes in contact with the stop, the arbor lever can continue to rotate and the arbor winds up the spring.

Fig. 7—Two-directional Override, Linear Motion. The previous mechanisms were overrides for rotary motion. The device in Fig. 7 is primarily a double override for small linear travel although it could be used on rotary motion. When a force is applied to the input lever, which pivots about point *C*, the motion is transmitted directly to the take-off lever through the two pivot posts *A* and *B*. The take-off lever is held against these posts by means of the spring. When the travel is such the take-off lever hits the adjustable stop *A*, the take-off lever revolves about pivot post *A*, pulling away from pivot post *B* and putting additional tension in the spring. When the force is diminished, the input lever moves in the opposite direction, until the take-off lever contacts the stop *B*. This causes the take-off lever to rotate about pivot post *B*, and pivot post *A* is moved away from the take-off lever.

# 17 ways of testing springs

C. J. McCLINTOCK, chief inspector, Hunter Spring Co.

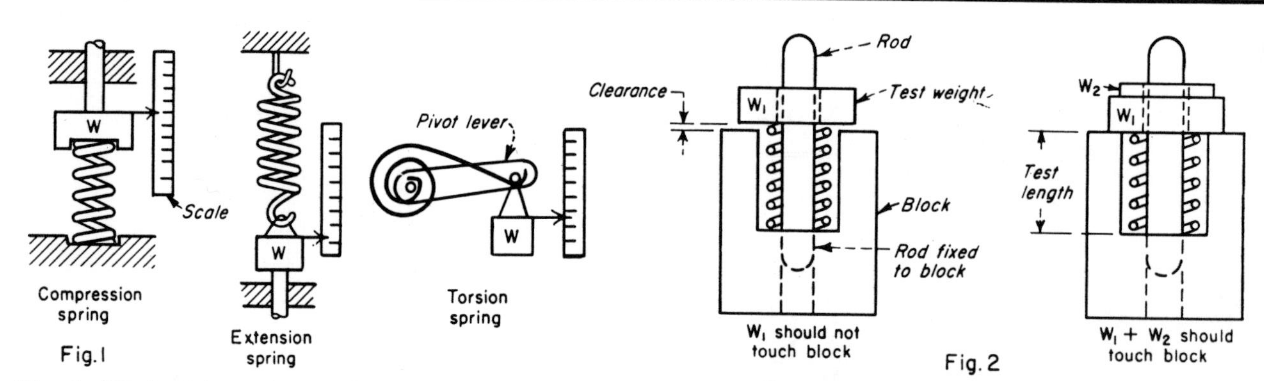


**Fig. 1—Dead-weight testing. Weights are directly applied to spring. In the compression spring and the extension spring testers, the test weights are guided in the fixture to prevent buckling. Instead of using a linear scale, the spring deflection can be measured with a dial indicator.**

**Fig. 2—Ordnance gage incorporates "Go-no-go" principle. Block is bored for specified test length L. Weight $W_1$ is slightly less than the minimum specified load at L and therefore should not touch block; $W_1$ plus load tolerance $W_2$ must touch block for the spring to be acceptable.**

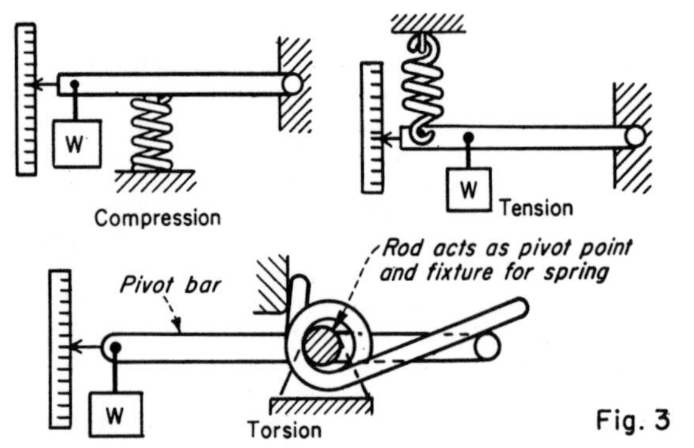


**Fig. 3—Pilot-beam testing. Fractional resistance offered to movement of parts is low. These testers are more sensitive than those in which the weight is guided in the fixture. Many of the commercial testers are based on this principle.**

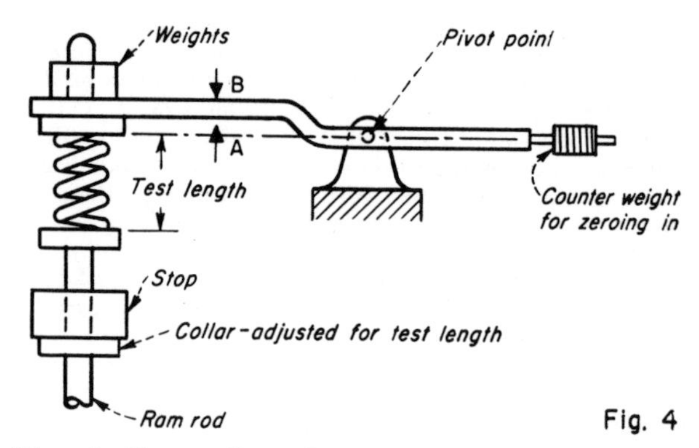


**Fig. 4—Zero-gradient beam. Uses refined pivot-beam principle. Ram rod is pushed up with pedal or air cylinder. Beam must not touch contacts A or B. Contacting A indicates spring too weak; B indicates spring too strong.**

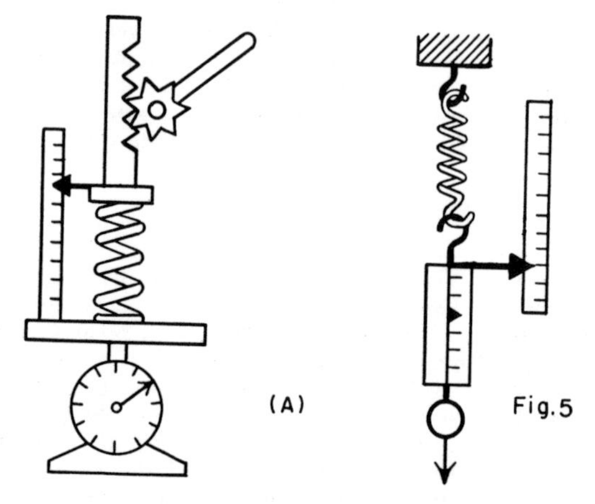


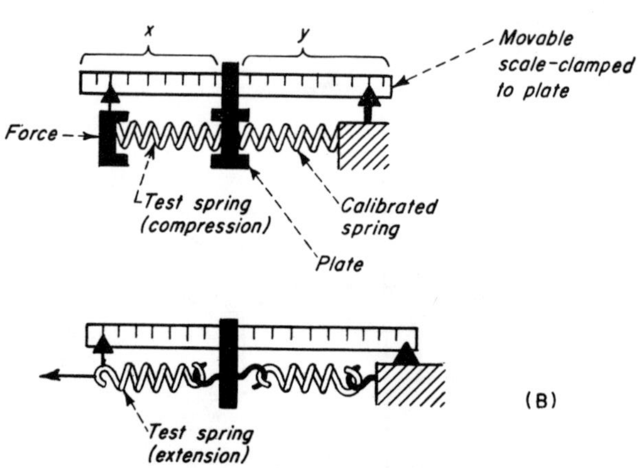


**Fig. 5—Spring against spring. (A) spring scales used in place of dead weights for testing short-run springs. (B) Similar results obtained by using calibrated springs. Section x calibrated for deflection readings; y for load.**

Spring dimensions are based on calculations using empirical-theoretical equations. In addition, allowances are made for material and manufacturing tolerances. Thus, the final product may deviate to an important degree from the original design criteria. By testing the springs: (1) Results can be entered on the spring drawing, thus including actual performance data; this leads to more realistic future designs. (2) Performance can be checked before assembling spring in a costly unit.

Shown below are 12 ways, Fig. 1 to 5, to quickly evaluate load-deflection characteristics; for more accurate or fully automatic testing, Figs. 6 and 7, describe 5 types of commercial testers.

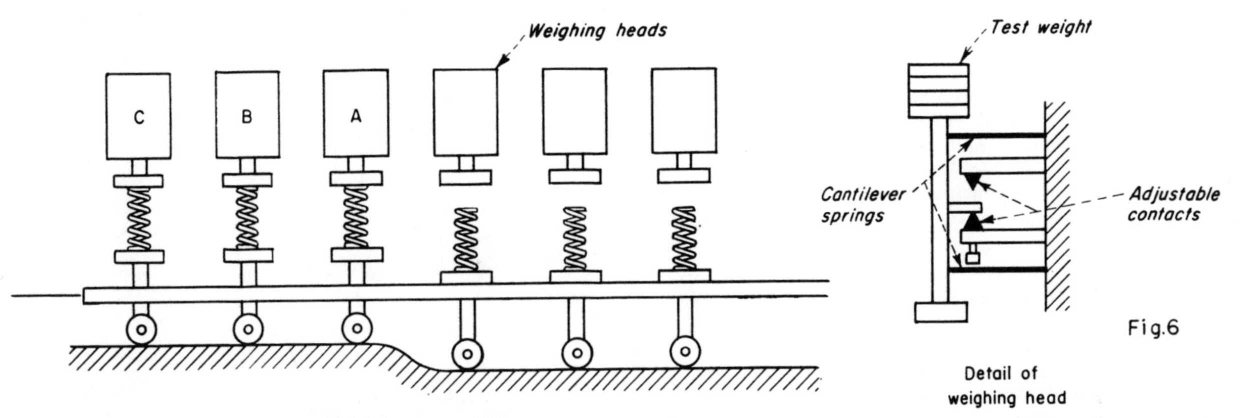


**Fig. 6—Fully-automatic testing. Continually moving rotary table with three testing positions. Springs are loaded manually but tested and ejected automatically. Weak springs that allow lower point to make contact are ejected at position A; springs too strong ejected at B. All springs reaching point C are ejected as acceptable.**

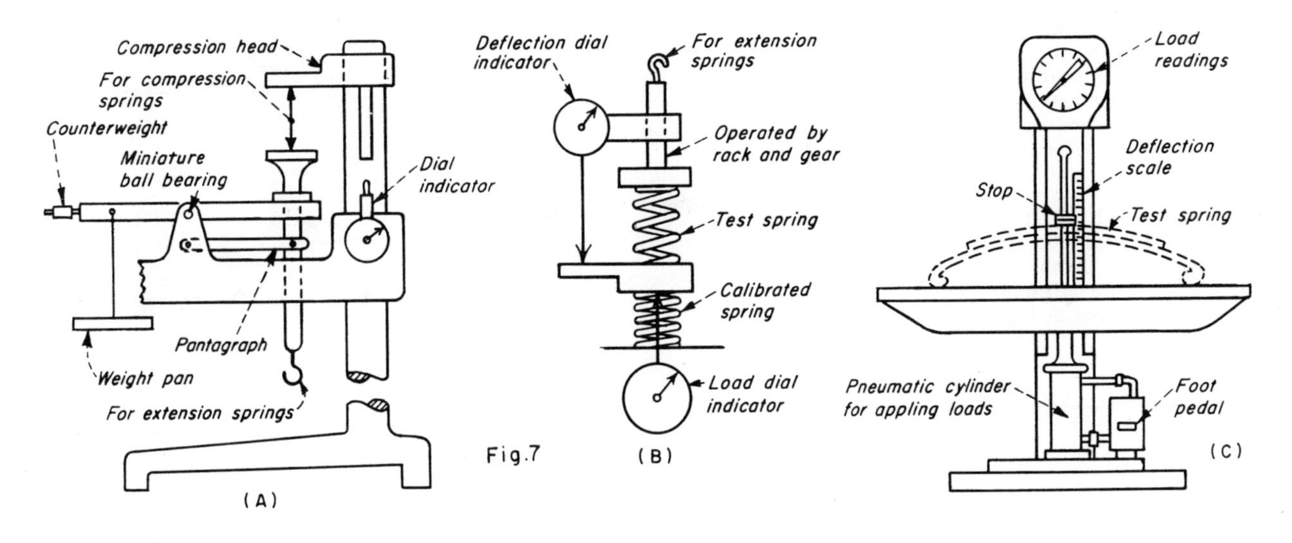


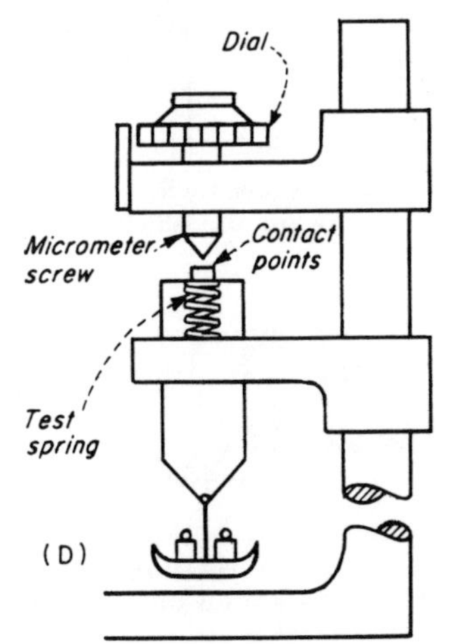


**Fig. 7—Commercial testers: (A) Balanced-beam tester uses dead weights for loads. Pantagraph linkage keeps weighing head vertical irrespective of beam movement. Load capacity: 10 lb. Baldwin-Lima-Hamilton Corp. (B) Calibrated spring tester available in several models for testing loads up to 1000 lb. Load applied manually through gear and rack; motor-driven units can be attained for applying heavier loads. Link Engineering Co. (C) Pneumatic-operated tester uses torque bar system for applying loads and a differential transformer for accurately measuring displacements. Wide table permits tests on leaf springs. Load capacity: 2000 lb. Tinius Olsen Testing Machine Co. (D) Electronic micrometer tester has sufficient sensitivity (0.0001 in.) to measure drift, hysteresis and creep as well as load deflection. Adjustments made by large micrometer dial; contact indicated by sensitive electronic circuit. Load capacity: 50 lb. J W Dice Co.**

# *Unusual Uses for* HELICAL WIRE SPRINGS

**SPRING BELTING (left).** For low power transmission at high speeds. Allows a certain amount of variation in the center distance and absorbs inertia forces. Spring ends can be joined smoothly by using a smaller internal spring as shown on the following page.

**ELECTRICAL FITTING (right).** An inexpensive lamp or fuse socket which insures proper contact even when subject to moderate vibration. Small threaded parts also can be joined in this same way.

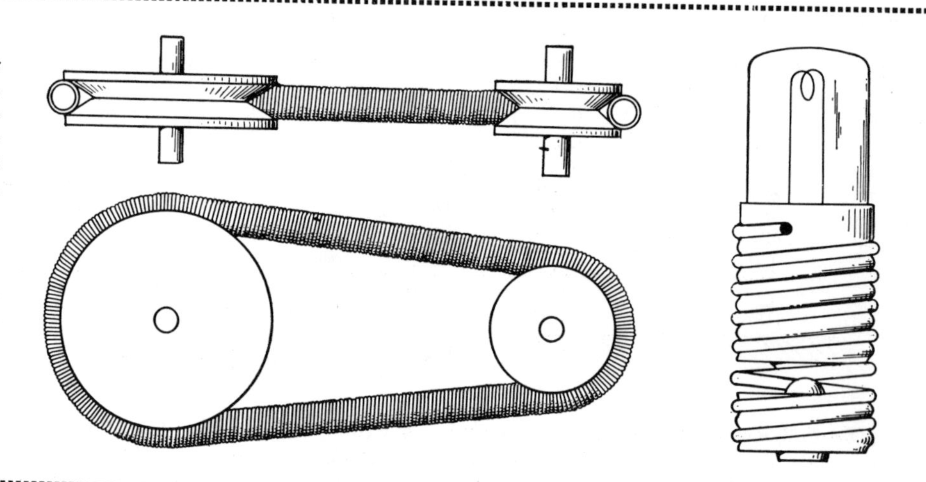

**SCREW THREAD INSERTS (left).** Wire with diamond cross section. For tapped holes in light alloys and plastics. Are made of stainless steel for corrosion-free threads. Can be used to renew worn-out tapped threads.

**ROTATING TYPE OIL SEAL (right).** Uses helical wire spring to exert a radial force on the packing. Friction is kept to a minimum and efficiency is high even at high shaft speeds.

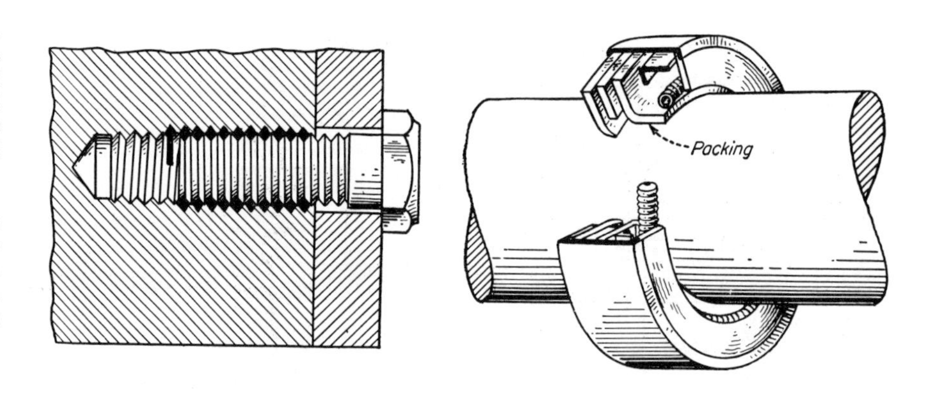


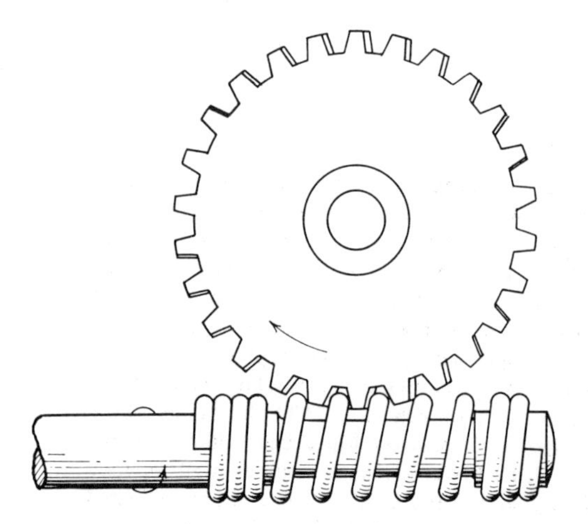

**WORM GEARING.** Used on low power transmissions. Allows a certain amount of misalignment between worm and wheel. Wheels are best made from laminated plastics.

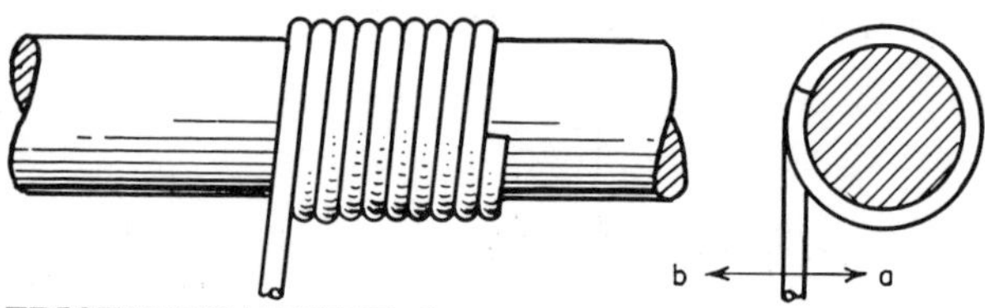


**FRICTION RATCHET.** Spring rotates shaft when pulled in the *a* direction, but turns freely on the shaft when pulled in the opposite or *b* direction.

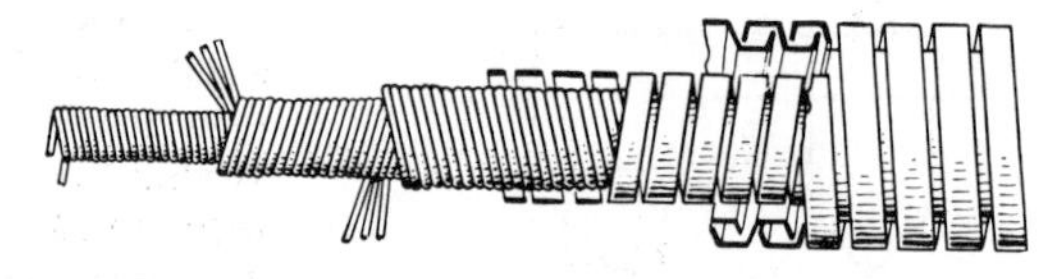

**FLEXIBLE SHAFT.** Inner springs serves as shaft, outer one as a casing. For single direction of rotation unless shaft consists of two or more springs wound in opposite directions.

A selection of practical applications that are characterized by the functions served in each case by the helical wire spring. The spring rate property is put to use in most cases, but not in the axial loading sense that represents the more common applications for which these types of springs are employed in industrial products.

HAIM MURRO
Bayonne, N. J.

**SENSITIVE STICK (left).** Round conductor bar is mounted within a spring fastened to insulators to serve as an electrical switch. Deflecting the spring laterally completes the circuit. Can operate relays or alarm and can be made with intermediate insulators where considerable length is required.

**THREAD MEASURING GAGE (right).** Dimension *d* allows calculating the effective dia of thread. Pressing the loops releases the bolt to be unscrewed. Can be used on fluted parts like thread taps.

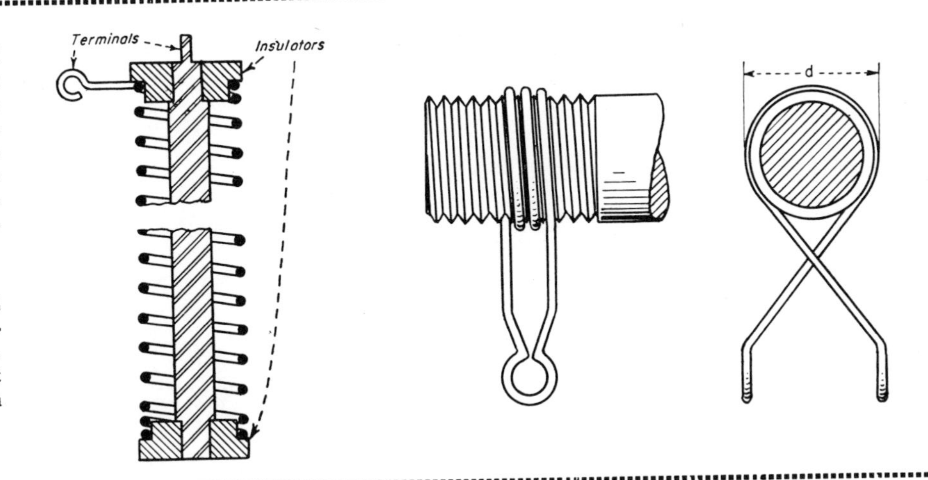


**FLEXIBLE CHUTE (left).** For feeding small articles from hoppers to automatic machines. Spring can be wound in different shapes as required by the articles being handled.

**TUBING REINFORCEMENT (right).** Gives plastic or rubber tubing added rigidity as well as protection against mechanical damage. Can be cast inside rubber as shown in lower sketch.

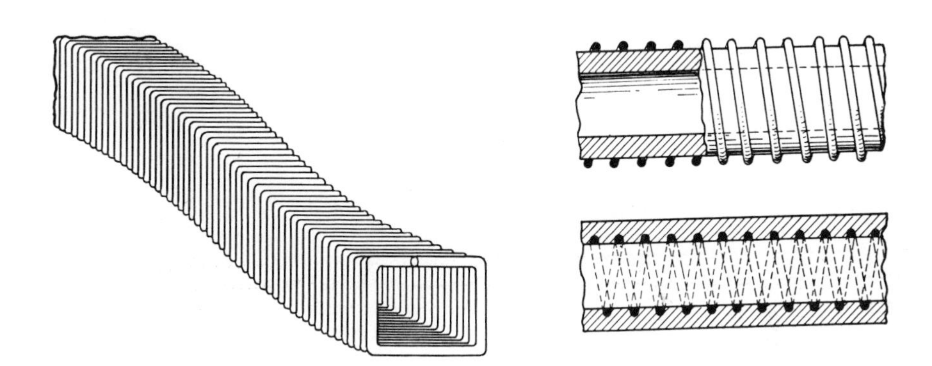

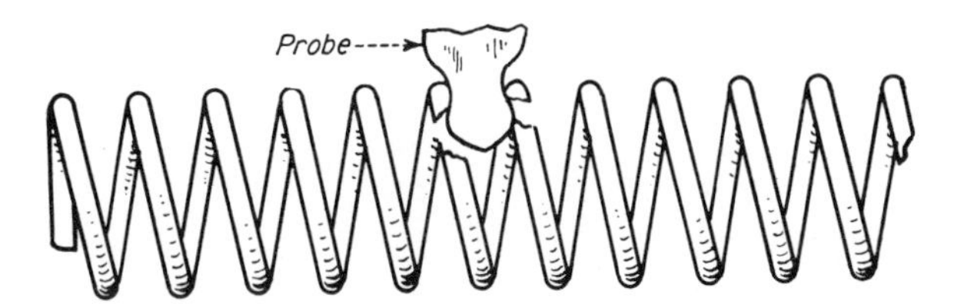


**ELECTRICAL CONNECTION** for small, light products like hearing aids uses a special probe that is easily inserted between coils of spring which is a conducting material.

**SMALL SPRING** connects ends of larger spring with a thread-like action. Useful where external projection cannot be tolerated, like the spring-belting on opposite page.

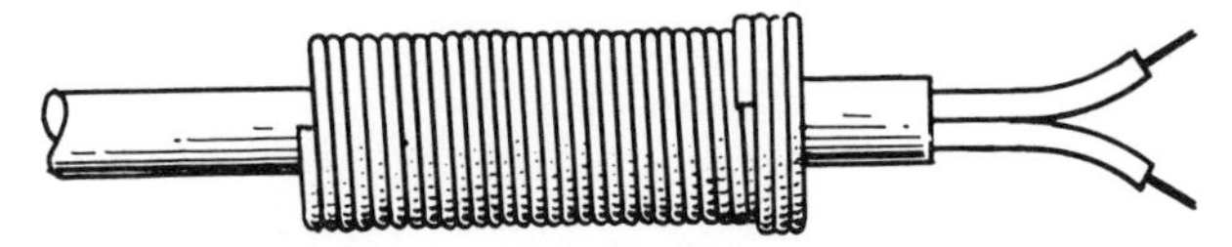

**SHIELD FOR ELECTRICAL WIRE AND CABLE.** Provides wear resistant covering for wires and protection against physical damage.

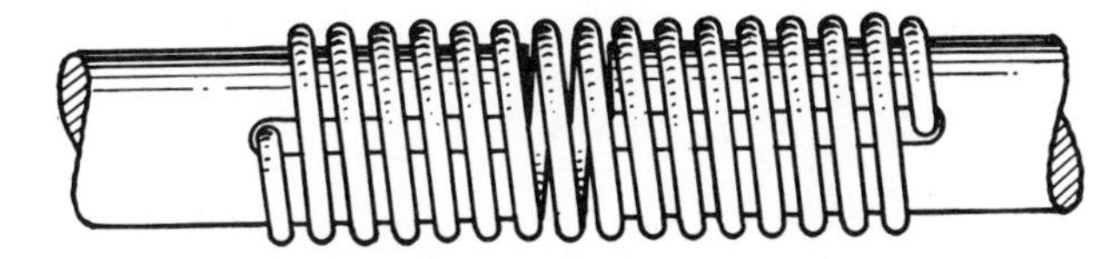

**SMALL DIAMETER SHAFT COUPLINGS.** Allows for some misalignment and can be used with shafts or unequal diameters. For single direction of rotation only.

# 14

# WELDING AND BRAZING

Methods for Placing Brazing Materials, and Vent Locations 332
Built-up Welded Constructions 334
Preparation of Materials for Resistance Welding 338

# METHODS FOR PLACING BRAZING

**Location and forms of copper used in brazing. Suggested positions for vents to prevent deformation of the final product. Two examples of joining dissimilar materials together by brazing. Sketches courtesy the General Electric Company.**

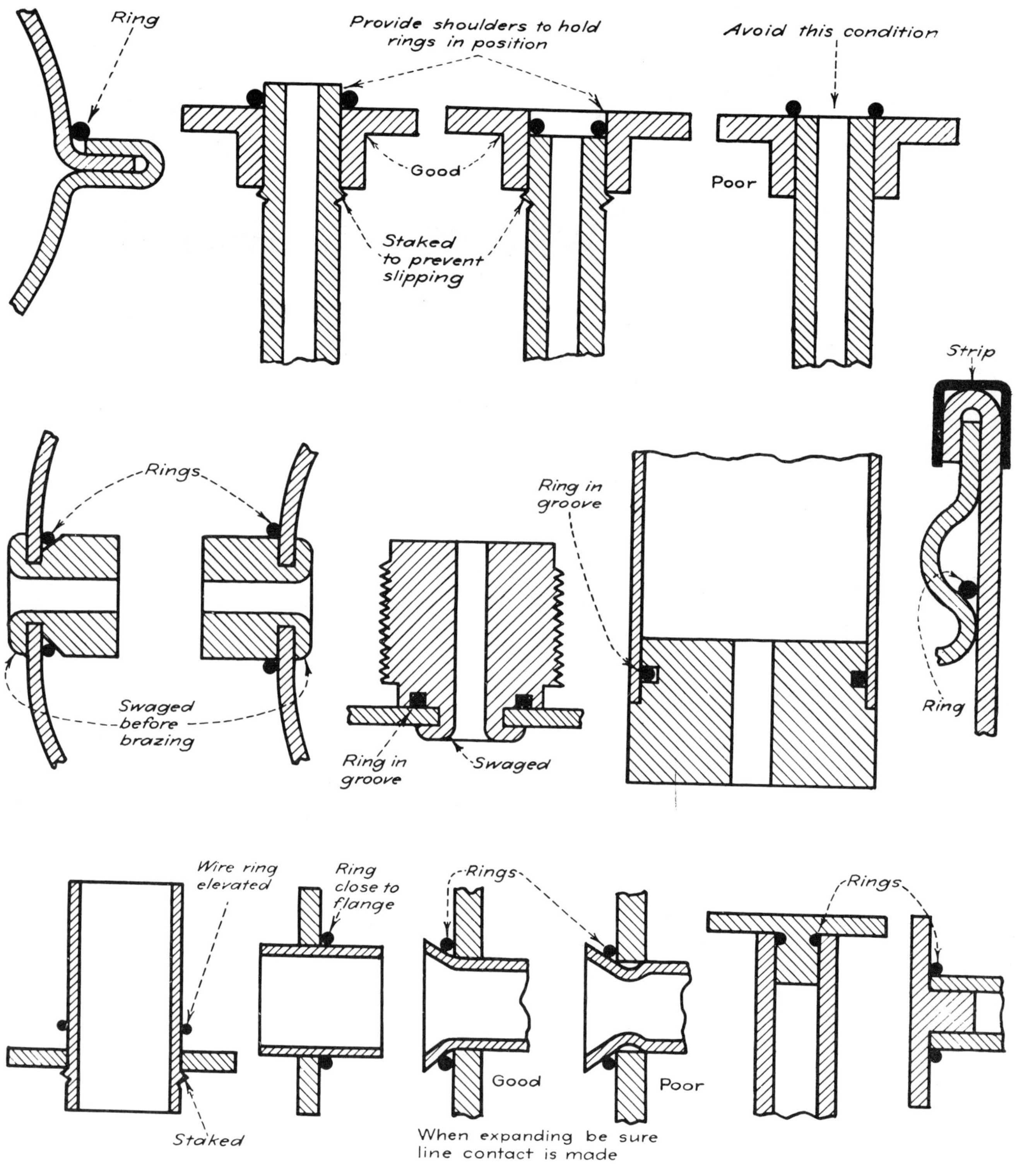


Placement of Copper Rings

# MATERIALS, AND VENT LOCATIONS —

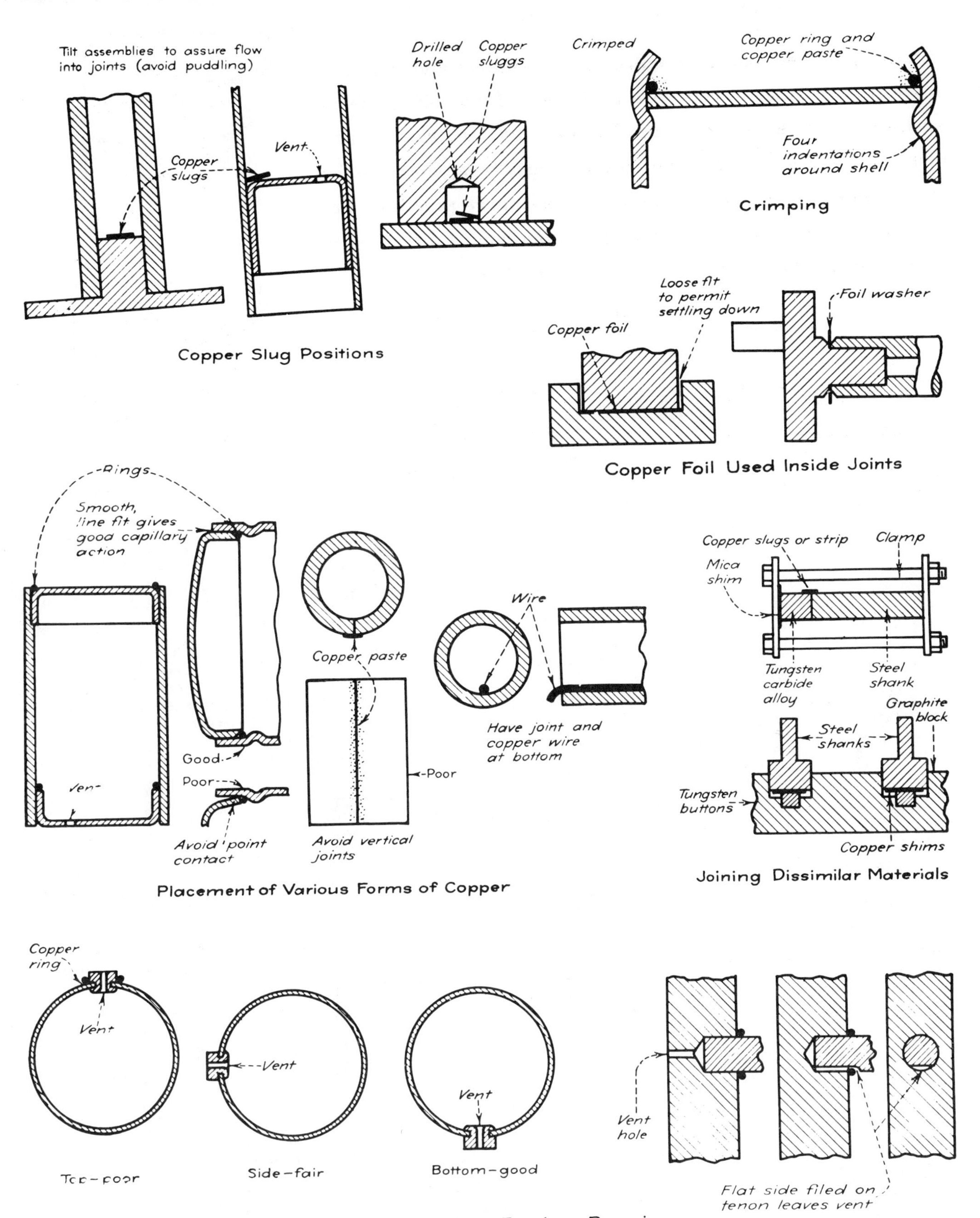

# BUILT-UP WELDED CONSTRUCTIONS

**Details illustrated, taken from designs for sanitary and chemical processing equipment, represent utilization of diversified metal-working equipment—bending rolls, power presses, flangers and such—to fabricate functionally correct parts from simple sheet and fittings. Commercial shapes are used where practical; but parts are flame-cut, forged or rolled when such fabrication is more economical or design requirements dictate**

O. W. FISHER
*Chief Engineer, L. O. Koven & Brother, Incorporated*

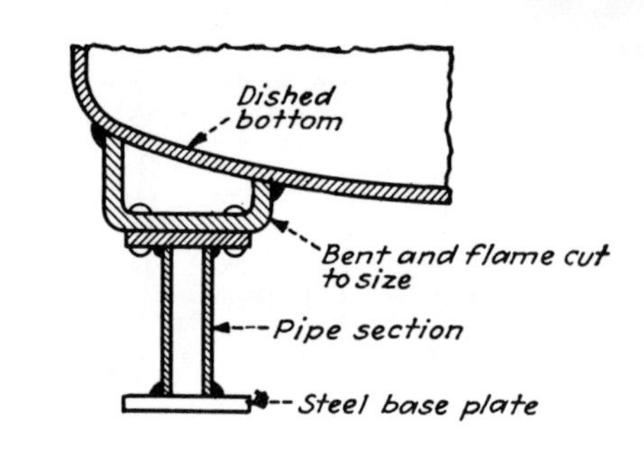


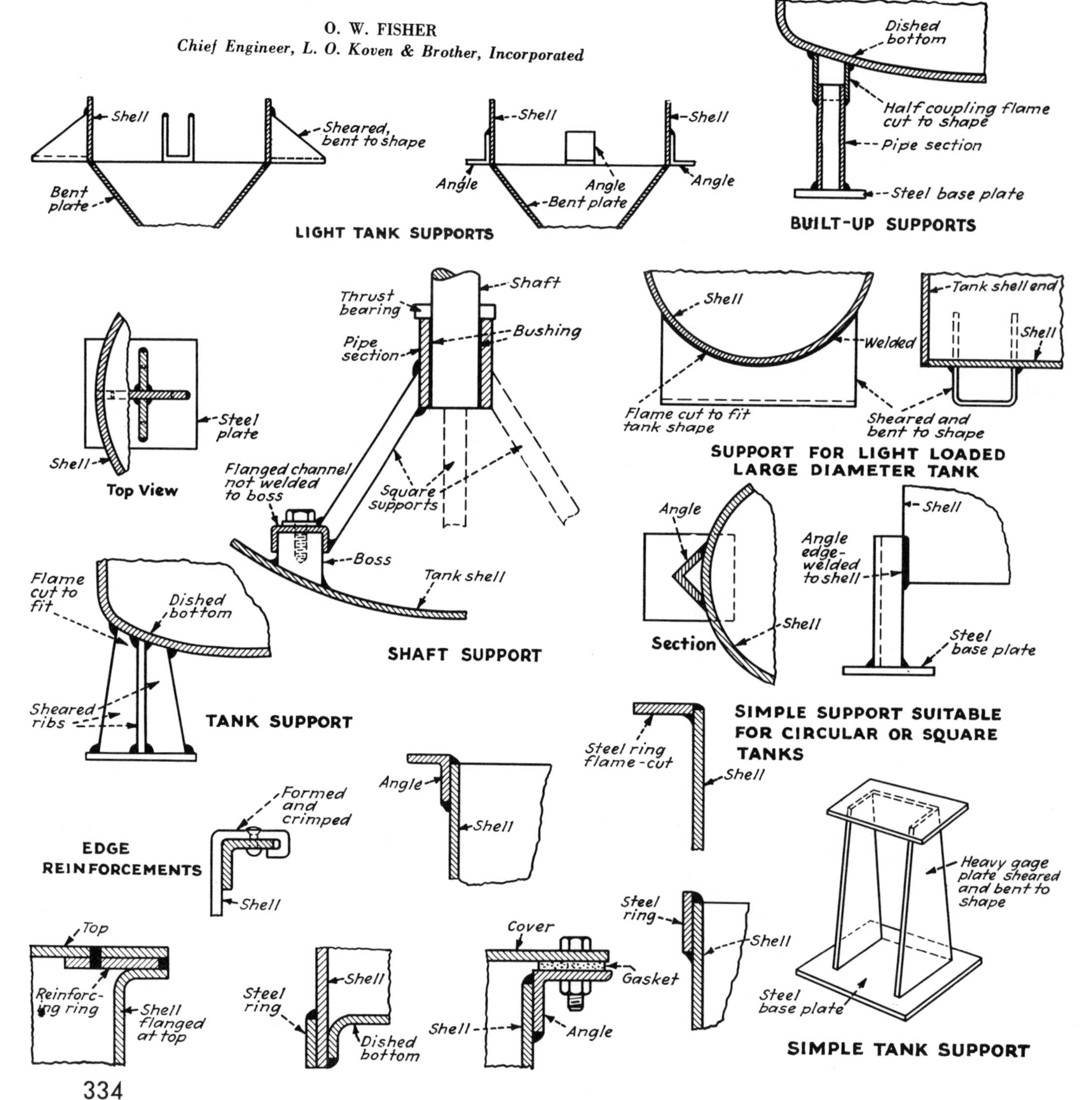

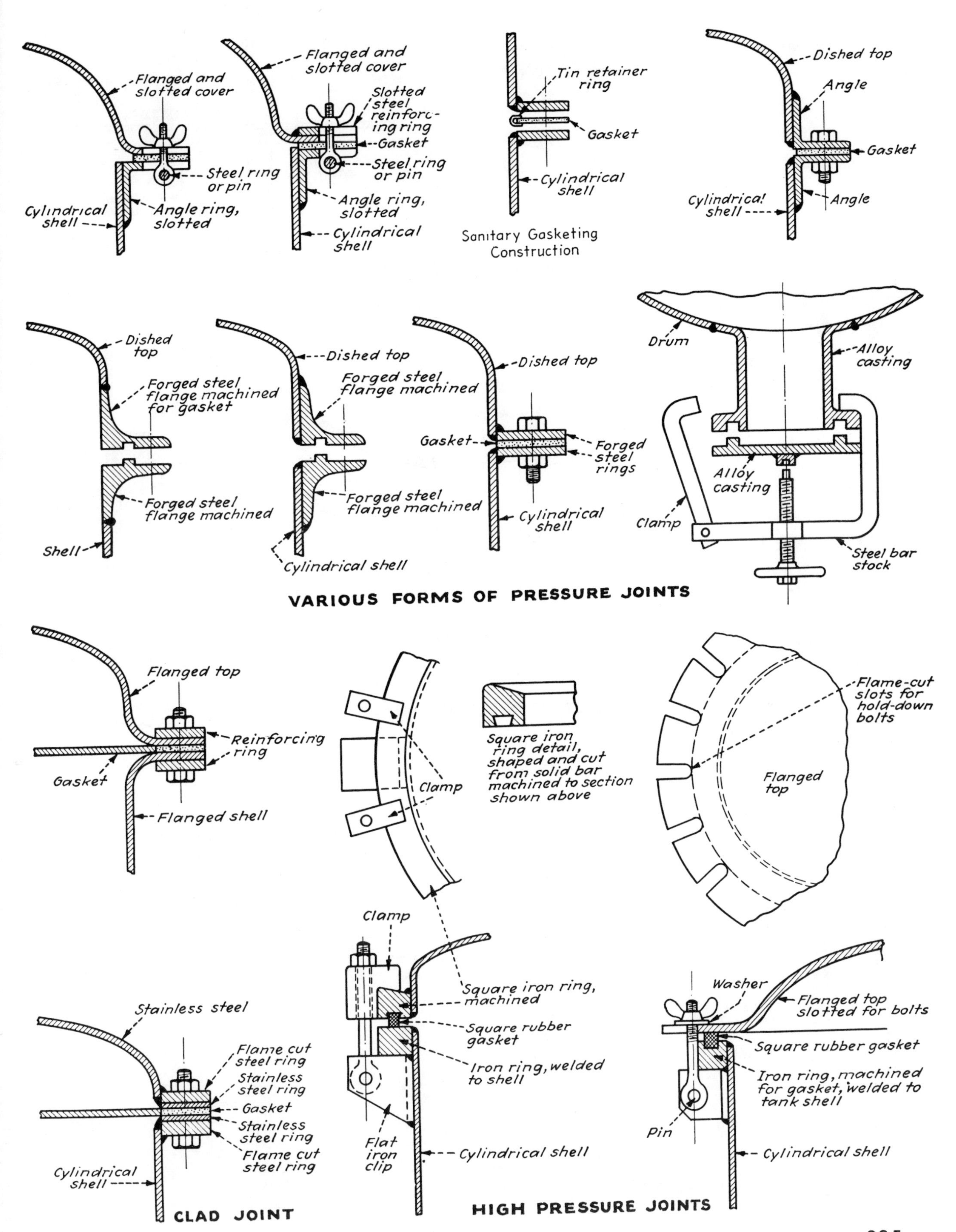
Flanged and slotted cover
Steel ring or pin
Cylindrical shell
Angle ring, slotted
Flanged and slotted cover
Slotted steel reinforcing ring
Gasket
Steel ring or pin
Angle ring, slotted
Cylindrical shell
Tin retainer ring
Gasket
Cylindrical shell
Sanitary Gasketing Construction
Dished top
Angle
Gasket
Cylindrical shell
Angle
Dished top
Forged steel flange machined for gasket
Forged steel flange machined
Shell
Dished top
Forged steel flange machined
Forged steel flange machined
Cylindrical shell
Dished top
Gasket
Forged steel rings
Cylindrical shell
Drum
Alloy casting
Alloy casting
Clamp
Steel bar stock
VARIOUS FORMS OF PRESSURE JOINTS
Flanged top
Reinforcing ring
Gasket
Flanged shell
Clamp
Square iron ring detail, shaped and cut from solid bar machined to section shown above
Flame-cut slots for hold-down bolts
Flanged top
Stainless steel
Flame cut steel ring
Stainless steel ring
Gasket
Stainless steel ring
Flame cut steel ring
Cylindrical shell
CLAD JOINT
Clamp
Square iron ring, machined
Square rubber gasket
Iron ring, welded to shell
Flat iron clip
Cylindrical shell
Washer
Flanged top slotted for bolts
Square rubber gasket
Iron ring, machined for gasket, welded to tank shell
Pin
Cylindrical shell
HIGH PRESSURE JOINTS

# BUILT-UP WELDED CONSTRUCTIONS

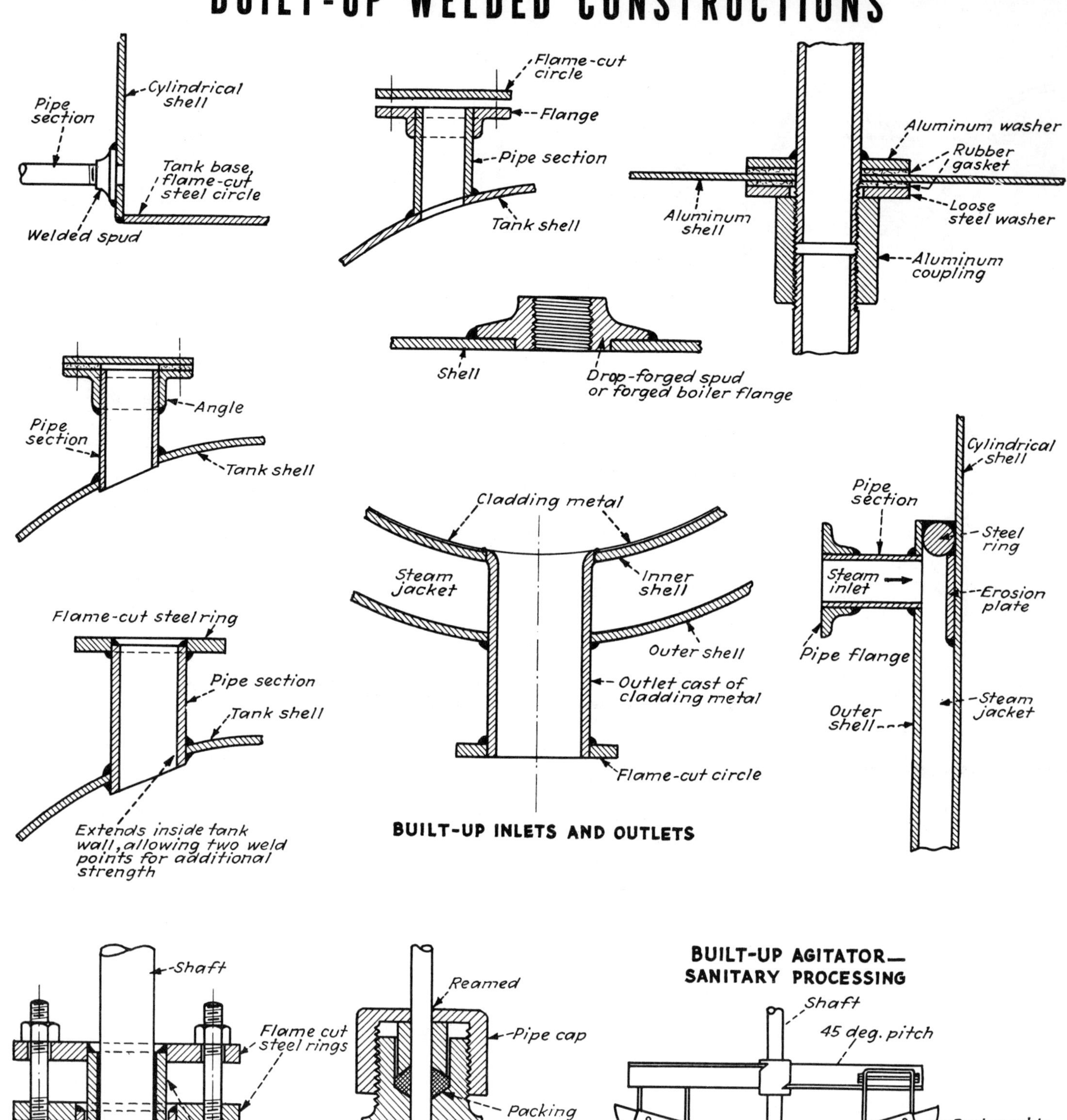


BUILT-UP INLETS AND OUTLETS

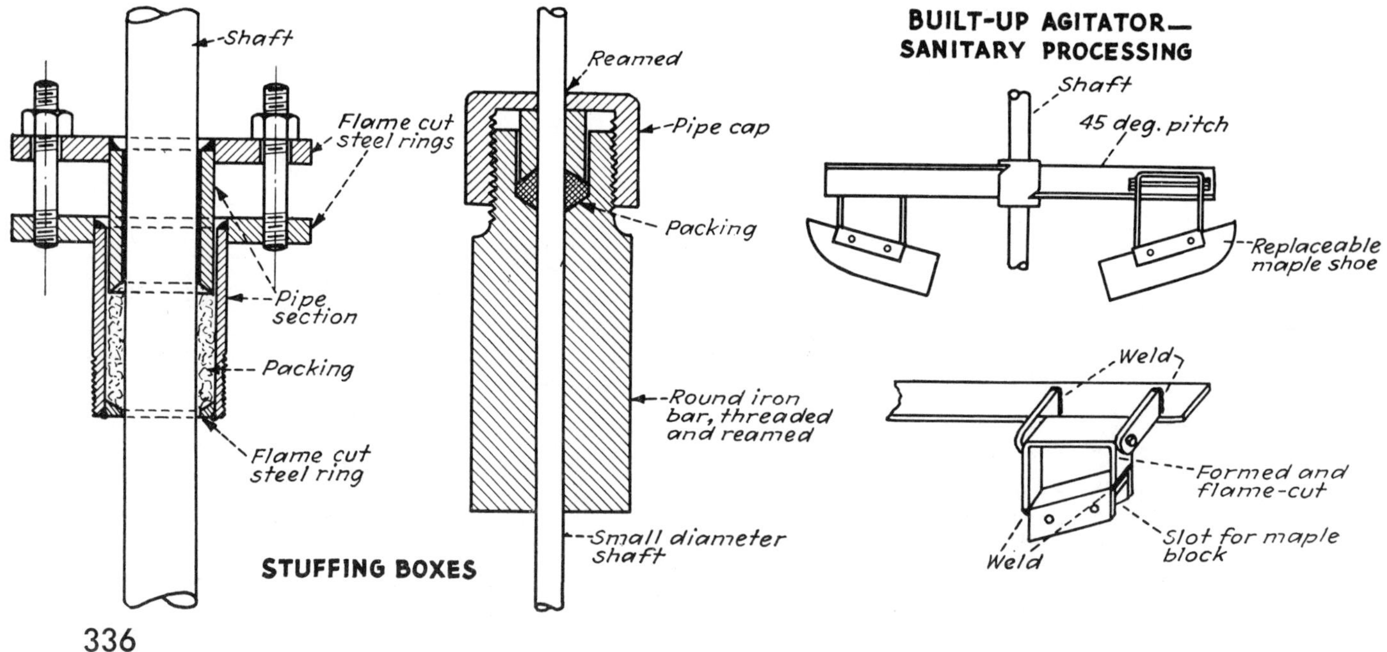


STUFFING BOXES

BUILT-UP AGITATOR—SANITARY PROCESSING

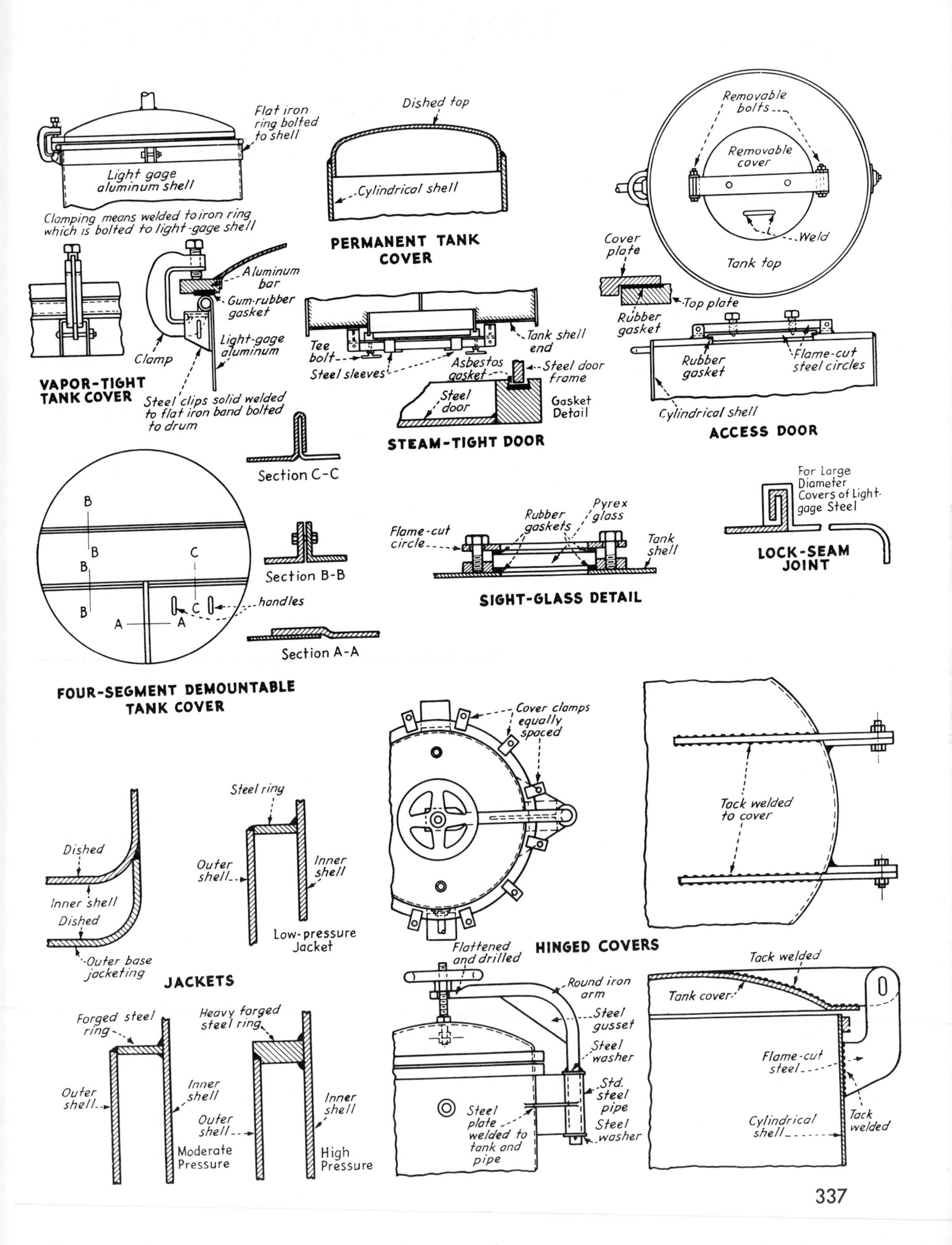
Flat iron ring bolted to shell
Light gage aluminum shell
Clamping means welded to iron ring which is bolted to light-gage shell
Aluminum bar
Gum-rubber gasket
Light-gage aluminum
Clamp
VAPOR-TIGHT TANK COVER
Steel clips solid welded to flat iron band bolted to drum
Dished top
Cylindrical shell
PERMANENT TANK COVER
Tee bolt
Steel sleeves
Asbestos gasket
Steel door frame
Tank shell end
Steel door
Gasket Detail
STEAM-TIGHT DOOR
Removable bolts
Removable cover
Weld
Tank top
Cover plate
Top plate
Rubber gasket
Rubber gasket
Flame-cut steel circles
Cylindrical shell
ACCESS DOOR
Section C-C
Section B-B
Section A-A
handles
FOUR-SEGMENT DEMOUNTABLE TANK COVER
Flame-cut circle
Rubber gaskets
Pyrex glass
Tank shell
SIGHT-GLASS DETAIL
For Large Diameter Covers of Light-gage Steel
LOCK-SEAM JOINT
Cover clamps equally spaced
Tack welded to cover
Dished
Inner shell
Dished
Outer base jacketing
Steel ring
Outer shell
Inner shell
Low-pressure Jacket
JACKETS
Forged steel ring
Outer shell
Inner shell
Outer shell
Moderate Pressure
Heavy forged steel ring
Inner shell
High Pressure
Flattened and drilled
HINGED COVERS
Round iron arm
Steel gusset
Steel washer
Std. steel pipe
Steel washer
Steel plate welded to tank and pipe
Tack welded
Tank cover
Flame-cut steel
Cylindrical shell
Tack welded

# PREPARATION OF RESISTANCE

J. M. COOPER

*Welding Engineer, General Electric Company*

FIG.1 FIG.2

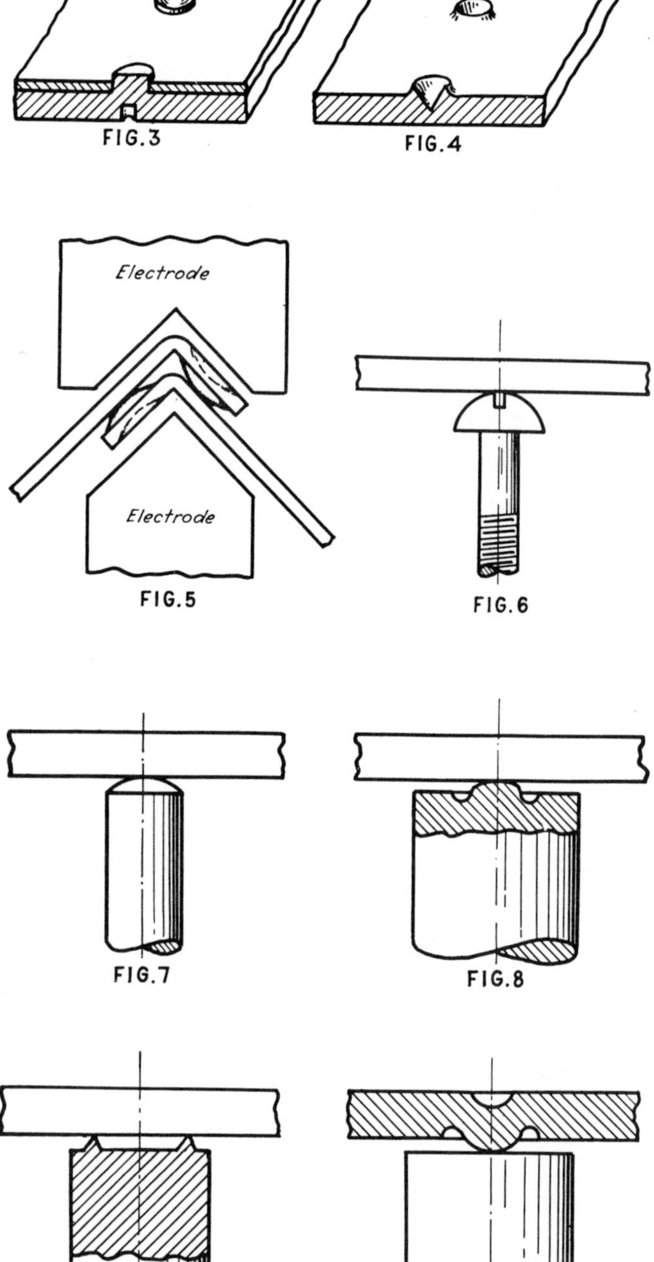


FIG.3 FIG.4 FIG.5 FIG.6 FIG.7 FIG.8 FIG.9 FIG.10

Fig. 1—Round, embossed projections make practical the use of flat electrodes. Several projection welds can be made simultaneously under the same large flat electrode. The larger area of electrode in contact with the work reduces current and pressure density on electrode, resulting in long electrode life.

Fig. 2—Typical pointed or cone-shaped projection, a form often used on light-gage materials such as 22 and 24 gage.

Fig. 3—Protruding punch-out may help to locate parts preparatory to welding. Where great strength is not required, the punch-out itself is sometimes electroforged down.

Fig. 4—Prick punch marks made with round punch (one blow) used in welding thick plates to light-gage sheets to throw up a crater which localizes welding heat and pressure.

Fig. 5—Design of embossed corners, formed over one another and welded between V-shaped electrodes. Self-aligning.

Fig. 6—Almost any form of screw, rivet or specially headed part can be projection welded.

Fig. 7—Slight radius on the end of the rod permits it to be welded to another part without throwing a fin or flash.

Fig. 8—The crater or ring-like cavity is filled with the heated metal of the round projection, resulting in close mechanical contact over the whole surface.

Fig. 9—A ring projection can be turned or coined on the shaft in order to localize the welding area.

Fig. 10—Construction similar to Fig. 8, except reversed, a crater projection combination being coined in the flat sheet.

Fig. 11—Projection swaged on the edges of a piece, a method of embossing thick plates or strap stock.

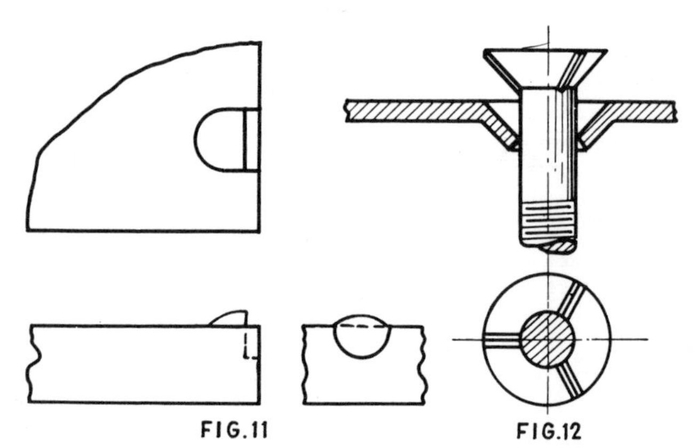

FIG.11 FIG.12

# MATERIALS FOR WELDING

Fig. 12—How specially headed screws or studs are prepared both to localize the weld and also to locate the screws without the necessity of using jigs or fixtures.

Fig. 13—Elongated projections that cross each other are recommended for the lightest gages and certain non-ferrous combinations, as well as for places where a good strong weld is imperative.

Fig. 14—Where possible, elongated projections or a pattern of ribs which cross one another, giving many points of small welded area, should be used for thin sheet metal.

Fig. 15—When projection welding to a curved surface, an elongated projection should be used. This is employed to assure ample contact surface in the direction in which movement is likely to occur.

Fig. 16—Upsetting a tube to form bulges can be done by heating and upsetting on a butt welder.

Fig. 17—Rods of almost any metal can be upset to provide increased sections or limiting rings.

Fig. 18—A method that helps to locate a spot-welded lap joint and also contributes to having one side of the object smooth.

Fig. 19—A method of welding both sides of a box form simultaneously, by employing a shunt block device.

Fig. 20—An example of "pry-bar" welding on a special machine. Done by prying against parts backed up by dies.

Fig. 21—Switch contacts of many different materials can be welded to their mounting members. This design shows a coined contact having three conical projections that nest in a ring groove stamped in the blade.

Fig. 22—When wire cables are "cut off" by clamping between the jaws of a butt welder and burning cable in two, a globule of metal, formed on the ends, holds the wires together.

Fig. 23—Pipe or heavy-walled tubes can be butted together or to other pieces. By chamfering pipe ends, the internal flash will be kept at the minimum.

Fig. 24—Rods or cables can be economically welded or swaged into sleeves or ferrules.

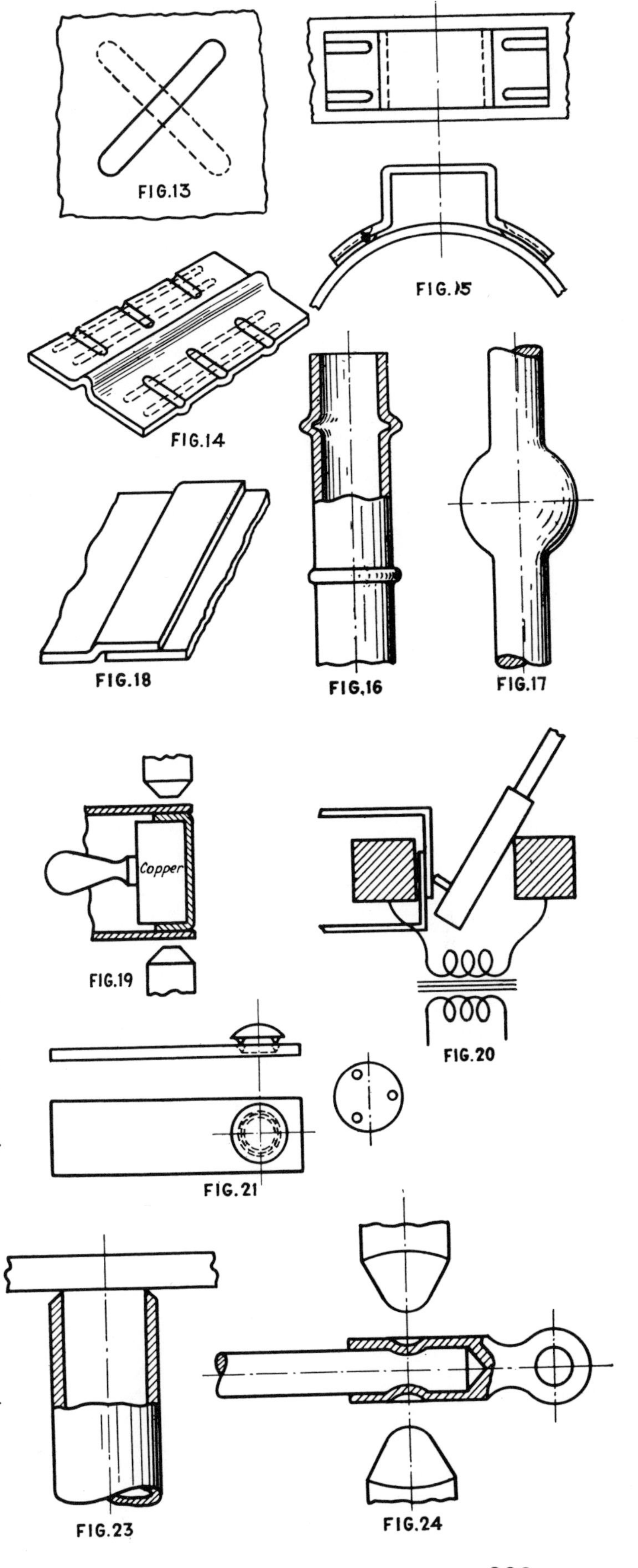

# INDEX

Adhesives, 28-33
  high-speed applicators for, 2, 3
  shopping list for, 33
  stresses in joints made with, 34-37
Adjustable parts, friction-held, 10, 11
Amplifiers, electronic circuits for, 218, 219
Applicators, high-speed, for adhesives, 2, 3
Automatic assembly, check-chart for, 150, 151
Automatic stop mechanisms, 130-135

Backlash in threaded parts, providing for, 292, 293
Ball bearings (*see* Bearings)
Ball slides, 112, 113
Bead chains, 4, 5
Bearings, miniature, applications of, 114, 115
  mounting units, 116, 117
  porous bronze, 110, 111
  ways to lubricate, ball, 108, 109
    small, 110, 111
Bellows, metal, 6, 7
Belt drives, 194-197
Bimetallic devices for temperature control, 142, 143
Brake, reel, magnetic, 235
Brazing, of dissimilar materials, 333
  methods of placing materials, 332, 333
Built-up welded constructions, 334-337

Calculations, charts and tables (*see* Formulas)
Cam, fast follower for, 285
Chains, bead, uses of, 4, 5
  reducing pulsations in, 198, 199
  roller, lubrication of, 300, 301
Clad steel, designing with, 158, 159
Clamp, magnetic, 235
  to retain electron tubes, 244, 245
Clutches, mechanical, basic types, 78, 79
  overriding, 80, 81
    ways to apply, 82, 83
  serrated, 86, 87
  small, for precision service, 84, 85
Compression springs, adjustment of, 316-319
Computing mechanisms, 250-253
Connections, electric terminal, 242, 243
  splined, 100, 101
Containers, sheet-metal, catches for lids of, 14, 15
    designs for, 60, 61
    top and bottom attachments for, 62, 63
Continuous beams, 152, 153
Control, automatic stop mechanisms for, 130-135
  automatic timers for, 146, 147
  electric, methods of, 120, 121
Control, limit-switching mechanisms for, 122, 123
  liquid-level, 124-127
  metal bellows used in, 6, 7
  photoelectric, 128, 129
  snap-action switches in, 240, 241
  temperature, 142, 143
  tubes for industrial, 220, 221
Copper used in brazing, 332, 333
Corrosion, reducing, design details for, 294, 295
  six types of, 156, 157
Couplings, magnetic, 232-234
  shaft, flexible, 90-95, 98
    parallel, 98, 99
    small, 96, 97
  slip, torque of, 208, 209
Covers, tank, 337
Crank and slot drive, 272-274
Cylinders, air and hydraulic, mechanisms actuated by, 248, 249

"D"-curve generating mechanism, 283
Detents, 8, 9, 86, 87
Disk-clutch torque, 207
Dissimilar materials, joining of, 333
  (*See also* Corrosion)
Door, access, in tank, 337
  check linkage for, 287
  latch, nonrattling, 235
Drives, bead chains used for, 4, 5
  belt, 194-197
    adjusting tension in, 196, 197
    horsepower loss in, 200
  chain, reducing pulsations in, 198, 199
  crank and slot, 272-274
  film, 283
  friction, 204-206
    for variable speed, 212, 213
  Geneva, 254-259, 262-271
  intermittent, 256-261, 282-285
  linear motion, 276, 277
  low torque, 324, 325
  power thrust linkages in, 278, 279
  problems in, solved by bead chains, 4, 5
  special-purpose, 282, 283
  for straight-line motion, etc., 284, 285
  toggle linkage, 286, 287
  winding machine, 288, 289

Electrical control methods, 120, 121
Electrical heating elements, design of, 222, 223
  industrial uses of, 224, 225
Electrical symbols, 216, 217
Electrical terminal connections, 242, 243
Electronic circuits, for industrial tubes, 220, 221
  power and voltage amplifiers, 218, 219
Electroplating, design hints in, 296-299
Extension springs, adjustment of, 320, 321
Fastening methods, brazing used in, 333, 334
  circular parts, 40, 41
  dissimilar materials, 333
  flexible sheet materials, 50, 51
  glass-to-metal structures, 52, 53
  liquid-tight, riveted joints, 68, 69
  porcelain enamel, 173
  quick-release fasteners used in, 58, 59
  sheet-metal parts, 64-67
  simplified with spring-steel fasteners, 70, 71
  small die-cast parts, 42-46
  tamper-proof, 72, 73
  twisting action used in, 74, 75
  wire inserts used in, 54, 55
Film drive, 283
Filter, magnetic, 232
Flexible chute, 329
Flexible joints for shafts (*see* Couplings, shaft)
Formulas, circular segments, 160-163
  continuous beams, 152, 153
  crank and slot drive, 272-274
  disk-clutch capacity, 207
  electric heating elements, 222, 223
  friction-wheel drives, 204-206
  Geneva drives, 254-259, 262-271
  horsepower to pipe liquids, 166, 167
  intermittent mechanisms (*see* Intermittent movements)
  magnet coil design, 226-229
  parallel-axis moment of inertia, 154
  press fits, 174-177
  ratchet layout, 178, 179
  shaft strength, 210, 211
  shrink fits, 182, 183
  slip couplings, 208, 209
  springs, sideways deflection of, 322, 323
  stretch-formed parts, 186, 187
  toothed components, 188, 189
Friction-held adjustable parts, 10, 11
Friction ratchet, 286, 328
Friction-wheel drives, 204-206, 260, 261

Gears, ball bearings as, 115
  Geneva (*see* Geneva drives)
  spur, capacity of, 201, 202
    index error in tooth, 275
    surface durability of, 203
  worm, made from spring, 328
  (*See also* Drives)
Geneva drives, 254-259, 262-271
Grommet, rubber, retains electron tubes, 244

Helical wire inserts, 54, 55
Helical wire springs, unusual uses for, 328, 329
High-speed applicators for adhesives, 2, 3
Holes, covers for, 16-19
Hoppers, basic types of, 12, 13

Index error in gear tooth, 275
Indicators, liquid-level, 124-127
Industrial tubes, circuits of, 220, 221
Instrument coupling, magnetic, 234
Intermittent movements, 256-261, 282-285
(*See also* Drives)

Joints, liquid-tight, 68, 69
pressure, 335

Knurls, how to specify, 164, 165

Latch, door, nonrattling, 235
locking, linkage for, 286
Lids, sheet-metal, catches for, 14, 15
Limit-switching mechanisms, 122, 123
Linear motion elements, 276, 277
Linkages, for multiplying short motions, 280, 281
power thrust, 278, 279
toggle, various, 286, 287
Liquid-level indicators, 124-127
Liquid-tight joints, 68, 69
Locking methods, detents used in, 8, 9
for threaded members, 56, 57
Locknuts, selection chart for, 47-49
Low-torque drives, 324, 325
Lubrication, of bearings, 108-111
of roller chains, 300, 301

Magnetron, principle of, 232
Magnets, design of coils for, 226-229
permanent, applications of, 232, 233
mechanisms using, 234, 235
types of, 230, 231
Miniature bearings, applications of, 114, 115
Motors, electrical fractional hp, 236, 237
synchronous, 232
"Mutilated tooth" drive, 282

Nonrubbing seals, 308, 309

O-ring seals, designing with, 306, 307
Oil retention, shaft seals for, 308-313
Overriding clutches, details of, 80, 81
ways to apply, 82, 83
Overriding spring mechanisms, 324, 325

Photoelectric control, 128, 129
Plastic parts, molded, 168-171
Plugs for holes, 16-19
Porcelain enamel, designing with, 172, 173
Press fits, 174-177
Pressure joints, 335

Quick-release fasteners, 58, 59

Ratchets, friction, 286, 328
how to lay out, 178, 179
Resistance welding, 338, 339
Retaining and locking detents, 8, 9
Riveted joints, liquid-tight, 68, 69
Rubber grommet retains electron tubes, 244
Rubber parts, design tips for, 180, 181
Rubbing seals for shafts, 310, 311

Seals, shaft, O-ring, 306, 307
nonrubbing, 308, 309
rubbing, 310, 311
typical, 312, 313
Sensitive stick, 329
Separator, magnetic, 233
Serrated clutches, 86, 87
Shafts, couplings for (*see* Couplings)
serrated, 100
skewed, 4
splined, 100, 101
Sheet metal, alignment of, 38, 39
boxes (*see* Containers)
fastenings (*see* Fastening methods)
Short motions, multiplying, 280, 281
Shrink fits, 182, 183
Slides, ball, 112, 113
Snap-action switch, magnetic, 235
applications of, 240
Special-purpose mechanisms, 282, 283
(*See also* Drives)
Speed, constant, methods of obtaining, 136-139
ways of measuring, 140, 141
Speedometer, principle of, 232
Splined connections, 100, 101
Spring belting, 328
Spring-steel fasteners, 70, 71
to retain electron tubes, 244
ways of testing, 326, 327
Springs, 316-329
adjusting, compression, 316-319
extension, 320, 321
for low-torque drives, 324, 325
helical, unusual uses for, 328, 329
sideways deflection, force for, 322, 323
Sprockets, misaligned, 4, 5
Stainless steel, finishes for, 184, 185
Stitching, wire, 190, 191
Stone crusher, linkage for, 286
Stop mechanisms, automatic, 130-135
Stress, in adhesive joints, 34-37
variable, designing for, 302, 303
(*See also* Formulas)
Stretch-formed parts, 186, 187
Stuffing boxes, welding in, 336
Synchronous motor, 232

Tamper-proof fastenings, 72, 73
Tank covers, 337
Tank supports, 334
Temperature, regulators for, 142, 143
thermocouple details for measurement, 144, 145
Tension devices, magnetic, 234
Terminals, electric, 242, 243
Testing methods for springs, 326, 327
Thread-measuring gage, 329
Threaded parts, providing for backlash in, 292, 293
Thrust linkages, 278, 279
Timers, automatic, 146, 147
Toaster-switch linkage, 287
Toggle linkages, 286, 287
(*See also* Drives)
Torque, disk clutches, 207
low, overriding mechanisms for, 324, 325
requirements of loads, 238, 239
slip couplings, 208, 209
Transducers, ultrasonic, 20, 21
Traversing mechanisms for winding machines, 288, 289
Tubes, electron, for industrial control, 220, 221
ways to retain, 244, 245
Tubing, how to connect, 102, 103
methods for attaching, 104, 105

Ultrasonics, applications of, 20, 21
Universal joints for shafts (*see* Couplings, shaft)

Valve, magnetic, 233
Variable-speed drives, 212, 213
Variable stress, designing for, 302, 303
Vents in brazing, 232, 233
Vibration isolation, mounts for, 24, 25
Vibration isolators, selecting, 22, 23

Welded constructions, 334-337
Welding, resistance, 338, 339
Winding machines, traversing mechanisms for, 288, 289
Wire retains electron tube, 244